TERRESTRIAL BIOSPHERIC CARBON FLUXES

TERRESTRIAL BIOSPHERIC CARBON FLUXES: QUANTIFICATION OF SINKS AND SOURCES OF CO$_2$

Bad Harzburg, Germany, 1-5 March 1993

Edited by

JOE WISNIEWSKI

and

R. NEIL SAMPSON

Reprinted from *Water, Air, and Soil Pollution,*
Volume 70, Nos. 1–4, 1993

SPRINGER-SCIENCE+BUSINESS MEDIA, B.V.

Library of Congress Cataloging-in-Publication Data

ISBN 978-94-010-4875-0   ISBN 978-94-011-1982-5 (eBook)
DOI 10.1007/978-94-011-1982-5

Printed on acid-free paper

TABLE OF CONTENTS

Section 2: Non-Forest Terrestrial Systems

Section 3: Land and Water

Section 4: Biomass and Energy

Section 5: Terrestrial Carbon Models

FOREWORD

Towards the Balance and Management of the Carbon Budget of the Biosphere

The current state of misunderstanding of the global C cycle and our failure to resolve an issue that has been debated for 100 years (Jones and Henderson-Sellers, 1990) speaks loudly about the limitations of modern science when faced with the complexity of the biosphere. Efforts to understand and balance the global C budget have gone through several phases. First was a holistic view of the C budget as part of efforts to understand the geochemistry of the Earth (e.g., Clarke, 1908). Next, came a period of data collection and sythesis which focused on the diversity of sectors of the biosphere. This phase culminated in the early 1970's with the realization that humans were greatly impacting the global C cycle as measured at the Mauna Loa Observatory (Keeling et al., 1973). New syntheses of the global C budget emerged at this time (Woodwell and Pacan, 1973; Bolin et al., 1979). The next phase was one of controversy and intense focus on particular sectors of the biosphere. The controversy rested on discrepancies about the role of the terrestrial biota in the global C cycle and the failure to account for sufficient C sinks to absorb all the C emitted by land-use change in the tropics (Woodwell et al., 1978, 1983; Houghton et al., 1983). The focus of investigators was on those sector of the biosphere believed to have changed since 1860, at the start of the industrial revolution, when it was assumed the atmosphere ceased to be in C steady state. Generally, most research progress (US Department of Energy, 1985) rested on the atmospheric steady state assumption, which allowed scientists to ignore the function of most of the biosphere. This period was characterized by its focus on atmospheric C sources in relation to human activity with less attention to C sinks and problem resolution. Eventually it was demonstrated that the atmosphere was not in C steady state prior to the industrial revolution (Houghton et al., 1990). In fact, CO_2 concentration in the atmosphere oscillated widely before the industrial revolution and it is very difficult to identify a time period when atmospheric CO_2 concentrations were not changing.

During the Palmas del Mar workshop (February 1992) that preceded this workshop in Germany, a group of scientists declared the C steady state assumption invalid (Wisniewski and Lugo, 1992; Lugo and Brown, 1992) and focused on a different path towards the balance of the terrestrial C budget. It was again recognized that all sectors of the biota need to be revisited to gain a global understanding of the balance between C sinks and sources. Greater attention was given to the identification of C sinks, the study of which had been shortchanged previously. At the Palmas del Mar workshop, emphasis shifted from stating problems to problem solution. In so doing, the door was opened for active management of the biota for its C sink function. This phase of the analysis of the global C cycle continues with this volume.

Unfortunately, full understanding of the C cycle still faces serious obstacles as is obvious from the various working group papers assembled in Germany. First is the enormous complexity of the problem; second is the perennial shortage of data; third is fundamental ignorance about the C cycle of mature ecosystems and lack of agreement on whether they are in C balance or not; and fourth is the still lingering vestigers of the steady state assumption which continues to retard progress. This is best illustrated by the report of the wetlands group in this volume.

The final statement on the C budget of the terrestrial biota is still to be made and it will require a significant leap of understanding of the C cycle of individual ecosystems before this occurs. However, the world cannot wait for scientists to balance the last Megagram of C before it acts and it is now time to do something about the problem. Participants in the Germany workshop recognized that the biosphere requires management to sustain human developments and regulate its C cycle. This volume contaeints practical suggestions for such actions and is another step towards gaining a better understanding and managing the global C budget.

Ariel E. Lugo
International Institute of Tropical Forestry
USDA Forest Service
Call Box 25000, Río Piedras, P.R. 00928-2500

References

Bolin, B., Degens, E.T., Kempe, S., and Ketner, P. editors: 1979, *The Global Carbon Cycle,* John Wiley and Sons, New York, 491 p.

Clarke, F.W.: 1908, *Data on Biogechemistry,* U.S. Geological Survey Bulletin No. 330.

Houghton, J.T., Jenkings, G.J., and Ephraums, J.J., editors: 1990, *Climate Change, the IPC Scientific Assessment,* Cambridge University Press, Cambridge, England. 365 p.

Houghton, R.A., Hobbie, J.E., Melillo, J.M., Moore, B., Peterson, B.J., Shaver, G.R., and Woodwell, G.M: 1983, Changes in the carbon content of terrestrial biota and soils between 1860 and 1980: a net release fo CO_2 to the atmosphere. *Ecological Monographs* 53:235-262.

Jones, M.D.H. and Henderson Sellers, A.: 1990, History of the green house effect. *Progress in Physical Geography* 14:1-18.

Keeling, C.D., Ekdahl, C.A., Guenther, P.R., Waterman, L.S., and Chin, J.F.S.: 1973, Atmospheric carbon dioxide variations at Mauna Loa observatory, Hawaii. *Tellus* 25 (5):

Lugo, A.E. and Brown, S.: 1992, Tropical forests as sinks of atmospheric carbon. *Forest Ecology and Management* 54:239-255.

U.S. Department of Energy: 1985, Atmospheric carbon dioxide and the global carbon cycle. J.R. Trabalka (editor). DOE/ER-0239.

Wisniewski, J. and Lugo, A.E., editors: 1992, *Natural sinks of CO_2,* Kluwer Academic Publishers, Dordrecht, The Netherlands, 466 p.

Woodwell, G.M. and Pecan, E.V., editors: 1973, *Carbon and the biosphere,* Conference 720510, National Technical Information Service, Springfiels, Va. 392 p.

Woodwell, G.M., Hobbie, J.E., Houghton, R.A., Melillo, J.M., Moore, B., Peterson, B.J., and Shaver, G.R.: 1983, Global deforestation: contribution to atmospheric carbon dioxide, *Science* 222:1081-1086.

Woodwell, G.M., Whittaker, R.H., Reiners, W.A., Likens, G.W., Delwiche, C.S., and Botkin, D.B.: 1978, The biota and the world carbon budget, *Science* 199:141-146.

PREFACE

Understanding the role of terrestrial ecosystems in the global carbon (C) cycle has become increasingly important as policymakers consider options to address the issues associated with global change, particularly climate change. Sound scientific theories are critical in predicting how these systems may respond in the future, both to climate change and human actions.

In March 1993, 60 scientists from 13 nations gathered in Bad Harzburg, Germany, to develop a state-of-the-science assessment of the present and likely future C fluxes associated with the major components of the earth's terrestrial biosphere. In the process, particular emphasis was placed on the potential for improving C sinks and managing long-term C sequestration.

The majority of the week's work was conducted in eight working groups which independently considered a particular biome or subject area. The working groups considered: the Global Carbon Cycle; Boreal Forests and Tundra; Temperate Forests; Tropical Forests; Grasslands, Savannas and Deserts; Land and Water Interface Zones; Agroecosystems; and Biomass Management. Each group spent long hours, often working late into the night, to prepare an overview paper on their particular subject. The workshop summary paper in Part I of this volume presents their major conclusions and findings. Furthermore, Table 1 of this paper brings together the best estimates from each group as to the current magnitude and estimated future direction of changes in the terrestrial C fluxes. The workshop summary paper is presented as brief conclusions, with minimal explanation of the calculations and assumptions that underlie the conclusions. These can be found in the eight overview papers in Part II, and the 38 individual papers contributed by the participants in Part III.

This workshop was co-sponsored by several governmental and private organizations. The findings and conclusions in this summary statement are those of the workshop participants, and not of the sponsoring agencies or organizations. None of the sponsors have officially reviewed or approved this statement. Therefore, they accept no responsibility for its accuracy, or its scientific or policy relevance.

Those of us who have coordinated the preparation of this volume, owe credit to the authors and workshop participants, whose input underlies the entire project. We can only hope that their contributions are adequately and fairly presented. We also hope that the considerable scientific uncertainty that remains on these questions is fairly portrayed. For many of the questions addressed here, basic methodologies still remain to be developed. The results of this workshop represent but one more step in the exciting and continuing search for understanding of the earth and our human influence upon it.

R. Neil Sampson   Joe Wisniewski

ACKNOWLEDGEMENTS

We acknowledge the individual contributions of the workshop participants, whose enthusiasm combined with an unflagging sense of duty resulted in this product. As the participants are responsible for providing the scientific substance of this report, they are due the final credit for producing the contents of this book. More specifically, we cite the efforts of the Co-Chairmen: Heinz-Detlef Gregor, Al Solomon and Peter Burschel.

Special thanks go to the U.S. Environmental Protection Agency (EPA), the Edison Electric Institute (EEI), the USDA Forest Service (FS), the German Federal Environmental Agency, Umweltbundesamt (UBA), the Ministry of Housing, Physical Planning and Environment of The Netherlands, the German Marshall Fund of the United States (GMF) and the ARCO Foundation for their financial support of the workshop.

From the funding organizations, we particularly thank: Courtney Riordan of EPA and Robert Beck of EEI for approving this effort and providing continued encouragement and guidance; Robert Dixon, Lowell Smith and Jack Durham of EPA who gave intellectual structure and substance to the workshop plan and guidance throughout its implementation; John Kinsman of EEI for continual support and technical assistance; Thomas Hamilton, Richard Birdsey and Linda Heath of the FS for overall encouragement and support of the effort; Heinz Gregor of the UBA for supporting and organizing the European participation, as well as serving as meeting host, together with Christoph Schluter, Katrina Kerber and Christa Morawa; and Bert Metz of The Netherlands Directorate-General for Environmental Protection, Marianne Lais Ginsburg of the GMF and Russell Sakaguchi of ARCO, who arranged for last-minute funds necessary to make the meeting successful.

The very able editorial guidance necessary to produce these publications has been provided by Billy McCormac, Editor-in-Chief of Water, Air and Soil Pollution (WASP) and Dee McCormac, WASP Copy Editor; and Janjaap Blom and Cynthia Feenstra of Kluwer Academic Publishers have offered invaluable publication assistance. Last, but not least, Lu Rose of American Forests provided essential administrative support for the project.

Joe Wisniewski   R. Neil Sampson

WORKSHOP SUMMARY STATEMENT

PART I

WORKSHOP SUMMARY STATEMENT: TERRESTRIAL BIOSHPERIC CARBON FLUXES-- QUANTIFICATION OF SINKS AND SOURCES OF CO$_2$

R. NEIL SAMPSON   *American Forests, 1516 P St. NW, Washington, DC 20005, USA.*

MICHAEL APPS   *Northern Forestry Centre, Forestry Canada, Northwest Region, Edmonton, Alta., T6H 3S5, Canada.*

SANDRA BROWN   *Department of Forestry, University of Illinois, W-503 Turner Hall, 1102 S. Goodwin, Urbana, IL 61801, USA.*

C. VERNON COLE   *USDA, Agricultural Research Service, Natural Resource Ecology Laboratory, Colorado State University, Fort Collins, CO 80523, USA.*

JOHN DOWNING   *Pacific Northwest Laboratories, Marine Sciences Laboratory, 1529 W. Sequim Bay Road, Sequim, WA 98362, USA.*

LINDA S. HEATH   *N.E. Forest Experiment Station, USDA Forest Service, 100 Matsonford Rd., #200, Radnor, PA 19087, USA.*

DENNIS S. OJIMA   *Natural Resource Ecology Laboratory, Colorado State University, Fort Collins, CO 80523, USA.*

THOMAS M. SMITH   *Department of Environmental Sciences, University of Virginia, Charlottesville, VA 22903, USA.*

ALLEN M. SOLOMON   *Environmental Research Laboratory, U.S. Environmental Protection Agency, Corvallis, OR 97331, USA.*

JOE WISNIEWSKI   *Wisniewski and Associates, Inc., 6862 McLean Province Circle, Falls Church, VA 22043, USA.*

Abstract

Understanding the role of terrestrial ecosystems in the global carbon (C) cycle has become increasingly important as policymakers consider options to address the issues associated with global change, particularly climate change. Sound scientific theories are critical in predicting how these systems may respond in the future, both to climate change and human actions.

In March 1993, 60 scientists from 13 nations gathered in Bad Harzburg, Germany, to develop a state-of-the-science assessment of the present and likely future C fluxes associated with the major components of the earth's terrestrial biosphere. In the process, particular emphasis was placed on the potential for improving C sinks and managing long-term C sequestration.

The majority of the week's work was conducted in eight working groups which independently considered a particular biome or subject area. The working groups considered: the Global Carbon Cycle; Boreal Forests and Tundra; Temperate Forests; Tropical Forests; Grasslands, Savannas and Deserts; Land and Water Interface Zones; Agroecosystems; and Biomass Management. This paper presents a brief overview of their major conclusions and findings. In addition, Table 1 brings together the best estimates from each group as to the current magnitude and estimated future direction of changes in the terrestrial C fluxes.

Water, Air, and Soil Pollution **70**: 3–15, 1993.
© 1993 *Kluwer Academic Publishers.*

1. Introduction

There is general scientific agreement that increasing atmospheric levels of CO_2 and other greenhouse gases will change climate and related components of the earth system. It is not clear whether these changes will enhance terrestrial C emissions further and thus lead to additional warming, or will enhance terrestrial C sequestration and lead to reduced warming rates.

In March 1993, 60 scientists from 13 nations gathered in Bad Harzburg, Germany, to develop a state-of-the-science assessment of the present and likely future C fluxes associated with the major components of the earth's terrestrial biosphere. In the process, particular emphasis was placed on the potential for improving C sinks and managing long-term C sequestration.

The challenge of the Bad Harzburg workshop was to gain a better understanding of the global C cycle and predict its response to and effect on future changes in climate and atmospheric chemistry. The critical importance of climate in controlling natural and managed biotic systems, upon which global food, fuel and fiber supplies depend, makes this a matter of highest priority.

Recent efforts to locate the "missing carbon" in the global C cycle have often focused on the complex terrestrial system. Portions of the terrestrial biosphere have undergone modifications by human societies over time, releasing C to the atmosphere at a quickened pace since the beginning of the industrial revolution. Some investigators have estimated that biospheric modification is responsible for up to half of the human-induced increase of atmospheric CO_2 experienced to date.

The primary objective of the Bad Harzburg workshop was to provide global change scientists and policymakers with a state-of-the-science assessment of: 1) present and future terrestrial biospheric C pools and fluxes associated with the major terrestrial biomes, and 2) the potential for managing the terrestrial biosphere to enhance long-term C sinks and reduce land-based CO_2 emissions.

This workshop builds on efforts and results of other recent related workshops. Of particular note are the Workshop on Natural Sinks of CO_2, held in Palmas Del Mar, Puerto Rico, during February 1992 (Wisniewski and Lugo, 1993), and the Workshop on Biotic Feedbacks in the Global Climate System, held in Woods Hole, Massachusetts, during October 1992. The former focused on articulating the issues related to biospheric sequestration of atmospheric carbon, while the latter focused on the cycle of cause/effect that a changing atmospheric composition and climate has on the biosphere and vice-versa.

Negotiations under the Framework Convention on Climate Change, signed by representatives of 160 nations at the UN Conference on Environment and Development in Rio de Janeiro in June 1992, are expected to begin in 1994. Their objective will be to reduce human-induced climate change by limiting emissions of CO_2 and other greenhouse gases, eventually stabilizing their atmospheric concentrations. The UN-sponsored Intergovernmental Panel on Climate Change (IPCC) is working with scientists from around the world to identify and assess important scientific and technical issues related to this objective.

The IPCC has begun development work on its 1995 Scientific Assessment and its Second Assessment Report on impacts and response to provide a scientifically sound consensual information base for the negotiators of the Framework Convention to use in their work.

This workshop was designed to complement IPCC assessment activities through the development of an updated quantification of the C flux associated with each of the major components of the terrestrial biosphere system. Additionally, it is hoped that the workshop results will be useful in assisting the design of management strategies for sustainable development in the future.

2. Conclusions of the Working Groups

The conclusions on the role of the terrestrial biosphere in global carbon cycling were as diverse as the participants backgrounds and interests, and the biomes and sectors analyzed. The most important of these are discussed below under the headings of the working groups. A few conclusions surfaced from most or all the analysis and these are described here.

The most consistent and surprising conclusion to emerge from the deliberations by all except the boreal group is that land use, not climate change or atmospheric chemistry (e.g., fertilization by CO_2 or N), has been and probably will continue to be the most important determinant of C storage, uptake and release in all terrestrial ecosystems. The conclusion is a surprise because the participants were focused on the roles of the atmosphere in terrestrial C cycling, not on land use. Characteristically, land use-induced losses of below-ground C stocks (soil C) appear to be of greatest concern in arid regions comprising savannas, grasslands, and deserts, and in wetlands, including boreal peatlands, lakes, rivers and coastal estuaries. Losses of above-ground C stocks are most evident in tropical and temperate forests. Only in boreal forest and tundra regions does climatic change appear to be of greatest concern during the next century.

Emphasis on the importance of land use appeared in several syntheses. The working group on agriculture pointed out that C loss to the atmosphere and aquatic systems from soils alone frequently reaches 50% of initial quantities after a half century or more. The biomass management working group saw equally important, but opposite changes in future C sequestration if programs could be implemented specifically to reduce land use effects on C stocks. The working group on land-water interface declared that land use of the past century or so has increased C delivery to streams, lakes, rivers and coastal estuaries many-fold. The grasslands working group suggested that the overall impact of land-use practices on C storage in grassland and dryland soils is substantially greater than that of climate change or increased CO_2 concentrations.

Similarly, the tropical forest group pointed to the massive amounts of C released from destruction of living forest resources (above-ground biomass) for crop and pasture land use as the source of much of the atmospheric C increases of the past century. They also expressed the expectation that land use will continue to dominate the tropical C source in the future. The temperate forest working group indicated that forests now cover less than 50% of areas that were previously 90% forested, primarily because these are prime farm and urban regions. Currently a C sink, the temperate regions are expected to be reduced to a C source in the future from continuing excessive land use. The boreal group concluded that land use in the form of management was unlikely to influence future C stocks, but poor management and ill-considered land use changes held great potential for decreased C storage capacity in boreal regions. Instead of land use, they identified rapidly changing climate as the chief risk to both above and below-ground C stocks.

As obvious as the emergent emphasis by working groups on effects of land use in today's and in a future warmed world, is the uncertainty regarding the direct influence of changing atmospheric chemistry on C sequestration. Definitive quantification of CO_2 "fertilization" effects is absent from assessments by the working groups and is considered only qualitatively if at all in their deliberations. Here, the inability to characterize CO_2 effects probably derives from a lack of definitive information demonstrating (1) that CO_2 fertilization actually occurs in the wildland ecosystems which control terrestrial C and (2) that there are systematic variations in CO_2 fertilization patterns with differences in biotic populations, ecosystems, soils and climate, which would permit calculating the consequences of CO_2 fertilization under shifting ecosystem and climate geography. Perhaps no other process has been at once so important to scientific and political conclusions on the functioning of the earth system and so mysteriously beyond our skill to measure in a way acceptable to the skeptical scientific mind.

Each working group evaluated the potential for humans to manage ecosystems to adapt to or reduce the impacts of climate change. Much can be done, and it all matters. However, the participants clearly stated their belief that vegetation-based management or mitigation measures cannot fully offset the existing anthropogenic disturbances of the global C cycle. The participants concluded that policies must be developed to reduce the root causes of greenhouse gas increases: fossil fuel use and non-sustainable land use practices, both driven by growing human populations and economic development.

Although terrestrial C management cannot solve the greenhouse gas problem in the absence of basic changes in land use and population dynamics, the participants concluded that it can play an important role in the solutions. In particular, they noted the need to reduce tropical deforestation, foster recovery of degraded lands, address the potential for increasing C storage by ecosystem management, and capture the potential benefits of increasingly efficient use of biomass to replace fossil fuel energy sources.

3. Estimates of Carbon Fluxes

The working groups attempted to quantify current C fluxes associated with the major terrestrial biomes, then to estimate future changes. No common scenario for future climate change was used; each group made its own assumptions regarding the timing of CO_2 doubling, the magnitude of associated warming and other climate effects, and the amount and degree of biome shift that might be associated with such changes. The individual work group efforts are summarized in Table 1, which presents C fluxes under current conditions, and under three future scenarios.

The three scenarios present the estimated impact of a climate changed by a doubling of atmospheric CO_2, but without human efforts to affect land use or ecosystem management; a climate changed by doubling of atmospheric CO_2, with optimum vegetation management aimed at maximizing C sequestration and improving terrestrial C pools; and, the impact of aggressive vegetation management aimed at energy conservation and production, which reduce fossil fuel emissions.

At present, the terrestrial biosphere may be a C source of as much as 1.4 Pg C yr-[1], but uncertainties also allow hypothesizing that much less C is released, and even that there may be a terrestrial C sink of as much a 0.9 Pg C yr-[1], based on data from individual biomes and sectors (Table 1, Column A). The largest contributor to the range of estimates was the tropical forest biome, apparently dominated by effects of deforestation.

The consideration of climate effects from doubling CO_2 widened the range of uncertainty, driven in large part by a wide range of possibilities which could change today's near neutral situation to either a large source or sink, depending on future actions which today are impossible to predict. In general, however, Table 1, Column B suggests that global systems, without effective management intervention, could become a much stronger C source under doubled CO_2 climate.

The global-scale working group noted that this uncertainty about future C fluxes could be reduced if data and models were available to generate estimates of transient responses as credible as those in Column A of Table 1. For example, current (insufficient) estimates indicate that transient processes during the next century could release 3-4 Pg C yr[-1] from the terrestrial biosphere. Under such a scenario, the transient C sources such as more intensive agriculture, climate-induced forest dieback, and C from exposed soils, would not be balanced by C sinks created when new trees immigrate to dieback areas and forests regrow in boreal, temperate and tropical regions. Others suggested that ecosystem management could significantly reduce these transient imbalances. What is the most likely outcome of these opposing effects on global carbon cycling? The answer is expected to emerge, but only with better data and models, and decreased scientific uncertainty.

Table 1. Current and future C fluxes in the terrestrial biosphere, with atmospheric CO_2 doubling, under different management scenarios.

Global Biotic System	Current C Flux	Future C Flux (Doubled CO_2 climate)	Future C Flux with Optimum Vegetation Management (Doubled CO_2 climate)	Fossil C Offset Potential from Biomass-Energy Management (Doubled CO_2 climate)
	(A)	(B)	(C)	(D)
	(Pg C yr[-1])			
Tundra/Boreal Forests	+0.5 to +0.7	-1.0 to -0.5[1]	0	
Temperate Forests	+0.2 to +0.5	-2.0 to +2.0	+0.3 to +2.0	+0.1 to +0.9
Tropical Forests[2]	-2.2 to -1.2	-1.0 to -0.5	-0.5 to 0	0 to +0.2
Grasslands, Savannas and Deserts	0 to +0.6	-0.3 to +0.1	+0.1 to +0.5	0 to +0.3
Agro-Ecosystems	-0.1 to +0.1	0.0 to +0.1	0.0 to +0.3	+0.4 to +2.4
Wetlands	+0.2	+0.1	+0.2	
TOTAL	-1.4 to +0.9	-4.2 to +1.3	-0.1 to +3.0	+0.5 to +3.8

Sink = (+); Source = (-).

[1] During transient (50-100 yr) response. In the long-term (200-1,000 yr), may revert to sink if climate stabilizes.
[2] From land use changes only

There was no consensus on how (or if) the estimates in columns B, C and D interact. Nor is this data sufficient to balance the global C budget. What was agreed, however, is that they illustrate the kinds of issues (and, hopefully, the range of magnitudes) facing scientists and policymakers as they consider the questions associated with global climate change.

The application of vegetation management (Column C) is predicted to generate positive effects on terrestrial C balance. These management opportunities include improving forest growth, building soil C through improved agricultural practices, and protecting forests and wetlands. Many of these techniques have been successfully demonstrated in regions prosperous enough to aim toward sustainable land use practice, although many are not currently in common use, either in developed or developing nations.

The work group on biomass management estimated the range of opportunities for management actions that utilize biomass to conserve energy or directly as an energy source to replace fossil fuels. These C "offsets" could reduce some of the net C emissions associated with fossil fuel combustion. The results of those estimates are shown as Column D.

4. Individual Working Group Reports

While the world can be viewed as a whole, the individual biomes and human sectors must be examined individually. Each possesses a set of unique attributes which disallow predictions of future global-scale trends in C dynamics based solely on the expectation of uniform effects of process and responses from one biome to another. The following sections examine the conclusions of the individual working groups.

4.1 THE GLOBAL CARBON CYCLE (Smith *et al.* 1993)

The global C cycle working group considered the earth as a whole, in light of estimates from current global models, without reference to the analyses of the other regions and sectors by the other working groups.

* The terrestrial biosphere is estimated to contain approximately 560 Pg C (Pg = petagram = 10^{15} g) in biomass and litter above ground, and 1,100 to 1,400 Pg C in roots and soil C below ground; the fluxes from the atmosphere into and out of the basic terrestrial C pools cannot be reconciled with what is known about the amount of CO_2 produced by land-use change and fossil fuel combustion, which is currently thought to be about 1 Pg greater per year than the known atmospheric, terrestrial and marine C sinks.

* Annual forest loss from land use change of 15 to 17 x 10^6 ha may contribute as much as 1.2 to 2.2 Pg C to the atmosphere. Uncertainties on C pool size, especially of forest systems, limits the ability to estimate the real impact of land use change on C flux with the atmosphere.

* Implementation of sustainable forest, grassland, and agro-ecosystem management options can be employed to conserve and sequester C in soils and vegetation over decades and centuries.

* The Framework Convention on Climate Change, especially provisions for joint implementation, provides the opportunity for C offsets in the terrestrial biosphere.

* Development of policy options may proceed ahead of the current scientific data base. Therefore, research is needed to: 1) define below-ground C cycling, 2) document impact of CO_2 enrichment on whole ecosystems, and 3) quantify and simulate the influence of land-use features and climate change on the global C cycle.

4.2 BOREAL FORESTS AND TUNDRA (Apps *et al.* 1993)

The boreal forests and tundra have been relatively free of direct land use impacts. The forests are continuous and much of forested and tundra regions have large areas of peatlands developed on internal drainages. Large scale disturbance by fire and insect infestation is probably more prominent in the cycling of C of boreal regions than in any other area. Equally critical is the expectation of great increases in winter temperatures, coupled with the unique limitation to biotic production of winter low temperatures. As a result, large uncertainties surround the potential release of vast amounts of C from oxidation of peats as well as from evaporation of CH_4 locked up as methane hydrates in permafrost.

* The boreal region (boreal forest and tundra) is estimated to contain 65 Pg C in living biomass, 270 Pg C in soil and detritus pools, 440 Pg C in boreal wetlands, and less than 3 Pg retained in wood products derived from the boreal region.
* The net C flux in the boreal region is estimated to currently be a sink of 0.4 to 0.6 Pg C yr^{-1}, including an estimated 0.08 Pg C yr^{-1} uptake through peat formation. The mechanisms responsible for this sink are believed to be continuing responses to: deglaciation; climatic perturbations (e.g., the little ice-age, 1250 to 1850); changes in disturbance regimes over the past two centuries (e.g., land-use change and wildfire); and, more recently, nutrient inputs from air pollution.
* The mechanisms responsible for the C sink will not likely continue at their present strength, even under the present climate. Reductions of the sink and increases in source mechanisms are probable.
* In a changing climate, regional shifts in boreal forest distribution and productivity will have significant future socio-economic impacts and will likely result in significant C pulses (transients). Forest resource policies can only partially mitigate changes in C storage but non-sustainable practices can adversely affect the C balance.
* Positive feedback (CO_2 release) mechanisms include increased disturbance regimes (e.g., wildfire), increased respiration, potential effects of higher UV-B radiation, and forest decline due to climatic shifts or in combination with other factors (e.g., soil acidification due to air pollution).
* Negative feedback factors include increased C sequestration due to a longer growing season and increased nutrient availability from faster detrital turnover. N deposition (pollution) could also favor C uptake in the short term.
* Other potential feedback processes for which there are great uncertainties include effects of CO_2 fertilization (C allocation to roots and other plant parts, net ecosystem productivity, water-use efficiency), melting of permafrost, and changes in the hydrological regime in peatlands (altering their distribution, productivity, and CH_4/CO_2 budgets).

4.3 TEMPERATE FORESTS (Heath *et al.* 1993)

Most of the temperate forests are located in the developed countries of the world. Here technology combined with climate and soils suitable for intensive agriculture has led to clearing and loss of forests to land use changes through the 19th century. Currently, these forests are a C sink because of considerable afforestation and forest regrowth. Pollutants from industrial activities are affecting these forests, with potential negative and positive growth effects. If current trends continue, temperate forests will become a C source because of increasing demand for wood products, and intensification of all land uses.

* Temperate zone forests are estimated to contain a C pool of 21 Pg in above-ground biomass and 70 to 100 Pg in below ground stocks.
* The current flux is estimated at about 0.2 to 0.5 Pg net increase each year in living biomass increment. The eventual changes in soil and detritus C are uncertain. The manufacture of wood products from these forests stores approximately 0.1 Pg C yr^{-1}.
* The temperate zone forests currently cover little of their potential growth area because intensive agriculture has replaced them, gradually in Europe and Asia, and within the past 200 yr in North America.
* Important policy considerations in temperate forests include the need to conserve these forests using sustainable management methods, protection from deforestation and forest degradation, and reduction of air pollution. Forest biomass can and should be increased by optimizing rotation periods, changing tree species and varieties, using appropriate harvesting methods, pursuing afforestation opportunities, and by soil conservation.
* Critical uncertainties affecting the future changes in the forest C cycle include the response of soil respiration to climate and atmospheric change and the net effect of CO_2 fertilization.

4.4 TROPICAL FORESTS (Brown *et al.* 1993)

During the past century, C cycling in tropical forests appear to have been affected almost entirely by forest destruction and deterioration by direct human actions. C release from tropical forests probably accounts for half or more of all land use-induced C emissions during this period. It has been suggested that tropical forests are also in a unique position to sequester large amounts of C as growth enhancement from C fertilization. This would result from the high density of photosynthetic surfaces coupled with the lack of severe limits to growth imposed elsewhere by low seasonal temperatures, limited soil moisture, or slow nutrient cycling. Even if the CO_2 fertilization process does function, it is expected that population growth and related land use will dominate C cycling in tropical forests for the foreseeable future, unless current trends are altered.

* The pre-industrial (1850) flux from changes in land use in tropical forest lands amounted to a small C source of 0.0 to 0.1 Pg yr^{-1} with a likely best estimate of 0.06 Pg yr^{-1}. The estimated C flux to the atmosphere from land use change for 1980 is from 1.0 to 1.3 Pg and for 1990, 1.2 to 2.2 Pg, with an average of 1.7 Pg. These estimates do not

include fluxes associated with a reduction in biomass caused by forest degradation nor biomass accumulation in mature forests, often assumed to be in steady state.

* The present (1990) C pools in tropical forests are estimated to be about 159 Pg in vegetation and about 216 Pg in soils and litter.
* No concrete evidence is available for predicting how tropical forest ecosystems are likely to respond to CO_2 enrichment and/or climate change, but C sources to the atmosphere from continuing deforestation will most likely overwhelm any environmentally-driven change in C fluxes (sources or sinks) unless land management efforts become more aggressive.
* Future changes in land use under a "business as usual" scenario could release 41 to 77 Pg of C over the next 60 yr; by 2050 the tropics could be a source of 0.5 to 1.0 Pg C yr^{-1} when deforestation has slowed significantly.
* Tropical forest losses to land use change may be lessened by aggressively pursued agricultural and forestry measures. These measures could reduce the magnitude of the tropical C source by 50 Pg by the year 2050.
* Policies to mitigate C losses must be multiple and concurrent, including reform of forestry, land tenure and agricultural policies, promotion of on-farm forestry, establishment of plantations, and forest protection. Policies should support improved agricultural productivity and general performance, especially in replacing slash and burn agriculture with more sustainable and appropriate approaches.
* The most important research needs for improving C flux estimates and developing policy options for C mitigation are: (a) map land use and cover over time, including disturbed and undisturbed forests, and produce corresponding maps of forest and soil C densities using remote sensing data and GIS technology; (b) establish a tropics-wide network of plots for continuous monitoring of the tropical landscape, including mature and various stages of secondary forests; (c) establish CO_2-enrichment and climate change studies at the ecosystem levels in tropical forests (mature and secondary, dry to humid), and (d) improve efforts to develop better estimates of lands technically suitable and available for management options for sequestering C.

4.5 GRASSLANDS, SAVANNAS AND DESERTS (Ojima *et al.* 1993)

Grasslands, savannas and deserts are unique in that C cycling is directly linked to amounts and seasonal distributions of precipitation and is only secondarily controlled by other climatic variables and atmospheric chemistry. The smaller amounts of above-ground C and the high densities of soil C equivalent to those in more moist temperate zones appear to be the basis for their vulnerability to human-induced land use changes. Permanent desertification of grasslands and savannas resulting from poor soil management is a serious consequence to C storage functions as well as to requirements for agricultural products in less developed nations where much of these biomes occur.

* The present C pool in grassland soils is estimated to be about 417 Pg.
* Grasslands may currently be a net C sink of 0.6 Pg yr^{-1}, mainly through gradual increase in soil C.
* Projections for optimal grassland management under a doubled CO_2 concentrations by

the year 2040, resulted in a net C sink ranging from +5.6 to +27.4 Pg, for three different land cover projections. The increased C storage resulted mainly from areal increases of the warm grassland biome (net increase of 280 x 10^6 ha.)

* Monitoring should be implemented to generate long-term assessments of grassland/dryland ecosystem productivity, above and below ground, in response to year-on-year variation in climate (temperature, precipitation, etc.).

* Specific investments that are cost effective in terms of C sequestration should be encouraged in dryland regions. These should also seek to achieve desertification program goals, with enhanced economic returns to inhabitants and preservation of dryland biodiversity.

* Models predict that during 100 yr, the difference in C emissions between the regressive and sustainable management scenarios will be 50 Pg (annual difference = 0.5 Pg) over the whole land base under consideration (4.5 x10^9 ha of grasslands and rangelands).

4.6 LAND AND WATER INTERFACE ZONES (Downing *et al.* 1993)

Ecosystems in the land-water interface cover only about 2% of the earth's surface, and yet they comprise a substantial C pool. Carbon is cycled in diverse ways at the land-water interface. Freshwater wetlands (peatlands, lakes, swamps, marshes) fix large amounts of atmospheric C photosynthetically, with very little CO_2 loss to heterotrophic respiration, and therefore are long-term C sinks. Carbon dynamics and storage in freshwater ecosystems, however, are sensitive to changes in land use (loss/gain of area; pollutant inputs), climate, hydrologic conditions and atmospheric chemistry. With the exception of changes in wetland area, the interaction of these factors and their influence on C storage on a global scale is currently unknown.

Rivers transport C weathered from the land to estuaries but take up little C via primary production directly from the atmosphere. Estuaries receive much more C from rivers than directly from the atmosphere by primary production. Both rivers and estuaries are strongly affected by land use. The coastal ocean (here meant to include tidal parts of rivers, estuaries, and continental shelves) exchanges CO_2 with the atmosphere, and at the shelf edge, dissolved CO_2 with the open ocean. The net CO_2 fluxes associated with these exchanges have yet to be determined. With currently available data, it is therefore impossible to estimate the portion of dissolved C fixed by coastal primary production that comes directly from the atmosphere. For the same reasons, it is not possible to predict the future C uptake by the coastal ocean.

* Estimates of temperate and boreal wetland C sinks suggest a decrease by about 50% since 1850, from about 0.2 to 0.1 Pg C/hr[1]. These estimates, however, do not include tropical wetlands, the area and carbon density of which are not well known.

* Even with doubled anthropogenic N and P loads, it is unlikely that rivers will fertilize more than about 20% of the net primary production on continental shelves. The balance of the net primary production is believed to be fertilized by nitrate transported from the open ocean. Whether this net primary production takes up atmospheric CO_2 or dissolved inorganic C from the open ocean, is not known. The fate of nutrients and C in coastal ecosystems is not well enough understood to predict their current, let along future roles, in the global C cycle.

* Land use management must consider the unique C sinks in coastal and alluvial wetland environments in order to minimize negative impacts of agriculture and urban development.
* Sea-level rise, changing hydrologic conditions, and sediment supply are the principal natural factors that will determine the future rate of C-storage in coastal wetlands. Conversion of wetlands to agri- and aquaculture, by expanding coastal populations in developing countries, will reduce their C-storage capacity and increase nutrient loads to the coastal ocean.
* Salt intrusion as a result of sea level rise and surface water diversion could switch respiration from methanogenesis to CO_2 production in some coastal ecosystems. Coincidentally, global warming could change the balance between methane production and oxidation. Although the severity of interactive effects is poorly known, water management could mitigate methane production to some degree.
* Regardless of the strategies used, long-term monitoring will be essential to establish the success, or failure, of management practices to sustain wetland resources in the future.

4.6 AGROECOSYSTEMS (Cole *et al.* 1993)

Agroecosystems are the land areas of the globe most subject to continuous anthropogenic disturbance. Land use changes involving major transformations from forests, grasslands and savannas have transformed large areas from relatively stable undisturbed ecosystems to agroecosystems under extensive and intensive management. The introduction of agriculture involving land clearing or breaking of sod, cultivation, replacement of perennial vegetation by annual crops, and nutrient subsidies in the form of fertilizers has had major impacts on C pools and fluxes of large regions of the globe. In the initial phases of these transformations, there have been major losses of CO_2 to the atmosphere as soil C pools adjusted to reduced C inputs and increased soil disturbance. In many areas this has also caused serious degradation by erosion and nutrient losses. These trends continue in many areas of the world. However, in countries able to provide subsidies of energy and technology, agricultural productivity has shown continuing increases, land degradation has slowed or reversed, and soil C pools have stabilized or increased. For analysis of C fluxes in agroecosystems the primary focus needs to be on current and predicted changes in land use and management.

* Existing agricultural lands are estimated to contain a C pool of 120 to 180 Pg C. They are neither a major source nor major sink for atmospheric C, with an estimated annual flux of -0.1 to +0.1 Pg yr^{-1}.
* Agro-ecosystems may be converted to a net C sink of up to 7 Pg C during the next 50 yr by use of appropriate soil management practices including enhanced use of crop residues, reduced tillage, and increased crop production with greater additions of organic C.
* Conversion of new lands into agricultural production, driven by increasing populations and land degradation, results in large C fluxes to the atmosphere(see Tropical Forests) on the order of one-fourth of emissions from fossil fuels.
* Policy challenges include the need to improve technical assistance in developing

countries to maintain and improve production on existing farm lands and to decrease land conversion.

* Governments need to encourage farmers to improve C sequestration in agroecosystems, through practices that improve crop residue management, reduce tillage, and increase biomass production, including the development of energy crops, preferably on marginal lands to avoid competition with food crop production.

4.7 BIOMASS MANAGEMENT (Sampson *et al.* 1993)

Biomass management is the essence of agriculture and forestry. Increasing emphasis recently has been placed upon managing resources to increase terrestrial C stocks and use that biomass as an energy source to replace fossil fuels. In addition, biomass can be used to help conserve energy (e.g., reducing electrical demand through shading and providing wind protection). These practices can reduce net CO_2 emissions. The major opportunities for biomass energy production exist in agricultural areas, but differ greatly between developed (located mainly in the temperate region) and developing nations (located mainly in the tropical or boreal regions). Energy conservation potential is greatest in temperate regions.

* The greatest potential for managing biomass to affect future energy-related C sinks and sources lies in the opportunities to utilize biomass for energy production. The combination of improved forest management, afforestation of appropriate lands, energy crops, improved agroforestry techniques, halophytes on saline lands, and the use of crop wastes for energy could result in an annual offset of C emissions ranging from 0.5 to 3.6 Pg yr^{-1}.
* There are opportunities to manage biomass for energy conservation--by substituting renewable materials like wood for high-energy materials like aluminum; using trees to shade homes and small buildings, reduce urban heat islands, and protect against winter winds--which, taken together, could reduce C emissions by 0.04 to 0.2 Pg C yr^{-1}.
* Forest and energy crop management must be performed in a way that is both ecologically and economically sustainable. Done well, biomass management for energy can offer multiple environmental benefits. Monitoring and reporting on those benefits is essential for public acceptance.
* National or international efforts to mitigate C emissions through expansion of forests and improved forest management must be coupled with incentives to develop an energy-related market for a significant portion of the new wood produced, or disruptive market dislocations could ensue within a few decades.
* Improving the science base for ecosystem management, with attention to the C cycle impacts of management actions, is needed for all ecosystems to help managers understand and incorporate C-related management factors into land use decisions.

5. References

Apps, M.J., Kurz, W.A., Luxmoore, R.J., Nilsson, L.O., Sedjo, R.A., Schmidt, R., Simpson, L.G., and Vinson, T.S.: 1993, The Changing Role of Circumpolar Boreal Forests and Tundra in the Global Carbon Cycle, this volume.

Brown, S., Hall, C.A.S., Knabe, W., Raich, J., Trexler, M.C. and Woomer, P.: 1993, Tropical Forests: Their Past, Present, and Potential Future Role in the Terrestrial Carbon Budget, this volume.

Cole, C.V., Flach, K., Lee, J., Saurbeck, D. and Stewart, B.: 1993, Agricultural Sources and Sinks of Carbon, this volume.

Downing, J.P., Twilley, R.R., Maybeck, M., Orr, J.C., Sharpenseel, H.W.: 1993, Land and Water Interface Zones, this volume.

Heath, L.S., Kauppi, P., Burschel, P., Gregor, H.D., Guderian, R., Kohlmaier, G.H., Lorenz, S., Overdieck, D., Scholz, F., Thomasius, H. and Weber, M.: 1993, Carbon Budget of Temperate Zone Forests, this volume.

Ojima, D.S., Dirks, B.O.M., Glenn, E.P., Owensby, C.E., and Scurlock, J.M.O.: 1993, Assessment of C Budget for Grasslands and Drylands of the World, this volume.

Sampson, R.N., Wright, L.L., Winjum, J.K., Kinsman, J.D., Benneman, J., Kürsten, E., and Scurlock, J.M.O.: 1993, Biomass Management and Energy, this volume.

Smith, T.M., Cramer, W.P., Dixon, R.K., Leemans, R., Neilson, R.P. and Solomon, A.M.: 1993, The Global Terrestrial Carbon Cycle, this volume.

Wisniewski, J. and Lugo, A.E. (eds): 1992, *Natural Sinks of CO₂* (Special Edition of *Water, Air and Soil Pollution*, **64**:1 & 2, Dordrecht, The Netherlands: Kluwer Academic Publishers, 466 pp.

WORKSHOP WORKING GROUP PAPERS

PART II

THE GLOBAL TERRESTRIAL CARBON CYCLE*

T.M. Smith
*Department of Environmental Sciences,
University of Virginia, Charlottesville, VA 22903 USA,*

W.P. Cramer
*Potsdam Institute for Climate Impact Research (PIK)
Telegrafen, 0-1500, Potsdam, Germany*

R.K. Dixon
*Environmental Research Laboratory
U.S. Environmental Protection Agency, Corvallis, OR 97331 USA,*

R. Leemans
*National Institute for Public Health and Environmental Protection (RIVM),
3720 Bilthoven, the Netherlands,*

R.P. Neilson
*USDA Forest Service-Pacific Northwest Research Station,
Corvallis , OR 97333 USA*

A. M. Solomon
*Environmental Research Laboratory
U.S. Environmental Protection Agency, Corvallis, OR 97331 USA*

Abstract. There is great uncertainty with regard to the future role of the terrestrial biosphere in the global carbon cycle. The uncertainty arises from both an inadequate understanding of current pools and fluxes as well as the potential effects of rising atmospheric concentrations of CO_2 on natural ecosystems. Despite these limitations, a number of studies have estimated current and future patterns of terrestrial carbon storage. Future estimates focus on the effects of a climate change associated with a doubled atmospheric concentration of CO_2. Available models for examining the dynamics of terrestrial carbon storage and the potential role of forest management and landuse practices on carbon conservation and sequestration are discussed.

*The information in document has been partially funded by the U.S. Environmental Protection Agency. It has been subjected to the Agency's peer and administrative review, and has been approved for publication as an EPA document. Mention of trade names or commercial products does not constitute endorsement or recommendation for use.

1. Introduction

Although representing only a small fraction of the oceanic C pool, the annual flux of C between the terrestrial surface and atmosphere is of the same order as the flux between the ocean and atmosphere (Figure 1). Previous work has addressed the impacts of changing landuse patterns on the storage and flux of C from the terrestrial biosphere (Houghton *et al.*, 1983; Emanuel *et al.*, 1984), however, the potential impacts of rising concentrations of atmospheric CO_2 (and other greenhouse gases) on patterns of terrestrial C storage and the consequent flux between terrestrial and atmospheric pools has yet to be quantified. In combination with future changes in landuse patterns, the increasing atmospheric concentrations of CO_2 represent a major uncertainty in projecting the future dynamics of the global C cycle, both in terms of the direct effects of CO_2 on ecosystems, as well as the potential impacts of rising atmospheric concentrations of greenhouse gases on the global climate system (i.e., global warming) (Houghton *et al.*, 1992).

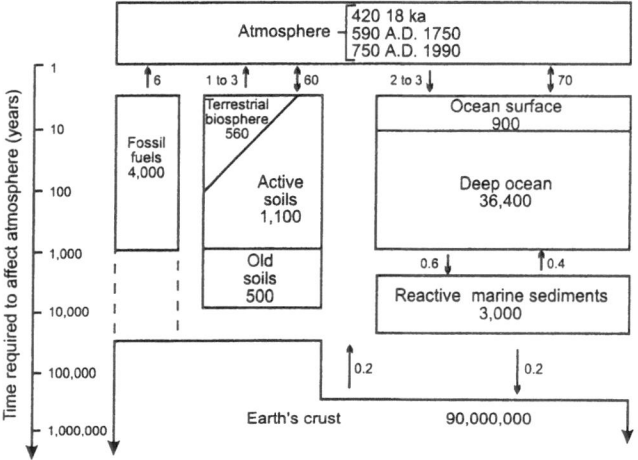

Figure 1. Principal reservoirs and fluxes (arrows) in the global carbon cycle. Vertical placements relative to scale on left show approximate time scales required for reservoirs and fluxes to affect atmospheric CO_2. Double arrows represent bidirectional exchange. Single arrows to and from the atmosphere are approximate estimates of anthropogenic fluxes for 1990. Terrestrial uptake of anthropogenic CO_2 is likely but not shown because of large uncertainties (from Sundquist, 1993).

The processes controlling C storage and release from the terrestrial biosphere are not documented well enough to accurately estimate and balance current exchanges (for discussion see: Sundquist, 1993). Measurements of C storage and release under present conditions are currently being undertaken in many places in the world, however, extrapolation of these estimates in time and space are dependent on a detailed understanding of the biological and physical processes involved in the flux of C between the atmosphere and biosphere. To predict the current and future patterns of C flux from the terrestrial surface will require the development of detailed models of these processes within a geographical framework as well as improved estimates of C pools.

The objectives of this manuscript are to: 1) examine estimates of past and present C fluxes from the terrestrial surface, 2) define the processes we feel are most important in controlling future patterns of C storage and net flux from the terrestrial surface, 3) review global-scale vegetation models which are currently being used to examine the dynamics of terrestrial C storage and exchange with the atmosphere, 4) assess the strengths and weakness of current and future estimates of terrestrial C dynamics, and research required to improve them.

2. Global Carbon Budget

2.1 PAST

The global budget of C pools and fluxes at the end of the Pleistocene is not well defined. The most complete historic record is atmospheric CO_2 concentrations derived from ice-core samples collected in Antarctica (Sundquist, 1993). Atmospheric CO_2 concentration climbed from 200 to 280 ppm from 18,000 to 10,000 BC. The terrestrial biosphere was probably a net sink during period but historical data describing vegetation and soil system C pools and flux are highly variable. During the past 1000 yrs the atmospheric CO_2 concentration was relatively stable, ranging from 270 to 290 ppm (Stauffer and Oescheger, 1985).

Human activities began to significantly influence atmospheric CO_2 in the late 1700s. Between 1750 and 1850, up to 40 Pg C was released due to deforestation and land cultivation (Sundquist, 1993). With the advent of the industrial revolution combustion of fossil fuels became a significant factor in the global C cycle. Fossil fuel emissions released over 200 Pg C during the past 250 years. The Antarctic ice core record revealed that atmospheric CO_2 concentrations rose from 280 to 290 ppm from 1750 to 1850. From 1850 to 1950 atmospheric CO_2 rose to 310 ppm.

2.2 PRESENT

In the latter half of the 20th century atmospheric CO_2 continued to rise exceeding 350 ppm in 1990. Currently, the global C budget can not be balanced as the amount of CO_2 produced by landuse change and fossil fuel combustion is greater than known terrestrial and marine sinks (Houghton *et al.*, 1992; Sundquist, 1993; Tans *et al.*, 1990). The budget imbalanced is due to the so-called "missing CO_2". Atmospheric CO_2 data and global simulation models suggest that the southern hemisphere is a weak C sink and the northern hemisphere is a major C source and sink. The enhanced northern hemisphere sink is attributed to both marine and terrestrial C accretion (Tans *et al.*, 1990; Keeling *et al.*, 1989). Despite significant release of CO_2 due to landuse change (eg., deforestation) most analyses suggest the biosphere is a net sink of CO_2. Research continues to define the magnitude of the terrestrial and marine C sinks of the northern hemisphere.

Estimates over the past 20 years suggest that terrestrial vegetation and soils contain 1500 to 2000 Pg C, with approximately two-thirds in soils (Dixon and Turner, 1991). Boreal, temperate and tropical forests are estimated to contain approximately two-thirds of the total, cycling 70 to 90 Pg C annually via photosynthesis, respiration and decomposition (Orr, 1993; Dixon *et al.,*1993). However, these estimates are based on a very course resolution definition of vegetation or ecosystem types (eg., boreal forest) and a very limited number of studies to determine C pools within each type. Recent regional-based estimates suggest global vegetation and soil C pools have been overestimated by 20 to 30% (Botkin and Simpson, 1993). These differences in estimates have largely been attributed to changes in forest area and forest destruction and degradation. Land-use changes, primarily deforestation within tropical latitudes, is currently releasing 1.2 to 3.2 Pg C annually (Brown, 1993). However, geographic variation do to climate, soils and species composition within any one vegetation or ecosystem type also introduces a large source of potential error. Research is needed to further define terrestrial C pools and fluxes within a geographic context, particularly in forest systems.

3. Estimating Future Trends in Terrestrial Carbon Dynamics : Processes Influencing Future Patterns of Terrestrial Carbon Storage

We can identify a number of critical processes which are likely to significantly modify the current distribution of above and below-ground C stocks. These processes include: 1) direct effects of increasing CO_2 on plants (eg., CO_2 fertilization), 2) response of plants to changing climate patterns, and 3) forest conservation and management including landuse practices such as afforestation and reforestation. We will briefly discuss these processes and then examine current attempts to incorporate these processes into a predictive framework for estimating future terrestrial C dynamics.

3.1 DIRECT EFFECTS OF CO_2

Numerous experimental studies have shown that enhanced atmospheric concentrations of CO_2 have the short-term effect of increasing both rates of photosynthesis and water-use efficiency (C fixed per unit of water transpired) at the level of the individual plant (Mooney *et al.,* 1991; Norby *et al.,* 1992; and see Bazzaz, 1990; Eamus and Jarvis, 1989 for reviews). If this short-term individual-level "fertilization" affect extrapolates to patterns of ecosystem net primary productivity, it would represent a large potential for increased terrestrial C storage. However, recent studies have shown that with prolonged exposure (eg., >1 year) some species acclimate to the higher CO_2 concentrations, with photosynthetic rates returning to values corresponding to those observed prior to treatment (eg., Oberbauer *et al.,* 1985; Smith *et al.,* 1987; Tissue and Oechel, 1987; Williams *et al.,* 1986). Secondly, data on ecosystem level response to elevated CO_2 from open-top chambers and free-air experiments are not consistent across different ecosystems (Curtis *et al.,* 1989; Ochel and Strain, 1985; Prudhomme *et al.,* 1984; see Bazzaz, 1990 for review), suggesting an interaction with other limiting factors (eg., N availability). Given the potential effects of elevated CO_2 on net primary productivity and terrestrial C storage, this area of research is critical to future projections and should represent a major focus in ecology over the next decade.

3.2 CLIMATE CHANGE

Rising atmospheric concentrations of CO_2 (and other greenhouse gases) over the next century have the potential to influence global climate patterns. Changes in global climate patterns as predicted by general circulation models (GCM's) for a doubled CO_2 atmosphere would have a major impact on the current global distribution of vegetation and soils (Emanuel et al., 1985; Smith et al., 1992a,b). Estimates of the changes in global patterns of vegetation and associated patterns of C storage for various GCM-derived climate change scenarios have been made and are discussed in detail in latter sections. These estimates represent equilibrium analyses, and a detailed analysis of the transient response of vegetation to changing climate patterns has yet to be undertaken. However, the qualitative nature of the transient response of terrestrial C storage can be estimated by examining the time scales associated with the basic ecological processes involved with the predicted shifts in vegetation distribution and associated patterns of C storage (Smith and Shugart, 1993a,b). Processes such as forest dieback resulting from increased aridity will occur on a time scale corresponding to the changes in climate conditions (decades to a century). Forest dieback and the associated increase in frequency of fire would represent a major net positive flux to the atmosphere (Neilson, 1993). In contrast, the increase in terrestrial C storage associated with natural forest expansion (eg., boreal forest into the current tundra region) will be limited by rates of species immigration into currently unforested areas. Estimates of forest migration rates from paleo-studies suggest a time-scale of centuries (Davis, 1984, 1989; Davis and Botkin 1985) for predicted shifts based on current analyses (eg., Emanuel et al., 1985; Smith et al., 1992a, b).

3.3 FOREST CONSERVATION AND MANAGEMENT FOR CARBON CONSERVATION AND SEQUESTRATION

Landuse changes, primarily deforestation within tropical latitudes, are currently releasing 1.2 to 3.2 Pg C annually (Brown, 1993). Future patterns of land-use, primarily forest clearing for agriculture in developing countries, could continue to be a major net positive flux of C to the atmosphere. However, these trends could be reversed through forest conservation and management for C conservation and sequestration. Forest ecosystems can be managed for the temporary storage of C (Harmon, 1990; Winjum et al., 1992; Brown et al., 1993). Forest and agroforest establishment and management practices can be grouped by three major functions: 1) maintain or 2) improve existing sinks and stores of C, and 3) expand forest areas that can serve as sinks of CO_2. We will briefly discuss these three groups of practices in a global context.

3.3.1 Maintenance of Existing Forest Carbon Sinks

Slowing deforestation within the tropical latitudes can conserve up to 1.6 Pg C annually. Although efforts to slow deforestation and degradation within the tropical latitudes have met with mixed success (Winterbottom, 1990), forest destruction in some regions of Brazil's Amazon basin dropped 20% in 1991 (Brown et al., 1992). Replacement of shifting agriculture by one hectare of sustainable agroforestry could potentially offset 5 to 20 ha of deforestation and consequently conserve existing C reservoirs (Sanchez and Benites, 1991). Based on the direct cost of providing economic incentives to practice sustainable forest management within tropical latitudes it has been estimated that C conservation can be achieved for $0.20 to 3.50 Mg C (Dixon *et al.*, 1993c; Winjum *et al.*, 1992; Schroeder and Ladd, 1991). For example, the Forest Village Project in Thailand, which provides incentives for shifting cultivators to establish agroforest systems and reforest degraded lands, has conserved or sequestered approximately 1 Pg C over the past 20 years.

Globally, temperate and boreal forest fires represent a significant source C to the atmosphere. Within Russia, 4 to 8 million ha of boreal forest burn annually, contributing 0.3 to 1.2 Pg of direct and indirect C emissions to the atmosphere (Krankina and Dixon, 1993). Over 40% of the boreal forests in the former Soviet Union have no fire monitoring or protection system. It is estimated that fire management and silvicultural practices could be employed to conserve C at a cost of $0.5 to 1.00 Mg.

Soils are significant reservoirs of C and conservation practices could be employed to reduce greenhouse gas emissions. Soil management can result in significant changes in soil C (Schlesinger, 1990). Management practices to conserve forest soil C include: 1) forestation to reduce erosion, 2) maintain or improve soil fertility using amendments, 3) concentrate tropical agriculture and reduce shifting agriculture, 4) remove marginal lands from agricultural production, and 5) retain forest litter and debris after silvicultural or logging activities.

3.3.2 Expansion of Forest Area and Potential Carbon Sinks

Several forest and agroforest establishment and management practices could be employed to expand forest C sinks in all latitudinal belts (Dixon *et al.*, 1993a,b,c). These practices include urban tree planting, implementing agroforestry and/or restoring forests on degraded lands and watersheds, establishing plantations on harvested or abandoned lands, and protection of existing forests from pathogens, insects and fires, or further degradation by humans. Within boreal, temperate and tropical latitudes C storage in plantations and natural forests range from approximately 25 to 250 Mg/ha, respectively. Carbon sequestration can initially be achieved via forestation and establishment of agroforestry systems for a median initial cost of $10 Mg C within tropical latitudes (Dixon *et al.*, 1993c). Agroforestry, a traditional land-use practices throughout the tropics, provides fuel, food, fiber, fodder, medicine and other goods and services. As a result of these benefits, the associated actual costs of C sequestration may be negative (Gregerson *et al.*, 1989). In boreal and temperate regions, establishment of forest plantations (poly or monocultures) can sequester C at $1 to 60 Mg C, with a median cost in Russia of $3 Mg C.

Expansion of the boreal, temperate and tropical forest systems is a large undertaking but national and international proposals have been developed (Winjum *et al.*, 1992). Various estimates have been made of land technically suitable for this effort, identifying its location and biophysical characteristics (Grainger, 1988; Houghton, 1990). An estimated 100 million ha of formerly harvested sites and abandoned lands are available for forestation in the former Soviet Union (Krankina and Dixon, 1993). Houghton (1990) identified 3,125 million ha available for forestation within the tropical latitudes. This land is climatically suitable, does not overlap with other productive land uses and includes: 625 million ha of degraded grazing land in Latin America, 625 million ha of degraded grasslands in south Asia and 1,875 million ha in Africa.

Grainger (1988) estimated that almost 5 billion ha of degraded tropical land is available for forest replenishment including: 544 million ha of deforested watershed, 2,068 million ha of degraded drylands, 1,258 million ha of forest fallows, 848 million ha of logged rainforests. Almost 75% of the 5 billion ha are suitable for plantation or agroforestry with the land being almost equally divided between Asia, Africa and Latin America. The land area available is 30 to 40 times the area of tropical forest plantations established in 1990. Proposals have been developed to implement a 50 yr forestation program on technically suitable lands where obstacles are minimal, while technical, social and economic constraints are confronted in other regions (Winjum et al., 1992).

4. Estimating Future Trends in Terrestrial Carbon Dynamics : Available Models for Examining The Dynamics of Terrestrial Carbon Storage

The flux of C between the terrestrial and atmospheric pools is largely a function of the uptake of CO_2 in photosynthesis by terrestrial plants and the release of CO_2 by plant and microbial respiration. Rates of photosynthesis and microbial decomposition are directly influenced by climate and substrata and therefore the rates of C sequestration and residence times differ geographically. These geographical relationships between climate and C storage form the fundamental basis for estimating current pools and predicting future C dynamics. It is our belief that predictive models of the terrestrial C cycle must begin with geographic specificity. We can then evaluate the impacts of various processes on the current patterns of storage.

There have been a wide variety of models developed to explore the response of vegetation and soils to environmental variation. These models range from purely statistical to models which simulate basic ecophysiological and demographic processes. Any number of these models have been used to examine patterns of C dynamics for a given site or region, but in this review we will focus solely on those models which have been used (or can easily be used) to provide global coverage.

4.1 BIOGEOGRAPHICAL MODELS

Perhaps the simplest of models for relating vegetation pattern to climate at a global scale is the approach of climate-vegetation classification. Assuming that the broad-scale patterns of vegetation (eg., biomes) are essentially at equilibrium with present climate conditions, one can relate the distribution of vegetation or plant types with biologically important features of the climate. Global bioclimatic classification schemes (von Humbolt, 1867; Grisebach, 1838; Koppen, 1900, 1918, 1936; Thornthwaite, 1931, 1933, 1948; Holdridge, 1947, 1959; Troll and Paffen, 1964; Box, 1981; Prentice, 1990) are essentially climate classifications defined by the large-scale distribution of vegetation. Although similar in concept, the wide variation in both the terminology used to describe categories of vegetation and the climate variables defined as important in influencing plant pattern make comparison among the models difficult. Bioclimatic classification models have a history of application in simulating the global distribution of vegetation under changed climate conditions, both past climatic conditions associated with the last glacial maximum (Manabe and Stouffer 1980; Hansen et al., 1984; Prentice and Fung, 1990) and predictions of future climate patterns under conditions of doubled CO_2 (Emanuel et al., 1985; Prentice and Fung, 1990; Smith et al., 1992a,b).

One of the most widely used of the bioclimatic classifications at a global scale is the model of Holdridge (1947, 1967). The Holdridge Classification is a bioclimatic classification relating the distribution of major ecosystem complexes (referred to as life zones) to the climate indices of biotemperature, annual precipitation and a ratio of potential evapotranspiration to annual precipitation (PET ratio). The classification has proven useful in estimating the potential impacts on vegetation distribution resulting from changes in global climate patterns as predicted by general circulation models of the earth's atmosphere for a doubling of CO_2 (Emanuel *et al.*, 1985; Smith *et al.*, 1992a,b). The model has also been combined with estimates of C storage in both soils and vegetation to estimate current patterns of potential C storage under both current and changed climate condition (Smith *et al.*, 1992a). Prentice and Fung (1990) used a modified version of the Holdridge model to estimate both past and future patterns of potential C storage based on global climate patterns simulated by the GCM developed at the Goddard Institute for Space Studies (Hansen *et al.*, 1988).

A difficulty with zonal concepts like the Holdridge Classification is that the vegetation is defined as an aggregate vegetation type or association. As such, the vegetation type responds to changes in climate patterns as a single unit. However, terrestrial ecosystems are composed of numerous species which can respond individualistically to changing environmental conditions (Davis, 1984) and whose distributions often cover more than one ecosystem or zone. Although it is currently impossible to construct bioclimatic models at a global scale where the species forms the basic unit of vegetation description, a number of efforts have been made to define classifications based on a functional scheme for aggregating species into groupings based on similarity in ecology and physiognomy.

Box (1981) developed a classification of "plant functional types" based on a combination of ecology and taxonomy. The classification defines 87 plant types whose potential global distribution is defined by correlating eight climatic indices to their current distributional limits. Using a rule based system which defines patterns of dominance for a given location (and climate), the model achieves a considerable detail in describing ecosystem structure as related to climate. However, the complexity of this scheme has also imposed a limit on its potential to be parameterized appropriately for all plant types and climatic indices. Some of these problems have been overcome by dramatically reducing the number of plant types defined and the selection of climatic variables whose influence on plant distribution have a more mechanistic interpretation. Two such models are BIOME (Prentice *et al.*, 1992) and MAPSS (Neilson, 1993).

Both BIOME and MAPSS rely on relatively simple methods to derive bioclimatic "envelopes" from long-term meteorological records. Both models define a set of plant types rather than vegetation complexes, allowing for the ecosystem or biome type to be defined by the set of plant types which can occur at any location. These models depend on calculating an index of plant moisture deficit based on a seasonal solution to potential and actual evapotranspiration. In addition they use a number of indices of seasonality, as well as values for absolute temperature tolerances for the plant types. Both models produce global patterns of vegetation cover which agree well statistically with the actual vegetation map of Olson *et al.* (1983). Like the Holdridge Classification, both BIOME and MAPSS have been used to estimate current C stocks for terrestrial vegetation and soils (Cramer and Solomon, in press; Prentice *et al.*, in press; Neilson, 1993)

Biogeographical models often have been, but do not need to be, limited to potential natural vegetation. There is a convincing body of data about the potential distribution of anthropogenically derived ecosystems, such as agronomic and forest crops, as a function of bioclimatic constraints. This type of information can be embedded into a predictive model of potential natural vegetation, thereby giving the potential for human land use as well (Cramer and Solomon, in press; Leemans, 1993). Ecosystems which are affected by slight or severe human impact cannot be described by biogeographical models alone, but need the inclusion of a specific landuse models which must be driven by socioeconomic variables such as population growth.

One major problem in the application of most bioclimatic models to assessing the potential response of vegetation to a CO_2-induced climate change is their inability to address the direct response of CO_2 on vegetation (Norby et al., 1992; Mooney et al., 1991) Models that do not simulate stomatal processes must find indirect, empirical approaches for incorporating the effects, as is done in BIOME. Models, such as MAPSS can incorporate increased WUE directly through manipulation of stomatal conductance. It is not yet certain which of these approaches produces accurate results.

The use of bioclimatic models to estimate terrestrial C dynamics is a two step process since the models do not directly simulate C pools. The models are used to define potential patterns of vegetation and associated soil properties based on simple climate indices. The calculation of C pools are done by multiplying the areal extent of each cover type (eg., vegetation type, ecosystem type, biome, life zone) by some estimate of C storage in vegetation and soils (see: Prentice and Fung, 1990, Smith et al., 1992a). Generally these estimates are solely dependent on the vegetation or biome type and do not vary geographically within any one type (eg., all tropical rain forests have the same value). This approach is in contrast to process-based ecosystem models which simulate patterns of net primary productivity for a given vegetation and climate.

4.2 MODELS OF NET PRIMARY PRODUCTIVITY

Several global productivity models have been developed and are being used to examine potential effects of both increasing atmospheric concentrations of CO_2 and associated prediction of climate change on global patterns of net primary productivity and biogeochemical cycles. Three examples are the Terrestrial Ecosystem Model (TEM) (Raich et al., 1991), the General Ecosystem Model (GEM) (Melillo et al., in press; Rastetter et al., 1991) and CENTURY (Parton et al., 1988). These models simulate the processes of plant energy and C balance at the canopy level. Given data on the site vegetation type (and characteristic physiological parameters), leaf area and soil water status, these models can predict vegetation growth and soil C and nutrient dynamics. The explicit consideration of ecosystem C dynamics in these models allows them to simulate the changes in net C flux for a given location, providing estimates of change in net primary productivity under changing climate conditions.

Melillo et al. (1993) recently employed GEM to estimate global patterns of net primary productivity for current and future climate conditions. GEM estimated that over half of global net primary productivity occurs in the tropical latitudes, mostly in tropical evergreen forests. Future responses of tropical and temperate biomes are predicted to be strongly influenced by CO_2, ambient temperature and nutrient availability (eg., N).

The explicit consideration of photosynthesis and transpiration allows this class of models to include estimates of the direct effects of CO_2 on net primary productivity and water use efficiency. However, they are not able to simulate changes in the composition and structure in response to changing environmental conditions. This is a major limitation to the use of these models for long term projections. Over the long term, changes in net primary productivity might be accompanied by a change in vegetation structure and composition comparable to those predicted by the biogeographical models.

4.3 OVERVIEW

In general, the two classes of models (i.e., biogeographical and net primary productivity) are addressing two sets of processes operating at different temporal scales. The models of net primary productivity require input defining the current cover of vegetation, the characteristics of which interact with climate and site conditions to determine the net flux of CO_2 with the atmosphere. The advantage of these models is that they are able to explicitly simulate the short-term dynamics of CO_2 flux from the terrestrial surface in response to CO_2 and changing climate patterns. The major constraint of these models is that they do not consider processes which define the longer-term response of vegetation structure and composition to these variables (CO_2 and climate). These longer-term changes in structure and composition are both a function of the changing patterns of net primary productivity as well as a constraint on future patterns of net CO_2 flux.

In contrast to the models of net primary productivity, the biogeographical models function as correlations between climate and vegetation/soils. The biogeographical models do not address the temporal dynamics of vegetation change or the processes by which those changes come about. In that respect they represent equilibrium solutions. Models which are able to examine both the short-term patterns of net primary productivity and the longer-term changes in vegetation and composition as a function of climate have been developed for a variety of ecosystems (eg., forest gap models: see Shugart, 1984). However, their utility to provide even region coverage is currently limited (Prentice *et al.*, 1992), because of the spatial scale at which they operate and problems with parameterization.

5. Estimating Potential Carbon Storage under Present and Future Climates

A critical question regarding the role of the terrestrial surface in the global C cycle is "How will patterns of terrestrial C storage respond to elevating atmospheric concentrations of CO_2 and associated changes in global climate patterns?" A number of studies have examined the sensitivity of terrestrial C storage to changes in global climate, both past and future (Lashof, 1987; Sedjo and Solomon, 1989; Prentice and Fung, 1990; Smith *et al.*, 1992a; King and Neilson, 1992; Prentice *et al.*, in press; Cramer and Solomon, in press; Smith and Shugart, 1993; Neilson, 1993). All of these studies share a common methodology, combining a mapping system for global patterns of vegetation with C density estimates for vegetation and soils relating to the classification units (i.e., vegetation or ecosystem types). The global distribution of vegetation is mapped by applying one of the biogeographical modelling approach outlined above to global databases of climate, soils and topography. By overlaying the predicted patterns of climate change as simulated by various global circulation models for past and future conditions (eg., $2XCO_2$ atmosphere) on the current global climate databases, the potential global distribution of vegetation and associated soils is mapped for the new climate conditions (eg., Emanuel *et al.,* 1985; Smith *et al.,* 1992a, b). Carbon density for each vegetation type and associated soil (i.e., classification unit) are estimated from published sources (eg., Olson *et al.,* 1983; Post *et al.,* 1982) and these densities are multiplied by the areal estimates for each of the vegetation types (classification units) as predicted by the biogeographical models.

5.1 CURRENT ESTIMATES

Current estimates of potential C storage in terrestrial vegetation and soils based on the approach discussed above are presented in Table 1. All three studies (Prentice and Fung, 1990; Smith et al., 1992; Cramer and Solomon, in press) use estimates of C in above-ground biomass from Olson *et al.* (1983) and estimates of soil C from Post *et al.* (1982). Although the three studies use different biogeographical models for mapping vegetation and soils distribution as a function of climate, the estimates of total C storage in vegetation and soils are in close agreement. Prentice and Fung (1990) used a modified version of the Holdridge Model (see Biogeographical Models) which defines 14 biome or ecosystem types (Prentice 1990). The study by Smith *et al.* (1992a) used the complete 39 class Holdridge Life Zone Classification. The estimates of Cramer and Solomon (in press) are based on the BIOME model of Prentice *et al.* (1992) which defines 14 biome or ecosystem types.

The studies by Prentice and Fung (1990) and Smith *et al.* (1992a) consider only potential C storage, in that the global vegetation maps which form the basis of the estimates (define areal coverage of ecosystems) consider only natural vegetation. These estimates therefore do not take into account lands in agricultural production, and in general overestimate C storage in terrestrial vegetation when compared to the values presented by Olson *et al.* (1983) (see Table 1). Cramer and Solomon (in press) provide estimates for conditions of natural vegetation only

Table 1. Estimates of current terrestrial carbon storage (Values in Gt).

Study	Vegetation	Soils	Total
[1]Olson et al. (1983)	561.3 (461-665)[3]		
[1]Post et al. (1982)		1308.6	
[2]Prentice and Fung (1990)	748	1143	1891
[2]Smith et al. (1992)	737.2	1158.5	1895.7
[2]Cramer and Solomon (in press)	754 (457-574)[4]	1367	2121

[1] Based on field based estimates from various ecosystem studies
[2] Based on estimates from Olson et al. (1983) and Post et al. (1982) combined with potential vegetation maps from biogeographical model (see text)
[3] Range for estimates
[4] With extensive and sparse agriculture

(i.e., potential cover) and with sparse and extensive agronomic cultivation. Their estimate of carbon storage without agronomic cultivation is similar to the other two studies. Estimates with agronomic cultivation agree well with the range of values presented by Olson *et al.* (1983).

Although the estimates of total terrestrial C storage are rather robust across mapping systems (i.e., biogeographical mode used), this does not speak for the quality or accuracy of the estimates, nor is landuse considered. As stated above, the estimates of C content for the various classification units used to describe vegetation and soils were all derived from Olson *et al.* (1984) and Post *et al.* (1982). Although these estimates represent some of the only data available at a global scale, they are based on a limited number of samples and geographical regions. A great deal of work is needed to provide more suitable estimates on a regional basis.

5.2 ESTIMATES UNDER FUTURE CLIMATES

A number of studies have used the approach outlined above to examine the potential impacts of a global climate change resulting from a doubling of atmospheric CO_2 on the potential of the terrestrial surface to store C. Lashof (1987) developed a classification system of 14 ecosystem types based on the vegetation classification of Olson *et al.* (1983). The global distribution of these types were then related to mean winter and summer temperature and precipitation. The resulting model was used to examine potential shifts in the distribution of ecosystems and associated changes in terrestrial C storage under climate change scenarios from the National Center for Atmospheric Research (NCAR) (Washington and Meehl, 1984), Geophysical Fluid Dynamics Laboratory (GFDL) (Wetherald and Manabe, 1986) and Goddard Institute for Space Studies (GISS) (Hansen *et al.*, 1984) general circulation models.

The GISS scenario predicted a 28 Gt increase in C storage. In contrast, C storage declined in both the GFDL (-64 Gt) and NCAR (-40 Gt) scenarios. Soil C declined under all three scenarios as a result of the transition from boreal forest to grassland. For the GISS scenario, this decrease was offset by a significant increase in forest cover in the tropics. Above-ground C decreased in the other two scenarios as a result of forest decline in the warm temperate, subtropical and tropical zones.

Prentice and Fung (1990) used a modified version of the Holdridge Classification (Prentice 1990) to examine the impacts of perturbed climates on the distribution of global vegetation and associated changes in terrestrial C. Using predicted climate patterns from the Goddard Institute for Space Studies (GISS) general circulation model (Hansen *et al.*, 1988) for conditions of $2XCO_2$, Prentice and Fung found a 235 Gt increase in terrestrial C storage, resulting in a strong negative feedback on elevated atmospheric CO_2 (128 ppm). The increase in C storage was primarily associated with a large increase (75%) in the extent of tropical rainforest under the changed climate conditions.

Smith *et al.* (1992a) examined the potential impacts of CO_2 induced climate change on terrestrial C storage using the Holdridge Life Zone Classification and four climate change scenarios derived from general circulation models: GISS (Hansen *et al.*, 1988), GFDL (Manabe and Wetherald, 1987), UKMO (Mitchell, 1983) and OSU (Schlesinger and Zhao, 1988). All four scenarios showed an increase in potential terrestrial C storage ranging from a 0.4% to 9.5%: GISS (147 Gt), GFDL (38 Gt), UKMO (8.5 Gt) and OSU (180 Gt). Carbon storage in above-ground biomass rose in all four scenarios, but soil C declined under the GFDL and UKMO scenarios. The increases in C storage were primarily a function of the poleward shift of the forested zones, with an increase in the areal extent of tropical forests and a shift of the boreal forest zone into the region currently occupied by tundra.

Neilson (1993) used the MAPSS model (Biogeographical Models above) to examine the potential changes in terrestrial C storage under four GCM scenarios based on $2XCO_2$ simulations (GISS, GFDL, UKMO and OSU). In contrast to the results of Prentice and Fung (1990) and Smith *et al.* (1992), Neilson found a decrease in potential C storage under the four scenarios investigated. This decline in C is a result of predicted patterns of forest decline, primarily in the subtropical and tropical regions.

Cramer and Solomon (in press) used the BIOME model (Prentice *et al*, 1992) discussed above (Biogeographical Models) to examine potential changes in terrestrial C storage under a climate change scenarios based on the GFDL and OSU general circulation models. Unlike the above analyses which examined only potential vegetation cover, Cramer and Solomon included agroecosystems in their classification. Their results were similar to Smith *et al.* (1992), predicting an increase in potential C storage under the two GCM scenarios (OSU 16 Gt, GFDL 76 Gt) when agricultural lands were excluded. In contrast to Smith *et al.*, Cramer and Solomon predict a decline in soil C under the OSU scenario. However, as with the GFDL scenario, increases in above-ground C more than offsets this decline giving a net positive gain in total C storage. In the case where agricultural lands are included in the classification, potential C storage declined under both scenarios.

With the exception of Neilson (1993), all of the analyses discussed above agree qualitatively in their predictions of an increase in potential terrestrial C storage under the climate change scenarios investigated. However, the quantitative predictions differ significantly. There are a number of sources of variation responsible for the differences in predicted future C stores in the above analyses. First, the differences among scenarios are a direct result of the differences in predicted global climate patterns and associated predictions for vegetation distribution. The largest difference among the scenarios is the greater degree of warming and associated mid-continental drying predicted by the GFDL and UKMO scenarios.

The second source of variation is in the differences among the models in their classification systems (for vegetation) and the associated estimates of C densities (above-ground biomass and soils). These two are interrelated in that the C densities have to be interpreted from the original sources (and classifications) for the specific vegetation classification system included in the model. As shown earlier (Table 1), despite the differences in the classifications associated with the different biogeographical models, the estimates of current potential C pools do not vary to a large degree. The differences arise from the predicted patterns of vegetation change under the climate change scenarios. The proportion and direction of vegetation change predicted for a given climate change will vary dependent on the classification system used and is particularly sensitive to the number of categories included in the classification (Smith *et al.*, 1992a). In general, the fewer the number of categories the greater the change in climate necessary to produce a shift in classification, and given a shift there will be a larger associated change in C density.

The analyses discussed above all represent equilibrium patterns of potential C storage resulting from a changed global climate pattern. As was noted in the discussion of biogeographical models, these models do not address the temporal or spatial dynamics of the transitions in vegetation and soils required to achieve the new equilibrium conditions. These transitional dynamics would be dependent on the rate of climate change and the rates of key ecological processes controlling the vegetation and soils dynamics associated with the predicted shift in cover (eg., tundra to boreal forest). These processes include vegetation dieback, successional replacement, species immigration, decomposition and soil formation. Despite the limitations presented by these equilibrium analyses, initial estimates of the transient dynamics of terrestrial C storage have been made by categorizing the shifts in vegetation and soils predicted by the equilibrium analyses and examining the rates associated with the ecological processes required for those shifts to occur (Smith and Shugart, 1993a,b). Initial results suggest that although the equilibrium analyses predict a net increase potential C storage, the transient dynamics necessary for those new equilibria to be achieved would result in a significant net flux of C from the terrestrial surface following a climate change.

5.3 ESTIMATES OF C FLUX UNDER FUTURE LANDUSE SCENARIOS

The timing and magnitude of future impacts of terrestrial C pools and flux will depend on both landuse and natural resource management policies, and environmental variables such as climate change and biogeochemical cycling. Future changes in landuse patterns could overwhelm any change in terrestrial C dynamics associated with CO_2 fertilization and climate change (eg., warming and drying). Demographic projections of human populations during the next few decades suggest forest and agroecosystem disturbance will be significant (Brown *et al.,* 1993).

The estimates of terrestrial C flux resulting from projected climate change do not consider the role of improved land management in mitigating C sources. If global climate change occurs, the boreal and temperate zone forest and agroecosystem vegetation will be in transition. The potential to manage and adapt terrestrial ecosystems to help minimize greenhouse gas emissions is significant (NAS, 1991; Sampson *et al.,* 1993). Simulation modes which consider landuse patterns, resource management options and climate change are needed to develop and test strategies for greenhouse gas mitigation options.

6. Integrating a Dynamic Terrestrial Surface Into Global Carbon Models

Traditionally, global C models have specified the dynamics between different compartments (atmosphere, oceans, and biosphere) and simulated the fluxes by balancing the different budgets globally (eg., Bolin *et al.,* 1981; Emanuel *et al.,* 1984; Goudriaan and Ketner, 1984). These C models are currently being used to project future trajectories of GHGs' emissions (eg., Rotmans, 1990). Recently, the major limitations of this traditional budget approach has been reviewed (Post *et al.,*1992; Solomon and Shugart, 1993). The major criticism of these highly aggregated budget models is that they do not account for significant regional differences within the terrestrial biosphere. The inclusion of geographic variation should yield better estimates of fluxes, sources and sinks, as well as improve our understanding of feedbacks. Recently several geographically explicit models have been developed (e.g. Raich *et al.,* 1991; Janecek *et al.,* 1989; Kohlmaier, 1993). However, these models do not take changes in landuse into account.

The earliest global C model to considered the dynamics of landuse was the Osnabrück model (Esser, 1991). Changes in landuse in the Osnabruck model are driven mainly by prescribed demographic processes. In contrast, in the more recent IMAGE 2.0 model (Alcamo *et al.,* 1993), landuse dynamics are lined to socio-economic developments with respect to energy, industrial and agriculture. The objective of the model is to evaluate a large array of policy and scientific global change issues. The model consist of three main components (Figure 1 of Vloedbeld and

Leemans, 1993): 1) the energy-industry component simulates regional specific energy and industrial demands, which results in the annual emissions of GHGs; 2) the terrestrial biosphere component simulates regional specific demands for agricultural products, like fibre, timber and food; and 3) the third component simulates the interactions between the atmosphere and oceans. The processes that are included are atmospheric chemistry, heat transport and C uptake by the oceans, and atmospheric circulation. It uses the emissions of GHGs from the other components and some specific land cover characteristics, like albedo, as inputs. It simulates a transient climate change, including several system feedbacks, but can also be run with climate-change results from different GCMs. The climatic change and final GHGs concentration immediate feedforward into the two other components and influence energy demand (heating/cooling) and land cover. Socio-economic impacts are therefore an intrinsic part of this model.

Initial results with the IMAGE 2.0 framework show the importance of combinations of different feedback mechanisms (Vloedbeld and Leemans, 1993). The magnitude and both positive and negative feedbacks are considered. The net result of all direct CO_2-effects (C fertilization, WUE) and climatic change (temperature response on growth and soil respiration, vegetation shifts) are significantly positive. The biosphere acts as a small sink under these conditions and until 2050 12.4 Gt can be sequestered. However, this additional C storage is not preserved, when changing land cover is taken in to account according to the IPPC business -as-usual scenario (Houghton, 1992), the biosphere act as a strong additional source.

7. Conclusions

The lack of consistent data on current patterns of terrestrial C storage (both pool sizes and current fluxes) presents a major limitation on our present understanding of the role of the terrestrial surface in the global C cycle. Landuse patterns and their current influence on terrestrial C pools and fluxes are difficult to estimate. The development and application of simulation models which can address the impacts of future landuse patterns on terrestrial C dynamics have only recently been initiated at a continental to global scale (Brown *et al.*, 1993; Hall *et al.*, 1993). Moreover, our limited understanding of the potential impacts of rising concentrations of atmospheric CO_2, both through direct effects on net primary productivity and indirectly through its influence on the global climate system, further limits our ability to make predictions of the future role of the terrestrial surface in the global C cycle. Given the overwhelming importance of understanding the global C cycle in the face of rising atmospheric CO_2 concentrations, research towards reducing these uncertainties is urgent.

Current analyses of the potential impacts of a global climate change resulting from increasing atmospheric concentrations of GHG's on patterns of terrestrial C storage suggest that in the equilibrium case, there is a potential for increased storage. This increase in potential C storage in the earth's vegetation and soils would represent an important negative feedback to rising atmospheric concentrations of CO_2. However, transient analyses designed to examine the dynamics of terrestrial C storage associated with achieving the new equilibrium conditions (i.e., changes in vegetation distribution and associated soils) suggest that the terrestrial surface may act as a major C source for a period following a climate change. A source of the same magnitude as that from current fossil fuel emissions. One of the major uncertainties with these analyses is that they do not address the possible direct effects of CO_2 on ecosystem dynamics. The potential for increased net primary productivity and water-use efficiency under rising levels of CO_2 could offset some of the predicted declines in forest distribution responsible for the net positive flux from the terrestrial surface in these transient analyses. An understanding of the direct effects of increasing landuse and CO_2 on natural ecosystems is essential to predicting the response of the earth's vegetation to possible changes in climate and the resulting future patterns of terrestrial C storage.

8.References

Alcamo, J.B., de Haan, B., de Vries, B., Klein-Goldewijk, K., Kreileman, E., Krol, M., Leeman, R., Vleoedbeld, M. and Zuidema, G.: 1993, IMAGE 2.0: A comprehensive tool to evaluate policy and scientific issues related to global change, *Water, Air and Soil Pollution* (in press).

Bazzaz, F.A.: 1990, *Ann. Rev. Ecol. Syst.* **21**, 167-196.

Bolin, B., Bjorkstrom, A., Keeling, C.D., Bacastow, R. and Siegentaler, U.: 1981, Carbon cycle modelling, in: Bolin, B. (ed), *Carbon Cycle Modelling*, Wiley and Sons, pp. 1-28.

Botkin, D.B. and Simpson, L.G.: 1993,*Water, Air and Soil Pollution* (this volume).

Box, E. O.: 1981, *Macroclimate and Plant Forms: An Introduction to Predictive Modeling in Phytogeography*, Junk Publishers, The Hague.

Brown, S., Lugo, A.E. and Iverson, L.R.: 1992, *Water, Air and Soil Pollution* **64**, 139-155.

Brown, I.F., Nepstad, D.C., Pires, I.O., Luz, L.M. and Alechandre, A.S.:, 1992, *Environmental Conservation* **19**, 307-315.

Brown, S.: 1993,*Water, Air and Soil Pollution* (this volume).

Cramer, W. & Leemans, R. (1993). Assessing impacts of climate change on vegetation using climate classification systems, in: Solomon, A.M. and Shugart, H.H. (eds.), *Vegetation Dynamics and Global Change*, Chapman and Hall, New York, pp. 190-217.

Cramer, W. and Solomon, A. M.: 1993, Climatic classifcation and future global redistribution of agricultural land, *Journal of Climate Research*, (in press).

Curtis, P.D., Drake, B.G. and Whigham, D.F.: 1989, *Oecologia* **78**, 297-301.

Curtis, P.D., Drake, B.G., Leadley, P., Arp, W. and Whigham, D.F.: 1989, *Oecologia* **78**, 20-26.

Davis, M.B.: 1984, Climatic instability, time lags and community disequilibrium, in:

Diamond, J. and Case, T.J. (eds) *Community Ecology,* Harper and Row, New York, pp. 269-284.

Davis, M.B.: 1989, *Climatic Change* **15**, 75-82.

Davis, M.B. and Botkin, D.B.: 1985, *Quat. Res.* **23**, 327-340.

Dixon, R.K. and Turner, D.P.: 1991, *Environmental Pollution* **72**, 245-262.

Dixon, R.K., Andrasco, K.J., Sussman, F.G., Lavin, M.A., Trexler, M.C. and Vinson, T.S.: 1993a, *Water, Air and Soil Pollution* (this volume).

Dixon, R.K., Winjum, J.K. and Schroeder, P.E.: 1993b, Conservation and sequestration of carbon: The potential of forest and agrosystem management practices. *Global Environmental Change* (in press).

Dixon, R.K., Winjum, J.K., Andrasko, K.J., Lee, J.J and Schroeder, P.E.: 1993c, Integrated systems: Assessment of promising agroforest and alternative land-use practices to enhance carbon conservation and sequestration, *Climatic Change* (in press).

Eamus, D. and Jarvis, P.G.: 1989, *Adv. Ecol. Res.* **19**, 1-57.

Emanuel, W.R. Killough, G.G., Post, W.M. and Shugart, H.H.: 1984, *Ecology* **65**, 970-983.

Emanuel, W. R., Shugart, H. H. & Stevenson, M. P.: 1985, *Climatic Change* **7**, 29-43.

Esser, G.: 1991, Osnabruck Biosphere model: structure, construction, results, in: Esser, G. and Overdieck (eds), *Modern Ecology: Basic and Applied Aspects*, Elsevier, pp. 679-709.

Goudriaan, J. and Ketner, P.: 1984,*Climatic Change* **6**, 167-192.

Grainger, A.: 1988, *International Tree Crops Journal* **5**, 31-61.

Gregerson, H., Draper, S. and Elz, D.: 1989, *People and trees: The role of social forestry in sustainable development*, EDI Seminar Series, World Bank, Washington, D.C.

Grisebach, A.: 1838, *Linnaea* **12**, 159-200.

Hall, C.: 1993, *Water, Air and Soil Pollution* (this volume).

Hansen, J., Lacis, A., Rind, D., Russel, G., Stone, P, Fung, I., Reudy, R., and Lerner, J.: 1984, Climate sensitivity: Analysis of feedback mechanisms', In: Hansen, J. and Thompson, R. (eds), *Climate Processes and Climate Sensitivity, Geophysical Monogr.* **29**, American Geophysical Union, Washington, D.C.

Hansen, J., Fung, I., Lacis, A., Rind, D., Russell, G., Lebedeff, S., Reudy, R. & Stone, P.: 1988, *J. Geophys. Res.* **93**, 9341-9364.

Harmon, M.E., Ferrell, W.K. and Franklin, J.F.: 1990, *Science* **247**, 699-702.

Holdridge, L. R.: 1947, *Life Zone Ecology,* Tropical Science Center, San José, Costa Rica.

Holdridge, L.R.: 1959, *Science* **130**, 572.

Houghton, R.A., Hobbie,J.E., Melillo, J.M., Moore, B., Peterson, B.J., Shaver, G.R. and Woodwell, G.M.: 1983, *Ecol. Monogr.* **53**, 235-262.

Houghton, J.T., Jenkins, G.J. and Ephraums (eds): 1990, *Climate Change - The IPCC Scientific Assessment*, Intergovernmental Panel on Climate Change, Cambridge University Press.

Houghton, J.T., Callander, B.A. and Varney, S.K. (eds): 1992, *Climate Change 1992 - The Supplimentary Report to the IPCC Scientific Assessment*, Intergovernmental Panel on Climate Change, Cambridge University Press.

Humbolt, A. von: 1867, *Ideen zu einem Geographie der Pflazen nebst einem naturgemalde der Tropenlander*, Tubingen, FRG.

Janecek, A., Benderoth, G., Ludeke, M.K.B., Kinderman, J. and Kohlmeier, G.H.: 1989, *Ecological Modelling* **49**, 101-124.

King, G.A. and Neilson, R.P.: 1992, *Water, Air and Soil Pollution* **64**, 365-383.

Kohlmaier, M.: 1993, *Water, Air and Soil Pollution*(this volume).

Koppen, W.: 1900, *Geogr. Z.* **6**, 593-611.

Koppen, W.: 1918, *Petermanns Geogr. Mitt.* **64**, 193-203.

Köppen, W.: 1936, Das geographische System der Klimate, in: Köppen, W. and Geiger, R. (eds.), *Handbuch der Klimatologie,* Gebrüder Borntraeger, Berlin, pp. 46.

Krankina, O.N. and Dixon, R.K.: 1993, *Forest management options to conserve and sequester terrestrial carbon in the Russian Federation,* World Resources Review (in press).

Lashof, D.A.: 1987, *The Role of the Biosphere in the Global Carbon Cycle: Evaluating Through Biospheric Modeling and Atmospheric Measurement,* Ph.D. dissertation, University of California, Berkley.

Manabe, S. and Stouffer, R.J.: 1980, *J. Geophy. Res.* **85**, 5529-5554.

Manabe, S. and Wetherald, R.T.: 1987, *J. Atm. Sci.* **44**, 1211-1235.

Melillo, J.M., McGuire, A.D., Kicklighter, D.W., Moore, B., Vorosmarty, C.J. and Schloss, A.L.: 1993, Global climate change and terrestrial net primary production, *Nature* (in press).

Mitchell, J.F.B.: 1983, *Q. J. Roy. Met. Soc.* **109**, 113-152.

Mooney, H.A., Drake, B.G., Luxmore, R.J., Ochel, W.C. and Pitelka, L.F.: 1991, *Bioscience* **41**, 96.

Neilson, R.P.: 1993, *Water, Air and Soil Pollution* (this volume).

Norby, R.J., Gunderson, C.A., Wallschleger, S.D., O'Neill, E.G. and McCracken, M.K.: 1992, *Nature* **357**, 322.

Oberbauer, S.F., Strain, B.R. and Fetcher, N.: 1985, *Physiologia Plantarum* **65**, 352-356.

Ochel, W.C. and Strain, B.R.: 1985, Native species responses to increased carbon dioxide concentration, in: Strain, B.R and Cure, J.D. (eds), *Direct Effects of Increasing Carbon Dioxide on Vegetation*, U.S. DOE NTIS, Springfield, Virginia.

Olson, J. and Watts, J. A.: 1982, *Major World Ecosystem Complexes Ranked by Carbon in Live Vegetation*, NDP-017, Carbon Dioxide Information Center, Oak Ridge.

Orr, J.:1993, *Water, Air and Soil Pollution* (this volume).

Parton, W.J., Stewart, J.W.B. and Cole, C.V.: 1988, *Biogeochemistry* **5**:109-131.

Post, W.M., Emanuel, W.R., Zinke, P.J., and Stangenberger, A.G.: 1982, *Nature* **298**, 156-159.

Post, W.M., Chavez, F., Mulholland, P.J., Pastor, J., Peng, T.-H., Prentice, K. and Webb III, T.: 1992, Climatic feedbacks in the global carbon cycle, in: Dunnette, D.A. and O'Brien, R.J. (eds), *The Science of Global Change: The Imact of Human Activities on the   Environment*, American Chemical Society, pp. 392-429.

Prentice, K.C. and Fung, I.Y.: 1990, *Nature* **346**, 48-51.

Prentice, K.C.: 1990, Bioclimatic distribution of vegetation for GCM studies, *J. Geophy. Res.* (in press).

Prentice, I. C., Cramer, W., Harrison, S. P., Leemans, R., Monserud, R. A. & Solomon, A. M.: 1992, *Journal of Biogeography* **19**, 117-134.

Prentice, I.C. and Sykes, M.T.: 1993, Vegetation geography and global carbon storage changes, in: Woodwell, G.M. (ed), *Woods Hole Workshop on Biotic Feedbacks in the Global Climate System*, Cambridge University Press, Cambridge.

Prentice, I.C., Monserud, R.A., Smith, T.M. and Emanuel, W.R.: 1992, Modeling large-scale vegetation dynamics, in: Solomon, A.M. and Shugart, H.H (eds), *Vegetation Dynamics and Global Change*, Chapman and Hall, New York.

Prudhomme, T.I., Ochel, W.C., Hastings, S.J. and Lawrence, W.T.: 1984, Net ecosystem gas exchange at ambient and elevated carbon dioxide concentration in tussock tundra at Toolik Lake, Alaska: an evaluation of methods and initial results, in: McBeath, J.H. (ed), *The Potential Effects of Carbon Dioxide-Induced Climatic Changes in Alaska: Proceedings of a Conference*, School of Agriculture and Land Resources Management, Univ. of Alaska, Fairbanks, Alaska.

Raich, J.W., Rastetter, E.B., Melillo, J.M., Kicklighter, D.W., Steudler, P.A., Peterson, B.J., Grace, A.L., Moore, B. and Vorosmarty, C.J.: 1991, *Ecological Applications* **1**:399-429.

Rastetter, E.B., Ryan, M.G., Shaver, G.R., Melillo, J.M., Nadelhoffer, K.J., Hobbie, J.E. and Aber, J.D.: 1991, *Tree Physiology* **9**:101-126.

Rotmans, J.: 1990, *IMAGE: An integrated Model to Assess the Greenhouse Effect*, Kluwer Academic Publishers.

Sampson, R.N.: 1993, *Water, Air and Soil Pollution* (this volume).

Sanchez, P.A. and Bennites, J.R.: 1987, *Science* **238**, 1521-1527.

Schlesinger, M. and Zhao, Z.: 1988, *Seasonal climatic changes induced by doubled CO₂ as simulated by the OSU atmospheric GCM/mixed layer ocean model*, Oregon St. U., Corvallis, OR, Climate Research Institute.

Schlessinger, W.H.: 1990, *Nature* **348**, 232-234.

Schroeder, P.S and Ladd, L.B.: 1991, *Climatic Change* **19**, 283-290.

Sedjo, R.A. and Solomon, A.M.: 1989, Climate and forests, in: Rosenberg, N.J., Easterling, W.E., Crosson, P.R. and Darmstadter, J. (eds.), *Greenhouse Warming: Abatement and Adaptation,* Resources For The Future, Washington, D.C.

Shugart, H.H.: 1984, *A Theory of Forest Dynamics*, Springer-Verlag, New York.

Smith, S.P., Strain, B.R. and Sharkey, T.D.: 1987, *Functional Ecology* **1**, 139-143.

Smith, T.M., Leemans, R. and Shugart, H.H.: 1991a, *Climatic Change* **21**, 367-384.

Smith, T.M., Shugart, H.H., Bonan, G.B., and Smith, J.B.: 1992b, *Advances in Ecological Research* **22**, 93-113.

Smith, T. M. and Shugart, H. H.: 1993a, *Nature* **361**:523-526.

Smith, T. M. and Shugart, H. H.: 1993b, *Water, Air and Soil Pollution* (this volume).

Solomon, AM. and Shugart, H.H. (eds): 1993, *Vegetation Dynamics and Global Change*, Chapman and Hall.

Sundquist, E.T.: 1993, The global carbon dioxide budget, *Science* **259**, 934-941.

Tans, P.P, Fung, I.F and Takahashi, T.: 1990, *Science* **247**, 1431-1438.

Thornthwaite, C.W.: 1931, *Geogr. Rev.* **21**, 633-655.

Thornthwaite, C.W.: 1933, *Geogr. Rev.* **23**, 433-440.

Thornthwaite, C.W.: 1948, *Geogr. Rev.* **38**, 5-89.

Tissue, D.T. and Oechel, W.C.: 1987, *Ecology* **68**, 401-410.

Troll, C. and Paffen, K.H.: 1964, *Erkund. Arch,. Wiss Geogr.* **18**, 5-28.

Vloedbeld, M. and Leemans, R.:1993, *Water, Air and Soil Pollution* (this volume).

Washington, W. and Meehl, J.: 1984, *J. Geophy. Res.* **89**, 9475-9503.

Wetherald, R. and Manabe, S.: 1986, *Climatic Change* **8**, 5-23.

Williams, W.E., Garbutt, K., Bazzaz, F.A. and Vitousek, P.M.: 1986, *Oecologia* **69**, 454-459.

Winjum, J.K., Dixon, R.K. and Schroeder, P.E.: 1992, *Water, Air and Soil Pollution* **64**, 213-228.

BOREAL FORESTS AND TUNDRA

M.J. APPS, W.A. KURZ, R. J. LUXMOORE, L.O. NILSSON,
R.A. SEDJO, R. SCHMIDT, L.G. SIMPSON, and T.S. VINSON

Abstract. The circumpolar boreal biomes cover *ca.* 2 10^9 ha of the northern hemisphere and contain *ca.* 800 Pg C in biomass, detritus, soil, and peat C pools. Current estimates indicate that the biomes are presently a net C sink of 0.54 Pg C yr^{-1}. Biomass, detritus and soil of forest ecosystems (including *ca.* 419 Pg peat) contain *ca.* 709 Pg C and sequester an estimated 0.7 Pg C yr^{-1}. Tundra and polar regions store 60-100 Pg C and may recently have become a net source of 0.17 Pg C yr^{-1}. Forest product C pools, including landfill C derived from forest biomass, store less than 3 Pg C but increase by 0.06 Pg C yr^{-1}. The mechanisms responsible for the present boreal forest net sink are believed to be continuing responses to past changes in the environment, notably recovery from the little ice-age, changes in forest disturbance regimes, and in some regions, nutrient inputs from air pollution. Even in the absence of climate change, the C sink strength will likely be reduced and the biome could switch to a C source. The transient response of terrestrial C storage to climate change over the next century will likely be accompanied by large C exchanges with the atmosphere, although the long-term (equilibrium) changes in terrestrial C storage in future vegetation complexes remains uncertain. This transient response results from the interaction of many (often non-linear) processes whose impacts on future C cycles remain poorly quantified. Only a small part of the boreal biome is directly affected by forest management and options for mitigating climate change impacts on C storage are therefore limited but the potential for accelerating the atmospheric C release are high.

1. Introduction

The boreal forest biome consists of a broad complex of forested and partially-forested ecosystems which form a circumpolar belt through northern Eurasia and North America. Its southern boundary is formed with temperate deciduous forests where oceanic influences moderate climate, and with arid steppe, prairie or semi-desert in continental regions. Climatically, these regions are characterized by short growing seasons and low mean temperatures resulting in a forest cover dominated by coniferous vegetation. The boundary between the boreal and temperate forests is not sharp, instead there is a transitional zone with mixed species stands or a mosaic of temperate deciduous species established on favorable soils and boreal conifers on colder sites. To the north, the subarctic woodlands contain a patchwork of treeless patches and stunted forest stands

Water, Air, and Soil Pollution **70**: 39–53, 1993.
© 1993 *Kluwer Academic Publishers.*

which form the boundary between the boreal forest and tundra zones. The term 'boreal forest biome', as used in this paper, includes the closed-canopy boreal forests, the open canopy sub-arctic wood lands and the mixed hardwood-softwood forests which form the southern transition zone. Tundra zones include the wet (coastal) and moist (tussock) tundra (Shaver *et al.*, 1992). The polar deserts and semi-desert regions, covered by ice and stone barrens, contain negligible quantities of active biogenic C.

The objectives of this paper are threefold. First, estimates of the current C-pool sizes and fluxes for the boreal regions will be summarized. This C budget synthesis includes recent literature and results presented by the authors in this volume. Secondly, the processes which may be responsible for the current state of the C budget will be discussed with an emphasis on those which may be important in an enhanced-greenhouse future climate. This discussion will be followed by an assessment of future C budgets for the boreal region, considering climatic change and the potential for human influences, both positive and negative.

2.   C in the Contemporary Boreal Region

2.1.   ESTIMATES OF CURRENT CARBON POOLS

Primary C pools in the boreal forest biome are forest biomass, forest soils including litter and coarse woody debris, and organic soils associated with peat formation. Additional C stores are retained in undecomposed wood products and landfill material derived from forest biomass. Table 1 provides a summary of the estimated present sizes of these pools, separated into biospheric components and geographical regions. More detailed breakdowns may be found in the individual works cited in the table footnotes. The western coastal regions of Canada (Cordilleran) and the entire forests of Scandinavia have been included in the boreal forest biome (Fig. 1) for this assessment to facilitate global accounting of C pools and fluxes. The Cordilleran regions are shown separately in Table 1 because of their ecologically different character and higher biomass C density.

Estimates of live belowground biomass are included in Table 1 for the North American and Russian boreal forests, but these data are not available for Scandinavia. Root biomass is an important contribution to the total biomass and the paucity of data underlying the estimates in Table 1, together with other uncertainties in plant C allocation, poses a significant scientific challenge for forecasting future C budgets in a changing climate.

Estimates of C stored in harvested forest biomass do not appear to have been previously estimated at a national scale for boreal systems except for Canada (Kurz *et al.*, 1992), although global estimates cited by Vitousek (1991) are consistent with the estimates in

Figure 1. A polar view of the circumpolar boreal regions.

pools, but the annual net storage in these pools plays a potentially important part in the annual forest sector C budget (Apps and Kurz, 1991).

The estimates for boreal peatlands shown in Table 1 for North America (see Kurz *et al.*, 1992) and Scandinavia were derived from Gorham's (1991) estimates, while the values for Russia are based on area estimates by Botch and Masing (see Kolchugina and Vinson, this volume). Gorham (1991), whose estimate of global northern peatlands (342 Mha and

Table 1:   Contemporary boreal forest biome areas and C pools. (Totals may not agree due to rounding errors)

| | AREA (Mha) | | CARBON POOLS (Pg C) | | | | | |
	Area	Peat-land	Plant Biomass	Plant Detritus	Forest Soil	Peat	Forest Products	Total
Alaska	52[1]	11[6]	2[10]	1	10	17[17]	<0.1	30
Canada (Boreal Forest Biome)	304[2]	89[7]	8[11]	N/A[14]	65	113[18]	0.2[21]	186
Canada (Cordilleran)	72[3]	3	6	N/A[14]	16	4[18]	0.3	27
Russia	760[4]	136[8]	46[12]	31[15]	100[16]	272[19]	2.9[22]	451
Scandinavia	61[5]	20[9]	2[13]			13[20]		15
Total	1249	260	64	32	199	419	3.4	709

1. From Birdsey (1992).
2. Areas in boreal, arctic, and subarctic ecoclimatic provinces that have biomass data in the national inventory (Kurz et al., 1992).
3. See text for explanation.
4. Forest area of Russia, ca. 95% of Former Soviet Union forests (Kolchugina and Vinson, this volume).
5. Area estimates from UN-ECE/FAO (1992), Table 11.
6. From Kivinen and Pakarinen (1981), including only thick peat soils, excluding 38.0 Mha of tundra bog soils.
7. Canadian Wetlands Working Group (1986) and Kurz et al. (1992).
8. From USSR State Forestry Committee, 1990 (See Kolchugina and Vinson, this volume).
9. From Kivinen and Pakarinen (1981).
10. From Birdsey (1992), above and belowground live biomass.
11. From Kurz et al. (1992); revised Kurz and Apps (in prep), above and belowground live biomass.
12. From Kolchugina and Vinson (this volume), above and belowground phytomass, understory vegetation and grasses.
13. Pekka Kauppi, pers. com. (1993), range 1.5-1.9 Pg C.
14. Plant detritus estimates are included in soil C pool.
15. From Kolchugina and Vinson, pers. com. (1993), litter (10.6 Pg C) plus necromass (20.2 Pg C).
16. From Kolchugina and Vinson (this volume). Forest ecosystems only, excluding peat.
17. Based on Gorham (1991), mean depth 2.5 m, mean bulk density 0.112 g cm^{-3}, 51.7% C content.
18. See note 17, mean depth 2.2 m.
19. Area, density (2,000 Mg C ha^{-1}), T. P. Kolchugina, pers. com.
20. See note 17, mean depth 1.1 m.
21. Residual C content of materials harvested from this region (1940-1980) (Kurz et al., 1992, Apps and Kurz, 1991).
22. From Sinitsin (1990).

455 Pg C) is comparable with the values shown in Table 1, discusses the major uncertainties associated with peatland C on a global scale. Broadly speaking, the largest uncertainties are associated with the areal extent of the different types of peatlands and their depth. Estimates of combined forest and peatland C pools introduce additional uncertainty because of the unknown degree of overlap between forest inventories and peatland inventories from which soil C pools and peatland pools are calculated. The potential double-accounting for the overlapping parts of the land-base is probably small relative to the other uncertainties in these pool estimates. The overlap, however, tends to confound separation of flux estimates from forest and peatland systems.

Two different sets of estimates of area and C pools for arctic tundra are provided in Table 2. The first is based on area estimates and average C densities following Kolchugina and Vinson (this volume). The second is based on the C pool estimate of Shaver et al., (1992) spatially distributed according to the BIOME model of Prentice et al. (1993) (36% in North America, 53% in Asian Russia, and 5.4% in western Europe, with the remaining 6% in scattered pockets). Because of the difficulties in scaling to the biome, both these estimates must be considered very uncertain. Previous estimates of C stores in the tundra and polar regions exhibit even higher ranges, from 55 Pg C (Post, 1990) to 256 Pg C (Prentice et al., 1993). The wide range of estimates arises partly from differences in definitions of the C pools involved (and their geographical distribution) and partly because of the way in which areal estimates are scaled up to the biome level.

Table 2:   Summary of estimates of contemporary tundra areas and C pools.

	AREA (Mha)	CARBON POOL (P_3 C)
North America	240^1 to 420^2	$41^{1'}$ to 22^3
Russia	226^4 to 624^2	59^4 to 32^3
Scandinavia	63^3	3^3
Other	70^3	4^3
Total	600 to 1180	107 to 61

1. Area of arctic regions, C density (17 ± 5 kg C m^{-2}) from Kurz et al. (1992).
2. Area estimates based on BIOME model (Prentice et al., 1993; Rik Leemans, pers. com.).
3. C density from Shaver et al. (1993), apportioned to areas from note 2.
4. From Kolchugina and Vinson (this volume) and Tatyana P. Kolchugina (pers. com.).

Adams *et al.* (1990) estimate an accumulation over the last 18 kyr of 240 Pg C in the closed-canopy boreal forest (soils and vegetation biomass), with an additional 190 Pg C uptake in open-canopy woodlands and tundra. In contrast, Prentice and Fung (1990) estimate an uptake over the same time frame of only 80 Pg C in boreal forests (50 Pg C in soils and 30 Pg C in biomass) and a decrease of 1-2% in tundra C storage (which they estimate presently to be 91 - 133 Pg C). Direct comparisons of these previous estimates with Tables 1 and 2, which are primarily based on synthesis of national-scale databases, are complicated by differences in definitions used to delineate boreal forest and boreal biome distributions. The discrepancies illustrate the divergence in different scaling assumptions - a challenge which also faces flux estimates.

2.2. ESTIMATES OF CURRENT CARBON FLUXES

The estimates of fluxes presented in Table 3 are inferred from estimated annual changes in the biospheric C pools. To be meaningful, this inference requires that all significant changes in these pools be accounted for (Apps and Kurz, 1993); bias and errors will result if significantly changing pools (or processes) are ignored. Comprehensive analyses of C fluxes have been performed for both the Russian and Canadian forest biomes which between them (including their peatlands) account for 91% of the circumpolar northern forest area, 90% of its C stores, and account for 90% of it's net atmospheric C exchange (Tables 1 and 3). Discussion of the forest sector portions are provided in the cited references (see also section 3 below) but peatland and tundra estimates require comment here.

Table 3 indicates that peatlands have been a significant sink of atmospheric C throughout the boreal region. These estimates are based on the historically-observed rate of peat accumulation (for references, see Kolchugina and Vinson & Kurz and Apps, this volume). These estimates have two sources of uncertainty; the areal extent of peatlands (which also affects the estimates of pool size) and the estimate of net ecosystem productivity (NEP) for these peatlands. The widely accepted value of 23 g C m^{-2}yr^{-1} (Gorham, 1991) falls in the middle of the range measured by Zoltai (1991) for boreal forest and subarctic peatlands of central Canada. The challenges of scaling to the biome with such an 'average' NEP have been previously noted (section 2.1, and Apps, 1993) and apply to both these sources of uncertainty. The apparent difference in peatland NEP between North American and Russian peatlands is almost certainly an artefact of the difference in procedures between the Russian and North American estimates; it is likely that the Russian forest soil estimates include some peat deposition fluxes which result in a decreased peat NEP and a corresponding increased forest ecosystem NEP (relative to the North American estimates in which a greater degree of separation has been attempted).

Table 3: Summary of present C fluxes in the circumpolar boreal regions. Net transfers from the atmosphere (+), net releases to the atmosphere (-).

	Ecosystem[1] (Tg C yr^{-1})	Peatland (Tg C yr^{-1})	Products (Tg C yr^{-1})	Net Flux (Tg C yr^{-1})
Boreal Forest Biome	+605	+45	+5̄7	+707
Alaska	+6[2]	+3[7]	<+1	+9
Canada (Boreal)	+62[3]	+25[8]	+8[10]	+95
Canada (Cordilleran)	+1[3]	+1[8]	+11[10]	+13
Russia	+493[4]	+11[9]	+26[1]	+530
Scandinavia	+43[5]	+5[7]	+12[12]	+60
Tundra				
moist (Tussock)	-140[6]			-140[6]
wet (coastal)	-30[6]			-30[6]
Total	+435	+45	+5̄7	+537

1. Annual net flux from ecosystem C pools includes all biomass and soil C pool dynamics, disturbance releases, and harvest removals. Where calculations were available, peatland fluxes are shown separately in column 3.
2. Flux estimate for Alaska assumes that the per ha fluxes on average are equal to those in the Canadian Yukon Territories (Kurz et al., 1992, revised in Kurz and Apps, in prep.).
3. Net change in ecosystem C (biomass plus soil and detritus C pools) from Kurz et al. (1992), revised in Kurz and Apps (in prep.).
4. From Kolchugina and Vinson (this volume and pers. com.), calculated from average NEP (1.05 Mg C ha^{-1} yr^{-1}) and area, minus fire release (199 Tg C yr^{-1}), harvest (152 Tg C yr^{-1}), and burned peat (100 Tg C^{-1}).
5. From Pekka Kauppi (pers. com.), range of estimates for ecosystem C sink, 29.7 to 56.6 Tg yr^{-1}.
6. From Oechel et al., 1993. Note, however, that Kolchugina and Vinson (this volume) and others consider tundra to be a sink, not a source. See text.
7. Assuming a net sink of 23 g C m^{-2} yr^{-1} (Gorham 1991).
8. Assuming a net sink of 28 g C m^{-2} yr^{-1} (Kurz et al., 1992; Gorham, 1991; Stephen Zoltai, pers. com.).
9. Assuming a net sink of 30 g C m^{-2} yr^{-1} and release of 30 Tg C yr^{-1} from peat burning (Tatyana P. Kolchugina, pers. com.).
10. Net balance between oxidation of stored forest products (including landfills), production emissions, and harvest input (from Kurz et al., 1992; Apps and Kurz, 1991).
11. Difference between forest product accumulation (statistical data 277 Mm^3 yr^{-1} and multiplier 0.26 to convert volume to C) and product decomposition (46 Tg C yr^{-1}) from Melillo et al. (1988).
12. Annual harvest from Pekka Kauppi (pers. com.), and assuming 50% C retention in forest products.

There have been several recent, but conflicting, estimates of atmospheric C exchange (CO_2 and CH_4) in the tundra and polar regions. The estimates in Table 3 are based on a 200 km transect study through moist and wet tundra of the North Slope of Alaska (Oechel et al., 1993) which indicate that a lowering of the water table and increased decomposition may already be occurring in response to warmer conditions. In contrast, however, Schell and Barnett (1993) have reported that production exceeds respiration; even where drainage may lead to aeration and increased respiration of peat, C continues to accumulate in valleys and low-lying foothill tundra of arctic Alaska at rates of up to 20 g C m^{-2}. Harden et al. (1992) also make the point that northern systems still continue to accumulate C; they are still responding to post glacial warming and although the response has slowed, they may not yet have reached a steady state. They have also shown that there are significant variations in the soil dynamics of northern continental North America following the retreat of the Laurentide ice-sheet. Because these deglaciated systems have a distinct spatial pattern, there is considerable danger in the extrapolation of site-specific average results to the global scale and for this reason the results shown for tundra in Table 3 must be considered very uncertain.

3.   C in the Future Boreal Region

The boreal forests are not static ecosystems and can not be assumed to be in equilibrium in terms of C exchange. In these systems, the current C cycle is strongly influenced by their past history over several time scales. The boreal forest is still evolving since the last glaciation on a time scale of millennia (e.g., Davis, 1969); species migration may still be occurring at the biome level. Recent evidence points at shorter-term disequilibria on the scale of decades and centuries which must be superimposed on these longer-term changes. Continuing responses to recent climate fluctuations such as the little ice-age (1250-1850 AD, Ian Campbell, pers. com.) and to changes in historical disturbance regimes have shaped the structure and function of the contemporary boreal forests on these shorter time and finer spatial scales (Apps, 1993; Kurz and Apps, this volume).

While the global boreal forests is presently estimated to be a sink for atmospheric C, can it be assumed that the mechanisms which are responsible for this sink will continue to function at their present strength? This does not appear likely. Even in the absence of climate change, the major sink mechanisms identified previously will become saturated over time and switches from C-sinks to C-sources are likely to occur at different parts of the biome over the coming decades. These transient phenomena will likely be amplified by changes in the global environment and in the discussion that follows, a distinction is made between long-term projections where equilibrium conditions (in both climate and ecosystem response) are assumed to have been achieved and short term (50 to 100 yr) C-cycle projections, where transient behavior is expected to dominate.

Kurz and Apps (this volume) argue that in the absence of other environmental changes (including changes in the disturbance regime) the forest ecosystem C sink associated with a shifting age-class structure will diminish, and eventually vanish, in the coming decades. Aging of the global boreal forest cannot continue indefinitely. If in addition, disturbance regimes (fire, insect-induced mortality, windthrow, and harvesting) increase in intensity and frequency, they will be accompanied by transient C-releases followed by longer-term response by younger forests whose spatial structure and growth characteristics will depend on the then-prevailing climatic and environmental conditions. Thus in the short-term, it is likely that the present C sink will vanish and be replaced by a transient C source even if the over the longer -term the new biome becomes an eventual sink for atmospheric C.

Kauppi *et al.* (1992) have suggested that several other factors may have also contributed to the net C uptake of European forests through enhancements of biomass growth at the tree and stand level. Site-specific studies in Europe have generally indicated a slight increase in tree growth which may reflect improved climate conditions and nitrogen deposition associated with air pollution. In particular they suggest that fertilization response to air pollution may have obscured detectable adverse effects of soil acidification. Experimental evidence for a fertilization mechanism has been provided by Nilsson (this volume). These authors are careful to point out that such increased forest productivity is unlikely to be sustainable in the medium-to-long term; soil acidification and decreasing availability of base cations will eventually lead to nutrient imbalances and site degradation. Nilsson also emphasizes the importance of soil water availability to the net C accumulation rates in such circumstances.

Because CO_2 doubling experiments with seedlings of boreal forest species have shown an average increase in growth of 38% (Wullschleger *et al.*, in press), some authors suggest that the 25% increase in atmospheric CO_2 over the last 100 years may have contributed to a forest C sink. There is no evidence, however, that trees growing under boreal field conditions have responded to this increase. Although some unexplained increases in tree growth have been reported, Graumlich (1991) has shown that similar unexplained increases in tree growth have occurred prior to industrialization, and, further, the increases often exceed the response that can be expected from CO_2 enrichment (Luxmoore *et al.*, this volume).

It is possible that boreal forests will respond to increasingly elevated CO_2 with increased root growth. This is expected on nutrient poor sites and under drier conditions; however, no observations of such root response to the historical rise in atmospheric CO_2 have yet been reported. Forecasting a boreal forest response to CO_2 fertilization suffers from difficulties of scaling. Small-scale physiological models generally predict increases in growth rate at elevated CO_2 levels, while plot-level succession models do not show much response to CO_2 after several decades, even when increased water-use efficiency is included. Biome-scale models generally show increased NEP with elevated CO_2, although the predicted response for the boreal biome can be small. One reason for the differing model predictions relates to model structure. In the case of physiological and biome models, the CO_2 response is not overridden by other processes, but this is not the case for succession models. The stochastic representation of mortality of old trees and ingrowth of new individuals introduces variability in stand biomass in succession models that masks detection of a CO_2 growth stimulation (Luxmoore et al., this volume).

Peat C accumulation rates are determined by the difference between net primary productivity (NPP) and decomposition, both of which are strongly influenced by site-specific hydrologic and nutrient conditions (Zoltai, 1991). Zoltai and Vitt (1990) have shown that large scale climatic and hydrologic changes - which vary considerably across the circumpolar boreal zone - can substantially influence the rate of peat accumulation. The range of accumulation rates reported by Zoltai (1991) reflects the role of permafrost and the age of deposition. Vegetation development plays a secondary role to hydrological and nutrient-cycling processes and is largely controlled by these other factors (Gignac et al., 1991).

The role of boreal peatlands under a warmer, but stable, future climate is also difficult to predict with certainty but these regions hold the potential for significant C-feedbacks to the climate system (Gorham, 1991). Distributions (and therefore C uptake) of boreal peatlands will shift northward in response to global warming with the consequent northward movement of the zones of continuous and discontinuous permafrost (Zoltai and Vitt, 1990; Zoltai, 1993a). Zoltai (1993b) has hypothesized that, when integrated across the entire expanse of Canadian peatlands, positive feedbacks (e.g., increased release of CH_4 in the warmer north and CO_2 in the drier south) may nearly balance negative feedbacks (e.g., increased C-uptake in the north and reduced CH_4 efflux in the south), thereby cancelling any significant changes. Kolchugina and Vinson (1993) have examined the future C cycle within the permafrost zone of Russia under a climate warming scenario and conclude that an additional 0.46 to 0.72 Pg C yr^{-1} may be gradually released to the atmosphere from increased decomposition of litter and coarse woody debris. They point out, however, that this efflux may be concurrently balanced by increased NPP and forest expansion into these regions thus moving towards a new equilibrium in the C cycle (albeit at a higher rate of C turnover); transient effects are likely to be more significant than long-term, equilibrium changes in C-storage.

3.1.   EQUILIBRIUM FUTURE BOREAL C-POOLS AND FLUXES

Some equilibrium projections (eg., Smith et al., this volume) suggest that future boreal forest regions will contain more C than they do today (i.e., when averaged over a long enough time, they will act as a net sink). These projections are based on potential vegetation and to be achieved, require that the forest ecosystems reach equilibrium with the new (and assumed stable) climate system - a process that will take millennia (Schlesinger, 1990). There are two principal reasons for expecting eventually enlarged boreal forest C pools: 1) increased temperatures result in increased forest productivity and 2) unimpeded migration of the boreal forest into the presently unforested tundra regions.

The equilibrium picture for boreal forests is, however, far from well-established. Prentice and Fung (1990) point out that the global C implications of these projections of potential vegetation are most sensitive to misclassifications in areas having a strong gradient in C density and cite the transition through the boreal forest, tundra and polar regions as a particular case in point. It is noteworthy that a slightly different approach by Nielson (this volume) suggests that increased drought stress will result in the boreal region losing a significant quantity of C - that is, the boreal forest will act as a long-term source of atmospheric C. Although C analyses have not yet been performed, the projections of Rizzo and Wiken (1992) for Canadian ecosystems suggest similar trends to those of Nielson - warmer and drier mid-continental conditions will favour encroachment of grasslands into the present boreal regions more strongly than migration of these forests into the present arctic tundra, resulting in a net loss of area having high C stores.

3.2. TRANSIENTS

Under projected climate change scenarios, the climate system is expected to undergo warming at a much faster rate than records show that species and ecosystems have had to respond in the past 20 kyr. Even if the climate conditions were to stabilize at the doubled-CO_2 scenario projections, the rapidity of these changes will produce transient and non-linear responses as ecosystems lose synchronization with their environment (Holling, 1992). Because of the stand-replacing role that disturbances play in boreal ecosystem dynamics, C storage and exchanges during such transients cannot be assumed to vary linearly between two steady-state or equilibrium states (Apps, 1993). Thus, projections based on observations of the relatively slow changes in prior times could be misleading and result in underestimates of short-term C fluxes.

On a regional scale, the transient C fluxes from forest ecosystems could be either positive or negative, depending upon differences in response times associated with NPP, respiration, disturbance regimes, and the changes in the regional environment (King and Nielson, 1992; Townsend et al., 1993). At the biome scale, the transient behavior (250 year period) of Canadian forests has been simulated by Nielson et al. (1993). In nearly every scenario that these authors examined, short-term (50-100 year) transients resulted in C release from forest ecosystems, even in cases where these same ecosystems showed net C uptake in longer term (100-250 year) projections.

Succession models suggest that large changes in ecosystem distribution could result from warming due to changes in soil water availability. Smith and Shugart (1993) used simulation of global warming effects on terrestrial ecosystems to show that transitions from one ecosystem to another occurring over a 50 to 100 year time frame could result in a significant release of CO_2 to the atmosphere. Much of the CO_2 release was due to transient changes in tundra and boreal ecosystems.

One of the most significant factors influencing the transient response of boreal regions to global change is the potential for dramatic changes in disturbance regime intensity and frequency (Apps, 1993; Overpeck et al., 1990). Disturbance regimes have an influence on both the immediate and the medium-term C cycle of boreal forests. A three-fold difference in Canadian wild fires between a high-fire year (1989) and a reference year (1986) resulted in an 86% reduction in the net ecosystem C sink (Kurz et al., 1992). Moreover, an additional 0.08 Pg C were transferred from aboveground biomass to the soil pools and will decompose in subsequent years. With increased decomposition rates accompanying a warming climate (Townsend et al., 1993), such transfers of aboveground biomass to forest-floor detrital pools may contribute to an increasingly important forest floor C efflux. Carbon transfers associated with disturbances must be balanced against the uptake from the regenerating forests and their potentially enhanced productivity.

Where soils permit, forests at the northern margin will expand into the present tundra and modest vegetative cover will develop as slumping and exposure of mineral soils accompanies permafrost melting. Losses of organic soil C in some areas will be balanced by increased aboveground C storage. Forests at the southern margin will likely suffer dieback as temperate forests and grasslands invade. The rate of dieback may exceed the rate of incursion in some areas, particularly where soil moisture is reduced. Forests in the middle of the boreal biome may exhibit small to modest increases in biomass in response to the generally more favorable growing conditions associated with the warming, provided that available soil water is not reduced.

Soil warming experiments have shown increased nutrient availability due to greater rates of organic matter decomposition and N mineralization (Van Cleve et al., 1990). Increased air and soil temperatures will result in enhanced soil respiration (Raich and Schlesinger, 1992) but it is uncertain if the enhanced growth from greater N mineralization will offset the loss of soil C through respiration as suggested by Townsend et al. (1993). Townsend et al. (1993) examined the transient behaviour of forest soils under global warming scenarios. They point out that if only warming occurs (i.e. without considering changes in soil water conditions), heterotrophic respiration (decomposition) will rise considerably more rapidly (exponential) than will NPP (linear) resulting in an approximately linear release of extant soil C. Townsend et al.'s analyses suggests that transient phenomena in soil processes will have a major impact on the C cycle of northern systems but the transient may play out over a relatively long time scale (100 years and beyond).

Transient behaviour in boreal peatlands may also be much more significant than the net changes in equilibrium storage might indicate (Gorham, 1991). While increased temperatures will likely enhance vegetation productivity, larger impacts are expected with changes in water table that will alter the balance between aerobic (CO_2 release) and anaerobic (CH_4 release) decomposition pathways. Hydrological and landscape changes in the patterns of bogs and fen formations with global change will be expected. In the north, melting of permafrost will likely favour an areal expansion of active wetlands, while drier conditions in southern regions will likely lower water tables changing the function and distribution of extant wetlands. Periodic severe droughts and increased incidence of forest fires that spread to the dry surfaces of peatland formations throughout the biome will accelerate reduction in C storage and increase the release of CO_2 (Post, 1990; Zoltai, 1993).

There have been few attempts to forecast climate-induced changes in the future tundra and polar regions. It has been estimated, however, that under a 5^0 C warmer climate the northern tundra and polar regions could release an additional 1.3 - 1.6 Pg C yr^{-1} as CO_2 and 0.1 Pg C yr^{-1} as CH_4, depending on the assumptions made about future moisture conditions (Post, 1990). The observations by Oechel et al. (1993) that arctic ecosystems may presently be a CO_2 source may already be an indication that transient C-releases associated with ecosystem readjustment to the slight warming trends of the last few decades are taking place.

4.   Conclusions

The circumpolar boreal region is one of the largest biomes in the world and contains significant pools of biogenic C that are sensitive to changes in the global environment. Although conflicting evidence exists, there are indications that the tundra regions may already act as a source of atmospheric C (0.17 Pg C yr^{-1}). The contemporary boreal forests are, however, believed to presently taking up ca. 0.7 Pg C yr^{-1}. The principal mechanisms responsible for this present sink are believed to be continuing responses to: post-glacial warming; climatic perturbations such as the little ice-age (1250-1850); changes in disturbance regimes over the past two centuries; and more recently, nutrient inputs associated with air pollution.

Although the circumpolar boreal forest is presently be a sink for atmospheric C, the mechanisms believed responsible for this sink are not likely to be sustained, even if there is no change in climate. One of the principal mechanisms appears to be an aging and aggrading boreal forest, primarily due to changes in past and present disturbance regimes, a structural phenomenon which cannot be maintained indefinitely. Forest protection measures, such as fire suppression for example, must take into account both the biological role of such disturbances and the geographical reality of lack of access to vast forest areas. Other possible mechanisms, such as nutrient inputs associated with low-level, but wide-spread air pollution, may yield short-term increased C uptake but may lead to future forest decline.

How the C cycle in the boreal biome will be affected by climate change (and whether it has already been) is a challenging problem. There has been no compelling evidence, at the biome level, for an increase in boreal forest productivity due to recent changes in climate (temperature and precipitation) or to ambient CO_2 levels. There seems to be a clear consensus, however, that both the structure and function of boreal regions will be drastically altered by the climate changes projected by present global circulation models. It is also expected that non-linear, transient ecosystem responses will dominate the C dynamics during a period of rapid climate change.

Can boreal forest nations do anything to mitigate against increasing atmospheric C? Apart from Northern Europe, much of the circumpolar boreal forest remains in a wilderness state in that silviculture and timber harvesting directly influence only a relatively small fraction of the area. Improved management practises have the potential for significantly increased biomass yields at the regional level but it is hard to see how these could be applied at the biome scale in North America and Russia. Non-sustainable development practises, on the other hand can adversely affect C-storage capacity and pools in a significant way. Even in the absence of such human intervention, however, transient response to climate change has the potential for significant positive feedbacks to the climate system by releasing large quantities of C to the atmosphere.

Acknowledgements: The authors would like to thank Tatyana P. Kolchugina, who gave freely of her time and energy to provide the estimates for Russian boreal systems in this manuscript. Two of the authors (M.J.A. and W.A.K.) gratefully acknowledge the support of the Canadian Federal Panel on Energy Research and Development (PERD) through the ENergy from the FORest program of Forestry Canada.

REFERENCES:

Adams J.M., Faure H., Faure-Denard L., McGlade J.M. and Woodward F.I.:1990, *Nature*, **348**, 711.

Apps M.J.: 1993, *World Resource Review*, **5**, 41.

Apps M.J. and Kurz W.A.:1991. *World Resource Review*, **3**, 333.

Birdsey R.A., 1992; Carbon Storage and Accumulation in US Forest Ecosystems, *General Technical Report WO-59*, USDA Forest Service, Washington, D.C., 51 pg.

Davis, M.B.: 1969, *Ecology*, **50**, 312.

Gignac L.D., Vitt D.H., Zoltai S.C., Bayley S.E.: 1991, *Nova Hedwigia*, **51**, 27.

Gorham E.: 1991, *Ecological Applications*, **1**, 182.

Graumlich L.: 1991, *Ecology*, **72**, 1.

Holling C.S.: 1992, *Ecological Monographs*, **62**, 447.

Kauppi P.E., Mielikäinen K., Kuusela K.: 1992, *Science*, **256**, 70.

Kivinen E. and Pakarinen P.;1981, *Annales Acad. Scientiarium Fennicae* A.III.**123**:1-28.

King G.A. and Nielson R.P.: 1992, *Water, Air and Soil Pollution*, **64**, 365.

Kolchugina T. P. and Vinson T. S.: 1993, *Permafrost and Peraglacial Processes*, **4**, 13.

Kurz W.A., Apps M.J., Webb T., and MacNamee P.: 1992, The Carbon Budget of the Canadian Forest Sector: Phase 1, *ENFOR Information Report NOR-X-326*, Forestry Canada Northwest Region, Edmonton, Alberta, 93 pg.

Nielson R.P., Lenihan J., and King G.A.: 1993, Global Biogeography and Biosphere Feedback During Climatic Change, in: Kanninen M. (ed), *Carbon balance of world's forested ecosystems: towards a global assessment*, Proceedings of Intergovernmental Panel on Climate Change Workshop, Joensuu, Finland, 11-15 May 1992, Publications of the Academy of Finland (in press).

Oechel W.C., Hastings S.J., Vourlitis G., Jenkins M., Riechers G., Grulke N.: 1993, *Nature*, **361**, 520.

Prentice I.C., Sykes M.T., Lautenschlager M., Harrison S.P., Denisenko O., and Bartlein P.J.: 1993, *Biogeographical Letters*, (in press).

Prentice K.C. and Fung I.Y.: 1990, *Nature*, **346**, 48.

Schlesinger W.H.:1990, *Nature*, **348**, 228.

Shaver G.R., Billings D.W., Chapin III F.S., Giblin A.E., Nadelhoffer K.J., Oechel W.C., and Rastetter E.B.: 1992, *BioScience*, **42**, 433.

Sinitsin, A.: 1990, *Forest Management*, **8**, 6.

Smith T.M. and Shugart H.H.: 1993, *Nature,* **361**, 523.

Townsend A.R., Vitousek P.M. and Holland E.A.: 1993, *Climatic Change*, **22**, 293.

UN-ECE/FAO : 1992, The forest resources of the temperate zones: major findings, United Nations Economic Commission for Europe, Geneva, 32 pg.

Vitousek P.M.:1991, *J. Environ. Qual.*, **20**, 348.

Wullschleger S.D., Post W.M., and King A.W.: 1993, On the Potential for a CO_2 Fertilization Effect in Forest Trees - An Assessment of 58 Controlled-Exposure Studies and Estimates of the Biotic Growth Factor, in: Woodwell G.M. (ed) *Biospheric Feedbacks in the Global Climate System: Will the Warming Speed the Warming?*, Oxford University Press (in press).

Zoltai S.C.: 1991, *Holocene*, **1**, 68.

Zoltai S.C.: 1993a, *Arctic and Alpine Research*, (in press).

Zoltai S.C.: 1993b, *5th International Mire Conservation Group Excursion Guide*, Swiss Federal Institute of Forestry Snow and Landscape Research, (in press)

Zoltai S.C. and Vitt D.H.: 1990, *Quaternary Research* **33**, 231.

CONTRIBUTION OF TEMPERATE FORESTS
TO THE WORLD'S CARBON BUDGET

LINDA S. HEATH   *USDA Forest Service, Global Change Research Program, Northeastern Forest Expt. Station, P. O. Box 6775, Radnor, PA, 19087, USA*

PEKKA E. KAUPPI   *Finnish Forest Research Institute, Unioninkatu 40 A, SF-00170 Helsinki, Finland*

PETER BURSCHEL   *Universität München, Lehrstuhl f. Waldbau und Forsteinrichtung, Hohenbachernstraße 22, W-8050 Freising, Germany*

HEINZ-DETLEV GREGOR   *Federal Environmental Agency of Germany, Bismarckplatz 1, W-1000 Berlin 33, Germany*

ROBERT GUDERIAN   *Institut für angewandte Botanik, Universität Essen, Universitätsstr. 5, W-4300 Essen, Germany*

GUNDOLF H. KOHLMAIER   *Institut für phys. Chemie der Universität Frankfurt, W-6000 Frankfurt 60, Germany*

SUSANNE LORENZ   *Institute for World Forestry, Federal Research Centre for Forestry and Forest Products, Leuschnerstr. 91, W-2050 Hamburg 80, Germany*

DIETER OVERDIECK   *TU Berlin, Institute of Ecology, Ecology of Woody Plants, Königin-Luise-Str. 22, W-1000 Berlin 33, Germany*

FLORIAN SCHOLZ   *BFA für Forst- und Holzwirtschaft, Institüt f. Forstgenetik, Sieker Landstr. 2, W-2070 Großhausdorf 2, Germany*

HARALD THOMASIUS   *Auf der Bismarckhöhe 24, O-8223 Tharandt, Germany*

MICHAEL WEBER   *Lehrstuhl f. Waldbau und Forsteinrichtung, Hohenbachernstr. 22, W-8050 Freising, Germany*

Abstract. Temperate forests currently cover about 600 MHa, about half of their potential. Almost all these forests have been directly impacted by humans. The total living biomass in trees (including roots) was estimated to contain 33.7 Gt C. The total C pool for the entire forest biome was estimated as 98.8 Gt. The current net sink flux of biomass was calculated at 205 Mt yr^{-1}, with a similar amount removed in harvests for manufacture into various products. The major cause of this C sink is forest regrowth. Forest regrowth is possible because fossil fuels are the major source of energy in temperate countries, instead of fuelwood. Future C in these forests will be greatly influenced by human activity. Options to sequester more C include conservation of forest resources, activities that increase forest productivity such as adopting rotation ages to optimize C production, afforestation, improvement of wood utilization, and waste management.

Water, Air, and Soil Pollution **70**: 55–69, 1993.
© 1993 *Kluwer Academic Publishers.*

1. Introduction

Temperate forests do not form a uniform belt around the globe, but exist in large blocks of discontinuous forest cover on five continents surrounded by extensive areas of prairie, steppe and desert (Figure 1). We define temperate forest as those forests in the mid-latitudes that are not included in the tropics or in boreal forests. The forest composition is diverse with deciduous broadleaved and mixed broadleaved/coniferous, evergreen and warm temperate mixed broadleaved, and cold-temperate coniferous types. The temperate forests in western Canada and Scandinavia were included in the boreal forest assessment for convenience.

North America currently contains 60% of the present area of temperate forest, Russia and Europe about 12% each, with the remainder scattered throughout the rest of Asia, Australia, New Zealand and South America. Virtually all temperate forests have been exploited and directly impacted by human beings, with the exception of those in major mountain systems. The forests share the landscape with agricultural land, pastures and urban areas, and seldom cover more than 40% of the land area in any one of the forest regions. Japan and some of the U.S. are exceptions with 50 to 70% of the area in forest.

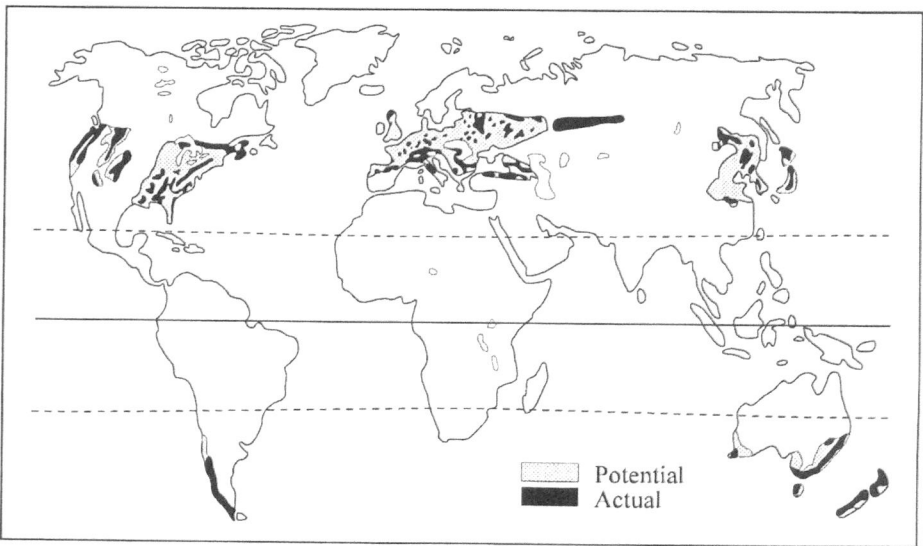

Figure 1. Potential and actual area of temperate forests (after Deutscher Bundestag [1990]).

2. Carbon Budget of Temperate Forests

2.1. METHODOLOGY OF CALCULATING

Figure 2 illustrates the C pools and fluxes in a forest. We distinguish one input flux, $F_{ab} = F_{npp}$, which is the annual net primary production and two output fluxes: F_{ha}, which is the flux from the decomposition of litter and soil, also called heterotrophic respiration, and F_{wa}, which in a natural system is equal to F_{bw}, the disturbance flux through catastrophic events, like fire, storm, insect infestation and also containing the component of herbivory feeding. As indicated in Figure 2, the three fluxes connecting the biota with the atmosphere determine whether the living biota and soils act as a source or a sink for atmospheric CO_2 or whether a steady state exists.

2.2. PRESENT POOLS AND FLUXES

Estimates for the pools and fluxes of C in aboveground biomass of temperate forests can be obtained from statistics on standing stock on forested land, and related wood increment and net primary productivity. For consistency, a simple methodology was chosen to calculate C pools and fluxes for most regions within the biome. The forest land areas were taken from UN-ECE/FAO statistics (1992) and other sources. Each hectare of temperate forest was assumed to contain 57.1 tons of C in living vegetation, which is the average C on a hectare from the largest individual region, the U.S. Soil was assumed to contain twice as much C as the living biomass. The net primary productivity was estimated corresponding to a net annual increment of 5 m^3 ha^{-1} yr^{-1} of stemwood. Stemwood was assumed to contribute 70% to the net primary productivity. Net storage in living biomass was estimated by assuming that fellings and natural losses account for 80% of the net primary production. Removal statistics were used to estimate the C transfer to forest products. Results are displayed in Table 1.

The total forested area assessed as temperate forests consisted of 600 Mha. This excluded forests in western Canada (130 Mha), an area considered by the boreal forest working group. The total living biomass in trees (above- and belowground) was estimated to contain 33.7 Gt C. The total C pool of temperate forest ecosystems was estimated as 98.8 Gt.

Net primary production was estimated as 892 Mt yr^{-1}. As 20% of it was assumed to accumulate in forests, a sink flux of 205 Mt yr^{-1} was sequestered in living trees. This estimate is similar to previous estimates by Armentano and Ralston (1980) and Sedjo (1992), after adjusting for differences in definitions of temperate forests. For those countries with available statistics, we calculated an additional 192 Mt C yr^{-1} was removed in harvests.

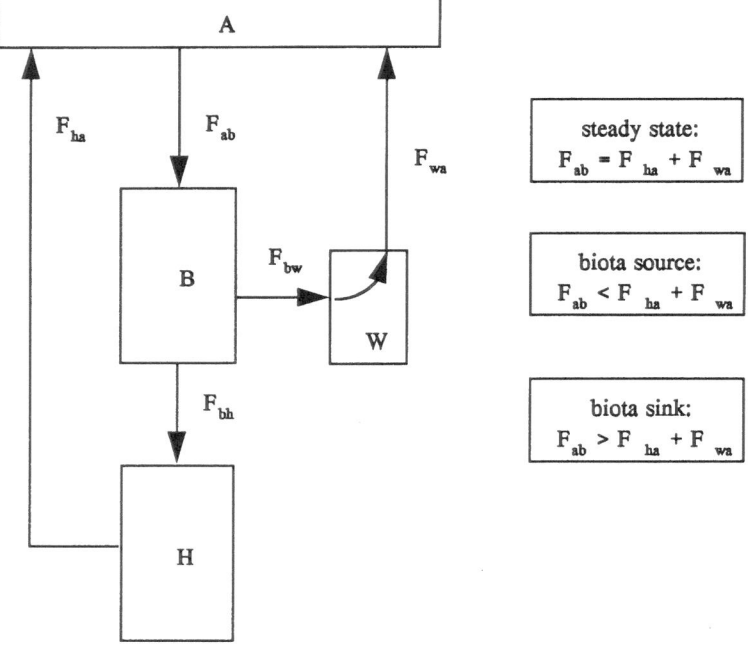

Pools:

A = C in atmosphere
B = C in biota (trees and understory)
H = C in litter and humus of soils
W = C in wood products of human society

Fluxes:

$F_{ab} = F_{NPP}$: net primary production (leaves, stems, branches, roots, fruits)
F_{bh} : litterfall
F_{ha} : heterotrophic (animal/microbial) respiration of litter and humus
$F_{bw} = F_{wa}$: with W = 0 in natural systems
      $= F_{disturb}$: disturbance flux (fire, insects, storms) in natural systems
$F_{bw} = F_{harvest}$: harvest in managed and natural forests, roundwood production
$F_{wa} = F_{decay}$: decay of wood products, burning of wood and biomass

Figure 2. Systems diagram for pools and fluxes in temperate forests including wood products in human society.

Table 1. C pools and above ground net C fluxes in temperate forests *ca.* 1990.

Region	Forested area (Mha)	Living Biomass (Gt C)	Total C pool (Gt C)	Net Primary Productivity (Mt yr⁻¹)	Net storage in living biomass (Mt yr⁻¹)	C removed in harvested wood (Mt yr⁻¹)
Australia	39.8	2.3	6.9	60	12	
Belarus	6.3	0.4	1.2	9	2	2
Canada (east)[a]	26.8	1.0	3.9	10[b]	2	5
Chile	7.5[c]	0.4	1.2	11	2	
China	45.0[c]	2.6	7.8	67	13	
Europe[d]	90.0[c]	5.1	15.3	135	27	25
Japan	24.7	1.4	4.2	37	7	
New Zealand	7.5	0.4	1.2	11	2	3
Russia	100.0[c]	5.7	17.1	150	30	14
Ukraine	9.2	0.5	1.5	14	3	3
USA[e]	243.2	13.9	38.5	388	105	140
Total	600.0	33.7	98.8	892	205	192

Source: UN-ECE/FAO (1992) unless noted.

a   Kurz *et al.* (1992), excludes 130 MHa of temperate forest in western Canada. This area is included in the boreal forest paper.

b   Estimate includes effects of disturbances.

c   Pekka Kauppi, personal communication (1993).

d   Excludes temperate forests in Scandanavian countries, which are included in boreal forest assessment. Also excludes former USSR.

e   Birdsey (1992), excludes Alaska and Hawaii.

3. Trends in Net Carbon Flux Over Time

The trends of net C flux over time may be estimated qualitatively. Past net C fluxes can be inferred from land use histories. We concentrate on Europe and the United States where we have more information. Currently, temperate forests are a C sink because of area expansion and forest regrowth. Future trends are uncertain even if no climate change is assumed because of heavy human demands on these forests.

3.1. PAST

3.1.1. History of European forests

Without human impact more than 90% of Central Europe would be covered by forests. The first clearing for pasture and primitive agriculture took place during the Bronze Age mostly on lowlands with fertile soils and mild climate. By the 12th century, widespread clearing of forests for agriculture and harvesting for fuelwood had occurred. Forest cover was reduced to 30% (Deutscher Bundestag, 1990). In the early 1800s, much natural forest was converted to even-aged coniferous monocultures in response to the industrial revolution and increased demand for wood. However, problems with this type of management have led recently to silvicultural practices which favor deciduous trees. Forest area has been increasing since the beginning of the 20th century. Thus, European forests were a weak CO_2 source in the 19th century, and a weak sink in the 20th century.

3.1.2. History of U.S. forests

By the early 1800s, only about 15% of pre-colonial forest in the continental U.S. had been cut, mainly in the East (Heath and Birdsey, this volume). Forests were rapidly cleared for agriculture and wood products, and by 1850, approximately 65% of the forests remained. Harvesting and land clearing continued at a rapid rate through the early 20th century as the U.S. land base and population grew. In the eastern U.S., the area of forestland has increased since the mid-1900s as land used for agriculture was abandoned and reverted to forests.

Harvesting was not the only influence on the forest during these two centuries. Wildfires have played a significant role in the landscape, unlike Europe, where fires are of only local importance. Wildfires annually consumed an estimated 8 to 20 Mha before 1930 (MacCleery, 1992). Repeated wildfires on the same areas left an estimated 32.4 MHa unstocked in the 1920s. After fires began to devastate lives and property in communities, strong fire programs were instituted. By 1960 and through the present, the area burned annually was reduced to about 10% of pre-1930 levels.

Based on this history, we can speculate that forests were small C sources in the early to mid-1800s. In the latter part of the 19th century, forests probably released great amounts of C as harvesting and land clearing increased throughout the end of the century, with huge wildfires unleashing CO_2 through the early to mid-20th century.

After this time, as forests became re-established on agricultural lands, fire suppression programs succeeded and fossil fuels took the place of fuelwood, the forests slowly became a sink of C.

3.2. USE OF FORESTS FOR FUELWOOD AND CURRENT NET C FLUX

The history of forests in the temperate zone is inextricably linked with energy production and fossil fuel use. Much of the wood harvested from forests in the past in both Europe and the United States was burned as fuelwood in the home. Over 90% of harvested wood was burned as fuel as late as 1850 in the U.S., while Europe used about 30%. Population growth continued, which led to local wood shortages that increasingly forced substitution of fossil fuels for fuelwood. Currently, fuelwood supplies under 10% of energy needs on these continents (Mather, 1990). That temperate forests are presently sequestering C is due directly to human preference of burning fossil fuels for energy (and thereby releasing CO_2 into the atmosphere) rather than fuelwood.

3.3. FUTURE

Projections of C pools and fluxes (Birdsey et al., 1993; Turner et al., 1993; Heath and Birdsey, this volume) assuming no changes in climate indicate that temperate forests in the U.S. will remain a weak sink over the next 50 yr. However, the rate of sequestration is projected to decrease as harvest levels increase to meet growing demands for wood. Figure 3 indicates the general shape of the net C trends in the temperate forests of Europe and the United States. No climate change is assumed. The future range of uncertainty in the future is due purely to human effects. Increasing demand for wood may change the forests into a C source, but adoption of policies to sequester C such as afforestation may keep the forests as sinks. Most temperate forests are managed, so that influencing net C flux by forestry activities is easier to realize that in the tropical or boreal zones.

4. Management Practices to Keep Temperate Forests a Sink

European nations such as Germany have been considering forestry activities to maintain forests as C sinks. We present some of Germany's suggested management practices in this section. A summary of the options and the duration of their effects are listed in Table 2.

4.1. CONSERVING FOREST (BIOMASS) RESOURCES

In comparison to tropical and subtropical regions, the conditions for sustainable management of temperate forests are favorable. All efforts must be made to practice sustainable management over the entire region.

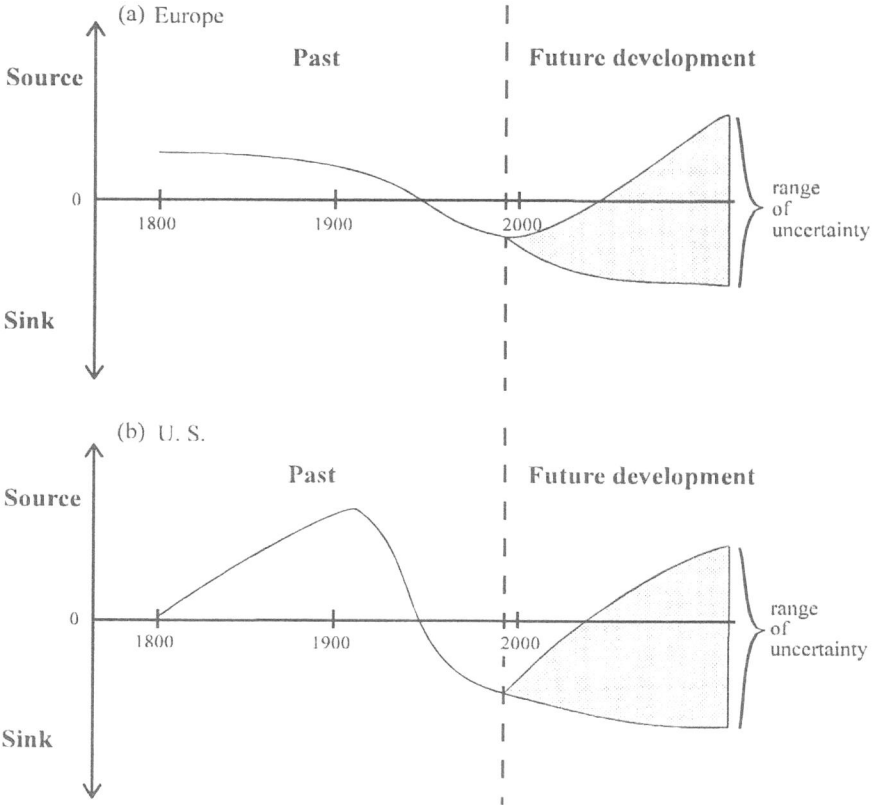

Figure 3.   Historical development and suggested future changes of C in temperate forests in (a) Europe and in the (b) United States.

Existing forest biomass in temperate zones of Europe and North America is not endangered by deforestation, but may be by degradation. In highly industrialized regions forests are stressed by anthropogenic air pollution, leading to forest dieback. This problem can only be solved by measures outside the forestry sector through reduction of emissions. Stand density and stand age are increasing on average in most countries in the temperate region, which tends to increase the risk of disturbances. Climate change may result in an increase of pests, windfalls and fires. Therefore intensive efforts must be undertaken to manage forests in such a way that these disturbances are minimized.

Option	short term 0-20 yr	mid term 21-60 yr	long term >60 yr
Conserving forest resources			
- sustainable management	x	x	x
- protection against deforestation and degradation	x	x	x
- reduction of air pollution		x	x
Increase of forest productivity			
use optimum rotation period	x	x	x
change of tree species		x	x
enrichment planting	x	x	x
use of appropriate harvesting practices	x	x	x
- soil properties			
fertilization, amelioration	x	x	x
erosion protection	x	x	x
no full tree utilization	x	x	x
less slash burning	x	x	x
- forest structures			
adapted and adaptable tree species		x	x
mixed stands		x	x
regulation of age and space structure		x	x
underplanting	x	x	x
- silvicultural systems			
transforming coppice into high forest systems		x	x
biomass conserving regeneration systems	x	x	x
Afforestation			
natural succession	x	x	x
high forests		x	x
energy plantations	x	x	x
Improvement of wood utilization			
- energy substitution	x	x	x
- material substitution (nonrenewable or energy intensive material substitution)	x	x	x
Waste management			
- energy production from used forest products	x	x	x
- improved landfill		x	x

Table 2. Options to manage the C budget using forestry related measures, and duration of effects.

4.2.   INCREASE OR RESTORATION OF FOREST PRODUCTIVITY

In terms of development dynamics, managed forests represent the aggradation phase of the forest ecosystem. Compared to mature natural forests, they are characterized by a lower amount of biomass but a high growth increment. Restricted harvesting would initially increase C stored in forests, but eventually growth would slow, as would C sequestration. In the long-term, harvesting forests at the optimum time and producing wood products and fuelwood as a substitute for fossil fuel might sequester a greater amount of total C.

Because of past devastative practices in some regions still persistent an improved accumulation of C can also be achieved by different measures. The main activities are include increasing stand productivity and biomass, improvement of soil properties, improvement of forest structures, and transformation of coppice systems to high forest systems.

4.3.   AFFORESTATION

One of the major options to increase C storage by forests is the afforestation of non-forested land. It is difficult to quantify the potential storage capacity for afforestation in temperate zones. Afforestation depends on the availability of agricultural land, which is related to the surplus of agricultural production, especially in Europe and North America. In Europe the total potential for conversion of farmland into forest land is estimated to be approximately 44 Mha. For the U.S., a biological potential of approximately 100 Mha is estimated. The potential for the entire temperate zone remains uncertain. Other areas that could be used for afforestation are degraded and marginal lands as well as areas susceptible to erosion. Socioeconomic and ecological aspects must be considered when estimating afforestation potential.

Afforestation programs should consider that establishment of high forest is more efficient than energy plantations in terms of C storage and ecological aspects. The latter are preferable when wood has to be produced as substitute for fossil fuels. In any case, afforestation is only useful if a corresponding demand for wood products and energy exists or can be stimulated.

4.4.   IMPROVED USE OF WOOD

The use of wood in the form of long-lived products including construction lumber and furniture, and the recycling of paper and paperboard are important options to increase C storage. In addition, using wood as an energy source contributes to the reduction of CO_2 emissions by substitution of fossil fuels.

4.5.   WASTE MANAGEMENT

In spite of efficient recycling, forest products will eventually become municipal waste. An optimal solution would be to use such waste in energy production as a substitute for fossil fuels. If that is not possible, waste could be stored in abandoned coal mines or in landfills in such a way that CO_2 and CH_4 emissions are minimized.

5. Future Carbon under Climate Change

Future C estimates for the temperate zone are quite uncertain if increasing atmospheric CO_2 concentration and climate change are considered. Species migrations have been forecast for the temperate forests of the U.S. within the next 100 years (Davis and Zabinski, 1992; also see IPCC, 1990) and species extinctions suggested (Peters and Darling, 1985). If these forecasts are accurate, the area we describe as temperate forest will have to be redefined. Because of uncertainty of identifying the onset of climate change, we concentrate on phenomena known to occur: increasing CO_2 concentration, air pollution, and evolution.

5.1. CO_2 EFFECTS ON THE CELL, LEAF, AND STAND LEVEL

Direct effects of CO_2 enrichment on plants are well documented. (E.g., see Eamus and Jarvis, 1989.) They include stimulation of photosynthesis, growth, increased fruit size and production, reduced transpiration and stomatal conductance, changes in inter- and intra-specific competition, and enhanced tolerance against air pollution. Such effects depend on light and sufficient nutrient supply and are utilized in "CO_2 fertilization" of greenhouse plant production.

On the cell level, the enzymatic capacity to fix atmospheric CO_2 will influence the total amount of C that can be incorporated per time unit. Temperate deciduous tree species differ in their ability to activate the responsible enzyme. The photosynthetic CO_2 uptake rates will be limited by the ability of most deciduous trees to translocate the assimilates to storage organs or to convert them into woody material or both.

On the leaf level, direct effects of CO_2 concentration on stomatal aperture have often been reported for deciduous trees. Water loss via stomata can be reduced under the predicted CO_2 concentration increase, resulting in increased water use efficiency on the leaf level. From this it follows that forests on drier sites under an unchanged precipitation regime will respond more favorably to CO_2 enhancement. Alterations of stomatal density have been reported as a consequence of CO_2-doubling, but there is still considerable debate on this subject. CO_2 concentration increases as well as temperature increases will influence respiration of leaves. Compensation of respiration by photosynthesis at low photon flux densities have been reported for elevated CO_2 concentrations. However, both possible effects are of minor importance if deciduous trees develop more leaves per individual, which may occur with increasing CO_2 concentrations.

Competition between the same species as well as that between different species will depend on CO_2 supply. Within a uniform single-species stand the capacity of the growing area will be reached earlier at high CO_2 concentrations. Competition between different deciduous tree species will be influenced due to the fact that pioneer species

respond differently from climax species. Our knowledge is restricted to the response
of a few deciduous tree species during the juvenile phase of development.

Preliminary results (Dieter Overdieck, pers. communication) of mineral analysis of
young maple (*Acer pseudoplatanus* L.) and beech (*Fagus sylvatica* L.) indicate that the
enhancing effect of increasing CO_2 concentration on plant growth and production will
not be limited in soils of medium fertility because mineral concentrations of the tissues
(C, N, P, K, Ca, Mg, Mn, Fe) can decrease at least to a certain degree without
obvious negative effects on growth. Two ecological consequences of these effects on
mineral contents should be noted: 1) with decreasing mineral concentration, the
nutritive value of the food for herbivores decreases, and 2) for the same reason,
microorganisms in the soil may decompose the litter more slowly and less effectively.
However, since the absolute mineral amounts in the whole vegetation or in the single
sampling are greater, it can be predicted that the flux rates of nutrients will increase in
the biogeochemical cycles if the tropospheric CO_2 concentration continues to increase
in the coming decades. This also means that more minerals will be taken up from the
soil. Therefore, nutrient-poor soils could become impoverished faster than before.

Considering all these uncertainties an enhancement factor for the effect of CO_2-
doubling on pool-size of deciduous forest can only be speculatively defined from case
studies. It might range from 1.05 to 1.3 at the time when the preindustrial CO_2
concentration is doubled.

5.2.   IMPACT OF AIR POLLUTION

Currently terrestrial ecosystems receive considerable amounts of N compounds (up to
60 kg of N ha^{-1} yr^{-1} in Central Europe) from air pollution by dry and wet deposition.
As humid ecosystems are naturally characterized by N deficiency, N deposition from
the atmosphere will initially improve plant growth. Later it leads to nutritional
imbalances and finally causes nutrient deficiency (e.g.: Mg, Ca or Zn) or
predisposition to drought, frost or pests. Oligotrophic or ombrotrophic systems are at
highest risk by nitrogen deposition. Persistent input of only 5 kg N ha^{-1} yr^{-1} will shift
interspecific competition in a way that certain plant species may become completely
suppressed.

A large number of growth observations have been made in individual stands. In
Europe they have indicated increased growth rates, which have often been attributed to
the high N deposition. Severe decline of tree stands due to air pollution has been
observed on relatively small areas so that they have only an insignificant effect on the
C pools and fluxes of the entire forest zone. However, there is concern that air
pollutants can cause nutrient imbalance in soils and adversely affect tree leaves and,
thereby, affect the C budget in the long term.

5.3.   POSSIBLE EVOLUTIONARY CONSEQUENCES

The shifts in the relation of CO_2 to O_2 that occurred during earth's history was an
important feature for the evolution of plant and animal life. The doubling of

tropospheric CO_2 predicted to occur in the upcoming century will therefore have marked ecological consequences. These will be stronger in autotrophic than in heterotrophic organisms. Those with shorter generations will adapt to the changes faster than others with longer generations. Changes in host-pathogen relations are likely to occur, but co-evolution of both hosts and symbiotic partners is less likely.

6. Research needs

1. Continue monitoring forest biomass growth, disturbances, mortality and removal to improve database and time series for the estimation of forest biomass in temperate forests, particularly for Russia and China, detect CO_2 fertilization effects in temperate forest ecosystems, obtain data and reduce uncertainties about C reservoirs and turnover, improve dynamic temperate forest ecosystem models, and better assess the fate of the sink potential and the change of ecological niches and distribution of species. The coordination of research efforts at all sites and on all forms of land use is a prerequisite for a reliable database that can be used by all working groups.

2. Questions about effects of increasing atmospheric CO_2 concentration on various topics need to be studied. The topics include interactions of CO_2 and temperature effects on water balance in forest ecosystems, effects of the nutritional conditions on the response to elevated CO_2, and problems and limitations in upscaling CO_2 enhancement experiments to natural conditions.

3. Investigate the effects of air pollution (including pollutants such as O_3, NO_x, CH_4) on forest growth and health and C/N interactions, along with other non-climate factors such as population growth.

4. Improve climate change scenarios and their application as modules for ecosystem impact models (including extreme events).

5. More research is needed on ecological processes that will affect C sequestration, including competition relations between different tree species by variation of site and succession stage, species adaptability (genetic potential), and host/symbiosis and host pathogen interactions.

7. Summary

Historically, temperate forests have been impacted greatly by human activity. The area covered by forest has decreased over the past millennia to about one half of its potential. Forests have been converted for agriculture, and for human habitation. During this conversion the forests were a C source.

New statistics from UN-ECE/FAO indicate that fellings and natural losses accounted for only 70 to 80% of the net annual increment in many of the temperate zone countries. Coupled with the historical outlook, it can be surmised that living vegetation of temperate forests at present is a sink of atmospheric CO_2. Data are less consistent regarding forest soils. The annual removal of C in wood for products was approximately equal to the net storage in living biomass, indicating that harvesting and wood production plays an important role in these forests.

A large number of management options are available to further increase the fluxes from the atmosphere into the biomass pools. In the short term, rotation ages could be adopted that optimize C production, and land can be afforested. In the longer term (70 to 150 yr), it would be more efficient to harvest forest biomass and store C in wood products or produce energy as a substitute for energy produced from fossil fuels.

It will be difficult to improve the estimates of the future development of temperate forests as sinks or sources of CO_2. The main constraints are related to the global change itself and to the negative and positive feedback mechanisms. Climate change has a potential of changing the pools and fluxes of C within the temperate forest system notably to increase decomposition rates, to cause forest decline and, thereby, to convert forests into a C source. The direct effect of CO_2 on photosynthesis and growth (CO_2-fertilization) can also affect the budget calculations. It has been difficult to estimate that effect at the present time, and it will be increasingly difficult to forecast its impact on the long term.

8. References

Armentano, T.V., and Ralston, C.W.: 1980, The role of temperate zone forests in the global carbon cycle, *Can. J. of Forest. Res.* **10**, 53-60.

Birdsey, R.A., Plantinga, A.J., and Heath, L.S.: 1993, Past and prospective carbon storage in United States forests, *For. Eco. Manage.* **58**, 33-40.

Birdsey, R.A.: 1992, *Carbon storage and accumulation in United States forest ecosystems*, Gen. Tech. Rep. WO-59, U.S. Department of Agriculture, Forest Service, Washington, DC, 51 p.

Davis, M.B., and Zabinski, C.: 1992, Changes in geographical range resulting from greenhouse warming: effects on biodiversity in forests, in R. Peters and T. Lovejoy (ed), *Global warming and biological diversity*, Yale University Press, p. 297-308.

Deutscher Bundestag (ed.): 1990, *Protecting the tropical forests: a high priority task*, translated by G. Woods-Schank, Bonn, Germany, 968 p.

Eamus, D., and Jarvis, P.G.: 1989, The direct effects of increase in the global atmospheric CO2 concentration on natural and commercial temperate trees and forests, *Adv. Ecol. Res.* **19**, 1-55.

Heath, L.S., and Birdsey, R.A.: 1993, Carbon trends of productive temperate forests of the coterminous United States. *J Air, Water, and Soil Pollution*, In press.

IPCC: 1990, *Climate change, the IPCC scientific assessment*, University Press, 365 p.

Kurz, W.A., Apps, M.J., Webb, T. M., and McNamee, P.J.: 1992, *The carbon budget of the Canadian forest sector: Phase I.*, Forestry Canada, Northwest Region, Northern Forestry Centre, Edmonton, Alberta, Inf. Rep. NOR-X-326, 93 p.

Mather, A.S.: 1990, *Global forest resources*, Timber Press, Portland, OR, 341 p.

MacCleery, D.W.: 1992, *American forests, a history of resiliency and recovery*, USDA Forest Service in cooperation with Forest History Society, SF-450, 59 p.

Peters, R.L, and Darling, J.D.S.: 1985, The greenhouse effect and nature reserves, *BioScience* **35**, 707-717.

Sedjo, R.A.: 1992, Temperate forest ecosystems in the global carbon cycle, *Ambio* **21**, 274-277.

Turner, D.P., Lee, J.P., Koerpner, G.J., and Barker, J.R. (eds): 1993, *The forest sector carbon budget of the United States: carbon pools and flux under alternative policy options*, U.S. Environmental Protection Agency, Office of Research and Development, Washington, DC, EPA/600/3-93/093, 202 p.

UN-ECE/FAO: 1992, *The forest resources of the temperate zones: general forest resource information*, Vol. 1, New York, 348 p.

TROPICAL FORESTS: THEIR PAST, PRESENT, AND POTENTIAL FUTURE ROLE IN THE TERRESTRIAL CARBON BUDGET

SANDRA BROWN
Department of Forestry, University of Illinois,
W-503 Turner Hall, 1102 S. Goodwin,
Urbana, IL 61801, USA

CHARLES A. S. HALL
Department of Environmental and Forest Biology,
College of Environmental Science and Forestry,
State University of New York, Syracuse, NY 13210, USA

WILHELM KNABE
Environmental Research and Consultancy,
Rumbachtal 69, d4330 Muelheim an der Ruhr, Germany

JAMES RAICH
Department of Botany, Iowa State University,
353 Bessey Hall, Ames, IA 50011, USA

MARK C. TREXLER
Trexler and Associates, 1131 SE River Forest Road,
Oak Grove, OR 97267, USA

PAUL WOOMER
TSBF, c/o UNESCO-ROSTA, UN Complex, Gigiri,
P.O. Box 30592, Nairobi, Kenya

Abstract. In this paper we review results of research to summarize the state-of-knowledge of the past, present, and potential future roles of tropical forests in the global C cycle. In the pre-industrial period (ca. 1850), the flux from changes in tropical land use amounted to a small C source of about 0.06 Pg yr^{-1}. By 1990, the C source had increased to 1.7 ± 0.5 Pg yr^{-1}. The C pools in forest vegetation and soils in 1990 was estimated to be 159 Pg and 216 Pg, respectively. No concrete evidence is available for predicting how tropical forest ecosystems are likely to respond to CO_2 enrichment and/or climate change. However, C sources from continuing deforestation are likely to overwhelm any change in C fluxes unless land management efforts become more aggressive. Future changes in land use under a "business as usual" scenario could release 41-77 Pg C over the next 60 yr. Carbon fluxes from losses in tropical forests may be lessened by aggressively pursued agricultural and forestry measures. These measures could reduce the magnitude of the tropical C source by 50 Pg by the year 2050. Policies to mitigate C losses must be multiple and concurrent, including reform of forestry, land tenure, and agricultural policies, forest protection, promotion of on-farm forestry, and establishment of plantations on non-

Water, Air, and Soil Pollution **70**: 71–94, 1993.
© 1993 *Kluwer Academic Publishers.*

forested lands. Policies should support improved agricultural productivity, especially replacing non-traditional slash-and-burn agriculture with more sustainable and appropriate approaches.

1. Introduction

Atmospheric CO_2 concentrations have increased by almost 30% during the last two centuries (Houghton et al., 1992: Raynaud et al., 1993). The causes of this increase are the burning of fossil fuels and changes in land use, particularly conversion of forests to agriculture (Houghton and Skole, 1990). The relative magnitudes and contribution to atmospheric CO_2 concentrations of these activities have changed over time.

Prior to this century CO_2 emissions from changes in land use, mainly caused by agricultural expansion in temperate countries, were higher than emissions from the combustion of fossil fuels (Houghton and Skole, 1990). From the turn of the century until about the 1930s, global CO_2 emissions from changes in land use were similar in magnitude to those from fossil fuel combustion. Since then, world-wide fossil fuel use has soared, land-use change in temperate regions has diminished, and deforestation in the tropics has accelerated. Thus biotic emissions from the temperate zone declined greatly and have approached a balance as forests expanded onto abandoned agricultural lands and as logged stands regrew (Houghton et al., 1987). After about the 1940s, CO_2 emissions from the changes in land use in the tropics have dominated the flux from the biota to the atmosphere, although they remain far below global fossil-fuel produced fluxes.

Although the trends in CO_2 additions to the atmosphere from burning fossil fuels and changes in land use are relatively well known, the C budget cannot be balanced at present (Houghton et al., 1992). The balance between all known C sources and sinks results in an amount of about 2.2 ± 2.5 Pg C yr^{-1} unaccounted for or "missing". Some research suggests that mid-latitude ecosystems are the repositories of this "missing" C (Tans et al., 1990) while others suggest that the tropics are (Taylor and Lloyd, 1992). However, the magnitude of the proposed mid-latitude sink depends on the magnitude of the tropical flux to set the bounds (Tans et al., 1990). It is clear, therefore, that progress in balancing the global C budget will be made only with improved understanding of the C fluxes from the tropics.

1.1. THE CARBON BUDGET FOR TROPICAL FORESTS

The important reservoirs and fluxes of C in tropical forest ecosystems are shown in Figure 1. The main reservoirs are the C in biomass, including above and belowground components, necromass (dead organic matter), and soil, and the much smaller quantities stored in wood products off-site. The main biotic flows of C between the atmosphere and forests are net C fixation via photosynthesis (net primary productivity) and heterotrophic respiration (decomposition of detritus [fine and coarse litter and dead roots] and soil C). Some C fixed by photosynthesis is also transferred to coastal systems via export in rivers as dissolved and particulate organic C, currently estimated at 0.1 Mg C $ha^{-1}.yr^{-1}$ for the humid tropics (Meybeck, this volume; Hall et al,. 1992).

A net CO_2 flux to the atmosphere is also brought about by changes in land use from high to low C density systems (e.g., from forests to agriculture). Net releases of C to the atmosphere by land-use change depend on the amount of C stored within native

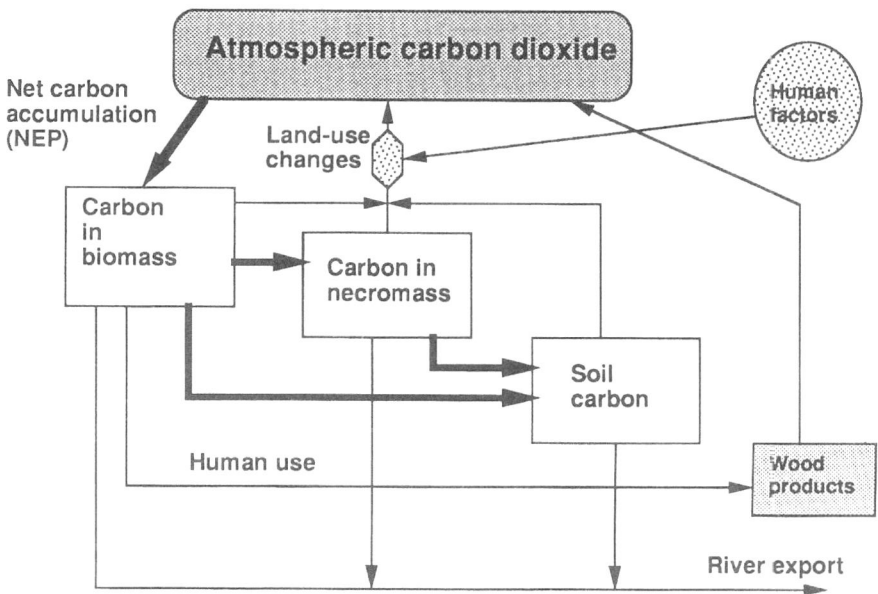

Figure 1. Basic model of the pools and flows of C within and between tropical forest ecosystems and the atmosphere caused by changes in land use or cover due to human and natural disturbances. The relative importance of the flows and stores varies over the course of time. The heavy arrows show the transfer of C from the atmosphere to the land during regrowth and recovery processes.

vegetation and soil, the rate and form of land clearing, and the intensity of soil disturbance. Rates of land-use change depend on social, economic, and political factors. A net accumulation of C occurs in forests recovering from disturbance via storage in biomass, necromass, and soil. The rate of C accumulation is a function of biotic and abiotic conditions and human activities, and can occur over decades to centuries (Lugo and Brown, 1992).

Tropical forests have undergone many changes in land use over the centuries. The changes usually follow certain pathways; the main paths of concern to the global C cycle are shown in Figure 2. The development process begins with mature tropical forests which may have once been primary forests, but most no longer are (Brown and Lugo, 1990). Mature tropical forests are converted to logged forests, permanent agriculture and pastures, or shifting cultivation. Logged forests, when managed or protected, regenerate and recover towards a mature state. However, they are also often converted to permanent or shifting cultivation because logging activities provide access to the forests (Knabe et al., 1990). Shifting cultivation and forest fallows include areas just cleared and burned for subsistence agriculture and areas in natural forest regrowth. Under long rotation periods, this system is relatively sustainable. However with ever-shortening cycles, particularly in areas with increasing population pressures, these lands often become badly degraded. Poor

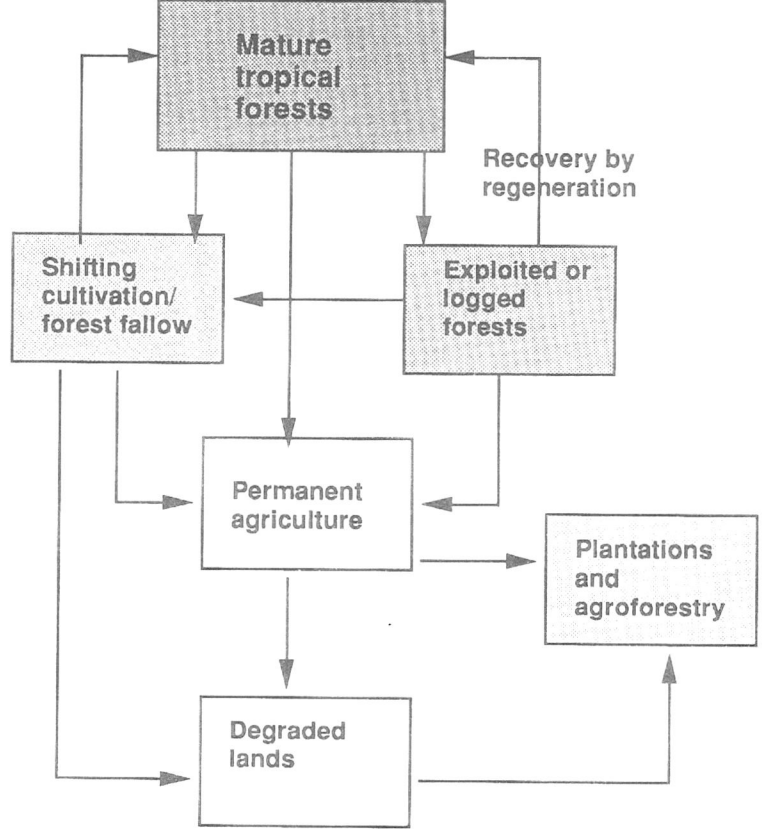

Figure 2. Land-use changes in the tropics of concern to the global C cycle.

agricultural practices or high grazing pressure also can lead to degraded lands. Re-establishment of vegetative cover through regeneration, plantations, or agroforestry or reduction in grazing would help to rehabilitate these degraded lands.

    To determine the role of tropical forests in the global C budget requires that the model in Figure 1 be quantified thoroughly for all the land uses in Figure 2, from the past to the present. This has generally been done for most land uses and land-use changes in Figure 2 as far back as the 1800s as well as for recent decades (Brown et al., 1986; Detwiler and Hall, 1988; Hall and Uhlig, 1991; Houghton et al., 1983). Agroforestry systems generally have been ignored because their rates of establishment are small and exert very little influence on the global C cycle at present. Rates of C accumulation or loss in mature forests have also been ignored in C models to date on the assumption that they are in steady state with the atmosphere, an assumption that is difficult to support today (Brown and Lugo, 1990; Lugo and Brown, 1986,1992). It is now time to relax this assumption and develop complete C budgets and their changes for all land uses (Lugo and Wisniewski, 1992).

During the last two centuries the role of tropical forests in the global C budget has been complex, and many data gaps and uncertainties still exist. The main purposes of this paper are to (1) present results of research that quantify the C model in Figure 1 for the land uses and land-use changes in Figure 2 for past (about 1850), present (1990), and future (ca. 2050) conditions as a means of summarizing the present state of knowledge concerning the role of tropical forests in the global C budget and (2) assess the alternatives for tropical land management for C conservation and sequestration. We also give recommendations for future research needs to improve C budgets for the tropics and to undertake effective mitigation options.

2. Past Carbon Budgets

2.1. PAST CARBON FLUX ESTIMATES FROM DECONVOLUTION

Estimates of C fluxes between the terrestrial biota and atmosphere can be derived from simulation models of land-use change (e.g., Detwiler and Hall, 1988; Houghton et al., 1983) or from another method referred to as deconvolution (e.g., Sarmiento et al., 1992). This method uses the historic record of atmospheric CO_2 concentrations (obtained from ice cores, and then direct measurements after 1958) and ocean models to calculate annual C emissions and accumulations required to generate the historic atmospheric record. Subtraction of only the fossil fuel emissions from the total yields a residual non-fossil flux of C which is usually assumed to be a terrestrial flux.

Results from the deconvolution suggest a steady net biotic source of about 0.4-0.6 Pg C yr[-1] to the atmosphere from 1850 to about 1935. After this time period, the flux became a net biotic sink (despite the apparent increase in C source from changes in land use in the tropics; cf. Figure 3) reaching a maximum of about 0.8 Pg C yr[-1] in the 1970s. Then it abruptly became zero by the 1980s, where it has hovered until the present time. The cumulative net flux for the whole time period was a net release of 25 Pg. This differs markedly from the results of Houghton (1992a) who suggests that the tropics alone have been a source of about 67 Pg C.

If the deconvolution results are valid, there has been a large biotic C sink operating in the terrestrial landscape that we know little about. The total sink is not only that suggested by the deconvolution analysis itself, but also that to overcome the net source caused by changes in land use (not considered in the deconvolution analysis). Clearly we are still far from understanding the past C budget which of course influences our understanding of the present and future budgets for the tropics.

2.2. WHOLE TROPICS

Past estimates of terrestrial C pool sizes in the tropics (latitudes 23.5 N to 23.5 S) are limited to biome-specific estimates of natural vegetation types only vaguely arranged by latitude (e.g., Whittaker and Likens, 1973). Estimates of net C fluxes from the tropics have been made by extrapolating changes as far back as to about 1750, under the assumption that prior to this time, the tropical forest landscape was in steady state with respect to C exchanges to and from the atmosphere.

Humans have been active components of the tropical landscape for millennia, and their populations have fluctuated widely in specific areas (e.g., Mayans and Incas of Central and South America, and the past great civilizations of Cambodia and Sri Lanka). Despite these fluctuations, there has been an overall increase in human populations through time and a parallel increase in human-induced forest disturbance. It is likely, therefore, that the tropical landscape has not been in C steady state in recent history. Carbon losses occur in areas of increased human activities but C gains also occur on previously disturbed and now abandoned lands. The magnitude of these gains and losses are likely to be different across different geographic regions and what their balance is in time and space at this scale is largely unknown. Factors to consider are the general increase in human populations and economies, implying greater overall C losses than gains in forests, and the increased substitution of fossil fuels for wood fuels in many tropical countries implying reduced pressure on forests. These factors make clear that the relationship between population growth and forest disturbance is not linear.

Since 1850, the estimated contribution to the total biotic CO_2 emissions from land-use change in the tropics varies by region for different time periods (Figure 3). Until the early 1900s, the emissions were estimated to be greatest from tropical Asia, after which time significant increases occurred in Latin America. This region became the dominant source until about the mid 1970s. From the 1970s to the present time, emissions from Latin America and Asia have been about equal to each other. Emissions from Africa appear to have been the lowest during the whole period of record.

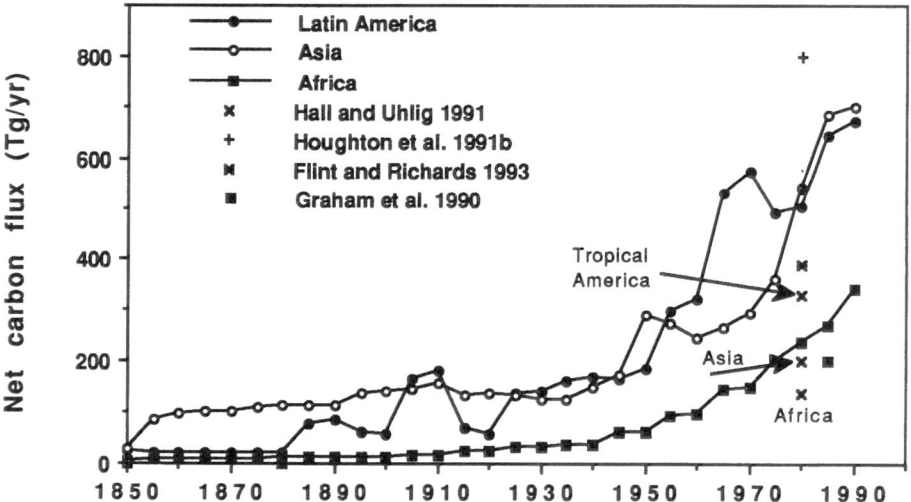

Figure 3. Net C flux from changes in tropical land use from 1850 to 1980 (data from Houghton, 1992a and indicated sources).

The high emissions from tropical Asia during the early part of the record reflect the accelerated pace of forest conversion and degradation caused by the introduction of plantations of spices, tea, rubber, and coffee after the Europeans arrived in the 19th century (Richards and Flint, 1993; Whitmore, 1984). This pace was accelerated further as mechanized logging was introduced and log extraction occurred over large areas (Whitmore, 1984). Large scale conversion of forests to croplands and pastures occurred after 1940 in Latin America (Houghton et al., 1991a) and still later in tropical Africa (Houghton, 1992a).

The total C flux from the tropics for the period 1850 to 1990 is estimated as 67.4 Pg (Houghton, 1992a). A best estimate for the annual flux for 1850 is 0.06 Pg C yr^{-1} (Houghton, 1992a), with an uncertainty range of 0.02-0.10 Pg C yr^{-1}.

2.3. REGIONAL ANALYSIS

Improvements in estimates of C emissions from the tropics are likely to be made when each region, with its unique history of land use and other ecological and socioeconomic information, is considered within the regional context rather than extrapolating a few data (e.g., forest C densities) across the whole tropics. Such improved analyses have been done to date for Latin America (tropical and non-tropical countries by Houghton et al., 1991a, b; selected countries by Hall et al., 1985), and tropical Asia (Flint and Richards, 1993; Houghton and Hackler, 1993; Iverson et al., 1993a; Richards and Flint, 1993). In addition, Hall and Uhlig (1991), using country-specific-data present results for all of the tropics. Although it is expected that these regional studies have improved estimates, it is important to realize that considerable uncertainties still remain, particularly the further back in time one goes.

2.3.1. Latin America

The analysis by Houghton et al. (1991a, b) for this region spanned the period 1850 to 1985. The major improvement in this analysis was a refinement of the data on land use and rates of land-use change in this region; data on C pools in vegetation and soils were basically the same as those used in other analyses by this group (e.g., Houghton et al. 1987). Up until about 1930, the simulated annual flux was less than 0.2 Pg C yr^{-1}, followed by a period of a slowly increasing flux to about 0.3 Pg C yr^{-1} by 1950. From 1950, the simulated flux increased rapidly to about 0.6 to 0.7 Pg yr^{-1} by 1965. By 1985, the net flux had increased to about 0.95 Pg yr^{-1}.

From a land use perspective, expansion of pasture lands and croplands was responsible for the greatest release of C (76% of total; Houghton et al., 1991a, b). From a process perspective, decay of plant material left on site at the time of harvest or clearing caused the largest release of C (43% of the total), yet this pathway is one of the most poorly understood (e.g., decomposition rates of slash, etc.). Changes in soil C contributed the least amount to the total flux (13%).

The estimated C flux from this region for the whole period was 30.7 Pg C, with an uncertainty range of 17.3 to 35.3 Pg C. An earlier analysis (Houghton et al., 1983), produced an estimate for this region of 42.5 Pg C for the period 1860 to 1980. Houghton et al. (1991b) believe that the lower estimate in their new analysis was due to improved estimates of soil C losses, more accurate assignment of lands cleared to ecosystem types, and presumably to C pools, and a more conservative approach to dealing with fuelwood. An even more recent simulation of the land-use change model by Houghton (1992a) results in an estimated flux of 25.1 Pg C for the period 1850 to 1985, a reduction from the

Houghton et al. (1991b) analysis. The work on Latin America demonstrates that more region-specific data can reduce C fluxes from the land and reduce the uncertainty range.

Table 1. Estimated carbon pools and net fluxes (by time interval; negative sign is a flux to the atmosphere; other values are a flux to the land) associated with changes in land use during different decades from 1880 to 1980 for 13 tropical Asian countries (Flint and Richards, 1993). The flux from soils is not included.

POOLS (Pg C):	1880	1920	1950	1980
Closed forests	46.39	39.18	31.16	20.63
Discontinuous forests	3.32	2.79	2.56	2.15
Grassland/shrubland	0.92	0.98	0.98	0.88
Barren/sparsely vegetated	0.06	0.06	0.06	0.05
Wetland forests	5.38	4.47	3.51	2.30
Wetland non-forests	0.24	0.18	0.14	0.10
Temporary crops	0.51	0.62	0.74	1.05
Permanent crops	0.17	0.30	0.59	0.93
Settled area, etc.	0.01	0.02	0.03	0.05
Total forest	55.08	46.44	37.24	25.09
Total non-forest	1.92	2.16	2.53	3.06
Total	57.00	48.60	39.77	28.15

FLUXES (Pg C yr^{-1}):	1880 -1920	1920 -1950	1950 -1980	1880 -1980
Closed forests	-7.21	-8.02	-10.53	-25.75
Discontinuous forests	-0.53	-0.23	-0.41	-1.16
Grassland/shrubland	0.06	0.00	-0.10	-0.04
Barren/sparsely vegetated	0.00	0.00	-0.01	-0.01
Wetland forests	-0.91	-0.96	-1.21	-3.07
Wetland non-forests	-0.06	-0.05	-0.04	-0.14
Temporary crops	0.11	0.12	0.31	0.54
Permanent crops	0.13	0.28	0.35	0.76
Settled area, etc.	0.01	0.01	0.02	0.03
Total forest	-8.64	-9.20	-12.14	-30.00
Total non-forest	0.25	0.37	0.53	1.14
Total net	-8.39	-8.84	-11.61	-28.84
Average annual	-0.21	-0.29	-0.39	-0.29

2.3.2. Tropical Asia

Flint and Richards (1993) approached their analysis for this region somewhat differently from the simulation models used for Latin America. Complete C inventories were developed for this region based on compiling data from the literature for more than 90 subnational units at different time periods (3 to 4 decade long periods) between 1880 to 1980. Differences between the C inventories through time were used to calculate the net flux. This approach accounted for conversion between land uses (forests to croplands and vice versa) and degradation of lands (mostly forests) remaining under the same use.

However, it did not account for soil C changes nor for time lags caused by decay of organic materials either left on site (e.g., slash) or at the user end (e.g., wood products).

A summary of their results of C pools and fluxes for the Asian region is given in Table 1. During the 100 yr period 1880 to 1980, the C pool in forest vegetation decreased by more than half, generating a net C flux of 28.8 Pg. Houghton and Hackler (1993), using a terrestrial C model and adding a new routine to account for forest biomass degradation, obtained a net C flux, including soils, of 19.2 to 32.6 Pg for the period 1850 to 1990. The estimate by Flint and Richards (1993) would be larger and closer to the upper limit of Houghton and Hackler's results if soil C losses (about 4.5 Pg C for the period; Houghton and Hackler, 1993) were added and the period extended to 1990.

2.3.3. Tropical Africa

To date, no specific historical analysis of this region has been done, other than the results shown in Figure 3.

3. Present (1980-90) Carbon Budgets

The C pool in living above and below ground vegetation of tropical forests (trees only, which accounts for about 95% of the total; Brown and Lugo, 1992) was estimated to be 159 Pg and that in soil (litter layer and mineral soil to 1 m depth) 216 Pg (R. Dixon and S. Brown, unpublished results from ongoing research; Eswaran et al., 1993).

Estimates of C emissions from the tropics caused by changes in land use have varied widely, from 0.4 to 2.5 Pg yr[-1] for 1980 and from 1.5 to 3.0 Pg yr[-1] for 1989/90 (Table 2). The range represents about 10 to 50% and 25 to 54% for 1980 and 1990, respectively, of the emissions from global fossil fuel burning. The uncertainty in these estimates has been large due to high uncertainties in rates of deforestation, rates of forest degradation, fate of deforested lands, and C pools of the forests being cleared (Brown and Iverson 1992).

Table 2. Recent estimates of the carbon flux, Pg C yr[-1], from the tropical landscape for 1980 and 1990.

Source	Year	Range	Average
Detwiler and Hall, 1988	1980	0.4-1.6	1.0
Hall and Uhlig, 1991	1980	0.52-0.64	0.58
Houghton et al., 1987	1980	0.9-2.5	1.7
Houghton 1992a	1980		1.3
Houghton 1991	1989	1.5-3.0	
Houghton 1992a	1990	1.2-2.2	1.7

To further compound this uncertainty, rates of tropical deforestation have increased in the last decade, from about 12 Mha yr[-1] in 1980 to 15.4 Mha yr[-1] in 1990 (Aldhous, 1993), and these new estimates of land-use change have not yet been incorporated into all the models. Furthermore, there is now evidence that many of the "undisturbed" forests, particularly in tropical Asia are undergoing degradation, that is their C pools are being reduced, often by illicit removals of wood for fuel or timber that does not get recorded by the usual channels (Brown et al., 1991, 1993; Iverson et al., 1993a; Richards and Flint, 1993). Attempts to incorporate such forest biomass degradation into the terrestrial C models have been made for tropical Asia (Houghton and Hackler, 1993) but not the rest of the tropics. Evidence also exists for the reverse process, that is forests assumed to be undisturbed are accumulating biomass (Brown et al., 1993; Lugo and Brown, 1992). However, to what degree the loss or gain of biomass in mature forests is pan-tropical still needs to be determined.

Our present best estimate for the net C flux to the atmosphere from tropical land-use change in 1980 is 1.0 to 1.3 Pg yr^{-1} and for 1990 1.2 to 2.2 Pg yr^{-1} (Table 2). The estimate from land use change in the Brazilian Amazon alone accounts for up to 58% of the 1990 flux (Fearnside 1992). These estimates do not include fluxes associated with reduction in biomass caused by degradation nor biomass accumulation in forests classed as "undisturbed" or assumed to be in steady state.

Estimates of the 1980 C flux, for regional and total tropics, based on the work of Houghton et al. (1991b) and Houghton (1992a) are higher in all cases than estimates made by others (Figure 3 and Table 2). The lowest estimates at the regional scale are those made by Hall and Uhlig (1991). However, compared to the range of flux estimates made by the different investigators about a 5 to 10 yr ago, present estimates are substantially closer in agreement.

Although all studies use essentially the same data sets, different results are produced because of differences in preferred rates of land-use change and C densities, differences in how certain modeling routines allocate C after clearing and burning, and differences in modeling shifting cultivation, forest logging, and forest regrowth. For example, the model used by Hall and his co-workers retains a portion of the original forest C as long term storage in charcoal and soot (not considered in the other models). While this process is logical and supported by the literature, the value of the parameter is very uncertain (3-19% of the initial biomass C) and the model output is very sensitive to it. For countries with extensive shifting cultivation, a significant portion of the biomass can be converted to charcoal each cycle, and with many cycles over large areas this will substantially reduce net C emissions.

4. Potential Future Carbon Budgets

Many variables will affect future C storage in, and fluxes to and from tropical forested regions. The most important of these, certainly in the short to medium term, is continued anthropogenic alteration of the landscape. Longer term influences include increasing atmospheric CO_2 concentrations, possible changes in temperatures and precipitation, and changes in patterns and magnitudes of chemical inputs from the atmosphere. We evaluate the potential impacts of these different variables and their possible influence on net C fluxes from the tropical forest regions for the year 2050. First, we consider probable land-use changes, followed by a brief evaluation of the potential impacts of higher CO_2 concentrations (doubling pre-industrial values) on net C fluxes in the altered landscape. Then we evaluate the possible changes associated with predicted climate changes. Global climate models are consistent in predicting a mean temperature rise of 1 to 2 C throughout the year in tropical regions, but vary in their predictions of the total amounts and seasonal distribution of precipitation. Therefore, both increases or decreases in precipitation will be considered as possible scenarios.

4.1. LAND-USE CHANGE

Continuing anthropogenic impacts on tropical forests are the most immediate source of future C fluxes from the tropical landscape into the atmosphere. Predictions of anthropogenic flux are dependent upon many variables, and consequently remain speculative. Despite the many factors that are important, human population growth and the demand for agricultural lands will dominate by driving land-use changes for the foreseeable future. Other related issues that will influence the level of future deforestation, forest degradation, forest recovery, and other C gains and losses include economic and social variables such as economic growth, agricultural productivity improvements, government policies ranging from land tenure to taxation, resource constraints, level of industrialization, and international terms of trade (Deutscher Bundestag, 1990). They also include public perceptions regarding the role of forests in furthering or impeding economic development. The relationship between C fluxes and these variables will not be linear, nor will it be consistent among nations.

With continued and complete deforestation of tropical forests, the maximum amount of C that could be released to the atmosphere would be nearly equal to the present "active" forest C pool, or about 200 Pg (160 Pg in vegetation and 40 Pg C from soil; see Section 3. above and using a soil C loss rate given in Houghton et al., 1991b). This represents an average annual C source of 3.3 Pg yr[-1] for the next 60 years. This value is approximately twice the net flux estimated for 1990 (Table 2). However, not all tropical forests will ever be cleared; even now rates of deforestation in several nations have slowed (e.g., Thailand and Brazil).

Not only are some tropical countries making efforts to curb their deforestation rates, but efforts by the international community are also being proposed and mounted to reduce rates. For example, the Enquete-Commission of the German Bundestag proposed an aggressive three-stage program to counteract deforestation (Deutscher Bundestag 1990). It first proposed a crash program within the scope of international negotiations to slow rates of deforestation and to protect undisturbed forests that are in great danger of being destroyed. Next, it proposed that deforestation be brought to a complete halt by 2010, followed by a program to restore forest cover in tropical countries to the level of 1990. Funding for such a program is to be accomplished from yearly contributions by developed, non-tropical countries. Certainly the outcome of the 1992 United Nations Conference on Environment and Development (particularly Agenda 21 and the Biodiversity and Climate Change Conventions) provides the groundwork for such a program proposed by the Enquete-Commission.

A recent study attempts to anticipate likely deforestation rates for 54 tropical countries encompassing most of the tropical forest region (Trexler and Haugen, 1993). This study starts with a 1990 baseline rate of tropical deforestation of 15 Mha yr[-1], with a corresponding C release of 0.95 to 18 Pg C yr[-1] under low and high biomass scenarios. These figures correspond well with other estimates of both the areal extent of deforestation (Aldhous, 1993) and associated C release (Table 2). Future deforestation rates were estimated by Trexler and Haugen for each country on a decade by decade basis, based on a subjective consideration of many physical, social and economic variables such as those mentioned above, and upon extensive interviews with country experts. They did not include natural processes of forest recovery and growth nor reduction in forest biomass density caused by degradation. The study concluded that 660 Mha are likely to be deforested in these 54 countries through 2050 (reducing the 1990 forest area by about a third), with net C emissions of 41 to 77 Pg under low and high biomass scenarios. By 2050, emissions level off towards 0.5 to 1.0 Pg C yr^{-1}, and would probably decline thereafter.

4.2. CHANGES IN ATMOSPHERIC CO_2 CONCENTRATIONS

By the year 2050 it is estimated that the atmospheric CO_2 concentration will be approximately 2.5 times the pre-industrial level of 280 ppm under the business-as-usual scenario (Houghton et al., 1992). It has been hypothesized that this change in C availability to plants will modify C exchange rates between the tropical biota and atmosphere (i.e., CO_2 fertilization), irrespective of any changes in climate.

A principal effect of increased atmospheric CO_2 concentrations is increased C fixation rates over the short term, as documented in a variety of field and laboratory experiments (e.g., Kimball et al., 1993). Increased rates of photosynthesis and plant growth favor C assimilation by plants and would thereby diminish the net C flux from the tropical region, if short-term responses of plants are predictive of their long-term responses, which is uncertain.

Recent work with humid tropical forest mesocosms in controlled-environment growth chambers (Korner and Arnone, 1992) indicated that leaf area and plant biomass were not significantly increased under high-CO_2 conditions. Furthermore, although C stores did not increase significantly, litter production, fine root growth, rates of mineral nutrient leaching, and rates of soil respiration all increased. Thus, doubling of atmospheric CO_2 concentrations had a small effect on C storage within the mesocosms, but stimulated rates of C and nutrient cycling substantially. However, the study was of a very short duration (about 3 mo). The lack of a significant increase in C stores in the mesocosms does not imply that humid tropical forests will respond this way in the longer term. Nor does it rule out that other tropical forests will be stimulated.

Plant growth is expected to be stimulated in tropical dry forests because they tend to be water-limited rather than nutrient-limited and CO_2 enrichment increases water-use efficiency (measured as the amount of water lost per unit of C assimilated) by C3 plants. Changes in the water-use efficiency of plants may have very important impacts in seasonally dry and arid regions. Under field conditions, it is possible that rates of C assimilation throughout the growing season will increase dramatically under high-CO_2 conditions (e.g., Idso, 1988), leading to a greater potential for agriculture, agroforestry, and forestry in these regions, and concomitant increases in net C storage. Furthermore, soils of the dry tropics tend to be more nutrient rich compared to the humid tropics as they have not been intensively leached. Although difficult to predict because of the lack of research on this topic, it is in the seasonally dry and semiarid lands of the tropics that we may expect to see the greatest changes in plant productivity and C sequestering. However, changes in climate could counteract this expectation if they become more arid.

Increases in plant C:N and root:shoot ratios are also observed under high-CO_2 conditions (Norby et al,. 1986; Overdieck this volume). The consequent change in the C:N ratio (and ratio of C to other plant nutrients) of litter produced by the vegetation has the potential to increase immobilization of plant nutrients by increasing competition between soil organisms and plants, and thereby decrease plant nutrient availability and plant production. This reduction may be balanced against the opposing effect of greater (potential) root production under high-CO_2 conditions. With a greater ability to exploit soil nutrient and water resources, plants may in fact benefit from high-CO_2 conditions. It is clear that further research is needed to shed light on these potential ecosystem responses.

All in all, there is no reason now to accept or reject the hypothesis that a change in the relative availability of CO_2 and plant nutrients caused by a change in the atmospheric CO_2 concentration will impact tropical vegetation dramatically. Experiments larger and longer than any yet conceived by ecologists would be necessary to begin to give definitive answers to this persistent question.

4.3. CLIMATE CHANGE IMPACTS

Climates have changed and will change from natural and anthropogenic phenomena. These changes in the past have changed vegetation patterns dramatically in the tropics. For example, much of the Amazon basin that is now forested was savanna during the Pleistocene. Natural climate changes are relatively slow, whereas human-induced climate change is expected to occur more rapidly. With respect to modeling the effects of greenhouse gas emission, global climate models are consistent in predicting a 1 to 2 C increases in temperature throughout the tropical region by 2050 (Houghton et al. 1992). These changes will lead to higher air and soil temperatures, greater water demand by the atmosphere, and, without compensating changes in precipitation, greater aridity during dry periods. Therefore, we expect direct temperature effects on plants, other organisms, and soil, and indirect effects on water and nutrient availability.

Not only will climate change affect processes within ecosystems, but it is likely to cause shifts in present distribution of vegetation types within and between biomes as well (Cramer and Solomon, 1993; Leemans, 1992; Neilson, this volume). For example, within the tropical latitudes, the area of dry forests and savannas is projected to increase by about 11% over present values under the future climate at double CO_2, whereas the areas of tropical seasonal and moist forests are projected to decrease by about 4% or less (R. Leemans, pers. comm.; results based on the BIOME model of Prentice et al., 1992). The changes are not uniform over all continents; tropical moist forests in South America are projected to increase in area over present day extent by about 5% whereas all other continents will experience a loss of up to 20% in Asia.

Increasing temperatures are expected to increase both net photosynthesis and plant respiration rates (e.g., Larcher, 1983). If rates of net photosynthesis increase more slowly than do rates of plant respiration (for example if nighttime temperatures increase more rapidly than daytime temperatures), it may be that net productivity at the stand level will diminish. However, statistical summaries show a positive correlation between total net primary productivity (NPP) and temperature (e.g., Box, 1978) and between actual evapotranspiration (AET) and NPP (Raich et al., 1991; Rosenzweig, 1968). Temperature and AET are closely related where water is abundant. At present, there are no stand-level empirical data which support a decrease in productivity at higher temperatures within the range of 20-30 C, despite the theoretical strengths of this argument.

Increasing soil temperatures increase rates of CO_2 efflux from soils, all other factors remaining the same (Raich and Schlesinger 1992). Under the assumptions that soil temperatures in the tropics will increase 1 to 2 C by change in climate and that rates of soil-CO_2 production double for every 10 C increase in temperature, the rate of soil-CO_2 efflux in tropical dry and moist forests could increase by 8% and 16% for a 1 and 2 C temperature rise, respectively. However, this potential C loss may be negated by greater C inputs to the soil via increased rates of plant productivity and changes in the chemical characteristics of plant litter (see Section 4.2). For example, an increased lignin:N ratio increases the proportion of litter inputs that become stabilized soil organic matter (Parton et al., 1987). Thus, C cycling rates could be stimulated with minimal impacts on mean net C pool sizes. Again, the actual effects of climate warming are not readily predicted from existing information.

Predicted changes in magnitudes and distribution of rainfall in the tropical forest regions of the world vary among the models from decreases to increases in different places. Plant production tends to increase with increasing precipitation in the tropics (Brown and Lugo, 1982; Raich et al., 1991). Decreasing precipitation and/or increased aridity due to increased evapotranspiration demands can be expected to decrease net C uptake. Furthermore, potential changes in water budgets will affect streamflow and export of C from the land by rivers. Without better information it is impossible to predict what the net effects of changes in precipitation will have on net C fluxes in tropical forests in 2050.

4.4. EFFECTS OF ALTERED PATTERNS OF CHEMICAL INPUTS TO THE ATMOSPHERE

Human activities stimulate the biogeochemical cycling rates of many materials, including the plant nutrients N, P, and S (Peterson and Melillo, 1985). This eutrophication of the biosphere could increase the forest C sink by about 0.25% of the annual forest net primary production only, and mostly in the temperate zone (Peterson and Melillo 1985).

Not only do human activities simulate the cycling of plant nutrients, but they also produce other gases and air pollutants that adversely affect forests. For example, the increase in tropical fires and biomass burning produces (1) methane and CO which interfere with stratospheric ozone, and (2) hydrocarbons, NOx, and SO_2 which are precursors of phytotoxic photo-oxidants (Andreae et al., 1988; Crutzen et al., 1989). These gases can affect forests by reducing photosynthesis and causing adverse effects on soils and root systems (Knabe 1983). Thus, there is the potential to damage tropical forests, but so far to an unknown degree.

4.5. INTERACTIVE EFFECTS OF CHANGES IN LAND USE, CLIMATE AND ATMOSPHERIC CO_2 ON NET CARBON FLUXES

The interactive effects of all future processes impacting tropical forests are extremely complex and are poorly understood as the forgoing single-effect discussions have illustrated. It is likely under current land management practices, however, that anthropogenic land-use change will overwhelm any future changes in C fluxes (sources or sinks) from CO_2-fertilization and climate change as less forest vegetation is left to respond. Even if C pools of forests increase due to CO_2-fertilization and/or climate change, then those forests being cleared in the future will have larger C pools and thus release larger quantities when they are deforested. Further increases in C fluxes from changes in land use could be slowed or reversed with different land management practices as will be discussed in Section 5.

Comprehensive models are needed to integrate all of the effects due to CO_2-fertilization, climate change, and other atmospheric chemical inputs within the next 5 to 10 yr. Information and sub models are now available to start this process. These models could give more accurate estimates of future C storage and fluxes, although the accuracy of the predictions would still be constrained by the broad uncertainties of social and political activities.

5. Potential for Management of Tropical Lands for Carbon Conservation and Sequestration

The "business-as-usual" scenario is seen above (Section 4.1) to result in the ongoing loss of tropical forests at a rapid rate. Anthropogenic deforestation will likely continue to be rooted in the same complex set of demographic, institutional, and policy variables that are operative today. This magnitude of forest loss and other forms of land degradation will impose tremendous economic, social, and environmental costs on the people and resource bases of many tropical countries. Past attempts to reduce global deforestation rates have met with limited success, partially caused by severe resource constraints for the types of programs that have been called for repeatedly, including forest protection, agricultural intensification, and natural forest management. The expanding interest in climate change mitigation may both be a lever to encourage policy reform in tropical forest nations, and to encourage international funding of measures to protect existing forests and to expand reforestation efforts (see Section 4.1 and Deutscher Bundestag, 1990). Such programs will be successful only if they are integrated with the social and economic needs of the local people.

Observation of the biotic C cycle suggests several techniques that would help slow the accumulation of C in the atmosphere. These are:

1. Slowing the loss and degradation of existing forests. This can be furthered through establishing protected areas, forest reserves and parks, managing natural forests for multiple use, improving the practices and efficiency of exploitative activities such as logging, and in particular working with shifting cultivators to implement different agricultural practices, thus decreasing pressure on remaining forests.

2. Expanding the size of existing C reservoirs in vegetation and in products. Over 1 Gha of previously forested land have been converted to agricultural, pastoral, and other uses in the tropics. In many cases this land has proved incapable of sustaining long-term uses, is continuously degrading, and may even be abandoned. Very large areas of land would be more productive and economic over time if returned to tree cover. This can be promoted by encouraging the regeneration or reforestation of previously forested areas, promoting sustainable agricultural practices to reduce slash-and-burn pressures, and assisting in the recovery of severely degraded forests. Such measures would often be ecologically and socially beneficial. Hundreds of millions of additional hectares could support in principle both higher agricultural yields and additional tree cover through agroforestry systems.

3. Conserving and expanding the size of soil C reservoirs. Forest clearing and continuous cropping has resulted in near universal declines of soil organic C. This C is lost directly into the atmosphere or transported via runoff and soil erosion. Present soil conservation measures are a start in the maintenance of existing soil C pools, but the

ability of degraded soils to re-accumulate C as a result of reforestation or other land management practices has significant long-term potential.

4. Substituting biomass for present and future fossil fuel use in energy production. Either waste products from other industrial processes may be used (e.g. sugar, rice, and timber production residues), or biomass can be grown specifically for energy production (see Sampson et al., this volume). Biomass could be grown in large-scale plantations on presently non-forested land or rely on a dispersed supply system using small-scale growers. Biomass could displace both the small-scale use of kerosene and gas-use in cities and the need for fossil-fuel electric plants.

The potential of these options to slow CO_2 accumulation in the atmosphere depends on previous experiences, costs, the response of the local population, and the will of governments and the international community to conserve forests and stabilize global climate (Deutscher Bundestag, 1990).

Several studies have assessed the availability of land on which additional C could be stored (Brown et al., 1992; Grainger, 1988, 1990; Houghton et al., 1991c; Iverson et al., 1993b). A summary of the literature concluded that land technically suitable in the tropics for expanded management and agroforestry is somewhere between 0.62 to more than 1 G ha (Dixon et al., 1991). These studies of land availability tend to define land potential differently, resulting in an inability to compare estimates and methodologies. In addition, almost all of the work has focused strictly on the physical availability of land, which is not likely to have much relationship to how much land can actually be made available for forestry interventions. Land regarded as degraded is often a vital base of subsistence for somebody and would be defended against any attempt to reforest without advantage to the traditional user. That is, very little of what might be considered degraded tropical lands is in practice likely to be abandoned land.

These estimates of land suitability must be viewed with care in that they generally omit consideration of the political, social, and economic forces influencing true land availability. Furthermore, very few projections exist of the potential for slowing deforestation or exploiting biomass resources in place of fossil fuels. An attempt to take these various variables into account for more than 50 tropical countries is presented by Trexler and Haugen (1993). Based on country-specific information and experts, the study estimates by decade how much deforestation could plausibly be reduced, how much land could be assisted to regenerate, how much land could be converted to agroforestry systems, and how much land could reasonably be converted to plantations in light of infrastructural and supply and demand variables (Table 3). The end-use of several hundred million hectares could probably be changed through concerted policy interventions between now and 2050.

Trexler and Haugen (1993) conclude that as much as 50 Pg C could be kept from or removed from the atmosphere by 2050 through the aggressive pursuit of the practices discussed in Table 3. Figure 4 illustrates continent by continent the gradual reduction in emissions resulting from the implementation of these practices. Slowing deforestation and natural forest regeneration account for the largest C gain (78%), followed by plantation establishment (19%), and agroforestry (3%) (Figure 5). This trend in C gains confirms previous experience and analysis that showed forest conservation and natural regeneration is cheaper, easier, and less likely to be resisted by the local population than reforestation (Deutscher Bundestag, 1990).

Table 3. Estimates of areas of tropical lands in 54 countries potentially and actually
available for C sequestration options (Trexler and Haugen, 1993).

Sequestration option	Potential area	Actual area
	(Mha)	
Slowing deforestation	647.3	509.2
Natural regeneration	314.8	216.7
Agroforestry	156.5	66.1
Plantation establishment	72.8	66.8

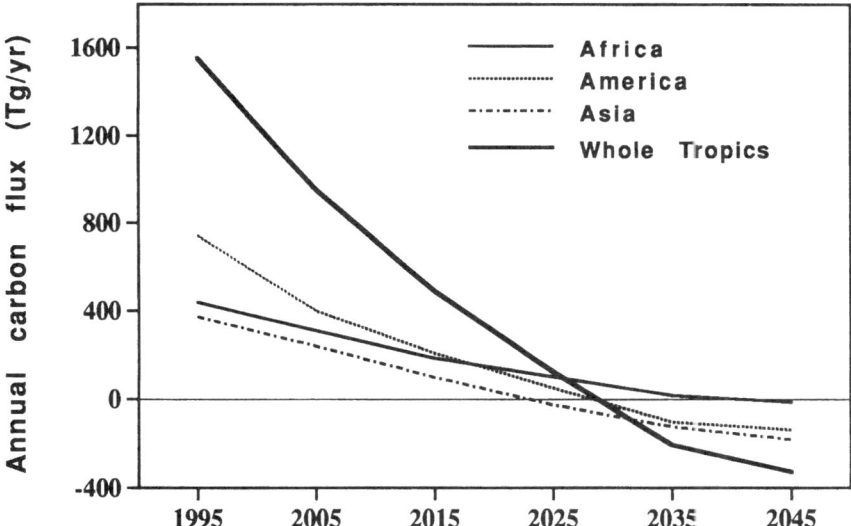

Figure 4. Gradual reduction in C emissions, continent by continent, resulting from
implementation of the C conservation/sequestration practices given in Table 3 (from
Trexler and Haugen, 1993).

Land availability per se is not the primary constraint to using forestry for climate
change mitigation in the near to mid-term (Trexler and Haugen, 1993). Social, political,
and infrastructural barriers are likely to keep possible reforestation rates modest for at least
a decade or so. Even then, the rate at which reforestation and other measures can be
accomplished realistically is modest compared to the plausible availability of land in many
countries of the tropics. Given the magnitude of ongoing deforestation, the continuing
pressures associated with rising human populations, and the necessarily incremental
impacts of measures to slow deforestation, Trexler and Haugen conclude that, even if such
a plan could be implemented, it would not be until 2030 before the 54 countries as a group
become a net sink for biotic C (Figure 4). It will be considerably longer, if ever, before

biotic C levels in these countries could return to the levels present today. With population
and other pressures building throughout the tropics, the scenarios presented by Trexler and
Haugen (1993) would represent dramatic increases in the effectiveness and magnitude of
forestry programs.

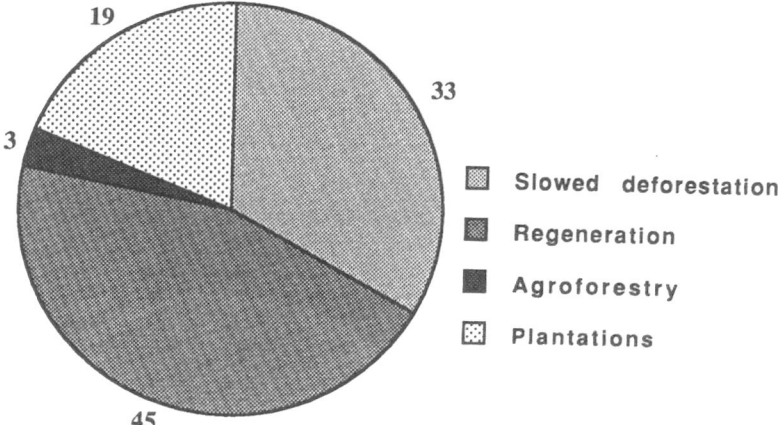

Figure 5. Relative contribution of each policy given in Table 3 to reduce C emissions to the
cumulative effect for 1990-2050 (from Trexler and Haugen, 1993).

        The analysis of Trexler and Haugen (1993) focuses on the ability to prevent
deforestatation and to return tree cover in some form to already deforested lands. It does
not include C losses or gains associated with further degradation or recovery of existing
forests, the potential recovery of biomass on land deforested between now and 2050, the
potential for increased forest management, or the land-use and C fluxes associated with
potential climate change itself. Neither do they include the potential for changes in soil
organic C (losses and gains) in their analysis. Activities that promote increases in C
sequestration in vegetation also cause increases in soil organic C (Lugo and Brown, 1993).
Judicious selection of plantation and agroforestry species can cause soil organic C to
accumulate faster than under secondary forests. The accumulation rate of soil C therefore,
need not be slow and there are alternatives for managers to increase or conserve the pool
(Lugo and Brown, 1993).
        Storage of additional C in soils will be an important mechanism for long term C
offset objectives. Soil organic C can be separated into different functional pools, active,
slow, and passive fractions, based on their residence times within soils (Parton et al., 1987;
Swift and Woomer, 1993). The potential to replenish different soil C pools is related to the
formation and turnover times of these pools. The passive pool is stable for hundreds to
thousands of years and is not subject to management. On the other hand, the active pool is
highly subject to recent organic additions to soils and subject to rapid loss when these
inputs are not longer available. The slow pool, derived from litter inputs, can form rapidly
depending upon the quantity and quality of the inputs, and is lost over decades. Because
this pool turns over slowly, it is likely to be a more secure C sequestering mechanism than
plant biomass (Lugo and Brown, 1993).

Trexler and Haugen (1993) did not include in their analysis the potential for large-scale commercial biomass utilization in the tropics. There are major impediments to the use of commercial biomass in the tropics, even on a more localized scale. Developing countries may not wish to be seen relying upon "primitive" biomass sources, preferring instead to emulate the energy technologies of more industrialized countries. Furthermore, the history of commercial plantations in the tropics is checkered at best, and indeed there is little precedent for the scale of planting that would be involved. Other important issues that would be faced include technology transfer barriers, the danger that natural forests would be cleared or relied upon for biomass supplies, and the sustainability of intensive biomass production on tropical soils. Nevertheless, the theoretical potential of biomass technologies to affect net C fluxes in the tropics is so large as to merit much more investigation.

Assessing the potential costs of measures such as those discussed here is difficult. Work by Dixon et al. (1991) is a first step in identifying ranges of potential costs, but the cost picture is complex. Some of the land-use changes proposed are likely to be cost-effective in their own right, requiring no "C subsidy". Others are not, and will require such a subsidy. It is also difficult to prepare a proper economic analysis as C storage is just one of the implications of many tropical forestry programs, and the many other costs and benefits (e.g. watershed protection, conservation of biological diversity, income generation) are equally difficult to quantify.

Overall, our ability to predict accurately future management opportunities in tropical forest areas is not well developed. Much more intensive country-specific assessments are called for. Shifting cultivators in particular will present very difficult problems calling for intensive extension and other services. Tropical forestry programs undertaken with global change mitigation in mind will need to be integrated into the social, environmental, and economic contexts and needs of the countries in which they are undertaken. Failure to understand this has brought about the failure of many tropical forestry efforts intended to solve fuelwood and other problems. The same could easily occur with forestry efforts intended to mitigate global climate change.

6. Research Needs

6.1. NEEDS FOR IMPROVING PAST, PRESENT AND FUTURE CARBON BUDGETS

1. The most effective way to reduce a large amount of the uncertainty in C flux and pool estimates in the tropics is to improve the mapping of area and area-change of land cover/land use using remote sensing technology, including the identification of all classes of disturbed forests (highly disturbed to mature). As the analysis requires several points in time, equal effort should be spent on earlier remote sensing imagery (1970s) as well as more recent imagery.

2. Efforts should be made to improve the estimates of C pool sizes in different types of tropical vegetation and soils. Recent work by Iverson et al. (1993a) and Brown and Iverson (research in progress) have demonstrated that producing maps of forest biomass (the tree component only) and soil C densities using a modeling approach in a geographic information system (GIS) is feasible. This approach needs to be expanded to the whole tropics, and improved to include C in other forest components (e.g., understory, woody debris, etc.). If coupled in a GIS with improved information and maps of land use over

time, a spatially detailed analysis of both magnitudes of C pools and fluxes will be possible. Thus changes in land use can then be matched with the appropriate C stocks, and present uncertainties in C pool and flux estimates can be reduced. This type of analysis could be coupled to other ecosystem or climate models to investigate future potential changes.

3. Establishment of a network of permanent, continuous forest inventory plots is needed to allow for detection of change in C densities of tropical forest lands. This should be coupled with remote sensing imagery to place the network of plots into a regional monitoring context. Such a network is the only way to determine rates of C accumulation or loss by different types of forests (i.e., mature and young, moist and dry, protected and unprotected). Until progress is made in establishing a continuous forest inventory network in conjunction with monitoring by remote sensing, our ability to balance the C budget in the tropics will remain elusive.

4. A more comprehensive analysis of the fate of burned or otherwise disturbed biomass-C such as charcoal is needed. This is particularly important in models of shifting cultivation as shown by Detwiler and Hall (1988) and Hall and Uhlig (1991); their simulated C flux is very sensitive to the proportion of the biomass that remains as charcoal.

5. Efforts should be made on developing relationships/models between changes in land use, forest degradation, and C densities and other variables such as population density, transportation network densities, logging, hardwood trade, agricultural production, etc. These empirical models are needed for reliably projecting future trends in land use and C fluxes in the tropics.

6. Comprehensive models need to be developed, using the spatially-oriented data bases generated by the above research items, to allow the synthesis and simulation of the complex interactions, including possible management scenarios, that are likely to influence tropical forests into the future.

7. Because past patterns of land use strongly influence the present C budget (pools and fluxes) by generating lands of different C pools and fluxes, it is imperative that the history of land use and corresponding C densities be well documented. The approach discussed above for tropical Asia (cf. Section 2.3.2) is a step in the right direction, however, the analysis needs to be expanded to the whole tropics.

8. Ecosystem-level studies to measure the effects of CO_2-enrichment and climate change on mature and secondary dry and humid tropical forests need to be established. These need to extend for several years as the effects may take this length of time to become fully apparent.

6.2. NEEDS FOR IMPROVING ESTIMATES OF CARBON SEQUESTRATION AND CONSERVATION FOR CLIMATE CHANGE MITIGATION

1. Improve research efforts to produce better estimates of areas and location of lands technically and actually available for C conservation and sequestration projects, and the likely quantities and rates of C that could be sequestered or conserved on these lands.

2. The management of soil organic C formation as a tool for C sequestration is poorly understood. Additional research on C allocation patterns (to aboveground litter or fine roots) in different plant species and how organic C enters into the different soil C pools could be particularly useful in selecting appropriate strategies for soil C management.

3. Develop and implement management schemes to learn which ones do and do not work for C conservation and sequestration over the long term. Emphasis should be placed on developing sustainable alternatives to the non-traditional forms of shifting cultivation to reduce deforestation.

7. References

Aldhous, P.: 1993, tropical deforestation: not just a problem in Amazonia, *Science* 259,1390.

Andreae, M. O., E. V. Browell, M. Garstang, G. L. Gregory, R. C. Harriss, G. F. Hill, D. J. Jacob, M. C. Periera, G. W. Sachse, A. W. Setzer, P. L. Silva Dias, R. W. Talbot, A. L. Torres, S. C. Wofsy.: 1988, Biomass-burning emissions and associated haze layers over Amazonia, *Journal of Geophysical Research* 93, 1509-1527.

Box, E.: 1978, Geographical dimensions of terrestrial net and gross primary productivity, *Radiation and Environmental Biophysics* 15, 305-322.

Brown, S., L. R. Iverson.: 1992, Biomass estimates for tropical forests, *World Resource Review* 4, 366-384.

Brown, S., A. E. Lugo.: 1982, Storage and production of organic matter in tropical forests and their role in the global carbon cycle, *Biotropica* 14, 161-187.

Brown, S., A. E. Lugo.: 1990, Tropical secondary forests, *Journal of Tropical Ecology* 6, 1-32.

Brown, S., A. E. Lugo.: 1992, Aboveground biomass estimates for tropical moist forests of the Brazilian Amazon, *Interciencia* 17, 8-18.

Brown, S., A. J. R. Gillespie, A. E. Lugo.: 1991, Biomass of tropical forests of south and southeast Asia, *Canadian Journal of Forest Research* 21, 111-117.

Brown, S., L. R. Iverson, A. E. Lugo.: 1993, Land use and biomass changes in Peninsular Malaysia during 1972-82: a GIS approach, in V. Dale (ed), *Effects of Land Use Change in Atmospheric CO2 Concentrations: Southeast Asia as a Case Study*, Springer-Verlag, in press.

Brown, S., A. E. Lugo, J. Chapman.: 1986, Biomass of tropical tree plantations and its implications for the global carbon budget, *Canadian Journal of Forest Research* 16, 390-394.

Brown, S., A. E. Lugo, L. R. Iverson.: 1992, Processes and lands for sequestering carbon in the tropical forest landscape, *Water, Air and Soil Pollution* 64, 139-155.

Cramer, W. P., A. M. Solomon.: 1993, Climatic classification and future global redistribution of agricultural land, *Climate Research*, in press.

Crutzen, P. J., W. M. Hao, M. H. Liu.: 1989, Estimates of annual and regional releases of CO2 and other trace gases to the atmosphere from fires in the tropics based on FAO statistics for the period 1975 to 1980, *Proceedings of the Third International Symposium on Fire Ecology*, Freiburg University, 16-20 May, 1989, Berlin.

Detwiler, R. P., C. A. S. Hall.: 1988, Tropical forests and the global carbon cycle, *Science* 239, 42-47.

Dixon, R. K., P. E. Schroeder, J. K. Winjum (eds).: 1991. *Assessment of Promising Forest Management Practices and Technologies for Enhancing the Conservation and Sequestration of Atmospheric Carbon and Their Costs at the Site Level*, EPA/600/3-91/067, US. Environmental Protection Agency, Environmental Research Laboratory, Corvallis, Oregon, USA.

Deutscher Bundestag (ed).: 1990, *Protecting the Tropical Forests a High-Priority International Task, 2nd Report of the Enquete-Commission*, Bonn, Germany.

Eswaran, H., E. Van Den Berg, P. Reich.: 1993, Organic carbon in soils of the world, *Soil Science Society of America Journal* 57, 192-194.

Fearnside, P. M.: 1992, *Carbon Emissions and Sequestration in Forests: Case Studies from Seven Developing Countries, Volume 2: Greenhouse Gas Emissions from Deforestation in the Brazilian Amazon*, LBL-32758 UC-402, Energy and Environment Division, Lawrence Berkeley Laboratory, Berkeley, CA, USA.

Flint, E. P., J. F. Richards.: 1993, Trends in carbon content of vegetation in South and Southeast Asia associated with changes in land use, in V. Dale (ed), *Effects of Land Use Change in Atmospheric CO2 Concentrations: Southeast Asia as a Case Study*, Springer-Verlag, in press.

Graham, R. L., R. D. Perlack, A. M. G. Prasad, J. W. Ranney, D. B. Waddle.: 1990, *Greenhouse Gas Emissions in Sub-Saharan Africa*, ORNL-6640, National Technical Information Service, Springfield, VA, USA.

Grainger, A.: 1988, Estimating areas of degraded tropical lands requiring replenishment of forest cover, *International Tree Crops Journal* 5, 31-61.

Grainger, A.: 1990, Modelling the impact of alternative afforestation strategies to reduce carbon dioxide emissions, in *Proceedings of the Conference on Tropical Forestry Response Options to Global Climate Change*, IPCC, Sao Paulo, Brazil, pp. 93-104.

Hall, C. A. S., R. P. Detwiler, P. Bogdonoff, P. S. Underhill.: 1985, Land use change and carbon exchange in the tropics: I. Detailed estimates for Costa Rica, Panama, Peru, and Bolivia, *Environmental Management* 9, 313-334.

Hall, C. A. S., M. R. Taylor, E. Everham.: 1992, A geographically-based ecosystem model and its application to the carbon balance of the Luquillo Forest, Puerto Rico, *Water, Air, and Soil Pollution* 64, 385-404.

Hall, C. A. S., Uhlig, J.: 1991, Refining estimates of carbon released from tropical land-use change, *Canadian Journal of Forest Research* 21, 118-131.

Houghton, J. T., B. A. Callander, S. K. Varney (eds).: 1992, *Climate Change 1992, The Supplementary Report to the IPCC Scientific Assessment*, Cambridge University Press.

Houghton, R. A.: 1991, Tropical deforestation and atmospheric carbon dioxide, *Climatic Change* 19, 99-118.

Houghton, R. A.: 1992a, Tropical forests and climate, Paper presented at the International Workshop Ecology, Conservation, and Management of Southeast Asian Rainforests, October 12-14, 1992, Kuching, Sarawak.

Houghton , R. A.: 1992b, Effects of land-use change, surface temperature, and CO2 concentration on terrestrial stores of C, Paper presented at IPCC meeting, October 1992, Woods Hole Research Center, MA, USA.

Houghton, R. A., J. L. Hackler.: 1993, The net flux of carbon from deforestation and degradation in South and Southeast Asia, in V. Dale (ed), *Effects of Land Use Change in Atmospheric CO2 Concentrations: Southeast Asia as a Case Study*, Springer-Verlag, in press.

Houghton, R. A., D. L. Skole. 1990.: Carbon, in: B. L. Turner, W. C. Clark, R. W. Kates, J. F. Richards, J. T. Matthews, and W. B. Meyer (eds), *The Earth as Transformed by Human Action*, Cambridge University Press, pp. 393-408.

Houghton, R. A., R. D. Boone, J. R. Fruci, J. E. Hobbie, J. M. Melillo, C. A. Palm, B. J. Peterson, G. R. Shaver, G. M. Woodwell, B. Moore, D. L. Skole, N. Myers.: 1987, The flux of carbon from terrestrial ecosystems to the atmosphere in 1980 due to changes in land use: geographic distribution of the global flux, *Tellus* 39B, 122-139.

Houghton, R. A., J. E. Hobbie, J. M. Melillo, B. Moore, B. J. Peterson, G. A. Shaver, G. M. Woodwell.: 1983, Changes in the carbon content of terrestrial biota and soils between 1860 and 1980: a net release of CO_2 to the atmosphere, *Ecological Monographs* 53, 235-262.

Houghton, R. A., D. S. Lefkowitz, D. L. Skole.: 1991a, Changes in the landscape of Latin America between 1850 and 1985, I. Progressive loss of forests, *Forest Ecology and Management* 38, 143-172.

Houghton, R. A., D. L. Skole, D. S. Lefkowitz.: 1991b, Changes in the landscape of Latin America between 1850 and 1985, II. Net release of CO_2 to the atmosphere, *Forest Ecology and Management* 38, 173-199.

Houghton, R. A., J. Unruh, P. A. Lefebvre.: 1991c, Current land use in the tropics and its potential for sequestering C, in *Proceedings: Technical Workshop to Explore Options for Global Forestry Management*, International Institute for Environment and Development, Bangkok, Thailand, pp. 297-310.

Idso, S. B.: 1988, Three phases of plant response to atmospheric CO_2 enrichment, *Plant Physiology* 87, 5-7.

Iverson, L. R., S. Brown, A. Prasad, H. Mitasova, A. J. R. Gillespie, A. E. Lugo.: 1993a, Use of GIS for estimating potential and actual biomass for continental South and Southeast Asia, in V. Dale (ed), *Effects of Land Use Change in Atmospheric CO_2 Concentrations: Southeast Asia as a Case Study*, Springer-Verlag, in press.

Iverson, L. R., S. Brown, A. Grainger, A. Prasad, D. Liu.: 1993b, Carbon sequestration in tropical Asia: an assessment of technically suitable forest lands using geographic information systems analysis, *Climate Research*, in press.

Kimball, B. A., J. R. Mauney, F. S. Nakayama, S. B. Idso.: 1993, Effects of increasing atmospheric CO_2 on vegetation, *Vegetatio* 104/105, 65-75.

Knabe, W.: 1985, Effects of chemical air pollution on forests and other vegetation, in G. B. Marini Bettolo (ed), *Chemical Events in the Atmosphere and their Impacts on the Environment, Pontificiae Academiae Scientiarum Scipta Varia* 56, Rome.

Knabe, W., P. Hennicke, W. Bach, M. Ganseforth, L. Hartenstein, V. Jung, M. Muller.: 1990, Supplementary opinion of the Enquete-Commission members on Part G, chapters 1-4, the causes of the destruction of tropical forests, in Deutscher Bundestag (ed), *Protecting the Tropical Forests a High-Priority International Task, 2nd Report of the Enquete-Commission*, Bonn, Germany, pp.393-407.

Korner, C., J. A. Arnone III.: 1992, Response of elevated carbon dioxide in artificial tropical ecosystems, *Science* 257,1672-1675.

Larcher, W.: 1980, *Physiological Plant Ecology, 2nd Edition*, Springer-Verlag.

Leemans, R.: 1992, Modelling ecological and agricultural impacts of global change on a global scale, *Journal of Scientific and Industrial Research* 51, 709-724.

Lugo, A. E., S. Brown.: 1986, Steady state terrestrial ecosystems and the global carbon cycle, *Vegetatio* 68, 83-90.

Lugo, A. E., S. Brown.: 1992, Tropical forests as sinks of atmospheric carbon, *Forest Ecology and Management* 54, 239-255.

Lugo, A. E., S. Brown.: 1993, Management of tropical soils as sinks of atmospheric carbon, *Plant and Soil*, in press.

Lugo, A. E., J. Wisniewski.: 1992, Natural sinks of CO_2, conclusions, key findings and research recommendations from the Palmas del Mar workshop, *Water, Air, and Soil Pollution* 64, 455-459.

Norby, R. J., J. Pastor, J. M. Melillo.: 1986, Carbon-nitrogen interactions in CO_2-enriched white oak: physiological and long-term perspectives, *Tree Physiology* 2, 233-241.

Parton, W. J., D. S. Schimel, C. V. Cole, D. S. Ojima.: 1987, Analysis of factors controlling soil organic matter levels in Great Plains grasslands, *Soil Science Society of America Journal* 51, 1173-1179.

Peterson, B. J., J. M. Melillo.: 1985, The potential storage of carbon caused by eutrophication of the biosphere, *Tellus* 37B, 117-127.

Prentice, I. C., W. Cramer, S. P. Harrison, R. Leemans, R. A. Monserud, A. M. Solomon.: 1992, A global biome model based on plant physiology and dominance, soil properties and climate, *Journal of Biogeography* 19, 117-134.

Raich, J. W., W. H. Schlesinger.: 1992, The global carbon dioxide flux in soil respiration and its relationship to vegetation and climate, *Tellus* 44B, 81-99.

Raich, J. W., E. B. Rastetter, J. M. Melillo, D. W. Kicklighter, P. A. Steudler, B. J. Peterson, A. L. Grace, B. Moore III, C. J. Vorosmarty.: 1991, Potential net primary production in South America, *Ecological Applications* 1, 399-429.

Raynaud, D., J. Jouzel, J. M. Barnola, J. Chappellaz, R. J. Delmas, C. Lorius.: 1993, The ice record of greenhouse gases, *Science* 259, 926-934.

Richards, J. F., E. P. Flint.: 1993, A century of land use change in South and Southeast Asia, in V. Dale (ed), *Effects of Land Use Change in Atmospheric CO_2 Concentrations: Southeast Asia as a Case Study*, Springer-Verlag, in press.

Rosenzweig, M. L.: 1968, Net primary productivity of terrestrial communities: prediction from climatological data, *American Naturalist* 102, 67-74.

Sarmiento, J. L., J. C. Orr, U. Siegenthaler.: 1992, A perturbation simulation of CO_2 uptake in an ocean general circulation model. *Journal of Geophysical Research* 97, 3621-3645.

Swift, M. J., P. Woomer.: 1993, Organic matter and the sustainability of agricultural systems: definition and measurement, in K. Mulongoy and R. Merckx (ed), *Dynamics of Organic Matter in Relation to Sustainability of Agricultural Systems*, J. Wiley and Sons, pp. 3-18.

Tans, P. P., I. Y. Fung, T. Takahashi.: 1990, Observational constraints on the global atmospheric carbon dioxide budget, *Science* 247, 1431-1438.

Taylor, J. A., J. Lloyd.: 1992, Sources and sinks of atmospheric CO_2, *Australian Journal of Botany* 40, 407-418.

Trexler, M. C., C. Haugen.: 1993, *Keeping it Green: Evaluating Tropical Forestry Strategies to Mitigate Global Warming*, World Resources Institute, Washington, DC, USA, in press.

Whitmore, T. C.: 1984, *Tropical Rain Forests of the Far East*, 2nd Edition, Oxford Science Publication, Clarendon Press.

Whittaker, R. H., G. E. Likens.: 1973, Carbon in the biota, in G. M. Woodwell and E. V. Pecan (eds), *Carbon and the Biosphere*, US Atomic Energy Commission, CONF-720510, National Technical Information Service, Springfield, VA, USA, pp. 281-302.

ASSESSMENT OF C BUDGET FOR GRASSLANDS AND DRYLANDS OF THE WORLD

DENNIS S. OJIMA — *Natural Resource Ecology Laboratory, Colorado State University, Fort Collins, CO 80523*

BJØRN O.M. DIRKS — *Department of Theoretical Production Ecology, Wageningen Agricultural University, P.O. Box 430, 6700 AK Wageningen. The Netherlands*

EDWARD P. GLENN — *Environmental Research Laboratory, University of Arizona, 2601 E. Airport Drive, Tucson, AZ 85706-6985*

CLENTON E. OWENSBY — *Department of Agronomy, Throckmorton Hall, Kansas State University, Manhattan, KS 66506-5501*

JONATHAN O. SCURLOCK — *Division of Life Sciences, King's College London, Campden Hill Road, London W8 7AH, United Kingdom*

Abstract. Intergovernmental Panel on Climate Change (IPCC) estimates indicate that potential changes in seasonal rainfall and temperature patterns in central North America and the African Sahel will have a greater impact on biological response (such as plant production and biogeochemical cycling) and feedback to climate than changes in the overall amount of annual rainfall. Simulation of grassland and dryland ecosystem responses to climate and CO_2 changes demonstrates the sensitivity of plant productivity and soil C storage to projected changes in precipitation, temperature and atmospheric CO_2. Using three different land cover projections, changes in C levels in the grassland and dryland regions from 1800 to 1990 were estimated to be -13.2, -25.5 and -14.7 Pg, i.e., a net source of C due to land cover removal resulting from cropland conversion. Projections into the future based on a double-CO_2 climate including climate-driven shifts in biome areas by the year 2040 resulted in a net sink of +5.6, +27.4 and +26.8 Pg, respectively, based upon sustainable grassland management. The increase in C storage resulted mainly from an increase in area for the warm grassland sub-biome, together with increased soil organic matter. Preliminary modeling estimates of soil C losses due to 50 yr of regressive land management in these grassland and dryland ecoregions result in a 11 Pg loss relative to current conditions, and a potential loss of 37 Pg during a 50 yr period relative to sustainable land-use practices, an average source of 0.7 Pg C yr^{-1}. Estimates of the cost of a 20 yr rehabilitation program are 5 to 8 x 10^9 US$ yr^{-}, for a C sequestering cost of approximately 10 US$ per tC.

Water, Air, and Soil Pollution **70**: 95–109, 1993.
© 1993 *Kluwer Academic Publishers.*

1. Introduction: Scope of the Problem

Grasslands and associated savanna and shrublands are clearly vulnerable to climate change. The sensitivity of grasslands to climate change has been documented by observations of past droughts in the semiarid and arid regions (Weaver and Albertson, 1943; Hare, 1977; Schlesinger *et al.*, 1990). Increased human activity has led to degradation of plant production and soil resources in many of these ecosystems leading to desertification in some regions (UNEP, 1991). The continuing pressure on these ecosystems and the projected modifications in regional climate indicate that further degradation of these lands will occur and that there is potential for still greater C losses.

Changes in seasonal rainfall and temperature patterns in central North America and the African Sahel will have a greater impact on biological response and feedback to climate than changes in the overall amount of annual rainfall (Houghton *et al.*, 1990; Ojima *et al.*, 1991). There are several ways in which changing climate and atmospheric CO_2 concentrations may affect grassland and semi-arid ecosystems. Productivity of these ecosystems is directly linked to precipitation (Le Houerou, 1984; Sala *et al.*, 1988; Parton *et al.*, in press), so changes in precipitation amounts will affect plant production. These changes in production can modify soil C storage. The soil C store in these ecosystems is a very important pool, since it represents a significant proportion of the total system C and is stabilized for hundreds to thousands of years.

In the following analysis, several important features of the grassland and dryland region will not be covered in our assessment at this workshop. First, the soil C stored as inorganic constituents in these arid environments is substantial, but the flux rates from the carbonates are relatively small. Second, the paleosols, in certain regions, can also be an important C pool; however, there is no systematic way to deal with the exposure of these soils and subsequent oxidation of their stored C. Third, invasion by or increase in woody species may be significant in grassland communities, and was not included in our climate change simulation. However, this is an indirect effect more closely linked to the frequency and intensity of burning and grazing (Archer, 1993, in press; Schlesinger *et al.*, 1990).

Our considerations will deal with those changes in plant production and decomposition due to global change or to management practices which will impact the level of stored soil C on a time scale of decades to centuries. The following sections will discuss the grassland and dryland ecoregions being considered in this analysis, key ecological and biological issues related to the uptake and storage of C in these ecosystems, estimate the range of the C source or sink of these ecoregions, and discuss the role of mitigation practices that would modify the magnitude of the source or the sink from the different grassland and dryland ecoregions.

2. Key Ecological Considerations for Assessment of Global Change in Grasslands and Drylands

Grasslands by their very nature are resource-limited, particularly for N and water. In the natural grassland ecosystems considered here, essentially all nutrient resources are supplied by the system through nutrient cycling and water through precipitation. The seasonal distribution of rainfall is a major determinant of plant production in many semiarid and arid regions. Simulation of ecosystem responses to climate change in grassland regions of the world demonstrate the sensitivity of soil C storage and grassland biogeochemistry processes to seasonal distribution of precipitation changes and to overall increases in temperature (Ojima et al., 1991). Modifications of resource use efficiency among various grassland and aridland communities are important to projecting how these ecosystems will respond to increased atmospheric CO_2, change in climate, or increases in atmospheric deposition of N.

In addition, these semiarid and arid lands are vulnerable to human-induced land use changes, and these land uses affect soil C storage, soil fertility, soil erosion rates, dust loading into the atmosphere, trace gas exchange, and water and energy balances. The overall impact of these management practices on C storage in grassland and dryland soils are potentially greater than that of climate change or increased atmospheric CO_2 concentrations.

In temperate grasslands, dominated by C_3 species, they are also limited by C due to high photorespiration rates. These resource limitations result in relatively low net primary and secondary productivities, especially in the arid regions. Plants with the C_3 photosynthetic pathway generally have increased C fixation rates when CO_2 levels are increased, while C_4 plants do not increase in C fixation to the degree that C_3 plants do (Kimball, 1983). Photosynthetic capacity of plants with the C_3 pathway is limited by current atmospheric CO_2 levels due to oxygenase activity of ribulose-1,5-bisphosphate carboxylase (Rubisco). Innumerable studies have shown increased C_3 photosynthesis with elevated CO_2 (Newton, 1991). C_4 photosynthesis is not considered C-limited, because C is initially fixed in the mesophyll by phosphoenolpyruvate carboxylase (PEPc) which does not have oxygenase activity (Edwards and Walker, 1983). It appears that the grassland ecosystems dominated by C_4 grasses will likely not experience increased C acquisition as a result of improved photosynthetic capacity (Knapp et al., 1993).

In ecosystems with frequent water stress, enhanced water-use efficiency due to partial stomatal closure in CO_2-enriched environments is likely more important than photosynthetic pathway (Gifford et al., 1990). Morrison (1985) reviewed the literature concerning CO_2 enrichment and water relations and reported a range of 60% to 160% increase in WUE for both C_3 and C_4 plants. For tallgrass prairie, increased above- and belowground biomass production under elevated CO_2 without input of additional water has been reported (Owensby et al., 1993a; Knapp et al., 1993). Owensby et al. (1993a) reported increased root production with CO_2

enrichment which would enhance water uptake. Changes in stomatal density may also impart water savings for plants under elevated CO_2. Woodward (1987) indicated that CO_2 enrichment over the past century has likely reduced stomatal density.

Any change in ecosystem function that improves productivity without additional N input will increase N use efficiency (NUE). NUE will almost certainly increase in most N-poor ecosystems under elevated CO_2. In natural grassland ecosystems, increased NUE was reported by Owensby et al. (1993b) in a CO_2 enriched tallgrass prairie over a 3-yr period. Reduced N requirement in N-limited systems may increase plant inputs into the soil system. The increased NUE will also result in a lower rate of decomposition due to changes in lower litter quality.

3. Ecoregion Definition and Analytical Methods

3.1.   GRASSLAND AND DRYLAND ECOREGION DEFINITION

Grasslands and drylands in this assessment are defined as: Natural grasslands and drylands determined by climate (e.g., an aridity index [the ratio of annual precipitation to potential evaporation] between 0.05 and 0.8). Some savanna grasslands in the tropics are similarly determined by the presence of a short, but intense dry season. This definition, therefore, excludes hyper-arid regions but incorporates both the tropical and temperate grasslands of the world, including the humid savannas and grasslands which are maintained primarily by fire and grazing. However, we exclude two types of grassland areas from this analysis, namely:

*   croplands previously converted from grasslands; and
*   grass-dominated areas in eco-regions or life zones normally classified
        as forest. Pastures in both tropical and temperate regions which
        require large inputs of fertilizers or other intensive management are,
        therefore, not included in this analysis.

Given this definition of grasslands and drylands, we estimated the areal extent of these regions from the BIOME model developed by Prentice et al. (1992). Two other estimates of grasslands and dryland area were made using the UNEP (1991) land cover estimates of arid, semiarid, and subhumid regions of the world, and a separate estimate based on Bailey's (1989) Ecoregion approach used by Ojima et al. (1993, this volume). We assessed the given land areas for the various classes (Table 1) defined by UNEP (1991) and by Bailey (1989) and grouped them as best as possible according to the BIOME map (Prentice et al., 1992). Potential grassland and dryland regions were taken to be the pre-industrial areas of grassland and dryland and used to estimate "past land cover" (Table 2). Current grassland and dryland areas were obtained after accounting for conversion of a fraction of the cool

Table 1. Grouping of grassland sub-regions as provided by Prentice *et al.* (1992), UNEP (1991) and Bailey (1989), according to the BIOME model of Prentice *et al.* (1992). Abbreviations: AA arid, SA semi-arid, SH sub-humid.

Prentice *et al.* (1992)	UNEP (1991)	Bailey (1989)
Semi-desert	AA Asia	dry continental
Cool grass and shrub	SA Asia (x 0.577) SA South-America (91 Mha)[1] SH Europe SH North-America (x 0.5)	dry temperate
Warm grass and shrub	AA Europe SA Africa SA Asia (x 0.423)[2] SA Australia SA Europe SA North-America SA South-America (174 Mha)[2] SH Africa SH Asia SH Australia SH North-America (0.5) SH South-America	humid temperate dry savanna savanna humid savanna mediterranean
Hot desert	AA Africa AA Australia AA North-America AA South-America	

[1]Based on Bailey (1989)
[2]Based on Ojima *et al.* (1993, this volume)

and warm grasslands to arable land (Prentice *et al.*, 1992). "Future" grassland and dryland extent was obtained from the 2xCO$_2$ GFDL CGM climate change estimate of altered land cover area (Cramer and Solomon 1993; in review). In addition we subtracted the fraction of area under cropland estimated for the "current" land cover case (Table 2).

Table 2. Changes in biome areas and carbon pools for grasslands and drylands for past, present and future.

BIO-REGION[2]	BIOME AREAS			CARBON POOLS[1]			
	POTENTIAL	CURRENT	FUTURE	PAST	CURRENT	FUTURE "REGRESSIVE"	FUTURE "SUSTAINABLE"
	M ha			Pg			
SEMI-DESERT							
[a]	499.5	499.5	298.9	27.620	27.620		15.740 (-11.9)
b)	626.0	626.0	374.6	34.615	34.615		19.726 (-14.9)
c)	137.1	137.1	82.0	7.581	7.581	5.2 (-2.38)	4.318 (-3.3)
COOL GRASS/SHRUB							
a)	571.3	272.5	121.3	11.642	5.553 (-6.1)[4]		1.789 (-3.7)
b)	791.0	377.3	167.9	16.120	7.689 (-7.6)		2.477 (-5.2)
c)	304.0	145.0	86.8	6.195	2.955 (-3.2)	2.26 (-6.95)	1.280 (-1.7)
WARM GRASS/SHRUB							
a)	1180.5	1016.4	1460.3	51.258	44.133 (-7.1)		63.725 (+19.6)
b)	2821.0	2428.9	3489.5	122.493	105.466 (-17.0)		152.282 (+46.8)
c)	1892.6	1629.5	2341.1	82.178	70.755 (-11.4)	63.00 (-7.75)	102.166 (+31.4)
HOT DESERT							
a)	1945.5	1945.5	2004.1	34.336	34.336		35.651 (+1.3)
b)	934.0	934.0	962.1	16.484	16.484		17.115 (+0.6)
c)	-	-					
TOTAL							
a)	4196.8	3733.9	3844.6	124.856	111.642 (-13.2)		116.905 (+5.6)
b)	5172.0	4366.2	4994.1	189.712	164.254 (-25.5)		191.600 (+27.0)
c)	2333.7	1911.6	1509.9	95.954	81.291 (-14.7)	70.46 (-10.82)	107.764 (+26.8)

[1]Carbon pools are total soil C to 20 cm depth as modeled using CENTURY (Ojima et al., 1993, this volume). An approximate value for soil C to 1.0 m depth may be obtained by multiplying these values by 3-4. Thus, current grassland and dryland soil carbon (range 96 to 190 Pg; average 119 Pg to a depth of 20 cm) is approximately 417 Pg to a depth of 1.0 m.

[2]Biomes as defined by Prentice et al. (1992).

[a] Areas for biomes defined by Prentice et al. 1992.

b) Areas derived from UNEP-based aridity index (UNEP, 1991).

c) Areas derived from Bailey Ecoregions (Bailey 1988).

[4]Values in parentheses indicate net flux in C pools from one period to the next.

3.2. ESTIMATING C STORAGE AND FLUXES RELATIVE TO GLOBAL CHANGE

Carbon storage and flux estimates from grassland and dryland regions were estimated under a prescribed climate change scenario (GFHI, IPCC Report, Houghton *et al.*, 1990) and land cover estimates described previously and given in Table 2. The grassland model of CENTURY (Parton *et al.*, 1987, 1992; Parton *et al.*, in press b), was used for the simulations of aboveground net primary productivity (NPP) and soil organic C (SOC) to a depth of 20 cm for 31 sites across the globe (Ojima *et al.*, 1993b, in press). Results from three sets of simulations were used.

* Current climate scenarios using acceptable land management practices;
* Combined effect of climate change and doubling of atmospheric CO_2 (+CC+CO_2+M) using acceptable or sustainable land management practices; and
* Combined effect of climate change and doubling of atmospheric CO_2 (+CC+CO_2-M) using unacceptable or regressive land management.

We modified the plant production parameters under a 2 x CO_2 climate by changing production relative to potential evapotranspiration (PET) and to N use efficiency (NUE). The magnitude of the effect is to cause a 20% increase in plant production with a change in atmospheric CO_2 concentration from 350 to 700 ppm (Ojima *et al.*, 1993, this volume).

In order to generate the double CO_2 climate, we spatially interpolated GCM grid values of projected 2 x CO_2 climate changes of monthly temperature (T) and precipitation (PPT) for each site based on the GCM output made for the 1990 IPCC report (Kittel, pers. comm., NCAR data retrieval system). We applied these projected monthly values in a linear fashion in a 50 yr ramp (see Ojima *et al.*, 1993, this volume).

4. Regional Ecosystem Modeling

For regional simulations, we selected a representative climate for each site within a region and simulated equilibrium ecosystem levels of soil C and plant production. Climate change simulations for a particular region was based on these representative sites and estimates of climate and CO_2 induced modifications to ecosystem dynamics were applied evenly across the region.

Past soil C pools under grassland were calculated using current soil organic matter content projected over a pre-industrial land use scenario (Table 2). Future soil C pools were calculated using both soil organic matter content occurring for a so-called 'regressive management' future scenario, accounting for removal of 50% of the aboveground biomass during grazed months, and an 'sustainable' scenario using the same moderate grazing and burning regimes specified by Ojima *et al.* (1993; this volume).

4.1.   CARBON CHANGES IN THE GRASSLAND AND ARID REGIONS IN THE PAST, PRESENT, AND FUTURE

The results from the "climate change scenario only" (i.e., using only projected changes in monthly rainfall and temperature levels and not CO_2 enhancement effects) indicate that soil C losses occur in all grassland regions (losses range from near 0 to 14% of current soil C levels for the surface 20 cm). Plant production varies according to modifications in rainfall amounts under the altered climate and to altered N mineralization rates. Soil decomposition rates responded most predictably to changes in temperature. CO_2 enhancement effects on plant production and the indirect effects on soil C loss tended to reduce the net impact of climate alterations in most of the regions, and actually resulted in net C sinks in the warm grasslands regions.

Using the sustainable land-management regime, results indicate that changes in soil C levels in the grassland and dryland regions from 1800 to 1990 are losses of -13.2, -25.5 and -14.7 Pg (based on land area estimates of Prentice, UNEP and Bailey, respectively, Table 2). The net loss of soil C is due primarily to land use conversion to cropland. Projections into the future based on a double-CO_2 climate including climate-driven shifts in biome areas by the year 2040 resulted in a net sink of +5.6, +27.4 and +26.8 Pg, respectively, based upon sustainable grassland management. The increase in C storage in this future projection resulted mainly from an increase in area for the warm grassland sub-biome (net increase of 280 M ha due to climate-change induced biome shifts) together with a net increase in soil organic C densities for the sites simulated in this sub-biome (Table 2).

We simulated "regressive" land management by increasing grazing levels from 30% to 50% removal for all of the sites analyzed. The impact of "regressive" land management resulted in a loss of soil C in all regions after 50 yrs (Table 2 "Future - regressive" column). Largest losses were evident in the warm grasslands. The total net loss relative to current condition is 10.8 Pg. When this regressive management is compared to a sustainable management system (i.e., light grazing) the net difference is 37.6 Pg gain of soil C. Modification of land use practices can greatly influence the net flux of soil C, changing grasslands from a source of C to a sink for atmospheric C.

4.2. AN ALTERNATIVE SOURCE/SINK ESTIMATE

Using the logic described by Gifford *et al.* (1990), estimates of annual C sequestration in grassland sinks can be alternatively derived. They hypothesized that given the continuous C exchange between live biomass, soil organic matter pools, and the atmosphere with turnover times in grassland ecosystems from 7 to 10 yr, changes in one pool would eventually be distributed to the other linked pools. Accordingly, they surmised that an increase in atmospheric CO_2 would be accompanied by an increase in C stored in live vegetation, the dead litter, and the soil C. The mechanisms involved are likely increased C fixation due to increased photosynthetic capacity and resource use efficiencies, and decreased decomposition rates. Basic assumptions relevant to C storage in grassland ecosystems include (i) turnover times for plant C are a decade or less; and (ii) changes in C:N ratios of plant biomass may reduce decomposition rates (Gifford et al., 1990).

Calculations based on those of Gifford et al. (1990) suggest that the present C flux in grassland soils (estimated at 417 Pg; Table 2), may be a net sink of 0.6 Pg yr^{-1} an upper limit. Owensby (1993; this volume) suggested that additional C storage in grassland soils could be as high as 25 to 30% with CO_2 doubling over the next 50 to 70 yrs, i.e., an annual increase of 0.4 to 0.5%. The grassland sink would be in the range of 1.7 to 1.8 Pg yr^{-1}, a number which we regard as an extreme upper limit. The of 0.6 Pg yr^{-1} sink should be regarded only as an alternative estimate of the upper limit for a possible grassland soil sink. The CENTURY model analysis of global grassland ecosystem sensitivity to climate change and CO_2 enhancement shows that changes in temperature and precipitation may more than offset the additional C storage produced by CO_2 enhancement (see Table 2).

5. Policy Issues

Semiarid and arid regions may be among the more sensitive terrestrial ecosystems with respect to global change effects, resulting in severe degradation of their potential to store C (OIES, 1991). Changes which decrease soil moisture and nutrient availability may result in large declines in both soil C and plant productivity. Under these conditions, human activities leading to desertification of certain ecosystems will be greatly accelerated. In the following section we outline an estimation of the level of C lost and the potential for conserving soil C.

5.1. COST OF SEQUESTERING C THROUGH ANTI-DESERTIFICATION PROGRAMS IN THE GRASSLANDS AND RANGELANDS

The CENTURY model estimates a potential soil C flux to the atmosphere from most grassland regions under scenarios of CO_2 enrichment and climate change. The scenarios assume a sustainable land management strategy with respect to fire and grazing pressure and preservation of native plant types. In contrast to this sustainable land management scenario, UNEP (Dregne et al., 1991) has estimated the extent of current land degradation and the projected rate given minimal anti-desertification measures. It has itemized the costs of a 20 yr program to arrest land degradation and restore already degraded lands (Kassas et al., 1991).

Soil C fluxes are driven in part by net primary productivity on the land surface. Desertification results in reductions in net primary productivity. Hence, it is theoretically possible to calculate the differences in soil C storage between scenarios of sustainable management versus current levels of regressive management projected into the future using the CENTURY model and UNEP data. The cost of anti-desertification measures, divided by the difference in annual C flux rates between the two scenarios yields a theoretical price per tonne of mitigating C losses into the atmosphere by restoring degraded drylands. Before performing the calculations it is necessary to examine some of the underlying assumptions and uncertainties.

5.2.   STATUS OF THE GRASSLANDS AND DRYLANDS UNDER PRESENT AND FUTURE MANAGEMENT SCENARIOS

The extent of present land degradation is documented in UNEP (1991). Their data were derived from two global data bases: GLASOD (International Soil Reference and Information Center, Wageningen, an index of soil degradation) and an ICASALS data base (Texas Tech University) on land degradation. The combined data were used to estimate the extent of land degradation in the rangelands of the world (Table 3). UNEP does not quantify percentage reduction in net primary productivity in each degradation class, but from the verbal descriptions we infer that moderate degradation is approximately 25 to 50%, severe 50 to 75% and very severe >75% loss of net primary productivity due to land use practices. At present, 70% of rangelands (3.32 Bha) are at least moderately degraded and UNEP estimates that desertification is increasing at an annual rate of 3.5%, defined by the percentage of the land base passing into a higher-degradation category from case-study data for eight well-studied countries. For the purpose of the calculations, we have assumed that the grazing removal rate is 50% using the site specific grazing management practices projected over 50 yr, compared to 30% removal rate for the sustainable management scenario. Again, we believe this to be a conservative projection from the UNEP data.

Differences in C flux between the two management scenarios were derived from the CENTURY model and are justified separately. They predict that over 50 yr the difference in C emissions between the regressive land use scenario and the progressive or optimal management scenario will be 37 Pg (annual difference = 0.7 Pg) over the whole land base under consideration (4.5 Bha of grasslands and rangelands). This is a significant rate of sequestration (ca. 12% of fossil fuel C emissions). It is important to note that the half-life of the 37 Pg (net) stored soil C is measured in hundreds of years. This is a much longer storage interval than can be achieved with tree plantations, forest preservation, or other sequestration scenarios that depend upon storing C in aboveground standing crop.

Table 3. Extent of desertification in rangelands within the drylands of the world (Dregne *et al.*, 1991).

Mha	Slight-none	Moderate	Severe-moderate	Very Severe	Total
Africa	347	274	716	5.3	995
Asia	384	485	692	10.8	1188
Australia	296	277	55	29.0	361
Europe	31	27	52	1.2	81
N. America	72	116	285	10.2	411
S. America	93	88	184	15.3	288
Total	1223	1267	1984	71.8	3323

5.3. COSTS OF ACHIEVING AN OPTIMUM MANAGEMENT SCENARIO

UNEP has developed a detailed series of regional plans for anti-desertification (Kassas *et al.*, 1991). The programs include afforestation, reforestation, planting of shrubs and grasses, control of grazing lands, planting halophytes on salinized land for animal feed and to sequester C (Glenn *et al.*, 1992), and numerous other remediation methods. Costs depend upon the severity of degradation and the intensity of land use (Table 4).

The urgency of initiating anti-desertification actions can be seen from Table 5 - as desertification proceeds, costs rise dramatically as land passes into higher degradation categories which require greater expense per unit of land. Our present concern is only with the rangelands (non-croplands) in Table 4. Annual costs to restore this land base are in the range 5.0 to 8.8 x 10^9 US$. Divided by the mean annual sequestration rate (i.e., 0.7 Pg per yr^{-1}), the cost per t C is about 10 US$. This estimate is within the range of estimates that have been calculated for forest sequestration schemes. However, as noted below, there are other policy considerations in regard to rehabilitating the drylands.

5.4. COSTS OF NOT CONDUCTING ANTI-DESERTIFICATION IN THE DRYLANDS

UNEP (1991) estimates the average annual income foregone due to degraded rangelands to be 23 x 10^9 US$ yr^{-1}. They calculate the cost:benefit ratio of restoring the rangelands as 1:3.5 on a global basis. However, the large sums of money required for investment in restoration have not been generally made available

Table 4. Global costs in 10^9 US\$ for a 20 yr program of direct anti-desertification measures (UNEP, 1991).

	Cropland			
	Irrigated	Rain-fed	Rangelands	Cost yr^{-1}
Preventive measures	10-31	12-36	6-18	0.3-.9
Corrective measures	17-50	18-55	13-38	0.7-1.9
Rehabilitation measures	21-41	22-59	80-120	4-6
Total			99-176	5-8.8

Table 5. Global average indicative costs for direct antidesertification measures in different land use systems (US\$) (UNEP, 1991).

Extent of Degradation	Cost of Restoration	US\$ ha^{-1}	
Slight/None	100-300	50-150	5-15
Moderate	500-1500	100-300	10-30
Severe	2000-4000	500-1500	40-60
Very Severe	3000-5000	2000-4000	30-70

(Kassas *et al.*, 1991). The funds required for restoration are available internally for the industrialized regions (N. America and Australia) and the oil-producing regions (the Middle East), but the developing countries in arid zones of Africa and Asia will require external support to conduct anti-desertification on a meaningful scale. The failure of past anti-desertification programs in these regions has been attributed directly to lack of funding (UNEP, 1991).

5.5. OTHER POLICY CONSIDERATIONS

The question of social equity needs to be considered. Should development and environmental projects be judged primarily on their net ability to sequester C in the future? If a massive transfer of payments from industrialized to developing countries for the purpose of C offsets occurs, will the decisions be based solely on storing the most C at the lowest price, even if there are no social benefits, or will priority be given to schemes which are necessary even in the absence of global warming?

A major concern is whether anti-desertification measures, even if fully funded, can be effective. Technical solutions to rehabilitating rangelands are available but if population increases put additional pressure on the landscape, desertification may continue despite external funding for anti-desertification programs.

6. Research Perspective

A number of research needs were identified at the workshop that would alleviate gaps in our current knowledge of grasslands and dryland ecosystems responses to changing atmospheric CO_2 and climate conditions. These research needs include:

Ecophysiology and Ecosystem Science - Response of natural ecosystems, above and below ground, across broad climate gradients to elevated CO_2 with particular emphasis on changes in relationships among environmental variables such as temperature, moisture, and nutrients. In mixed C_3-C_4 ecosystems, determination of differential responses among species that affect productivity and interspecific competition leading to composition shifts. Particular emphasis should be placed on carbon pool quantification over time and space for selected ecosystems.

Land Use Management - Response of natural ecosystems to elevated CO_2 under traditional and anticipated management, including sustainable, rehabilitative, and regressive strategies utilizing, for example, grazing, fire, and conversion strategies.

Modeling of Ecosystem and Ecophysiological Responses - Using data obtained from the ecophysiology and management research, models should simulate responses across all scales. Modeling efforts are currently constrained by inadequate databases of ecosystem responses to elevated CO_2 for essentially all ecosystems. In addition, collaboration with development and application of data bases that include land cover, land use, and soils. This would facilitate projections of future scenarios for assessment purposes.

Monitoring and Coordination of Research Efforts - Long-term assessments of grassland and dryland ecosystem productivity, above and below ground, in response to year-on-year variation in climate (temperature, precipitation, etc.), concentrated at a series of study sites covering the whole range of these ecosystems. These measurements should be combined with more extensive monitoring over large areas using techniques such as satellite and aircraft remote sensing. Research efforts at all sites should be coordinated to provide relatively uniform data collection and synthesis methodologies required to produce databases for efficient distribution to varied uses.

7. Summary

The potential effects of management on carbon storage in grassland and dryland soils are substantially greater than that of climate change or CO_2 enhancement. Projections into the future based on a double-CO_2 climate including climate-driven shifts in biome and accounting for cropland areas by the year 2040 resulted in a net C sink ranging from +5.6 Gt to +27.4 Gt, for three different land cover projections respectively, based upon optimal grassland management. The increase in C storage in this future projection resulted mainly from an increase in area for the warm grassland sub-biome (net increase of 280 M ha due to climate-change induced biome shifts) together with a net increase soil organic C density resulting from net ecosystem response to climate changes and enhanced CO2 concentrations.

Differences in C flux between sustainable and regressive management scenarios were derived from the CENTURY model and are justified separately. They predict that over 50 yr the difference in carbon emissions between the regressive scenario and the sustainable management scenario will be 37 Gt (annual difference = 0.7 Gt) over the whole land base under consideration (4.5 B ha of grasslands and rangelands). The cost per t C is around US$10. This estimate is within the range of estimates that have been calculated for forest sequestration schemes.

The soil store of C in these ecosystems is a very important pool, since it is stabilized for hundreds to thousands of years, and forms the bulk of the grassland C pool. Calculations based on those of Gifford *et al.* (1990) suggest that the present C pool in grassland soils (estimated at 417 Gt; Table 2), may be a net sink of 0.6 Gt per annum.

8. References

Archer S.: 1993, in: Varra M., W. Laycock and R. Pieper (eds), *Ecological implications of livestock herbivory in the West*, Society for Range Management (in press).

Bailey, R.G.: 1989, *Environmental Conservation* **16**, 307-309.

Cramer, W.P. and Solomon, A.M.: 1993, *Climatic Research* (in review).

Dregne, H., Kassas, M. and Rosanov, B.: 1991, *Desertification Control Bulletin* **20**, 6-18. United Nations Environment Programme, P.O. Box 30552, Nairobi, Kenya.

Edwards, G., and Walker, G.: 1983, C_3, C_4: *Mechanisms, and Cellular and Environmental Regulation, of Photosynthesis*. 542 p. Blackwell, Oxford.

Gifford, R.M., Cheney, N.P., Noble, J.C., Russell, J.S., Wellington, A.B. and Zammit,C.: 1990, in: *Australia's Renewable Resources: Sustainability and Global Change*, Gifford, R.M. and M.M. Barson (eds) Bureau of Rural Resources Proceedings No. 14. Resource Assessment Commission, Queen Victoria Terrace, Parkes ACT 2600. pp. 151-187.

Glenn, E.P., Hodges, C., Leith, H., Pielke, R. and Pitelka, L.: 1992, *Environment* **34**, 40-43.

Hare F.K.: 1977, in: *Desertification: its causes and consequences*, United Nations Conference on Desertification, Nairobi, Kenya, Pergamon Press, Oxford UK.

Houghton, J.T., Jenkins, G.J. and Ephraums, J.J. (eds): 1990, *Climate Change. The IPCC Scientific Assessment*, World Meteorological Organization (WMO), Cambridge University Press, Cambridge.

Kassas, M., Ahmad, Y. and Rosanov, B.: 1991, *Desertification Control Bulletin* **20**, 19-29.

Kimball, B.A.: 1983, *Agron. J.* **75**, 779-788.

Knapp, A.K., Hamerlynck, E.P. and Owensby, C.E.: 1993, *Environmental and Experimental Botany* (in press)

Le Houerou, H.N.: 1984, *J. Arid Environments* **7**, 1-35.

Morrison, J.I.L.: 1985, *Plant, Cell and Environment* **8**, 467-474.

Newton, P.C.D.: 1991, *New Zealand J. Agr. Res.* **34**, 1-24.

OIES: 1991, *Arid and semi-arid regions: response to climate change*, Office of Interdisciplinary Earth Science, Boulder, Colorado, USA.

Ojima, D.S., Kittel, T.G.F., Rosswall, T. and Walker, B.H.: 1991, *Ecol. Appl.* **1**, 316-325.

Ojima, D.S., Parton, W.J., Schimel, D.S., Scurlock, J.M.O. and Kittel, T.G.F.: 1993, *Water, Air, and Soil Pollution* (this volume).

Owensby, C.E., Coyne, P.I., Ham, J.M., Auen, L.M. and Knapp, A.K.: 1993a, *Ecological Applications* (in press)

Owensby, C.E., Coyne, P.I. and Auen, L.M.: 1993b, *Plant, Cell and Environment*.

Owensby, C.E., 1993, *Water, Air, and Soil Pollution* (this volume).

Parton, W.J., Schimel, D.S., Cole, C.V. and Ojima, D.S.: 1987, *Soil Science Society of America Journal* **51**, 1173-1179.

Parton, W.J., McKeown, B., Kirchner, V. and Ojima, D.: 1992, *CENTURY Users' Manual*, Natural Resource Ecology Laboratory, Colorado State University, Fort Collins CO 80523, USA.

Parton, W.J., Coughenour, M.B., Scurlock, J.M.O., Ojima, D.S., Gilmanov, T.G., Scholes, R.J., Schimel, D.S., Kirchner, T., Menaut, J-C., Seastedt, T., Garcia, E. Moya, Kamnalrut, A., Kinyamario, J.L. and Hall, D.O.: 1993a, *Global Biogeochemical Cycles*.

Parton, W.J., Schimel, D.S., Ojima, D.S. and Cole, C.V.: 1993b, submitted, *Soil Science Society of America Journal* (in press b).

Prentice, I.C., Cramer, W., Harrison, S.P., Leemans, R., Monserud, R.A. and Solomon, A.M.: 1992, *Journal of Biogeography* **19**, 117-134.

Sala, O.E., Parton, W.J., Joyce, L.A. and Lauenroth, W.K.: 1988, *Ecology* **69**, 40-45.

Schlesinger, W.H., Reynolds, J.F., Cunningham, G.L., Huenneke, L.F., Jarrell, W. M., Virginia, R.A. and Whitford, W.G.: 1990, *Science* **247**, 1043-1048.

UNEP: 1991, UNEP Governing Council decision - desertification, *Desertification Control Bulletin* **20**, 3-5.

Weaver, J.E. and Albertson, F.W.: 1943, *Ecological Monographs* **13**, 63-118.

Woodward, F.I.: 1987, *Nature* **327**, 617-618.

AGRICULTURAL SOURCES AND SINKS OF CARBON

C. VERNON COLE

USDA, Agricultural Research Service
Natural Resource Ecology Laboratory
Colorado State University
Fort Collins, CO 80523 USA

KLAUS FLACH

12601 Builders Road
Herndon, VA 22070 USA

JEFFREY LEE

U.S. EPA
Environmental Research Laboratory
Corvallis, OR 97330 USA

DIETER SAUERBECK

Bonhoeffer Weg 6
W-3300 Braunschweig, GERMANY

BOBBY STEWART

USDA, Agricultural Research Service
P.O. Drawer 10
Bushland, TX 79012 USA

Abstract. Most existing agricultural lands have been in production for sufficiently long periods that C inputs and outputs are nearly balanced and they are neither a major source nor sink of atmospheric C. As population increases, food requirements and the need for more crop land increase accordingly. An annual conversion of previously uncultivated lands up to 1.5×10^7 hectares may be expected. It is this new agricultural land which suffers the greatest losses of C during and subsequent to its conversion. The primary focus for analysis of future C fluxes in agroecosystems needs to be on current changes in land use and management as well as on direct effects of CO_2 and climate change. A valid assessment of C pools and fluxes in agroecosystems requires a global soils data base and comprehensive information on land use and management practices. A comprehensive effort to assemble and analyze this information is urgently needed.

Water, Air, and Soil Pollution **70**: 111–122, 1993.

1. Introduction

Of all the major biomes of the terrestrial biosphere, agroecosystems represent the land areas of the globe most subject to continuous anthropogenic disturbance. Land use changes involving major transformations from forests, grasslands and savannas have converted large areas from relatively stable, undisturbed ecosystems to agroecosystems under extensive and intensive management. The introduction of agriculture involving land clearing or breaking of sod, cultivation, replacement of perennial vegetation by annual crops, and nutrient subsidies in the form of fertilizers has had major impacts on C pools and fluxes in large regions of the globe. In the initial phases of these transformations there have been major losses of CO_2 to the atmosphere as soil C pools adjusted to increased soil disturbance and reduced C inputs. In many areas under intense pressure for production, this has led to serious soil degradation by erosion and nutrient losses. These trends continue in many areas of the world. On the other hand, in countries able to provide subsidies of energy and technology, agricultural productivity has shown continuing increases, land degradation has slowed or reversed, and soil C pools have stabilized or slightly increased. For analysis of future C fluxes in agroecosystems the primary focus needs to be on current changes in land use and management as well as future adjustments in management when agricultural communities respond to climatic change.

2. Assessment of Sources and Sinks

2.1.   INVENTORY OF AGRICULTURAL SOILS

In order to assess the involvement of agriculture as a source or sink in the overall C budget, data on the amounts of land actually and potentially used for crop production are required. According to FAO statistics and their evaluation by Bouwman et al. (1990a), the total area of crop land at the present time amounts to about $1.5 \cdot 10^9$ ha. Roughly half of this area is located in the temperate climatic zones and the other half is in the tropical and subtropical areas of the world (Table 1, Sauerbeck, 1992).

As population increases, especially in the developing countries, food requirements and the need for more crop land increase accordingly. Estimates for the period until 2025 assume an additional requirement of only about +5% in the temperate zone, but more than 60% in the tropical/subtropical zone. This would add up to an overall increase of 36% from $1.5 \cdot 10^9$ ha at present to more than $2 \cdot 10^9$ ha within less than 40 yr (Sauerbeck 1992).

Considering the amount and quality of the soils which are still available for potential agricultural uses, it is questionable whether these projected land requirements can in fact be met. For the time being, however, an annual conversion of somewhere between 1 and 1.5×10^7 ha of formerly virgin land must be assumed. It is this new agricultural land which suffers the greatest losses of C during and subsequent to its conversion.

Table 1: Projections of human population, arable land requirement, and N-fertilizer consumption for 1990-2025, for temperate or tropical and subtropical climates (Bouwman 1990b, Sauerbeck 1992).

Region	Population 1990	Population 2025	Arable land 1990	Arable land 2025	N-Fert. Use 1990	N-Fert. Use 2025
	x 10^6		10^6 ha		kton N yr^{-1}	
Temperate Areas	1,156	1,335	689	720	40,427	45,289
% of total	22	16	47	36	51	38
% change till 2025		+15		+ 6		+12
Tropical and subtropical	4,084	7,042	787	1,286	39,137	73,081
% of total	78	84	53	64	49	62
% change till 2025		+72		+63		+87
Total	5,240	8,407	1,476	2,006	79,564	118,370
% change till 2025		+63		+36		+49

2.2. QUANTIFICATION OF C POOLS AND FLUXES

Different approaches have been used to assess the amounts of soil organic C (SOC) which are presently stored in agricultural soils (Eswaran, 1993). Schlesinger (1984) assumed 7.9 kg m^{-2} and arrived at a global total of 111 Gt for 1.4 billion ha of cultivated land. Buringh (1984) assumed 9.5 kg m^{-2} for 1.5 billion ha of cropland for a total of 142 Gt and 11.6 kg m^{-2} for 3.04 billion ha of grassland for a total of 353 Gt. Buringh also gave estimates for the C content of the various soil orders, which, for cropland, ranged from 2.0 kg m^{-2} for Aridisols to 13 kg m^{-2} for Mollisols and 10 kg m^{-2} for Oxisols. The high value for Oxisols is probably realistic, although a low SOC content for these soils had been assumed in the past. As most of the increase in cultivated land in the next 35 yr will be in the tropics and subtropics (Table 1) a realistic assessment of the SOC pool in these soils is important.

These pool size estimates are rather rough, the variation with soil orders and C contents at different soil depths are large, and it will be difficult to arrive at more reliable figures soon. However, soil mapping has become an international priority and improved soil C estimates will become available if supported by adequate analytical work.

To understand the turnover of and the CO_2 fluxes from these overall soil C pools, one needs to realize that soil organic matter consists of several sub-pools of exceedingly different accumulation rates and residence times. Soil organic components are often characterized on the basis of their density, size, and chemical composition. These components consist of a continuum from labile compounds that mineralize rapidly to more recalcitrant residues that accumulate as they are deposited during advanced stages of decomposition. The turnover rate of the labile and stable pools of soil organic C vary from a few months to several thousand years. The most resistant components of SOC are highly polymerized humic substances. The resistance of humic substances to microbial degradation results from both physical configuration and chemical structure. These substances become complexed with clays and mineral colloids so they are greatly influenced by soil texture and structure.

The organic C most rapidly lost within just a few years after land conversion is - apart from the native standing plant biomass - the more recent and labile, but quantitatively rather small soil humus fraction. This is then followed by a more gradual but long-lasting loss of stabilized SOC due to continuous disturbance by soil tillage. This is true for plowed grassland and forest soils in temperate zones as well as for cleared tropical forests.

3. Managing Soils for C Storage

3.1. MANAGEMENT OPTIONS

It is well documented that soil organic C levels decline when land is converted from grassland or forest ecosystems to cropland. This decline is most rapid in the first few years following conversion and then continues at slower rates until a new steady state is reached. After 50 to 100 yr SOC levels are often 50 to 60% lower than the initial levels.

Most existing agricultural lands have been in production for sufficiently long periods that they are approximately in steady state, and thus are neither a major source nor major sink of atmospheric C. However, conversion of previously uncultivated lands (Watson et al., 1990; Leggett et al., 1992) into agricultural production, driven by increasing population and land degradation, results in large C fluxes to the atmosphere of approximately one fourth of the emissions from fossil fuels (Brown et al., 1993).

Soil management practices have significant effects on both the rate and extent of SOC decline and on restoration of SOC levels. Improved management of crop residues, reduced tillage and inputs of more biomass with higher crop production offer the greatest potential for reducing the decline and for storing some additional C in soils. At the same time, reduced tillage will generally require lower inputs of fossil fuel.

3.2. LIMITATIONS FOR C STORAGE

The extent to which soil management can influence gains or losses of soil C is highly variable and difficult to predict. Stewart (1993) points out that the maintenance of SOC becomes more difficult as temperatures increase and the amounts of precipitation decrease. The reasons are many, but are dominated by the fact that organic matter decomposition rates are accelerated with rising temperatures, and the production of biomass to replenish the SOC reserves becomes less as water becomes more limiting.

Sauerbeck (1993) postulates that under optimum soil management it might be feasible to increase the C level of existing arable soils in the temperate zones (690 x 10^6 ha) by up to 1 kg m^{-2}. This soil C increase would represent about a 10% increase of SOC in these soils. Sauerbeck estimates that it would take 50 to 100 yr to reach this new level of SOC and that little additional SOC could be stored unless a new set of management practices was initiated. Such a change in SOC would sequester in the order of 6.2 Gt of C over the 50 to 100 yr period.

Similar increases are not likely in all soils. Aridisols cover large regions of the world and are characterized by hot dry climates. These soils are inherently low in SOC. Kimble (1990) analyzed 98 pedons as part of a global assessment and found an average SOC of 4.2 kg C m^{-2}. It is unlikely that the SOC in most of these soils could be increased by more than a small fraction. Also, many of the Aridisols are located in developing countries where populations are great and continue to increase at a rapid rate (Table 1). Therefore, even the small amounts of crop residues that are produced in these water deficient regions are often utilized as fodder for animals or fuel for cooking. Under such conditions, these soils do not have a significant potential as a sink for C, but neither will they be a major source because of their low content of SOC.

The likely potential for soil management to sequester C ranges from 0 to about 1 kg m^{-2} for important soil groups. Assuming an average of 0.5 kg m^{-2} as an achievable goal for agricultural cropland, about 7 Gt of C sequestration could be achieved. However, this increase would take at least 50 yr and could be achieved only once.

4. Tradeoffs for Management of Biomass Production and Soil Organic C

Managing croplands to increase SOC has implications for the emission of several greenhouse gases (GHG) and for environmental quality in general. Some of these effects occur off-site.

4.1. CARBON COSTS OF FERTILIZER MANUFACTURE

The potential for increasing SOC in agricultural systems is strongly influenced by the levels of C inputs. These, in turn, largely depend on the supply and improved utilization of inorganic and organic N sources. On average, the energy required to produce 1.0 ton of nitrogen in fertilizer releases approximately 1.5 tC to the atmosphere. This constitutes the largest share of the fossil C use of agriculture. The production of 80 Mt of fertilizer N in the world (Table 1) results in the release of about 120 MtC to the atmosphere each year. Thus, the C sink associated with increased SOC is partially offset by the fossil C used for production of the additional N. As pointed out by Flach et al. (1993), the need for continuous N inputs to maintain an increased equilibrium SOC value can, eventually, totally offset the C credits for the increased SOC.

Some options for increasing SOC do not require additional N input. For example, Lee et al. (1993) project that the adoption of no-till and winter cover crops can increase SOC in the U.S. cornbelt, while keeping N inputs constant. In this case, because tillage has been reduced, there might actually be a reduction in energy required for agricultural management. If similar increases could be obtained for large regions, as suggested by Sauerbeck, potentially, most of the fossil C cost of fertilizer production could be offset by increasing SOC.

A permanent C benefit can be achieved by increasing the efficiency of N utilization. For example, if increased N efficiency were to result in a reduction of 10% in the use of N fertilizer, world-wide fossil C emissions would be decreased by about 12 Mt yr^{-1}.

4.2. INFLUENCE ON EMISSION OF OTHER GREENHOUSE GASES

Of the fertilizer N applied to agricultural soils, about 1.1% may be emitted from the soil into the atmosphere during a cropping season as N_2O-N (CAST, 1992). A considerable fraction of the fertilizer N is removed from the field to which it was applied through NH_3 volatilization, erosion and nitrate leaching. An unknown fraction of this N is eventually converted to N_2O and emitted to the atmosphere (Duxbury et al., 1993). Over the course of about 50 yr more than 80% of the N applied to a field is returned to the atmosphere (about 60% in 1 to 10 yr) through denitrification (McElroy et al., 1977) after it has been processed through the food chain. Generally, greater than 95% of this N returns to the atmosphere as N_2 but some unknown amount is released as N_2O (Duxbury et al., 1993). Thus, to the extent that increases in SOC depend on increased N inputs, the GHG benefits of sequestering atmospheric C will be partially offset by increased N_2O emissions. Since N_2O has about 270 times the global warming potential of CO_2 and a relatively long (100 to 200 yr) atmospheric residence time (Isaksen et al., 1992) the magnitude of this offset is not well quantified.

Changes in land use practices and related N inputs modify the magnitude of soil CH_4 oxidation (Mosier et al., 1991). The net effect of land cover changes and increased N deposition to temperate ecosystems have resulted in about 30% reduction in the CH_4 sink relative to the soil sink assuming no disturbance to any of the temperate ecosystems (Ojima et al., 1993).

Adoption of no-till and cover crops will, in many cases, cause an increase in soil water content. If this results in increased occurrence or duration of anaerobic soil conditions, there is the potential for increased production and release of CH_4 and N_2O from soils. Methane has a global warming potential of about 11 relative to CO_2 (Isaksen et al., 1992), so increased production of CH_4 and N_2O through changes in management could partially offset the GHG benefits of increasing SOC.

4.3 WATER QUALITY

Excessive application of fertilizers and organic wastes to croplands results in increased N and P concentrations in surface waters and in groundwater. These cause eutrophication of surface waters, with concomitant degradation of both aquatic life and water quality. Increased N (especially NO_3) concentrations in groundwater can make well-water unfit for human consumption. Thus, it is critical that fertilizer and manure applications for SOC sequestration be made in ways that do not result in excessive N loading.

Adoption of reduced tillage systems will, in many cases, lead to a substantial decrease in soil erosion. The protection of the soil is, in itself, a positive environmental contribution. Decreased erosion also reduces the delivery of sediments to surface waters. However, it is possible that reduced tillage might, under certain conditions, contribute to water pollution. Under reduced tillage systems, use of herbicides for weed control might be increased. Another consideration is that by increasing infiltration of water into and through the soil, agrochemicals might be more easily transported to aquifers.

4.4. BIOMASS PRODUCTION FOR C OFFSETS

The production of perennial crops (trees and grasses) as dedicated biomass energy feedstocks provides an opportunity for agricultural systems to reduce the demand for fossil fuels (Wright and Hughes, 1993). It also provides an alternative source of income to landowners in areas of excess food production. While economic availability of land is difficult to determine, there are in the U.S. and the E.C. currently about 40 to 60 M ha of land in set-aside or cropland removal programs. Larger amounts of excess cropland are predicted for the future.

Oilseeds and grain crops which can be converted to liquid transportation fuels are likely to occupy some portion of this available land, but there would be little overall reduction of C emissions due to the large energy inputs required for production and conversion to liquid fuels. Short rotation woody crops (such as hybrid poplars) grown on croplands with good moisture retention capacity and/or croplands with moderate wetness limitations are anticipated to yield 10 to 15 Mg biomass ha^{-1} yr^{-1} now (1990-2000), but have the potential of yielding 20 to 25 Mg biomass ha^{-1} yr^{-1} (Wright and Hughes, 1993). Perennial grasses (such as switchgrass) which can reduce erosion losses on croplands with moderate erosion susceptibility appear to have similar yield potentials. Both are suitable for conversion to electricity or liquid fuels through a wide variety of conversion processes. Fossil fuel inputs to short rotation woody crops, including diesel fuel for tractors and harvesters, pesticides and herbicides and N-fertilizers (at average rates of 50 kg ha^{-1} yr^{-1}) for trees, will have an approximate C cost of 0.3 Mg C ha^{-1} yr^{-1} at current yields and 0.4 Mg C ha^{-1} yr^{-1} at high yields. Perennial grasses will be slightly higher due to higher fertilizer additions and annual harvest. The fossil fuel offset benefits will depend largely on assumed yields and on the conversion process used (Sampson et al., 1993).

5. Competition for Limited Soil Resources

5.1. HISTORY OF LAND USE CHANGES

The need for food and fibers has grown with the growth of the world's population since the beginning of time. Until the late 19th century, almost all increase in food and fiber production was achieved through increases in the land area used for farming at the expense of forest and grass land. In recent times, food production has been increased to a large degree through gains in the production per unit land area, at least in the developed countries.

The conversion of forest land to farmland involved a large initial release of CO_2 from the destruction of vegetation, often through fire, and a smaller annual release of C from the soil organic matter over the next 100 to 150 yr until soil organic matter reached a new steady state, usually about one half of the original SOC content. In grassland soils the initial loss of C from the vegetation was much smaller, but the loss of soil organic matter was of similar magnitude as in forest soils.

Houghton (1977) estimated the annual C emission from land use changes at about 0.6 Gt in 1860 and at about 2.5 Gt in the 1980's. Carbon emission from the burning of fossil fuels exceeded emission from land use changes for the first time in the early 1960's. Wilson et al. (1978) documented significant decreases in values of $\delta^{13}C$ of cellulose in bristlecone pine in California with influx of soil organic matter-derived CO_2 into the atmosphere between 1850 and 1890 due to extensive land clearing and pioneer agriculture. This was before major inputs from fossil fuels.
The change of forest and grassland to agricultural land has been mostly irreversible with the exception of parts in the eastern United States where large areas that were converted to farmland in the 19th century were found unsuitable for sustained farming and reverted to forest.

5.2. ALTERNATIVE LAND USES: FOOD, FIBER AND ENERGY

At the present time some 80% of the potentially arable land of the world is being farmed. Much of the remaining land that might be converted to farmland is in developing countries and of poor quality. Conversion to farmland may require greater investments in capital energy and skill than are available, and may be associated with a major release of C from vegetation and soils.

Currently, the annual increase in world-wide agricultural production is about keeping pace with the increasing worldwide demand (CAST, 1992). There have been no major catastrophic reductions in world-wide production in recent years, but the world's food reserves are small (about 20% of the annual consumption), the world's population is growing rapidly and is expected to double by the year 2030, and large parts of the world's population are still living on an inadequate diet. Crosson (1992) has estimated that the demand for U.S. grains and soybeans will grow by about 1.4% annually in the next 20 yr, and that yields will increase at about the same rate. Hence, even in the United States, the areas of productive cropland available for release for fuel wood production are likely to be small. Yields may be increased more, and land may be released, if there are financial incentives.

Large areas of the world have soils that present major difficulties for agricultural production, but we have no breakdown as to the land areas that are subject to specific problems. The major problems are excess salinity, excess wetness, excess acidity caused by S in the soil, shallow and/or highly erodible soils, steep topography, and excess stoniness. Some of these problem soils may be advantageously used for fuel wood production if tolerant fast growing wood species are available. Real energy offset requires that the fuel wood be harvested economically and that there is a local market with an energy demand that would otherwise be satisfied with fossil fuels.

6. Uncertainties and Research Needs

6.1. COLLATION AND ORGANIZATION OF DATABASES

This analysis is based on information that has been accumulated by a small number of investigators. Although we are confident that this is the best information available and that these investigators did excellent work within their constraints of time and money, we have noted significant inconsistencies among the various sources. For better analyses of the impact of global warming, much better data bases are needed.

An international soils data base is being developed by ISRIC, the International Soil Resource Information Center in Wageningen (Netherlands) in cooperation with IGBP (International Geosphere-Biosphere Programme), FAO (Food and Agriculture Organization of the United Nations) and USDA (United States Department of Agriculture) (Scholes and Skole, in prep.). A realistic assessment of the SOC pools and dynamics requires comprehensive information on land use and management practices in addition to the soil data. A comprehensive effort to assemble this information is urgently needed.

7. Summary

7.1. SCIENTIFIC CONSIDERATIONS

* Existing agricultural lands are neither a major source nor major sink for atmospheric C.

* Agroecosystems in temperate regions may be converted to a net C sink up to a total of 7 Gt C during 50-100 years by use of appropriate soil management practices including enhanced use of crop residues, reduced tillage, and increased crop production with greater additions of organic C.

* Production of perennial grasses or trees as energy sources can offset significant C fluxes, but will be constrained by competition for limited land resources.

* Conversion of new lands into agricultural production, driven by increasing populations and land degradation, results in large C fluxes to the atmosphere (see Tropical forests) of approximately one-fourth of emissions from fossil fuels.

7.2. POLICY CONSIDERATIONS

* Policies should encourage technical assistance in developing countries to maintain and improve production on existing farm lands and to decrease land conversion.

* Policies should enable and encourage farmers to improve C sequestration in agroecosystems. These should support practices that improve crop residue management, reduced tillage and increased biomass production.

* Policies should encourage the development of energy crops, preferably on marginal lands to avoid competition with food crop production.

8. References

Bouwman, A.F.: 1990a, Global distribution of the major soils and land cover types, in: Bouwman, A.F. (ed), *Soils and the Greenhouse Effect*, John Wiley & Sons, New York, pp. 33-59.

Bouwman, A.F.: 1990b, Exchange of greenhouse gases between terrestrial ecosystems and the atmosphere, in: Bouwman, A.F. (ed), *Soils and the Greenhouse Effect*, John Wiley & Sons, New York, pp. 61-127.

Brown, S., Hall, C.A.S., Knabe, W., Raich, J., Trexler, M.C., and Woomer, P.: (this volume). *Tropical forests: Their past, present, and potential future role in the terrestrial carbon budget.*

Buringh, P.: 1984, Organic carbon in soils of the world, in: Woodwell, G.M. (ed), *The Role of Terrestrial Vegetation in the Global Carbon Cycle, SCOPE 23*, John Wiley & Sons, 247 p.

CAST: 1992, *Task Force Report No. 119, Council for Agricultural Science and Technology*, Ames, Iowa. 96 p.

Crosson, P.R.: 1992, *United States Agriculture and Environment: Perspective in the next 20 years*. U.S. Environmental Protection Agency (in press).

Duxbury, J.M., Harper, L.A. and Mosier, A.R.: 1993, Contributions of agroecosystems to global climate change, in: Harper, L.A., Mosier, A.R., Duxbury, J.M. and Rolston, D.E. (eds), *Agricultural Ecosystem Effects on Trace Gases and Global Climate Change*. ASA Special Publication Number 55, American Society of Agronomy, Inc., Madison, Wisconsin, pp. 1-18.

Eswaran, H., Van Den Berg, E. and Reich, P.: 1993, *Soil Sci. Soc. Am. J.* **57**, 192-194.

Flach, K., Barnwell, T.O. and Crosson, P.: (in press), Impacts of agriculture on soil organic matter in the United States, in: Paul, E.A. and Elliott E.T. (eds), *Soil Organic Matter in Temperate Agroecosystems*, Lewis Press.

Houghton, R.A., Skole, D.L. and Lefkowitz, D.S.: 1977, *Forest Ecology and Management* **38**, 173-199.

Isaksen, I.S.A, Ramaswamy V., Rodhe H. and Wigley T.M.L: 1992 Radiative Forcing of Climate, in: Houghton J.T., Callander, B.A. and Varney S.K. (eds), *Climate Change 1992*, Cambridge University Press, pp. 51-67.

Kimble, J., Cook, T. and Eswaran, H.: 1990, Organic matter in soils of the tropics, in: Proc. Symp. Characterization and role of organic matter in different soils. Int. Congr. Soil Sci. 14th, Kyoto, Japan. 12-18 Aug. 1990. ISSS, Wageningen, the Netherlands.

Lee, J., Phillips, D. and Lin, R.: 1993, *The effects of trends in tillage practices on erosion and carbon content of soils in the U.S. corn belt*.

Leggett, J., Pepper, W.J. and Swart R.J.: 1992, Emissions Scenarios for the IPCC: an Update, in: Houghton, J.T., Callander, B.A. and Varney, S.K. (eds), *Climate Change 1992*, Cambridge University Press, pp. 73-95.

McElroy, M.B., Wolfsy, S.C. and Yung, Y.L.: 1977, *Philosophical Transactions of the Royal Society of London*, **277**, 159-181.

Mosier, A.R., Schimel, D.S., Valentine, D., Bronson, K. and Parton, W.J.: 1991, *Nature* **350**, 330-332.

Ojima, D.S., Valentine, D.W., Mosier, A.R., Parton, W.J. and Schimel, D.S.: 1993, *Chemosphere* **26**, 675-685.

Sampson, N.: 1993, *Biomass management and energy*.

Sauerbeck, D.: 1993, *CO_2-Emission from Agriculture: Sources and Mitigation Potentials*.

Sauerbeck, D.: 1992, *IPCC Update WG III AFOS Section 2. Temperate Agricultural Systems*.

Schlesinger, W.H.: 1984, Soil organic matter: a source of atmospheric CO_2, in: Woodwell, G.M. (ed), *The Role of Terrestrial Vegetation in the Global Carbon Cycle*, John Wiley & Sons, 247 p.

Scholes, R.J. and Skole D.: (in prep.) Global soils data: a proposal for a synthesis
    task. *Global Change Report No. 27, IGBP, Stockholm.*
Stewart, B.A.: 1993, *Managing crop residues for the retention of carbon.*
Watson, R.T., Rodhe, H., Oeschger, H. and Siegenthaler, U.: 1990, Greenhouse
    Gases and Aerosols, in: Houghton, J.T., Jenkins G.J. and Ephraums, J.J. (eds),
    Climate Change: The IPCC Scientific Assessment, Cambridge University Press, pp.
    5-40.
Wilson, A.T.: 1978, *Nature* **273**, 40-41.
Wright, L. and Hughes, F.E. 1993, *U.S. carbon offset potential using biomass energy
    systems.*

LAND AND WATER INTERFACE ZONES

JOHN P. DOWNING — *Pacific Northwest Laboratories, Marine Sciences Laboratory, 1529 West Sequim Bay Road, Sequim, WA 98382, USA*

MICHEL MEYBECK — *Laboratoire de Géologie Appliquée, C.N.R.S., Place Jussieu, 75257 Paris Cedex 05, France*

JAMES C. ORR — *Laboratoire de Modélisation du Climat et de l'Environnement, DMS/CEN Saclay/CEA, L'Orme des Merisiers, Bât. 709, F-91191 Gif-sur-Yvette, France*

R.R. TWILLEY — *Department of Biology, University of Southwestern Louisiana, P.O. Box 42451, Lafayette, LA 70504 ,USA*

H-W. SCHARPENSEEL — *Institute fur Bodenkunde, Universitat Hamburg, Allendeplatz 2, D-2000 HAMBURG 13, Germany*

Abstract. This paper reports analyses of C pools and fluxes in land-water interface zones completed at the International Workshop: Terrestrial Biospheric Carbon Fluxes; Quantification of Sinks and Sources of CO_2 (Bad Harzburg, Germany, March 1-5, 1993). The objective was to determine the role of these zones as global sinks of atmospheric CO_2 as part of a larger effort to quantify global C sinks and sources in the past (ca. 1850), the present, and the foreseeable future (ca. 2050). Assuming the world population doubles by the year 2050, storage of atmospheric C in reservoirs will also double, as will river loads of atmospheric C and nutrients. It is estimated that C sinks in temperate and boreal wetlands have decreased by about 50%, from 0.2 to 0.1 Gt C yr^{-1}, since 1850. The total decrease for wetlands may be considerably larger when tropical wetlands are taken into account, however, the area and C density of tropical wetlands are not well known at this time. Changes in cultivation practices and improved sampling of methaneogenesis have caused estimates of CH_4 emissions from ricelands to drop substantially from 150 to 60 Tg yr^{-1}. Even with doubled N and P loads, rivers are unlikely to fertilize more than about 20% of the new primary production in the coastal ocean. The source of C for this new production may not be the atmosphere, however, because the coastal ocean exchanges large quantities of DIC with the open ocean. Until the C fluxes from air-sea exchange of CO_2 and DIC are better quantified, the C-sink potential of the coastal ocean will remain a major uncertainty in the global C cycle. Analysis of model simulations of oceanic C uptake reconfirmed that the open ocean appears to take up about 2.0 Gt C yr^{-1} from the atmosphere and that model estimates are in better accord now, ± 0.5 Gt C yr^{-1}, than ever before. Land use management must consider the unique C sinks in coastal and alluvial wetlands in order to minimize the future negative impacts of agriculture and urban development. Long-term monitoring will be essential to prove the success, or failure, of management practices to sustain wetlands in the future. Relative to the other systems examined at the workshop, the C-sink capacity of the ocean (excluding estuaries) is not likely to be measurably affected in the foreseeable future by the management scenarios considered at the workshop.

Water, Air, and Soil Pollution **70**: 123–137, 1993.
© 1993 *Kluwer Academic Publishers.*

1. Introduction

This paper reports data and analyses completed by a working group of the authors at the International Workshop: Terrestrial Biospheric Carbon Fluxes; Quantification of Sinks and Sources of CO_2 (Bad Harzburg, Germany, March 1-5, 1993). Using available data, the group set out to quantify C pools and fluxes in rivers, wetlands, estuaries, and the ocean in order to determine their role in and potential for sequestration of atmospheric CO_2. Like the other working groups, our charge was to evaluate the current C sink and source potential of these systems at the present time, a time in the historical past when they were nearly pristine (ca. 1850), and in the foreseeable future (ca. 2050). The overall goal of the exercise was to evaluate the benefits and effectiveness of management scenarios described by Sampson et al. (1993) on the global, terrestrial C-sink capacity.

It was understood at the outset that major components of the systems that we examined, notably the coastal ocean, exchange C with the atmosphere and adjacent systems at unknown net rates (Downing and Cataldo, 1992). Also, we discuss C pools in ricelands and boreal wetlands that were analyzed by other working groups (eg. Cole et al., 1993; Apps et al., 1993). We did this for the reader's convenience, but to avoid double accounting of C, we did not include C fluxes between these pools and the atmosphere in our estimates for the workshop statement (Sampson et al., 1993).

For the 1850 scenario, we assumed that CO_2 emissions from fossil fuel combustion and land use were low ($\sim$0.1-0.2 Gt C yr-1), the CO_2 concentration in the atmosphere was 280 ppmv, and that sea level was about 0.3 m lower and average air temperature was 0.5°C cooler than today. The present scenario consists of today's climate, sea level, and atmosphere (355 ppmv CO_2), with total CO_2 emissions of 7.0 Gt C yr-1, and a landscape drastically modified by infrastructure, agriculture, and the harvesting of wood. For the future scenario, we assumed the IPCC Scenario A ("business as usual") projections for 2050 when sea level, mean temperature, and atmospheric CO_2 content may be: 0.6 m, +2.5°C, and 550 ppmv, respectively (Houghton et al, 1990).

The paper is organized by environments. It traces atmospheric C in the hydrological cycle from rock weathering to rivers, through lakes, rivers, reservoirs, and estuaries, and finally to the open ocean. It summarizes information in papers by Meybeck (1993), Twilley et al. (submitted), Orr (1993), and Scharpenseel (1993), provides new analyses based on their data, and concludes with scientific and policy considerations that managers of resources in land-water interaction zones will need to address in the future.

1.1.   LAND-WATER INTERACTIONS

Land-water interactions occur in narrow zones at the edges of rivers, lakes, and continents. Although the total area of these zones is only about 8% of the total earth surface, the C pools in soils and biomass (excluding marine sediments) total more than 600 Gt C. Ecosystems where land and water interact are among the most productive and biologically active known. They strongly influence loads of C, nutrients, and pollutants carried to the sea by rivers as well as the storage of C in sediment in the coastal zone (Wollast, 1983, 1991). For these reasons, land-water interactions are very important in understanding global carbon fluxes.

2. Lakes, Reservoirs, and Rivers

This section presents analyses of the present and future capacities of freshwater systems to store atmospheric C based on data and nomenclature presented in Meybeck (1993). To interpret them, it is essential to understand what is meant by atmospheric C in these systems. For convenience, the definition in Meybeck (1993) is summarized here. Atmospheric C in rivers includes: atmospheric DIC from rock-weathering, DOC from soils, and POC from primary production. Together, they comprise the total riverine load of atmospheric C. The proportion of these three constituents varies from one river to another depending on precipitation, bedrock, land use in the drainage area, and a variety of other factors discussed elsewhere. A summary of the analyses described in this section is given in Table 1.

2.1. STORAGE OF C IN LAKES

Lakes are abundant in Canada, Scandinavia, and parts of Russia and store particulate matter very efficiently, upstream lakes storing the greatest amount. The total area of lakes larger than 10 ha is at least 2.5×10^6 km^2 (Meybeck in preparation). Small lakes (0.1 to 1 km^2) have an average depth of 4.3 m. Very small ones (1 to 10 ha) have an average depth less than 3 m and can be considered as wetlands. The global storage estimate presented here is based on the total area draining to lakes larger than 100 km^2. Smaller lakes are assumed to be unlike these watersheds and should not be counted twice. The total area of larger lakes is estimated to be 1.57×10^6 km^2 and the average basin area/lake area ratio to be 15. Therefore, the total basin area upstream of lakes likely to produce particulate material is about 23.5×10^6 km^2.

Internal drainage is considered as a special sink in our C budget and must be subtracted from this figure. It corresponds to lake basins of the Caspian and Aral seas, Titicaca, Issyk-Kul, Balkash, Great Salt Lake, Eyre, and Chad and is roughly estimated to be 3.5×10^6 km^2, leaving 20×10^6 km^2 for the external drainage of upstream lakes. It means that as much as 20% of the drainage transit passes through lakes on the way to the ocean.

The efficiency at which suspended-matter is trapped in lakes frequently exceeds 95% and therefore we assume that all POC carried to lakes is stored there for at least 10^3 years. The average POC export rate in world rivers has been estimated to be 1.8 g m^{-2} yr^{-1}, of which 1.0 g m^{-2} yr^{-1} originate in soil. Therefore, the total POC storage in lakes that drain to oceans is about 36×10^{12} g C yr^{-1}. Probably some additional DOC precipitates and is stored in lakes, although this has not been well documented. Total organic C, 90% of which is DOC, greatly decreases in Finnish rivers as the area of lakes in the drainages increases. Also, up to 60% of DOC can be stored or oxidized to CO_2. If a precipitation rate of 30% DOC in lakes is applied here for 20% of river discharge, this would correspond to 15×10^{12} g DOC yr^{-1} stored, or approximately to a TOC retention of 51×10^{12} g yr^{-1} in lakes globally, of which 35×10^{12} g yr^{-1} is of atmospheric origin. This figure corresponds to a TOC retention of 32 g C m^{-2} yr^{-1}, a value close to the estimate of 29 g m^{-2} yr^{-1}, by Mulholland and Elwood (1982).

Calcium Carbonate precipitation in lakes is significant in drainages with carbonate bedrock. The total area of lakes located in carbonate terrains is about 0.18×10^6 km^2 (10% of external continental drainage). Average rates of DIC retention in lakes (estimated from Ca budgets) are 10^2 g DIC m^{-2} yr^{-1} for carbonate lakes and 5 g C m^{-2} yr^{-1} for non-carbonate lakes. Applying these rates to the corresponding lake areas, the total DIC

Table 1. Summary of global transport and storage of atmospheric C. Entries are in 10^{12} g yr^{-1}.

System	ca. 1850	Present	2050
River inputs to oceans			
Atmospheric C	540-600	550-610	550-610
TN	35.6	44.5	51.0
DP + POP	9.4	11.0	12.5
Storage of atmospheric C [1] in:			
Lakes	51	53	---[2]
Reservoirs	~0	100	200
Internal Drainages	20	25	30
Direct anthropogenic inputs to oceans			
DIC	~0	~0	~0
TOC	<20	40	>80
TN	<3	3	>6
DP	<0.3	0.3	>0.6

[1] See text and Meybeck (1993) for explanation.
[2] No estimate made.

retention in lakes is estimated to be about 26 x 10^{12} g C yr^{-1} (at least 70% originates in the atmosphere); the total organic and inorganic is about 77 x 10^{12} g C yr^{-1}. To this figure must be added the detrital carbonate particulate matter originating from land erosion and redeposited in lakes that does not play any role in the present CO_2 cycle.

2.2. STORAGE OF C IN RESERVOIRS

Reservoirs are built or being built on most major rivers (Columbia, Colorado, Missouri, Tocantins, Panama, Nile, Zambezi, Orange, Volta, Senegal, Indus, Chiang Ziang, Ob, etc.). In the 1970's, the total reservoir area was about 0.4 x 10^6 km^{-2}. The basin area/reservoir area ratio is much greater than for lakes, about 15. The C-retention rate in reservoirs is also much higher, about 500 g m^{-2} yr^{-1} (Mulholland and Elwood, 1982) of which only 50% originates in soil. Assuming these estimates, the total amount of C stored in reservoirs is estimated to be about 100 x 10^{12} g C yr^{-1} in 1970 and is projected to double, 200 x 10^{12} g m^{-2}, by 2050.

2.2.1. Eutrophication of Lakes and Reservoirs

Eutrophication is potentially an extra sink of atmospheric carbon. However, as for river eutrophication, it is not likely that it will play a major role in the global C cycle. About 60% of present lake area is located in formerly glaciated areas which are still not densely populated in Canada, Scandinavia, and Russia. If 10% of world lakes are considered to be eutrophied, and if the organic carbon long-term storage in sediments is assumed to be

1% of the new production (NP) of lakes, the total organic carbon storage in lakes would be 1.25×10^{12} g C yr-1 based on an average NP productivity of 500 g C m-2 yr-1. Including existing reservoirs does not change this figure much.

2.3. POLLUTANTS IN RIVERS

Untreated wastes comprise direct release of C, N, and P to the coastal oceans. At least 10^9 people discharge sanitary waste to rivers without modern treatment (primary at best). The C, N, and P discharge associated with these wastes are not accounted for in estimates of river loads because sampling monitoring stations are usually located upstream from waste discharges. The estimates of OC, TN, and DP fluxes in these waste streams are: 40×10^{12} g yr-1, 3×10^{12} g yr-1, and 0.3×10^{12} g yr-1, respectively. They are based on 40, 3, and 0.3 kg yr-1 of C, N, and P per capita, respectively (Meybeck, 1982).

Assessing the present influence of man on global riverine TOC is difficult because of the lack of appropriate measurements. Whereas BOD and COD are regularly surveyed in about 25% of river-to-oceans discharge, TOC has been regularly monitored in Northern America and Western European rivers only for about the last decade. Generally, BOD and COD levels in rivers of industrialized regions (Mississippi, Rhine, etc.) have stabilized or declined since the mid 1970's as a result of secondary waste treatment. Total organic C levels have increased in very few rivers as a result of human activities. This is thought to result from self- purification, whereby TOC is converted to CO_2.

By assuming a 2 mg l-1 increase of TOC in half of the rivers of the temperate zone (i.e., 5×10^3 km³ yr-1), the present human influence is believed to increase TOC to 10^{13} g C yr-1. Instead of 6.4 mg l-1 the average TOC concentration for temperate rivers (1.03×10^4 km³ yr-1) would drop to 5.4 mg l-1 which is well beyond uncertainties. This amount would correspond to a net input of 20 kg C yr-1 per capita from 0.5×10^9 people, after self-purification. Another 18×10^{12} g C yr-1 is likely from the less-developed world (i.e. 5 kg C yr-1 per capita x 3.5×10^9 people, living inland). The rest of the world population (10^9 people) is assumed to live in the coastal zone and directly releases its waste to the coastal ocean. Per capita release of TOC could not be much higher than these values when densely populated rivers are compared to less populated rivers. For instance, a TOC release of 5 kg C yr-1 per capita in China would roughly correspond to an increase of TOC in the Chiang Ziang (Yangtze-Kiang) by 4 mg l-1, that is about 30% of the present average TOC concentration. Altogether about 30×10^{12} g TOC yr-1 could presently originate from domestic wastes, or about 8% of global TOC load.

Inorganic C has been fairly constant in the few rivers that have been monitored over a 30-to-50-year period. The average annual changes of DIC export rates in kg C km-2 yr-1 in the last 25 years are the following: Volga +13, Kura (Azerbaijan) -24, Don (FSU) -5, St. Lawrence +17. The world-average DIC export rate is 3.9×10^3 kg km-2 yr-1. The DIC trends in these rivers therefore can be up or down but are less than 1% of the average export rate. Therefore, the changes are not yet detectable with current methods.

There is much better documentation of the changes in N and P loads (Meybeck, in preparation). Nitrate and phosphate loads have increased rapidly since the 1950's in most North American and Western European rivers (Rhine, Seine, Thames). Ammonia-N is about 10% of nitrate-N in natural rivers. In industrialized regions, ammonia rose to levels approaching $N-NO_3$ during the 1970's, then dropped due to secondary treatment of domestic and industrial sewage. But nitrate from fertilizers and groundwater is increasing in those rivers and will likely continue to do so for decades because of its long residence time in groundwater and the slow response of water quality to current mitigative countermeasures. Phosphorous, anthropogenic, and natural sources are increasing in many rivers. Countermeasures, particularly P-detergent banning, may drastically reduce P inputs to rivers by as much as 25% within 1 or 2 years if methods developed in Switzerland are applied worldwide.

Meybeck (1982) has estimated the transport of $N-NO_3$ and $P-PO_4$ in pristine rivers to be 37 and 3.7 kg km^{-2} yr^{-1}, respectively. Globally, the additional N and P inputs to oceans were estimated to be 7 x 10^{12} g N yr^{-1} and 10^{12} g P yr^{-1} for the 1970's, compared to 14.5 x 10^{12} g N yr^{-1} of natural transport to oceans for total dissolved N, and 10^{12} g P yr for total dissolved P. Other authors provide more alarming estimates, for example, Van Bennekom and Salomons 1981 (see Meybeck 1982) stated that "man's activities have increased the P load of rivers by a factor 5 and N load 3 or 4 times."

The global budget of N and P pollutants is very difficult to make because most rivers are under-sampled and probably have very low $N-NO_3$ and $P-PO_4$ concentrations (about 100 and 10 μg l^{-1} respectively); Western European rivers, on the other hand, have enormous N and P levels (2 to 8 mg N l^{-1}, and 0.05 to 0.25 mg P l^{-1}). If we consider an average concentration increment of 2 mg N l^{-1} and 0.3 mg P l^{-1} (the average N:P molar ratio in polluted waters is about 15) for half of the rivers of the temperate zone, which seems to be an upper estimate of the present situations, the excess N and P loads are respectively 10^{13} g N yr^{-1} and 1.5 x 10^{12} g P yr^{-1}, which corresponds to a nitrate increase by a factor of 3, and phosphate increase by a factor 4, which is consistent with Van Bennekom and Salomons (1981). When the organic forms of N and P are taken into consideration these factors increase by 0.6 and 1.5 for total dissolved N and total dissolved P respectively.

2.3.1. Eutrophication of Rivers

Most rivers in Western Europe and in the United States are not becoming eutrophic despite the tremendous increase in total N and P loads. In rivers like the Rhine, Loire, and Thames, the average annual chlorophyll A content may exceed 50 μg l^{-1} with peaks up to 250 μg l^{-1}, while for non-eutrophic rivers in tropical regions, or in low densely populated temperate regions the areal chlorophyll is less than 5 μg l^{-1}. The POC flux in such rivers is not important on a global scale. If all Western European rivers are assumed to have an average chlorophyll content of 50 μg l^{-1}, further assuming that POC:chlorophyll equals 30, it would correspond to an extra load of 0.27 x 10^{12} g C y^{-1}, which is a highly conservative estimate. This is highly labile and may cause severe anoxia in turbid estuaries or secondary eutrophication of coastal waters, as in the North Sea and in the Baltic.

3. Wetlands

The analysis of wetlands builds on the assessment by Twilley et al. (1992). The objective was to place existing estimates of the C pools and fluxes in an historical perspective. Carbon dioxide fertilization of wetlands was not addressed, even though Drake (1992) and Thom (submitted) demonstrate that it occurs, because the long-term fate of the fixed C is not known. Recent estimates of C pools and fluxes for wetlands and mangroves are in Table 2.

Table 2. Pools and accumulation rates of C in wetlands.

Biome	Pool (Gt C)	Sink (Gt C yr[-1])
Temperate and Boreal Wetlands	455[1]	0.096
Tropical Wetlands	---[2]	0.18[3]

[1] Armentano and Menges (1986); modified by Gorham (1991), includes zones from 40° N/S to 65° N/S.
[2] No data.
[3] Twilley (1992); mangrove wood and soil.

Biogeochemical processing of C and nutrients by wetland ecosystems changes both the quantity and composition of C and nutrient loads in rivers. Net ecosystem production in wetlands results in the accumulation of large C pools in wetland soils and the utilization of nutrient loads in surface waters. By converting wetlands to agri- and aquacultural uses, the quantity and composition of C and N compounds transported from land to sea is changed. In the conversion process, much of the C stored in wetlands soils over the last 5,000 years of slow sea level rise has been released to the atmosphere. The nutrient loads produced by agri- and aquacultural uses in converted wetlands contribute to the sanitary waste loads discussed in the previous section. The magnitude of these loads and their ecological effects are only known for a few intensely studied estuaries in industrialized regions of Europe and North America (Wollast, 1983). The global ramifications of these modifications, however, remain a major uncertainty worthy of systematic research.

Thus, in the past century, wetlands have been replaced with systems that contribute carbon to the atmosphere and nutrients to lakes and the coastal ocean. Losses of wetlands, mainly peatlands, have been particularly severe in the temperate zone of the northern hemisphere. Shoreline hardening, flood control, sediment management, and development in densely populated coastal regions may have already put a cap on future wetland losses to sea level rise (Edgerton, 1990) because there is little of the resource left. In some regions, such as the Mississippi River delta, diversion of freshwater to control flooding of low lying areas has lead to unprecedented loss of coastal wetlands. In the tropics, harvesting of mangrove forests for forest products and charcoal production has seriously diminished their areal extent. The largest peatlands in the world, located in the boreal regions, have not been as severely impacted by human activities as those in the temperate zone and remain a sink of C at a rate of about 0.096 Pg per yr. Most of the decreased C-storage capacity of wetlands is associated with losses of temperate wetlands.

Twilley et al. (submitted) estimated that the C flux from the atmosphere to temperate and boreal wetlands was about 0.2 Gt C y[-1] in 1850, and is only about half that now, 0.096 Gt C y[-1]. To put these fluxes into perspective globally, it is useful to consider Sundquist's (1993) estimate of the cumulative C imbalance calculated from ocean model simulations, ice-core and atmospheric CO_2 data, and land use estimates. For the period from 1750 to 1990, the range of the cumulative imbalance is 64 to 101 Gt C. Assuming that wetlands lost C at the average rate of 0.05 Gt C yr[-1] since 1750, wetlands have lost 12 Gt C to the atmosphere; therefore, the imbalance term of the global anthropogenic C budget would become 12% to 19% higher. This simple calculation underscores the potential significance of small fluxes integrated over long periods of time and the importance of a better understanding of the role of wetlands in the global C cycle.

It is our assessment that the C-management scenarios presented in Sampson et al. (1993) would not significantly affect the C sink capacity of wetlands by the year 2050. The main reason for this is that the majority of remaining wetlands are in sparsely populated, high-latitude regions and are unlikely to be targeted for C management in the near term. We must emphasize, however, that our assessment does not include tropical wetlands which may change the picture quite a bit. In order to preserve the C- and nutrient-sinking capacity of wetlands, future management practices will have to consider the processes we have summarized here.

4. Ricelands

Ricelands are affected by climate in ways similar to freshwater wetlands, and like wetlands, they produce CH_4, a potent greenhouse gas. For these reasons, they were on the agenda of the working group. The area of ricelands is presently about 140×10^6 ha and is rapidly expanding in West Africa, Egypt, Madagascar, and Brazil. Assuming a soil depth of 0.30 m, and C density of 8 kg C m^{-2}, the C pool in ricelands is estimated to be 11.2 Gt. Irrigated rice crops are expanding steadily and presently comprise more than 50% of total production. The grain yields range from about 1 t ha^{-1} for subsistence agriculture and upland rice up to 10 t ha^{-1} for irrigated high-yield varieties grown with mineral fertilizers. Irrigated riceland, with fast-growing varieties, can produce 2 to 4 crops per year.

A report on methane production by ricelands (Neue et al., 1990) reviews closed-box studies of methaneogenesis in ricelands conducted over a ten-year period at the International Rice Research Institute (IRRI). The IRRI data indicate that global CH_4 emissions from ricelands were 150 Tg CH_4 yr^{-1} at the time of the study. More recent measurements suggest that CH_4 emissions may now be only about 60 Tg CH_4 yr^{-1} (10% of the total CH_4 from all sources). Gradual abandonment of marginally productive paddies may also be measurably reducing the total area of ricelands and CH_4 emissions. Also, increased irrigation and multiple annual crops have decreased the amount of straw returned to paddies. Based on historical trends in CH_4 emissions from China, Khalil and Rasmussen (1993a) suggest that increasing substitution of mineral for organic fertilizers, such as straw and manure, may be part of the reason for declining CH_4 emissions from ricelands. The atmospheric CH_4 concentration is currently 1.73 ppmv and is increasing about 12 ppbv yr^{-1}, or about 0.7% yr^{-1} (Khalil and Rasmussen, 1993b). The rate of increase of atmospheric CH_4 in the past four years has declined by about 6 ppbv yr^{-1}. Changes in rice cultivation practice may have influenced these trends. The IRRI studies further indicated that ricelands are a small C sink for soil temperatures up to 28°C; at higher soil temperatures, they become a small C source. Neue (1990) determined from studies with varying SOM that the optimum yields are obtained when SOM content is 4 to 4.5%; yields slowly decrease as SOM content rises above 4.5%. Fertilizers increase the optimum SOM content slightly.

5. Oceanic C Fluxes

To be consistent with the discussion and analyses presented in Downing and Cataldo (1992), we define the coastal ocean to include saline wetlands, estuaries, bay-lagoon complexes, and continental shelves out to the 200-m isobath. This section discusses gaps in current knowledge of C cycling in wetlands and estuaries, recent hypotheses about NP and metabolism of organic carbon in the coastal ocean, and the status of model estimates of C uptake by the open ocean.

5.1. ESTUARIES

River loads of C, N, and P loads pass through estuaries on the way to the continental shelf. While in estuaries, fresh and saline waters are mixed by tidal-, wind-, and density-driven circulation. The fate of river C and N in an estuary involves the complex interaction of hydrodynamics, biogeochemistry, and morphological characteristics (Wollast, 1983, 1991; Kjerfve, 1988). Extreme combinations of these characteristics range from highly stratified, anoxic estuaries in deep glacial fjords with entrance sills to shallow estuaries dominated by strong discharge of freshwater like the Amazon for example.

Only a few large estuaries in densely populated temperate regions of the northern hemisphere have been studied long enough and in sufficient detail to adequately quantify material fluxes in time and space. Urban estuaries in the United States (Chesapeake and San Francisco Bays, Puget Sound, the Hudson Estuary, and Galveston Harbor, for example) and others in Western Europe have been the object of comprehensive research and long-term observations driven by fisheries, maritime commerce, and environmental issues. Because of the regional and political nature of these issues, studies in one estuary are rarely directly comparable to those in another with respect to biogeochemical cycling of C, N, and P (Thom, submitted); therefore, global conclusions drawn from data scattered in space and time are tentative at best. The main limitations of the data sets we considered include: 1) scarce measurements of air-sea gas fluxes, 2) incomplete knowledge as to whether estuaries are net autotrophic or heterotrophic, and 3) semi-quantification, at best, of estuarine-shelf exchanges of C, N, P, DIC, and particulate matter. All three limitations make it extremely difficult to assess the role of estuaries as a C sink, or source, on a global basis.

5.2. CONTINENTAL SHELVES

The working session focused on research reported in the last year and on the possible fate of river nutrient loads in the coastal ocean. Although aquaculture of macroalgae has drawn some attention in the past (Ritschard, 1992), the projected costs to sequester C is the highest of all ecological options evaluated by Spencer (1993) and the analysis of Orr and Sarmiento (1992) showed that enhanced algal growth would have only a small cumulative effect on atmospheric CO_2 levels. For these reasons, the subject is not discussed further. Also, because historical data on physical processes and C cycling is sparse on continental shelves, it was not possible to analyze the carbon budgets for the period from 1850 to 2050.

As with estuaries, C cycling on continental shelves have not been examined in a consistent way (Walsh, 1988). Data are clustered unevenly near regions of strategic military interest, mineral and fisheries resources, and commerce routes to major ports. With the exception of arctic shelves exploited for oil and gas, notably in the Alaskan

Bering, Chukchi, and Beaufort Seas and the Canadian arctic, little is known of C storage and transport, or biogeochemical processes in vast shelf areas bordering the former Soviet Union, Greenland, eastern Canada, and Antarctica.

Sea ice drastically reduces light levels, circulation, and mixing most of the year, but the biological consequences of these harsh conditions are difficult to observe on arctic shelves because they are inaccessible most of the year. Low light levels and sluggish recycling of nutrients result in low primary production. For example, primary production at a site in the Alaskan Beaufort Sea is about 20 g C $m^{-2} yr^{-1}$ (Dunton, 1984). This is less than 10% of average primary production, 274 g C $m^{-2} yr^{-1}$, at 22 sites reported by Smith and Hollibaugh (1993).

5.2.1. Primary Production on Continental Shelves

Two recent papers, one by Smith and Hollibaugh (1993) and the other by Walsh (1991), provide divergent views of the role of the coastal ocean in the global C cycle. The general arguments presented in both papers are quite plausible, however, the steps and assumptions leading to their results are very different. The idea proposed by Smith and Hollibaugh is that net organic metabolism (NOM) in the coastal ocean may be an important process in the global carbon cycle. They define NOM as the difference between primary production and respiration of organic material. The product of NOM, they say, is about 84 x 10^{12} g C yr^{-1}, and they further speculate that this C is "a source of CO_2 released to the atmosphere".

Walsh's hypothesis is that substantial NP is fertilized on continental shelves by river inputs of N (0.6 x 10^{14} g yr^{-1}) and onshore transport of NO_3 in slope water by coastal and eddy-induced upwelling (5.6 x 10^{14} g yr^{-1}). After correction for denitrification losses (0.5 x 10^{14} g yr^{-1}), the available N is (5.7 x 10^{14} g yr^{-1}). Stoichiometric conversion of this N, using a C:N ratio of 5, yields a maximum global C export of 2.9 Gt C yr^{-1} from continental shelves. Walsh goes on to speculate that this new production is a potential sink for atmospheric CO_2, provided the partial pressure of CO_2 (pCO_2) of surface water on continental shelves remains below the pCO_2 of the atmosphere (~360 µatm).

It is unlikely that either of these C fluxes will ever be confirmed by direct observation and it will be a long time before numerical models will be able to simulate the coupled shelf-ocean system, with realistic biology, at a global scale. Both these estimates lead to open-ended questions about gas exchange and cross-shelf transport of C. A check on how reasonable it is for air-sea exchange of CO_2 to provide the C fluxes to balance NP and NOM can be made by calculating the required CO_2 partial pressure difference between the ocean and atmosphere (ΔpCO_2) with accepted gas exchange coefficients. We assumed a range of coefficients, 0.04 to 0.1 mol C $m^{-2} yr^{-1}$ µatm^{-1} from Sarmiento et al. (1992), a total shelf area of 27 x 10^{12} m^2, and calculated ΔpCO_2 using the approach of Tans et al. (1990). The range of ΔpCO_2 values required to balance Walsh's NP value is 90 to 223 µatm, and the range to balance the NOM of Smith and Hollibaugh is 2.6 to 6.5 µatm. To get all of the C to support Walsh's NP from the atmosphere would require an unreasonably large ΔpCO_2 over the global shelf; the pressure difference needed to outgas the NOM, however, is not unreasonable.

Recent measurements off the coast of Ireland, by Watson et al. (1991) and in the North Sea by Kempe and Pegler (1991), indicate that average ΔpCO_2 levels are not as high as those required to support Walsh's estimate of NP directly from the atmosphere. Kempe and Pegler's calculated pCO_2 levels range from about 50 to 150 µatm above atmospheric level, making the North Sea a strong CO_2 source at the time of the measurements. The Watson et al. data range from 0 to 45 µatm below the atmospheric level, making the survey area a moderately strong sink at the time of the measurements. These limited data sets indicate that the coastal ocean has highly variable CO_2 fluxes, both in and out of the ocean. To explain the new production hypothesized by Walsh, we think better estimates of both gas exchange and mixing between shelf and slope waters are required.

Walsh's new production is due almost entirely to natural oceanic fluxes of N, not river delivery. Furthermore, unless natural processes have changed substantially since preindustrial times (clearly not the case for nutrient delivery by rivers), these processes play virtually no role in sequestering anthropogenic CO_2. Another perspective on NP and potential C fluxes in the coastal ocean is obtained with the nutrient inputs from rivers given in Section 2. We calculated the NP for these loads and the associated changes in atmospheric C (Table 3). The estimates in Table 3 indicate that river loads of nutrients do not significantly impact the budget of atmospheric CO_2. A final conclusion from our brief review and simple calculations is that we are a long way from understanding C cycling in the coastal ocean well enough to put quantitative estimates of natural CO_2 fluxes associated with this C cycling in the global C budget.

Table 3. Estimates of potential NP[1] in the coastal ocean and changes in atmospheric C[2] (since 1850) resulting from global river inputs of N and P. Entries are in 10^{12} g C yr[-1].

Nutrient	Ca. 1850		Present		2050	
	NP[1]	Δ atm. C[2]	NP	Δ atm. C	NP	Δ atm. C
N	204	0	256	-20.6	290	-34.2
P	386	0	452	-28.9	514	-27.1

[1] Potential NP determined from river nutrient loads (Table 1) and the Redfield ratio (C:N:P = 106:16:1).
[2] Losses of C from the atmosphere (calculated with the model-derived relationship of Orr and Sarmiento (1992)) associated with changes in NP.

6. Open-Ocean C Sinks

Attempts to directly measure anthropogenic CO_2 in limited regions of the ocean have been moderately successful (Takahashi et al., 1986). At global scales, however, the measurements are difficult because the change in DIC from man's input is minute compared to the huge DIC pool in the ocean and DIC varies spatially and temporally so much that a global-scale sampling program would be required to detect a meaningful trend. Brewer (1978) and Chen et al. (1985, 1986) estimate the amount of DIC from anthropogenic sources, but their approach was not well accepted (Shiller, 1982; Broecker et al., 1985). Currently, the most promising approach is to measure changes in oceanic del [13]C (Quay et al., 1992), although better global coverage and more surveys will be necessary before reliable estimates are obtained.

At present, ocean models provide the best estimates of oceanic uptake of anthropogenic CO_2. With updated results from three GCMs and three box models, and some additional corrections, Orr (1993) shows that the ocean absorbed 2.0 ± 0.5 Gt yr[-1] of anthropogenic CO_2 during the period 1980-1989. Simulations with the Princeton model (Sarmiento and Orr, 1991) predict that in 2050, the ocean will absorb 5.5 GT C yr[-1] under the IPCC emission scenario A. All models assume that the ocean absorbed no anthropogenic CO_2 in 1750 and that the ocean's carbon cycle has not changed substantially due to variations in ocean biology and physics since the start of the industrial revolution. Sarmiento (1991) and Sarmiento and Siegenthaler (1991) provide convincing evidence that supports the latter assumption.

Even though model estimates agree with each other on the net global flux, the picture they and the observations give of the latitudinal distribution of sources and sinks is inconsistent. Most anthropogenic CO_2 enters three-dimensional ocean models in the southern hemisphere because most of the ocean surface is there. The analysis of Tans et al. (1990), using an atmospheric model and CO_2 data, suggests that a large terrestrial CO_2 sink exists in the northern hemisphere. In contrast, Keeling et al. (1989), using atmospheric ^{13}C data, also identify a northern hemisphere sink, but suggest that it is the ocean. These inconsistencies most likely will only be resolved when CO_2, and perhaps river C fluxes, are added to coupled models of the atmosphere and oceans. In addition, Tans et al. calculated an oceanic C sink about half as large as is simulated in ocean models. However, after correcting for ocean skin temperature, atmospheric CO transport, and river C loads (Sarmiento and Sundquist, 1992), the adjusted Tans et al. C sink agrees well with the ocean models.

7. Conclusions and Recommendations

Carbon dynamics and storage in freshwater systems are sensitive to changes in land use, pollutant loads, climate, hydrologic conditions and atmospheric chemistry. With the exception of changes in wetland area, the interaction of these factors and their influence on C storage on a global scale is currently unknown. Relative to the other systems examined at the workshop, the C-sink capacity of the ocean (excluding estuaries) is not likely to be measurably affected in the foreseeable future by the management scenarios considered at the workshop. Assuming the world population doubles by the year 2050, storage of atmospheric C in reservoirs and river loads of anthropogenic nutrients and TOC will also double. Changes in cultivation practices and improved sampling of methaneogenesis have caused estimates of CH_4 emissions from ricelands to drop substantially from 150 to 60 Tg yr^{-1}. The coastal ocean exchanges CO_2 with the atmosphere and dissolved CO_2 with the open ocean, but the net CO_2 fluxes associated with these exchanges have yet to be determined. With currently available data, it is therefore impossible to estimate the portion of dissolved C fixed by coastal primary production that comes directly from the atmosphere. For the same reasons, it is not possible to predict the future C uptake by the coastal ocean.

Current model estimates of the uptake of CO_2 by the ocean appear to be converging and the mean CO_2 absorption rates for the ocean have not changed much since the 1990 IPCC report. More importantly, the range of estimates has decreased by nearly 0.6 Gt C y^{-1}. As long as ocean currents and mixing continue to operate in their present mode, it is not likely that the C-sink capacity of the open ocean and the distribution of sources and sinks for atmospheric CO_2 could change drastically before the year 2050.

7.1   SCIENTIFIC CONSIDERATIONS

- Estimates of temperate and boreal wetland C sinks suggest a decrease by about 50% (from 0.2 to 0.1 Pg C yr^{-1}) since 1850. The main reason for the decline is the conversion of wetlands to agricultural and urban lands. In the future, the C sink capacity of temperate wetlands may decline further or remain stable. Management practices and climate changes will determine what their actual fate will be. These conclusions, however, do not include tropical wetlands, the area and carbon density of which are not well known.

- Even with doubled anthropogenic N and P loads, it is unlikely that rivers will fertilize more than about 20% of the new primary production in the coastal ocean. The balance of the new production is believed to result from landward transport of nitrate from the open ocean. Whether this C is atmospheric CO_2 or DIC from the open ocean is not known. The fate of nutrients and C in coastal ecosystems is not understood well enough to predict their current or future roles in the global C cycle.

- Sea-level rise, changing hydrologic conditions, and sediment supply are the principal natural factors that will determine the future rate of C storage in coastal wetlands. Conversion of wetlands to agri- and aquacultural uses by expanding coastal populations in developing countries will reduce their C-storage capacity and increase nutrient loads to the coastal ocean.

7.2   POLICY CONSIDERATIONS

- Land use management must consider the unique C sinks in coastal and alluvial wetland environments in order to minimize negative impacts of agriculture and urban development.

- Salt intrusion as a result of sea-level rise and surface water diversion could switch respiration from methaneogenesis to CO_2 production in some coastal ecosystems. Coincidentally, global warming could change the balance between CH_4 production and oxidation. Although the severity of interactive effects is poorly known, water management could mitigate CH_4 production to some degree.

- Regardless of the strategies used, long-term monitoring will be essential to establish the success, or failure, of management practices to sustain wetlands in the future.

8.   Acknowledgements

The senior author wishes to thank the U.S. Department of Energy for financial support while this paper was in preparation and Ron Thom for his contribution on coastal ecosystems. PNL is operated for DOE under contract DE-AC06-76RLO-1830 by Battelle Memorial Institute. We thank the workshop participants in other working groups who provided results, information, and comments on our analysis and Eileen Stoppani and Michele Hemp for turning the manuscript into photo-ready copy.

9.   References

Apps, M.J., Kurz, W.A., Luxmoore, R., Nilsson, L.-O, Sedjo, R.A., Schmidt, R., Simpson, L.G., and Vinson, T.S.: this volume.
Armentano, T.V. and Menges, E.S.: 1986, *Journal of Ecology* **74**. 755.

Brewer, P.G.: 1978, *Geophys. Res. Lett.* **5**, 997.

Broecker, W.S, Takahashi, T., Peng, T.-H.: 1985, *Reconstruction of past atmospheric CO₂ contents from the chemistry of the contemporary ocean, an evaluation*, DOE Tech. Rep., DOE/OR-857, U.S. Department of Energy, Washington, D.C., pp. 79.

Chen C.T., Poission, A., and Goyet, C.: 1986, *Preliminary data report for the INDIVAT 1 and INDO1/INDIVAT 3 cruises in the Indian Ocean*, U.S. Department of Energy, pp. 106.

Chen, C.T.A.: 1985, *Cont. Shelf Res.* **4**, 465.

Cole, C.V., Flach, K., Lee, J., Sauerbeck, D., and Stewart, B.: this volume.

Downing, J.P. and Cataldo, D.A.: 1992, *Water, Air, and Soil Pollution* **64**, 439.

Drake, B.G.: 1992, *Water, Air, and Soil Pollution* **64**, 25.

Dunton, K.H.: 1984, An Annual Carbon Budget for an Arctic kelp Community, in: Barnes, P.W., Schell, D.M., and Reimnitz, E. (eds), *The Alaskan Beaufort Sea: Ecosystems and Environments*, Academic Press, Inc., pp. 311-323.

Edgerton, L.T.: 1990, *The Rising Tide, Gobal Warming and World Sea Levels*, Island Press, Washington, D.C.

Gorham E.: 1991, *Ecological Applications* **1**, 182.

Houghton, J.T., Jenkins, G.J and Ephraums, J.J.: 1990, *Climate Change, the IPCC Scientific Assessment*, Cambridge University Press, New York, pp. 365.

Keeling, C.D., Piper, S.C, Heimann, M.: 1989, in: Peterson H, (ed), *Aspects of Climate Variability in the Pacific and the Western Americas*, American Geophysical Union, Washington D.C., pp. 165-236.

Kempe, S and Pegler, K.: 1991, *Tellus* **43B**, 224.

Khalil, M.A.K. and Rasmussen, R.A.: 1993a, *Chemosphere* *1-4*, 803.

Khalil, M.A.K. and Rasmussen, R.A.: 1993b, *Chemosphere* *1-4*, 127.

Kjerfve, B.: 1988, *Hydrodynamics of estuaries*, CRC Press, Inc. Boca Raton, Florida.

Meybeck, M.: 1982, *Amer. J. Sci.* **282**, pp. 401.

Mulholland, P.J. and Elwood, J.W.: 1982, *Tellus* **34**, 490.

Meybeck, M.: 1993, C, N, P and S in Rivers: From Sources to Global Inputs, in: Wollast R., Mackenzie F.T. and Chou L. (eds), *Interaction of C, N, P and S Biogeochemical Cycles and Global Change*, Springer-Verlag, pp. 163-193.

Neue, N.U., Becker-Heidmann, P. and Scharpenseel, H.W.: 1990, A review of Rice Soil Agronomy and Chemistry, in: Bouwman A.F. (ed), *Soils and the Greenhouse Effect*, John Wiley & Sons, pp. 61-127.

Orr, J.C.: this volume.

Orr, J.C. and Sarmiento, J.L.: 1992, *Water, Air, and Soil Pollution* **64**, 405.

Quay, P.D., Tilbrook, B, Wong, C.S.: 1992, *Science* **256**, 74.

Ritschard, R.L.: 1992, *Water, Air, and Soil Pollution* **64**, pp. 289-303.

Sampson, R.N., Apps. M.J., Brown, S., Cole, C.V., Downing, J.P., Heath, L., Ojima, D.S., Smith, T.M., Solomon, A.M., Wisniewski, J.: this volume.

Sarmiento, J.L.: 1992, Biogeochemical Ocean Models, in: Trenberth, K. (ed), *Climate Systems Modeling*, Cambridge University Press, Cambridge.

Sarmiento, J.L.: 1991, *Global Biogeochem. Cycles* **5**, 309.

Sarmiento, J.L. and Orr, J.C.: 1991, *Limnol Oceanogr.* **36**, 1928.

Sarmiento, J.L, Orr, J.C., and Siegenthaler, U.: 1992, *J. Geophys. Res* **97** No. C3, 3621.

Sarmiento, J.L. and Siegenthaler, U.: 1991, in: Falkowski P. and Woodhead A. (eds), *New Production and the Global Carbon Cycle*, Plenum, New York.

Sarmiento, J.L. and Sundquist, E.T.: 1992, *Nature* **356**, 589.

Scharpenseel, H-W.: this volume.

Shiller, A.M.: 1981, *J. Geophys. Res.*, **86**, 11,083.

Smith, S.V. and Hollibaugh, J.T.: 1993, Coastal Metabolism and the Oceanic Organic Carbon Balance, *Reviews of Geophysics* **31**, 75.

Spencer, D.F.: 1993, An Overview of Energy Technology Options for Carbon Dioxide Mitigation, in: Rosen L. and Glasser R. (eds), *Climate Change and Energy Policy*, American Institute of Physics, New York, pp. 176-191.

Sundquist, E.T.: 1993, *Science* **259**, 934.

Takahashi, T., Goddard, J., Sutherland, S., Chipman, D.W. and Breeze, C.S.: 1986, *Seasonal and Geographic Variability of Carbon Dioxide Sink/Source in the Oceanic Areas: Observations in the North and Equatorial Pacific Ocean, 1984-1986 and Global Survey*, Technical Report to U.S. DOE Carbon Dioxide Research Division, Washington, D.C., pp. 65.

Tans, P.P., Fung, I.Y, and Takahashi, T.: 1990, *Science* **247**, 1431.

Thom, R.M.: submitted, *Water, Air and Soil Pollution*.

Twilley, R.R., Chen, R.H., Hargis, T.: 1992, *Water, Air, and Soil Pollution* **64**, 265.

Twilley, R.R., Chen, R.H., Bouergois, J.: submitted, *Water, Air and Soil Pollution*.

Van Bennekom, A.J. and Salomons, W.: 1981, Pathways of nutrients and organic matter from land to ocean through rivers, in: SCORE, *Proceedings of the Workshop on River Inputs to Ocean Systems (RIOS)*, New York, United Nations, pp. 33.

Walsh, J.J.: 1991, *Nature* **350**, 53.

Walsh, J.J.: 1988, *On the Nature of Continental Shelves*, Academic Press, New York, pp. 521.

Watson, A.J., Robinson, C., Robertson, J.E., Williams, P.J.IeB.. and Fasham, M.J.R.: 1991, *Nature* **350, No. 6313**, 50.

Wollast, R.: 1983, Interactions in Estuaries and Coastal Waters, in: Bolin B. and Book R.B. (eds), *The Major Biogeochemical Cycles and Their Interactions*, John Wiley & Sons, New York, pp. 385-407.

Wollast, R.: 1991, The coastal carbon cycle: fluxes, sources, and sinks, in: Mantoura, R.F.C., Martin, J.M. and Wollast, R. (eds), *Ocean Margin Processes in Global Change*, Plymouth Marine Lab, Plymouth, U.K., pp. 365-381.

BIOMASS MANAGEMENT AND ENERGY

R. NEIL SAMPSON	*American Forests, 1516 P St. NW, Washington, DC 20005 USA*
LYNN L. WRIGHT	*Environmental Sciences Division, Oak Ridge National Laboratory, Oak Ridge, TN 37831-6331 USA*
JACK K. WINJUM	*National Council on Air and Stream Improvement at Environmental Protection Agency, Environmental Research Laboratory, 200 SW 35th St., Corvallis, OR 97333 USA*
JOHN D. KINSMAN	*Environmental Affairs, Edison Electric Institute, 701 Pennsylvania Avenue NW, Washington, DC 20004-2696 USA*
JOHN BENNEMAN	*907 Tropic Drive, Vero Beach, FL 32963 USA*
ERNST KÜRSTEN	*PRIMA KLIMA, Ikenstasse 1B, D-4000 Dusseldorf 12, Germany*
J.M.O. SCURLOCK	*Division of Biosphere Sciences, King's College London, Campden Hill Road, London W8 7AH, Great Britain*

Abstract.

The impact of managing biomass specifically for the conservation or production of energy can become a significant factor in the global management of atmopsheric CO_2 over the next century. This paper evaluates the global potential for: (1) conserving energy by using trees and wood for shading, shelterbelts, windbreaks, and construction material; and (2) increasing the use of biomass and improving its conversion efficiency for producing heat, electricity, and liquid biofuels. The potential reduction in CO_2 emissions possible by the anticipated time of atmospheric CO_2 doubling was estimated to be up to 50×10^6 t C yr^{-1} for energy conservation and as high as 4×10^9 t C yr^{-1} for energy production. Of the many opportunities, two stand out. Through afforestation of degraded and deforested lands, biomass energy production offers the potential of 0.36 to 1.9×10^9 t C yr^{-1} emission reduction. Dedicated energy crops, which include short-rotation woody crops, herbaceous energy crops, halophytes, some annual crops, and oilseeds, offer the potential of 0.2 to 1.0×10^9 t C yr^{-1} emission reduction. Also addressed in the paper, but not quantified, were establishment of new forests, increasing the productivity of existing forests, or protecting forests to sequester C as an offset against CO_2 emissions from burning fossil fuels or forest destruction. Also addressed are uncertainties, gaps in scientific knowledge about ecosystems and their management, and policy considerations at the international and national levels.

Water, Air, and Soil Pollution **70**: 139–159, 1993.
© 1993 *Kluwer Academic Publishers.*

1. Introduction

In today's world, humankind utilizes about 400×10^{18} joules (J) of energy annually to provide for 5.6×10^9 people (Scurlock and Hall, 1990). Future energy needs will grow as population expands and as goods and services become more equitably available.

Currently, the primary means of supplying energy is combustion of fossil and wood fuels. The anthropogenic oxidation of fossil fuels is estimated to emit 5.5 to 6.5×10^9 t C to the atmosphere annually (Houghton $et\ al.$ 1992). Such emissions strongly contribute to the annual net gain of 3.5×10^9 t C as atmospheric CO_2, the major greenhouse gas projected to cause global warming (Houghton $et\ al.$ 1992).

Through photosynthesis, terrestrial and marine plants capture energy from sunlight to power the conversion of atmospheric CO_2 into biomass. For land areas, human technology in the form of agriculture and forest management can increase the amount of useable biomass. A portion of this biomass is currently used to produce 55×10^{18} J yr^{-1} of energy globally (Scurlock and Hall, 1990). If forested land areas are increased, management is intensified, and productivity is maintained on a sustainable basis, more biomass could be produced and utilized as bioenergy without contributing to anthropogenic emissions of CO_2, while, at the same time, more C can be captured within terrestrial ecoystems.

This paper summarizes information from a workshop held in Bad Harzburg, Germany, in March 1993, and the discussions of this working group regarding:

* Energy conservation from using trees and wood for shading, shelterbelts, wind-breaks, and construction material;
* Increasing the use of biomass to produce heat, electricity and liquid biofuels; and
* Establishing new areas of forest, increasing the productivity of existing forests, or protecting forests to sequester C as an offset against CO_2 emissions from burning fossil fuels or forest destruction.

For these topics, the paper addresses their current status, future projections, and key information gaps and research priorities.

2. Energy Conservation

Energy conservation is one of the major means available to developed countries to reduce fossil fuel emissions. Trees can be grown and utilized in ways that reduce the need for fossil fuel use, while simultaneously increasing terrestrial C storage in both above-ground biomass and soils. In addition, significant benefits are achievable in terms of C storage offsite in long-term wooden structures, given appropriate management and policy attention.

2.1 CONSERVATION TREES IN AGROECOSYSTEMS

Windbreaks and shelterbelts associated with croplands provide multiple benefits, including: increased C storage in woody plants and soils; reduced energy needs for farming the affected croplands; lowered rates of soil erosion, which reduces loss of soil organic matter and stored C; improved air quality due to reduction of windborne dust; and fuelwood that can replace fossil fuel or dung as an energy source.

On U.S. soils subject to wind erosion, it is estimated that windbreaks covering 6% of the cropland are needed to provide adequate erosion control (Brandle *et al.* 1992). In Germany, Burschel *et al.* (1993) concluded that it is ecologically desirable to have 4% of the agricultural land in hedges and shelterbelts, but the current cover is closer to 1 to 2%.

Kürsten and Burschel (1993) have shown that, in agroforestry systems where trees and crops are grown together or in rotation on the same land, trees can make mineral fertilization more effective or even unnecessary by catching nutrients that are being transported out of the crop root zone into the groundwater and recycling them back to the upper soil horizons in litter; by increasing the cation retention capacity of the soil by elevating soil humus content; by adding additional nitrogen to the system in the case of N-fixing tree species; and, by reducing the need for herbicides through mulching with leaves in alley-cropping systems.

Incorporation of trees within agroecosystems is beneficial in all types of agricultural systems, but the C benefit of conservation trees is likely to be higher in areas using mechanized agriculture. Shelterbelts protecting land under developed-nation mechanized agriculture in the temperate zones could reduce C emissions by 8 to 16 x 10^6 t C yr^{-1} if 2 to 4% of cropland was converted to shelterbelts. The C benefit of tropical agroforestry trees is lower since fuel and fertilizer use is presumed to be lower. However, in the case of maize production in the wet tropics, the CO_2-mitigation effect of adding agroforestry trees was estimated to be 0.024 to 0.1 t C ha^{-1} yr^{-1} (Kürsten and Burschel, 1993).

Although incorporating more trees in the agricultural landscape shows potential benefits, the reality is that trees are diminishing in many regions. In areas where intensive mechanized agriculture is practiced, windbreaks, shelterbelts and hedgerows are destroyed to accommodate large machinery, irrigation systems, and to capture maximum land for cultivation (Sampson, 1981). Also, in non-mechanized agriculture, the pressures of human populations, along with the increasing need for fuelwood and fodder, often means that trees are diminishing, not increasing.

2.2 URBAN AND COMMUNITY FORESTS

Properly placed urban trees can have a significant impact on atmospheric CO_2 buildup through energy conservation, reducing air conditioning and winter heating requirements (Sampson *et al.* 1992). From a C perspective, a properly-placed urban tree can be 4 to 15 times as effective at reducing atmospheric CO_2 as a rural tree, which primarily is only involved in C sequestration (U.S. EPA, 1992). Several considerations for urban tree planting should be noted. First, little fossil energy is used for cooling in many parts of the world. In such areas, urban trees can make the quality of human life far more healthy and pleasant, but energy savings will not be significant. Proper maintenance of urban trees is critical. Urban trees live notoriously short lives, succumbing to many different stresses such as drought, vandalism, and urban air pollution. Planting must be associated with a program for tree care. In areas with power lines, the choice of species and tree location is critical to avoid unnecessary damage during storm events. Despite these concerns, urban tree planting makes environmental and economic sense in many situations.

With energy conservation as a major goal, tree planting programs should focus first upon those trees that provide direct energy benefits to buildings. The next priority would be trees that provide maximum shade for parking lots, streets, and other dark-surfaced areas. The lowest priority would be to "fill in" the open spaces that, while not directly shading or protecting buildings, would help reduce the urban heat island effect by modifying albedo and wind patterns as part of the total urban forest. Sampson *et al.* (1992) proposed a 10-yr program aimed at increasing the canopy cover by 10% on residential lands, and 5 to 20% on other urban lands in the U.S. They estimate that the effect of such a program could result in a 7 to 29 x 10^6 t C yr^{-1} reduction in C emissions due to energy conservation from improved shading, increased evapo-transpiration, and reduction of the urban heat island, along with wintertime heat savings. Although these U.S. estimates are impossible to extrapolate to a global estimate, it would appear that urban and community trees could play a significant role in helping manage global fossil fuel use and CO_2 emissions.

2.3 MATERIALS SUBSTITUTION

Wood can be used as a raw material instead of aluminum, steel, concrete and plastics. A recent study on construction options estimated the CO_2-mitigation effect of using 1 m^3 of wood instead of steel and concrete to be 0.28 t C (Burschel *et al.* 1993).

Another field of substitution could be production of chemicals from resin, bark and leaves instead of coal and oil. Production of bio-based chemicals would reduce energy consumption and CO_2 emissions, as well as eliminating chemical byproducts.

3. Energy Production

Estimating biomass energy production potential requires making assumptions about available land, biomass productivity, percent used for energy, conversion efficiencies, and fuel substitution factors. The assumptions used in this paper are summarized in Table 1 and briefly discussed in the following sections.

Table 1. Assumptions for estimating the C emission reduction potential of biomass energy systems.

Region and Resource	Land Area[1]		Net C Yield[2]		% Energy Use	Substitution Factors		Higher Heat Value
	Low	High	Low	High		Low	High	
	(Mha)		(t C ha⁻¹ yr⁻¹)					(J x 10⁹)
Forest Management								
Boreal	80	120	0.5	2	20	0.9	0.95	20
Temperate	80	100	2	5	20	0.9	0.95	20
Tropical	200	500	2	6	20	0.5	0.7	20
Grass/Desert Conversion								
Irrigated Halophytes	10	43	3	7	100	0.5	0.7	20
Boreal/Temperate Afforestation	100	200	3	7	50	0.9	0.95	20
Tropical Afforestation	220	440	4	8	50	0.5	0.7	20
Cropland Conversion								
Temperate Shelterbelts	13	26	2	4	75	0.5	0.7	20
Temp. Energy Crops	26	73	5	9	100	0.65	0.75	18.5
Tropical Agroforestry	41	65	3	6	75	0.5	0.7	20
Tropical Energy Crops	41	57	6	12	100	0.65	0.75	18.5
Agricultural Residues					100	0.6	0.7	17.5

---

[1]It is assumed that 10 to 20% of forested and grassland/degraded areas and 10 to 15% of cropland areas could be available for production of biomass energy resources. The 10% est mate agrees with Hall *et al.* (1993).

[2]Yields are based on information in Farnum *et al.* (1983), Hall *et al.* (1993) Graham *et al.* (1992), and he experience and judgement of the working group.

3.1 BIOMASS ENERGY CONVERSION AND UTILIZATION TECHNOLOGIES

Biomass can be converted to a number of energy forms through a variety of conversion technologies. In the context of CO_2 emission reduction, the most important consideration is efficiency of the conversion process. This is because the major C benefit of bioenergy systems is based on the amount of fossil fuel C that is not used (or delayed in use).

About 14% of total energy used worldwide is bioenergy, with most of that use occurring in developing countries (Scurlock and Hall, 1990). The primary bioenergy technology used worldwide today is the direct combustion of solid biomass fuels for space heating, cooking, and industrial processes. Although these are generally very low intensity and low efficiency conversion processes (5 to 20%), they will always have a very important role to play in the total world economy. Major improvements can be made in biomass and charcoal cookstoves, and well as home heating stoves, zone heating systems, and industrial furnaces, to improve efficiency. Nearly all biomass not currently used for heat is converted to electricity via direct combustion.

Utilization of the potentially available biomass energy resources for C offsets will only have significant impact if conversion efficiencies become much higher than those commonly obtained today. Conversion efficiencies tend to be in the 15 to 25% range. Wright and Hughes (1993) estimated that if average conversion efficiencies of 42% could be obtained in 170,000 MW of biomass electrical production capacity, then 20% of 1990 U.S. fossil fuel emissions could be offset. However it would likely require at least 35 yr to implement this level of change in the power industry assuming that biomass energy efficiency levels of 33% were available by 1996. Steam cycle biomass energy efficiencies of 30 to 35% are anticipated to be available now or within the next 5 yr (Wright and Hughes, 1993). These efficiencies are achieved in part by use of conversion facilities of 100 MW or larger, allowing use of high-temperature, high-pressure steam cycles. Co-firing of biomass with coal provides the opportunity to take advantage of high-efficiency (30 to 40%) coal conversion technology available now.

To reach average efficiency levels of 40 to 50% for conversion of biomass to electricity by 2030 will require both successful research and development and strong market forces to stimulate the development and installation of technologies such as integrated gasification combined cycle and fuel cells.

Fuel substitution factors greatly influence the amount of CO_2 emission reduction that can be obtained by use of biomass energy. Biomass can most effectively reduce CO_2 emissions when substituted for coal at similar conversion efficiency. A recent analysis by Graham *et al.* (1992) revealed that substituting ethanol for gasoline provided only about half the benefit of substituting wood for coal. Use of biomass for liquid fuels (ethanol, methanol, reformulated gasoline components and biodiesel) is already occurring and likely to continue. For instance, the French government is encouraging the development of biofuels primarily as a quick means of finding an alternative crop for excess cropland. Thus our analysis assumes that some energy crop production will be converted to liquid fuels or biofuels, even though the C offset potential is less.

The column showing fuel substitution factors in Table 1 incorporates arbitrary assumptions about which fossil fuel is being substituted for by the biomass and the relative conversion effeciency of the fuels. The high fuel substitution factor allocated to wood resources in temperate regions (0.9 to 0.95) was believed conceivable since coal is the major fuel for electricity production and the technology for converting biomass to electricity at similar efficiencies as coal to electricity is likely to be available. In tropical regions, it was assumed that wood resources may be substituted for coal or oil at lower relative conversion efficiencies. Energy crops, whether produced in the temperate or tropical areas, were assumed to be used for both electricity and liquid fuels production. It was assumed that halophytes would be used only for electricity production but the energy losses due to irrigation requirements and boiler derating as a result of the high salt content lead to lower fuel substitution factors. In the case of wood for shelterbelts, it was assumed that a portion of the wood would be used for on farm or household needs at low conversion effeciency.

3.2   BIOMASS ENERGY FEEDSTOCK RESOURCES

3.2.1   Existing Forests

The world's forests, closed plus open, occupy 3.6×10^9 ha or 26% of the earth's total land area (excluding Antarctica). Actively managed forests total approximately 355 Mha or 10% of the total worldwide (Table 2). Within these managed lands are about 30 Mha of forest plantations established within the last three decades (WRI, 1992).

Table 2. Forests of the world, by region and management.

Region	Unmanaged	Managed	Total
	(Mha)		
Boreal	835	85	920
Temperate	577	190	767
Tropical	1,857	80	1,937
Total	3,269	355	3,624
Source: WRI, 1992			

Annual harvests of wood from the world's forests total about 3.4×10^9 m^3. Half of the volume is removed as logs, i.e. roundwood and pulpwood, which are manufactured into solid wood and paper products. The other half, 1.7×10^9 m³, or about 0.85 $\times 10^9$ t biomass is used for fuelwood, both domestically and commercially. By regions, approximately 70% of the fuelwood is utilized in the tropics, 20% in the temperate, and 10% in the boreal regions (WRI, 1992). Thus, forests currently contribute a significant amount of biomass energy today.

Hall (1991) notes that the global potential for using wood for bioenergy is uncertain. However, the potential to grow additional woody biomass appears high either through increasing the amount of existing forests under active management or expanding the amount of lands with forest cover by afforestation practices.

If the amount of forest land under active management was doubled during a 50-yr period, it would add another 360 Mha of actively managed lands or a total of 20% under management. This does not seem unreasonable under a doubled CO_2 situation. In fact, in a world scenario where reducing fossil C emission has high value, it does not seem unreasonable to estimate that 30% or more of the world's forests would be managed, since the temperate region of the world already has about 25% of its forested area under management. The greatest change would have to occur in the tropical region where currently less than 4% of the forested area is managed (Winjum *et al.* 1992)

Bringing a forest under management does not necessarily imply that natural forest stands have to be converted to plantations. For instance, in Sweden addition of N has resulted in a 31% increase in dry matter production of 25 yr old Norway spruce stands, and irrigation with liquid fertilizer resulted in a 57% increase in dry matter (Nilsson and Wiklund, 1992). The costs of such treatments would be very high, however. It is estimated that the mean annual yields of temperate forest stands under active management could range from about 2 to 5 t C ha^{-1} yr^{-1} with both silvicultural treatments and genetic improvements (derived from Farnum *et al.* 1983). However, this approach would imply a conversion from natural regeneration to forests planted with genetically improved materials. This is already done in most European forests. Since tropical forests exist under a wide range of rainfall conditions, it is very difficult to characterize what the average C yields might be in the future. It is likely, however, that more intensive management would likely occur in the moister, more productive forests. Consequently, the working group estimated that managed tropical forests would show a average annual C yield range of 2 to 6 t C ha^{-1} yr^{-1}, with the upper end being considerably less than the maximum possible.

The range in C emission reduction that can be achieved by managing existing forest resources will depend both on the area of forest managed, and the proportion of the managed forest that is dedicated to biomass production. The assumptions summarized in Table 1 result in estimated C emission reduction levels ranging from 36 to 141 $\times 10^6$ t C yr^{-1} in boreal plus temperate forests and 40 to 420 $\times 10^6$ t C yr^{-1} in tropical forests.

3.2.2   Afforestation of Non-Forested Lands

Afforestation on lands capable of supporting trees such as some grasslands, pasture lands and lands degraded or abandoned following other land uses such as grazing or cropping would significantly expand global forest resources and increase the level of terrestrial C storage (Dixon *et al.* 1993).

Estimates of the amount of land in the world that is technically suitable and socially-politically available for establishing new forests is believed to range from about 10 to 20% of current grasslands and degraded areas, or 320 to 640 Mha. These estimates bound a recent worldwide estimate of about 500×10^6 ha (Winjum *et al.* 1992). The largest part of the afforestation would be assumed to occur in the tropics. Average annual yields could be quite high as it would be anticipated that these afforested areas would be plantations with the best silvicultural techniques and genetic materials utilized (Table 1).

Afforested areas could become new sources of wood products at maturity. It is important to note that wood yields from this level of new plantations could disrupt world wood product markets, if it were only to be made available through traditional markets. Thus, a significant portion of this new production will likely be used for biomass energy, both because of its availability, and because of the continuing pressure to turn from fossil to renewable sources of energy. Use of the wood for energy may occur either directly as a feedstock into energy-producing facilities, or as waste from other wood processing operations.

If half of the new wood resource available for afforestation were used for energy under the assumptions given in Table 1, then the C emission reduction could range from 0.3 to 1.9×10^9 t C yr^{-1} for tropical and temperate afforestation combined.

3.2.3   Halophyte Production in Deserts

Since irrigation is an energy-intensive operation, crops requiring irrigation are generally not considered to be an energy resource. An exception to this generalization are halophytes which can be grown on saline land, using brackish or sea water. Glenn *et al.* (1993) concluded that a maximum of 43 Mha are suitable for producing plants with a sufficiently low salt level to be utilized for combustion to produce electricity and that yields could range from 3 to 7 t C ha^{-1} yr^{-1}. They furthermore suggest that irrigation requires only 20 to 30% of the final energy produced. If 100% of the biomass is dedicated to electricity production in lieu of coal, under the fuel substitution scenario given in Table 1, the C emission reduction potential is estimated to range from 15 t C yr^{-1} (if only 10 Mha are planted) to a high of 211 t C yr^{-1}.

Algae which can grow in saline water offer an opportunity to produce an energy product, a biodiesel fuel. Desert conditions would be required to obtain sufficient sunlight intensity for high algae yields. The need for high CO_2 levels essentially requires that algae ponds be associated with industrial processes or power production facilities emitting CO_2, thereby limiting it's potential as an energy resource.

3.2.4   Cropland Conversion to Biomass Production

Cropland can contribute biomass energy feedstocks in several ways: (1) by using agroforestry techniques to integrate production of food and energy resources on cropland; (2) by dedicating large blocks of land to long-term production of cellulosic energy crops; and, (3) by using starches, sugars and oils from plants as energy feedstocks. It was presumed that energy crops would share the available land with agroforestry plantings, shelterbelts, and filter strips.

Agroforestry, through which trees can be grown for energy conservation as described in section 2.1, can also result directly in the production of biomass energy feedstocks. In temperate regions where mechanized agriculture may preclude integrating trees and food crops in the same fields, it still is reasonable to suggest that 2 to 4% of the total cropland area could be dedicated to production of shelterbelts and filter strips. In tropical areas, it is presumed that integration of energy trees and food crops will be a common practice occupying 5 to 8% of the total cropland. Given the assumptions in Table 1, C emission reduction potential would range from 10 to 55 x 10^6 t C yr^{-1} in temperate areas and 46 to 205 x 10^6 t C yr^{-1} in tropical areas.

Dedicated Energy Crops, which include short-rotation woody crops (SRWC) and perennial herbaceous energy crops (HEC), offer an opportunity to significantly reduce fossil fuel usage in temperate zones (Wright et al. 1992). It is assumed that the ellulose in the plants will be used for production of electricity or liquid fuels. Sustainably grown perennial energy crops could provide an alternative use for 8 to 11% of the excess marginal to good cropland in the temperate zone. In the tropics, a lower percentage of land is likely to be fully dedicated to energy crops so a reasonable range of estimates may be 5 to 7%.

In the major crop growing regions of the U.S and Europe, delivered energy crop yields of about 5 t C ha^{-1} yr^{-1} are believed to be currently achievable on selected lands while delivered yields around 9 t C ha^{-1} yr^{-1} are believed possible by 2030 on a larger land base. Such productivity increases, averaging about 1.5% per yr, have been achieved with several major agricultural crops in the past 50 yr. They have occurred, however, only with major research and development efforts in plant genetics and with increased management inputs. In humid tropical regions such as Brazil, SRWC yields in the range of 20 to 30 t C ha^{-1} yr^{-1} are sometimes achieved (Betters et al. 1992), representing a 100% increase over average yields achieved 30 yr ago. It is anticipated that average tropical energy crop yields will be much lower, ranging from 6 to 12 t C ha^{-1} yr^{-1}.

Given the assumptions in Table 1, the range of C emission reduction from energy crops in the tropical region is estimated to be 160 to 513 x 10^6 t C yr^{-1}. In the temperate regions C emission reduction potential is estimated to range from 85 to 493 x 10^6 t C yr^{-1}.

Sugar, Starch or Oilseed Crops include annual or perennial crops in which only a portion of the crop is used for energy, primarily liquid fuel production. The best example is in Brazil, where 17 Mha are used to grow sugarcane for conversion to 9 x 10^9 l of alcohol fuels, sufficient to supply 20% of Brazilian transportation fuels (Hall, 1991). In the U.S. about 3 x 10^9 L of ethanol fuel are produced from corn, although no specific corn acreage is devoted to fuel production (Keim and Venkatasubramanian, 1989).

Recently, the production of vegetable oil crops for production of "biodiesel" has attracted considerable attention in the U.S. and Europe, as a method of utilizing surplus agricultural lands currently devoted to these highly-subsidized markets. These crops may occupy substantial amounts of the land considered to be "available" for dedicated energy crops. Unfortunately, the use of crops from which only the sugar, starch or oil is used may be of limited value in reducing CO_2 emissions (Marland and Turhollow, 1991). The C emission reduction potential of this category of crops was believed to be very low and thus was not estimated for this analysis.

3.2.5 Bio-wastes

Forestry, agricultural, and municipal wastes and residues represent a major current and potential biomass energy resource. (Wastes are those materials already collected and brought to a central location, for which no specific use currently exists, such as sawdust or mill ends, while residues are the materials left in the field after harvesting of timber or crops. Utilization of residues requires their collection and transportation, but no specific production costs are incurred.)

The major current biomass energy source in developed countries is forestry wastes - bark, sawdust, waste lumber, etc. - generated at lumber and paper mills. In the U.S. about 6,000 MW of installed capacity is dedicated to burning forestry wastes (Kinsman and Trexler, 1993).

The other sources of biomass wastes currently used for conversion to energy are municipal solid wastes and animal and agricultural wastes. In addition to reducing fossil C emissions, the use of these wastes for energy also reduces CH_4 (methane) emissions, another greenhouse gas of significant interest. The potential for additional energy recovery from wastes is significant, and many schemes could be implemented in the near term, given the right regulatory and economic incentives. However, from a global perspective, bioenergy from wastes is a small fraction of the potential of biofuels from agricultural and forestry residues, or energy crops.

Agricultural residues could become a major future energy resource. In many countries, agricultural residues are left on the fields and, in some cases, burned to control disease before replanting. Restrictions on field burning being implemented in many countries are making residue removal and use as fuel more attractive. In developing countries, agricultural residues are already used in many cases as animal feeds and for fuel. Excessive removal of agricultural residues is not desirable, as it will result in loss of nutrients and reduced soil organic C. As a general rule, about 50% of agricultural residues can be removed from fields without affecting future crop productivity and perhaps 25% should be recoverable. Hall *et al.* (1993) suggest that for the major crops alone (wheat, rice, corn, barley and sugarcane), with residue production assumed to be similar to U.S. agriculture and a 25% residue recovery rate, the residue resource is equivalent to 12.5×10^{18} J. This converts to a net C amount of 357×10^6 t. With the assumptions provided in Table 1, the C emission reduction potential is estimated to be 214 to 319×10^6 t C yr^{-1}. It is likely that increased utilization efficiency and recovery of other crop residues could significantly increase the potential of agricultural residues for CO_2 mitigation.

4. Energy Offset Programs

4.1   BIOMASS MANAGEMENT FOR CO_2 OFFSETS

Carbon offset forestry differs from energy conservation and production since it refers primarily to methods of increasing the terrestrial store of C in trees and forests, whereas energy conservation and production aim to reduce the emission of fossil C. CO_2 is different from a pollutant such as SO_2 in that it can be photosynthetically removed from the atmosphere after being emitted. It also is long-lived in the atmosphere, mixing globally, and thus can be offset anywhere.

Several types of forestry-based C offsets are available:

*   Reducing C emissions and retaining C sinks by protecting or managing existing forests that would otherwise be lost through deforestation, poor management, and stresses from rapid climate change; and,
*   Increasing C stores by encouraging natural regeneration or reforestation efforts.

Tropical forests were being cleared at a rate of 17 Mha per year during the 1980's (Houghton *et al.* 1992). Protecting and/or managing standing forests can be an attractive means of implementing a C offset program and can provide a series of ancillary environmental benefits such as maintaining biodiversity and reducing soil erosion.

Tree plantations can offer rapid growth rates over large areas of land, uniform management, and quantifiable costs and benefits (Sedjo and Lyon, 1990). Storage of C in soils and biomass of properly managed plantations may occur at more than double the rate of C storage in naturally regenerating forests (Farnum *et al.* 1983). The history of plantation forestry in the tropics is checkered at best (Trexler and Haugen, 1993), but over 30 Mha of forest plantations have been established worldwide over the last three decades (WRI, 1992).

There are hundreds of millions of hectares around the world that previously supported tree cover and that could support tree cover again (Winjum *et al.* 1992). Lands have become degraded due to slash and burn agricultural systems, salinization, etc., with C storage on the land declining. Removing barriers to natural regeneration often provides good results with low operating expenses. In many areas the control of wildfire by itself would result in the regeneration of forest cover over large areas.

The incorporation of tree cultivation with agricultural and other practices can also play a significant role in tropical C offset projects. Since only a portion of area is planted in trees, large areas could be required to achieve significant long-term C offset credits.

Tree planting and management programs work best if they are successfully integrated into the local economy. If regeneration projects are designed so that local people are given an incentive to allow the regeneration to take place, the chances of success are good (Gregersen *et al.* 1989). Plantings designed for C offset can include energy production, as well as other wood uses, of potential high value to local economies.

4.2   OFFSET POLICY DEVELOPMENT AND PROJECT EXPERIENCE

The recent United Nations Conference on Environment and Development (UNCED) featured the signing of a Framework Convention on Climate Change by over 150 countries. The Framework Convention commits all developed country parties to adopt policies and implement measures to mitigate climate change by limiting emissions of greenhouse gases and enhancing sinks and reservoirs. The "joint implementation" provision allows industrial countries to receive credit towards their own emissions reduction and stabilization objectives by reducing emissions in other countries. The U.S. federal government's new "Forests for the Future" program, developed as part of the U.S. response to UNCED, involves the negotiation and implementation of C offset projects in a number of countries (U.S. EPA, 1993).

The U.S. Energy Policy Act of 1992 requires establishment of a program for the voluntary collection and reporting of information on greenhouse gas emissions and reductions, with the latter including fuel switching, forest management practices, tree planting, use of renewable energy and other actions.

Also in the U.S., the Bonneville Power Administration is beginning to require CO_2 mitigation as part of its power purchase contracts. One power supplier agreed to spend $1 x 10^6 on C offset as part of an agreement for the construction of a 250 MW natural gas fired facility in Washington state.

Kinsman and Trexler (1993) and Dixon *et al.* (1993) describe the above programs in more detail and discuss pilot C offset forestry programs being sponsored by power producers in several countries.

The cost of forestry offsets depends on the amount of land required for the offset, and the cost per unit of land area. High-quality C offsets can probably be obtained for under $5 t^{-1} on a non-discounted basis (Trexler, 1993). There is a tradeoff between cost and risk, with the cheaper but riskier projects often being found in the tropics, and the more secure but expensive projects often found in temperate areas.

Carbon offset forestry holds great promise as one method to manage atmospheric CO_2, with key advantages being cost-effectiveness, secondary environmental benefits, and benefits to local peoples in developing nations. The major limitation is the competition for land from other uses. Local people must be involved in planning and implementation if these projects are to succeed. A long-term strategy might be to link C offsets and biomass energy by sequestering C to be used later in producing fuels on a sustainable basis. For electric utilities, forestry projects offer an approach to CO_2 mitigation during development of more energy-efficient and/or lower CO_2-emitting technologies, such as renewables, integrated gasification combined cycle, fuel cells, and magnetohydrodynamics.

It is difficult to predict the future extent of C offset projects and their CO_2 implications, because regulatory consideration of many key questions is only now beginning. One can only hazard a guess at the potential "market" for C offset forestry projects. For example, the North American Electric Reliability Council predicts a 19% increase in U.S. electric production between 1990 and 1999. Assuming a CO_2 emission rate equivalent to the current utility average, an additional 80×10^6 t C would be emitted in 1999. To sequester these emissions would require roughly 10 Mha of new area of a fast-growing species taking up 8 t C ha^{-1} yr^{-1}.

5. Implications for Global CO_2 Management

The impact of managing biomass specifically for the conservation, production, or offset of energy can become a significant factor in the global balance of CO_2 emissions over the next century. It is important to recognize, however, that biological systems alone, no matter how well managed, cannot overcome increasing anthropogenic greenhouse emissions. The changes in land use patterns, energy usage, and population of the 20th century appear to be unsustainable. Slowing or reversing those trends is a major challenge that must be addressed in a broad array of policies. Table 3 illustrates our judgment regarding the potential implications for the global C balance of the broad array of energy-related biomass management strategies discussed in Section 3. It is important to note that no single strategy will, by itself, be fully adequate in altering global C balances. Taken together, however, they can play an important role.

Perhaps as important will be the underlying indirect effects of social, cultural, and economic changes inherent in their application. If societies adopt biomass management strategies such as these, coupled to associated policies of aggressive energy conservation, land use management and conservation, and sustainable development, the total effect will be much more important than Table 3 suggests because of the synergistic accumulation of associated and indirect positive effects.

In constructing Table 3, we limited consideration to activities and products directly related to the production of energy. Table 3 shows the primary energy that would be added as a result of new biomass production and does not reflect the 55×10^{18} J of biomass energy production believed to be utilized at present. It represents our best estimate of the potential increase in biomass energy impact that could be experienced by the time of CO_2 doubling. It does not attempt to factor in the impact of changing climate, nor the ability of land managers to adapt to such climate changes.

Table 3. High and low estimates of biomass production for energy, and associated C feedstocks and C emission offsets, under doubled CO_2 conditions.

	Net C Stock[3]		Primary Energy[4]		C Emission Reduction[5]	
	Low	High	Low	High	Low	High
	(t C yr^{-1})		(J x 10^{18})		(10^6 t C yr^{-1})	
Forest Management						
Boreal	40	240	0.3	1.9	7	46
Temperate	160	500	1.3	4.0	29	95
Tropical	400	3000	3.2	24.0	40	420
Grass/Desert Conversion						
Irrigated Halophytes	30	301	1.2	12.0	15	211
Bor./Temp. Afforest.	300	1400	6.0	28.0	135	665
Tropical Afforestation	880	3520	17.6	70.4	220	1232
Cropland Conversion						
Temperate Shelterbelts	26	104	0.8	2.2	10	55
Temp. Energy Crops	130	657	4.8	18.2	85	493
Tropical Agroforestry	123	390	3.7	8.2	46	205
Tropical Energy Crops	246	684	9.1	20.5	160	513
Bio-wastes	357	457	12.5	16.0	214	319
TOTAL	2692	11253	60.5	205.5	982	4,254

Table 3 also does not attempt to document the effect of biomass management on increasing the standing volume of woody biomass (and, thus, C storage), nor the increase in C storage associated with additions to the stable fraction of soil organic matter. Those effects can be significant, and are relevant to calculating the CO_2 offsets discussed in the paper, but since they were taken into account by the

---

[3]Based on land area and yield assumptions provided in Table 1.

[4]Estimate of primary energy equivalence of the total net C amounts in columns 1 and 2, based on the higher heat value factors in Table 1.

[5]Numbers were calculated from the net C amounts in columns 1 and 2 and the fuel substitution factors, and percentages of use for each resource and region in Table 1.

calculations of the work groups associated with each of the basic biomes, they were omitted here to prevent double-counting of effects.

In order to make such judgments, of course, we were required to make a significant set of assumptions about the most likely scenarios for the future. Those assumptions were summarized in Table 1 and described in the various subsections of section 3 of this paper.

6. Major Uncertainties and Research Needs

Major uncertainties and research needs exist in the field of biomass energy. Some are clearly gaps in science that can be remedied by research. Others are likely to remain uncertainties since future events and conditions are subject to hypothesis and speculation, but not scientific proof.

The following list attempts to separate uncertainties and research needs as they relate to our current knowledge and future needs for technology in energy-producing biomass management. First, the uncertainties:

* The impact of climate change on ecosystems and their production is uncertain, and almost impossible to predict. As climate changes, both scientists and managers will be tested to meet those changes with effective adaptive management techniques.
* Sustainable management requires thoughtful and constructive human interaction with natural resources in ways that protect ecosystem integrity indefinitely. However, the ability of human societies to carry out such management is highly uncertain. Most societies have histories more noted for exploitation than conservation, and the rapid population growth and land use changes currently underway in much of the world foretell that human populations may become increasingly destructive to natural and managed ecosystems.
* The future of fossil energy is uncertain. Depletion of nonrenewable stocks, or disruptive world events, can have great effect, and those disruptions are unpredictable. What is certain is that major changes in fossil fuel prices or availability can send shocks through natural resource management, particularly in the developed countries whose agriculture and forestry technologies are dependant on fossil fuel inputs.

Uncertainties such as these, while they can never be accurately predicted, can be built into modeling exercises to evaluate the sensitivity of systems to such changes, and to test the effectiveness of policies to dampen some of the extreme effects.

In addition to uncertainties, there are clear gaps in scientific knowledge about ecosystems and their management that research, monitoring, and testing can help resolve. Some of the research needs identified include:

* Improved information on the actual C feedstock value of forest, agroforestry, and

agricultural management systems.

* Improved ecosystem management methods that optimize energy production, conservation, or offset while achieving other necessary production or social goals.
* Better data on energy inputs for the production of wooden goods and tree-derived chemicals and their substitutes.
* Additional data on soil C, and the potential of soils to sequester C under afforestation.
* Better data on land availability, including cultural, social, and political factors that may preclude some lands from being available for afforestation or C offset projects.
* Better documentation of the impacts and effects of tree planting projects in terms of their C sequestration, likelihood of survival, and final product utilization as basic tools to build effective economic and regulatory cases for C offset projects.
* Improved analytic tools and models to predict the C sequestration and CO_2 emission reduction benefits of C offset projects and methodologies to monitor and verify project results.
* Better data, including economic analyses, for the use and efficiency of biomass in energy applications, particularly where that usage is conducted outside markets.
* Forest and energy crop management techniques and systems that can be demonstrated to be both ecologically and economically sustainable. Monitoring and reporting on the full range of energy and environmental effects of biomass energy systems is essential for public acceptance.
* An improved science base for ecosystem management, with attention to the C cycle impacts of management actions, on all ecosystems, to help resource managers understand and incorporate C-related factors into management decisions.
* Additional research on institutional approaches, successes, and failures in order to help afforestation projects deal successfully with the social and cultural issues involved.

7. Conclusions

The management of biomass for the conservation or production of energy can be an effective way of reducing the burning of fossil fuels. In addition, forests planted or managed to sequester or store C can offset the C emissions from energy generation or other C-emitting facilities. There are many opportunities, but two stand out as being of major importance:

* Large-scale afforestation (tree planting) projects can produce biomass on lands which have been degraded or abandoned. A major problem lies in avoiding the social, cultural, and political disruptions that can occur with significant land use

changes of this type. The land that is technically suitable and may be socially and politically available ranges from 320 to 640 Mha. Through afforestation, these lands could produce biomass for energy production that could be used to offset fossil fuel emissions by 0.4 to 1.9 x 10^9 t C yr^{-1}.

* Dedicated energy crops, which include short-rotation woody crops (SRWC) and herbaceous energy crops (HEC), annual crops and oilseeds, offer an opportunity to convert 67 to 130 Mha to crops that can significantly reduce fossil fuel usage. In the temperate region, energy crop yields are expected to range from 5 to 9 t C ha^{-1} yr^{-1}, while in the tropics, energy crop yields are expected to range from 6 to 12 t C ha^{-1} yr^{-1}. Use of energy crops for production of electricity and liquid fuels offers the potential of 0.2 to 1.0 x 10^9 t C yr^{-1} offset of fossil fuel emissions.

On a global scale, the opportunities to reduce C emissions by using biomass for energy production (0.9 to 4.3 x 10^9 t C yr^{-1}) greatly outweigh the energy conservation opportunities (50 x 10^6 t C yr^{-1}) for C mitigation. Although the opportunity to sequester C by forest protection and tree planting for non-energy use could initially equal the C emission reduction gained by biomass energy, the latter would eventually outweigh non-energy C offsets due to the cumulative effect of biomass energy use in keeping fossil C stocks from being converted into long-lasting atmospheric CO_2. Marland and Marland (1992) suggest that high-productivity land is best used for energy crops, and lower-productivity land for reforestation/C storage.

The availability of biomass as an energy feedstock also determines its value, and concurrently, the value of other products such as wood products which are alternate uses of the biomass. If the nations of the world decide to undertake large-scale improvements and expansions of forests, the resulting production could swamp traditional markets for wood, paper and fuel products. The most obvious outlet for this material is new capacity to convert these materials to energy as part of a global shift away from non-renewable to renewable energy sources.

This new capacity will only emerge in a favorable situation created by forward-looking government policies, aggressive research to improve both feedstock production and energy conversion, and economic incentives (or, at the least, a minimum of disincentives). Under the right conditions, market incentives will draw a growing number of land managers and industrial investors to programs that produce and manage biomass for energy-related reasons.

There are important policy considerations at both the international and national levels related to ecosystem management and biomass production as an element in the global effort to address climate change issues, and it is important for the science community to remain engaged in that effort. The current situation and some of the relevant issues include:

* The international Framework Convention on Climate Change and the U.S. Energy Policy Act of 1992 both address C sinks, and provide impetus to the

small but growing number of private and public-private programs testing C offset forestry. It is difficult to predict the future extent of C offset projects and their CO_2 implications, because regulatory consideration of many key questions is only now beginning.

* The use of biomass for energy applications is heavily dependent on government policies, research priorities, and economic incentives and/or disincentives. Policies can remove impediments or provide incentives, and speed or slow the transition to a renewable energy future.
* National or international efforts to mitigate C emissions through expansion of forests must be coupled with incentives to develop an energy-related market for a significant portion of the new wood produced, or disruptive market dislocations could ensue.
* The electric power industry can be a major participant in the expansion of biomass-energy markets, but it will need support from its public regulatory bodies at the national and sub-national levels.
* Advanced systems of biomass feedstock and energy production may be key to providing electrical power in developing nations and can serve as a exportable technology for developed nations.
* Offset projects such as afforestation of degraded lands and protection of existing forest are technically viable and appear to be cost-effective. Policymakers should consider them seriously as options for managing C.
* Trees are an important means to conserve energy and reduce CO_2 emissions in urban environments, and improved tree planting and care programs are needed.
* Management of woody biomass in agriculture is critical for both resource-conserving and economic reasons. Government policies at every level should focus financial and technical assistance to encourage farmers to adopt appropriate agroforestry, soil conservation, and energy-related management practices.
* Environmental impacts of increased biomass production, both positive and negative, must be considered and understood from the outset. This is essential for the successful expansion of biomass usage, especially in the long term.

8. References

Betters, D. Wright, L.L., and Couto, L.: 1992, Short Rotation Woody Crop Plantations in Brazil and the United States, *Biomass and Bioenergy* **1(6)**, 305-316.

Brandle, J.R., Wardle, T.D., and Bratton, G.F.: 1992, Opportunities to Increase Tree Planting in Shelterbelts and the Potential Impacts on Storage and Conservation, in Sampson, R.N. and Hair, D., eds., *Forests and Global Change, Volume 1: Opportunities for Increasing Forest Cover,* Washington, DC: AMERICAN FORESTS, pp. 157-176.

Burschel, P., Kürsten, E., and Larson, B.C.: 1993, Die Rolle von Wald und Forstwirtschaft im Kohlenstoffhaushalt - Eine Betrachtung für die Bundesrepublik Deutschland, Forstl. Forschungsberichte München, Nr. 126, (short version in

English).

Dixon, R.K., Andrasko, K.J., Sussman, F.A., Trexler, M.C. and Vinson, T.S.: 1993, this volume.

Farnum, P., Timmis, R. and Kulp, J.L.: 1983, Biotechnology of Forest Yield, *Science* **219**, 694-702.

Glenn, E.P., Pitelka, L.F. and Olsen, M.W.: 1993, this volume.

Graham, R.L., Wright, L.L., and Turhollow, A.: 1992, The Potential for Short Rotation Woody Crops to Reduce CO_2 Emissions, *Climatic Change* **22**, 223-238.

Gregersen, H., Draper, S., and Elz, D.: 1989, People and Trees: The Role of Social Forestry in Sustainable Development, Washington, DC: World Bank, 273 pp.

Hall, D.O., 1991, Biomass Energy, *Energy Policy* **19(8)**, 711-737.

Hall, D.O., Rosillo-Calle, F., Williams, R.H., and Woods, J.: 1993, Biomass for Energy Supply Prospects, in: Johansson, T.B., Kelly, H., Reddy, A.K.N., and Williams, R.H, eds., *Renewable Energy*, Washington, DC: Island Press.

Houghton, J.T., Callander, B.A. and Varney, S.K. (eds): 1992, *Climatic Change 1992 -- the Supplementary Report to the IPCC Scientific Assessment, Intergovernmental Panel on Climate Change, Report prepared for IPCC by Working Group I,* New York: Cambridge University Press.

Keim, C.R. and Venkatasubramanian, K.: 1989, Economics of Current Biotechnological Methods of Producing Ethanol, *Trends in Biotech*, **7**, 22-29.

Kinsman, J.D. and Trexler, M.C.: 1993, this volume.

Kürsten, E. and Burschel, P.: 1993, this volume.

Marland, G. and Marland, S.: 1992, Should We Store Carbon in Trees, *Water, Air & Soil Pollution,* **64**, 181-196.

Marland, G. and Turhollow, A.F.: 1991, CO_2 Emissions from the Production and Combustion of Fuel Ethanol from Corn, *Energy* **16**, 1307-1316.

Nilsson, L. and Wiklund, K.: 1992, Influence of Nutrient and Water Stress on Norway Spruce Production in South Sweden -- The Role of Air Pollutants, *Plant and Soil* **147**, 251-265.

Sampson, R.N., Moll, G.A., and Kielbaso, J.J.: 1992, Opportunities to Increase Urban Forests and the Potential Impacts on Storage and Conservation, in: Sampson, R.N. and Hair, D., eds., *Forests and Global Change, Volume 1: Opportunities for Increasing Forest Cover*, Washington DC: American Forests, pp. 51-72.

Sampson, R.N.: 1981, *Farmland or Wasteland: A Time to Choose*, Emmaus, PA: Rodale Press. 465 pp.

Scurlock, J.M.O. and Hall, D.O.: 1990, The Contribution of Biomass to Global Energy Use (1987), *Biomass* **21**, 75-81.

Sedjo, R.A. and Lyon, K.S.: 1990, *The Long-Term Adequacy of World Timber Supply*, Resources for the Future, Washingtonm, DC: John Hopkins University Press, pp. 30-31.

Trexler, M.C.: 1993, this volume.

Trexler, M.C. and Haugen, C.A.: 1993 (forthcoming), Keeping it Green: Global Warming Mitigation Through Tropical Forestry, Washington, DC: World Resources Institute.

U.S. Environmental Protection Agency: 1992. *Cooling our Communities: A Guidebook on Tree Planting and Light-Colored Surfacing*, 22P-2001, Washington, D.C.

US Environmental Protection Agency: 1993, *Forests for the Future: Launching Initial Partnerships*, Report of an Interagency Task Force, January 15.

Winjum, J.K., Dixon, R.K, and Schroeder, P.E.: 1992, Estimating the Global Potential of Forest and Agroforest Management Practices to Sequester Carbon, *Water, Air & Soil Pollution* **64**, 213-227.

WRI: 1992, *World Resources 1992-1993*, World Resource Institute, New York: Oxford University Press, 383 pp.

Wright, L.L. and Hughes, E.: 1993, this volume.

Wright, L.L., Graham, R.L., Turhollow, A.F. and English, B.C.: 1992, The Potential Impacts of Short-Rotation Woody Crops on Carbon Conservation, in: Sampson, R.N. and Hair, D., eds., *Forests and Global Change, Volume 1: Opportunities for Increasing Forest Cover*, Washington DC: American Forests, pp. 123-156.

WORKSHOP RESEARCH PRESENTATIONS

PART III

CONTRIBUTION OF NORTHERN FORESTS TO THE GLOBAL C CYCLE: CANADA AS A CASE STUDY

WERNER A. KURZ[1] and MICHAEL J. APPS[2]

[1] *ESSA Environmental and Social Systems Analysts Ltd., 1765 West 8th Avenue, Vancouver, B.C. Canada, V6J 5C6*

[2] *Forestry Canada, Northwest Region, 5320 - 122 Street, Edmonton Alberta, Canada, T6H 3S5*

Abstract. Boreal forests are exposed to periodic stand-replacing disturbances such as wildfire. Unchanging disturbance regimes in unmanaged forests result in an age-class structure in which the proportion of forest area is largest in the youngest age class and decreases exponentially in older age classes. The current (*ca.* 1970) age-class structure of Canadian forests contains a much smaller proportion of the forest area in each of the two youngest 20-yr age classes than in each of the next three age classes (i.e., the 40 to 99-yr age-classes). We hypothesize that more intensive disturbance regimes in the late 19th and early 20th centuries, compared to disturbances in the period 1920 to 1969, have resulted in this unusual age-class structure. The reduction in disturbance regimes has resulted in an increase of the average forest age and therefore an increase in total forest biomass carbon (C). This C sink is obtained without altering age-dependent growth or decomposition rates. If the average forest age of Canadian forests continues to increase, additional C sequestration of forests, (i.e., the C sink strength) will diminish. This result of a C sink in Canadian forest ecosystems is supported by more detailed C budget calculations for the year 1986.

1. Introduction

The discrepancy between the observed annual increase in atmospheric CO_2 and the increase expected from the estimates of global C sources and sinks has triggered a search for the missing C. Recent efforts to better quantify the global C cycle have led to improved estimates of C sources and sinks in oceans and terrestrial systems (Wisniewski and Lugo, 1992; Boden *et al.*, 1991). Tans *et al.* (1990) suggested that a northern hemispheric C sink was required to maintain the observed atmospheric CO_2 gradients. Quay *et al.* (1992) suggested a stronger ocean C uptake, and a near zero net contribution from the biosphere. Because of C sources in equatorial and tropical forests (Hall and Uhlig, 1991; Houghton, 1991), a near-zero net contribution of the biosphere can only be achieved through a sink outside the regions of active deforestation. Apps and Kurz

Water, Air, and Soil Pollution **70**: 163–176, 1993.
© 1993 *Kluwer Academic Publishers.*

(1991), Kauppi *et al.* (1992), Sedjo (1992), and Vinson and Kolchugina (1993) documented C sinks in northern temperate and boreal forests.

The current lack of understanding of the missing C sink is unacceptable for useful policy analysis (Edmonds, 1992). Unless the mechanisms behind the missing C sinks are understood, assumptions about their future contributions to the global C budget must be speculative. The great concern, of course, is that the sinks cease to function, or worse that the processes involved cause them to become C sources. In this paper, we will discuss a structural mechanism that, in the case of the Canadian forest sector, has contributed to the observed C sink in northern terrestrial ecosystems, and suggest that this sink mechanism will diminish in the future.

2. The Role of Forest Age-class Structures

It has been demonstrated that forest-level C storage in managed forests is lower than the maximum possible C storage in an old-growth landscape (Cooper, 1983; Harmon *et al.*, 1990; Dewar, 1991). These analyses, however, assumed that the old-growth landscapes are free of natural disturbances, (Harmon *et al.* 1990 included a comparison with a 450-yr disturbance regime), and that biomass C in old-growth forest stands never declines. Most Canadian temperate and boreal forests are affected by periodic stand-replacing disturbances with return frequencies in the range of 50 to 200 years (Bonan and Shugart, 1989; Payette 1992; Rowe 1983). Moreover, stand biomass dynamics often result in stand breakup and biomass decline in 80 to 120-yr old stands (Whynot and Penner 1990; Alban and Perala, 1992). We therefore suggest that an assessment of C storage and dynamics should be based on analyses of forest age-class structures and C storage in each age class.

Under unmanaged conditions, it is generally observed that with increasing forest age the forest area in each age class follows a negative exponential decline. The expected proportion of the area in each age class is a function of the average disturbance probability (p) and can be calculated from the cumulative proportion f(x) of all stands up to a given age (x) using the equation (Van Wagner, 1978):

$$f(x) = 1 - e^{-px}$$

This relationship between forest age-class structure and disturbance regime has long been used to infer historic disturbance regimes from age-class structure data (Yarie, 1981; Van Wagner, 1978; Johnson and Fryer, 1987). As the disturbance frequency increases, a greater proportion of the total area is in younger forest age classes and the average age of the forest decreases. A reduction in the area annually disturbed (e.g., through forest protection against fire and pest) shifts forest age-class structures to the right (i.e., increases the average age). Forest harvesting at rotation lengths that are shorter than the average period between natural stand-replacing disturbances can shift the forest age-class structure to the left. Fire disturbances affect stands of all ages nearly equally. If disturbances were restricted to stands above a certain age (e.g., successful forest protection and harvesting), the forest age-class structure would be greatly altered. The analysis of forest management impacts on regional C storage must therefore include a comparison of age-class distributions under managed and unmanaged conditions.

3. Current Age-class Structure in Canadian Forests

The current age-class structure of Canadian forests is derived from the National Forest Biomass Inventory (Bonnor, 1985) that forms the basis for the carbon budget model of the Canadian forest sector (CBM-CFS; Apps and Kurz, 1991; Kurz et al., 1992). Figure 1 shows the area distribution in 20-yr age classes. Derivation of the data and the underlying assumptions are described in Kurz et al. (1992), although additional data (Forestry Canada, 1988) have been used to derive the distribution in Figure 1. The age-class structure is based on the national forest inventory (Bonnor, 1982) in which the average age of records was approximately 10 yr. The forest age-class structure thus represents the conditions at ca. 1970. Less than 1% of the forest area in Canada is classified as uneven-aged forest, and in the CBM-CFS this area is distributed equally across all age classes (Kurz et al., 1992).

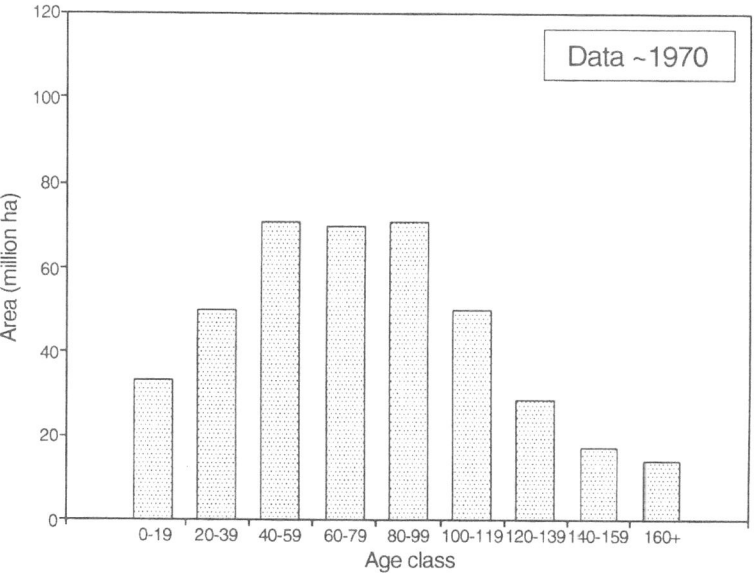

Figure 1:   The age-class structure of Canadian forests in ca. 1970 (Bonnor, 1985; Forestry Canada, 1986; Kurz et al., 1992).

In Figure 1, the right half of the age-class structure (i.e., > 80 yr) resembles the negative exponential decline that results from an unchanging disturbance regime (Van Wagner, 1978). The notable characteristic of the age-class distribution is that the proportion of forest area in the 0 to 19 and 20 to 39-yr age classes is smaller than the area in each of the next three 20-yr age classes. A reduction in the area in younger age classes compared with older age classes can also be observed in more recent forest inventories at the national (J. Lowe, pers. comm., 1992), the provincial (Ontario Ministry of Natural Resources, 1992), and the regional level (Dellert, 1991).

Two explainations can be given for the reduction in forest area in the youngest two age classes in the forest inventory (Figure 1). This reduction can only be achieved if either the annual rate of forest disturbance has been reduced during the 40 to 50 yr prior to 1970 or the areas that have been disturbed have been reclassified as non-stocked or non-forest land (and thus effectively removed from the age-class distribution). The two explanations are discussed in more detail below.

Statistics on annual disturbances provide an independent estimate of the expected forest age-class structure because all stands that have been affected by stand-replacing disturbances during a time period are transferred to the youngest forest age class. The areal sum of stand-replacing disturbances in the periods 1930 to 1949 and 1950 to 1969 should therefore be close to the area in the 20 to 39 and 0 to 19-yr age classes in the forest inventory, respectively. We have recently compiled Canadian forest disturbance statistics for the period 1920 to 1989. Canadian provincial and federal resource management agencies have been collecting disturbance statistics since the early part of this century.

The area disturbed in the period 1950 to 1969 exceeds the area in the 0 to 19-yr age class by 23%. The area in the 20 to 39-yr age class is, however, nearly identical (2% larger) with the area disturbed in the period 1930 to 1949. This discrepancy between the area in an age-class and the area disturbed during the 20-yr period in which the age class was initiated may be explained by examining the influence of recurring disturbances. Some of the areas in the youngest age class will be disturbed more than once in the 20-yr period. Any stand disturbed between 1950 and 1969 will be added to the youngest age class in the 1970 inventory and if this stand is disturbed a second time during the same period, the statistic for disturbed area will increase but the area in the youngest age class will not be affected. Any disturbed area in the second (or older) age class will be removed from that age class and transferred to the youngest age class. With increasing age a decreasing proportion of the initial area will be retained in any given age class.

The second reason for the decrease in the area in the youngest age-class in the inventory could be a problem with data classification. Some disturbed areas may not have been assigned a forest age at the time of the inventory, that is they were classified as non-stocked or non-forest areas. Some years after the disturbance, forest regrowth will have occurred on nearly all sites, and in a subsequent forest inventory, areas that were previously classified as non-stocked will re-enter the inventory, albeit at an advanced age. This re-entry into the forest inventory will contribute to the closer match between the area disturbed and the area in the inventory in the second age class.

Our estimates of the sum of the area disturbed in the two 20-yr periods and the area in the inventory in the youngest two age classes are in close agreement. We therefore believe that the decline in the area in the youngest two 20-yr age classes in the national inventory (Figure 1) reflects primarily a reduction in the area disturbed annually, and not a data classification problem.

4. C-budget Implication of the Age-class Structure

The aforementioned observation that the area in the two youngest 20-yr age classes in the Canadian forest inventory is considerably smaller than the area in the older age classes (Figure 1) leads to the hypothesis that a reduction in disturbance regimes relative to the disturbance frequency of the late 19th and early 20th centuries has resulted in an increase in the average forest age in Canadian forests. Implications of this hypothesis for the forest sector C budget can be calculated.

The biomass C dynamics of boreal (and some temperate) forest ecosystems are age dependent and can be characterized by four phases of stand dynamics: a regeneration phase immediately following disturbance; a logistic growth phase with the highest rates of biomass C accumulation; a mature phase during which biomass C accumulation is reduced; and a stand breakup phase with a high transfer rate of biomass C to soil and detritus C pools (Alban and Perala, 1992; Whynot and Penner, 1990; Kurz and Apps, 1992). Differences in energy flux to the forest floor, water-balance, and input of litter fall and coarse-woody debris in each of these four stand dynamics phases also cause changes in soil and detritus C dynamics. To assess C storage at the regional scale, it is therefore necessary to know the biomass and soil C density (i.e., C ha^{-1}) in each forest age class as well as other strata such as ecosystem type, species, and site class (Apps, 1993).

It is possible to compile the average biomass and soil and detritus C density of each age class of all Canadian forest ecosystems. Figure 2 represents the area-weighted average biomass, soil and detritus, and total ecosystem C density as compiled from CBM-CFS2 (Kurz and Apps, 1993). Average ecosystem C storage (C density) differs between the 20-yr age classes represented in the age-class distribution because of age-dependent C dynamics and because the area of ecosystem types that contribute to each age class varies. Carbon density is highest in the 120 to 139-yr age class. Regional C storage, the sum of the products of ecosystem C density and the area in each age class, depends on the regional forest age-class structure. Furthermore, the characteristics of the ecosystem C dynamics are such that regional C storage can be maximized by allocating forest area to the age classes with the highest ecosystem C storage. Unfortunately, this maximum C storage will be a transient stage because forest ecosystems continue to age and will eventually enter the stand breakup stage or be affected by disturbance.

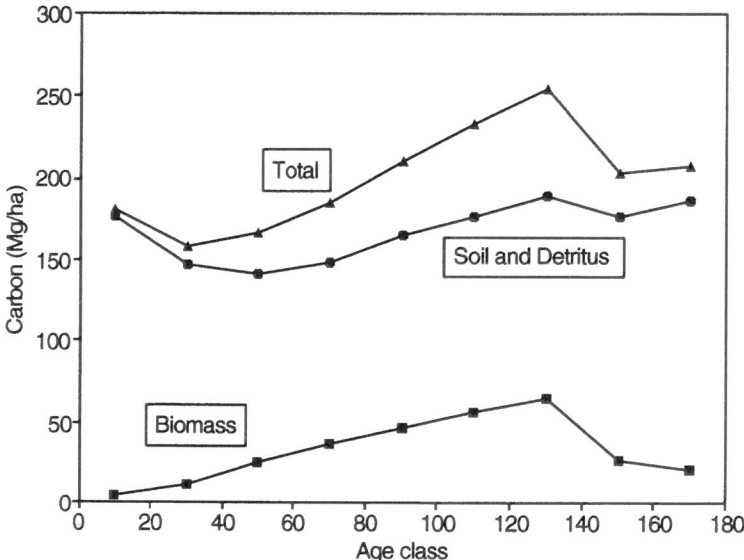

Figure 2:   Area-weighted average biomass (live above and below ground), soil and detritus, and total ecosystem C density in each forest age class. Data compiled from the Carbon Budget Model of the Canadian Forest Sector (CBM-CFS2) (Kurz and Apps, 1993).

This simple example also shows that the current forest age-class structure of Canadian forest ecosystems results in a larger ecosystem C storage than an age-class distribution that has the largest proportions of the area in the youngest age class, i.e., a distribution that resembles the negative exponentially declining distribution that will likely have been the precursor of the current distribution. If we assume that (1) the forest age-class structure in 1920 resembled the negative exponentially declining age-class structure (Figure 3); (2) around 1920 the total forest area was the same as at the time of the 1981 forest inventory (Bonnor, 1982) which represents forest conditions around 1970; and (3) the forest ecosystem C density in each age class in 1920 was similar to that in 1970 (Figure 2), we can calculate the biomass and soil C storage in 1920. A simple analysis, based on these assumptions and the data shown in Figures 1-3, suggests that from ca. 1920 to ca. 1970 forest ecosystem C storage in Canadian forests has increased by about 4.33 Pg C, averaging 86.6 Tg C yr^{-1}. Most of this increase (4.28 Pg C) occurred in the biomass C pools. This estimate does not include any additional accumulation of C that may have occurred in peatlands and in forest products.

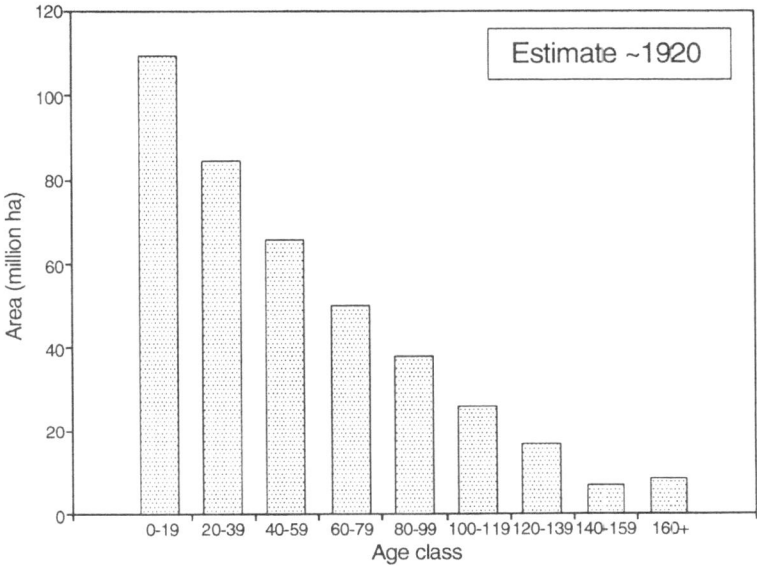

Figure 3:   Hypothetical age-class structure of Canadian forests in *ca.* 1920. Age-
             class structure based on assumptions of a 100-yr fire cycle and age-
             dependent stand mortality (e.g., insect disturbances) similar to those
             described in Kurz *et al.* (1993).

We must emphasize that this analysis is only an estimate and that the result is sensitive
to the assumed age-class structure for 1920. For this analysis, this age-class structure is
derived by combining a 100-yr fire cycle with a second disturbance that represents stand
breakup and insect-induced stand mortality (similar to the assumptions used in Kurz *et
al.*, 1993). We know, however, that the conditions in 1920 were somewhat different
because a simulation of the disturbance regimes for the period 1920 to 1969 only
approximates the forest age-class structure observed in *ca.* 1970. It appears that in 1920
an even greater proportion of the total forest area must have been in the youngest age
classes. Alternative initial age-class distributions are currently being explored.

The second assumption (i.e., constant forest area) is not valid if some forest disturbances
result in non-forest or non-stocked conditions that are not included in the forest age-class
distribution. Honer *et al.* (1991) estimated that, in the period 1977 to 1986, $0.47 \cdot 10^6$ ha
yr^{-1} of productive, non-reserved forest land in Canada were reclassified as non-stocked or
not satisfactorily restocked (NSR) land. Definitions of NSR standards used by Canadian
resource management agencies are based on commercial criteria, however, and much NSR

land supports rapidly growing, but noncommercial, hardwood species that are sequestering C. Estimates of conversion of forest land to non-stocked land for the period 1920 to 1970 are not available, but the close match between disturbed areas reported for the period 1930 to 1949 and the area in the forest age class that was initiated during this period suggests that the net change in forest area was small. (The net change must be considered because some areas are classified as non-stocked while other areas, previously classified as non-stocked, re-enter the forest inventory as a result of natural forest regeneration following disturbance.)

The third assumption (i.e., similar C density in each age class in *ca.* 1920 and *ca.* 1970) cannot be tested directly because no national data on biomass or soil C densities exist for 1920. If, however, disturbance frequency in Canadian forest ecosystems in the late 19th and early 20th centuries was indeed higher than in the period 1920 to 1969, then soil and detritus C density will likely have been smaller than in *ca.* 1970 because higher disturbance frequencies reduce soil and detritus C pools. This suggests that an assumption of similar biomass and soil C densities in *ca.* 1920 and *ca.* 1970 could yield conservative estimates of terrestrial C sinks.

We emphasize again that our main conclusion from this analysis is that the reduction in disturbance regimes that must have occurred to generate the known age-class structure (*ca.* 1970) will lead to a terrestrial C sink in Canadian forest ecosystems. The numerical estimate of the sink strength presented here is preliminary. We are currently in the process of calculating the annual C budget of the Canadian forest sector for the period 1920 to 1989 and our initial modeling results support the conclusions from the simple analysis presented here.

5. Current C-budget of Canadian Forests: an Update

The Carbon Budget Model of the Canadian Forest Sector (CBM-CFS) (Apps and Kurz, 1991; Kurz *et al.*, 1992; Apps and Kurz, 1993) has been developed to estimate the current contribution of the Canadian forest sector to the global atmospheric C budget. The model simulates the C dynamics of the biomass, soil and detritus, peatland, and the forest product sector C pools. It makes extensive use of the biomass data compiled in the National Forest Biomass Inventory (Bonnor, 1985). The model simulates the impacts of 5 disturbance types, (wildfire, insect-induced stand mortality, clear-cut logging, clear-cut logging with slash burning, and partial cutting). The simulation of C dynamics in soil C pools (which include detritus, coarse woody debris, and soil organic matter) is coupled with the dynamics of the biomass C pools. The fate of the harvested biomass is simulated in a forest product sector module. The entire model is described in detail elsewhere (Kurz *et al.*, 1992; Apps and Kurz 1993).

We have recently enhanced the model by adding the simulation of root biomass dynamics (Kurz and Apps, in prep. a). In addition, we made a number of minor modifications to the assumptions used for the derivation of biomass growth curves. In parallel, we have developed a dynamic version of the C budget model (Phase 2) that simulates forest sector C dynamics over several decades (Kurz and Apps, 1993).

Our revisions to the Phase 1 model resulted in an increase in estimates of the annual forest ecosystem C sink, because of the additional C sink in root biomass and the slightly higher growth rates (resulting from an improved parameterization). The revised Phase 1 model results indicate that in 1986, the reference year for the analysis, forest ecosystems were a C sink of about 78 Tg C (Kurz and Apps, in prep. b). Additional C sinks are estimated for peatlands (26.2 Tg C) and for the forest product sector (21.1 Tg C) (Kurz et al., 1992).

The CBM-CFS incorporates five important design criteria: (1) all major systems components are analyzed (Kurz et al., 1992); (2) belowground biomass C dynamics are simulated (Kurz and Apps, in prep. a); (3) soils dynamics are directly linked to biomass dynamics (Kurz et al., 1993); (4) disturbance regimes and their effects on C pools are simulated (Apps, 1993); and (5) age-class structures are explicitly recognized (Kurz et al., 1992). The model does not rely on assumptions of steady state for unmanaged forests. It does not invoke changes in forest growth rates or ecosystem productivity. The observed C sink in 1986 is primarily a function of changes in the forest age-class structure.

6. Comparison with other C-budget Assessments

Kauppi et al. (1992) suggest that in European forests, improved site conditions due to, for example, the nutrient import associated with air pollution has increased forest growth and contributed to the increase of forest growing stock between 1970 and 1990 by 25%. The results of our comparison of age-class structures in ca. 1920 and ca. 1970 suggest that in Canadian forests, the forest growing stock, and therefore the biomass C pool, has increased but this increase is independent of any assumptions about changes in growth rates. In fact, the increase in biomass C pools in Canadian forests would be greater than suggested in this analysis if growth rates had increased and enhanced net ecosystem productivity. There is, however, no unequivocal evidence that average forest growth has increased across the wide range of forest ecosystem types and environmental conditions encountered in Canada.

As Rastetter and Houghton (1992) correctly pointed out, the C sink in European forests (Kauppi et al., 1992) should not entirely be accounted towards the missing C in the global C budgets: only the difference between earlier estimates of C sinks and the recent, larger estimates of C sinks in European forests should be considered. In this context, it is important to point out that earlier estimates of the C budget of north American forests suggested that these are C sources (Houghton *et al.*, 1983; 1987). The significance of our results therefore is not merely the numerical value but more importantly the reversal of the direction of C exchange with the atmosphere. This conclusion is supported by recent studies about C dynamics in US forests (Birdsey, 1992; Heath and Birdsey, 1993).

The analyses by Houghton *et al.* (1983; 1987) indicated that north American forests are net C sources. Their analyses include an estimate of the contribution of croplands that are C sources (i.e., cropland newly created through land clearing and some land currently under cultivation) or C sinks (i.e., abandoned cropland now accumulating biomass). The estimated net contribution of north American (USA and Canada combined) croplands was a sink of 3 Tg C yr^{-1} (Houghton *et al.*, 1987). The C budget of the Canadian forest sector is based on that protion of the Canadian land base, which is included in the forest inventory. This does not include land in agricultural use but it may include some forest land that was previously used for agricultural purpose. Even if some forested, abandoned-farm land is included in our budget, the area involved will be a very small fraction of the total area in the inventory and its contribution to the observed C sink would be negligible.

7. Shifts in Forest Age-class Structure: A Saturating C Sink

The forest aging process that results in the observed C sink in Canadian forest ecosystems cannot continue indefinitely. Could this portion of the terrestrial C sink therefore become saturated?

The aging and declining phase of forest dynamics is poorly researched and quantified. Most traditional yield curves reach a maximum volume or biomass and stop rather than indicating the dynamics of stand breakup. Forest biomass in some boreal and temperate forest ecosystems often declines in the late stages of stand development (Cogbill, 1985; Whynot and Penner, 1990; Alban and Perala, 1992). As the average age of Canadian forests increases, this decline in forest biomass will become an increasingly important contribution to the declining sink strength in the C budget. Perhaps even more importantly, as stand age increases, vigour and productivity decrease, fuels accumulate, and the stands become increasingly susceptible to forest insect and fire disturbances (Blais, 1983; Kurz *et al.*, 1993). The CBM-CFS explicitly simulates the possible decline in living biomass in many older Canadian forest ecosystems. This biomass is transferred to the soil and detritus pools (including coarse woody debris) for decomposition in subsequent years. Preliminary sensitivity analyses that incorporate the assumption that disturbance regimes remain constant, show that the forest ecosystem C sink will diminish in the coming decades, unless changes in net ecosystem productivity compensate for the effects of the changing age-class structure (Kurz and Apps, 1993).

8. Conclusions

The results of the Phase 1 analyses of the C budget of the Canadian forest sector indicate that Canadian forests were a C sink in 1986 (Apps and Kurz, 1991; Kurz *et al.*, 1992). The result of the simple analysis presented here further suggest that Canadian forest ecosystems have been C sinks for some decades because a change in disturbance frequency must have occurred to generate the age-class distribution in the national forest inventory. This estimate of a C sink is largely the result of shifts in the forest age-class structure (increase in average forest age) and was obtained without invoking changes in growth rates (e.g., due to climate change or CO_2-fertilization effect). This structural sink mechanism will become saturated as more stands enter the breakup phase of stand development. Our results emphasize the need to abandon assumptions of steady state conditions for forests that are not immediately affected by land-use: these assumptions are only valid if natural disturbance regimes remain constant over many decades. The age-class structure of Canadian forests suggests that disturbance regimes in the period 1920 to 1970 have declined relative to those in the late 19th and early 20th century. We are conducting a retrospective analysis of the Canadian forest sector C budget for the period 1920 to 1989 and of forest policy options for maintaining a C sink in the coming decades.

9. Acknowledgements

Work for this project was funded in part by the Canadian Federal Panel on Energy Research and Development (PERD) through the ENFOR (ENergy from the FORest) program of Forestry Canada. The results reported here are based on an ongoing study that has received input from scientists and resource managers in many Canadian federal and provincial agencies. Their continuing support is greatly appreciated. We thank Sarah Beukema, Tamara Lekstrum, and Ralph Mair for programming and research assistance.

10. References

Alban, D.H. and D.A. Perala: 1992. Carbon storage in Lake States aspen ecosystems. *Can. J. For. Res.* **22:** 1107-1110.

Apps, M.J.: 1993. NBIOME: a biome-level study of biospheric response and feedback to potential climate changes. *World Resourc. Rev.* (in press).

Apps, M.J. and W.A. Kurz: 1991. Assessing the role of Canadian forests and forest sector activities in the global carbon balance. *World Resourc. Rev.* **3:** 333-344.

Apps, M.J. and W.A. Kurz: 1993. The role of Canadian forests in the global carbon budget. In: *Carbon Balance of the World's Forested Ecosystems: Towards a Global Assessment,* Publications of the Academy of Finland, in press.

Birdsey, R.A.: 1992. Carbon storage and accumulation on United States Forest Ecosystems. USDA For. Serv., Gen. Tech. Rep. WO-59, 51 pp.

Blais, J.R.: 1983. Trends in the frequency, extent, and severity of spruce budworm outbreaks in eastern Canada. *Can. J. For. Res.* **13:** 539-547.

Boden, T.A., R.J. Sepanski, and F.W. Stoss, (eds): 1991. Trends '91: A Compendium of Data on Global Change. Carbon Dioxide Information Analysis Center, Oak Ridge National Laboratory, Oak Ridge, Tennessee, U.S.A., ORNL/CDIAC-46. 665 pp. and appendices.

Bonan, G.B., and H.H. Shugart: 1989. Environmental factors and ecological processes in boreal forests. *Ann. Rev. Ecol. Syst.* **20:** 1-28.

Bonnor, G.M.: 1982. Canada's Forest Inventory 1981. Canadian Forestry Service, Environment Canada, Forestry Statistics and Systems Branch. 79 pp.

Bonnor, G.M.: 1985. Inventory of Forest Biomass in Canada. Canadian Forestry Service, Petawawa National Forestry Institute. 63 pp.

Cogbill, C.V.: 1985. Dynamics of the boreal forests of the Laurentian Highlands, Canada. *Can. J. For. Res.* **15:** 252-261.

Cooper, C.F.: 1983. Carbon storage in managed forests. *Can. J. For. Res.* **13:** 155-166.

Dellert, L.H.: 1991. What is British Columbia's timber supply forecast to the year 2050? In: Brand, D.G. (ed). *Canada's Timber Resources,* Forestry Canada, Petawawa National Forestry Institute, Chalk River, Ontario, Inf. Rep. PI-X-101. pp. 157-163.

Dewar, R.C.: 1991. Analytical model of carbon storage in the trees, soils, and wood products of managed forests. *Tree Physiol.* **8:** 239-258.

Edmonds, J.: 1992. Why understanding the natural sinks and sources of CO_2 is important: a policy analysis perspective. *Water Air Soil Poll.* **64:** 11-21.

Forestry Canada: 1988. Canada's Forest Inventory 1986. Forestry Canada, Ottawa, Ont. 60 pp.

Hall, C.A.S. and J. Uhlig: 1991. Refining estimates of carbon released from tropical land-use change. *Can. J. For. Res.* **21:** 118-131.

Harmon, M.E., W.K. Ferrell, and J.F. Franklin: 1990. Effects on carbon storage of conversion of old-growth forests to young forests. *Science* **247:** 699-702.

Heath, L.S. and R.A. Birdsey: 1993. Carbon trends of productive temperate forests of the coterminous United States, this volume.

Honer, T.G., W.R. Clark, and S.L. Gray: 1991. Determining Canada's forest area and wood volume balance, 1977-1986. In: Brand, D.G. (ed). *Canada's Timber Resources.* Forestry Canada. Petawawa National Forestry Institute, Chalk River, Ontario, Inf. Rep. PI-X-101, pp. 17-25.

Houghton, R.A.: 1991. Tropical deforestation and atmospheric carbon dioxide. *Climatic Change* **19:** 99-118.

Houghton, R.A., J.E. Hobbie, J.M. Melillo, B. Moore, B.J. Peterson, G.R. Shaves, and G.M. Woodwell: 1983. Changes in the carbon content of terrestrial biota and soils between 1860 and 1980: a net release of CO_2 to the atmosphere. *Ecol. Monogr.* **53:** 235-262.

Houghton, R.A., R.D. Boone, J.R. Fruci, J.E. Hobbie, J.M. Melillo, C.A. Palm, B.J. Peterson, C.R. Shaver, G.M. Woodwell, B. Moore, D.L. Skole, and N. Myers: 1987. The flux of carbon from terrestrial ecosystems to the atmosphere in 1980 due to changes in land use: geographic distribution of the global flux. *Tellus B.* **39:** 122-139.

Johnson, E.A. and G.I. Fryer: 1987. Historical vegetation change in the Kananaskis Valley, Canadian Rockies. *Can. J. Bot.* **65:** 853-858.

Kauppi, P.E., K. Mielikainen, and K. Kuusela: 1992. Biomass and carbon budget of European forests, 1971 to 1990. *Science* **256:** 70-74.

Kurz, W.A. and M.J. Apps: (in prep.a). Estimation of root biomass and dynamics for the Carbon Budget Model of the Canadian forest sector. (manuscript in preparation).

Kurz, W.A. and M.J. Apps: (in prep.b). Canada's forests: a net carbon sink in 1986. (manuscript in preparation).

Kurz, W.A. and M.J. Apps: 1992. Atmospheric carbon and Pacific Northwest forests. In: Wall, G. (ed). *Implications of Climate Change for Pacific Northwest Forest Management.* Department of Geography, University of Waterloo, Waterloo, Ontario, Dept. of Geography Publ. Ser. Occassional Paper No. 15, pp. 69-80.

Kurz, W.A. and M.J. Apps: 1993. The carbon budget of Canadian forests: a sensitivity analysis of changes in disturbance regimes, growth rates, and decomposition rates. *Environ. Poll.* (in press).

Kurz, W.A., M.J. Apps, T.M. Webb, and P.J. McNamee: 1992. The Carbon Budget of the Canadian Forest Sector: Phase I. Forestry Canada, Northwest Region, Northern Forestry Centre, Edmonton, Alberta, Inf. Rep. NOR-X-326. 93 pp.

Kurz, W.A., M.J. Apps, B.J. Stocks, and W.J.A. Volney: 1993. Global climate change: disturbance regimes and biospheric feedbacks of temperate and boreal forests. In: Woodwell, G. (ed). *Biotic Feedbacks in the Global Climate System: Will the Warming Speed the Warming?* Oxford University Press, Oxford, UK, (in press).

Ontario Ministry of Natural Resources: 1992. Ontario Forest Products and Timber Resource Analysis. Volume I and Volume II -- October 1992. Joint Study by: Resource Information Systems, Inc. Bedford, MA and Resource Economics, Inc. Corvallis, OR. Published by Ontario Ministry of Natural Resources, Sault Ste. Marie, Ontario. 139 pp.

Payette, S.: 1992. Fire as a controlling process in the North American boreal forest. In: Shugart, H.H., R. Leemans, and G.B. Bonan (eds), *A Systems Analysis of the Global Boreal Forest*, Cambridge: Cambridge University Press, pp. 144-169.

Quay, P.D., B. Tilbrook, and C.S. Wong: 1992. Oceanic uptake of fossil fuel CO_2: Carbon-13 evidence. *Science* **256:** 74-79.

Rastetter, E.B. and R.A. Houghton: 1992. Carbon budget estimates. *Science* **258:** 382.

Rowe, J.S.: 1983. Concepts of fire effects on plant individuals and species. In: Wein, R.W. and D.A. MacLean (eds), *The Role of Fire in Northern Circumpolar Ecosystems*, SCOPE 18, Chichester: John Wiley & Sons. pp. 135-154.

Sedjo, R.A.: 1992. Temperate forest ecosystems in the global carbon cycle. *Ambio* **21** 274-277.

Tans, P.P., I.Y. Fung, and T. Takahashi: 1990. Observational constraints on the global atmospheric CO_2 budget. *Science* **247:** 1431-1438.

Van Wagner, C.E.: 1978. Age-class distribution and the forest fire cycle. *Can. J. For.*

Res. **8:** 220-227.

Vinson, T.S. and T.P. Kolchugina: 1993. Pools and Fluxes of Biogenic Carbon in the Former Soviet Union, this volume.

Whynot, T.W. and M. Penner: 1990. Growth and yield of black spruce ecosystems in the Ontario Clay Belt: implications for forest management. Forestry Canada, Petawawa National Forestry Institute, Chalk River, Ontario, Inf. Rep. PI-X-99. 81 pp.

Wisniewski, J. and A.E. Lugo (eds): 1992. Natural Sinks of CO_2, Kluwer Academic Publishers, Dordrecht, The Netherlands, 466 pp.

Yarie, J.: 1981. Forest fire cycles and life tables: a case study from interior Alaska. *Can. J. For. Res.* **11:** 554-562.

CARBON SEQUESTRATION IN NORWAY SPRUCE IN SOUTH SWEDEN AS INFLUENCED BY AIR POLLUTION, WATER AVAILABILITY, AND FERTILIZATION

Lars Owe Nilsson

Swedish University of Agricultural Sciences, Department of Ecology and Environmental Research, Box 7072, 750 07 Uppsala, Sweden

(Received Mars 1, 1993; revised May 3, 1993)

Abstract. Carbon sequestration in 30 yr old Norway spruce in south Sweden following manipulation of nutrient and water availability is presented. The site has an annual precipitation of 1100 mm and a deposition of about 20 kg N and 25 kg S per ha^{-1} yr^{-1}. The soil type is a poorly developed podzol. Treatment include irrigation; artificial drought; ammonium sulphate addition; nitrogen-free-fertilization and irrigation with liquid fertilizers including a complete set of nutrients. The experiment has a randomized block design with four replicates per treatment. A comprehensive investigation of the above ground C storage on an areal basis was made at the start of the experiment and after 3 yr of treatment. After 3 yr of treatment with simulated N-S deposition using ammonium sulphate (100 kg N, 114 kg S ha^{-1} yr^{-1}), C accumulation rates in the above ground compartments had increased by 37%. Similarly, irrigation caused increased C accumulation rates by 25%, whereas simulated drought during the vegetation period during 2 yr followed by 1 yr of recovery caused a 15% reduction of the C accumulation rates. Irrigation combined with liquid fertilization (100 kg N ha^{-1} yr^{-1}), including all important nutrient elements, led to 65% increase in C accumulation rates compared to the control. The C sequestration of the latter treatment gradually increased and, during yr 5 of treatment, 8.6 Mg C ha^{-1} accumulated in stems and branches, compared to 3.6 Mg ha^{-1} for the control. It is concluded that there is a strong interaction between N-deposition and C accumulation rates in Norway spruce in south Sweden. The C accumulation rates are also sensitive to water availability. The study indicates a great potential to cultivate Norway spruce in south Sweden as a renewable energy source. A shift in energy source from fossil fuels to renewable energy sources will directly reduce the net emissions of CO_2 to the atmosphere.

1. Introduction

The increased emissions of green house gases to the atmosphere are expected to result in future undesirable global temperature increase and changes in precipitation patterns. The most important gas in this context is CO_2. Emphasis has been placed to quantify the different sources and sinks of CO_2 on a global scale (Detwiler and Hall, 1988; Tans *et al.*, 1990). A net imbalance in the form of an unknown sink of 1.6 PgC yr^{-1} has been found (IPCC, 1992).

Recent studies indicate increased forest production during the last 50 yr in several European countries (Kenk, 1990; Kauppi *et al.*, 1992). According to the Swedish Forest Survey, forest productivity in Sweden has increased considerably during the last decades and amounts today to about 100 million m³ yr^{-1} compared to an annual harvesting of 70-75 mil m³ yr^{-1} (REF). There are several explanations for the widespread production increases; improved forestry practices (e.g., choice of provenances, ditching, fertilization etc.), changed land utilization (Johnson *et al.*, 1991); but also environmental factors like increased concentration of atmospheric CO_2.

Wisniewski and Lugo (1992"a" and "b") concluded that "a vast number of natural and managed ecosystems are currently accreting carbon, and the quantity may be large enough to account for the so-called 'missing carbon'". However, today there are still considerable uncertainties in the global flows of CO_2 between the atmosphere and the terrestrial ecosystems and how those flows are affected by anthropogenically-induced environmental change.

The aim of this study is to clarify the influence of N-deposition and water availability on the sequestration of CO_2 in a managed Norway spruce stand in south Sweden. The aim is further to indicate the potential to cultivate spruce as an energy source.

2. Materials and Methods

2.1. SITE

The site is situated at Skogaby (Lat 56°33'N; long 13°13'E; alt ca. 100 m above sea level) about 25 km from Kattegat in the south-western part of Sweden. The bedrock belongs to the Southwest Swedish gneiss area covered by more than a 2 m thick till layer. The soil type is a poorly developed podzol with a 6.7 cm thick humus layer (O) and a 2.1 cm thick leached layer (E). The parent material is a sandy loamy till. The clay content varies between 4-7% in the upper 50 cm of the mineral soil. The average pH value (H_2O; soil:solution=1:2) in 1987 was 3.9 in the humus layer and increased about 0.5 pH units down to 20 cm depth and another 0.1 pH unit down to 50 cm depth of the mineral soil. The effective base saturation was 30% in the humus layer and 7-14% in the mineral soil. Aluminium saturation increased from 16% in the humus layer to 66-77% in the mineral soil, indicating a prevailing buffer system of aluminium.

2.2. STAND

The stand was planted in 1966 with Norway spruce (*Picea abies* (L.) Karst). The previous generation forest was a planted pine forest (*Pinus sylvestris* (L.)) replacing grazed heath land 1916. Before starting the experiment in 1987/88 the stand volume was 144 m³ ha⁻¹, the number of trees 2285 ha⁻¹, breast height diameter 11.3 cm and a basal area 24.4 m² ha⁻¹ (Nilsson and Wiklund, 1992). The stand was homogeneous and without visible damage. The area consists of two provenances; Istebna (I) originating from S. Poland (lat 49 °34' N, long 18° 56'E, 5-700 m altitude) and Augustow (A) from N. Poland (lat 54°N, long 23°; 2-300 m altitude).

2.3. DEPOSITION AND CLIMATE

On an annual basis, the climate is characterized by high precipitation, 1139, 974, 1164, 1096 and 1220 mm during 1988-1992, respectively. Marked presummer droughts were noted 1989 and 1992. The drought during 1992 was unusual with an almost two month long period with no precipitation and much higher air temperatures than normal. Annual mean air temperatures

TABLE I

Treatments within the Skogaby Project. Each treatment consist of 4 replicates.

Symbol	Treatment	Description
C	Control	No treatment
D1	Drought 1988-89	A roof prevents 2/3 of the throughfall from reaching the ground during the growing season April 1 - September 30. The roof is located at 0.5-2 m above ground. During winter all precipitation can reach the ground.
I	Irrigation	Irrigation is done using sprinkler technique which gives an even distribution of water on the soil surface. Changes in water storage are calculated once a week using actual weather data and irrigation is performed when a 20 mm storage deficit has developed. The potential evapotranspiration is calculated with the Penman-Montieth combination formula.
NS	NS-addition	Ammonium sulphate is added manually three times a year (100 kg N ha^{-1} yr^{-1}).
V	N-free-fertilization	Altogether 1000 kg ha^{-1} of "Skog-Vital" was added during 1988-89. 1988 the fertilizer addition was divided between two occasions. In 1989 there was only one addition. "Skog-Vital" is a commercial fertilizer (manufactured by Supra, Sweden) without any N but including other elements of importance for forest growth (Table II).
IF	Optimum fert-ilization with irrigation	100 kg of N ha^{-1} yr^{-1} is added together with a complete set of nutrients essential for an optimum forest yield according to the Ingestad principle. The fertilizer solution is sprayed evenly above the ground with the same technique as for the I-treatment. Fertilization is done from late May until late August, with peaks in late June.

TABLE II.

Composition and amounts of macro nutrients in three treatments (kg ha^{-1}).

Treatment	Element						Period of application
	N	P	K	Ca	Mg	S	
NS	100	0	0	0	0	114	Yearly
IF	100	17	48	6[a]	6	9	Yearly
V	0	48	43	218	46	75	Total during 1988-89

[a] was given as a single dose prior the start of the irrigation as ground limestone.

were 7.4, 8.2, 8.3, 7.1 and 7.7 °C during 1988-92, respectively. The annual throughfall at Skogaby was 1989/90; 11 kg NO_3-N and 30 kg SO_4-S ha^{-1}.

2.4. EXPERIMENTAL DESIGN AND TREATMENT

The general approach of the treatments was either to increase or to decrease the availability of water and nutrients separately or together. The experiment has a randomized block design with six treatments and 4 replicates. Each plot was approximately 2000 m². Availability of water was manipulated by irrigation (I) or artificial drought using a roof (D1, Table I). As the roof is transparent and only about 2% of the incident light (400-700 nm) reaches the ground the roof has a negligible effect on the light level that reaches the ground. The experimental approach enables us to quantify effects of natural drought on the control. The nutrient balance and the availability of nutrients other than N were manipulated by ammonium sulphate addition (NS) or by application of nitrogen-free-fertilizer "SkogVital" (V). Combined improvement of both water and nutrient availability was achieved by irrigation with liquid fertilizers (IF) with a composition of nutrients as indicated by Ingestad (1979). A detailed description of the different treatments is given in Tables I and II. Treatment started in 1988.

2.5 CALCULATION OF CARBON ACCUMULATION RATES

The treatment effect on different biomass compartments was studied in two steps: 1) First 16 trees from each treatment were harvested both before the start of treatment 1987 and after 3 yr of treatment. The trees were used for determining allometric relationships so that biomass of different tree compartments (stem, bark, needles, branches, cones) could be calculated as a function of breast height diameter (D), tree height (H) and/or crown length (CL). 2) Using those equations the biomass per areal unit of a subplot including about 40 growing trees per plot (E-unit) was calculated for the two different points in time. The procedure has earlier been described by Nilsson and Wiklund (1992).

During individual years (1988-92) biomass of stem and branches was determined using breast height diameter only (c.f. Marklund, 1987; Whittaker and Marks, 1975). Due to the strong influence of the different treatments on stem form, the allometric relationships used for prediction of biomass based on only breast height diameter, undergoes change as a function of time. The biomass 1990 was determined using the available allometric equation 1987 and was compared with measured values. In this way a correction factor for 1990 for the use of equations from 1987 could be established. Using the equations from 1987 and assuming a linear change in the correction factors over time for the different treatments and, using the annual measurements of the trees within the E-unit, biomass during 1987, 1988, 1989, 1990, 1991 and 1992 was then calculated for each plot.

Carbon content was determined in the various compartments of the tree in connection with the destructive samplings 1987 and 1990. Emphasis was then made to establish samples that represent the whole stand in terms of stems, needles and branches for each treatment. Total carbon concentration was determined with an elemental analyzer (Perkin Elmer 2400) according to the routines of the Faculty of Forestry, Swedish University of Agricultural Sciences, Umeå. Data on C content were combined with biomass data for calculation of total quantity of C in the stand during the different years. The acccumulation rates of C over time were then calculated on each plot within the four replicates of each treatment.

2.6. STATISTICAL METHODS

When analyzing treatment effects on C accumulation rates in a certain compartment initial H or D was used as covariate. That is, the experiment was analyzed as a randomized block design with one covariate. The analysis of variance was done using the procedure GLM in the SAS package (SAS, 1989). The significant F-values when testing the over-all treatment effect and significant t-values in comparing the differences between any two treatments are marked when occurring. All significances including the F-values are on the 5% level.

3. Results

The C density showed not significant variation over time or as a result of the different treatments in the various compartments of the tree (data not shown). Carbon density generally was 50%. Increased C accumulation rates above ground during 1988-90 was noted for treatments IF (+65%), I (+25%) and NS (+37%) compared to C (Table III). Drought treatment during 2 yr followed by 1 yr of recovery, however, resulted in a 15% lower C accumulation rate compared to C. Treatment V showed a minor increase in the accumulation rate of C (+13% vs C). During the first 3 yr of treatment the C flow as litter fall varied between 2.53 Mg C ha^{-1} for treatment I and 3.13 Mg C ha^{-1} for treatment IF.

C accumulation rates as a function of time for the different treatments are given in Figure 1. There is an annual variation of C accumulation rate for control trees. Increased rates of C accumulation vs control are noted also during 1991-92 for trees treated with ammonium sulphate. Trees treated with irrigation with liquid fertilization (IF) showed a gradual sequestration of C and C-accumulation rates 1992 were 141% higher than control (+65% during 1988-90). Irrigation led to 35% increased C accumulation rates as a mean value during all 5 seasons, whereas drought treated trees for 2 yr followed by no treatment for 3 yr resulted in decreased C accumulation rates up to 1991.

4. Discussion

Simulated fivefold higher N-S deposition resulted in about 40% increased C accumulation rates over the first 5 yr of treatment. The results indicate an interesting relationship between N deposition and C fixation. The observations support findings by Kenk (1990) and Kauppi et al. (1992) who have indicated a possible positive correlation between some air pollutants and forest production. If higher N-deposition rates than today result in production increases it is quite likely that the increased N-deposition in this region also has previously resulted in increased forest production. This is supported by a recent study within the same area by Erikson and Johansson (1993). The authors have compared the production of the first and

TABLE III

Pools of C at the start of the experiment, 1987, and accumulation rates of carbon ($Mg\ C\ ha^{-1}$) following 3 yr of treatment in various above ground compartments of the tree, and, litter fall for the different treatments. The values are corrected for initial variation with basal area as covariate. Pair of mean values in the same row marked with different letters differ significantly. A star (*) means that the F-value in testing the over-all treatment effects in a row is significant.

	1987	C		D1		I		IF		NS		V
Stem (*)	26.00	8.86	a	8.21	a	11.03	b	13.16	c	10.99	b	9.61
wood (*)	23.10	8.24	a	7.66	a	10.24	b	12.26	c	10.28	b	9.01
bark (*)	2.90	0.62	ad	0.55	a	0.79	bd	0.90	bc	0.71	d	0.60
Branches (*)	9.84	1.68	ac	1.19	a	1.46	a	3.30	b	3.35	b	2.45
alive	7.30	0.99	a	0.68	a	1.35	ab	2.20	b	1.40	ab	0.98
dead (*)	2.54	0.69	ac	0.51	ab	0.11	b	1.10	ce	1.95	d	1.47
Needle (*)	7.28	1.05	a	0.15	b	2.05	cd	2.62	c	1.44	ad	1.03
Cone	0.00	0.00	a	0.34	b	0.00	a	0.00	a	0.05	a	0.03
Total (*)	43.12	11.59	ad	9.89	a	14.54	be	19.08	c	15.83	b	13.12
Litter-fall (*)		2.72	a	3.10	a	2.53	a	3.13	a	3.01	a	2.55
C-uptake (*)		14.31	ad	12.99	a	17.07	be	22.21	c	18.84	b	15.67

second generation of Norway spruce on the same site with similar provenances and forest management between 1880-1919 and 1950-1989. During the two compared periods the mean annual air temperature was similar whereas precipitation was 6% higher during the latter period. The results indicate 40% increased production of Norway spruce during the latter period. The authors conclude that the most probable cause of the increased production is the increased N-deposition which has fertilized the forests. It thus seems obvious that air pollutants have enhanced a CO_2 sink. In my view it seems likely that this effect will continue as N-deposition rates hardly will be reduced to a greater extent during the coming decades. On a long-term perspectice the effects of air pollutants may be the opposite. Decreasing availability of base cations may induce nutrient imbalances in the trees which may result in decreased C-accumulation rates, meaning that a C-source will develop.

Nitrogen-free-fertilization increased C accumulation rates by 22% (however, not statistically significant) compared to control over the 5 yr experimental period. The V-

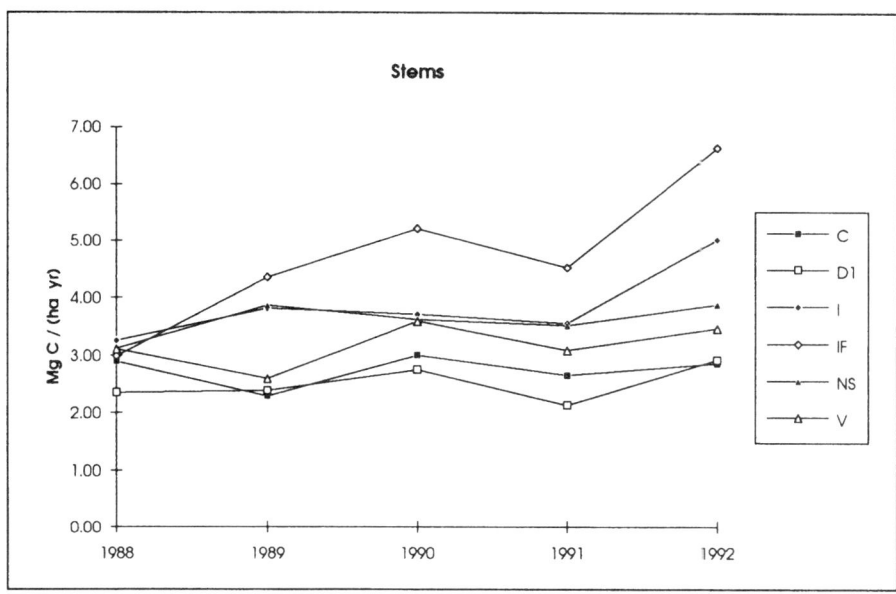

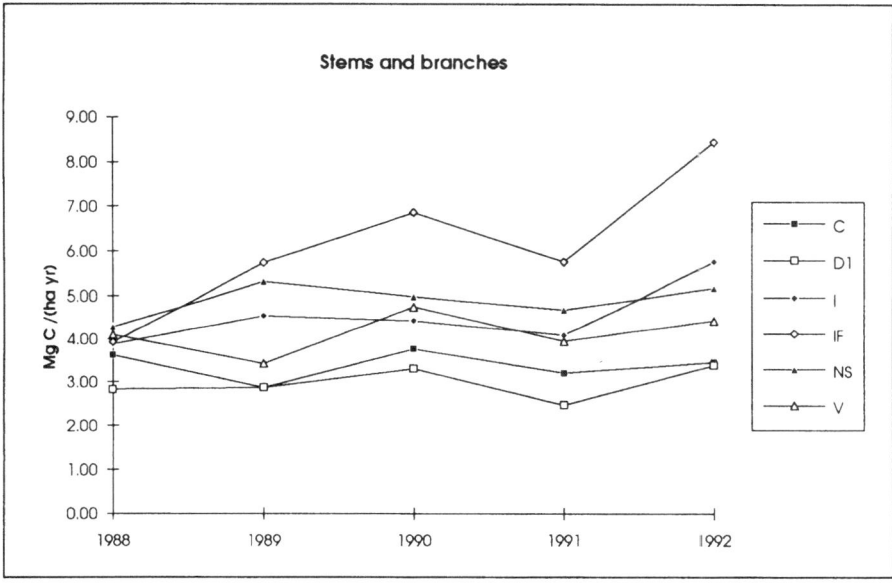

Figure 1. Carbon accumulation rates in stems or stems and branches of Norway spruce in the Skogaby site following different treatments (see Table I).

treatment included addition of all necessary nutrients but N. The response in growth probably is caused by improved N-availability in the soil. After 3 yr of treatment the N-uptake had increased significantly compared to the control (data not shown).

Carbon accumulation rates increased significantly with greater water availability. Although the site has an annual precipitation of about 1100 mm, C accumulation rates increased by about 35% due to irrigation as a mean value over the 5 yr experimental period. This is partly explained by an increased N-uptake (data not shown) following increased N-mineralization. Simulated seasonal drought led to lower (non-significant) C fixation compared to untreated trees. The order of reduction, however, was somewhat lower than the stimulation due to irrigation. The experimental results show that any changes in the precipitation pattern for instance as a consequence of the green house effect will interact dramatically in the C cycle of Norway spruce.

Liquid fertilization combined with irrigation resulted in significantly increased C accumulation rates in the stand. During the first 3 yr of treatment, C acccumulation rates above ground increased by 65%. However, accumulation rates in stems and branches increased over time and during the 5th yr of treatment (1992) it reached 8.4 Mg C
ha^{-1} y^{-1} (141% higher than control). This equals about 17 ton dry mass ha^{-1} yr^{-1}. The production increase during the first 3 yr of treatment was correlated with a dramatic increase of the N-uptake in the stand. N-fertilization equalled 300 kg N per ha^{-1} during 1987-90 (about 355 kg ha^{-1} including N-deposition). This resulted in a N-storage increase above ground of 210 kg N ha^{-1} during the same period (data not shown).

Several examples of growth response in Norway spruce in South Sweden due to N-fertilization have been reported by Möller (1986) and Linder (1992). The dramatic growth response clearly shows that unfertilized Norway spruce today grows very far from its maximum potential. Growth is highly restricted by nutrient availability, in particular by N.
Several studies have indicated an increased C storage in the soil following N-fertilization (eg. Nohrstedt et al., 1989; Johnson, 1992). Fertilization leads to long-term lower rates of litter decomposition as indicated by Berg and Ekbohm (1991). The overall CO_2 evolution is also reduced as indicated from preliminary investigations of the IF-treatment in this study (Tryggve Persson 1993). Thus cultivation of Norway spruce with fertilizers most probably would lead to sequestration of C in the soil at least on a short time perspective.

During the last decades, the use of forest products for combustion has increased dramatically in Sweden. If this energy replaces the use of fossil fuels the implication is a net reduction of the emissions of CO_2 to the atmosphere. The enormous potential to sequester CO_2 following fertilization and irrigation, as indicated in this study, opens up a possibility to cultivate Norway spruce for energy use. This renewable energy source would actively reduce the net emissions of CO_2 to the atmosphere and consequently be of great environmental value. This of course requires that leaching of N is kept on a minimal level by highly controlled additions of fertilizers. In a situation with increased recycling of forest products the demands of, raw material for the forest industry are likely to decrease or increase in lower rates. An alternative use of forest biomass, as, for example, energy, will be welcomed today by the producers.

5. Conclusions

Unfertilized Norway spruce in southern Sweden grows far from its maximum potential. The most limiting factor for production is the availability of N.

There is a strong positive correlation between N deposition and C accumulation rates in Norway spruce in southern Sweden.

Changes in the precipitation pattern for instance as a consequence of the green house effect will interact with the C cycle of Norway spruce.

6. Acknowledgements

The Skogaby Project is financed by the National Swedish Environmental Protection Agency and the Foundation of Forestry Research (Stiftelsen Skogsbrukets Forskningsfond). Instrumentation and constructions in the field were financed by the Swedish bank, Nordbanken. I want to thank Karin Wiklund for having recalculated data on biomass into C.

7. References

Berg, B. and Ekbohm, G. : 1991, *Can. J. Bot.* **69**, 1449.

Detwiler, R.P. and Hall, C.A.S.: 1988, *Science.* **239**, 42.

Eriksson, H. and Johansson, U.: 1993, Yield of Norway spruce (*Picea abies* (L.) Karst.) in two consecutive rotations in south-western Sweden. *Plant and Soil.* (accepted).

Ingestad, T.: 1979, *Physiol. Plant.* **45**, 373.

IPCC (International Panel on Climate Change), 1992. IPCC 1992 Science Assessment. Report to IPCC from Working Group 1, prepared by the IPCC Group at the meteorological Office, Bracknell, UK.

Johnson, D.E.:1992, *Water, Air, and Soil Pollution.* **64**, 83.

Johnson, D.W., Cresser, M.S., Nilsson, S.I., Turner, J., Ulrich, B., Binkley, D. and Cole, D.W.: 1991, Soil changes in forest ecosystems: evidence for and probable causes. International Conference on Acid Deposition. Its Nature and Impacts, Glasgow, 1990, Proceedings of the Royal Society of Edinburgh, 97B, pp. 81-116.

Kauppi, P. Mielikäinen, K., Kuusela, K.: 1992, *Science.* **26**, 70.

Kenk, G.:1990, Effects of air pollution on forest growth in south-western Germany - hunting for a phantom? Proc. Div. 2 XIX IUFRO World Congress, Montreal, Canada.

Linder, S.:1992, private communication.

Marklund, L.G.: 1987, Biomass functions for Norway spruce (*Picea abies* (L.) Karst) in Sweden. Dept. For. Survey, Swed. Univ. Agric. Sci., Rep. 43, 127 pp.

Möller, G.:1986, Results from forest fertilization trials in areas with high nitrogen input, in *Nitrogen saturation, abstracts from a workshop*. National Swedish Environmental Protection Board. Report 3153, pp 49-61.

Nilsson, L.O. and Wiklund, K.: 1992, *Plant and Soil*. **147**, 251.

Nohrstedt, H.-Ö., Arnebrant, E. Bååth, E. and Söderström, B.: 1989, *Can. J. For. Res.* 19, 323.

Persson, T.:1993, private communication.

SAS Institute Inc. 1989 SAS/STAT User's Guide: Basics, Version 6, Fourth Edition, Volume 2, Cary. NC:SAS Institute INC. 846 p.

Tans, P.P., Fung, I.Y. and Takashi, T.: 1990, *Science*. **247**, 1431.

Whittaker, R.H. and Marks, P.L.: 1975, Methods of assessing terrestrial productivity, in: Lieth H. and Whittaker R.H. (eds) *Primary Productivity of the Biosphere* , Springer-Verlag, New York Inc, pp 55-118.

Wisniewski, J. and Lugo, A.E.: 1992a, Workshop Statement. Natural sinks of CO_2. Palmas Del Mar, Puerto Rico, 24-27 February 1992, Kluwer Academic Publishers, 6 pp.

Wisniewski, J. and Lugo, A.E.: 1992b, Natural sinks of CO_2. Palmas Del Mar, Puerto Rico, 24-27 February 1992, Kluwer Academic Publishers, 466 pp.

IMPACT OF FORESTS ON NET NATIONAL EMISSIONS OF CARBON DIOXIDE IN WEST EUROPE

P.E. KAUPPI and E. TOMPPO

Finnish Forest Research Institute, Unioninkatu 40 A, SF-00170 Helsinki, Finland

Abstract. The amount of forest biomass increased and thereby reduced the net national emissions of CO_2 in Europe, in some countries more than in others. Estimates of the annual C fluxes through forests in 17 west European countries are presented, based on recent statistics. The flux in each country is subdivided into components referring to removal, change in the remaining growing stock, and detritus formation. The relative contribution of forests to the national C budget varied by two orders of magnitude between the upper extremes (Sweden and Finland) and the lower extremes (Netherlands, Belgium and the UK). Such large differences between countries must have an impact on policies and strategies for controling the net C emissions. Because the capacity of forests to carry biomass is limited, the net flux of C from the atmosphere into the forests is expected to decrease. Afforestation and improved management of the removed forest biomass could compensate for this development.

1. Introduction

The United Nations Climate Convention, in order "...to achieve stabilization of greenhouse gas concentrations in the atmosphere ... at a level that would prevent dangerous anthropogenic interference with the climate system", engages all Parties to formulate national and, where appropriate, regional programs including measures to mitigate climate change (UN Framework Convention..., 1992). Forestry affects the fluxes of C between the atmosphere, vegetation, soils and forest products, and is a potential agent both in controlling and in contributing to national net emissions. Forestry measures, on a country-by-country basis, could be developed to contribute to the policies of meeting national emission goals. Both direct and indirect evidence suggests that boreal and temperate forests have acted as a stronger sink of atmospheric C than was estimated earlier (Sedjo, 1992; Tans *et al*, 1990). Detailed analyses have been carried out for Canada, indicating net transfer of C from the atmosphere into forest ecosystems and forest products (Apps and Kurz, 1991; Kurz and Apps, this volume). In Europe, forest biomass has increased and additional C has been sequestered in forest products (Kauppi *et al*, 1992). However, tropical deforestation has continued at a rapid rate (Tolba *et al*, 1992) and the World's forests are currently estimated to be a source term in the global C budget (Sedjo, 1992).

Water, Air, and Soil Pollution **70**: 187–196, 1993.
© 1993 *Kluwer Academic Publishers.*

When old-growth forest is cleared and converted into forest which is managed for maximum sustained yield, standing biomass is reduced by about two thirds (Cooper, 1983). Although such conversion and biomass development has taken place in many regions, as in north western U.S. (Harmon *et al*, 1990), biomass development has been the reverse in other areas. Large parts of the old-growth forests in central Europe were cleared in the Middle Ages (Kandler, 1992) and almost all of the contemporary harvest is from secondary forests. New statistics from ECE/FAO are used in this paper to analyze C fluxes and the capacity for C sequestration in forestry on a country by country basis (The Forest Resources of the Temperate Zone..., 1992). The statistics refer to the mid and late 1980s.

The focus of this paper is on five member countries of the European Free Trade Association (EFTA): Austria, Finland, Norway, Sweden and Switzerland; and on the 12 member countries of the European Community (EC): Belgium, Denmark, France, Germany, Greece, Ireland, Italy, Luxenbourg, Netherlands, Portugal, Spain and the United Kingdom. The total forest resources of these 17 countries expanded in the 1980s in terms of both the growing stock and the area of forest and other wooded land (The Forest Resources of the Temperate Zone..., 1992; and The Forest Resources of the ECE Region..., 1985).

2. Carbon Fluxes and Storages

We denote the total annual flux of C in forestry as C_{fr} (Figure 1) and divide it into three flux components: 1) C in removal, C_{rm}, 2) C in detritus formation, C_{dt}, and 3) C in the net change of living biomass, C_{st} (the storage rate). The description of fluxes and storages is similar to that presented by Apps and Kurz (1991) and Dewar and Cannell (1992).

Statistics on gross annual increment (The Forest Resources of the Temperate Zone..., 1992) were converted into estimates of the total annual flux of C in forestry, C_{fr}, by accounting for branches, roots and foliage, and for the C content in biomass. The lower and upper estimates of the biomass in roots, branches and foliage were obtained by multiplying stemwood biomass by 0.4 and 1.1, respectively (Brown and Lugo, 1984). Bulk density of 400 g dry matter per liter wood, and a C content of 50 % were assumed (Hakkila, 1989). Carbon in removal, C_{rm}, was calculated in the same way from the statistics (The Forest Resources of the Temperate Zone..., 1992), taking only stemwood (overbark) into account. Carbon in the net storage of biomass, C_{st}, was estimated as the difference between net annual increment and felling, in terms of whole tree biomass. Carbon in detritus formation, C_{dt}, was calculated by subtracting C_{st} and C_{rm} from the total flux, C_{fr}.

The net change of biomass, C_{st}, can be negative as a result of wild fire, logging, deforestation, pollution damage, etc. The net change is positive when biomass builds up, *i.e*, when increment exceeds the sum of losses and removals. Net changes of the C pools in detritus and in products were not included in this analysis but have been estimated in a similar study for Canada (Apps and Kurz, 1991) and the UK (Dewar and Cannell, 1992).

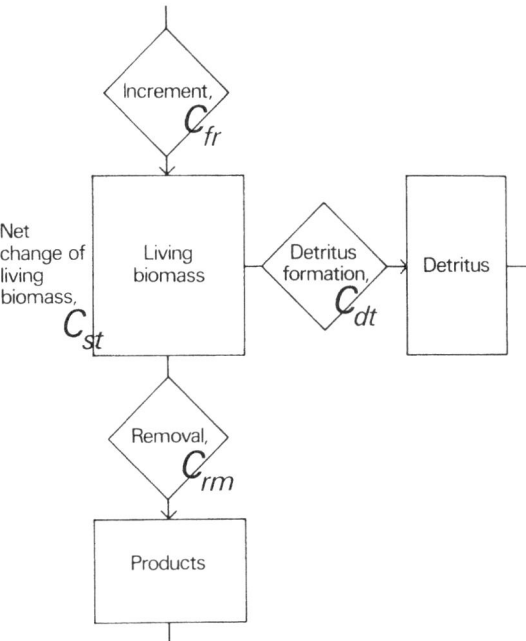

Fig 1. Fluxes of carbon in forestry. The total forestry flux (C_{fr}) is divided into three compoments: 1) Removal (C_{rm}), 2) Detritus formation (C_{dc}), and 3) Net change of living biomass (C_{st}).

2.1. CARBON IN REMOVAL (C_{rm})

The statistics on wood removal indicate a harvest of 143 and 155×10^6 m^3 in EFTA and EC, respectively. This amount of wood, used for various products, contains a total of 60 Tg C (Tg = 10^{12} g). The fate and the decay rate of the different products are less well known. Soft products like newsprint have a short life span, whereas construction wood can store carbon for decades or centuries. Carbon in soft products can also remain intact if disposed of in landfill sites.

2.2. CARBON IN DETRITUS FORMATION (C_{dt})

The removal of 60 Tg C in stemwood implies an annual transfer of 24 to 66 Tg C from the pool of living biomass into the detritus pool in the form of the roots, branches and foliage of felled trees. This estimate refers to the whole region (EFTA+EC). Natural losses and stem residues account for additional detritus fluxes, and the total C flux in detritus formation, C_{dt}, was estimated at 31 to 76 Tg. An additional detritus flux, not included in this estimate, is due to the annual turnover of foliage and fine roots.

2.3. CARBON IN THE NET STORAGE OF LIVING BIOMASS (C_{st})

Felling comprised a modest 72 and 68 % of the net annual increment in EFTA and EC, respectively (The Forest Resources of the Temperate Zone..., 1992). A positive net change of living biomass was reported by each of the 17 countries; the gross annual increment exceeded the sum of natural losses, removal and harvest residues. The growing stock increased annually in the late 1980s by 68 and 72×10^6 m^3 in EFTA and EC, respectively. This corresponds to a net increase of 39 to 58 Tg C in living biomass in EFTA+EC.

2.4. TOTAL CARBON FLUX IN FORESTS AND FORESTRY (C_{fr})

The three subfluxes within forestry, as described in Figure 1, were of the same order of magnitude in many of the European countries (Table 1). However, there ware large differences between countries in terms of the impact of forest on net national emissions. Forest is an important component of the national C budget of EFTA countries, in fact decisive in Sweden and Finland, but less important in the EC being almost negligible in countries like Belgium, the Netherlands and the UK. The relative contribution of forests to C budget was an order of magnitude larger in EFTA than in the EC.

The total flux of C (C_{fr}) was 62 to 94 Tg in EFTA, about the same as the emissions of C from the combustion of fossil fuels (C_{fos} as documented in IEA/OECD, 1992). EFTA countries export large quantities of wood products, whereas the EC countries are net importers. It can be estimated that a flux of 5 to 10 Tg C — a fraction of the removal flux — enters from EFTA to the European Community. Forest products from other parts of the world are also imported to the EC, where the total C flux in forest sector is thus larger than the domestic flux estimated in Table 1.

3. Discussion

3.1. UNCERTAINTY OF THE ESTIMATES

The accuracy and precision of forest statistics vary considerably from country to country. Applying well established methods (Lindeberg, 1923; Langsaeter, 1926; Loetsch *et al*, 1973), growing stock and annual increment were estimated over large areas, e.g. on a national scale. Temporary sampling plots were located in a grid covering the study area, and accurate measurements were made on those plots. Currently, at least one national forest inventory has been carried out in most European countries, though increment measurements have not been taken in all inventories.

Permanent sample plots have allowed reliable estimates of mortality (natural loss) to be made. Both temporary and permanent plots have been included in inventories, e.g. in Austria, Finland, Sweden and Switzerland. The use of sampling with partial replacement (Ware and Cunia, 1960) has thus been possible.

As national inventories have provided information relating to large areas, another inventory system, standwise inventory, has been maintained in many European countries for operational management of forestry. This system is costly and involves subjective

Table 1.   Annual fluxes of carbon in removal (C_{rm}), into detritus (C_{dt}), into living biomass (C_{st}), within forestry in total (C_{fr}), and in energy-related CO_2 emissions (C_{fos}).

Country/Region	C_{rm}	C_{dt}	C_{st}	C_{fr}	C_{fos}
			10^{12} g a^{-1}		
Austria	3.3	1.5–4.0	1.9–2.8	6.7–10.1	15.6
Finland	10.4	5.6–13.6	4.5–6.7	20.5–30.7	16.0
Norway	2.4	1.3–3.2	1.9–2.8	5.6–8.4	8.7
Sweden	11.4	5.8–14.4	10.6–15.9	27.8–41.7	15.1
Switzerland	1.1	0.5–1.3	0.2–0.3	1.8–2.7	12.1
EFTA	29	15–36	19–28	62–94	68
Belgium	0.7[*]	0.3–0.8[*]	0.3–0.5[*]	1.3–2.0[*]	33.8
Denmark	0.4	0.3–0.6	0.3–0.5	1.0–1.6	15.3
France	10.4	3.9–11.1	5.5–8.3	19.8–29.7	104.7
Germany	9.4[*]	5.4–12.9[*]	4.9–7.3[*]	19.7–29.6[*]	283.4
Greece	0.6	0.7–1.2	0.1	1.3–2.0	22.1
Ireland	0.3	0.1–0.3	0.5–0.8	0.9–1.4	9.0
Italy	1.6[*]	1.2–2.6[*]	2.6–3.9[*]	5.4–8.1[*]	112.1
Luxembourg	0.06[*]	0.05–0.1[*]	0.1[*]	0.2–0.3[*]	2.8
Netherlands	0.3	0.09–0.3	0.3–0.5	0.7–1.1	49.9
Portugal	2.2	0.9–2.5	0.3–0.4	3.3–5.0	11.7
Spain	3.7	2.2–5.2	4.1–6.3	10.1–15.1	61.2
United Kingdom	1.5	0.9–2.0	0.8–1.2	3.1–4.7	160.6
EC	31	16–40	20–30	67–100	870
EFTA+EC	60	31–76	39–58	130–194	930

[*] Gross annual increment was estimated comparable to that reported in the neighbouring countries in cases when data were missing.

assessment and field work which can result in large errors when the data are aggregated to describe large areas. New technology has recently been introduced in order to combine the two inventory systems and to provide both statistically reliable data and relevant information within small regions. The Finnish National Forest Inventory, for example, employs satellite image data and digital map data in addition to ground measurements on sample plots (Tomppo, 1991). Figure 2 shows an example of an output thematic map. The technology, including space- and airborne remote sensing, both in optical and micro-wave regions, is developing rapidly and also shows promise for estimating C storages. Forthcoming, more precise and accurate surveys may alter some of the figures in Table 1, notably those for countries like Germany and Italy, where the estimates of gross annual increment have not been derived from a statistically representative sample.

Soil C, a large reservoir, was assumed stable in this analysis. In other words, it was assumed that the decomposition of organic matter in the humus layer and in soil fully compensated for the formation of dead organic matter. Lugo (1992) has criticized such steady state assumptions. In Europe, it is possible that the pool of C in forest soils is increasing rather than decreasing. Although prescribed burning, silvicultural site prepara-tion and drainage of peatlands (Gorham, 1991) tend to decrease C storage in soils, other and probably more powerful trends, such as improved fire control and the reduction of cattle grazing in forest have the opposite effect. Differences can occur between countries in terms of changes of C pool in soils. In Finland, for example, a large program has been carried out to drain peatlands for timber production. Decomposition of organic soil on drained peatlands can have a significant effect on net national CO_2 emissions (however, see Laine et al, 1992).

3.2. POLICY IMPLICATIONS

Given the work of international programs and conferences such as IPCC and UNCED, the concept of net national emisssions is becoming increasingly important in developing and evaluating policy alternatives. Net national emissions determine the impact of each individual country on the concentration of CO_2 in the atmosphere. Each country is com-mitted according to the Climate Convention to report and periodically update the "emis-sions by sources and removals by sinks". National emission targets are negotiated. Each country has the freedom to choose cost-efficient strategies for emission control and to take into account the country-specific infrastructure affecting "emissions and removals".

Using forests to sequester carbon dioxide from the atmosphere has long been discussed (Dyson, 1977). In western Europe in the 1980s, the total annual flux of C in forests and forest products was 14 to 20 % of the C flux from the use of fossil fuels. Although forests affect the C budget of the region, measures within the forest sector cannot be used as the main solution eventually to reduce C emissions. The C flux through forests is simply too small. However, in Sweden and Finland, taken together, the C flux within the forest sector is larger than the flux from fossil fuels by a factor of 1.6 to 2.3. In the Netherlands and the UK the domestic C flux within forest sector is only 2 to 3 % of the C releases from fossil fuels. The potential of the forest sector to affect the net national emissions of CO_2 thus varies by as much as two orders of magnitude between countries in industrial western Europe. Such differences between countries can have profound im-pacts on national strategies for controling CO_2 emisssions.

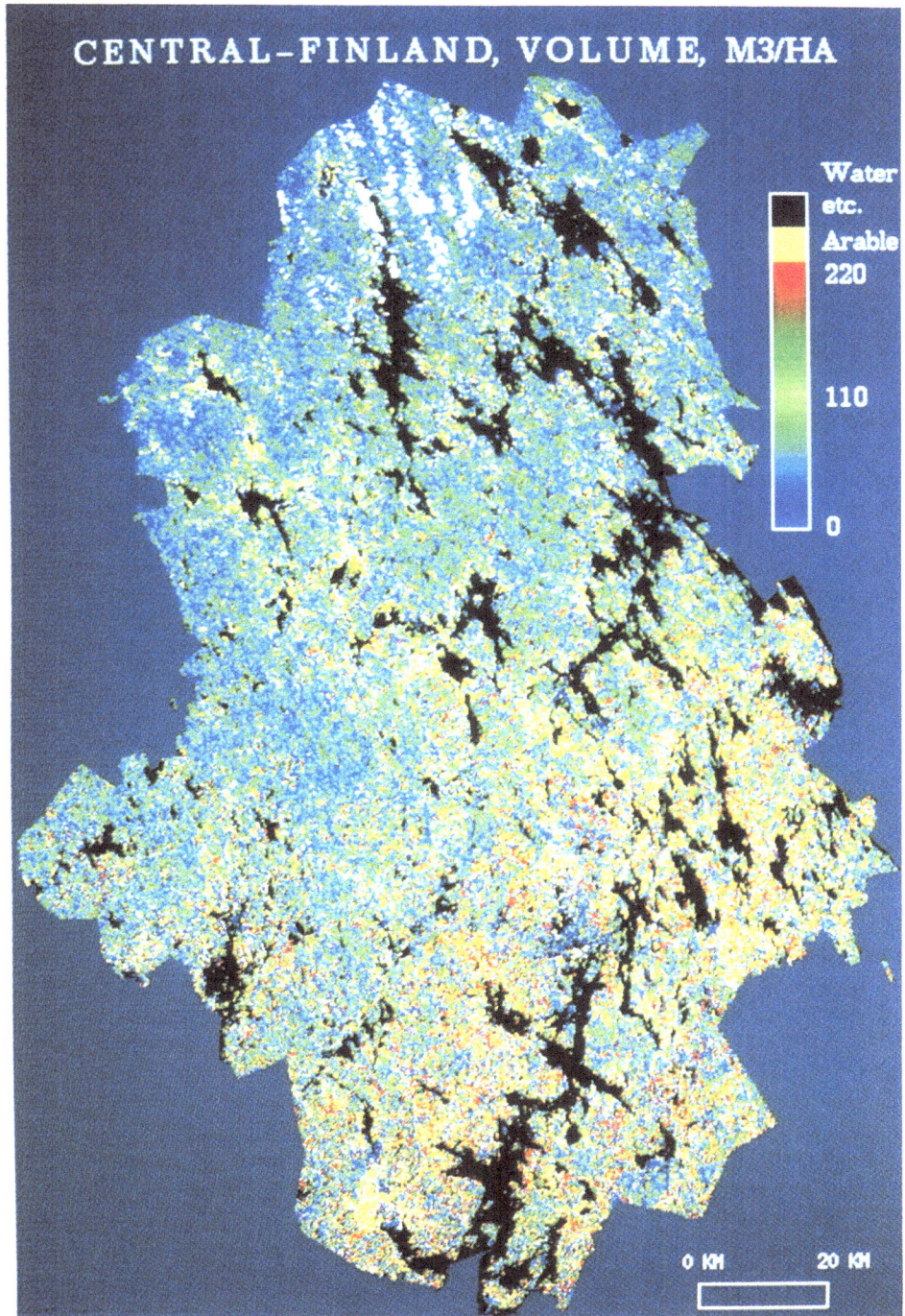

Fig 2. Thematic map of forest inventory results.

Many forestry policies are available for controlling the net national emissions of CO_2. Fire control, drainage, site preparation practices, and game management affect C pools and fluxes in forest soils, but the quantification of these effects is difficult at the present time. It is easier to evaluate measures that affect C pools and fluxes in living biomass and in the biomass harvested and removed.

First, it is possible to plant new forests on agricultural or other land, which would increase the C pool in living biomass and soil. On the global scale, it has been calculated that afforestation of 4.56×10^6 km^2 could stabilize the CO_2 concentration in the atmosphere over a time period of several decades (Sedjo, 1989). Afforestation can be an attractive policy in central and southern Europe where large areas of potential forest land have historically been cleared for agriculture. In northern Europe, afforestation has less potential because most of the land is forest anyway.

Secondly, growing stock can be increased on a given area of forest land. Growing stock per unit of forest land has increased in most industrial countries. This development has been widely ignored although in many countries it has affected the national C budget more strongly than afforestation. Underlying factor has been the low harvesting rate. Only 65 to 80 % of the net annual increment has been harvested (The Forest Resources of the Temperate Zone..., 1992). The growing stock has increased in all west European countries, including countries such as Austria, Finland and Sweden where the data are reliable and there has been very little afforestation. Natural losses (mortality of trees) have been insignificant, and the difference between increment and harvest has accumulated in forests almost entirely as living biomass. This development has been unplanned and undesired. The intention in Austria, Finland and Sweden has been to maintain the amount of living biomass approximately at the present level, and to match harvest with the net increment. However, low wood demand has prevented the full utilization of the sustained yield.

It would be possible actively to stimulate the trend and further to increase the C pool in living biomass. This policy could be effective in areas where the present growing stock is low compared to the site potential. For example, relatively short rotation cycles are applied in the existing plantations in the UK and in Ireland. There is potential in these forests for building up the growing stock. In contrast, there is only a small potential of increasing the growing stock per hectare of forest land in Switzerland and Germany. The average growing stock in these countries is approaching the potential upper limit.

Building up the C pool of existing forests is beneficial in certain cases. There is a need in western Europe to reserve new forest areas for nature protection. Such areas would support higher biomass levels than areas managed for maximum sustained yield (Cooper, 1983). However, it is questionable to aim at increasing biomass levels as a general policy in all forests. An increase of standing stock will lead to a reduction of net annual increment in the long term, implying less potential for developing the use of biomass as a renewable source of raw material and energy. Building up living biomass in current European forests can decrease the national net emissions of CO_2 in the short term (<50 yr), but tends to increase the emissions in the long term (>50 yr).

A *third* option for controling net national emissions is the management of removed biomass. Changes in favor of long-lasting products and recycling would increase the C storage bound in products. Bioenergy applications such as using waste paper to substitute fossil fuels can contribute to a decrease of net CO_2 emissions. The disposal of waste paper and other used products in landfill sites or in abandoned mines can delay or even prevent decomposition and contribute to C sequestration. Due consideration is needed for the potential increase of NH_3 emissions.

Proper management of the removed biomass should be recognized as an important policy alternative for controling the net national emissions of CO_2. In EFTA+EC, the present annual C flux associated with timber removal is 60 Tg C yr^{-1}. It determines the present upper limit which can be sequestered with the management of the removed biomass. A theoretical upper limit is higher. The present harvest rate has been low compared to the sustainable yield. New plantations, in addition to increasing the C pool in biomass and soils, would add to the removal. Assuming an increase of forest area by 5 to 30 %, slight improvement of the present average yield, and the full use of sustained yield, the upper limit in EFTA+EC is 100 to 200 Tg yr^{-1} of sequestering C by means of the management of removed biomass. This is 10 to 20 % of the present emisions from fossil fuels.

A combination of afforestation, forest protection and the managemant of removed biomass could be developed to control net national emissions of CO_2. According to our estimate, the sustainable potential to be achieved ranges between 150 and 250 Tg C annually in western Europe in the long term. This is higher than the present sequestration rate by a factor of 2 to 4. The management of the removed biomass would account for more than half of the estimated potential. An eventual climate change, the CO_2 stimulation of phosysthesis and growth, and other factors not included in this analysis generate large uncertainty to these projections.

The accumulation of living biomass accounts for a sequestration of 39 to 58 Tg C yr^{-1} in western Europe at the present time. This flux is mainly due to the build-up of biomass in existing middle-aged and mature stands. Accumulation of biomass can continue temporarily but is unsustainable in the long term. Therefore, this sink is expected to disappear. Afforestation and controlling the C flux associated with removal could substitute this temporary sink and have a more long-term impact.

4. References

Apps, M. and Kurz, W.: 1991, World Resource Review 3, 333.

Brown, S. and Lugo, A.E.: 1984, *Science* **223**, 1290.

Cooper, C.F.: 1983, *Can. J. For. Res.* **13**, 155.

Dewar, R.C. and Cannell,M.G.R.: 1992, *Tree Physiol.* **11**, 49.

Dyson, F.J.: 1977, *Energy* **2**, 287.

Gorham, E.: 1991, *Ecological Applications* **1**, 182.

Hakkila, P.: 1989,*Utilization of Residual Forest Biomass,* Springer-Verlag, Berlin, Heidelberg, New York.

Harmon, M.E., Ferrell, W.K. and Franklin J.F.: 1990, *Science* **247**, 699.

IEA/OECD Energy and the Environment Series: 1992, Climate Change Policy Initiatives.

Kandler, O.: 1992. *Environmental Toxicology and Chemistry* **11**, 1077.

Kauppi, P.E., Mielikäinen, and Kuusela, K.: 1992, *Science* **256**, 70.

Kurz, W. and Apps, M.: 1993, This volume.

Laine, J, Wasander, H. and Puhalainen, A.: 1992, Effect of Forest Drainage on the Carbon Balance of Mire Ecosystems, in :Proceedings of the 9th International Peat Congress.

Langsaeter, A.: 1926, *Medd. Norske Skogsforsöksv.* **27**, 5.

Lindeberg, J.: 1923, *Acta Forestalia Fennica* **25**, 1.

Loetsch, F., Zöhrer, F. and Haller, K.E.: 1973, *Forest Inventory*, BLV Verlagsgesellschaft, München, Bern, Wien, vol. 1–2.

Lugo, A.E.: 1992, The Search for Carbon Sinks in the Tropics, in Wisniewski J. and Lugo A.E. (eds), Natural Sinks of CO2, Kluwer Academic Publishers, pp. 3–9.

Sedjo, R.A.: 1989, *J. Forestry* **87** (7), 12 .

Sedjo, R.A.: 1992, *Ambio* **21**, 274.

Tans, P.P., Fung, I.Y., and Takahashi, T.: 1990, *Science* **247**, 1431.

Tolba, M.K., El-Kholy, O.A., El-Hinnawi. E., Holdgate, M.W., McMichael, D.F., and Munn, R.E. (eds) : 1992, The World Environment 1972–1992. UNEP Chapman & Hall, London.

Tomppo, E.: 1991, *Int. Archives of Photogrammetry and Remote Sensing* **28**, 419.

The Forest Resources of the Temperate Zones, Main Findings of the UN-ECE/FAO 1990 Forest Resource Assessment, ECE/TIM/60, UN Publications Sales No. E.92.II.E.23 (1992); The Forest Resources of the Temperate Zones, The UN- ECE/FAO 1990 Forest Resource Assessment, Vol 1, General Forest Resource Information, ECE/TIM/62, UN Publication Sales No. E.92.II.E.27 (1992).

The Forest Resources of the ECE region (Europe, the USSR, North America), UN-FAO/ECE, ECE/TIM/27 (1985).

UN Framework Convention on Climate Change. United Nations (1992).

Ware, K.D. and Cunia, T.: 1960, *For. Science-Monograph* **3**, 1.

Acknowledgments:

We thank J. Alcamo, P. Hari, K. Kuusela, P. Nöjd and P. Stenberg for comments, and M. S. Jarvis for editing the language.

THE POTENTIAL ABOVEGROUND CARBON STORAGE
OF NORTH AMERICAN FORESTS

Lloyd G. Simpson
Department of Biological Sciences, University of California
Santa Barbara, CA, 93106-9610, USA

Daniel B. Botkin
Program on Global Change, George Mason University
Fairfax, VA, 22030, USA

and

Robert A. Nisbet
Department of Biological Sciences, University of California
Santa Barbara, CA, 93106-9610, USA

Abstract. To assess the possibility of using C offset as a method of sequestering CO_2 produced by the burning of fossil fuels, it is necessary to have accurate estimates of C reservoirs and fluxes. Recent studies have shown that estimates of C commonly used in the past are too large, and this may lead to confusion about the global C budget. Field data used in recent estimates of present C storage for the North American boreal and eastern deciduous forest biomes were reanalyzed to estimate their maximum potential C storage. The original data were collected using a stratified two-stage cluster survey sampling design. The reanalysis suggests that the boreal forest and eastern deciduous forest could sequester possibly as little as 13.4% (3.0 Pg) and 18.5% (1.5 Pg), respectively, more C than they presently store. These estimates represent the potential increase in C storage under present conditions, if the study areas were allowed to revert back into forests.

1. Introduction

There is much discussion today about the possibility of using the technique of "carbon offset" as a partial solution to the rapid buildup of CO_2 in the atmosphere, and the possibility of global warming. Carbon offset means that producers of CO_2 (e.g., nations dependent on fossil fuels; industries producing automobiles) would pay a fee to be used to promote forest growth, thus using trees to sequester C from the atmosphere. While the idea of C offset is

Water, Air, and Soil Pollution **70**: 197–205, 1993.
© 1993 *Kluwer Academic Publishers.*

appealing, there are several problems and gaps in our knowledge that we must deal with before leaping into large scale expenditures for C offset in forests. Rosenfeld and Botkin (1990) have suggested that it is not cost-effective to invest in planting small trees if large-scale deforestation of mature forests is proceeding unchecked. They estimate that, to a first approximation, one would have to plant 100 ha of new forest to offset the destruction of every ha of mature forests. A program in C offset that looked only at planting and ignored net C storage change would likely fail.

A C offset project must be soundly based on valid information so that a legitimate economic value can be established. It is necessary to know how much C is likely to be sequestered per year for each ha of forest land set aside. Only with this information can acceptable prices be established for a C offset program. If economic value is placed on a unit area of land without this information, then a false valuation will be made, most likely overestimating the value of the land for C storage. This will lead to false expectations that will not be realized, and may unnecessarily discourage future efforts to sequester C. There are three major questions that must be answered before success can be assured: (1) How much C can be stored in a given forest even under the best of conditions? (2) What is the rate of C uptake associated with that forest? (3) What is the likely fate of stored C under human-induced global change, such as might occur from global warming and is occurring now from the deforestation of large areas?

Recognizing the need for information that will answer these questions, we initiated a program to obtain statistically reliable estimates of aboveground C storage on a large scale several years ago. In the past, analyses of the global C cycle and C budget used estimates of C density found in the literature that were not designed to represent large areas. These were the only numbers available, and necessity dictated their use. However, times have changed, and there is genuine concern about the long term effects of increased CO_2 accumulations in the atmosphere. To truly understand the threat and develop methods of alleviating it, new more rigorous measures that have a reasonable degree of certainty associated with them are needed.

Our early work in the North American boreal forest showed that the C stored in above ground vegetation was 3 to 4 times lower that previously thought (Botkin and Simpson, 1990a,b). These initial results lead to the hypothesis that the above ground C storage of forests worldwide is overestimated. We tested this hypothesis in the eastern deciduous forest of North America, one of the most highly studied forest ecosystems in the world, and found that our estimates were again lower than those in common use today (Botkin et al., 1993). In this paper we present results that begin to answer the first question listed above by using our recent estimates of present C storage in the North American boreal and eastern deciduous forests to estimate their potential C storage.

2. Methods

2.1   SAMPLE DESIGN AND SELECTION

The sample design for this study is presented in detail in Botkin and Simpson (1990a,b) and Botkin *et al*. (1993). Briefly, we used a stratified two-stage cluster design of survey sampling to estimate total C storage and C density. Survey sampling has been used for many years to estimate the crop yields and forest volumes. It permits an accurate estimation of parameters using an extremely small sample size. Study areas were defined using environmental parameters, and the areas were then stratified for sampling as shown in Figure 1.

A map of the strata was input into Earth Resources Data Analysis System (ERDAS) geographic information system (GIS), from which locations of sample clusters were selected at random. Once the strata map was entered into the GIS, it was converted to an Alber's equal area projection, and primary sampling units (PSUs), 24 x 24 km in size, were selected from the computer screen within each stratum using a table of randomly generated screen coordinates. Once selected, the longitude and latitude of the northwest corner of each PSU were recorded. The number of PSUs allocated to a stratum was proportional to its size, with at least two PSUs selected from each stratum to obtain an unbiased estimate of the variance. PSUs were then located and marked on Canadian and USGS topographic maps. Four secondary sampling units (SSUs) were selected and marked on the map using a table of randomly generated Universal Transverse Mercator (UTM) coordinates. Each SSU consisted of five 20 m diameter subplots, with one subplot located in the center and four located tangentially in the cardinal directions. A 2 m diameter understory plot was established at the center of each subplot. An SSU was omitted if it occurred over a lake or river. No other consideration such as a clearing or exposed bedrock eliminated an SSU, since our goal was to estimate the present C content for entire strata.

2.2   FIELD METHODS

Each SSU was located on the ground and visited using topographic maps and aerial photographs. At each SSU, the following data were measured: 1.) For all trees ≥ 2 cm dbh (diameter at breast height -- 1.37 m above the ground) - dbh, total height, and species were recorded in each 20 m subplot. 2.) For all tree seedlings and saplings > 2 cm dbh and shrubs in the understory plots - stem diameter at the base, stem diameter at 15 cm above the ground, and species were recorded in each shrub plot. 3.) Data about site conditions were also collected at each plot including: slope, aspect, topographic position, and type and degree of disturbance.

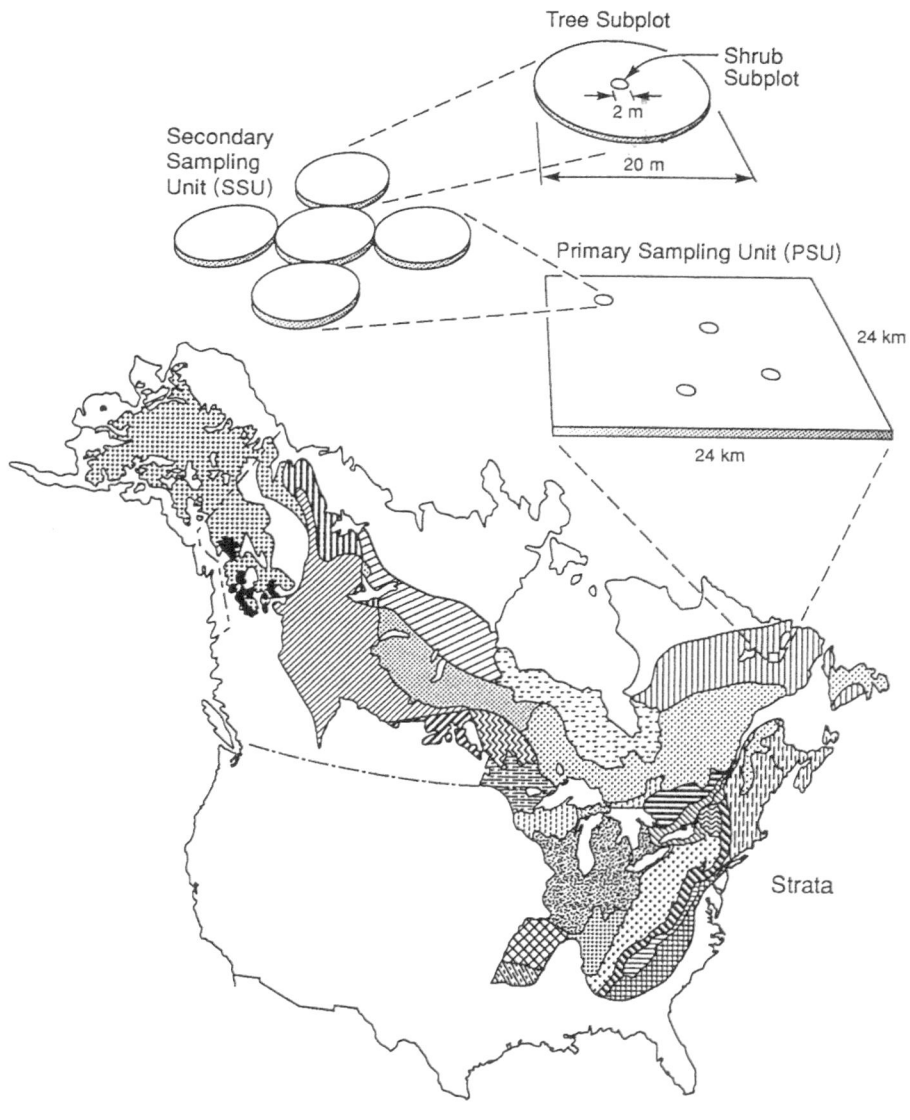

Figure 1. Sampling design used in the estimation of biomass and C storage for the boreal and eastern deciduous forests of North America.

2.3   DATA ANALYSIS

Total aboveground, ovendry woody biomass was calculated for each secondary sampling unit using species specific dimension-analysis equations developed by the US Forest Service and Forestry Canada for most trees (Taras and Clark, 1974, 1977; Taras and Phillips, 1978; Saucier and Boyd, 1982; Evert, 1985; Standish *et al.*, 1985; Clark *et al.*, 1986a, b; Clark and Schroeder, 1986) and most shrubs (Stanek and State, 1978; Ribe, 1979; Smith and Brand, 1983). When no equation was available for a species, an equation from a suitable analog species was used.

Carbon density for an SSU was calculated by dividing the sum of the ovendry biomass of all the trees or shrubs in a plot by the plot area for both the overstory and the understory, and multiplying the result by 0.45. Mean C density (t C ha^{-1}), total C, and their 95% error bounds were calculated with a set of survey-sampling equations derived for the sample design (Yamane, 1967; Botkin *et al.*, 1993).

The potential additional C storage of the boreal and deciduous forests was estimated using two methods:

1.) Potential maximum C storage was calculated by eliminating SSUs with recent anthropogenic disturbance (within the last 10 to 15 yr) or SSUs that had been permanently cleared from the data set. This approach assumes that forests were subject to natural disturbances from fire, windstorms, disease, insect outbreaks, etc. It also assumes that there would be spatial variation in soils, topography, and microclimate that would lead to a mosaic of stands similar to those of today, with a variety C storage at any one time distributed over the landscape. Less than 10% of the SSUs in the boreal forest and 30% of the SSUs in the deciduous forest had no woody vegetation on them and were located in areas that had been recently logged or permanently cleared for agriculture.

2.) Potential maximum C storage was estimated by selecting the SSU with the highest C density from each PSU, and estimating biomass and its error bound using equations based on a stratified random design (Yamane, 1967: pp. 105-113). This approach assumes that, within each PSU, forests were at a classical ecological climax stage -- the forest was homogeneous and at maximum biomass. This assumption is embedded in many global scale estimates of C storage, though few if any scientists believe it to be true today, and there is ample evidence against it (Botkin, 1990; Botkin and Simpson, 1990a,b; Botkin and Sobel, 1975; Hall *et al.*, 1991).

3. Results

The estimate of present aboveground C storage for the boreal forest and deciduous forest combined (735,609,600 ha) is 17.6 ± 2.7 Pg (Table 1). The mean C density for the combined areas is 24.1 ± 3.7 t C ha^{-1}. The estimate of potential maximum C storage using method 1 is 11.0 ± 2.5 Pg for the boreal forest, and 9.6 ± 1.4 Pg for the deciduous forest, and for the combined areas is 20.6 Pg. These values are not significantly different from the estimate of present carbon storage. The potential maximum C storage using method 2 for the boreal forest is estimated to be 18.9 ± 4.6 Pg, and for the deciduous forest, 16.1 ± 2.6 Pg. The estimate for the combined forest areas using method 2 is 34.1 ± 5.3 Pg. The potential increase in C storage for the combined areas is between 3.0 and 16.5 Pg based on the difference between present C storage and the two methods of estimating maximum potential C storage.

Table 1. Present and Potential Maximum C storage in the North American boreal and eastern deciduous forests estimated by two different methods.

| | Present | | Potential Maximum | | | |
| | | | Method 1 | | Method 2 | |
Forest Biome	Density (t C ha^{-1})	Total C (Pg)	Density (t C ha^{-1})	Total C (Pg)	Density (t C ha^{-1})	Total C (Pg)
Boreal[§]	18.8 ± 4.5	9.7 ± 2.3	21.2 ± 4.6	11.0 ± 2.5	37.0 ± 9.1	18.9 ± 4.6
Deciduous[†]	36.2 ± 6.2	8.1 ± 1.4	43.2 ± 7.5	9.6 ± 1.4	67.7 ± 11.0	15.1 ± 2.5
Combined[‡]	24.1 ± 3.7	17.6 ± 2.7	28.0[¥]	20.6[¥]	46.3 ± 7.1	34.1 ± 5.3

[§] 512,642,700 ha
[†] 223,027,200 ha
[‡] 735,669,900 ha
[¥] This estimate does not include an error estimate because the data sets were analyzed using different computer programs with different data structures. It was not possible to combine the data for an estimate of the error.

4. Discussion

The estimate of present C storage calculated from field data, has been shown to be dramatically smaller than previously thought (Apps et al., this volume; Bonnor, 1985; Botkin and Simpson, 1990a,b; Botkin et al., 1993; Kurz et al., 1992). It follows that the maximum potential is also smaller than was previously thought. Previous studies suggest that total C stored at present in these two North American biomes is 57 Pg, calculated from earlier C storage densities (e.g. Houghton et al., 1983; 135 t C ha^{-1} for the deciduous forest and 90 t C ha^{-1} for the boreal forest), applied to the areas of our studies (735,669,900 ha).

The estimate based on method 1 is more appropriate for estimating the potential maximum C storage of the boreal and deciduous forests. This means that the maximum potential C storage for the combined area is 20.6 Pg or an increase of 2.8 Pg over the present C storage. The reason for the use of method 1 for the two North American biomes is based largely on the natural heterogeneous state of both forests. We are reluctant to use the results of method 2, because it implies unrealistically that the forests are homogeneous. The utility of method 2 is simply that it gives the largest values possible based on our sample.

The boreal forest is the largest on the continent, and the vast majority of what we have defined as boreal forest is unaccessible to human exploitation (Bonnor, 1985). The estimate of maximum C storage of the boreal forest based on method 2 is not significantly different from the value that can be calculated using values for "mature" forests from data presented by Forestry Canada (Bonnor, 1985; Kurz et al., 1992). The term "mature" carries with it the implicit assumption that the forest is at an undisturbed maximum. However, the boreal forest is in fact a large and heterogeneous mosaic, composed of a patchwork of vegetation stands in various successional stages (Hall et al., 1991). This has always been the natural state of the forest. It is therefore, inappropriate to use an estimate based on the highest amount of biomass found in a given PSU. Results of a recently developed C budget model using an extensive data set suggest that the boreal forest of Canada may be a net C sink (Kurz et al., 1992). This is due primarily to the fact that the vast majority of disturbed forest in the boreal region is subject to natural disturbances and is composed of younger, faster growing stands. Kurz et al. (1992) suggest that the general structure of the boreal forest is skewed toward younger more vigorous stands because of past disturbance, both natural and anthropogenic. In the data set used in this and our earlier studies of the boreal forest, 10.5% of the SSUs had been recently (within the last 10 yr) burned (Botkin and Simpson, 1990a). These fires were all natural fires located, for the most part, in areas away from human settlement. Only 8.5% of the SSUs had signs of recent human disturbance or permanent clearing.

The temperate deciduous forest region of eastern North America is also heterogeneous, and there is evidence to suggest that it always has been so. Day (1953) suggests that fires were commonly set by Native Americans to aid in hunting, management wildlife habitat, and to make travel easier. These fires were in addition to natural fires and hurricanes that resulted in large areas of blow down, especially near the coasts. The present heterogeneity of the deciduous forest region is due primarily to human land clearing over the last 200 yr, coupled with human induced fire suppression. Windstorms, intrinsic edaphic variation, and natural forest dynamics are also important. Over 30% of the SSUs in our data set were devoid of woody vegetation due to human impacts. Thus anthropogenic effects are more significant in this biome than in the boreal biome. All the forest area in this biome is accessible to exploitation, and therefore it is reasonable to assume that virtually all of it has been affected by human activities.

The estimates of maximum C storage presented in this paper suggest that, if conditions remain the same and the region is allowed to revert to forest, the boreal forest would increase C storage by 13.4%, and the deciduous forest would increase C storage by 18.5%. These estimates can serve as an upper limit of C storage that could be expected from these areas for use in studies of the global C cycle and budget. However, because of land use changes that have caused decreased soil productivity, especially in the deciduous forest biome, the maximum C storage might be less than would have occurred in presettlement times or might occur at some time in the future if soils recover. An intensive program of soil restoration coupled with tree planting and forest conservation programs could lead to higher C storage than is predicted by this study.

5. Conclusions

There has been much speculation about the location of the "missing C" in the global C budget, but there are few data to support any theory. Indeed, it is the lack of accurate data about C reservoirs and fluxes that has resulted in the "missing C" phenomenon. Accurate estimates of actual and potential C storage may help to resolve the question about the location of the "missing C" and other questions related to the global C budget and the global C cycle.

6. References

Bonnor G. M.: 1985, Inventory of forest biomass in Canada, Petawawa Nat. For. Inst., Can. For. Serv. Petawawa, Ontario, Canada.

Botkin D. B. 1990. *Discordant Harmonies: A New Ecology for the 21st Century*. Oxford University Press, New York.

Botkin D. B. and Simpson L. G.: 1990a, *Biogeochem.* **9**, 161.

Botkin D. B. and Simpson L. G.: 1990b, The distribution of biomass in the North American boreal forest. in: *Global Natural Resource Monitoring and Assessments: preparing for the 21st century*, Cini F. G. (ed), Proceedings of the international conference and workshop, Am. Soc. for Photogram. and Remote Sens., Washington D. C. pp. 1036-1045.

Botkin D. B., Simpson L. G., and Nisbet R. A.: 1993, *Biogeochem.*, In Press.

Botkin D. B. and Sobel M. J.: 1975, *American Naturalist* **109**, 625.

Clark III A., Phillips D. R., and Frederick D. J.: 1986a, Weight , volume, and physical properties of major hardwood species in the Piedmont, USDA Forest Service Tech. Rpt. SE-255

Clark III A., Phillips D. R., and Frederick D. J.: 1986b, Weight , volume, and physical properties of major hardwood species in the upland-south, USDA

Forest Service Tech. Rpt. SE-257

Clark III A. and Schroeder J. G.: 1986, Weight , volume, and physical properties of major hardwood species in the southern Appalachian Mountains, USDA Forest Service Tech. Rpt. SE-253

Day G. M.: 1953, *Ecology* **34**,329.

Evert F.: 1985, Systems of equations for estimating ovendry mass of 18 Canadian tree species Petawawa Nat. For. Institute, Can. For. Serv., Info. Rpt. Pl-X-59.

Hall F. G., Botkin D. B., Strebel D. E., Woods K. D., and Goetz S. J.: 1991, *Ecology* **72**, 628.

Houghton R. A., Hobbie J. E., Melillo J. M., Moore B., Peterson B. J., Shaver G. R., and Woodwell G. M.: 1983, *Ecol. Mono.* **53**, 235.

Kurz W. A., Apps M. J., Webb T. M. , and McNamee P. J.: 1992, The carbon budget of the Canadian forest sector: Phase I, Forestry Canada Infor. Rpt. NOR-X-326.

Ribe J. H.: 1979, A study of multi-stage sampling and dimensional analysis of puckerbrush stands, The Complete Tree Institute, Univ. of Maine, Orono Bull. 1.

Rosenfeld A. H., and Botkin D.B.: 1990, *Physics and Society* **19**, 4.

Saucier J. R. and Boyd J. A.: 1982, Above ground biomass cf Virginia pine in north Georgia, USDA Forest Service Res Pap. SE-323.

Smith, W. B. and Brand G. J.: 1983, Allometric biomass equations for 98 species od herbs, shrubs and small trees, N. Central For. Exp. Sta., St. Paul, Minn. Res. Note NC-299.

Standish J. T., Manning G. H., and Demaerschalk J. P.: 1985, Development of biomass equations for British Columbia tree species, Forestry Canada infor. Rpt. NOR-X-264.

Stanek W. and State D.: 1978, Equations predicting primary productivity (biomass) of trees, shrubs, and lesser vegetation based on current literature, Can. For. Serv., Pacific For Res. Cen. Victoria, B.C. Infor. Rep. BC-X-183.

Taras M. A. and Clark III, A.: 1974, *Tappi* **58**, 103.

Taras M. A. and Clark III A.: 1977, Aboveground biomass of longleaf pine in a natural sawtimber stand in southern Alabama, USDA Forest Service Res. Pap. SE-162.

Taras M. A. and Phillips D. R.: 1978, Aboveground biomass of slash pine in a natural sawtimber stand in southern Alabama, USDA Forest Service Res. Pap. SE-188.

Yamane T.: 1967, Elementary Sampling Theory,. Prentice Hall, Englewood, N.J.

COMPARISON OF TWO METHODS TO ASSESS THE CARBON BUDGET OF FOREST BIOMES IN THE FORMER SOVIET UNION

TATYANA P. KOLCHUGINA and TED S. VINSON
Oregon State University, Corvallis, Oregon, 97331
U.S.A

Abstract. The sink of CO_2 and the C budget of forest biomes of the Former Soviet Union (FSU) were assessed with two distinct methods: (1) ecosystem/ecoregional, and (2) forest statistical data. The ecosystem/ecoregional method was based on the integration of ecoregions (defined with a GIS analysis of several maps) with soil/vegetation C data bases. The forest statistical approach was based on data on growing stock, annual increment of timber, and FSU yield tables.

Applying the ecosystem/ecoregional method, the area of forest biomes in the FSU was estimated at 1426.1 Mha (10^6 ha); forest ecosystems comprised 799.9 Mha, non-forest ecosystems and arable land comprised 506.1 and 119.9 Mha, respectively. The FSU forested area was 28% of the global area of closed forests. Forest phytomass (i.e., live plant mass), mortmass (i.e., coarse woody debris), total forest plant mass, and net increment in vegetation (NIV) were estimated at 57.9 t C ha^{-1}, 15.5 t C ha^{-1}, 73.4 t C ha^{-1}, and 1.0 t C ha^{-1} yr^{-1}, respectively. The 799.9 Mha area of forest ecosystems calculated in the ecosystem/ecoregional method was close to the 814.2 Mha reported in the FSU forest statistical data. Based on forest statistical data forest phytomass was estimated at 62.7 t C ha^{-1}, mortmass at 37.6 t C ha^{-1}; thus the total forest plant mass C pool was 100.3 t C ha^{-1}. The NIV was estimated at 1.1 t C ha^{-1} yr^{-1}. These estimates compared well with the estimates for phytomass, total forest plant mass, and NIV obtained from the ecosystem/ecoregional method. Mortmass estimated from the forest statistical data method exceeded the estimate based on the ecosystem/ecoregional method by a factor of 2.4. The ecosystem/ecoregional method allowed the estimation of litter, soil organic matter, NPP (net primary productivity), foliage formation, total and stable soil organic matter accumulation, and peat accumulation (13.9 t C ha^{-1}, 125.0 t C ha^{-1}, 3.1 t C ha^{-1} yr^{-1}, 1.4 t C ha^{-1} yr^{-1}, 0.11, and 0.056 t C ha^{-1} yr^{-1}, respectively). Based on an average value of NEP (net ecosystem productivity) from the two methods, and following a consideration of anthropogenic influences, FSU forests were estimated to be a net sink of approximately 0.5 Gt C yr^{-1} of atmospheric C.

Water, Air, and Soil Pollution **70**: 207–221, 1993.
© 1993 *Kluwer Academic Publishers.*

1. Introduction

Forests play an important role in the global C cycle. Forest vegetation contains more than 80% of the C stored in terrestrial vegetation and forest soils contain more than 70% of the world's soil C pool (Post *et al.*, 1982; Olson *et al.*, 1983). The role of forests in the global C cycle may be assessed only if existing C pools and fluxes are quantified, i.e., their C budget is established. Carbon sinks in forest ecosystems are associated with C sequestration in growing vegetation, soils, and peatlands. Sources of C in forest biomes are primarily associated with forest fires and anthropogenic disturbances (mainly wood harvesting).

The forests of the FSU account for approximately 30% of the world's total forested area; growing stock amounts to 24% of the world resources (Alimov *et al.*, 1989). Most of the forests (95%) are in Russia (USSR State Forestry Committee, 1990). The character and magnitude of C sources and sinks which reflect the contributions of both natural ecosystems and the industrial sector remain uncertain.

Two distinct methods may be used to assess C accumulation parameters of forest ecosystems. The ecosystem/ecoregional method is based on information on C cycling in different ecosystems and the use of maps that allows one to estimate the areas of forest ecosystems. The forest statistical data method is based on information collected under a national forest inventory. The advantage of the ecosystem/ecoregional method is that it is based on data collected for all components of an ecosystem ,e.g., tree stems, leaves, branches, roots, understory, grass, soils, etc. In contrast, forest statistical data relate mainly to that part of the ecosystem which is associated with wood products. Brown *et al.* (1989) state that the advantage of the forest statistical method is that the data are based on "an abundant sample area," whereas ecosystem data result from a "limited number of experimental plots".

The purpose of the study presented herein is to assess the C budget of forest biomes in the FSU using two distinct methods: 1) ecosystem/ecoregional, and 2) forest statistical data.

2. Ecosystem/Ecoregional Method

2.1. METHODOLOGY

The ecosystem/ecoregional method was implemented in two stages. In the first stage average densities of C accumulation parameters in vegetation and soil were estimated based on the assumption that C uptake by forest ecosystems was balanced by the heterotrophic respiration. In the second stage the present rate of net C sequestration in forest ecosystems was determined by considering the successional stage of FSU forest ecosystems. Further, the area estimates of forest ecosystems and the vegetation C pool were improved by incorporating data on actual forest coverage and the relative phytomass densities of young and established forest ecosystems.

2.1.1. First stage

The C budget of FSU forest biomes was assessed based on the methodology described in Vinson and Kolchugina (1993, this volume). Specifically, spatially distributed data on maps were combined within a geographic information system (GIS) to isolate ecoregions. The soil-vegetation complexes for the ecoregions were linked to FSU data bases of soil and vegetation C accumulation parameters. The C budget for an ecoregion was established by multiplying the area of the ecoregion (in hectares) by the C content(s) or rate(s) associated with the soil-vegetation complex for the ecoregion. The C contents and fluxes for all the ecoregions within the four forest biomes of the FSU were summed to estimate the biogenic C budget. A detailed discussion of the activities (maps and data sources) conducted is also presented in Kolchugina and Vinson (1991, 1993a) and Vinson *et al.* (1992).

Average densities of C accumulation parameters estimated for the four forest biomes in the FSU taken as a whole (weighted by the area of the ecosystems that comprise a specific forest biome) are as follows: phytomass (i.e., above and belowground live plant mass)- 67.1 t C ha^{-1}; mortmass (i.e., *above and belowground coarse woody debris*) - 18.0 t C ha^{-1}; litter (i.e., forest floor) - 13.9 t C ha^{-1}; soil organic matter (excluding peat) - 125.0 t C ha^{-1}; NPP (i.e., *net primary productivity*) - 3.1 t C ha^{-1} yr^{-1}; foliage formation - 1.4 t C ha^{-1} yr^{-1}, formation of total soil organic matter; 0.11 t C ha^{-1} yr^{-1}, and formation of stable organic matter - 0.056 t C ha^{-1} yr^{-1}.

The maps and data sources that were used in the analyses were not designed specifically for the purpose of estimating the C budget. As an example, at this stage of work it was possible to isolate only arable and non-arable land in the FSU. The exact area under natural vegetation within the forest biomes in the FSU remained uncertain. Further, data on C accumulation in forest plant mass given by Bazilevich (1986) reflect C accumulation in mature (established) forest ecosystems but not young forest ecosystems.

2.1.2. Second stage

Net ecosystem productivity (NEP) is the net C increment in soil and vegetation after C is expended for autotrophic and heterotrophic respiration. The NPP represents C sequestration in an ecosystem after autotrophic respiration. The NEP equals the difference between NPP and C loss resulting from heterotrophic respiration. Carbon accumulates in soils as stable organic matter. Stable organic matter decomposes at slower rates than the unprotected (labile) organic reservoir (Kobak, 1988).

The NEP of a forest ecosystem depends on its successional stage (Odum, 1953; Vorobyov, 1986). The NEP may be zero or negative at the beginning of succession because of the combination of reduced NPP and enhanced soil respiration. Then NEP increases with time, reaching a maximum in the productive phase. In a climax ecosystem the rate of autotrophic and heterotrophic respiration is high and NEP may approach zero. Thus, maximum NEP occurs in middle age forest stands (20 to 60 years).

Data on NEP of FSU forests were not available at the present stage of work. Gucinski *et al.* (1992) following Birdsey (1991) estimated site specific values of forest NEP for the United States. The data that can be related to the FSU forests (under the State Management excluding forest set aside) are presented in Table 1. The application of North American forest NEP data to FSU forests may be disputed. A comparison of data on growing stock of North American (Birdsey, 1991, the same sites that were used in the present study) and FSU (Kozlovski and Pavlov, 1967) revealed an apparent similarity (Kolchugina and Vinson, 1993b). At present NEP of mature/overmature forest stands was assumed to be zero. The *net increment in vegetation* (NIV) was obtained by subtracting the rate of stable soil organic matter formation in forest ecosystems from total NEP.

For the purpose of the present study the term *actual forest coverage* is defined as the percentage of land within a forest biome actually covered by forest ecosystems. The largest area of forested land in the FSU is the taiga biome, where 50 to 80% (65% on average) of the land is covered by forest ecosystems (Vorobyov, 1985). Forest ecosystems occupy from 30 to 45% of the area of the mixed-deciduous forest biome, and from 10 to 20% of the area of the forest-steppe biome. Specific data for the forest-tundra/sparse-taiga biome were lacking. Data on actual forest coverage for the taiga biome were extrapolated to the forest-tundra/sparse taiga biome. The area of forest ecosystems was estimated based on data on actual forest coverage as a percentage of the total area of the forest biomes in the FSU defined from the GIS analysis.

Table 1 presents data on the age-class distribution of forest stands in the FSU. *Growing stock* (GS) (i.e., above-ground woody phytomass excluding leaves and branches) is approximately four times greater in middle-age, premature, and mature/overmature forest stands compared to young forest stands. By analogy, total phytomass was also assumed to be approximately four times greater. The differences in GS between middle-age, premature, and mature/overmature forest stands were not significant. Average densities of vegetation C accumulation parameters (total phytomass and, by association, mortmass, NPP, and the rate of foliage formation) derived at the first stage of work were decreased by a factor of four to receive C accumulation parameters of young stands. For the present, litter, soil organic matter, and the formation of soil organic matter were assumed to be independent of the age of the stand. The area of young forest ecosystems was derived from the total area of forest ecosystems and the ratio of the areas of young to middle age - overmature forest stands (1:4.5) (Table 1).

The area of peatlands in the FSU is 83 Mha (Botch and Masing, 1983). The area of peatlands within the forest biomes was 77 Mha (Kolchugina and Vinson, 1993b). Carbon accumulation in peatlands within the forested zone in the FSU was estimated from data on the annual rate of peat accumulation in Sweden (Eriksson, 1991), Finland (Laine and Paivanen, 1992) and Canada (Apps *et al.*, 1991).

Table 1 Age-class Distribution, Growing Stock, and NEP of Forests in the Former Soviet Union (USSR State Forestry Committee, 1990; Birdsey, 1991; Gucinski, 1992)

| Age-class | Area | Growing Stock | | NEP |
Main forest species	(Mha)	(Mm³)	(m³/ha)	C t/ha
Young 1st Class; total including:	61.3	822.00	13.4	
Larch	20.7			0.75
Spruce/fir	8.7			1.64
Pines	18.5			2.23
Hardwoods	13.4			2.90
Young 2nd Class; total including:	56.6	2,968.00	52.5	
Larch	19.8			2.70
Spruce/fir	5.1			2.03
Pines	16.9			2.80
Hardwoods	14.8			3.62
Middle-Age; total including:	146.9	16,406.00	111.5	
Larch	53.5			3.40
Spruce/fir	10.2			1.95
Pines	37.1			2.49
Hardwoods	46.1			2.38
Premature; total including:	65.9	9,742.00	147.8	
Larch	20.8			3.00
Spruce/fir	7.6			1.48
Pines	21.3			2.00
Hardwoods	16.2			1.79
Mature-overmature; total including:	323.2	44,948.00	139.1	
Larch	152.2			-
Spruce/fir	61.8			-
Pines	63.8			-
Hardwoods	45.4			-
Total:	653.9	74,886.00	114.5	

2.2. RESULTS

The area of natural ecosystems within the forest biomes in the FSU was estimated at 1,306.2 Mha (10^6 ha) with 799.9 Mha of forested lands. The phytomass C pool of forest ecosystems was 46,354 Mt C, the mortmass C pool was 12,434 Mt C, the litter C pool was 11,118 Mt C, NPP was 2,145 Mt C yr^{-1}, and foliage formation (i.e., production of green parts) was 974 Mt C yr^{-1}. Formation of total soil organic matter was estimated at 88 Mt C yr^{-1}, formation of stable soil organic matter was estimated at 45 Mt C yr^{-1}. Carbon accumulation in peatlands within the forested zone in the FSU was estimated at 23 Mt C yr^{-1}. Annual peat extraction in the FSU reaches 200 Mt (Botch and Masing, 1983). A significant portion of peat (60 Mt yr^{-1}) is used as fuel and undoubtedly represents a C efflux to the atmosphere, which corresponds to about 30 Mt C yr^{-1}. Overall peat accumulation/combustion was estimated to be a net efflux of 7 Mt C. The NEP of the forest biomes in the FSU was estimated at 825 Mt C yr^{-1}. The NIV, established as a difference between NEP and formation of stable organic matter (45 Mt C yr^{-1}), was 780 Mt C yr^{-1}.

3. Forest Statistical Data Method

3.1. METHODOLOGY

FSU forest statistical data are reported under two area related terms: forested land and forest land. Forested land identifies the area where forests presently grow; forest land identifies the area where forests can potentially grow or are presently growing. Three major problems must be resolved when converting data on the volume of merchantable timber to total forest plant mass, specifically, one should account for (1) parts of trees other than stems , (2) ecosystem components other than trees , and (3) mortmass accumulation. Sampson (1992) suggests the use of a multiplier of 0.53 to convert timber volume (m^3) to total forest phytomass (t C) (including roots, branches, and understory). This multiplier was applied in the present study.

The USSR yield tables (Kozlovski and Pavlov, 1967) were used to assess the mortmass accumulation in FSU forests. According to this data source, the ratio of plant mass (phytomass and mortmass) to phytomass depends on the forest type (i.e. species composition) and the age of the forest stands (Table 2). Total forest plant mass was obtained by multiplying the estimates of the total forest phytomass by a corresponding plant mass to phytomass coefficient.

The GS data were used to estimate the total forest plant mass and annual increment of timber data were used to estimate NIV. The maximum possible annual C flux associated with wood harvesting was estimated by converting data on the annual cut (AC) of timber in the FSU to an estimate of the total forest plant mass that may be extracted from the logging area. Considering the fact that wood harvesting in the FSU has remained at approximately the same level during the last several decades this amount of C may represent the maximum possible annual C flux from wood harvesting.

Table 2. Forest Statistical Data (Alimov et al., 1989)

Economical Region	Forest Type	Area	Growing Stock	Annual Increment	Annual Cut	Ave. Age	Ratio: Total Plant Mass[2] to Phytomass
		(Mha)	(Mm³)	(m³/ha/yr of timber)	(Mm³)	yrs	
European-	coniferous	96.0		1.8		93.0	1.6
Ural	broadleaf	11.0		4.1	235.1	71.0	2.0
	small leaf	43.8		2.9		42.0	1.2
	total	150.8[1]	19139.4[1]	2.2			
West Siberia	coniferous	51.8		1.1		117.0	1.7
	broadleaf	0.0			34.4		0.0
	small leaf	21.4		2.1		58.0	1.4
	total	73.2[1]	9180.9[1]	1.4			
East Siberia	coniferous	180.1		1.1		132.0	1.8
	broadleaf	0.0			72.3		
	small leaf	31.4		1.8		50.0	1.3
	total	211.5[1]	27641.2[1]	1.2			
Far-East	coniferous	188.2		0.8		104.0	1.7
	broadleaf	3.4		0.9	38.2	104.0	2.1
	small leaf	3.1		1.7		41.0	1.2
	total	194.7[1]	18561.5[1]	0.9			
FSU	coniferous	516.1		1.2		113.0	1.7
	broadleaf	14.4		2.2	380.0	79.0	2.1
	small leaf	99.7		2.3		147.0	1.3
	total	630.2[1]	74523.0[1]	1.4			

[1]Estimated as the sum of coniferous, broadleaf, and small leaf forests under USSR State Management

[2]Phytomass and mortmass

3.2. RESULTS

The area of forests in the FSU has not changed appreciably during the last 20 yrs (1966 - 1988) (Alimov *et al.*, 1989; USSR State Forestry Committee, 1990). Forested land is approximately 814.2 Mha, or 64.9% of the total area of the forest fund of the FSU (1254.2 Mha). Forests of the European-Ural economical region comprise 23.9% of the forested land of the FSU (Table 2). The West, East Siberian, and Far-East economical regions have 11.6, 33.6, and 30.9%, respectively.

The phytomass density of the FSU forests was estimated at 62.7 t C ha^{-1}, the mortmass density was estimated at 37.6 t C ha^{-1}, thus total forest plant mass density was estimated at 100.3 t C ha^{-1}. The NIV was estimated at 1.1 t C ha^{-1} yr^{-1} (including phytomass and mortmass accumulation). For the same area as the area of forest ecosystems identified from the ecosystem/ecoregional method (799.9 Mha), the phytomass, mortmass, and the total plant mass C pools would be 50.2, 30.0, and 80.2 Gt C, respectively. The NIV would be 879.9 Mt C yr^{-1}.

The AC in the FSU comprises approximately 380 Mm^3 of timber per year (Table 2). Most of the harvested timber comes from Russia (91.2%). Logging activities in the European-Ural economical region account for 61.9% of AC. Logging activities in the West, East Siberian economical regions, and the Far-East economical region amount to 9.1, 19.0, and 10.1%, respectively. The amount of plant mass that may be extracted from the area of logging was estimated at 317.6 Mt C yr^{-1}.

4. Comparative Analysis of Carbon Budget Components

The area of forest ecosystems defined with the help of the ecosystem/ecoregional method (799.9) corresponded well with the forested area reported in the forest statistical data (814 Mha) (USSR State Forestry Committee, 1990). The area of natural ecosystems within FSU forest biomes based on the ecosystem/ecoregional method (1306.2) was in good agreement with the forest area from the forest statistical data (1254.2 Mha). The FSU forest area (i.e.,"closed forests and other woodlands") reported by the World Resources Institute (WRI, 1992) is 929.6 Mha and the forested area (i.e.,"closed forests") is 739.9 Mha. The FSU forested area was three times greater than the forested area of Canada (264.1 Mha) (WRI, 1992); it comprises 28% of the world's forested area (2,823 Mha).

A comparison of the C pools and the NIV is presented in Table 3. The estimates of phytomass, total plant mass, and NIV are in good agreement, while mortmass estimated from the forest statistical data exceeded mortmass estimated from the ecosystem/ ecoregional method by a factor of 2.4. Based on the two methods presented, NEP was estimated at 875 Mt C yr^{-1} on average (assuming 45 Mt C yr^{-1} accumulating in forest soils). The net accumulation of phytomass (NAPh) was estimated at 593.5 Mt C yr^{-1}, considering AI of 1.4 m^3 ha^{-1} yr^{-1}, a 0.53 conversion factor, and 799.9 Mha of forest ecosystems (Table 2). The net accumulation of mortmass (NAM) was estimated as the difference between NIV and NAPh at 236.5 Mt C yr^{-1}. Considering the distribution of

Table 3.   Comparison of Estimates for Carbon Pools and Net Annual Increment
of Forest Vegetation

Approach	Phytomass		Mortmass		Total Plant Mass[2]		Net Increment in Vegetation (NIV)	
	Gt C	t C/ha	Gt C	t C/ha	Gt C	t C/ha	Mt C/yr	t C/ha/yr
Ecosystem/ Ecoregional	46.3	57.9	12.4	15.5	58.7	73.4	780	1
Forest statistical	50.2[1]	62.7	30.0[1]	37.6	80.2[1]	100.3	879.9[1]	1.1
Average	48.3±2.0	60.3±2.4	21.3±8.8	26.6±11.0	62.5±10.7	86.9±13.4	830.0±50.0	1.05±0.05

[1] Estimate adjusted for the same area (799.9 Mha) as in the ecosystem/ecoregional approach

[2] Phytomass and mortmass

forested land in the FSU, forests of the European-Ural economical region sequestered
209.1 Mt C yr^{-1}, forests of the West, East Siberian, and the Far-East economical regions
sequestered 101.5, 294.0, and 270.4 Mt C yr^{-1}, respectively. Russian forests sequestered
872.4 Mt C yr^{-1}.

The landscape-average NEP of the forest biomes in the FSU was estimated at 67 g
C m^{-2} (total NEP 875 Mt C for a forest area 1,306.2 Mha). This value is in a good
agreement with the landscape-average net C sink of 74 g C m^{-2} reported by Bonan
(1991) for Alaskan (Fairbanks) boreal forests.

The C budget of undisturbed forest ecosystems is presented in Figure 1. The
vegetation C pool was estimated at approximately 58.7 Gt C; the soil C pool exceeded the
vegetation C pool by a factor of 1.7. About one-half of NPP was accumulating in forest
plant mass. Carbon effluxes were estimated as described in Vinson and Kolchugina (1993,
this volume). Mortmass decomposition was estimated as the difference between NPP and
sum of foliage formation, NAPh, and NAM. Mortmass formation from phytomass was
estimated as the sum of NAM and mortmass decomposition. Mortmass decomposition
was approximately 11% of mortmass accumulation. It is difficult to accurately assess the
amount of carbon transformed from mortmass to soil organic matter (Kobak, 1988).
Consideration of this process would increase the value for soil organic matter
accumulation and, as a result, the NEP estimate.

5. Disturbance Effects in FSU Forests

The assessment of the C flux due to forest fires was based on the following assumptions:
(1) there is no post fire effect (i.e., all C is burned directly during the forest fire), (2) on

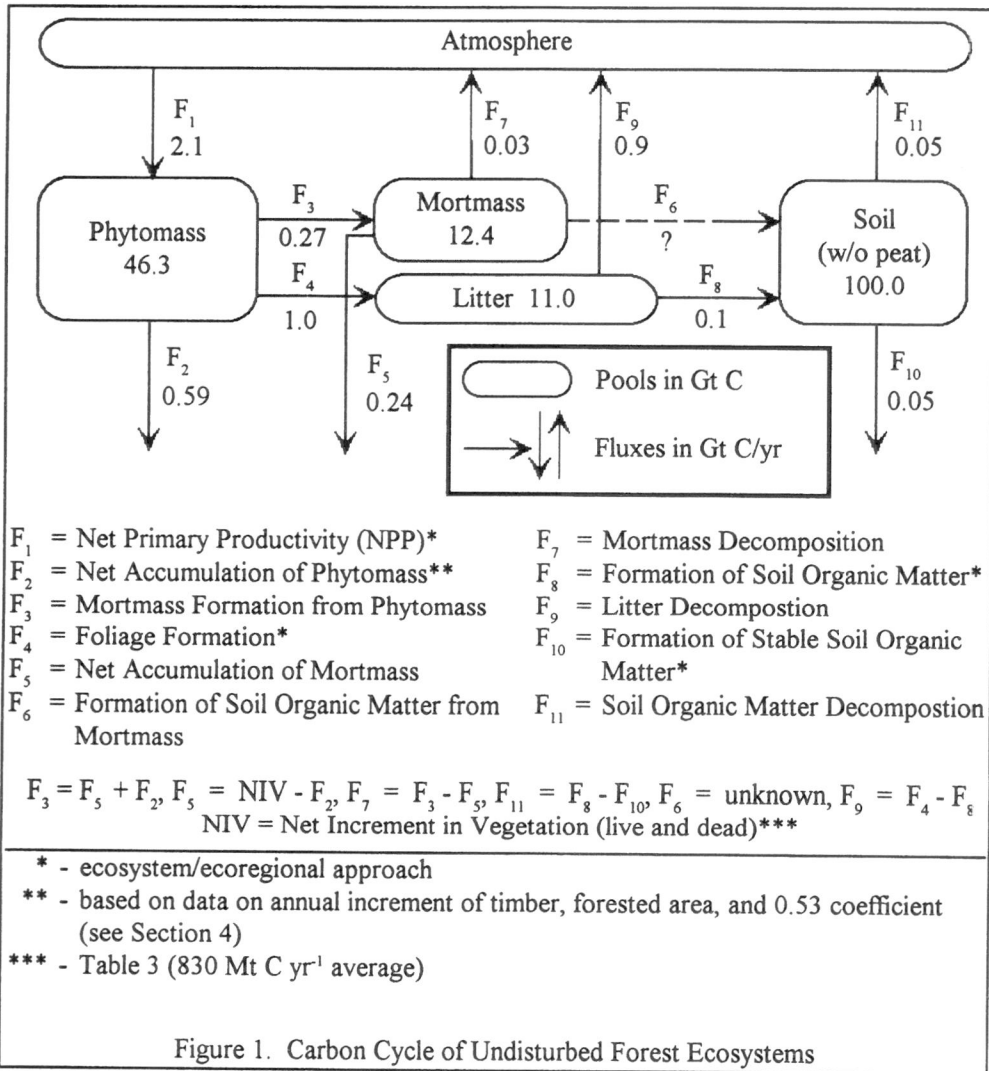

F_1 = Net Primary Productivity (NPP)*
F_2 = Net Accumulation of Phytomass**
F_3 = Mortmass Formation from Phytomass
F_4 = Foliage Formation*
F_5 = Net Accumulation of Mortmass
F_6 = Formation of Soil Organic Matter from Mortmass

F_7 = Mortmass Decomposition
F_8 = Formation of Soil Organic Matter*
F_9 = Litter Decompostion
F_{10} = Formation of Stable Soil Organic Matter*
F_{11} = Soil Organic Matter Decompostion

$F_3 = F_5 + F_2$, F_5 = NIV - F_2, F_7 = $F_3 - F_5$, F_{11} = $F_8 - F_{10}$, F_6 = unknown, $F_9 = F_4 - F_8$
NIV = Net Increment in Vegetation (live and dead)***

* - ecosystem/ecoregional approach
** - based on data on annual increment of timber, forested area, and 0.53 coefficient (see Section 4)
*** - Table 3 (830 Mt C yr^{-1} average)

Figure 1. Carbon Cycle of Undisturbed Forest Ecosystems

average 8% of aboveground plant mass is burned (Auclair, 1985), (3) litter is entirely burned during a fire, and (4) 20 to 30% of soil C is lost (Raison et al., 1985; Johnson, 1992). The average aboveground plant mass (phytomass and mortmass) density was estimated at 58.0 t C ha^{-1} (Kolchugina and Vinson, 1993b). The average densities of the litter and soil C (including peatlands) (13.9 t C ha^{-1} and 244.1 t C ha^{-1}, respectively) were derived for the forest biomes during the first stage of the ecosystem/ecoregional method. Based on the historical records, Krankina (1991) estimates that annually about 2.5 Mha of forests are subjected to fires in the FSU. This is equal to the area of forest fires in Canada (Kurz et al., 1992). The FSU forested area is three times greater than in Canada. On average, there is 0.003 ha per one ha of the forested area in the FSU burned annually. The same parameter for Canada is three times greater (0.009 ha).

Considering plant mass, litter, and a 25% (average) soil C loss, each hectare of burned forest would result in a C efflux of 79.5 t C yr^{-1}. For 2.5 Mha of burned forest the annual C efflux from forest fires would be 199 Mt C yr^{-1}. This estimate is 31% less than the average estimate made by Krankina (1991) (287.5 Mt C yr^{-1}). Krankina assumed a post fire C efflux which was 2.8 times the direct C emissions. Considering the distribution of areas between the economical regions in the FSU, forest fires in the European-Ural economical region represent a C efflux of 47.5 Mt C yr^{-1}. Forest fires in the West, East Siberian, and the Far-East economical regions represent a C efflux of 23.1, 66.8, and 61.6 Mt C yr^{-1}, respectively.

Melillo *et al.* (1988) assessed the C exchange between FSU harvested forests and the atmosphere and estimated C effluxes resulting from slash decay (82 Mt C yr^{-1}), burned material (24 Mt C yr^{-1}), and the decay of 10- to 100-yr wood products (46 Mt C yr^{-1}). These estimates were used to evaluate the contribution of wood harvesting to the C budget. Thus, the total C efflux from wood harvesting was estimated at 152 Mt C yr^{-1}. The estimate of C efflux due to wood harvesting based on results presented by Melillo *et al.* (1988) is probably more realistic than the estimate obtained in the present study from AC data (317.6 Mt C yr^{-1}).

Wood harvesting and forest fires are the main factors that contribute to forest age-class distribution and, therefore, they have a direct influence on NEP. The NEP estimated from the data on age-class distribution of forest stands accounted for C sequestration in regrowing vegetation after wood harvesting and forest fires.

Taking into account the relative harvest intensity in the different economical regions in the FSU (AC by economical region), the distribution of the C efflux due to wood harvesting was as follows: the European-Ural economical region contributed 94.1 Mt C yr^{-1}, the West, East Siberian, and the Far-East economical regions contributed 13.8, 28.8, and 15.3 Mt C yr^{-1}, respectively.

In the mid 80s, the FSU forests were a net sink for atmospheric C (Table 4). This sink was estimated at 517 Mt C yr^{-1}, considering (1) net C increment in forest vegetation and soils (830 + 45 Mt C yr^{-1}), (2) C efflux resulting from peat accumulation/burning (7 Mt C yr^{-1}), and (3) C efflux resulting from forest fires and wood harvesting (199 and 152 Mt C yr^{-1}, respectively). This value corresponded well with the net C sink estimated for FSU forests by Sedjo (1992) (416 Mt C yr^{-1}, 1992). Forests of the European-Ural economical region were a sink for 65.9 Mt C yr^{-1}. Forests of the West, East Siberian, and the Far-East economical regions were net sinks for 63.8, 196.0, and 191.3 Mt C yr^{-1}. The distribution of the net sink of C estimated per unit of forested area in the different economical regions in the FSU within the forested area is as follows: the European-Ural economical region sequestered only 0.4 t C ha^{-1} yr^{-1}, while forests of the West, East Siberian economical regions, and the Far-East economical regions sequestered 0.8 - 1.0 t C ha^{-1} yr^{-1}. Approximately 62% of the timber harvested in the FSU is from the European-Ural economical region; however, forests in this region occupy only 24% of the total FSU forested area.

Table 4. Budget of the Carbon Fluxes in the FSU Forests

Economical Region	Influxes	Effluxes		Net Sink	
	Integrated NEP and peat accumulation/ burning (Mt C/yr)	Forest fires (Mt C/yr)	Wood Harvesting (Mt C/yr)	Total (Mt C/yr)	(t C/ha/yr)
FSU	868.0	199.0	152.0	517.0	0.8
European-Ural	207.5[1]	47.5	94.1	65.9	0.4
West Siberia	100.7[1]	23.1	13.8	63.8	0.8
East Siberia	291.6[1]	66.8	28.8	196.0	0.9
Far-East	268.2[1]	61.6	15.3	191.3	1.0

[1] In accordance with the distrubtion of forested land in the FSU (see Section 3.2)

The forested area in the FSU (799.9 Mha) comprised 28% of the world forested area (2,823 Mha, i.e., "closed forest" (WRI, 1992)). The estimated net sink of 517 Mt C yr^{-1} was approximately 25% of the C absorbed by the terrestrial ecosystems of the Northern Hemisphere based on the low estimate of 2.0 Gt C by Tans et al. (1990). It is equivalent to one-half of the fossil fuel emissions in the FSU (Makarov and Bashmakov, 1990). The reasons for the net sink that presently exists in the FSU are: (1) vast forested area, (2) specific age structure, and (3) relatively low level of disturbances.

6. Summary and Conclusions

The C budget of FSU forest biomes was established based on data for undisturbed forest ecosystems and forest statistical data. The estimates of vegetation component obtained with the help of two methods agreed well. The landscape average net C sink estimated herein corresponded well with the estimate for North American boreal forests. FSU forests are an important component of the global C cycle. They store and sequester a significant amount of C. FSU forest biomes were a net sink of approximately 0.5 Gt C yr^{-1} or approximately 25% of the C absorbed by the terrestrial ecosystems of the Northern Hemisphere. The net sink of C within the FSU forest biomes was equivalent to one half of the fossil fuel emissions in the FSU.

7. Acknowledgments

The work presented herein was funded by the U.S. Environmental Protection Agency (EPA) - Environmental Research Laboratory, Corvallis, Oregon, under Cooperative

Agreement CR820239 to Oregon State University. Jeffrey J. Lee is the Project Officer for the project entitled "Carbon Cycling in Terrestrial Ecosystems of the Former Soviet Union." The work presented is a component of the U.S. EPA Global Climate Research Program, Global Mitigation and Adaptation Program, Robert K. Dixon, Program Leader. This paper has not been subjected to the EPA's review and, therefore, does not necessarily reflect the views of the EPA, and no official endorsement should be inferred. Anatoly Shvidenko provided resource materials and many suggestions that were extremely valuable to the research work. Shane Trenary diligently produced the camera ready copy of the manuscript.

8. References

Alimov, Y.P., I.V. Golovikhin, L.B. Zdanevich, and I.V. Yunov (eds.). 1989. Dynamics of forests under forest management organization regarding the main forest forming species in 1966-1988, U.S.S.R State Forestry Committee,159 pp., Moscow.

Apps, M.J., A.W. Kurz, and T.D. Price. 1991. Estimating carbon budgets of Canadian forest ecosystems using a national scale model. In: T. Kolchugina and T. Vinson, (eds.) *Proceedings of the Workshop on Carbon Cycling in Boreal Forests and Subarctic Ecosystems.* Corvallis, Oregon, Sept. 9-14 , pp. 241-250.

Bazilevich, N.I. 1986. Biological productivity of soil-vegetation formations in the USSR. Bulletin of the Academy of Sciences of the USSR. *Geographical Series*, 2:49-66.

Birdsey, R.A. 1991. Prospective changes in forest carbon storage from increasing forest area and timber growth. USDA Forest Service, Washington DC. (Data on the file with the USDA Forest Service).

Bonan, G.B. 1991. Boreal forests, the carbon cycle and global change: a challenge for ecologists. In: T. Kolchugina and T. Vinson, (eds.) *Proceedings of the Workshop on Carbon Cycling in Boreal Forests and Subarctic Ecosystems.* Corvallis, Oregon, Sept. 9-14 , pp. 139-153.

Botch, M.S. and V.V. Masing. 1983. Mire ecosystems in the USSR. In: A.J.P. Gore (ed.) Ecosystems of the World: Mires, Swamp, Bog, Fen and Moor, 4B, Regional Studies. Elsevier, New York, pp. 95-152.

Brown, S., A.G.R. Gillespie, and A.E. Lugo. 1989. Biomass estimation methods for tropical forests with application to forest inventory data. *Forest Science*, 35(4):881-902.

Eriksson, H. 1991. Sources and sinks of carbon dioxide in Sweden. *AMBIO*, 20(3-4):146-150.

Gucinski, H., D.P. Turner, and G. Koerper. 1992. The carbon flux on forested lands of the United States. In: *Proceedings of the IUFRO Conference on Integrating Forest Information over Space and Time.* Canberra, Australia.

Johnson, D.W. 1992. Soil carbon storage. *Water, Air, and Soil Pollution* 64:83-120

Kobak, K.I. 1988. *Biotical compounds of carbon cycle.* Hydrometeoizdat Press, Leningrad, 248 pp.

Kolchugina, T.P. and T.S. Vinson. 1991. Framework to quantify the natural terrestrial carbon cycle of the former Soviet Union. In: T. Kolchugina and T. Vinson, (eds.) *Proceedings of the Workshop on Carbon Cycling in Boreal Forests and Subarctic Ecosystems.* Corvallis, Oregon, Sept. 9-14, pp. 257 - 273.

Kolchugina, T.P. and T.S. Vinson. 1993a. Equilibrium analysis of carbon pools and fluxes of forest biomes in the former Soviet Union. *Canadian Journal of Forest Research* 23:81-88.

Kolchugina, T.P. and T.S. Vinson. 1993b. Carbon sources and sinks in forest biomes of the former Soviet Union. *Global Biogeochemical Cycles.*

Kolchugina, T.P., A.Z. Shwidenko, T.S. Vinson, R.K. Dixon, K.I. Kobak, and M.S. Botch. 1992. Carbon balance of forest biomes in the former Soviet Union. *Proceedings of the IPCC AFOS Workshop.* University of Joensuu, Joensuu, Finland, May 11-12.

Kozlovski, V.B. and V.M. Pavlov. 1967. Growth Rates of the Main Forest Forming Species in the USSR. Lesnaya Promyshlennost Press, Moscow, 327 pp.

Krankina, O.N. 1991. Forest fires in the USSR: past, present and future greenhouse gases contributions to the atmosphere. In: T. Kolchugina and T. Vinson (eds.) *Proceedings of the Workshop on Carbon Cycling in Boreal Forests and Subarctic Ecosystems.* Corvallis, Oregon, Sept. 9 - 14, pp. 177 - 184.

Kurz, W.A., M.J. Apps, T.M. Webb, and P.J. McNamee. 1992. The carbon budget of the Canadian forest sector: Phase 1. *Information Report NOR-X-326,* Forestry Canada, Northwest Region, Northern Forestry Centre, 93 pp.

Laine, J. and J. Paivanen. 1992. Carbon balance of peatlands and global climatic change: summary. In: *The Finnish Research Program on Climate Change Progress Report.* Publications of Academy of Finland, March, VAPK - Publishing, Helsinki, pp. 189 - 192.

Makarov, A.A. and I. Bashmakov. 1990. The Soviet Union. In: W.U. Chandler (ed.) *Carbon Emission Control Strategies: Case Studies in International Cooperation,* 263 pp.

Melillo, J.M., J.R. Furry, R.A. Houghton, B. Moor III, and D.L. Scole. 1988. Land-use change in the Soviet Union between 1850-1980: causes of a net release of CO_2 to the atmosphere. *Tellus,* 40B:116-128.

Odum, E.P. 1953. *Ecology.* W.B. Saunders Co., Philadelphia-London, 384 pp.

Olson, J.S., J.A. Watts, and L.J. Allison. 1983. Carbon in live vegetation of major world ecosystem, ORNL-5862, Oak Ridge.

Post, W.M., W.R. Emanuel, P.J. Zinker, and A.G. Strangenberger. 1982. Soil carbon pools and world life zones. *Nature,* July 8, pp.156 - 159.

Raison, R J., P.K. Khanna, and P.V. Woods. 1985. Mechanisms of element transfer to the atmosphere during vegetation fires. *Canadian Journal of Forest Research* 15:132-140.

Sampson, R.N. 1992. Forestry opportunities in the United States to mitigate the effects of global warming. *Water, Air, and Soil Pollution* 64:83-120

Sedjo, R. A. 1992. Temperate forest ecosystems in the global carbon cycle. *Ambio* 21(4):274-277.

Stocks, B.J. 1991. The extent and impact of forest fires in northern circumpolar countries. In: J.S. Levine (ed.) *Global Biomass Burning: Atmospheric, Climatic and Biogenic Implications.* MIT press, Cambridge, MA, pp. 197-202.

Tans, P.P., I.Y. Fung, and T. Takahashi. 1990. Observational constraints on the global atmospheric CO_2 budget. *Science* 247:1431-1438

USSR State Forestry Committee. 1990. *Forest Fund of the USSR,* Vol 1,1005 pp., Moscow.

Vinson T. S, T.P. Kolchugina, R.K. Dixon, P.M. Bradley, and G.G. Gaston. 1992. Framework to assess the natural component of the carbon budget of the former Soviet Union. In: *Proceedings of the IPCC Workshop.* University of Joensuu, Finland, May 11-15 (in press).

Vinson T. S and T.P. Kolchugina. 1993. Pools and fluxes of biogenic carbon in the former Soviet Union (this volume).

Vorobyov, G.I., (ed.). 1985. *Forest encyclopedia,* 1, Sovetskaya Encyclopedia Press, Moscow, 563 pp.

Vorobyov, G.I., (ed.) 1986. *Forest encyclopedia,* 2, Sovetskaya Encyclopedia Press, Moscow, 631 pp.

World Resources Institute (WRI). 1992. World Resources 1992-1993. Oxford University Press, New York - Oxford.

POOLS AND FLUXES OF BIOGENIC CARBON
IN THE FORMER SOVIET UNION

TED S. VINSON and TATYANA P. KOLCHUGINA,

Oregon State University, Corvallis, Oregon, 97331
U.S.A.

Abstract. The Former Soviet Union (FSU) was the largest country in the world. It occupied one-sixth of the land surface of the Earth. An understanding of the pools and fluxes of biogenic C in the FSU is essential to the development of international strategies aimed at mitigation of the negative impacts of global climate change. The territory of the FSU is represented by a variety of climate conditions. The major part of the FSU territory is in the boreal and temperate climatic zones. The climate in the FSU changes from arctic and subarctic in the North to subtropical and desert in the South. From west to east, the climate makes a transition from maritime to continental to monsoon. The vegetation of the FSU includes the following principal types: forest, woodland, shrubland, grassland, tundra, desert, peatlands and cultivated land. Arctic deserts and tundra formations are found in the northern part of the FSU; deserts and semi-deserts are found in the southern part.

A framework was created to assess pools and fluxes of biogenic C in the FSU. Under the framework spatially distributed data were analyzed with a geographic information system to isolate ecoregions. The soil-vegetation complexes for the ecoregions were linked to FSU data bases of soil and vegetation C pools and fluxes. The C budget for an ecoregion was established by multiplying the area of the ecoregion by the unit area C content(s) or rate(s) associated with the soil-vegetation complex for the ecoregion. The C pools and fluxes for all the ecoregions were summed to arrive at an initial estimate of the pools and fluxes of biogenic C for 95% of the territory of the FSU. Based on the framework, net primary productivity (NPP) for the FSU was estimated at 6.17 ± 1.65 Gt C yr^{-1}, the vegetation C pool (live plant mass and coarse woody debris) at 118.1 ± 28.5 Gt C, the litter C pool at 18.9 ± 4.4 Gt C, and total soil C pool at 404.0 ± 38.0 Gt C. The phytomass pool of the FSU was 16% of the global biomass pool. The soil and litter pools of the FSU were 20 and 23% of the global soil and detritus pools, respectively. The NPP of the FSU was 10% of the global NPP. The phytomass, soil and litter densities of the FSU were greater than the world average. The productivity of terrestrial ecosystems in the FSU was slightly lower than the world average.

Water, Air, and Soil Pollution **70**: 223–237, 1993.
© 1993 *Kluwer Academic Publishers.*

1. Introduction

The long-term ecological consequences of the change in the chemical composition of the atmosphere are not fully understood; however, a warmer global climate is highly probable (PPIGW, 1992). If CO_2 concentrations were to double, the earth's temperatures may rise between 1 and 5° C (Schneider, 1990). Climatic changes may be more pronounced in the Northern Hemisphere (Etkin, 1990). Global warming may accelerate the rates of plant respiration (Keeling *et al.*, 1989) and decay of organic matter (Dixon and Turner, 1991).

It may be necessary to manage terrestrial C stores to offset increased amounts of atmospheric CO_2 (PPIGW, 1992). Before any global strategy can be formulated, policies aimed at maintaining a desirable C balance within national boundaries would be required. The determination of a C balance includes the quantification of the biogenic and anthropogenic contributions to the C cycle within national boundaries. The quantification of the C cycle following an assessment of C pools and fluxes is generally referred to as the *carbon budget*. Biogenic C budgets recently have been established for Sweden (Eriksson, 1991) and the forest sectors of Canada (Apps *et al.*, 1991; Kurz *et al.*, 1992)) and the United States (Gucinski *et al.*, 1992).

The Former Soviet Union (FSU) was the largest country in the world. It occupied one-sixth of the land surface of the earth. An understanding of the pools and fluxes of biogenic C of the FSU is essential to the development of a global strategy aimed at mitigation of the negative impacts of climate change.

The territory of the FSU is represented by a variety of climate conditions. The major part of the territory is in the boreal and temperate climatic zones. The climate in the FSU changes from arctic and subarctic in the North to subtropical and desert in the South. From west to east, the climate makes a transition from maritime to continental to monsoon.

Matthews (1983) identified eight principal types of vegetation in her global data base: forest, woodland, shrubland, grassland, tundra, desert, peatlands and cultivated land. The vegetation of the FSU is represented by a variety of formations, including all of these major types. Arctic deserts and tundra formations are found in the northern regions of the FSU; deserts and semideserts occur in southern regions. A vast area, the largest of any country in the world, is occupied by forests and grasslands. The total area of forest zone under State supervision (in 1983) is 1,259 Mha (Vorobyov, 1985), which is approximately 56.5% of the territory of the country. About 95% of the forest area is in Russia. Tundra and boreal forests store a significant amount of organic matter. Twenty-seven percent of the approximately 80% of terrestrial organic matter stored in soil is found in boreal ecosystems (Billings, 1987). Grasslands are also an important component of the terrestrial C cycle. Despite the fact that grasslands do not accumulate large quantities of plant mass (compared with forest ecosystems), they exhibit high *net primary productivity* (NPP) and, therefore, may influence the terrestrial C cycle.

Peatlands, which are wetlands where peat is accumulating, store a significant amount of C. Organic soil C content reaches 2,000 t ha^{-1} (Bohn, 1982). The FSU has the greatest expanse of peatlands in the world (Tyuremnov, 1976). Wetlands are known to be a source of CH_4 to the atmosphere (Bartlett *et al.*, 1985a,b; Harriss *et al.*, 1985). Although the atmospheric concentration of CH_4 is much lower than the concentration of CO_2, CH_4 is 20 times more effective (per molecule) than CO_2 as a greenhouse gas (Blake and Rowland, 1988).

There is an abundance of Soviet data on C-cycle parameters in desert, tundra, forest and grassland ecosystems (Bazilevich *et al.*, 1970; Kazimirov and Morozova, 1973; Aleksandrova, 1977; Tytlyanova, 1977; Vatkovskyi, 1976; Dylis and Nosova, 1977; Kazantseva, 1980; Bazilevich, 1986; Bazilevich *et al.*, 1986; and others). The purpose of the study presented herein is to estimate the pools and fluxes of biogenic C in the FSU.

2. Pools and Fluxes of Biogenic Carbon

The biogenic C cycle consists of a combination of pools and fluxes. The pools are C stores in soil and vegetation, including living vegetation (i.e. *phytomass*) and plant detritus (i.e. *mortmass* and *litter*). In the present study, the term *mortmass* was used to describe coarse above-ground and below-ground woody debris. The term *litter* was used to define the upper soil layer comprised of fine woody debris and leaves that are not completely decomposed. The effluxes are C emissions resulting from plant respiration and decomposition of organic matter. The processes of formation of new organic matter in soil and vegetation (i.e. *humus, foliage formation* and *NPP*) represent C influxes. The NPP equals the difference between *gross photosynthesis* (GPP) and *respiration of autotrophic organisms* (R_A). The R_A amounts to 44 to 52% (48% on average) (Kobak, 1988) of GPP. *Root respiration* (R_{Ar}) comprises one-third of the R_A.

Carbon fluxes can be measured or calculated. However, when C effluxes are measured, the contribution from different processes cannot be distinguished. For example, when soil C efflux is measured, it is difficult to distinguish between effluxes resulting from R_{Ar} and R_H (decomposition of litter, below-ground mortmass and soil organic matter). The quantitative method allows one to separate fluxes.

3. Methodology

To calculate pools and fluxes of biogenic C the geographic area within which C may be quantified must be isolated. The term *ecoregion* was applied to the boundaries and areal extent of the geographic area. The term *ecosystem* was applied to the combination of certain soil-vegetation formations within an ecoregion. The concept of an ecosystem is a broad one; its function is to emphasize obligatory relationships and interdependence (Odum, 1953). The term *biome* was applied to the complex of ecosystems within a climatic belt or subbelt. Nine biomes within the FSU were identified: polar desert, tundra, forest-tundra/sparse taiga, taiga, mixed-deciduous forest, forest-steppe, steppe, desert-semidesert and subtropical woodlands.

Carbon cycle parameters have been quantified by soil, agricultural and forest scientists, ecologists and botanists for several decades. The C cycle parameters may be expressed in terms of C content (for pools) or rate (for influxes or effluxes)/ha for a variety of soil-vegetation complexes. If the soil-vegetation complexes which comprise an ecosystem are related to the natural attributes identified on maps which are used to isolate ecoregions, then the C budget for an ecoregion can be established simply by multiplying the area of the ecoregion (in hectares) by the C content(s) and flux(es). The C contents and fluxes for all the ecoregions may be summed to arrive at the C budget for a larger region, biome or nation.

Based on the preceding discussion, the framework shown in Figure 1 was created to assess pools and fluxes of biogenic C in the FSU. Initially, maps were used to isolate ecoregions (Frames 1 and 2), and data bases which contain natural C cycle parameters were compiled (Frame 3). The areal coverage of the ecoregions was integrated with the C content and flux data bases to establish the C budget within the ecoregion (Frame 4). The organization of the C cycle parameter data base, the hectare data for the ecoregions, and the calculations that were required to establish the C budget, were performed with personal computer hardware and commercially available spreadsheet software (Microsoft Corporation, 1991) in a Windows™ environment. The C budgets for the ecoregions were summed to establish the C budget for a biome or the entire territory of the FSU (Frames 5 and 6). The specific activities related to the execution of these steps are discussed in the following paragraphs.

Isolation of Ecoregions (Frames 1 and 2): About 95% of the territory of the FSU, including Russia, Ukraine, Belorussia, Kazakhstan and the Baltic states, was categorized by the soil-vegetation type of the ecosystem, the presence of peatlands and cultivation intensity. Maps containing information on the distribution of zonal soil-vegetation associations within the FSU (Ryabchikov, 1988), distribution of peatlands (Isachenko, 1988) and cultivation intensity of arable lands (Cherdantsev, 1961) were digitized and computer-superimposed with a geographical information system (GIS) (Burrough, 1986). The map with the distribution of soil-vegetation associations (Ryabchikov, 1988) provided the basis for ecoregion isolation. In addition, eight georegions: Near Ocean; Eastern, Middle and Western Siberia; Eastern, Central and Western Europe; and Kazakhstan were defined to accommodate geodependence of C accumulation (Gerasimov, 1933; Bazilevich, 1986). These georegions were mapped, digitized and computer-superimposed with the GIS. After computer superimposition of the four maps previously described, more than 70 ecoregions related to different ecosystems (i.e. soil-vegetation associations) were identified. Peatland, floodplain, mountain ecosystems and arable land were isolated within these ecoregions.

The ecosystems presented by Ryabchikov (1988) were aggregated into nine biomes. The *polar desert* biome included areas covered by ice and stony barrens. The *tundra* biome included herbaceous and shrub tundra formations of polar and subpolar belts on dry cryic, arctic, peat, turf, gleyic, "podbur" and podzolized soils. The *forest-tundra/sparse taiga* biome included forest ecosystems within the subpolar climatic belt and northern areas of the boreal climatic belts. This biome unified ecosystems with sparse forest cover on peat-turf, podzol and "podbur" soils. The *taiga* biome included ecosystems within the boreal climatic belt with light- or dark-crown coniferous forest vegetation of northern, middle and southern subzones, mainly on podzol and podzolic

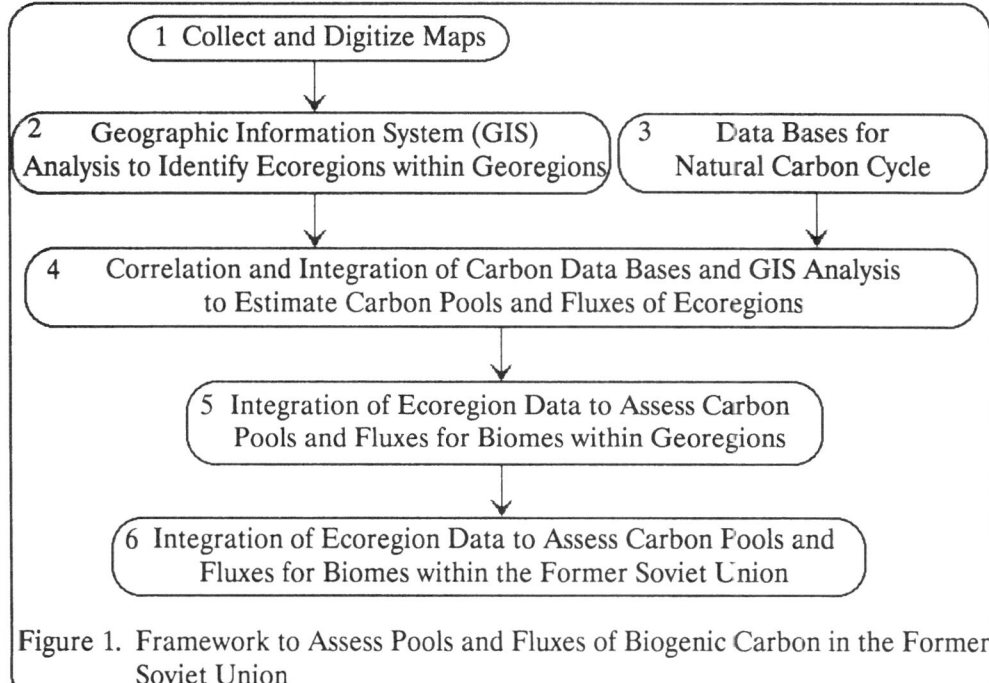

Figure 1. Framework to Assess Pools and Fluxes of Biogenic Carbon in the Former Soviet Union

soils. The *mixed-deciduous forest* biome unified mixed (coniferous - broadleaf or small-leaf) or broadleaf forests on podzolic or gray soils within the subboreal climatic belt. The *forest-steppe* biome included mixed coniferous-deciduous forests and grasslands of the subboreal climatic belt. Gray soils are characteristic for the northern part of the forest-steppe biome and chernozem soils for the southern part. The *steppe* biome included grasslands on chernozem and castanozem (i.e. chestnut) soils. The *desert-semidesert* biome included shrub-grass and shrub-tree desert formations of subboreal and subtropical belts on castanozems, yermosols, sands and primitive desert soils. The *subtropical* biome included mountainous formations of the subtropical belt.

Mountain ecosystems identified by Ryabchikov were described by a combination of different vegetation formations. The vegetation of mountain ecosystems makes the transition from zonal type at low altitudes to meadow, tundra, or polar desert type at high altitudes.

The map used to isolate peatlands (Isachenko, 1988) allows one to determine the total area of peatlands, but does not allow different peatland landscapes to be distinguished. There are at least three main classifications of peatland systems in the FSU (Botch and Masing, 1983). According to the trophic conditions and the developmental stage, peatlands can be classified as eutrophic, mesotrophic or oligotrophic. According to the hydrological conditions, peatlands may be divided into minerotrophic (both ground and rainwater supply) and ombrotrophic (rainwater supply). According to the main layer of plant communities, peatlands can be divided into moss, graminoid, dwarf-shrub and shrub. Carbon accumulation differs depending on peatland type. Future studies will require the incorporation of maps in which specific types of peatlands are identified. However, for the present, the superimposition of peatland and soil-vegetation maps allowed the determination of peatland type within a given ecosystem and the specification of C accumulation parameters.

Data Bases for Natural C Cycle Parameters (Frame 3): Bazilevich (1986) compiled a data base on C accumulation in vegetation from studies of 1,500 vegetation complexes in the FSU. The data base is a comprehensive source of information on all vegetation formations in the FSU, namely, 13 polar desert and tundra, 40 forest, 57 grassland, 20 mire ecosystems and more than 50 desert-semidesert formations. These vegetation complexes were correlated to the ecosystems presented by Ryabchikov (1988). The data base provides site-specific values for total phytomass content and phytomass productivity for all vegetation formations in the FSU. The data base allows the assessment of phytomass and phytomass increment allocation. Phytomass was categorized as green-assimilating, woody above- (stems and branches) and below-ground (roots and buried stems) parts of plants. Mortmass was categorized as woody above- (dead stems, branches, grass, windfall) and below-ground parts of plants and litter. Productivity of phytomass or NPP was categorized in the same manner as phytomass. The net C content of plant mass was assumed to be 50% (Kobak, 1988). This percentage was used to calculate the net C storage and rates of C accumulation in vegetation.

Bazilevich (1986) provided descriptions of vegetation formations for the eight georegions considered in the present study. Polar deserts and tundra formations include arctic and subarctic deserts and tundra. Forest-tundra and taiga forests include forests composed of spruce, larch, fir, pine and cedar (*Pinus sibirica, P. koraiensis*). Mixed-deciduous forests, forests of the forest-steppe zone and subtropical woodlands are represented mainly by pine, oak, birch aspen and subtropical broadleaf forests. Steppe formations are represented by meadow-grasslands, moderately dry and dry steppes. Desert-semidesert formations include steppified semideserts and semishrub and succulent deserts; zonal and psammophytic deserts and semideserts, including solonchak formations. Carbon accumulation data are also provided for peatlands and continental and floodplain meadows of tundra, forest-tundra, taiga, mixed and broadleaf forests, steppes, deserts and semideserts. Data for meadow-grasslands and moderately dry grasslands reported by Bazilevich (1986) were used to quantify the C cycle in the forest-steppe biome.

Kobak's (1988) data base was used to characterize the soil component of the C cycle. The data base resulted from the analysis of the published soil data. About 70 different Soviet and foreign sources were included (Kononova, 1963; Schlesinger, 1977; Bohn, 1982; Post *et al.*, 1982; Kobak and Kondrashova, 1986; and others). Soil C cycle parameters for more than 40 soil types identified in the polar, boreal and tropical belts represent average values from the data presented in different sources. However, in some cases, specific data were selected. For example, only Soviet data were used to characterize the C contents of podzol and chernozem soils. The data base includes: (1) carbon contents of soil (total and stable portion) to a depth of 1 m in inorganic soils and for the entire depth of deposit of peatland organic soils, (2) the annual rate of foliage and humus formation (total and stable portion), and (3) CO_2 efflux from soils. Kobak (1988) reports the following names of soil types that were related to the polar desert and tundra biomes: arctic, tundra gleyic and peatland soil. Soils of the forest biomes were represented by cryic-taiga and podzolic soils, brown and gray forest soils, chernozems, peatland soils, mountain-meadow, and floodplain soils. Soils of the steppe biome were chernozem, chestnut, floodplain, peatland and solonetz soils. Chestnut (castanczem) and gray-brown desert soils were related to the desert-semidesert biome. The combinations of mountain-meadow, podzolic, gray forest and chestnut soils were used to describe soils of mountainous formations.

Matthews and Fung (1987) compiled a data base from published sources for typical CH_4 emissions from natural wetlands. Data on CH_4 emissions from wetlands were correlated to the data on peatland distribution within the nine biomes in the FSU.

Correlation of Data Bases to Mapped Ecosystems (Frame 4): The maps and data bases are not specifically designed for C cycle quantification. However, the names of soil-vegetation associations reported by Ryabchikov (1988) corresponded well with the descriptions of vegetation formations given by Bazilevich (1986) and soil types given by Kobak (1988). For example, for the ecosystem named "Lichen-moss tundra on gleyic and podbur soils" isolated in the Central European georegion in accordance with the Ryabchikov (1988) map, the following name from Bazilevich's (1986) data base was used: "Southern subarctic tundra (West and East Europe);" the soil type was described as "Tundra-gleyic soil." For the ecosystem named "Dark- and light-crown coniferous moderately humid taiga, middle subzone, on podzolic soils and ferric podzols," the name "Pine and spruce forests of northern taiga; fir forests of middle taiga, and larch forests of southern taiga (Central Europe)" was used; the soil type was described as "Podzolic soil." For the ecosystem named "Broadleaf-coniferous moderately humid forests on turf-podzolic soils," the name "Coniferous-broadleaf forests and oak-lime *(Tilia)* forests (Europe)" was used; the soil type was described as "Podzolic soil." For the ecosystem in the Kazakhstan georegion named "Floodplain formations of grass steppes on orthic and southern chernozems," the name "Floodplain solonchak meadows and mesophytic meadows of steppe zone (West Siberia, Kazakhstan)" was used; the soil type was described as "Chernozem and solonetz soils." For the peatland formation identified in the ecosystem named "Broadleaf moderately humid forests on turf podzolic soils, gray luvisols and acid cambisols," the name "Peatlands of broadleaf-coniferous forests" was used. When a direct correspondence did not exist, certain extrapolations and interpolations were made. For example, in the West Siberian georegion for the ecosystem called "Mountainous semidesert/steppe/desert," isolated with the help of the Ryabchikov (1988) map, the combined name from Bazilevich's (1986) data base "Subboreal semideserts; zonal meadows of subboreal steppified semideserts; northern semishrub subboreal deserts" was used, the soil type was described as "Mountain-meadow and gray-brown desert soils."

Integration of Ecoregion Areas and C Data Bases (Frame 5 and 6): Carbon pools and fluxes for natural ecosystems in the FSU were estimated by integrating the C data bases and the GIS analysis results (hectare data). Productivity of green-assimilating parts of plants was used to characterize the rate of foliage formation. Data on C content in peatlands were integrated with hectare data on the extent of peatlands within biomes. Low, mean and high estimates were made for each of the eight georegions by summing the contributions from each ecoregion. The C pools and influxes for the biomes in the FSU were obtained by summing the georegion totals for the nine biomes. Initially, it was assumed that (1) natural ecosystems are presently in a state of equilibrium (NPP equals R_H); (2) forests totally cover the area of ecosystems (excluding arable land) within forest-tundra/sparse taiga, taiga and mixed-deciduous forest biomes; and (3) forests occupy one-half of the area (excluding arable land) of the forest-steppe biome.

     Carbon effluxes were calculated from the influxes assuming that all ecosystems were initially in an equilibrium state. Mortmass decomposition was assumed to be equal to mortmass production. In turn, mortmass production was assumed to be equal to phytomass production (NPP and production of different parts of plants). Carbon efflux from litter decomposition was calculated as the difference between foliage formation (green-assimilating parts production) and the sum of total humus formation and peat accumulation. The carbon efflux from soil organic matter decomposition was calculated as the difference between total and stable humus formation. The carbon efflux from R_{Ar} was calculated from the NPP, assuming that R_{Ar} comprises one-third of the total R_A, and R_A comprises 48% (on average) of the GPP; NPP equals the difference between GPP and R_A. The sum of R_{Ar} and R_H (below-ground mortmass, litter, and soil organic matter decomposition) was compared with field measurements of the surface soil C efflux (Kobak, 1988).

     The estimates of C cycle parameters obtained in the present study were compared, where possible, with estimates available in other data bases. The global data base for the NPP of terrestrial ecosystems, compiled with the help of advanced very high resolution radiometer (AVHRR) data (Fung et al., 1987), was incorporated in the study. The NPP for natural (non-arable) ecosystems was calculated by dividing NPP totals for the ecoregion by the total number of hectares and multiplying by the number of hectares of natural ecosystems within the ecoregion.

4. Biogenic Carbon Pools and Fluxes in the FSU

The vegetation C pool of natural terrestrial ecosystems in the FSU was estimated at 118.1±28.5 Gt C. The vegetation C pool included phytomass (91.0±22.0 Gt C) in green assimilating, above-ground woody, and below-ground woody parts (6.6 ± 1.7, 59.2 ± 15.6, and 25.2 ± 4.8 Gt C, respectively) and mortmass (27.1±6.5 Gt C) in above-ground and below-ground parts (8.2 ± 1.9 and 18.9 ± 3.8 Gt C, respectively). The litter C pool was estimated at 18.9±4.4 Gt C. The soil C pool was estimated at 404.0±38.0 Gt C (including peatlands), with 269.0±25.0 in stable form. Peatlands accumulated 148.0 Gt C.

     The total productivity of phytomass of the nine biomes was estimated at 6.17±1.65 Gt C yr^{-1}. The productivity of green assimilating parts based on Bazilevich's (1986) data was estimated at 2.51±0.58 Gt C yr^{-1}. Humus formation was estimated at 257.0±94.0

Mt C yr^{-1} (87.8±15.1 Mt C yr^{-1} in stable form). CH_4 production from peatlands (77 Mha) was estimated at 7.57 ±1.6 Mt C yr^{-1}.

FSU biomes may be arranged in four groups based on the distribution of plant mass (phytomass and mortmass): (1) polar-desert biome; (2) tundra and forest-tundra/sparse taiga biome; (3) taiga, mixed-deciduous forest, forest-steppe and mountainous subtropical woodlands biome; and (4) steppe and desert-semidesert biome (Figure 2). In the polar desert biome, mortmass was greater than phytomass and was distributed equally above and below ground. Above-ground phytomass was significantly greater than below-ground phytomass. In the tundra and the forest-tundra/sparse taiga biomes, mortmass also exceeded phytomass. In the tundra biome, below-ground biomass was more developed. In the forest-tundra/sparse taiga biome, phytomass and mortmass were distributed in equal proportions above and below ground.

In the forest biomes, phytomass was greater than mortmass. Above-ground parts were greater than below-ground parts in both pools (Figure 2). In the steppe and desert-semidesert biomes, phytomass and mortmass did not differ substantially. Below-ground parts were greater than above-ground parts in both pools . In the steppe biome, above-ground mortmass was greater than above-ground phytomass, while in the desert-semidesert biome, more living parts of plants were found above ground.

The NPP of natural ecosystems estimated with the data base from Fung et al. (1987) was compared with the estimate of phytomass productivity based on data given by Bazilevich (1986) (Figure 3). Total phytomass productivity (6.17 Gt C yr^{-1}) estimated from Bazilevich (1986) was about 2.1 times higher than NPP (2.98 Gt C yr^{-1}) obtained from Fung et al. (1987) (the desert biomes represented an exception: NPP obtained from Fung et al. (1987) was greater than phytomass productivity obtained from Bazilevich (1986)). The NPP obtained from Fung et al. (1987) corresponded well only with the productivity of green-assimilating parts given by Bazilevich (1986). The discrepancy may be due to the fact that either root productivity or both root and above-ground woody phytomass productivity are underestimated by Fung et al. (1987). Further, a discrepancy in the results could be related to the use of different methods to identify C-quantifiable regions.

The C efflux from litter decomposition estimated from the equilibrium analysis was 2.25 Gt yr^{-1}. Soil organic matter decomposition (excluding peatlands) was 0.17 Gt C yr^{-1}. The R_{Ar} was estimated at 1.84 Gt C yr^{-1}. Below-ground mortmass decomposition was estimated at 2.47 Gt C yr^{-1}.

The C efflux from litter and soil organic matter decomposition (2.42 Gt C yr^{-1}) was less than the soil surface CO_2 efflux (3.55± 1.13 Gt C yr^{-1}) estimated for the nine biomes based on data reported by Kobak (1988) (Figure 4). The aggregated efflux from R_{Ar} and average efflux from the decomposition of below-ground mortmass, litter, and soil organic matter was estimated at 6.73 Gt C yr^{-1}. This value was 30% greater than the maximum CO_2 efflux (4.68 Gt C yr^{-1}) estimated from Kobak's (1988) data.

Based on Avogadro's number, CH_4 emissions from wetlands were equivalent to 3.76 x 10^{35} molecules of CH_4. The total efflux from below-ground mortmass, litter and soil organic matter decomposition was equivalent to 3,437 x 10^{35} molecules of CO_2.

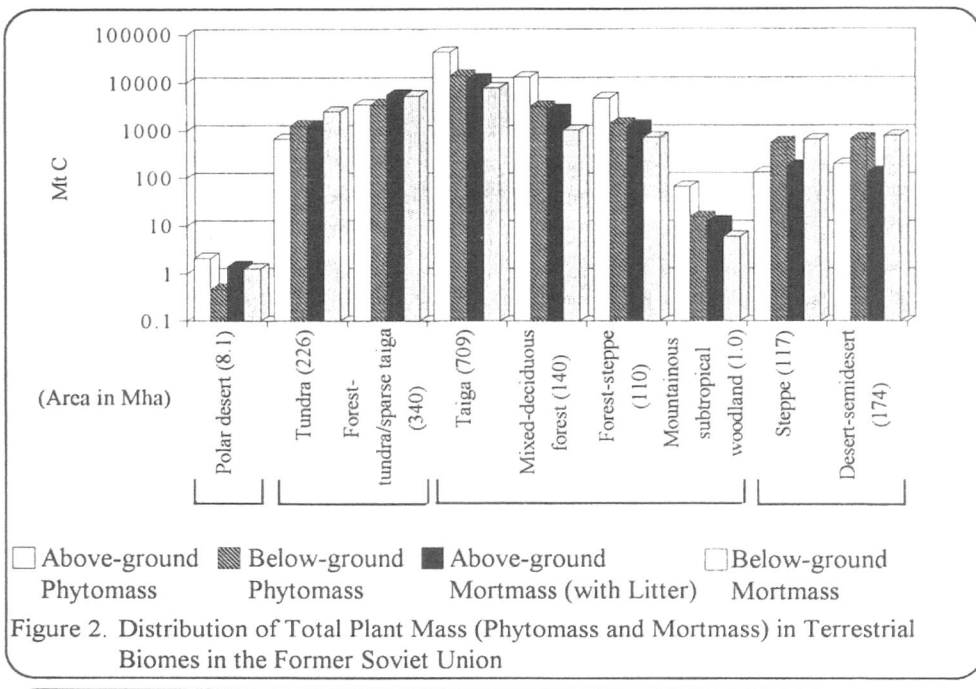

Figure 2. Distribution of Total Plant Mass (Phytomass and Mortmass) in Terrestrial
Biomes in the Former Soviet Union

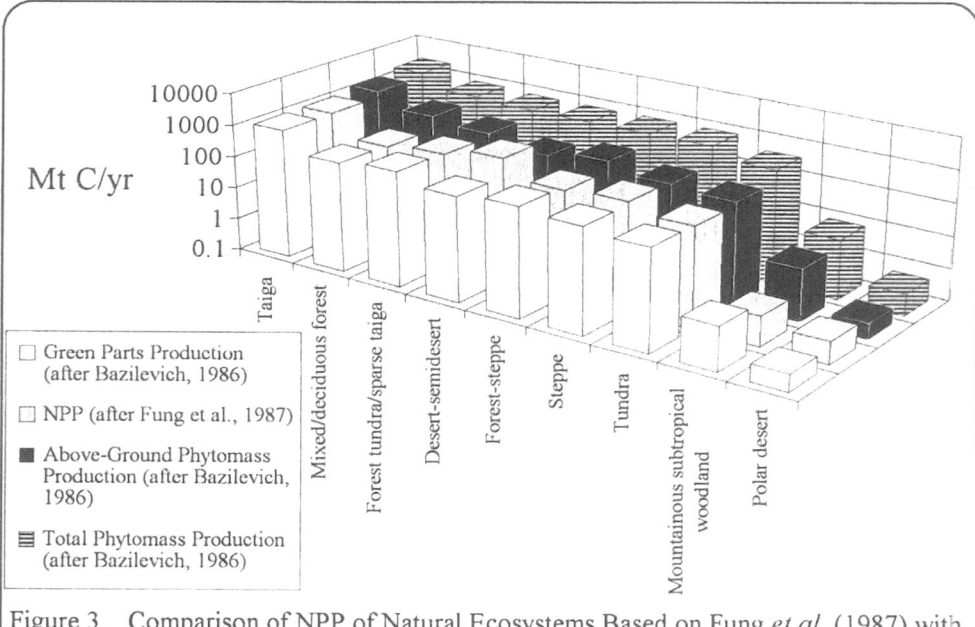

Figure 3. Comparison of NPP of Natural Ecosystems Based on Fung *et al.* (1987) with
Phytomass Productivity Based on Bazilevich (1986).

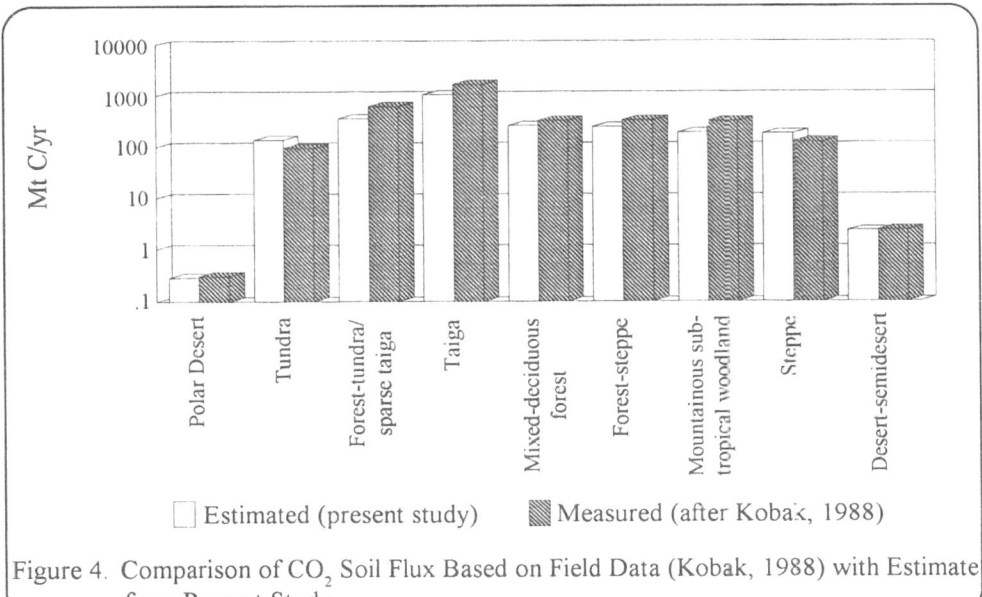

Figure 4. Comparison of CO_2 Soil Flux Based on Field Data (Kobak, 1988) with Estimate from Present Study

Assuming CH_4 is 20-fold more active than CO_2 as a greenhouse gas, CH_4 emissions represented approximately 2.2% of the soil surface efflux estimated in the present study. However, CH_4 emissions from peatlands could be underestimated because many peatland areas were not accurately mapped (Botch, 1991).

The total area of terrestrial ecosystems (with arable land) analyzed in the present study (2,020 Mha) was 14% of the area of the world terrestrial ecosystems (14,917 Mha). The area of natural ecosystems defined in the present study (1,830 Mha) was 12% of the world terrestrial area. Phytomass estimated in the present study was 16% of the global biomass (Figure 5) estimated by Kobak (1988). Soil and litter estimated in the present study were 20 and 23% of the global soil and detritus pools, respectively. Above- and below-ground mortmass was 32% of the global detritus pool. The phytomass, soil, litter, and mortmass densities estimated for the natural terrestrial ecosystems of the FSU were 49.7, 220.8, 10.3, and 14.8 t C ha^{-1}, respectively. The phytomass, soil, and detritus densities estimated for the world terrestrial ecosystems were 37.5, 135.4, and 5.6 t C ha^{-1}, respectively. The NPP and humus formation in the FSU each were 10% of the comparable global fluxes. The NPP and humus formation densities estimated for the FSU were 3.4 and 0.14 t C ha^{-1}yr^{-1}, respectively. The NPP and humus formation densities estimated for the globe were 4.0 and 0.17 t C ha^{-1}yr^{-1}, respectively. Based on this comparison it is reasonable to conclude that: (1) the phytomass, soil and litter densities of the FSU were greater than the world average (the FSU has the greatest expanse of forests, peatlands, and grasslands with rich phytomass, soils, and litter), (2) it is possible that mortmass (coarse above- and below-ground woody debris) was omitted in the global

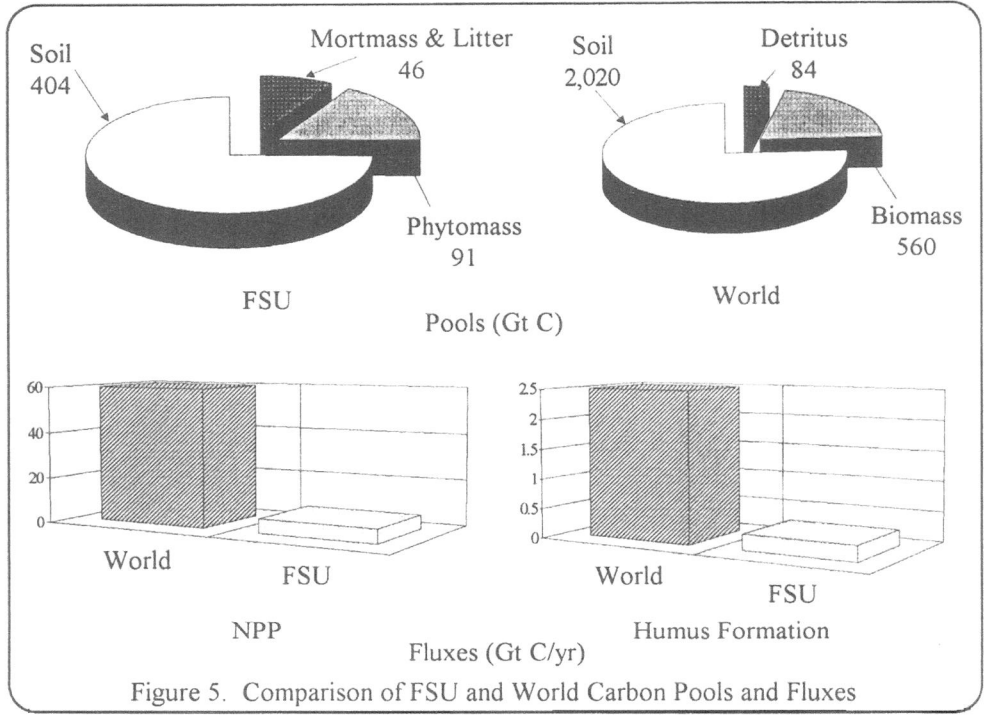

Figure 5. Comparison of FSU and World Carbon Pools and Fluxes

assessment, (3) the productivity of terrestrial ecosystems in the FSU was slightly lower than the world average. This is a reasonable observation since a significant part of the FSU is in the boreal climatic zone; ecosystems of temperate and tropical climatic zones are more productive than ecosystems of boreal climatic zone.

5. Summary and Conclusions

Based on the framework, net primary productivity (NPP) for the FSU was estimated at 6.17 ± 1.65 Gt C yr^{-1}, the vegetation C pool at 118.1 ± 28.5 Gt C, the litter C pool at 18.9 ± 4.4 Gt C, and total soil C pool at 404.0 ± 38.0 Gt C. The area of natural terrestrial ecosystems in the FSU considered herein was 12% of the land surface of the Earth. The phytomass pool of the FSU was 16% of the global biomass pool. The soil and litter pools of the FSU were 20 and 23% of the global soil and detritus pools, respectively. The NPP of the FSU was 10% of the global NPP. The FSU has the greatest expanse of forests, peatlands, and grasslands with rich phytomass, soils, and litter. Thus, the phytomass, soil and litter densities of the FSU were greater than the world average. The productivity of terrestrial ecosystems in the FSU was slightly lower than the world average. A significant part of the FSU is in the boreal climatic zone; ecosystems of temperate and tropical climatic zones are more productive than ecosystems of the boreal climatic zone.

6. Acknowledgments

The work presented herein was funded by the U.S. Environmental Protection Agency (EPA) - Environmental Research Laboratory, Corvallis, Oregon, under Cooperative Agreement CR820239 to Oregon State University. Jeffrey J. Lee is the Project Officer for the project entitled "Carbon Cycling in Terrestrial Ecosystems of the Former Soviet Union." The work presented is a component of the U.S. EPA Global Climate Research Program, Global Mitigation and Adaptation Program, Robert K. Dixon, Program Leader. This paper has not been subjected to the EPA's review and, therefore, does not necessarily reflect the views of the EPA, and no official endorsement should be inferred. Shane Trenary diligently produced the camera ready copy of the manuscript.

7. References

Aleksandrova, V.D. 1977. Geobotanical Regionalization of Arctic and Antarctic. Nauka Press, Leningrad.

Apps, M.J., A.W. Kurz, and T.D. Price. 1991. Application of a carbon budget model to strategic planning for the effects of climate change on the Canadian forest sector. In: T. Kolchugina and T. Vinson, (eds.) *Proceedings of the Workshop on Carbon Cycling in Boreal Forests and Subarctic Ecosystems*. Corvallis, Oregon, Sept. 9-14 , pp. 243-252.

Bartlett, K.B., D.S. Bartlett, and D.I. Sebacher. 1985a. CH_4 flux from coastal salt marshes. *Journal of Geophysical Research* 90:5710-5720.

Bartlett, K.B., D.S. Bartlett, D.I. Sebacher, R.C. Harris, and D.P. Brannon. 1985b. Sources of atmospheric CH_4 from peatlands. Paper presented at the 36th Congress of the International Austronautic Federation, Stockholm, Sweden, Oct. 7-12.

Bazilevich, N.I. 1986. Biological productivity of soil-vegetation formations in the U.S.S.R. Bulletin of Academy of Sciences of the U.S.S.R. *Geographical Series* 2:49-66.

Bazilevich, N.I., O.S. Grebenshikov, and A.A. Tyshkov. 1986. Natural Association of Geographical Patterns with the Structure and Functioning of Ecosystems. Nauka Press, Moscow.

Bazilevich, N.I., L.Ye. Rodin, and N.N. Rozov. 1970. Geographical aspects of biological productivity studies. In: *Proceedings of 5th Congress of Geographical Society of USSR, Leningrad.*

Billings, W.D. 1987. Carbon balance of Alaskan tundra and taiga ecosystems: past, present and future. *Quaternary Science Review* 6:165-177.

Blake, D.R. and F.S. Rowland. 1988. Continuing worldwide increase in traposheric CH_4, 1978-1987. *Science* 239:1129-1131.

Bohn, H.L. 1982. Estimates of organic carbon in world soils: II. *Journal of Soil Science Society of Arizona* 4:1118-1119.

Botch, M.S. 1991. Carbon storage in peat based on regionality of Russian peatlands. In T. Kolchugina and T. Vinson, (eds.) *Proceedings of the Workshop on Carbon Cycling, in: Boreal Forests and Subarctic Ecosystems.* Corvallis, Oregon, Sept. 9-14, pp. 101-108

Botch, M.S. and V.V. Masing. 1983. Mire ecosystems in the U.S.S.R. In: A.J.P.Gore (ed.) Ecosystems of the World Peatlands: Swamp, Peatland, Fen, and Moor. Vol. 4B, Regional Studies. Elsevier, New York, pp. 95-152.

Burrough, P.A. 1986. Principles of Geographical Information Systems for Land Resources Assessment. Clarendon Press, Oxford, New York, p. 193.

Cherdantsev, G N. 1961. Map: Arable land in the U.S.S.R. in 1954. In: J.P. Cole (ed.) A Geography of the U.S.S.R. - Background to a Planned Economy. F.C. German, p. 290.

Dixon, R.K. and D.P. Turner. 1991. The global carbon cycle and climate change: responses and feedbacks from below-ground systems. *Environmental Pollution* 73: 245-262.

Dylis, N.V. and L.M. Nosova. 1977. Phytomass of Forest Biocenosis. Nauka Press, Moscow.

Eriksson, H. 1991. Sources and sinks of carbon dioxide in Sweden. *AMBIO* 20(3-4):146-150.

Etkin, D. 1990. Greenhouse warning: consequences for Arctic climate. *Journal of Cold Regions Engineering*, Amer. Soc. of Civil Engineers 4(1):54-66.

Fung, I.Y., C.J.Tucker, and K.C. Prentice. 1987. Application of advanced very high resolution radiometer vegetation index to study atmosphere - biosphere exchange of CO_2. *Journal of Geophysical Research* 92(D3):2999 - 3015.

Gerasimov, I.P. 1933. About soil-climatical faces of the flatlands of the U.S.S.R. and adjacent countries. *Proceedings of the V.V. Docuchayev Soil Institute*, Vol. 8, No. 5, U.S.S.R. Academy of Sciences Press, Leningrad.

Gucinski, H., D.P. Turner, and G. Koerper. 1992. The carbon flux on forested lands of the United States. *Proceedings of the IUFRO Conference on Integrating Forest Information over Space and Time.* Canberra, Australia.

Harriss, R.C., E. Gorham, D.I. Sebacher, K.B. Bartlett, and P.A. Flebbe. 1985. CH_4 flux from northern peatlands. *Nature* 315:652-653.

Isachenko, A.G., ed. 1988. Landscape Map of the U.S.S.R. Institute of Geography, Leningrad State University.

Kazantseva, T.I. 1980. Productivity and dynamics of above-ground phytomass of deserts. *Problems of Desert Utilization* 2:76-83.

Kazimirov, N.I. and R.M. Morosova. 1973. Biological Turnover of Substances in Spruce Forests of Karelia. Nauka Press, Leningrad.

Keeling, C.D., R.B. Bacastow, A.F. Carter, S.C. Piper, T.P. Whorf, M. Heimann, W.G. Mook, and H. Roeloffzen. 1989. A three-dimensional model of atmospheric CO_2 transport based on observed winds: 1. Analysis of observational data. Geophysical Monographs, Vol. 55, pp. 165 - 236.

Kobak, K.I. 1988. Biotical Compounds of Carbon Cycle. Hydrometeoizdat, Leningrad.

Kobak, K.I. and N.Yu. Kondrashova. 1986. Distribution of organic carbon in soils of the globe. *Trudy* GGI 320: 61-76.

Kononova, M.M. 1963. Soil Organic Matter. Nauka Press, Moscow.

Kurz, W.A., M.J. Apps, T.M. Webb, and P.J. McNamee. 1992. The carbon budget of the Canadian forest sector: Phase 1. *Information Report NOR-X-326*, Forestry Canada, Northwest Region, Northern Forestry Centre, 93 pp.

Matthews, E. 1983. Global vegetation and land use: new high - resolution data bases for climate studies. 22:474-487.

Matthews, E. and I.Y. Fung. 1987. CH_4 emission from natural peatlands: global distribution, area, and environmental characteristics of sources. *Global Biogeochemical Cycles* 1(1):61-86.

Microsoft Corporation. 1991. Microsoft Excel, Redmond, Washington.

Odum, E.P. 1953. Fundamentals of Ecology. W.B. Saunders C., Philadelphia-London.

Panel on Policy Implications of Greenhouse Warming (PPIGW). 1992. Policy Implications of Greenhouse Warming - Mitigation, Adaptation, and the Science Base, Committee on Science, Engineering, and Public Policy, Natl. Academy of Sciences, Natl. Academy of Engineering, Inst. of Medicine, National Academy Press, Washington, D.C.,

Post, W.M., W.R. Emanuel, P.I. Zinke, and A.G. Stagenberger. 1982. Soil carbon pools and world life zones. *Nature* 298(5870): 156-159.

Ryabchikov, A.M., ed. 1988. Map: Geographical Belts and Zonal Types of Landscapes of the World. School of Geography, Moscow State University, Moscow.

Schlesinger,W.H. 1977. Carbon balance in terrestrial detritus. *Annual Review of Ecological Systems*. 8:51-81.

Schneider, S.H. 1990. Global Warming. Vintage Books, New York..

Tytlyanova, A.A. 1977. Biological Cycle of Carbon in Herbaceous Formations. Nauka Press, Novosibirsk.

Tyuremnov, S.N. 1976. Torfyanye Mestorozhdeniya, 3rd ed. Nedra Press, Moscow.

Vatkovskyi, O.S. 1976. The Analysis of Primary Production of Forest Formations. Nauka Press, Moscow.

Vorobyov, G.I., ed. 1985. Forest Encyclopedia. Sovetskaya Ecyclopedia Press, Moscow.

FOREST MANAGEMENT AND CARBON STORAGE:
AN ANALYSIS OF 12 KEY FOREST NATIONS

JACK K. WINJUM
National Council for Air and Stream Improvement
USEPA Environmental Research Laboratory
200 SW 35th Street
Corvallis, OR 97333
USA

ROBERT K. DIXON
US Environmental Protection Agency
USEPA Environmental Research Laboratory
200 SW 35th Street
Corvallis, OR 97333
USA

PAUL E. SCHROEDER
ManTech Environmental Technology, Inc.
USEPA Environmental Research Laboratory
200 SW 35th Street
Corvallis, OR 97333
USA

Abstract. Forests of the world sequester and conserve more C than all other terrestrial ecosystems and account for 90% of the annual C flux between the atmosphere and the Earth's land surface. Preliminary estimates indicate that forest and agroforest management practices throughout the world can enhance the capability of forests to sequester C and reduce accumulation of greenhouse gases in the atmosphere. Yet of the 3600×10^6 ha of forests in the world today, only about 10% (350×10^6 ha) are actively managed. The impetus to expand lands managed for forestry or agroforestry purposes lies primarily with nations having forest resources. In late 1990, an assessment was initiated to evaluate the biological potential and initial site costs of managed forest and agroforest systems to sequester C. Within the assessment, 12 key forested nations were the focus of a special analysis: Argentina, Australia, Brazil, Canada, China, Germany, India, Malaysia, Mexico, South Africa, former USSR, and USA. These nations contain 59% of the world's natural forests and are representative of the world's boreal, temperate, and tropical forest biomes. Assessment results indicate that though the world's forests are contained in 138 nations, a subset of key nations, such as the 12 selected for this analysis, can significantly contribute to the global capability to sequester C through managed tree crops. Collectively, the 12 nations are estimated to have the potential to store 25.7 Pg C, once expanded levels of practices such as reforestation, afforestation, natural regeneration and agroforestry are implemented and maintained. Initial site costs based upon establishment costs for management practices are less than US$33/Mg C.

Water, Air, and Soil Pollution **70**: 239–257, 1993.
© 1993 *Kluwer Academic Publishers.*

1. Introduction

International interest is growing to explore all potential means to reduce the accumulation of greenhouse gases in the atmosphere (IPCC, 1992). A recent analysis shows that expanding forest management within global forest regions has high potential for sequestering atmospheric C (Winjum *et al*, 1992). Such an effort over the next half century would help to offset the increase in atmospheric CO_2, a major greenhouse gas (Dixon *et al*, 1991).

Managed forests in the world total about 350 x 10^6 ha in area or approximately 10% of the forests worldwide (Table 1). Forest and agroforest management practices can contribute to the capability of forests to sequester C (Winjum *et al* , 1992). Examples of practices include: reforestation in the boreal, temperate, and tropical zones; afforestation in the temperate zone; and agroforestry in the tropical zone (Schroeder *et al*, 1993).

The impetus to expand managed forest and agroforestry systems lies primarily with the nations of the world that contain forests. Out of approximately 200 states and territories in the world, 138 are reported to have some closed forests, woodlands, or both (WRI, 1992). Of these, about 100 nations have forest areas that are actively managed (WRI, 1992), i.e., with the objective of enhancing biomass productivity. This paper describes an analysis of 12 key forest nations to determine if such a subset could, as a start, significantly contribute to increased C storage through expanding existing levels of forest management. The 12 nations are: Argentina, Australia, Brazil, Canada, China (i.e., The Peoples Republic of China), Germany, India, Malaysia, Mexico, South Africa, former USSR, and the United States.

As noted above, the paper suggests a starting approach to a global effort to expand forest management to increase C sequestration. It is assumed that the implementation phase of such an effort would begin in the next few decades so that climatic considerations would not be greatly different than at present. Should the predictions of rapid climate change occur over succeeding decades, however, such changes would have to be considered in further planning and executing of the effort. Examples of such considerations would include changes in land suitability for forests, selection of appropriate tree species and genetic stock, and utilization of silvicultural practices that would adapt forests to changing climatic conditions.

2. Background

Forests play a prominent role in the global C cycle (Tans *et al*, 1990). About 26% of the world's land area or 3600 x 10^6 ha is occupied by forests (Postel and Heise, 1988; WRI,

Table 1. Land and forest area statistics for 12 key forest nations compared to world totals. Area numbers are ha x 10⁶.

	Nation's land area (1)				Forest and Woodland (2)							Managed Closed Forest 1980
	Boreal	Temperate	Tropical	Total	Natural Forest			Plantation	Wooded Area	Total	Total as % nation	
Nation					Closed	Open	Sub-total					
1	2	3	4	5	6	7	8	9	10	11	12	13
Argentina		241	36	277	44		44	(3)	16	60	22	x (4)
Australia		622	146	768	42		42		64	106	14	x (3)
Brazil		43	808	851	357	157	514	4	162	680	80	
Canada	769	223		992	264		264		172	436	44	x
China, PR		924	36	960	98	17	115	13	28	156	16	x
Germany		36		36	19		19		1	10	28	7
India			317	317	52	5	57	2	15	74	23	32
Malaysia			33	33	21		21	(3)	5	26	79	3
Mexico		4	193	197	46	2	48	(3)	86	134	68	x
So. Africa		118		118	1		1		3	4	3	10
USSR, former	1344	896		2240	739		739		190	929	42	96 (5)
USA	118	817		935	226		226	(6)	72	298	32	102
Total	2231	3924	1569	7724	1899	181	2090	19	814	2913		249
% World	91	65	31	56	67	24	59	66	48	55		71
World	2461	6074	5009	13812	2822	742	3564	29	1695	5288		350

(1) Values from New York Times, Atlas of the World (1988).
(2) Values from WRI (1992).
(3) Values are less than 1 x 10⁶ ha.
(4) x denotes data not currently available (WRI, 1992).
(5) Estimated from Krankina and Dixon (1992).
(6) In 1985, the USA had approximately 5 x 10⁶ ha of plantations included in the 226 x 10⁶ ha of closed forests (Laarman and Sedjo, 1992).

1992). Moreover annually, forests account for some 90% of the C flux between the atmosphere and terrestrial ecosystems through their photosynthetic uptake and respiratory release of CO_2. This flux amounts to about 90 Pg of C (Schneider, 1989). Further, the estimate of the C stored in global forest systems is approximately 1400 Pg (500 Pg above ground and 900 Pg below ground) or about 77% of the world's total terrestrial C of 1800 Pg (Sedjo, 1992).

At the UN Conference on Environment and Development (UNCED, i.e., the Earth Summit) in Rio de Janeiro during June 1992, over 100 nations of the world recognized the prominent role of forest biomes in global ecology and the global C cycle. This recognition led to the promulgation of a set of 51 Forest Principles. The Principles address a broad array of concerns about global forest resources, but primary objectives are to: 1) promote sustained forest management and development; 2) curb deforestation; 3) protect biodiversity; and 4) identify threats to the world's forests (Heiner, 1992). Timetables and specific targets for action will be determined in the future, but key forest nations like those analyzed here must ultimately provide the leadership for successful implementation of the Principles on a world scale.

3. Methods

The 12 key nations were selected by three criteria: those having 1) a large amount of forest land, existing or potential; 2) credible biologic and economic data on the nation's forest resources and management practices; and 3) locations that collectively make a cross section of the boreal, temperate, and tropical latitudes in a manner that provided coverage of all forested continents or major island groups of the world (Howlett and Sargent, 1991).

The analyses of key nations are based upon a global database assembled from the technical literature on managed forest and agroforest systems. Typical forest practices covered in the database are reforestation, afforestation, and natural regeneration. Agroforest management was considered as one practice in this analysis though within this practice there is a wide variety of agroforestry approaches (MacDicken and Vergara, 1990).

Within nations where these practices are utilized and described in the literature, values were entered into the database on two variables: annual stemwood growth of trees per hectare and implementation costs. Use of the values for these variables is briefly described below and in detail by Dixon *et al*, (1991).

3.1 MEAN CARBON STORAGE

To evaluate forestation practices from place to place in the world, this paper deals with the average amount of C stored above ground over an indefinite number of rotations. The concept is called mean C storage or MCS. It was described and used by Schroeder (1992) and the same caveats apply here. That is, it is assumed that once a practice is implemented, the forest system is sustainable, and there is no yield reduction in later rotations. Thus the result is the same as the average amount of C site over one full rotation. Because any number of biological, climatic, or social events could contribute to some level of yield reduction that cannot be predicted (Smith, 1986), the approach presented here may represent an upper bound. This calculation can be made by summing the C standing crop for every year in the rotation and dividing by the rotation length -- i.e., by calculating the Mean C Storage (MCS) as follows:

$$MCS = \frac{\sum_{i=1}^{n} (C\ ha^{-1})_{standing\ crop}}{(rotation\ length)}$$

where: n = rotation length, and C ha^{-1} = stem wood volume ha^{-1} x 1.6 (i.e. conversion factor to get whole-tree biomass ha^{-1}) x wood density x 0.5 (i.e., assuming whole-tree biomass is 50% C) (Brown and Lugo, 1982; Sedjo and Solomon, 1989; Schroeder and Ladd, 1991)

Not included are data for C stored below ground or in durable wood products manufactured from tree harvests. C storage for these items would add to the base MCS calculated for each newly-established forest systems.

This approach assumes that at harvest or shortly thereafter, all stored C returns to the atmosphere. Further, the levels of C sequestration computed by formula 1 for each practice are gross and not net increases of C sequestered. That is, the C sequestered if the land was left unmanaged with some natural revegetation which is not accounted for in this analysis. That is, where land is left unmanaged, the C in natural revegetation occurring in these lands is not accounted for in this analysis.

3.2   IMPLEMENTATION COSTS

The implementation costs of forest management practices were collected concomitantly with stemwood growth data, though not all references contained data for both costs and forest productivity. Implementation costs are reported in various ways, but they generally include site preparation, stock costs (i.e., seed/cuttings, nursery/greenhouse propagation, packing, storage, and transportation), and planting labor plus supervision. Other cost considerations, not included here, will be added in later analyses as data become available; examples are: 1) the cost of land; 2) annual or tending costs; and 3) nonfinancial costs associated with maintaining the forest stock for the duration of the crop rotation. Similarly, revenues from forest resource benefits are not credited in this analysis. Though reliable data on all costs and revenues for world forests are not generally available at present, they are very important and will undoubtedly affect future conclusions regarding the cost effectiveness of forest management for C sequestration (Winjum and Lewis, 1993).

Economic data are reported in U.S. dollars adjusted to 1990 accounting for the inflation and exchange rates for individual nations (Dixon *et al*, 1991). When forests are considered as renewable resources (Smith, 1986), the costs of initiating forest management or establishing plantations are recurring costs. In estimating costs, it is important to account for these additional costs that will occur at more or less periodic intervals in the future (Davis and Johnson, 1987). Therefore, the present value of a series of successive future costs over a 50-yr period was computed for expressing the implementation costs for each practice (Dixon *et al*, 1991).

3.3   CALCULATION OF CARBON COSTS

Cost per Mg of C was calculated as the present value of implementation costs over a 50 year period divided by mean C storage. It is important to emphasize that costs computed in this manner do not account for any financial benefits that result from the initial investment and the production of useful products (Gregersen *et al*, 1989).

3.4   STATISTICAL METHODS

The data assembled for this analysis were from a very wide variety of sources and were not selected by random sampling. Plots of the data showed that distributions were not normal but unimodal with a positive skew, i.e., highest frequency for the low numbers (Dixon *et al*, 1991). Therefore, in the following presentations, results are more properly presented (Table 2) as sample medians with variation indicated by the interquartile ranges (middle 50% of observations)(Devore and Peck, 1986). The sample sizes (n) are also shown.

3.5   DATA QUALITY

In a broad review involving large amounts of data from many sources, data quality will vary. Referenced technical data for this analysis were considered the best available. When data on forest productivity or initial costs were encountered that were clearly outside reported ranges and were not adequately explained, they were not used in the analysis.

4. Results

Overall, analysis results show that a subset of 12 key nations appear representative of the world's 138 nations or territories that have forest resources. That is, the total forest and woodland area within the 12 key nations is 2900 x 10^6 ha, slightly over half (55%) of the world total (Table 1). Of the total 2900 x 10^6 ha in the 12 key nations, 1900 x 10^6 ha (65%) are closed forests, that is those forests where tree crowns cover more than 20% of the ground area (WRI, 1992). Managed closed forests cover 249 x 10^6 ha, or 71% of the 350 x 10^6 ha managed worldwide (Table 1).

Table 2. Values on forest management practices for 12 key nations based upon a database assembled from the technical literature and calculations of statistics releve to carbon storage such as mean carbon storage (MCS in Mg C ha^{-1}), implementation costs ($ ha^{-1}), and cost of carbon stored ($/Mg C) (Winjum et al, 1992).

Nation/Practice	Mg C ha^{-1}				$ ha^{-1}				$/Mg C
	Lower Quartile	Median	Upper Quartile	n	Lower Quartile	Median	Upper Quartile	n	Column 6 ÷ Column 2
Column Number	1	2	3	4	5	6	7	8	9
Argentina									
Reforestation	33	54	88	114	662	1684	1684	7	31
Afforestation	51	55	63	13	988	988	988	4	18
Australia									
Reforestation	46	74	118	16	306	347	740	16	5
Brazil									
Reforestation	53	65	101	57	293	637	1207	52	10
Canada									
Reforestation	39	39	41	8	335	417	513	10	11
Natural regen.	12	19	23	12	93	121	170	12	6
China, Peoples Republic									
Reforestation	28	41	95	43	329	393	410	8	10
Germany									
Reforestation	38	48	78	6	442	1391	3662	6	29
India									
Reforestation	8	31	70	33	220	477	1845	11	15
Malaysia									
Reforestation	60	66	115	14	285	303	309	5	5
Mexico									
Reforestation	59	105	143	8	354	402	526	4	4
So. Africa									
Reforestation	105	110	118	8	910	952	993	8	9
USSR, former									
Reforestation	13	15	17	6	69	83	171	6	6
Natural regen.	12	17	18	7	83	83	83	7	5
USA									
Reforestation	19	56	98	21	53	256	346	33	5
Afforestation	98	126	194	104	39	255	373	104	2
Total				470				293	

In addition, across latitudinal zones (Table 1), 91% of the boreal forest region is contained in Canada, USSR (former), and the United States (i.e., Alaska). Approximately 65% of the world's temperate forests are contained in Argentina, Australia, Brazil, Canada, China, Germany, Mexico, South Africa, USSR (former), and the United States; and 31% of all tropical forests lie in Argentina, Australia, Brazil, China, India, Malaysia, and Mexico (Table 1).

In each nation's forest land management program, data were found in the literature for the practice of reforestation, and in some cases for afforestation and natural regeneration (Table 2). Sufficient data to evaluate agroforest management were not found. In the database, therefore, the sample size (n) for C storage across the forestation practices in the 12 nations summed to 470 data points; for implementation costs, the total sample size (n) reached 293 data points (Table 2).

4.1   NATIONAL HIGHLIGHTS

These data are sufficient to provide useful insights on the role of forest management in global C storage now, and its potential for expansion. National highlights, given below, serve as an illustration.

4.1.1. Argentina

Approximately 22% of Argentina or 60 x 10⁶ ha is occupied by closed forests or woodlands (Table 1). Though the amount of land under active management is not reported, the annual reforestation rates in the 1980s were reported to be about 50,000 ha. Reforestation of formerly forested areas and afforestation of pastures appears to be viable options (Houghton *et al,* 1991). Several species of pine (Pinus) and Eucalyptus are planted. Mean C storage (MCS) levels for reforestation and afforestation in Argentina are 54 and 55 Mg C ha⁻¹, respectively (Table 2). Costs of C sequestration through reforestation and afforestation practices are US$31 and US$18/Mg C (Table 2).

4.1.2   Australia

With 6% of world's land area, Australia is a large nation, though much of the land is too arid for forest growth. However, 14% of the nation is occupied by closed forests and woodlands. In addition, agroforestry is practiced by 25% of the farmers and ranchers in this nation where trees are planted as windbreaks, shade, and shelter for agricultural crops (Prinsley, 1991). The humid temperate lands (12%) and the humid tropical lands (19%) are among the most productive forest lands in the world. For reforestation, species of pine (Pinus) and Eucalyptus are planted. The mean C storage potential of plantations has a median of 74 Mg C ha⁻¹ (Table 2). Costs of C storage are US$5/Mg C for reforestation (Table 2). The Australian government has committed to a national reforestation program (1 x 10⁹ trees) that will eventually encompass 10 x 10⁶ ha (Howlett *et al,* 1991). Compared to the 42 x 10[6] ha of closed forested reported for Australia (WRI, 1992), achieving such a reforestation goal would significantly increase the nation's forest area.

4.1.3 Brazil

Deforestation in Brazil is a significant source of CO_2 emissions (Houghton *et al*, 1991). A large body of literature reviews the current deforestation and forest degradation patterns in Brazil (Fearnside, 1989). Approximately 80% of Brazil, or 680 x 10[6] ha, is covered with closed and open forests or woodlands (Table 1). Given the rich biodiversity and large stock of standing C in forest systems, the most cost efficient option to store C may be to conserve the remaining forests in Amazonia (IPCC, 1992). Large-scale forestation projects, e.g., FLORAM, have been proposed, but demographic, economic, and political factors are significant barriers to implementation (Andrasko *et al*, 1991).

In the 1980s, the annual reforestation rate in Brazil was about 450,000 ha in which several species of Araucaria, Eucalyptus, and Pinus were planted. Mean C storage (MCS) for these plantations is about 65 Mg C ha-1. Agroforestry systems Brazil appear to have high potential, but definitive data to compute MCS were not available. For reforestation practices, costs of C sequestration based upon establishment costs are estimated to be US$10/Mg C (Table 2).

4.1.4. Canada

Three-fourths of Canada's almost 1000 x 10[6] ha lie in the boreal region (Table 1). Approximately 44% of Canada is occupied by natural forests and woodlands (Table 1). The forest sector of Canada is estimated to store 88 Pg C with about 40% contained in the boreal forests (Kurz *et al*, 1992). Annual reforestation rates in the 1980s were reported to be 720,000 ha (WRI, 1992) using species of spruce (Picea) and some Douglas-fir (Pseudotsuga menziesii) (Lavender, 1991). Because of limiting edaphic and climatic conditions of the boreal region, however, forest growth and mean C storage (MCS) are relatively low in much of Canada, e.g., 19 Mg C ha-1 for naturally regenerated forests (Table 2). Yet, a recent analysis by Apps and Kurz (1991) indicates that the Canadian forest sector in 1986 was a net C sink of 0.05 Pg, including changes in forest biomass, soil C, and C storage in forest products. MCS can be as high as 39 Mg C ha-1 for reforested lands (Table 2). Based on implementation costs, reforestation and natural regeneration can sequester C at US$11 and US$6/Mg C, respectively (Table 2).

4.1.5. China (The Peoples Republic of China)

Approximately 16% of China is covered with natural forests or woodlands for a total of 156 x 10[6] ha (Table 1). China is phytogeographically transitional between humid temperate and tropical regions.

The Green China program has a goal of establishing 10 to 15 x 10^6 ha of forests annually (OCAC, 1986). An earlier national goal was to increase the total forest area from 115 x 10^6 ha to 192 x 10^6 ha (20% of the nation) by the year 2000 (FAO, 1982). These programs are aimed at restoring abandoned or under-utilized land, as well as providing employment, income, and other social benefits.

Numerous temperate and tropical trees are used for plantations in China, and among the most common are species of Eucalyptus, larch (Larix), pine (Pinus), and poplar (Populus). Established plantations can produce MCS levels of 41 Mg C ha^{-1} (Table 2). Based upon implementation costs, reforestation C sequestration costs are about US$10/Mg C (Table 2). Agroforestry systems are widely used in China, and are a vital part of the nations wood resources. Typical use of trees with agricultural operations includes windbreaks, amenity plantings around farm buildings and roads; dune forest plantations to stabilize encroaching sands; shade intercrops with the growing of some cereals and vegetables (Richardson, 1990). Amounts of MCS by the agroforestry systems in China are uncertain (Richardson, 1990).

4.1.6. Germany

As a historically important nation regarding forest management practices, Germany has approximately 9 x 10^6 ha of closed and open forests and 1 x 10^6 ha of woodlands totaling 28% of the national land area (Table 1). Further, more than 0.7 x 10^6 ha of land have been identified as suitable for afforestation within pine (Pinus) and spruce (Picea) species (Volz et al, 1991). Currently Central and Western European forests are expanding in volume and are a net C sink (Kauppi et al, 1992). MCS in plantations resulting from reforestation in Germany was found to be 48 Mg C ha^{-1} at US$29/Mg C (Table 2) based upon implementation costs. These costs escalate significantly if mid-rotation management costs and subsidies to land owners are considered. Urban forestry as commonly practiced in Germany, may also provide opportunity for C sequestration in Central and Eastern Europe (Smith, 1991).

4.1.7. India

The forests of India have been degraded and harvested due to extreme demographic and environmental pressures (Jain et al, 1989). At the beginning of the 20th century, more than one-half of India was forested. Today less than 20% of India (52 x 10^6 ha) is occupied by closed forests (Table 1). Demographic factors and industrialization have created a large demand for fuelwood. Over 800 x 10^6 cattle, sheep, and goats browse commercial forests and woodlands. Nevertheless, current plans call for implementing forest and agroforest practices to establish tree cover on at least 33% of India's land area (Pandeya, 1991).

Tree species commonly planted are species of Acacia, Albizzia, Eucalyptus, Leucaena, and Pinus (Dixon et al, 1991). Plantations typically sequester 31 Mg C ha^{-1} at US$15/Mg C

(Table 2). Implementing forest management and agroforestry systems in India would also provide numerous social benefits such as employment and wood building materials (Pandeya, 1991). Further, wood stock for a biofuels program would greatly supplement national energy resources. Constraints to large-scale forest programs include limited training, poor forest management infrastructure, and a lack of funding (Sharma *et al*, 1989).

4.1.8 Malaysia

Lying in the tropical region, Malaysia has 21×10^6 ha of closed forests occupying 64% of the nation's land area (Table 1). The deforestation rate in Malaysia as of 1989 was 480,000 ha yr[-1] (Tho, 1991). Malaysia is very active in the international forest-product markets. The states of Sabah and Sarawak, in particular, are major exporters of tropical hardwoods.

Reforestation rates during the 1980s in Malaysia were 20,000 ha yr[-1] (WRI, 1992). Plantations include species of Acacia, Albizzia, Gmelina, Pinus, and teak (Tectonia) (Dixon *et al*, 1991). C storage in the plantations of Malaysia has a median of 66 Mg C ha[-1] and a sequestration cost of US$5/Mg C (Table 2).

4.1.9. Mexico

Demographic and environmental factors have significantly contributed to deforestation in Mexico (Masera *et al*, 1992). Greenhouse gas emissions resulting from deforestation are estimated to be 45×10^6 Mg annually. Today, less than 25% of the total land area of Mexico is covered with forests (Table 1). The potential land technically suitable for reforestation, agroforestry, and fuelwood systems is approximately 120×10^6 ha.

Reports indicate that annually about 20,000 ha of plantations were established in the 1980s (WRI, 1992) mostly with species of pine (Pinus) though species of Casuarina and teak (Tectona) are also utilized. The MCS potential for reforestation in Mexico is about 105 Mg C ha[-1] (Table 2). Costs of C sequestration for reforestation practices were found consistently low in Mexico at US$2/Mg C (Table 2).

4.1.10. South Africa

A large proportion of the land area in South Africa is dedicated to pasture or other non-forest uses. Livestock grazing is a prominent industry. Silvopastoral systems which integrate livestock and forests may be an appropriate management option in the future. South Africa in the 1980s had about 1×10^6 ha of closed forests and 3×10^6 ha of woodlands (Table 1). The closed forests of the South Cape area were extensively exploited by European settlers beginning in the 1700s. Today, even-aged stands of introduced species of Acacia, Eucalyptus, and Pinus are planted and grown in this area for timber production, particularly after fire disturbance (Geldenhuys, 1982). Reforestation here can sequester C at US\$9/Mg C for MCS levels of approximately 110 Mg C ha-1 (Table 2).

4.1.11. USSR (former)

The former USSR covers 16% of the earth's land area or 2200 x 10^6 ha. Russia contains 26% of world's closed forest area (Table 1) or approximately 739 x 10^6 ha. Phytomass, litter, and soil C pools of the forest sector of former USSR are 91, 12, and 319 Pg, respectively (Kolchugina and Vinson, 1993). This C is about one-sixth of the C in the terrestrial biosphere of the world. Currently, the forest sector of the former USSR is a net sink of 0.5 Pg C annually (Kolchugina and Vinson, 1993). Boreal forest fires in Russia are currently a major source of annual greenhouse gas emissions (Dixon and Krankina, 1993).

Although a complete forest inventory of the former USSR has not been completed, the opportunity for reforestation, stand improvement with silvicultural treatments, and forest protection appears significant (Krankina and Dixon, 1993). The area of mature and over-mature forests, however, is greater than 200 x 10^6 ha so that the opportunity to stimulate C sequestration in these existing forests is low. During the decades of the 1970s and 80s, tree planting and seeding were practiced on nearly 1 x 10^6 ha yr-1 (Burdin, 1991). Plantations are commonly stocked with species of pine (Pinus), poplar (Populus), and spruce (Picea) (Dixon et al, 1991). C sequestration potential for reforestation and natural regeneration practices are 15 and 17 Mg C ha-1, respectively (Table 2). Costs of C sequestration using initial costs for these practices are US\$6 and US\$5/Mg C, respectively (Table 2).

4.1.12. United States

About 32% of the United States is occupied by closed forests (226 x 10^6 ha) and open woodland (72 x 10^6 ha; Table 1). In the 1980s, the average reforestation rate in the U.S. was 1.8 x 10^6 ha yr-1 (WRI, 1992). This level of annual reforestation has grown from small programs sponsored by the federal government in the 1930s. As the programs evolved, sponsors have included various federal, state, and private organizations, and the scope encompasses tree planting on recently harvested forest lands, marginal agricultural lands, parks and even urban settings (USDA, 1988; Skiera and Moll, 1992). Plantation species are numerous, but commonly include the following groups: alder (Alnus), fir (Abies and Pseudotsuga), cottonwood (Populus), pine (Pinus), spruce (Picea), and walnut (Juglans) (Dixon et al, 1991).

The most comprehensive analysis to date of C storage by forest management in the United States was done by Moulton and Richards (1990). They estimated that an extensive tree planting and forest management program on private lands could store up to 730 x 10^6 Mg C. The total cost of this program, including land rental cost, would be US\$19.5 x 10^9.

The mean C storage (MCS) for reforestation and afforestation across the nation's temperate regions are 56 and 126 Mg C ha[-1], respectively (Table 2). Based on implementation costs, C can be sequestered by these practices at less than US$5/Mg C (Table 2). Using annualized data for land rent, treatment costs, and C sequestration, Moulton and Richards (1990) calculated that the cost of C sequestered by active forest management across U.S. forest regions ranged from US$10 to US$30/Mg C; the range for tree planting alone is US$6 to US$24/Mg C.

4.2   POTENTIAL OF THE "12"

Globally, what contribution could be made if the 12 nations in this analysis expanded their forest and agroforest management? In part, the answer to the question would also depend upon the amount of land that could be reforested, afforested, naturally regenerated, or utilized for agroforestry. A challenging world target suggested at the Noordwijk Ministerial Conference (1989) was to increase forest management on 12×10^6 ha yr[-1] as a net increase over global deforestation.

Deforestation rates in the 1980s varied by forest region. In the boreal and temperate regions, reforestation and some national afforestation projects have roughly stabilized the overall area in forest cover during the last decade (Allan and Lanly, 1991). In the tropical regions, deforestation rates in the 1980s were estimated to be about 17×10^6 ha yr[-1] (Allan and Lanly, 1991; WRI, 1992). During the same period, annual reforestation in the tropics was about 1×10^6 ha (WRI, 1992). Thus the global deforestation rate is estimated to be approximately 16×10^6 ha yr[-1].

If the Noordwijk target is to be achieved by the year 2000 and maintained until mid-century, then approximately 28×10^6 ha would need to be added annually to world's total land area through forestation practices (i.e. in 10^6 ha, $16 + 12 = 28$). Assuming these new forests remain under management over 50 yr, about 1400×10^6 ha would be added to the 350×10^6 ha now reported (Table 1). Since the 16×10^6 ha yr[-1] makes up for global deforestation, the net gain in forest area by these estimates would be 600×10^6 ha by 2050 (12×10^6 ha x 50 yr). The total area of closed plus open forests would then be 4200×10^6 ha (Table 1). If this projection were realized, 42% of the total would then be actively managed (i.e., [{$1400 + 350 \times 10^6$ ha} ÷ 4200×10^6 ha] x 100), a seemingly reasonable target for mid-21st century.

The 12 key forest nations in this analysis contain 59% of the 3600×10^6 ha of natural forests in the world (Table 1). Taking 59% of the increased area by forestation practices as a fair share for the 12 key forest nations, the target becomes 826×10^6 ha by the year 2050 (i.e., $0.59 \times 1400 \times 10^6$ ha). Using total natural forest areas for individual nations (Column 8, Table 1) to weight medians of C storage (MCS) for practices within nations (Column 2, Table 2), the weighted average for C storage for the 12 nations is 47.4 Mg C ha^{-1}. Thus multiplying 47.4 Mg C ha^{-1} times 826 x 10^6 ha gives 40 Pg C, the amount stored by all the practices listed in Table 2 for the 12 nations once implemented and maintained on a continuous basis.

It was shown in a similar analysis of promising forestry practices implemented on socially and politically available lands worldwide (i.e., across most of the 138 nations with forests) that 50 to 100 Pg of C could be stored (Winjum *et al,* 1992). Preliminary projections here are that just 12 key forest nations could conceivably make a significant contribution to increased C storage in the world via managed forest ecosystems. That is, perhaps as much as 80% (40 of 50 Pg) or as little as 40% (40 of 100 Pg) of the world potential for enhancing forest C storage could be attained through expanded forest and agroforest management by this subset of the world's nations with forests.

For costs, a weighted average of the medians for US$/Mg C (Column 9, Table 2) is US$7.30/Mg C, when computed in the same manner as above. This cost times 40 Pg is US$292 x 10^9 over 50 yr, or US$5 to US$6 x 10^9 yr^{-1} for the world to invest in increasing the production of basic forest goods and services including C storage. It should be remembered that the full lists of costs for these forest management practices are not yet included in this analysis, but neither are revenues from the benefits.

5. Discussion and Conclusions

The results indicate that the 12 forest nations evaluated in this analysis could significantly enhance global storage of C through increased utilization of forest and agroforest practices. The range of medians for mean C storage (MCS) is 15 to 126 Mg C ha^{-1} (Table 2). These levels compare favorably to similar long-term storage values reported for other ecosystems. For example in the tropics, Brown and Lugo (1982) calculated that primary closed forests contain 90 Mg C ha^{-1} and primary open forests have about 31 Mg C ha^{-1}. Waring and Schlesinger (1985) show that storage values for temperate grassland and tropical savanna to be 10 and 20 Mg C ha^{-1}, respectively.

Considering only implementation or initial site costs, the costs per Mg C range from US$2 to US$31/Mg C. If financial data were available on land and crop maintenance for these practices, the investments or costs in US$/Mg C could double. Likewise, however, not considered in the analysis are potential credits of revenues realized from forest benefits -- this would help offset and in some cases cover the higher costs. But as they are, these approximate costs compare favorably with alternative options for C sequestration and conservation presented by the U.S. National Academy of Sciences (NAS, 1991). The NAS analysis considered such options as improved energy efficiency and alternatives to fossil fuels. The low cost category was less than US$33/Mg C; the moderate ranged from US$34 to US$363/Mg C (NAS, 1991).

It is concluded, therefore, that since more than 90 of the 138 forest nations are reported to have ongoing programs of forest and agroforest management (Dixon *et al,* 1991), a considerable near-term contribution toward C sequestration could be made in the world by expanding management practices within a subset of capable nations. In the meantime, efforts to find solutions could begin in nations wishing to make contributions to global C sequestration but who have constraints (biological, social, economic, etc.) to developing their management programs. Such constraints are very significant barriers in many instances and will likely require intensive and innovative efforts to achieve resolutions.

This "easy-first" approach, however, has been successful in accomplishing large forestation projects with many uncertainties at the outset (Winjum *et al,* 1986). It means getting started with what is known and easiest to do in the early years while simultaneously solving constraints as a need to allow progress in later years. Here, the suggestion is to consider the approach on a nation-to-nation basis to harness and expand the global potential of forest and agroforest management to sequester C over the next half century. Expansion could begin with a subset of nations most able to contribute (e.g., the 12 analyzed in this paper) while international efforts are at work to resolve constraints within other nations that could contribute in later years.

6. Caveats and Follow-up Research

The results and conclusions presented are subject to several caveats and cautions. These also serve as guidelines to follow-up research, such as:

- The effect of projected climate change on the productivity of forest systems, particularly those under active management;
- The role of other forested nations;
- Contributions by additional forest practices such as use of silvicultural tending treatments, genetic improvement, and protection from catastrophic wildfires;
- Consideration of all costs besides just those for implementation such as for land and crop maintenance;
- Inclusion of dollar values for benefits so that net costs can be computed;
- Improved estimates of land suitable and available;
- Capability of management practices to contribute to:
  - sustainable forest productivity; and
  - adaptations to global climate change.

7. Disclaimer

The research described in this paper has been funded by the US Environmental Protection Agency . The paper has been prepared at the EPA Environmental Research Laboratory in Corvallis Oregon, USA, through Interagency Agreement No. DW12934530/Grant PNW 91-0051. It has been subjected to the Agency's peer and administrative review process and approved for publication. Mention of trade names or commercial products does not constitute endorsement or recommendation for use.

8. References

Allan, T. and J.P. Lanly. 1991. Overview of status and trends of world forests. p. 17-39. In: Technical workshop to explore options for global forestry management. D. Howlett and C. Sargent (eds.). International Institute for Environment and Development, London. 349 pp.

Andrasko, K., K. Heaton, and S. Winnett. 1991. Evaluating the costs and efficiency of options to manage global forests: a cost curve approach. p. 216-233. In: Technical workshop to explore options for global forestry management. D. Howlett and C. Sargent (eds.). International Institute for Environment and Development, London. 349 pp.

Apps, M.J. and W.A. Kurz. 1991. Assessing the role of Canadian forests and forest sector activities in the global carbon balance. World Resource Review, 3:333-344.

Brown, S. and A.E. Lugo. 1982. The storage and production of organic matter in tropical forests and their role in the global carbon cycle. Biotropica, 14:161-187.

Burdin, N.A. 1991. Trends and prospects for the forest sector of the USSR: a view from inside. Unasylva, 165(42):43-50.

Davis, L.S. and K.M. Johnson. 1987. Forest Management. McGraw-Hill, New York, NY. 790 pp.

Devore, J. and R. Peck. 1986. Statistics, The Exploration and Analysis of Data. West Publishing Co., St. Paul. MN. 699 pp.

Dixon, R.K. and O.N. Krankina. 1993. Forest fires in Russia: contribution of carbon dioxide to the atmosphere. Canadian Journal of Forest Research. (In press).

Dixon, R.K., P.E. Schroeder, and J.K. Winjum (eds.). 1991. Assessment of Promising Forest Management Practices and Technologies for Enhancing the Conservation and Sequestration of Atmospheric Carbon and Their Costs at the Site Level (USEPA/600/3-91/067). US Environmental Protection Agency Environmental Research Laboratory, Corvallis, OR. 138 pp.

Fearnside, P.M. 1989. Contribution to the greenhouse effect from deforestation in Brazilian Amazonia. Paper presented to the International conference on soils and the Greenhouse Effect, Wageningen, The Netherlands, 14-18 August, 1989. 43 pp.

Food and Agricultural Organization (FAO). 1982. Forestry in China. Forestry Paper 35. UN Food and Agricultural Organization, Rome. 308 pp.

Geldenhuys, C.J. 1982. Management of the Southern Cape forests. South Africa Forestry
    Journal, 121:4-10.
Gregersen, H., S. Draper, and D. Elz. 1989. People and trees: The role of social forestry
    in sustainable development. EDI Seminar Series, World Bank, Washington, D.C.,
    273 pp.
Heiner, H. 1992. Report from UNCED: The challenge of global forest management.
    Journal of Forestry, 90(9):28-31.
Houghton, R.A., J. Unruh, and P.A. Lefebvre. 1991. Current land use in the tropics and
    its potential for sequestering carbon. p. 297-310. In: Technical workshop to
    explore options for global forestry management. D. Howlett and C. Sargent (eds.).
    International Institute for Environment and Development, London. 349 pp.
Howlett, D. and Sargent C.(eds.), 1991. Technical Workshop to Explore Options for
    Global Forestry Management. International Institute for Environment and
    Development, London. 349 pp.
Intergovernmental Panel on Climate Change (IPCC). 1992. Agriculture, Forestry, and
    Other Human Activities: Working Group III Response Strategies, Supplement.
    IPCC, UNEP-WMO, Geneva. 65 pp.
Jain, R.K., K. Paliwal, R.K. Dixon, and D.H. Gjerstad. 1989. Improving productivity of
    multipurpose trees growing on sub-standard soils. Journal of Forestry, 87:38-42.
Kauppi, P.E., K. Mielikäinen, and K. Kuusels. 1992. Biomass and carbon budget of
    European forests, 1971 to 1990. Science, 256:70-78.
Kolchugina, T.P. and T.S. Vinson. 1993. Carbon sources and sinks in forest biomes of
    the Soviet Union. Global Biogeochemical Cycles. (In press).
Krankina, O.N. and Dixon R.K.. 1992. Forest management in Russia: Challenges and
    opportunities in the era of perestroika. Journal of Forestry, 90(6):29-34.
Kurz, W.A., M.J. Apps, T.M. Webb, and P.J. McNamee. 1992. The Carbon Budget of
    the Canadian Forest Sector. Information Report NOR-X-326, Forestry Canada,
    Northern Forestry Centre, Edmonton, Alberta. 93 pp.
Laarman, J.G. and R.A. Sedjo. 1992. Global Forests: Issues for Six Billion People.
    McGraw-Hill, Inc., New York, NY. 335 pp.
Lavender, D.P. 1991. Reforestation in British Columbia. In: Proceedings of the
    International Workshop on Large-Scale Reforestation. J.K. Winjum, P.E.
    Schroeder, and M.J. Kenady (eds.). No. EPA/600/9-91/014, USEPA Washington
    D.C. pp. 29-40.
MacDicken, K.C. and N.T. Vergara. 1990. Agroforestry: Classification and Management.
    Wiley & Sons, New York, NY. 381 pp.
Maini, J.S., 1991. Towards an international instrument on forests. p. 278-285. In:
    Technical workshop to explore options for global forestry management. D. Howlett
    and C. Sargent (eds.). International Institute for Environment and Development,
    London. 349 pp.
Masera, O., M. de Jesús Ordòñez, R. Dirzo. 1992. Carbon emissions from deforestation
    in Mexico: Current situation and long-term scenarios. Lawrence-Berkeley
    Laboratory, Berkeley, CA. Report No. 32665. 45 pp.

Moulton, R.J. and K.R. Richards. 1990. Costs of sequestering carbon through tree
    planting and forest management in the United States (General Technical Report
    WO-58). US Department of Agriculture Forest Service, Washington, D.C., 48 pp.
National Academy of Science (NAS). 1991. Policy Implications of Greenhouse Warming.
    US National Academy Press, Washington, D.C., 127 pp.
New York Times (NYT). 1988. Atlas of the World. Times Books of Random House, New
    York, NY. 241 pp.
Noordwijk Ministerial Conference (NMC). 1989. The Noordwijk Declaration on Climate
    Change. Ministerial Conference on Atmospheric Pollution and Climate Change,
    Noordwijk, The Netherlands. 15 pp.
Office of Central Afforestation Committee (OCAC). 1986. Green China. The Great Wall
    Publishing House, Beijing. 32 pp.
Pandeya, C. 1991. Forests of India. p. 117. In: Technical workshop to explore options
    for global forestry management. D. Howlett and C. Sargent (eds.). International
    Institute for Environment and Development, London. 349 pp.
Postel, S. and L. Heise. 1988. Reforesting the Earth. Worldwatch Paper 83. Worldwatch
    Institute, Washington, D.C. 66 pp.
Prinsley, R.T. 1991. Australian Agroforestry. Rural Industries Research and Development
    Corporation, Canberra. 90 pp.
Richardson, S.D. 1990. Forests and Forestry in China. Island Press, Washington, D.C.
    352 pp.
Schneider S.H. 1989. The changing climate. Scientific American, 261(3):70-79.
Schroeder, P.E. 1992. Carbon storage potential of short rotation tropical tree plantations.
    Forest Ecology and Management, 50(1992):31-41.
Schroeder, P.E., R.K. Dixon, and J.K. Winjum. 1993. An assessment of forest
    management and agroforestry to sequester and conserve atmospheric CO_2.
    UNASYLVA. (In press).
Schroeder, P.E. and L. Ladd. 1991. Slowing the increase of atmospheric carbon dioxide:
    A biological approach. Climate Change, 19:283-290.
Sedjo, R.A. 1992. Temperate forest ecosystems in the global carbon cycle. Ambio.
    21(4):274-277.
Sedjo, R.A. and A.M. Solomon. 1989. Climate and forests. In: Greehouse Warming:
    Abatement and Adaptation, Proceedings of a Workshop. N.S. Rosenburg, W.E.
    Easterling, P.R. Crossen, and J. Dormstadter (eds.). Resources for the Future
    Washington D.C., 1988:105-109.
Sharma, R.N., H.L. Sharma, and H. S. Garcha. 1989. Fuelwood from Wasteland.
    Bioenergy Society of India, New Delhi. 486 pp.
Skiera, B. and G. Moll. 1992. Urban forests: The sad state of city trees. American Forests,
    98(3&4):61-64.
Smith, D.M. 1986. The Practice of Silviculture, Eighth Edition. Wiley, New York, NY.
    527 pp.
Smith, D. 1991. Launching Hungarian ReLeaf. American Forests, 97:33-36.
Tans, P.P., I.Y. Fung, and T. Takahashi T. 1990. Observational constraints on the global
    atmospheric CO_2 budget. Science, 247:1431-1438.

Tho, Y.P. 1991. Tropical moist forests - facts and issues. p. 43-62. In: Technical workshop to explore options for global forestry management. D. Howlett and C. Sargent (eds.). International Institute for Environment and Development, London. 349 pp.

U.S. Department of Agriculture (USDA). 1988. The South's Fourth Forest: Alternatives for the Future. USDA Forest Service, Forest Resource Report No. 24, Washington D.C. 512 pp.

Volz, H.A., W.U. Kriebitzsch, and T.W. Schneider. 1991. Assessment of Potential, Feasibility and Costs of Forestry Options in the Temperate and Boreal Zones. p. 124-158. In: Technical workshop to explore options for global forestry management. D. Howlett and C. Sargent (eds.). International Institute for Environment and Development, London. 349 pp.

Waring, R.H. and W.H. Schlesinger. 1985. Forest Ecosystems, Concepts and Management. Academic Press, Orlando, FL. 340 pp.

Winjum, J.K., R.K. Dixon, and P.E. Schroeder. 1992. Estimating the global potential of forest and agroforest management practices to sequester carbon. Water, Air, and Soil Pollution, 64:213-227.

Winjum, J.K., J.E. Keatley, R.G. Stevens, and J.R. Gutzwiler. 1986. Regenerating the blast zone of Mount St. Helens. Journal of Forestry, 84(5):28-35.

Winjum, J.K. and D.K. Lewis. 1993. Forest management and the economics of carbon storage: The nonfinancial component. Climate Research. (In press).

World Resources Institute (WRI). 1992. World Resources 1992-93. A Guide to the Global Environment. Oxford University Press, Oxford. 386 pp.

EFFECTS OF ATMOSPHERIC CO₂ ENRICHMENT ON CO₂ EXCHANGE RATES OF BEECH STANDS IN SMALL MODEL ECOSYSTEMS

D. OVERDIECK

Institut für Ökologie, Fachgebiet Ökologie der Gehölze
TU Berlin, Königin-Luise-Str. 22, D-1000 Berlin 33, Germany

Abstract. CO_2 enrichment experiments were performed during two vegetation periods on young beech stands in four closed mini-greenhouses. The houses were climatized according to the outside microclimate ($\pm$ 0,5 °C, $\pm$ 15 % rel. air humidity, wind speed approximately to outside in the range of 0.5 - 2.5 m s^{-1}, max. 17 % PAR reduction). The model ecosystems – consisting of 36 young beech (2.5 yr-old) in a soil block of 0.38 m^3 and an air volume of 0.64 m^3 – were exposed to CO_2 concentrations of the unchanged ambient air (350 $\pm$ 34 ppmv, control) and of 700 ppmv (698 $\pm$ 10 ppmv). Plant growth parameters were measured non distructively and at the end of the 1st season samples were taken for weighing the phytomass. CO_2 gas exchange of the stands taken as a whole were continuously measured with two entire mini-greenhouses and, in addition, a compact mini-cuvette system (CMS 400, Walz) was used for measuring dark respiration and CO_2 net assimilation rates of single leaves in both stands. Under the influence of the additional CO_2 supply stem diameter (2 cm above the first lateral roots) was increased by 13.5 %, stem height by 27.4 %, and the number of leaves/tree by 33 % at the end of the 2nd season. The number of buds was not significantly different and the effect on mean area per leaf was insignificant. Leaf area index was by 1.4 units greater. All dry weights of the main organs were increased after the 1st season: leaf 60 %, stem 34 %, bud 54 %. Roots < 2 mm ϕ weighed 1.5-fold more and roots > 2 mm ϕ 1.7-fold more under elevated CO_2. CO_2 gas exchange of two systems was measured. Whole system CO_2 losses during night as well as photosynthetic CO_2 gains during days were greater at 700 ppmv than in the control system. However, if one balances CO_2 gains with CO_2 losses over a period of five days in August both model-ecosystems taken as a whole were sinks for CO_2. During this selected time period of 5 days at the peak of the season the beech stand at 350 ppmv was the greater sink. At 350 ppmv CO_2 (control) the average leaf respiration for 20 °C amounted to 0.31 $\pm$ 0.18 and at 700 ppmv to 0.57 $\pm$ 0.42 μmol CO_2 m^{-2} s^{-1} (n = 35/40, t = 3.48, α < 0.05), and correlated positively with leaf temperature. At light saturation the mean net assimilation rate was 4.48 μmol m^{-2} (leaf area) s^{-1} in the control and 6.21 μmol m^{-2} s^{-1} at

Water, Air, and Soil Pollution **70**: 259–277, 1993.
© 1993 *Kluwer Academic Publishers.*

the high CO_2 concentration corresponding with an enhancement factor of 1.39 for the selected time period.

Results from the whole stand and from single leaf measurements are compared by means of mathematical modelling procedures in order to quantify CO_2 enrichment effects on beech model ecosystems.

1. Introduction

The productivity of agriculturally used plants can be enhanced if more CO_2 is available. "CO_2-fertilization" of cultures in greenhouses has already been known since the last century (Godlewski, 1873; Kreusler, 1885). Kimball published literature reviews (1983a, b) which are based on 430 and 770 single observations, mainly on agriculturally or horticulturally used plant species. According to these data, the yield will be increased by about 36 % on the overall average if today's CO_2 concentration is doubled. Consequently, in many cases the actual CO_2 content of the troposphere seems to be suboptimal for plant growth. Therefore, Botkin (1977) concluded that the total phytomass of the terrestrial vegetation would act as a sink for the anthropogenic CO_2, and that its mass would increase in the future. In spite of that knowledge and this founded hypothesis – from the ecological point of view – we are, however, still confronted with unsolved problems concerning direct CO_2 effects on vegetation.

Studies on plant communities exposed for long periods, i.e. for months and years, to realistic CO_2 concentrations under field-like conditions are rather rare. Investigations on the terrestrial plant-community level are necessary because, from the results of the various CO_2 enrichment experiments with single individuals it is not possible to know if and when units of the upper soil layers and vegetation cover will be sources or sinks for additionally emitted CO_2. For these reasons, we continuously exposed mixtures (1:1) of white clover and perennial ryegrass (*Trifolium repens / Lolium perenne*) and red clover and meadow fescue (*Trifolium pratense / Festuca pratensis*) for more than 2 yr to elevated CO_2 concentrations under outdoor microclimatic conditions (Overdieck, 1991). Comparable investigations on the long-term effects of CO_2 on entire plant communities were conducted in the Alaskan Tundra and on a salt marsh of the eastern North America (Hilbert *et al*, 1987; Curtis *et al* 1989; Drake *et al*, 1989).

Only about 10 % of the publications available to us deal with the effects of elevated CO_2 concentrations on woody plants which play the most important role in the terrestrial regional and global C cycle and could form a large sink for additional CO_2. The few recent long-term and autecological investigations on trees refer – just like the studies on herbs – to enhanced growth and increasing CO_2 net assimilation (Rogers *et al*, 1983; Dahlmann *et al*, 1985; Hollinger, 1987; Gaudillère and Mousseau, 1989). Research about the effects on whole forest ecosystems has been based on experiments with single tree species only. Therefore, our team tried to broaden our understanding about the direct CO_2 effects on woody vegetation by means of long-term CO_2 enrichment experiments on juvenile beech stands. First results are reported in this paper.

2. Material and Methods

2.1.   MODEL ECOSYSTEMS

Four cubical acrylic glass (permeable for wavelengths > 300 nm; DIN 16957, RHÖM) boxes consisting of closed bottom parts for soil and roots and closed top-parts for the shoot parts of small stands of juvenile beech (3rd - 4th year) were constructed for continuous CO_2 gas exchange and water vapor exchange measurements on the model-ecosystem level (modified according to Overdieck and Bossemeyer, 1985; Forstreuter, 1991). The ground area was 0.64 m2, the soil volume 0.38 m3, the air volume 0.6 m3, and top and bottom parts were sealed together. After having filled the bottom parts with homogenized garden soil of medium fertility (loamy sand, bulk density: 1.1 g cm-3) in two of the mini-greenhouses the CO_2 concentration levels were permanently kept at 350 and in the two others at 700 ppmv throughout two complete vegetation periods (except: Dec, 1 - April, 1) by means of electronic regulators and injection of pure CO_2 from gas bottles via thermic mass flow meters (Brooks-Instruments). An air stream of 2,000 l h-1 was pumped through the systems (pumps: KF Neuberger). The microclimatic conditions inside were continuously regulated electronically according to the outside using an air conditioning system attached to each northern side-wall of the aerial part of the mini-greenhouse (workshop of the University of Osnabrück, Rathmann). Copper-constantan thermocouples were used for temperature control. Each air conditioning system consists of a car-radiator which is connected with an external cooling aggregate. From there the cooling liquid (water + ethylenglycol) was pumped towards the radiator according to regulative demand. In addition, the mini-greenhouse air is pressed through the radiator by means of a fan in case of temperature deviations inside from outside and streamed back into the mini-greenhouse passively. Temperature deviations from outside amounted to ± 0.5 °C and rel. air humidity to ± 15 %. Photon flux density of photosynthetically active radiation (PAR) was continously measured inside and outside with a Quantum-Sensor (Li-190 SB, LICOR, Lincoln, Neb., USA) showing decreased densities up to 17 % at maximum dependent on the angle of solar radiation. The wind way was measured at 4 m above-ground with an anemometer (Typ 1469, W. Lamprecht) and relatively approximated inside with a second, controllable fan in the range 0.5 - 2.5 m s-1.

2.2   REGISTRATION OF DATA

In the beginning of the first season each parameter (Table 1) was registered every 48 s with a Polycomp 2 (Hartmann and Braun) on paper stripes. Parallely, the row data were collected with a data-logger and transfered via a analog-digital-converter (workshop of the University of Osnabrück) to two portable mini-PC's (Olivetti M10) and stored as half-hourly means with their standard deviations on a PC.

During the first season this equipment was replaced by on-line storage of data at every 45 s per sensor and data reduction to half-hourly means using a special PC for data storage that was simultaneously used for regulation of the microclimate inside and CO_2 concentration control. In addition to the automatic measuring of CO_2 concentrations and microclimatic parameters, five further parameters (compare Table 1) were registered by separate instruments once a day.

2.3   CHOICE OF BEECH SAPLINGS

Almost 2-yr-old beech saplings of the same provenance were bought from a tree nursery. Individuals for the first two mini-greenhouses and the two adjacent plots were selected from the whole population of 2,000 saplings after having determined the height of all and taking those from the medium classes of the normal distribution of stem height. Thus, 48 saplings of mean size and mean developmental stage were planted into 1.5 l-black-plastic-sacs in May, 1991 and distributed to two houses and two open plots at random.

In addition, in August the smallest and the most developed saplings of the population were selected and distributed in groups of 36 to two additional mini-greenhouses and the two other neighbouring plots outside. In the beginning of winter, the number of saplings was reduced to 36 in the first two mini-greenhouses and the adjacent open plots, and all saplings were planted into the soil of the mini-greenhouse and the open basins directly.

Tab. 1:   Measured parameters inside and outside of closed mini-greenhouses for CO_2 enrichment studies

parameter (48s or 45s interval)	Sensor or instrument	mini-greenhouse 1	2	3	4	open-plot	number of measuring channels	unit
air temperature	Pt-100, HARTMANN and BRAUN	1	1	1	1	2	6	[°C]
soil temperature	Pt-100, HARTMANN und BRAUN	1	-	-	-	-	1	[°C]
relative air humidity	capacitive sensor TESTOTHERM, type 9763	1	1	1	1	1	5	[%]
wind velocity	cap-star anemometer No.1469, W. LAMPRECHT	-	-	-	-	1	1	[km h^{-1}]
radiation (PAR)	quantum-sensor, LICOR (400-700 nm) LI 190 SB	1	-	-	-	1	2	[µE m^{-2} s^{-1}]
CO_2-concentration	URAS 2T, 3G, HARTMANN and BRAUN	1	1	1	1	1	5	[ppm]
CO_2-injection rate	thermal mass flow meter modell 5850 TR BROOKS INSTRUMENTS	1	1	1	1	-	4	[ml min^{-1}]
parameter per day	Sensor or instrument	mini-greenhouse 1	2	3	4	open-plot	number of measuring channels	unit
air flow	volumetric flow meter BROOKS INSTRUMENT gasometer, KROMSCHRÖDER OSNABRÜCK	1	1	1	1	-	4	[m^3 h^{-1}]
air pressure	precision barometer THIES	-	-	-	-	1	1	[mbar]
precipitation	rain meter by Prof. Hellmann, W.LAMPRECHT	-	-	-	-	1	1	[mm]
condense water	calibrated glass cylinder balance	1	1	1	1	-	4	[ml]

During both seasons all eight plots were nearly kept constant to 20 vol.% of soil water content. The four plots in the mini-greenhouses were rewatered manually with tap water according to the values of evapotranspiration (condense water + water content of air stream). Direct measurements of soil water content of all eight plots were made on soil samples (100 cm³) by oven drying at 105 °C to constant weight. The conversion of water content per weight unit into water content per volume was done by multiplying the weight percentage by the bulk density of soil.

2.4   CO₂ GAS EXCHANGE

The CO_2 gas exchange of the four entire model-ecosystems having 48 beech saplings in season 1 and 36 beech saplings in season 2 was continuously measured. The mini-greenhouses were considered as half-open systems in which the CO_2 concentration levels were maintained compensatively. The CO_2 gas exchange of the systems (plants + soil) was calculated from the difference in concentration between the instreaming air and probes at the air outlet and the amounts of injected CO_2 following equation (1.0) (Forstreuter, 1991, 1993). The values were expressed on soil area basis according to Overdieck and Bossemeyer (1985). The unit [g m⁻² h⁻¹] was converted to [mol m⁻² h⁻¹] by multiplying with 22.7.

$$F_{CO_2} = \frac{(J_v \cdot \Delta C_{CO_2} + J_{vCO_2}\frac{T_v}{T_c})\frac{T_0}{T}\frac{P}{P_0}\frac{M_r}{V_0}}{A} \tag{1}$$

A :	ground area	$[m^2]$
ΔC_{CO_2} :	difference in CO_2 concentration of airstream	$[ppmv]$
F_{CO_2} :	CO_2 gas exchange	$[g\ m^{-2}h^{-1}]$
J_v :	air flow rate	$[m^3 h^{-1}]$
J_{vCO_2} :	injected pure CO_2	$[m^3 h^{-1}]$
M_r :	molar mass of CO_2	44 $[g\ mol^{-1}]$
P :	barometric pressure	$[Pa]$
P_0 :	standard pressure	10.13 $[Pa]$
T:	air temperature	$[K]$
T_0 :	standard temperature	273.15 $[K]$
T_c :	calibration temperature of thermal mass flowmeter	293.15 $[K]$
T_v :	ambient temperature of thermal mass flowmeter	293.15 $[K]$
V_0 :	molar volume of an ideal gas at standard temperature and pressure	22.414 10⁻³ $[m^3 mol^{-1}]$

2.5   TEST OF CO₂ GAS EXCHANGE MEASUREMENTS

The data of CO_2 gas exchange of the completely empty mini-greenhouses I and II were evaluated in the beginning for both situations, ambient and elevated CO_2 concentration level (700 ppmv). Slight differences from the expected zero gas exchange rate occured which amounted to values about 2 % of the final CO_2 gas exchange rates. After having filled both bottom parts with soil the CO_2, efflux from the soil alone (without plants) was measured at 350 ppmv. The differences between the two mini-greenhouses were in the still tolerable range from magnitude of 2.6 mmol CO_2 m^{-2} · h^{-1}. The rates of soil together with saplings still having closed buds differed less evidently (~ 1.0 mmol CO_2 m^{-2} h^{-1}).

2.6   GROWTH PARAMETERS

The following growth parameters were measured non distructrively during the first and the second vegetation period: stem diameter (2 cm above the first lateral roots) [mm], stem height [cm], number of leaves per tree, mean area per leaf [cm^2]. The leaf area index (LAI) was calculated using these data.

The leaf area per leaf was calculated by means of the following equation for each leaf at five representative trees per plot:

$$y = C_0 + C_1 x \qquad\qquad\qquad \begin{aligned} C_0 &= 0.26 \\ C_1 &= 0.67 \\ x &= L_1 L_2 \end{aligned}$$

L_1 = length of the leaf [cm]
L_2 = width of the leaf at its greatest expansion [cm].

The coefficient for the correlation between the leaf area measured with an areameter (Li-Cor, Model Li-3000, Lincoln, Nebraska, USA) and the calculated area using the formula was r = 0.989 (n = 44).

During the first and the second week of Oct., 1991, the six smallest and the six greatest saplings of each of the eight plots were harvested and leaves, buds, stems, roots ϕ > 2 mm and roots ϕ < 2 mm were dried (85 °C) and weighed.

Two experiments were conducted. Experiment I: May 17, 1991 - Dec 1, 1992; Experiment II: July 31, 1991 - Dec 1, 1992.

3. Results

3.1   STEM DIAMETER

By the end of the first vegetation period (Oct.) the stem diameter of beeches grown at 700 ppmv CO_2 had increased by 18 % (n = 48, t = 4.89, α < 0.05) compared to the control mini-greenhouse in experiment I (Figure 1). At the end of the 2nd vegetation period (end of Sept.) the difference amounted to 14 % (n = 36, t = 3.6, α << 0.05).

In experiment II (Figure 2) the difference was small at the end of the first vegetation period (Oct.): 2.2 % (n = 36, t = 0.69, α >> 0.05). At the end of the 2nd period, however, the CO_2 effect was obvious and amounted to 13 % (n = 31/27, t = 2.77, α << 0.05). In the open plots the mean stem diameters were nearly as great as in the control-mini-greenhouse and did not differ significantly from each other (compare Figure 1).

3.2   STEM HEIGHT

A tendency to increase stem length developed already during the first vegetation period in
experiment I. In Oct. the stems were 7.6 % significantly longer at the high CO_2 concentration level (n = 48, t = 2.09, α < 0.05). In Sept. of the 2nd yr the increase amounted to
27.4 % (n = 36, t = 3.78, α << 0.05).

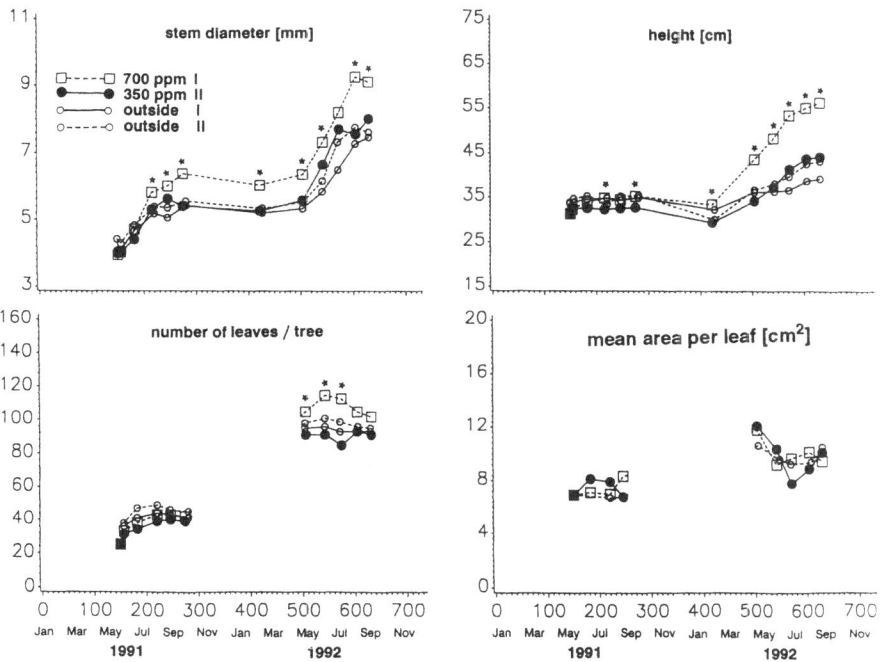

Fig. 1:   Experiment I, growth of beech at 350 and 700 ppmv CO_2 during their 3rd and 4th yr
            of development on natural, homogenized soil in mini-greenhouses climatized accord-
            ing to outside conditions and on outside plots; *: significant on the α < 0.05-level.

In experiment II (Figure 2), which was started later in the first yr, no significant difference in stem height could be found at the end of the first or second vegetation period
comparing the two CO_2 concentrations.
        The mean stem heights in the outside plots I and II did not differ significantly
from each other and were almost at the same level as in the control-mini-greenhouse during the course of both vegetation periods (compare Figure 1).

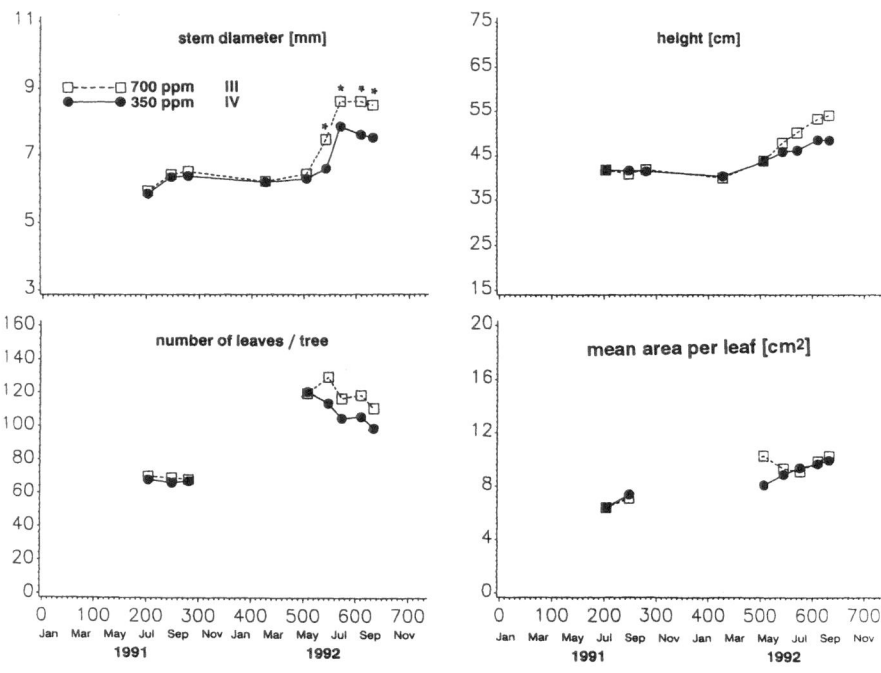

Fig. 2:   Experiment II, growth of beech at 350 and 700 ppmv CO_2 during their 3rd and 4th yr
            of development on natural, homogenized soil in mini-greenhouses climatized accord-
            ing to outside conditions; *: significant on the $\alpha < 0.05$-level.

3.3   NUMBER OF LEAVES / TREE

At the end of the first vegetation period the number of leaves per tree was not significantly higher at 700 ppmv although the averages differed by 9 %, and the tendency to increase the number was obvious during the whole vegetation period in experiment I (n =
47/48, t = 1.17, $\alpha > 0.05$). In contrast to the first vegetation period, the mean number of
leaves/tree was elevated by 33 % significantly in the middle of the 2nd vegetation period
(n = 36, t = 4.74, $\alpha \ll 0.05$).

This result is corroborated by those from experiment II. Again, there was a tendency to increase the leaf number/tree (Figure 2) during the 2nd vegetation period (June,
1992: n = 31/27, t = 1.72, $\alpha > 0.05$).

On the average, the mean number of leaves/tree was slightly higher outside (I,
II) than in the control-mini-greenhouse, but not reaching the same elevated level as at 700
ppmv during the 2nd vegetation period (compare Figure 1).

3.4   NUMBER OF BUDS

At the end of February in 1992 the buds were counted. The mean numbers/tree were not significantly different: 23 - 25 in the mini-greenhouse I and II, 27 - 28 on the outside plots I and II, 32 - 36 in the mini-greenhouses III and IV and 38 - 39 on the outside plots III and IV.

3.5   MEAN AREA PER LEAF

A tendency to increase the mean leaf area could be detected.

The leaf area values were not normally distributed. Therefore the differences were not tested by means of the Student t-test. At the first of July, 1991 the mean leaf area amounted to 8.1 cm^2 under 350 ppmv and to 7.1 cm^2 at high CO_2 in experiment I. In August it was 7.9 and 7.0 cm^2. In September, however, the leaves were greater at high CO_2 on the average, reaching 8.4 cm^2 whereas the mean leaf area at 350 ppmv was 6.8 cm^2.

In 1992, in total, the CO_2 effect on leaf area was negligible (Figure 1). The greatest differences between the two CO_2 treatments occured in July: 9.7 cm^2 at 700 ppmv and 7.8 cm^2 in the control.

In experiment II, there were also no clear differences between the mean values at the end of the vegetation period (Figure 2) although they started from different levels (10.3 cm^2 at 700 ppmv and 8.1 cm^2 at 350 ppmv).

The data from one outside plot (II) were evaluated. They were on the same level as those from the two different CO_2 treatments (Figure 1).

3.6   LEAF AREA INDEX

During the first 3 mo of our measuring period the leaf area index (LAI) was almost on the same level at both CO_2 concentration levels of experiment I (Figure 3). In September, the first change could be detected (control: 2.0, 700 ppmv: 2.6).

In the middle of the 2nd vegetation period – due to the increased number of leaves – a clearly higher LAI was reached at the elevated CO_2 concentration (control: 4.7, 700 ppmv: 6.0).

This result was verified in experiment II where no clear difference could be found during the first vegetation period and a distinct increase developed during 1992, again due to the greater number of leaves at 700 ppmv (control: 4.2, 700 ppmv: 5.6). The outside plot II had a higher LAI than the control-mini-greenhouse II (in experiment I) but was lower than at 700 ppmv (1992: 5.3; compare 4.7 at 350 ppmv and 6.0 at 700 ppmv).

3.7   DRY MATTER ACCUMULATION

At the end of the first season the dry phytomass of all organs was increased significantly under the influence of elevated CO_2 concentration in comparison to the control: bud 54 %, leaf 60 %, stem 34 %, root $\phi > 2$ mm 67 %, root $\phi < 2$ mm 48 % (Table 2).

The total dry matter per tree was increased by 53 % on the average, the above-ground phytomass by 41 % and the below-ground phytomass by 63 %, at the end of the 1st vegetation period.

Tab. 2:   Mean dry mass of main organs of young beech (3rd yr, 1st experimental yr) and total weights after one vegetation period (harvest 1st - 2nd week of Oct.) at the different CO_2 concentration levels 350 and 700 ppmv in mini-greenhouses climatized according to the outside conditions and on open plots with the same natural soil; n.s.: insignificant, (*): significant on the $\alpha < 0.1$-level; *: on the $\alpha < 0.05$-level; **: on the $\alpha < 0.01$-level; n = number of trees (Student t-test).

dry matter [g]								
		CO_2-treatment [ppm]						
organ		350	α	700	outside I	α	outside II	
bud	n	6		6	6		6	
	x̄	0.35	<0.05	0.54	0.27	>0.1	0.36	
	±s	0.07	*	0.15	0.05	n.s.	0.15	
leaf	n	11		9	9		10	
	x̄	1.27	<0.1	2.03	1.38	>0.1	1.41	
	±s	0.65	(*)	0.98	0.63	n.s.	0.99	
stem	n	11		9	13		12	
	x̄	4.22	<0.05	5.66	4.01	>0.1	4.22	
	±s	0.82	*	1.68	0.56	n.s.	0.88	
root ⊘ >2 mm	n	11		10	13		12	
	x̄	5.28	<0.01	8.84	4.86	>0.1	6.65	
	±s	1.36	**	2.69	1.08	n.s.	2.55	
root ⊘ <2 mm	n	11		10	6		6	
	x̄	1.50	<0.01	2.22	2.03	<0.05	1.27	
	±s	0.27	**	0.52	0.54	*	0.53	

CO_2 level [ppm]	total	above-ground	below-ground
350	12.62	5.84	6.78
700	19.29	8.23	11.06
outside I	12.55	5.66	6.89
outside II	13.91	5.99	7.92

In Oct. of the first year the specific leaf area was decreased by 9 % under the influence of elevated CO_2. There was 12 % less leaf area per unit total weight. The leaf dry weight was slightly increased in relationship to the total plant dry weight (4 %).

The leaf weight was increased by 17 % more, relative to the stem weight. It was slightly decreased (5 %) in relation to the root weight. On the average, the root/ shoot relationship increased after one yr at elevated CO_2 concentration.

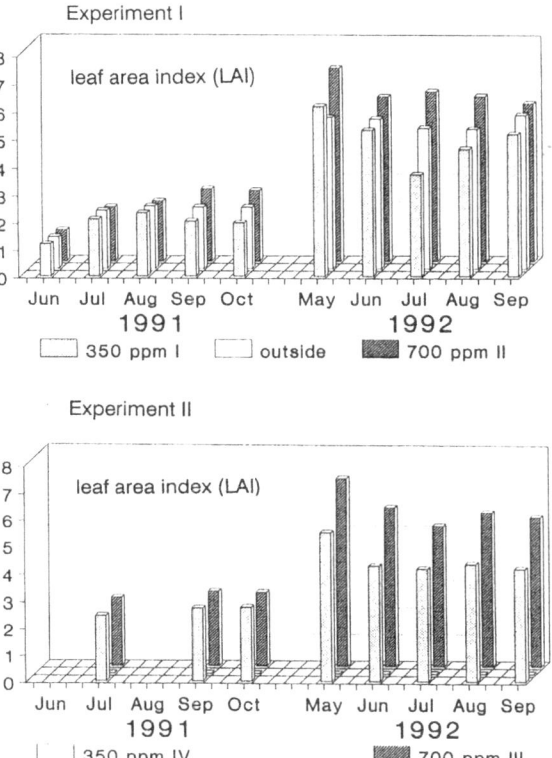

Fig. 3:   Leaf area index (LAI) of small stands of beech (3rd and 4th yr) at 350 and 700 ppmv
            CO_2 on natural, homogenized soil in mini-greenhouses climatized according to
            outside microclimate (I, II, III, IV) and on an open plot outside in two parallel experi-
            ments (48 and 36 trees).

3.8 CO_2 GAS EXCHANGE OF STANDS DURING THE VEGETATION
            PERIODS

During the first few weeks of the first experimental yr, respiration was higher than photosynthesis at both CO_2 concentration levels (mini-greenhouse I and II). Only on some days
in July and August, CO_2 uptake by photosynthesis exceeded respirational CO_2 losses at
midday. On the other hand, photosynthetical CO_2 gains during the days were also greater
at 700 ppmv. At midday of August the 5th for instance (1st year) having photon flux densities around 1,000 mmol m-2 s-1, the net CO_2 uptake of the high CO_2-mini-greenhouse
was 3-times as large as in the control. The daily balance of this day amounted to a CO_2

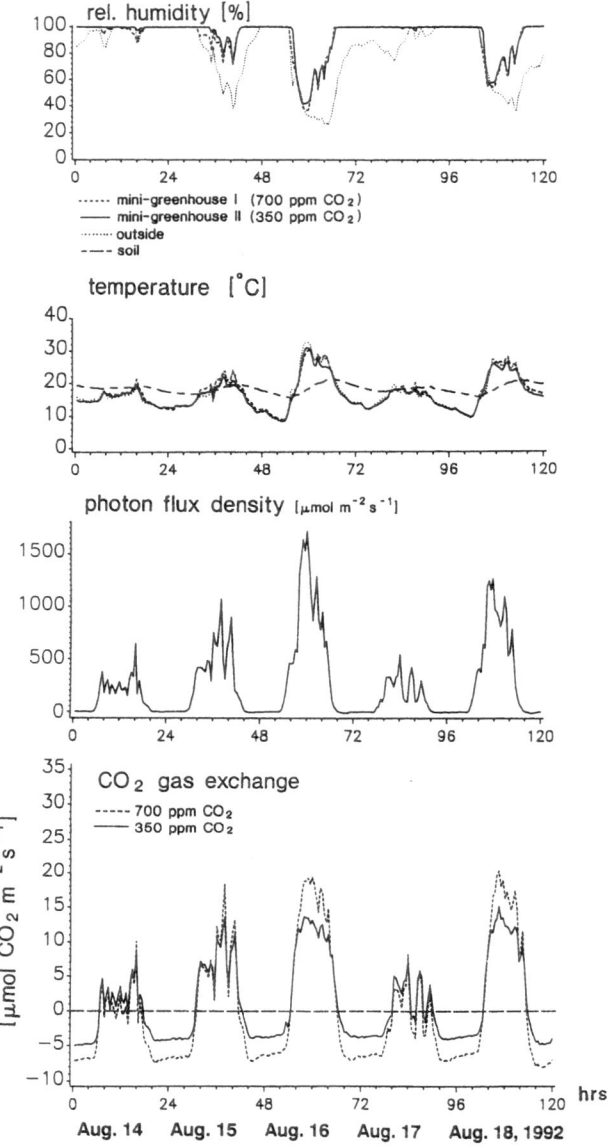

Fig. 4:   Example for daily courses (Aug. 1992) of CO_2 gas exchange (based on half-hourly
            means of CO_2 net uptake) of small stands of beech saplings (n = 36) on natural, ho-
            mogenized soil permanently exposed to 350 and 700 ppmv for two vegetation periods
            in two mini-greenhouses climatized according to the outside conditions (evaluated by
            M. Forstreuter).

loss of 7.3 g d^{-1} m^{-2} ground area in the case of 700 ppmv CO_2 and to a loss of 13.7 g d^{-1} m^{-2} at 350 ppmv. So, at that developmental stage (LAI around 2.0) the young stands (3rd yr after germination) together with their soil bed reacted as CO_2 source. The source, however, was by about 53 % smaller at the high CO_2 concentration taking the 5th of August as representative example for the peak of the vegetation period 199:.

In the 2nd vegetation period respiration rates of both systems are still great (Figure 4). Again respirational CO_2 losses during night time as well as photosynthetical CO_2 gains during days are greater at 700 ppmv than in the control. However, if one balances CO_2 gains with CO_2 losses for instance over a period of five days (Aug. 14 - 18) on the basis of half-hourly means both mini-greenhouses then are sinks for CO_2. Contrary to our expection, during this selected time period of 5 days the control was the greater sink (Table 3).

The half-hourly means of total dark respiration from a period of about three weeks in August were plotted against soil temperature and that relationship was described by functions using a mathematical iteration procedure (Figure 5). The function was used to calculate in half-hourly steps at each given temperature respiration and canopy gross photosynthesis for the time period Aug. 14 - 18, 1992 following this general equation:

$$P_{G\ canopy} = F_{CO2\ system} + R_{D\ system} \tag{2}$$

$P_{G\ canopy}$: canopy gross photosynthesis
$F_{CO2\ system}$: CO_2 gas exchange of the whole system
$R_{D\ system}$: dark respiration of the whole system
[all : μmol m^{-2*} s^{-1}]

*: ground area

The result is given in Figure 6.

Tab. 3: CO_2 gas exchange rates (dark respiration and CO_2 net uptake) balanced over five days in Aug. 1992 for small stands of beech saplings (n = 36) growing on natural homogenized soil at 350 and 700 ppmv CO_2 in mini-greenhouses climatized according to outside conditions. Positive values indicate not CO_2 uptake in the beech stands (compare Figure 4).

time	light [mol m^{-2}]	350 ppm [g m^{-2}]	350 ppm [mol m^{-2}]	700 ppm [g m^{-2}]	700 ppm [mol m^{-2}]
Aug. 14/92	11.14	-3.75	-0.09	-10.75	-0.24
Aug. 15/92	21.69	5.85	0.13	0.55	0.01
Aug. 16/92	38.52	12.86	0.29	13.04	0.30
Aug. 17/92	10.63	-2.48	-0.06	-10.77	-0 24
Aug. 18/92	30.97	10.68	0.24	10.08	0 23
total	112.97	23.16	0.51	2.15	0.06

Accumulated CO_2 gas exchange ($F_{CO2system}$)

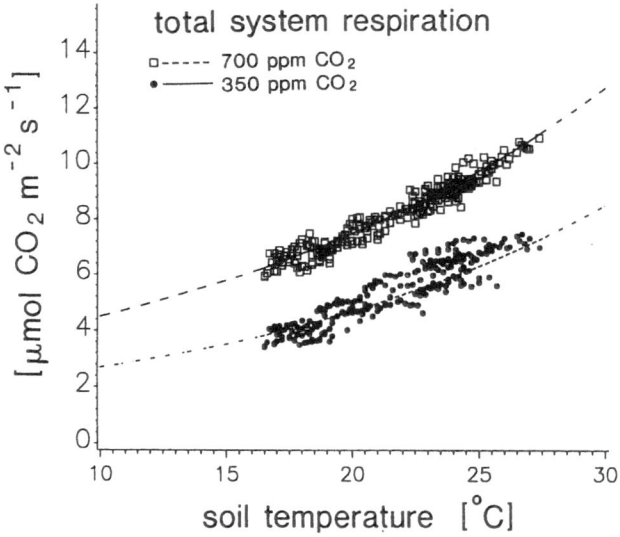

Fig. 5:   Correlation between soil temperature and dark respiration rates of small stands of
            beech saplings on homogenized, natural soil at 350 and 700 ppmv CO_2 in mini-green-
            houses climatized according to outside conditions (evaluated by M. Forstreuter).

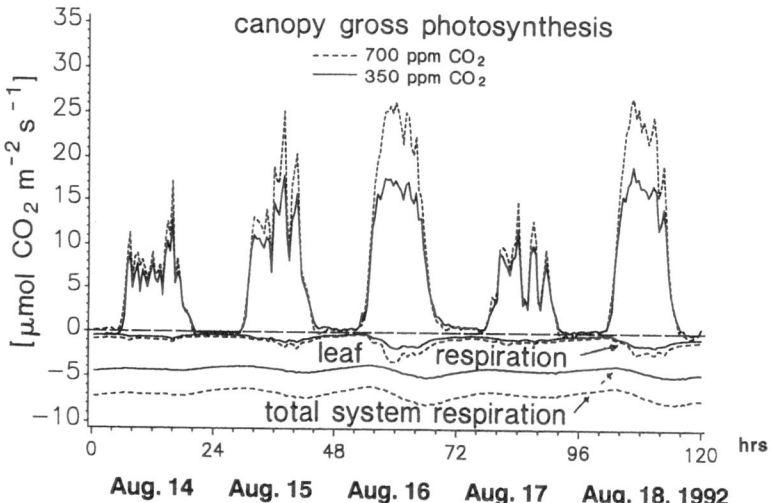

Fig. 6:   Daily courses (Aug. 1992) of respiration and canopy gross photosynthesis rates based
            on half-hourly means for small stands of beech saplings (n = 36) at 350 and 700 ppmv
            CO_2 on natural, homogenized soil in mini-greenhouses climatized according to out-
            side conditions (evaluated by M. Forstreuter); leaf respiration data are derived from
            single leaf measurements (compare 3.9)..

Thus after having separated respiration from photosynthesis it was now possible to plot canopy gross photosynthesis against photon flux density and approximate exponential functions to the data (Figure 7).

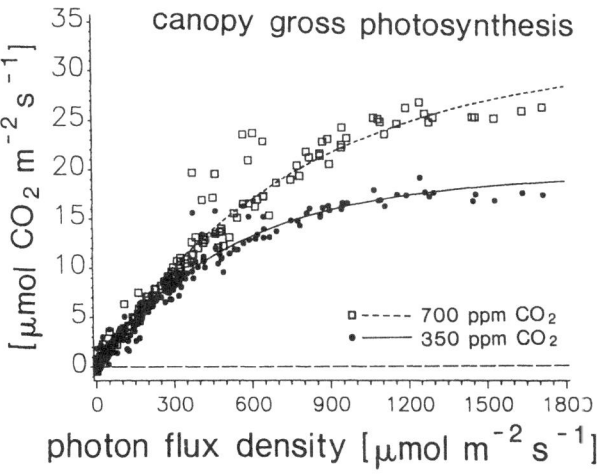

Fig. 7:   Correlation between photon flux density and canopy gross photosynthesis rates of
           small stands of beech saplings (n = 36) at 350 and 700 ppmv CO_2 on natural, homo-
           genized soil in mini-greenhouses climatized according to outside conditions (evaluat-
           ed by M. Forstreuter).

3.9 CO_2 GAS EXCHANGE OF SINGLE LEAVES

The first results from the leaf gas exchange measurements (June 1991) showed a small difference in dark respiration rates between 350 ppmv and the elevated CO_2 concentration. However, this difference was not significant.

In July, 1992 for 20°C on 350 ppmv CO_2, the average leaf dark respiration amounted to 0.31 ± 0.18 and at 700 ppmv to 0.57 ± 0.42 µmol CO_2 m^{-2} s^{-1} (n = 35/40, t = 3.48, a < 0.05). The correlation with leaf temperature is shown in Figure 8.

During the first measuring campaign in 1991 it was found that CO_2 net assimilation was clearly enhanced by 700 ppmv CO_2. The mean net assimilation rate at light saturation amounted to 2.97 µmol CO_2 m^{-2} (leaf area) s^{-1} at 350 ppmv and to 6.47 µmol CO_2 m^{-2} s^{-1} at 700 ppmv CO_2.

In 1992, the mean net assimilation rate at light saturation was 4.48 µmol m^{-2} (leaf area) s^{-1} in the control and 6.21 µmol m^{-2} s^{-1} at the high CO_2 concentration corresponding to an enhancement factor of 1.39.

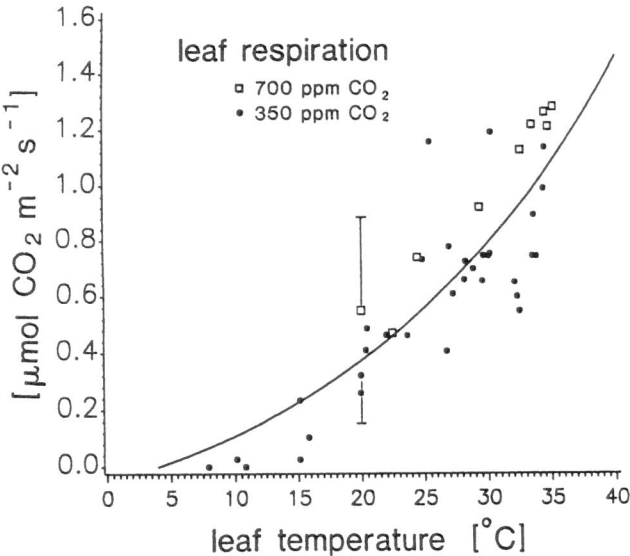

Fig. 8:   Correlation between leaf temperature and leaf dark respiration rates of beech leaves
            having developed at 350 and 700 ppmv CO_2 in mini-greenhouses climatized accord-
            ing to the outside conditions (measured in July; 2nd yr; evaluated by M. Forstreuter).

The leaf temperature response of leaf dark respiration (compare Figure 8) was
used to calculate leaf gross photosynthesis by means of the following equation:

$$P_{G\ leaf} = P_{N\ leaf} + R_{D\ leaf} \tag{3}$$

$P_{G\ leaf}$: leaf gross photosynthesis
$P_{N\ leaf}$: leaf net photosynthesis
$R_{D\ leaf}$: leaf dark respiration
[all: $\mu mol\ m^{-2**}\ s^{-1}$]

**: leaf area

Knowing the LAI (leaf area index) of the whole stands (compare Figure 3) it was now
possible to calculate the total gross photosynthesis of the whole stand in response to
photon flux density on the basis of the single leaf measurements following this relationship:

$$P_{G\ canopy} = (P_{N\ leaf} + R_{D\ leaf})\ LAI\ k \tag{4}$$

$P_{G\ canopy}$: canopy gross photosynthesis
LAI : leaf area index
k : light penetration coefficient.

The results from the two experimental ways in the search of the CO_2 enrichment effect on CO_2 gas exchange rates of small stands of beeches having reached LAI's between 4 and 6 can now be compared (Figure 9). In this first approach to solve the problem k was set to 1 in equation (4).

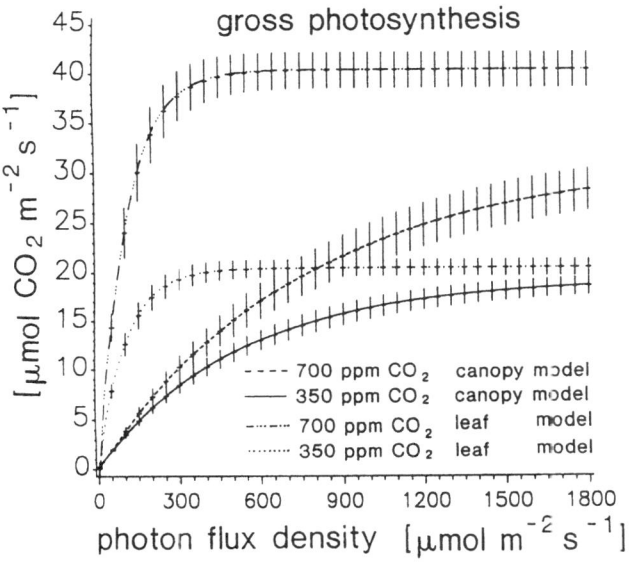

Fig. 9:   Comparison of functions derived from stand and leaf measurements describing the relationship between photon flux density and gross photosynthesis rates (gross CO_2 exchange rates) of small stands of beech saplings on natural, homogenized soil permanently exposed to 350 and 700 ppmv for two vegetation periods in mini-greenhouses climatized according to outside conditions (data base: 2nd vegetation period, July - Aug.; evaluated by M. Forstreuter).

4. Discussion

In total, therefore, the working hypothesis that young woody plants are stimulated to grow more at elevated CO_2 concentration levels could be fully verified for beech. The effect on stem diameter is obviously positive whereas the stem heights varied so much that only a certain tendency to increase shoot growth could be detected. One major reason for the great variation in this parameter is the ability of beech to develop additional flushes during the season irregularly. The probability that a second flushing is enhanced by additional CO_2 supply is great.

In many former investigations on plants, among those also a few woody, long-living species like *Vitis vinifera*, the increase of leaf area per plant was reported (Lindstrom, 1965; Kriedemann *et al*, 1976; Imai and Murata, 1976; Raper and Peedin, 1978; Wong, 1979; Carlson and Bazzaz, 1982; Sionit *et al*, 1982; Wulff and Strain, 1982; Overdieck, 1986). This increase was also evident in our recent investigations. The reason for that was the increased number of leaves per tree and not so much the area increase per single leaf. Hence it follows that the leaf area index (LAI) of a stand of young beeches is increased at elevated CO_2 concentration.

After 24 weeks O'Neill *et al* (1987a) detected an enhancement of dry weight by 72.5 % at 692 ppmv CO_2. Norby *et al* (1986) and O'Neill *et al* (1987b) weighed seedlings of *Quercus alba* after 30 and 40 weeks under 690 ppmv CO_2 and found a total dry weight increase of 71 and 85 %. Mousseau and Enoch (1989) found a strong effect on the roots of *Castanea sativa* and total dry weight increase of 43 %. The 53 % increase of beech total dry weight which could be determined in our experiment after one year corresponds with the effects on deciduous trees found in literature. In other words, the probability is great that the pool 'living phytomass' in young stands of deciduous trees will increase faster under the influence of tropospheric CO_2-enrichment.

Whereas, the probability that the fluxes remain the same at elevated CO_2 concentrations is low because at greater phytomass accumulation per unit soil area also the respirational CO_2 release rates will increase. Our first data evaluations indicate that at least during summer a small stand of juvenile beech (3 to 4-yr-old trees) is a smaller sink at 700 ppmv CO_2 than at the current CO_2 concentration level.

Causes for the increased respiration could be searched for above- and below-ground. Above-ground: In denser stands more leaves will be shaded and the compensation of respiration by photosynthesis-rates could demand for higher PAR-intensities which can then be restricted to a few summer days. Below-ground: Root exudation might increase and in consequence the microbial activity could be stimulated.

In general, also the amounts of litter from roots and shoots will increase at greater CO_2 supply and its remineralization by the edaphon might add greater amounts of CO_2 to the atmosphere than before.

5. References

- Botkin, D.B.: 1977, Bioscience **27**, 325-331.

- Carlson, R.W.; Bazzaz, F.A.: 1982, Oecologia **54**, 50-54.

- Curtis, P.S.; Drake, B.G.; Leadley, P.W.; Arp, W.J.; Wigham, D.F.: 1989, Oecologia **78**, 20-26.

- Dahlmann, R.C.; Strain, B.R.; Rogers, H.H.: 1985, Journal of Environmental Quality **14**, 1-8.

- Drake, B.G.; Leadley, P.W.; Arp, W.J.; Nassiry, D.; Curtis, P.S.: 1989, Functional Ecology **3**, 363-371.

- Forstreuter, M.: 1991, Verh. Ges. Ökol. (Osnabrück 1989), **IXX, III**, 265-279.

- Forstreuter, M.: 1993, Dissertation, Osnabrück, pp. 31-33.
- Gaudillère, J.P.; Mousseau, M.: 1989, Acta Oecologia. Oecol. Plant **10**, 95-105.
- Godlewski, E.: 1873, Arbeiten des Botanischen Instituts in Würzburg, Leipzig, 343-370.
- Hilbert, D.W.; Prudhomme, T.I.; Oechel, W.C.: 1987, Oecologia **72**, 466 - 472.
- Hollinger, D.J.: 1987, Tree Physiology **3**, 193-202.
- Imai, K.; Murata, Y.: 1976, Japan. J. Crop Sci. **45**, 598-606.
- Kimball, B.A.: 1983a, Agronomy Journal **75**, 779-788.
- Kimball, B.A.: 1983b, WCL Report 14, U.S. Dept. of Agric., Agric. Res. Serv., pp. 71.
- Kreusler, U.: 1885, Landwirtschaftliche Jahrbücher **14**, 913-965.
- Kriedemann, P.E.; Sward, R.J.; Downtown, W.J.S.: 1976, Aust. J. Plant Physiol. **3**, 605-618.
- Lindstrom, R.S.: 1965, Proc. Am. Soc. Hort. Sci. **87**, 521-524.
- Mousseau, M.; Enoch, H.Z.: 1989, Plant, Cell and Environment **12**, 927-934.
- Norby, R.J.; O'Neill, E.G.; Luxmoore, R.J.: 1986, Plant Physiology **82**, 83-89.
- O'Neill, E.G.; Luxmoore, R.J.; Norby, R.J.: 1987a: Plant and Soil **104**, 3-11.
- O'Neill, E.G.; Luxmoore, R.J.; Norby, R.J.: 1987b, Canadian Journal of Forest Research **17**, 878-883.
- Overdieck, D.: 1986, Int. J. Biometeor. **30**, 323-332.
- Overdieck, D.: 1991, Carbon dioxide effects on vegetation, in: Esser, G. and Overdieck, D. (eds.), Modern Ecology: Basic and Applied Aspects, Elsevier, pp. 623-657.
- Overdieck, D.; Bossemeyer, D.: 1985, Angewandte Botanik **59**, 179-198.
- Raper, C.D.; Peedin, G.F.: 1978, Bot. Gaz. **139** (2), 147-149.
- Rogers, H.H.; Bingham, G.E.; Cure, J.D.; Smith, J.U.; Surano, K.A.: 1983, Journal of Environmental Quality **12**, 569-574.
- Sionit, N.; Hellmers, H.; Strain, B.R.: 1982, Agron. J. **74**, 721-725.
- Wong, S.C.: 1979, Oecologia **44**, 68-74.
- Wulff, S.C.; Strain, B.R.: 1982, Canadian Journal of Botany **60**, 1084-1091.

CARBON TRENDS OF PRODUCTIVE TEMPERATE FORESTS
OF THE COTERMINOUS UNITED STATES

LINDA S. HEATH AND RICHARD A. BIRDSEY

USDA Forest Service
Global Change Research Program
PO Box 6775
Radnor, PA 19087 USA

Abstract. Carbon trends of U. S. timberlands reflect past and current harvesting patterns and forest growth. Using periodic forest inventory data coupled with the Carbon Budget Model, we estimate C inventory from 1952 to the present, and project future trends through 2070. Two sets of projections are presented, one based on economically derived harvest levels and the other assuming no harvests after 1990. Productive forests sequester an average of 250 Tg C yr^{-1} from 1952-1987, but projections under expected harvests assuming no changes in growing conditions indicate this rate will fall to 60 Tg C yr^{-1} from 1987 to at least 2050, and then become a C source by 2070. Carbon sequestered in products and landfills over the projection period average 75 Tg C yr^{-1}. An estimated 328 Tg C yr^{-1} would be sequestered if harvesting ceased.

1. Introduction

The proposition of a large unidentified C sink in the Northern Hemisphere, containing both terrestrial and oceanic components, has prompted researchers to estimate current magnitudes of C sources and sinks. One terrestrial process that is likely to contribute to this large sink is forest regrowth (Watson *et al.*, 1992). Kauppi *et al.* (1992) estimated forest regrowth in European forests accumulated 85 to 120 Tg C yr^{-1} in the 1970s and 1980s. About 30 Tg C yr^{-1} are being sequestered in Canadian forests, not including peatlands (Kurz and Apps, 1993). Much of the forest land of the United States is also in a period of regrowth, and this should be reflected in its C pools. Recent estimates of net C sequestration for all temperate forests of the entire U. S. include 152 Tg yr^{-1} (Sedjo, 1992) and 230 Tg yr^{-1} (Birdsey *et al.*, 1993) for productive temperate forests. Furthermore, C is sequestered in products that were manufactured from wood harvested from these forests. Approximately 36 Tg C yr^{-1} was stored in products over the period 1952-1987 in the U. S. (Birdsey *et al.*, 1993), and an estimated 21 Tg C yr^{-1} in Canada in 1986 (Kurz *et al.*, 1992).

In growing forests such as these, the magnitudes of C pools may vary considerably, depending on the time frame considered. The potential trends of C pools and net C fluxes in U. S. forests and forest products are of interest to determine how long and to what extent they will continue to act as net C sinks.

Water, Air, and Soil Pollution **70**: 279–293, 1993.
© 1993 *Kluwer Academic Publishers*.

The objectives of this paper are to (1) describe the past history of land use and current conditions of productive U. S. forests excluding Alaska and Hawaii on a regional basis, and discuss its importance in estimating C storage, (2) provide past estimates and future projections of regional C sequestration using historical growth data and projections of C storage in wood products manufactured or processed from harvested wood, and (3) investigate the hypothesis that the C pool in forest ecosystems and from harvesting timber and manufacturing wood products would be larger over time than the C pool in forest ecosystems with no harvesting. The estimates are based on all timberlands of the coterminous U. S., which is currently 201 Mha, or 80% of all forest land. Timberland is defined as forest land that is capable of producing more than 1.4 m^3 ha^{-1} yr^{-1} of wood in natural stands.

2. Forest Resource Trends in the Continental United States

The history of the evolution of the original forest and its conversion to its current status is important for understanding trends and modeling projections of C storage. In 1630, the continental U. S. had about 333 Mha of forest land, excluding slow-growing dry forests such as pinyon-juniper and chaparral (United States Department of Agriculture [USDA], 1928). The forests were characterized by a high proportion of large trees and in approximate equilibrium with respect to growth and mortality. Most of the pre-settlement forest has now been cleared, except for pockets in the western U. S., and replaced with agricultural land, successional forest, or developed land use. The continental U.S. can be split into three regions, North, South, and West, which broadly reflect differences in the use and management of the pre-settlement forest lands (Figure 1).

Figure 1. Regions of the United States with similar histories of forest land use.

The pattern of total area of timberlands is illustrated by region over time in Figure 2. Starting in 1630, the West shows little change in total area over the 350-yr period. The North and South show similar trends with only 50% of the total area in 1630 recognized as timberland in 1920. The South features a distinct increase from 1920-1952 as marginal agricultural lands reverted to forest; this increase is obscured in the North as agricultural lands reverted to forests in northeastern states, but forests continue to convert to other uses in north central states. These long-term trends of total area mask the true extent of forest disturbance in the regions. The pattern of forest land area use is reflected in the percentage of area by age-class for private timberlands for each region (Figure 3).

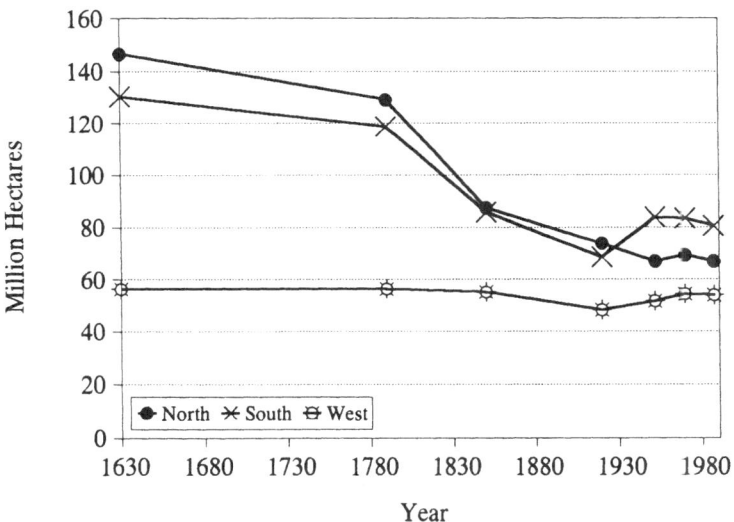

Figure 2. Area of productive forest land by region and year for the continental United States.
Sources: Reynolds and Pierson, 1941; USDA, 1928, 1958, 1974; Waddell *et al*., 1989

2.1. NORTHERN REGION

By the end of the 19th century, much of the northern forest had been cleared for agricultural use or heavily logged for timber products, with the exception of extreme northern areas of Maine and the Great Lakes States and some inaccessible areas in the Appalachian mountains. Beginning in the mid-nineteenth century and accelerating in the 20th century, marginal agricultural land has reverted to forest, producing a large proportion of forests in the 25 to 65 yr age-classes (Figure 3). These forest lands of mixed species are in the middle of a period of rapid growth that can be sustained for several more decades before reaching a period of declining net growth (Gansner *et al.*, 1991). In the near future, harvesting is not expected to increase enough to offset continued accretion of biomass (Haynes, 1990).

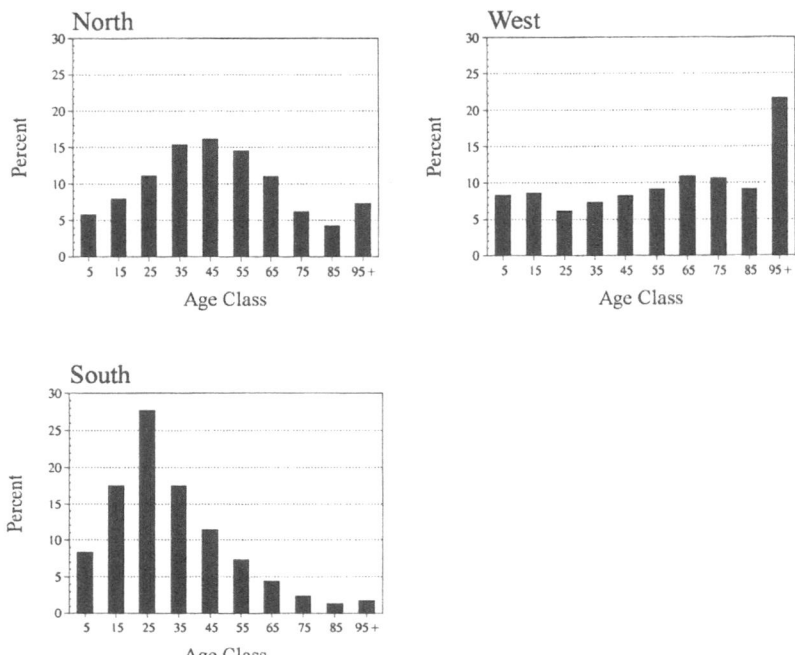

Figure 3. Percent timberland by age-class by region for private ownership in the continental United States, 1987.

2.2. SOUTHERN REGION

As in the northern U. S., only scattered fragments of pre-settlement forest remain in the South. The current forest has been regenerated either naturally or artificially on marginal cropland, pasture, or cutover forest land, and many areas are being cut for the third time and regenerated to a fourth timber stand (USDA, 1988). The intensively managed southern forests have the youngest average age in the U.S. with a large proportion in the 15 to 35 yr age classes (Figure 3). Because of intensive management and a continuing shift of industrial timber supply to the South, most of these forests are not expected to reach biological maturity before further harvest and regeneration (Haynes, 1990). Forest biomass is expected to remain constant or decline slightly.

2.3. WESTERN REGION

Although the West has not had major land use shifts characteristic of the other regions (Figure 2), forest disturbance nonetheless has dominated the landscape as the original forests have been logged and converted to second-growth forest. The remaining old-growth in the West comprises less than 10% of the pre-settlement forest. Most of the logged areas have been regenerated at a constant rate. Private western forests have a relatively flat age-class distribution and a large component of older forest classes (Figure 3). Public forest lands show a similar age-class distribution. It is expected that western public forest lands will be increasingly reserved (protected from timber cutting), producing a long-term net gain in forest biomass on public lands. As harvesting shifts from public to private lands, forest biomass will show a corresponding decrease on private lands (Haynes, 1990).

3. Methods of Estimating Past, Current, and Future Carbon Storage

Separate analyses were conducted for major forest types in 8 geographic regions, and then aggregated to the appropriate level for analysis. Projections include estimates of C that is harvested and retained in processed wood. The methods used to calculate the past, current, and future estimates are discussed in the following section. Additional details of the assumptions, estimation methods, and models can be found in Birdsey (1992a,b) and Plantinga and Birdsey (1993).

3.1. ESTIMATES FOR BASE YEAR, 1987

Carbon storage in forest ecosystems was estimated in four separate components: trees, soil, forest floor, and understory vegetation. The definitions of these components are broad enough to include all sources of organic C in the forest ecosystem. The tree portion includes all above- and belowground portions of all live and dead trees, including the merchantable stem, limbs, tops, and cull sections, stump, foliage, bark and rootbark, and coarse tree roots (greater than 2 mm). The soil component includes all organic C in mineral horizons to a depth of 1 m, excluding coarse tree roots. The forest floor includes all dead organic matter above the mineral soil horizons except standing dead trees: litter, humus, and other woody debris. Understory vegetation includes all live vegetation besides live trees.

Estimates of C storage in trees were based on periodic forest inventories designed to provide statistically valid estimates of timber volume, growth, removals and mortality (Waddell et al., 1989). Timber volume included merchantable live tree, 12.8 cm and larger at diameter breast height. Aboveground tree biomass was calculated by multiplying timber volume by conversion factors derived from the national biomass inventory (Cost et al., 1990). Belowground tree biomass was similarly calculated using conversion factors that range from 0.155 in northern hardwoods to 0.197 in southern hardwoods (Koch, 1989).

Equations were devised to estimate C storage in the forest floor and understory vegetation, based on the compilation by Vogt *et al.* (1986) and reviews of numerous intensive-site ecosystem studies (see Birdsey, 1992a). It was assumed that understory biomass peaked at age 5 and declined to 1 or 2% of the tree C by age 50 in the South and age 55 elsewhere. Forest floor estimates from forest ecosystems studies were applied to related forest types. Soil C for individual states and forest types was related to mean annual temperature and precipitation using a model similar to Burke *et al.* (1989) with coefficients derived from data in Post *et al.* (1982), and modified to account for forest growth development as estimated by average stand age.

3.2   HISTORICAL ESTIMATES

Estimates of past C storage in privately owned forests were derived from periodic assessments of forest resource conditions, each including a compilation of national inventory statistics (USDA Forest Service, 1958, 1965, 1974, 1982). Conversion factors used to estimate C storage for the base year (1987) were applied retroactively to previous estimates of growing stock volume by region and species group (softwoods and hardwoods).

3.3.   PROJECTIONS

3.3.1.   *Carbon in forest ecosystems*

For private timberlands, profiles of average C storage by age of forest stands were composed for each ecosystem component for forest classes defined by region, forest type, and land use history. The profiles were developed using methods similar to those used in estimating C in the base year (1987). However, additional assumptions were required to estimate soil C over time. A review of the literature indicated that a major forest disturbance, such as a clearcut harvest, can increase coarse litter and oxidation of soil organic matter. The balance of these two processes can result in a net loss of 20% of the initial C over a 10 to 15 yr period following harvest (Woodwell *et al.*, 1984; Pastor and Post, 1986). After the initial loss, C begins to accumulate in the soil unless the harvest is followed by conversion to agricultural use, in which case loss of soil C could reach 60% under intensive cultivation (Anderson, 1992; Johnson, 1992). The model was based on a net loss of 20% of initial soil C on the site over a 10 to 15 yr period, depending on forest class, with soil C returning to initial levels by age 50 in the South and 55 elsewhere.

If the forest land had reverted from agriculture, soil C was assumed to accumulate over time to levels similar to forest land that had never been cultivated. This assumption was based on the knowledge that tree plantations established on agricultural land with depleted C stores can cause a substantial buildup in soil organic matter, depending on species, soil characteristics, and climate (Johnson, 1992). For example, *Populus* spp. established on sandy soils showed large increases in soil and forest floor C due to high litter production (Dewar and Cannell, 1992). Expected changes in C storage for soil and forest floor components were derived by assuming a linear transition from average nonforest to average forest conditions.

The profiles form the basis of the Carbon Budget Model (Plantinga and Birdsey, 1993), and are linked with projected inventory estimates from a forest sector model, TAMM/ATLAS (Adams and Haynes, 1980; Haynes and Adams, 1985; Alig, 1985; Mills and Kincaid, 1992). The main disturbance in U. S. forests is harvesting, and the forest sector model provides an economic framework through which price, consumption, and production of timber and wood products and land use area change are projected. Linkage with a forest sector model is critical for making valid projections because interactions with expected market responses will affect estimates of future forest harvest levels.

For public lands, changes in forest inventories are estimated using an inventory model not linked to TAMM/ATLAS. Planned harvest levels are assumed to be realized on lands not reserved from timber harvest. Changes in forest inventories are then converted to C estimates using the current (1987) conversion factors listed in Birdsey (1992a). For lands reserved from timber harvest, a model was developed using average age-class distributions, stand profiles, and total area in reserved status.

3.3.2. Carbon in wood products

The C pools of wood from projected harvests on both private and public lands were estimated with a model based on the work of Row and Phelps (1991). Only C from forest harvests after 1980 is considered, which means the estimates do not include emissions from products manufactured in the past. There are four disposition categories: products, landfills, energy, and emissions. Products are goods manufactured or processed from wood including lumber and plywood for housing and furniture, and paper for packaging and newsprint. Landfills store C as discarded products that eventually decompose and are released as emissions. Emissions also include C from wood burned without generation of usable energy, or from decomposing wood. Energy is a separate category from emissions because wood used for energy may be a substitute for fossil fuels.

3.3.3. Description of scenarios

Two sets of projections were made over the years 1980 to 2070. One set was based on harvests determined by economic criteria using the forest sector model for years 1980 to 2040, with harvest levels in the years 2050 to 2070 held constant at 2040 levels. The assumptions underlying the projections can be found in Haynes (1990). The other set is projected under the assumption of no harvesting after 1987. Together, the projections provide the probable extremes for future C pool estimates under the assumption that future growing conditions will be identical to past or current growing conditions. A comparison of these scenarios will indicate the effect of harvesting on the C budget.

An important assumption implicit in the projections is that future growing conditions will be identical to past or current growing conditions. Potential effects of climate change, increasing atmospheric CO_2 concentration, changing forest productivity, and changing disturbance regimes (other than harvesting) are not included. Considering the uncertainty of these effects, the introduction of these factors would probably do more to obfuscate than to clarify the results.

5. Results and Discussion

5.1   ESTIMATES OF CARBON IN FORESTS

The historical, current, and projected estimates of C pools and changes in pools are displayed by region for the harvesting scenario in Table 1, and the scenario of no harvesting after 1987 in Table 2. The historical estimates are the same on both tables; they are included in both to facilitate comparisons over time. The live portion of organic matter is the sum of tree and understory vegetation C, and soil and forest floor C for the dead portion.

The net change estimates in the tables reveal that increments in C pool can vary considerably in a relatively short time. The fluctuations would be even greater if we had reported results by ten-year period instead of twenty-year period. Balancing the U. S. C budget will be difficult over time with variable net C fluxes.

The total C sequestered in productive forest ecosystems under the harvesting scenario over the 83-yr period from 1987 to 2070 is 2.57 Pg C, which is about 45% of the amount of C currently released globally to the atmosphere in anthropogenic CO_2 emissions in one year. Over ten times as much C is projected to be sequestered in these timberlands with no harvests. However, the results of the no harvesting scenario should be interpreted accordingly because the model at this time does not allow for natural species composition changes. Without harvesting, by 2070 average stand age in the North rises to 132 yr, 143 yr in the West and 84 yr in the South. The volume yield equations were developed through 175 yr in the North and West, and 80 yr in the South, so some stands may begin to exhibit behavior not expressed in the model. Furthermore, we have assumed no change in the probability of disturbances such as fire, insect attack, and disease as stands age.

5.2   CHANGES IN LAND USE EFFECTS

The C estimates are a consequence of the land use history of the regions. Historical estimates indicate that 60% of the accumulation of C in forests occurred in the North and about 40% occurred in the South, with negligible changes in the West. Accumulation in the dead C pool continues through the projections of the harvesting scenarios in the North, as C accumulates on lands that reverted from agriculture to

Table 1. Estimates of C pools (Pg) in live and dead organic matter and changes in pools (Pg yr^{-1}) for U. S. timberlands under expected harvest levels by year. Net changes are annual estimates, averaged over the period.

Region	Component	Historical estimates				Projections[a]			
		1952	1970	1987	2010	2030	2050	2070	
North	Alive	2.001	2.824	3.663	4.415	4.673	4.714	4.659	
	Dead	4.825	6.816	8.809	10.042	11.070	12.007	12.831	
	Total	6.826	9.640	12.472	14.457	15.743	16.721	17.490	
	Net Change		0.156	0.167	0.086	0.064	0.049	0.038	
South	Alive	2.675	3.432	4.288	4.118	3.707	3.435	3.390	
	Dead	3.493	4.619	5.726	6.185	6.396	6.004	3.547	
	Total	6.168	8.051	10.014	10.303	10.103	9.438	6.937	
	Net Change		0.105	0.116	0.013	-0.010	-0.033	-0.125	
West	Alive	3.716	3.698	3.482	3.293	3.422	3.700	3.818	
	Dead	6.375	6.319	5.919	5.486	5.695	5.951	6.212	
	Total	10.091	10.017	9.401	8.779	9.117	9.551	10.030	
	Net Change		-0.004	-0.036	-0.027	0.017	0.022	0.024	
U. S.	Alive	8.392	9.954	11.433	11.826	11.802	11.748	11.866	
	Dead	14.693	17.754	20.454	21.713	23.161	23.961	22.590	
	Total	23.085	27.708	31.887	33.539	34.963	35.709	34.456	
	Net Change		0.257	0.246	0.073	0.071	0.037	-0.063	

[a]Projections are based on economically derived harvests through 2040. Harvests for 2050-2070 are held constant at 2040 level.
Note: Projections are based on the assumption that future growing conditions will be similar to past growing conditions.

Table 2. Estimates of C pools (Pg) in live and dead organic matter and changes in pools (Pg yr^{-1}) for U. S. timberlands with no harvest after 1987 by year. Net changes are annual estimates, averaged over the period.

Region	Component	Historical estimates			Projections[a]			
		1952	1970	1987	2010	2030	2050	2070
North	Alive	2.001	2.824	3.663	5.730	6.801	7.764	8.642
	Dead	4.825	6.816	8.809	10.760	11.539	12.204	12.784
	Total	6.826	9.640	12.472	16.490	18.340	19.968	21.426
	Net Change		0.156	0.167	0.175	0.093	0.081	0.072
South	Alive	2.675	3.432	4.288	7.254	8.909	10.087	10.754
	Dead	3.493	4.619	5.726	7.050	7.722	8.817	9.250
	Total	6.168	8.051	10.014	14.304	16.631	18.904	20.004
	Net Change		0.105	0.115	0.187	0.116	0.114	0.055
West	Alive	3.716	3.698	3.482	5.605	6.566	7.319	7.819
	Dead	6.375	6.319	5.919	8.206	9.011	9.569	9.881
	Total	10.091	10.017	9.401	13.811	15.577	16.888	17.700
	Net Change		-0.004	-0.036	0.192	0.088	0.066	0.041
U. S.	Alive	8.392	9.954	11.433	18.589	22.276	25.170	27.214
	Dead	14.693	17.754	20.454	26.018	28.272	30.589	31.915
	Total	23.085	27.708	31.887	44.605	50.548	55.759	59.129
	Net Change		0.257	0.246	0.552	0.297	0.261	0.169

[a]Projections are based on no harvesting after 1987.

Note: Projections are based on the assumption that future growing conditions will be similar to past growing conditions.

forest. Increasing harvest levels eventually result in a decline in the dead C pool in the young forests of the South in the harvesting scenario, as average stand age decreases from 30 yr in 1980 to 18 yr in 2070. In the West, decreased harvesting leads to increases in both live and dead C pools.

5.3 CONTRIBUTIONS OF PRODUCTS TO FOREST CARBON POOL

In the harvesting scenario, a substantial amount of C is projected to be removed from U. S. timberlands through harvests over the period 1980 to 2070. Table 3 provides estimates of the C pool in wood removed from forests over the projection period, and shows the disposition of the C into component pools. Although the C in both the energy and emissions components are now part of the atmospheric C pool, they are tracked separately, as biomass energy may be considered as a substitute for energy from fossil fuels. Only C from wood harvested after 1980 is considered.

Table 3. Projected C pools (Pg) of uses of harvested wood, and net change of C (Pg yr^{-1}) in use.

Component	1990	2010	2030	2050	2070
Products	0.741	1.837	2.770	3.535	4.011
Landfill	0.000	0.462	1.163	1.979	2.818
Energy	0.469	1.699	3.213	4.908	6.613
Emissions	0.188	0.878	1.902	3.218	4.748
Removals	1.398	4.867	9.048	13.640	18.190
In Use net change[a]		0.078	0.082	0.079	0.066

[a] In Use means C in product and landfill components.

These products pools and the forest C pools were summed and are displayed in Figure 4 for comparison. Over the period 1952 to 2070, the general trend is for C pools in U. S. forests to increase, except for the period 2050 to 2070, in which pool size in the harvesting scenarios declines. Carbon from harvested wood is presented two ways. The C in forests plus all the C in removals is the line marked by asterisks. The difference between this line and the top line is due mainly to C in the dead C pool, and indicates that when all C from forests is accounted for, there is less than 7 Pg C difference by 2070 between the harvesting and no harvesting scenarios. However, some of the C removed has been released to the atmosphere in 2070. The line marked by the squares is an estimate of the C pool which remains sequestered. This estimate includes products, landfills, and that part of the energy pool which

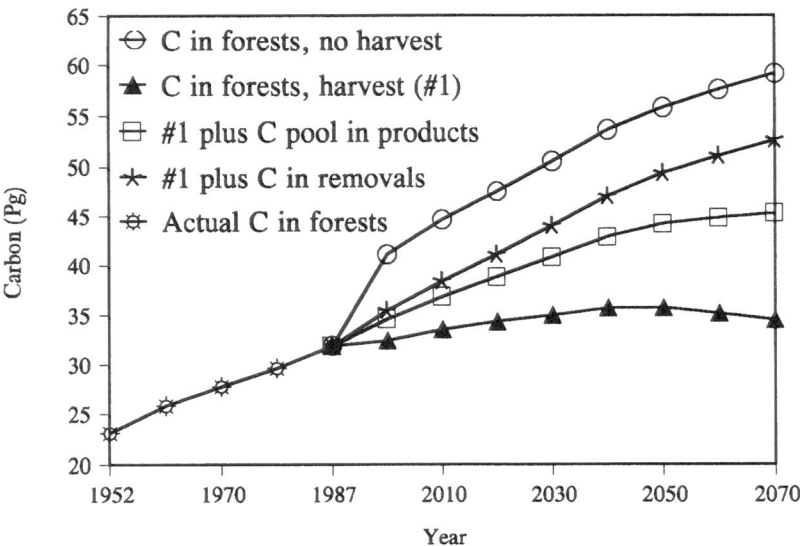

Figure 4. Actual (1952 to 1987) and projected C pools (Pg) in U. S. timberlands for two scenarios, and projections for harvested C.

may be considered a substitute for energy from fossil fuels. (This C would remain part of the fossil C pool.) Following Marland and Marland (1992), we assume that 60% of the C in energy from wood would be conserved in fossil fuels. The forest C pool under harvesting plus the C pool from these products is about 15 Pg less in 2070 than the forest C pool under no harvesting.

6. Conclusions

Forests in the United States exhibit regional growth patterns that are a consequence of their past use by humans. In the recent past, these forests have been sequestering an average of 250 Tg C yr^{-1}, but under projected harvests and growth rates the net change in forest C pools is expected to fall in the near future to an average of 60 Tg yr^{-1} through at least 2050, and then begin releasing C. Carbon sequestered in products and landfills over the projection period adds 75 Tg C yr^{-1}. If all harvesting ceased, the forest C pool would increase at a rate of 328 Tg C yr^{-1}, but this sequestration rate is declining near the end of the period.

We originally hypothesized that harvesting trees and manufacturing wood products would eventually lead to larger C pools in the forest and products than the C pool in the forest from no harvesting. This does not occur within the time frame of this study. We tracked products that were harvested over the 90 yr period from 1980 to 2070 and still find that the no harvesting scenario features a larger C pool. Further research should concentrate on analysis over a longer time period.

7. References

Adams, D.M. and Haynes, R.W.: 1980, *Forest Science* **26,** Monograph 22, 64 p.

Alig, R.J.: 1985, *Forest Science* **32**, 119-134.

Anderson, J.M.: 1992, *Advances in Ecological Research* **22**, 163-210.

Birdsey, R.A.: 1992a, *Carbon storage and accumulation in United States forest ecosystems,* Gen. Tech. Rep. WO-59, U.S. Department of Agriculture, Forest Service, Washington, DC, 51 p.

Birdsey, R.A.: 1992b, Changes in forest carbon storage from increasing forest area and timber growth, in: Sampson, R.N. and Hair, D. (eds), *Forests and Global Change, Vol. 1: Opportunities for Increasing Forest Cover*, American Forests, Washington, DC, pp. 23-39 and App. 2.

Birdsey, R.A., Plantinga, A.J., and Heath, L.S.: 1993, *Forest Ecology and Management* **58**, 33-40.

Burke, I.C., Yonker, C.M., Parton, W.J., Cole, C.V., Flach, K., and Schimel, D.S: 1989, *Soil Sci. Soc. Am. J.* **53**, 800-805.

Cost, N.D., Howard, J., Mead, B., McWilliams, W.H., Smith. W.B., Van Hooser, D.D., and Wharton, E.H.: 1990, *The biomass resource of the United States*. Gen. Tech. Rep. WO-57, U.S. Department of Agriculture, Forest Service, Washington, DC, 21 p.

Dewar, R.C. and Cannell, M.G.R.: 1992, *Tree Physiology* **11**, 49-71.

Gansner, D.A., Birch, T.W., and Lacy, S.E.: 1991, *American Forests* **97**, 48-49.

Haynes, R.W. (coord.): 1990, *An Analysis of the Timber Situation in the United States: 1989-2040*, General Technical Report RM-199, U.S. Department of Agriculture, Forest Service, Rocky Mtn. Forest and Range Expt. Stn., Fort Collins, CO.

Haynes, R.W. and Adams, D.M.: 1985, *Simulations of the Effect of Alternative Assumptions of Demand-Supply Determinants on the Timber Situation in the United States*, USDA Forest Service, Forest Economics Research, Washington, DC.

Johnson, D.W.: 1992, *Water, Air, and Soil Pollution* **64**, 83-120.

Kauppi, P.E., Mielikäinen, K., and Kuusela, K.: 1992, *Science* **256**, 70-74.

Koch, P.: 1989, Estimates by species group and region in the USA of: I. Below-ground root weight as a percentage of ovendry complete-tree weight; and II. carbon content of tree portions, Consulting report, 23 p.

Kurz, W.A. and Apps, M.J.: 1993, *Environmental Pollution*, **In press**.

Kurz, W.A., Apps, M.J., Webb, T.M., and McNamee, P.J.: 1992, *The Carbon Budget of the Canadian Forest Sector: Phase I.*, Forestry Canada, Northwest Region, Northern Forestry Centre, Edmonton, Alberta, Inf. Rep. NOR-X-326. 93 p.

Marland, G. and Marland, S.: 1992, *Water, Air, and Soil Pollution* **64**, 181-195.

Mills, J.R. and Kincaid, J.C.: 1992, *The Aggregate Timberland Assessment System-ATLAS: A comprehensive timber projection model.* Gen. Tech. Rep. PNW-GTR-281, U.S. Department of Agriculture, Forest Service, Pacific Northwest Research Station, Portland, OR, 160 p.

Pastor, J. and Post, W.M.: 1986, *Biogeochemistry* **2**, 3-27.

Plantinga, A. J. and Birdsey, R.A.: 1993, *Climatic Change* **23**, 37-53.

Post, W.M., Emanuel, W.R., Zinke, P.J., and Stangenberger, A.G.: 1982, *Nature* **298**, 156-159.

Reynolds, R.V., and Pierson, A.H.: 1941, *The sawtimber resource of the United States, 1630-1930*, Forest Survey Release, U.S. Department of Agriculture, Forest Service, Washington, DC, 21 p.

Row, C. and Phelps, R.B.: 1991, Carbon cycle impacts of future forest products utilization and recycling trends, in: *Agriculture in a World of Change*, Proceedings of Outlook '91, 67th Annual Outlook Conference, U.S. Department of Agriculture, Washington, DC.

Sedjo, R.A.: 1992, *Ambio* **21**, 274-277.

U.S. Department of Agriculture: 1928, *American forests and forest products*, Statistical Bulletin No. 21, U.S. Government Printing Office, Washington, DC, 17 p.

U.S. Department of Agriculture, Forest Service: 1958, *Timber resources for America's future*, For. Resour. Rep. 14., U.S. Government Printing Office, Washington, DC, 713 p.

U.S. Department of Agriculture, Forest Service: 1965, *Timber trends in the United States*, For. Resour. Rep. 17, U.S. Government Printing Office, Washington, DC, 235 p.

U.S. Department of Agriculture, Forest Service: 1974, *The outlook for timber in the United States*, For. Resour. Rep. 20, U.S. Government Printing Office, Washington DC, 367 p.

U.S. Department of Agriculture, Forest Service: 1982, *An analysis of the timber situation in the United States: 1952-2030,* For. Resour. Rep. 23, U.S. Department of Agriculture, Forest Service, Washington, DC, 499 p.

U.S. Department of Agriculture, Forest Service: 1988, *The South's fourth forest: alternatives for the future*, For. Resour. Rep. 24, U.S. Department of Agriculture, Forest Service, Washington, DC, 512 p.

Vogt, K.A., Grier, C.C., and Vogt, D.J.: 1986, *Advances in Ecological Research* **15**, 303-377.

Waddell, K.L., Oswald, D.D., and Powell, D.S.: 1989, *Forest statistics of the United States, 1987*, Resour. Bull. PNW-RB-168, U.S. Department of Agriculture, Forest Service, Pacific Northwest Research Station, Portland, OR, 106 p.

Watson, R.T., Meira Filho, L.G., Sanhueza, E., and Janetos, A.: 1992, Greenhouse gases: sources and sinks, in: Houghton, J.T., Callander, B.A, and Varney, S.K., *Climate Change 1992: The Supplementary Report to the IPCC Scientific Assessment*, Intergovernmental Panel on Climate Change, World Meteorological Organization /U.N. Environment Program, Cambridge University Press, Cambridge, Great Britain.

Woodwell, G.M., Hobbie, J.E., Houghton, R.A., Melillo, J.M., Moore, B., Park, A.B., Peterson, B.J., and Shaver, G.R.: 1984, Measurement of changes in the vegetation of the earth by satellite imagery, in: G.M. Woodwell (ed), *The Role of Terrestrial Vegetation in the Global Carbon Cycle: Measurement by Remote Sensing*, SCOPE No. 23, John Wiley and Sons, Ltd, 272 p.

The Carbon Cycle and Global Forest Ecosystem

ROGER A. SEDJO
Senior Fellow, Resources for the Future,
1616 P St NW, Washington, D.C. 20036 USA

ABSTRACT. Attempts to account for the fluxes by quantifying C sources and sinks have provided evidence of a missing C sink (Detwiler and Hall, 1988), which may be located somewhere in the temperate region of the northern hemisphere (Tans et al., 1990). Until recently, most estimates have concluded that the temperate forest is a small C source. Two recent papers (Sedjo, 1992; Kauppi et al., 1992) provided evidence that the temperate forests are substantial C sinks. This paper combines these earlier findings on temeperate forest carbon sequestration with a new estimate of the annual C releases due to tropical deforestation, 1.7 Gt, which is obtained using the FAO estimates of the rate of deforestation in the tropics over the decade of the 1980s and conservative estimates of C releases associated with this deforestation. Finally, to this is added the crude estimate of C export by the global river system found in Hall et al. (1992). Applying these estimates of the C sink function of both temperate and tropical forests to Detwiler and Hall's alternative C budgets largely eliminates the "missing C" hypothesized by Detwiler and Hall, and Tans et al.

Key Words: carbon, carbon fluxes, carbon sink, carbon cycle, tropical and temperate forests, deforestation, global warming

Forests dominate the dynamics of the terrestrial C cycle. They contain 483 of the 562 Gt (86%) of the globe's above-ground C (Olson et al. 1983), and an estimated 73% of the C in the world's soil (927 of 1,272 Gt) is in forest soils (Post et al., 1982). It is estimated that of the 31% of the world's land area that is forested, some 22% is closed forest (closed canopy) (Persson, 1986).

1. Forests and Carbon: The Debate

The global C cycle has been examined by researchers for clues to the likelihood and timing of global warming. Rising atmospheric CO_2 and other greenhouse gases have lead to the prediction of global warming. Atmospheric CO_2 levels have now been observed to have been rising for 30 yr, and other evidence indicates that they almost certainly have been increasing for well over 100 yr. Rising CO_2 levels reflect a global C cycle in which more C is released into the atmosphere (from sources) than is absorbed (in sinks). However, the weakness in our current understanding of the global C cycle is demonstrated by the difficulties that researchers have encountered when trying to balance the C budget.

Water, Air, and Soil Pollution **70**: 295–307, 1993.
© 1993 *Kluwer Academic Publishers.*

C in the atmosphere has been increasing at about 2.9 x 10^9 t annually. The major source of atmospheric C is fossil fuel burning, which is estimated to have released about 5.3 x 10^9 t yr^{-1} in the 1980s. A second source of atmospheric C is believed to be in land-use changes in terrestrial ecosystems, which have been estimated as having net C releases of 0.4 to 4.7 x 10^9 yr. In the process of growth, forest plant life captures C. An expanding forest is able to build-up its inventories of C in its increasing biomass and dead organic material. The oceans are known to be a sink but the extent is uncertain, with various studies estimating that the oceans are absorbing between 26 and 44% of the fossil C.

Until recently the conventional wisdom held that forest ecosystems, both tropical and temperate, were net sources (e.g., see Houghton et al., 1983; Houghton, 1990). Although the early work of Armentano and Ralston (1980) estimated that net forest growth throughout the north temperate zone to be a C sink for 1.0 to 1.2 Gt of C yr^{-1}, other early studies estimated the annual C releases from land-use changes in the nontropical ecosystems to be between 0.5 and 0.8 x 10^9 t yr^{-1} (e.g., Woodwell et al., 1978; Houghton et al., 1983). More recent work has put the nontropical terrestrial biosphere roughly in C balance as reflected in Table 1 (e.g., Melillo et al., 1988). The major differences between the results of the early study of Houghton et al. (1983) and the later study of Melillo et al. (1988) are due largely to different assumptions about the rates of C loss from the soil: assumptions that reflect the variability in empirically estimated rates of soil C losses. In addition, these studies treat the land area in forest as essentially stable in the post World War II period.

In their assessment of the global C budget, Detwiler and Hall accepted the view of the nontropical terrestrial biosphere C balance being roughly balanced and ignored other possibilities. However, improving data on global forest extent and volume have provided considerable evidence, both anecdotal and empirical, of the widespread expansion of forest area and forest volumes in the temperate northern hemisphere, thereby placing into question their assumption of temperate forest C balance.

Detwiler and Hall (1988) report estimates of C release in the tropics from various sources as ranging from 1.0 to 4.2 Gt yr^{-1}. In their simulation study they estimate the feasible range of tropical forest C releases to be between 0.4 and 1.6 Gt yr^{-1}. The recent assessment by Detwiler and Hall (1988) of the implications of their estimates of deforestation on the global C budget is presented in Table 1.

The recent work of Tans et al. (1990) has presented further evidence that calls into question the conventional estimates of sinks and the magnitude of the fluxes. Using a general circulation model (GCM) and information on the observed north-south atmospheric CO_2 concentration gradient by latitude, they simulate the implications of various source-sink distributions in an attempt to try to explain the divergence between predicted and measured atmospheric CO_2 buildup. The paper concludes that "the total

Table 1. Global C Budget, 1980.

C	Flux ($\times 10^{15}$ g C yr^{-1})		
	Extreme	Median	Extreme
Released			
Fossil fuel combustion; cement production	4.8	5.3	5.8
Tropical forest clearing	0.4	1.0	1.6
Nontropical forest clearing	-0.1	0.0	0.1
Accounted for			
Atmospheric increase	-2.9	-2.9	-2.9
Ocean uptake	-2.5	-2.2	-1.8
"Missing"*	-0.3	1.2	2.8

* A minus indicates the need for a source of the size shown; a plus the need for a sink.

Source: As presented by Detwiler and Hall (1988).

CO_2 uptake by the oceans is considerably less than uptake by terrestrial systems." They continue, "there must be a terrestrial sink at temperate latitudes to balance the C budget..." Their C budgeting implies that the unknown sink's magnitude could be as large as 2.0 to 3.4 x 10^9 t of atmospheric C annually--a result apparently inconsistent with high C releases from both fossil fuels and from tropical forest biosystems. Their paper then calls for a reanalysis of the contribution of mid-latitude reforestation, as well as studies of the feedbacks between ecosystem function, climate, and atmospheric composition.

In summary, the work of Detwiler and Hall suggests that the unexplained C sink could range from -0.3 to 2.8 Gt yr^{-1} while that of Tans et al. put the sink's size in the range of 2.0 to 3.4 Gt yr^{-1} and suggest the existence of a large sink in the northern hemisphere. This wide range of estimates demonstrates the lack of preciseness in our understanding of the relative size of the various fluxes and indicates the extent of uncertainty of each.

2. Some Recent Findings

Recent work (Kauppi et al., 1992; Sedjo, 1992) has suggested that the forest ecosystem, especially the temperate northern forests (including the boreal forests), may be a significant C sink. This work utilized a wide variety of data sources that demonstrate that areas of temperate forests have experienced considerable expansion over recent decades in much of the northern hemisphere. Kauppi et al. found that measurements from a number of countries within Europe "show a general increase of forest resources" and that biomass was built up in the 1970s and 1980s in European forests because of a relaxation of earlier rates of exploitation. They suggest that similar developments in other continents could account for a large portion of the missing C. Sedjo, relying upon data recently compiled by the ECE/FAO, finds that forest inventories are estimated to have expanded throughout the northern hemisphere. Using forest inventory data and common C coefficients, he estimates that the temperate forests have sequestered large volumes of C in recent decades and that the levels of captured C are consistent with the hypothesis of Tans et al. regarding a "missing sink" in the northern hemisphere. Sundquist (1993) has accepted the view that "...northern nontropical forests are currently accumulating CO_2 due to regrowth and afforestation..." but notes that the magnitude of the net sink includes considerations other than simply forest growth.

In general, these estimates of the C fluxes which have shown the northern temperate forest ecosystem to be a C source may be inconsistent with the evidence of substantial growth being experienced by these forests. The evidence is of two types. First, is anecdotal evidence that reveals that in many regions of the world the forest biosystem has expanded considerably in recent decades. Second, is empirical evidence based on newly developed data by the ECE/FAO and new C conversion factors developed by the US Forest Service.

In many of the temperate regions of the world a strong trend toward reforestation, largely via natural regeneration, began as early as the 19th century. For example, in New England the area of forest was much more extensive in 1980 than it was in the mid-1800s. Furthermore, this phenomenon was not confined to the New England States. The area of forestland in the U.S. reached its nadir about 1920 (Clawson, 1979) and rebounded sharply thereafter continuing into the post-World War II period. A contributing factor to forest expansion in the U.S. was the decline of agriculture on marginal lands as agricultural productivity rose. This phenomenon was accelerated by the Depression of the 1930s where, especially in the eastern U.S., abandonment of agricultural lands was followed by reforestation, again largely by natural regeneration. Also contributing to the rebirth of the forest was the gradual control of forest fires (Sedjo, 1991). In the 1930s wildfires affected an average of almost 12.5 Mha yr^{-1}; by the 1980s the area affected was reduced to one tenth, 1.25 Mha yr^{-1} (USDA-FS 1987). Although the postwar period has seen the stabilization, and even a modest decline, of the U.S. forested area, forest biomass has continued to expand as the forests that regenerated in the 1930s and 1940s continue to experience rapid growth. The first complete inventory of U.S. forests was released in 1952. Four subsequent inventories of U.S. timberlands have followed, the most recent released in 1987. Each one reported an increase in timber stock with the total increase being 29% over the 35-yr period (USDA-FS 1989).

Forest expansion was not restricted to North America. Parts of Europe began their forest expansion as early as the mid-1800s (e.g., see Johann, 1990). Recently forest volumes have been increasing substantially. For example, data for Sweden (National Board of Forestry 1990, p. 60) indicate an increase of forest growing stock of about 25% over that same period. Furthermore, the Swedish data which begin in the mid-1920s show a continuing increase in growing stock to the extent that the 1986 stock is over 50% greater than the 1926 stock. In addition, recent data indicate that even the vast forest lands of the Soviet Union, while exhibiting deterioration in some regions, nevertheless have experienced large overall net reforestation as the total area of the forests increased from 738 Mha in 1961 to 811 Mha in 1984 (Holcwacz, 1985). This general tendency within the former Soviet Union is supported by newly emerging data from some of the former Soviet states. For example, data recently available from Estonia estimates that the forested area in 1992 was over twice that forested in 1940. The reasons given include the reversions of large area of former meadow and pasturelands to forests (Estonian Forest Survey Centre, 1992).

Finally, even Canada, which has had large harvesting pressures on its forests, shows a modest increase in forest biomass in the past two decades, while exhibiting some decline in forested area (Honer et al., 1990). Given this recent well documented experience of the temperate climate forests, it is difficult to envisage the temperate forests not being a significant C sink.

3. Empirical Evidence: New Data

Although regional data have shown the northern temperate forests to be expanding in many regions, it was a recent effort by the ECE/FAO (1986,

Table 2

Country	Timber Volume Change $(10^9 \text{ m}^3 \text{ yr}^{-1})$	Time Period
USSR[a]	0.4	1973-84 (p. 14)
Europe[b]	0.1	1985
US[c]	0.168	1986 (p.20)
Canada[c]	0.025	1985 (p.21)
TOTAL	0.693	

Source: Sedjo (1992) as drawn from a)ECE/FAO 1989; b) 1986; c) 1990.

1989, 1990) that documented the pervasiveness of the expansion in recent years of forest area and forest biomass for all the temperate regions of the northern hemisphere. Table 2 is drawn from the findings of that effort. Although the data presented is for the late 1970s into the mid 1980s, the data sources reveal that this phenomenon has been occurring for several decades. All four of the major regions have been experiencing substantial increases in timber volume and, therefore, in the associated C sequestered by the timber. Using the common conversion factor of 0.26 t C m^3 of woody biomass (see Marland 1988, p.35) gives an estimate of C sequestered by the timber of the temperate forest in recent years of 0.180 Gt yr^{-1}. However, the amount of C estimated to be captured in the timber volume is only one component of the total C sequestered by the forest ecosystem. For the U.S. forest, Birdsey (1990) has estimated that 31% of the C is in live trees including tree roots, 59% in the forest soils, 9% in the forest litter, and 1% in other vegetation. These other components are also assessed.

A number of conversion factors for converting timber volumes into woody biomass have been estimated that adjust commercial volume to total wood biomass, including the limbs and root systems (e.g., Brown et al., 1989; Johnson and Sharp, 1984).

Table 3 estimates total annual C sequestered by temperate forest ecosystems as developed by Sedjo (1992).

Table 3: C Sequestered by Temperate Forest Ecosystems

Region/ Country	Timber C (Gt)	C conversion factor	Total annual C (Gt)
(1)	(2)	(3)	(4)
USSR	0.104	4.0	0.416
Europe	0.026	3.5	0.091
US	0.044	3.5	0.153
Canada	0.001	4.0	0.026
			0.686

Source: Column (2) from table 3; Column (3) drawn from Birdsey (1990).

This estimate, approximately 0.7 Gt., is similar to that of Armentano and Ralston (1980), and is also consistent with both the anecdotal, and more recently, the systematic evidence of the gradual but continuous buildup of the northern hemisphere temperate forest biomass. The estimate of the annual amount of C sequestered by the forest ecosystem, however, is inconsistent with the a number of studies that have estimated, or assumed, that temperate climate forests are either a C source, or at most a very modest sink, for atmospheric C.

4. Implications of Temperate Forest Expansion for the Global Carbon Budget

Table 4 applies the revised estimates of temperate forests as a C sink to the C budgets of Detwiler and Hall. The temperate forest C sink of this size substantially reduces the problems raised by Detwiler and Hall as the range and the asymmetry of the missing component are reduced substantially.

Table 4. Global C Budget Revised Temperate Forest, 1980.

C	Flux ($\times 10^{15}$ g C yr^{-1})		
	Extreme	Median	Extreme
Released			
Fossil fuel combustion; cement production	4.8	5.3	5.8
Tropical forest clearing	(0.4)	(1.0)	(1.6)
Nontropical forest clearing	(-0.1)	(0.0)	(0.1)
Revised	**-0.7**	**-0.7**	**-0.7**
Accounted for			
Atmospheric increase	-2.9	-2.9	-2.9
Ocean uptake	-2.5	-2.2	-1.8
"Missing"*	(-0.3)	(1.2)	(2.8)
Revised	**-0.9**	**0.5**	**2.0**

    * A minus indicates the need for a source of the size shown; a plus the need for a sink.

    Source: Sedjo (1992), Adapted from Detwiler and Hall (1988)

5. Tropical Forests and Carbon

Tropical forests are widely believed to be a relatively large C source. While citing a range of studies estimating tropical forest C releases of 1.0 to 4.2 Gt yr^{-1}, Detwiler and Hall derived the range 0.4 to 1.6 Gt yr^{-1} C release from tropical forests in their analysis.

Tropical deforestation is estimated at 15.4 Mha yr^{-1} for the decade of the 1980s (FAO, 1993). Assuming the same proportion of closed and open forest as the 1980 estimate provided by Lanly (1982), about 9.6 Mha of closed forest and 5.8 Mha of open forest would have been deforested annually over the decade. Using an estimate for C release of 150 t ha^{-1} from biomass, soil, and litter from closed forests and 50 t ha^{-1} from open forest yield an estimate of gross C release, gives gross C releases from tropical deforestation of about 1.7 Gt annually.

This figure is slightly above the range provided by Detwiler and Hall, but it is calculated based upon the higher preliminary estimates as provided by FAO of the deforestation rates of the full decade of the 1980s. Hence, the estimates are essentially within the range of Detwiler and Hall (Table 1). Additionally, although arrived independently, this estimate is identical to that arrived at by the tropical forest group in this workshop (see the report of the tropical forest group in this volume).

6. Total Forest Implications for the Carbon Cycle

These estimates of carbon releases from forests do not account for the C exported to the oceans through rivers. If we accept the crude estimate of 0.8 Gt annually, reported by Hall et al. (1992), and assign it to forests, the net release of the "forest system" is reduced to a modest 0.2 Gt annually, (e.g., 1.7 - 0.8 - 0.7). These values are introduced to table 5 to assess the carbon balance adjusted for the influences of tropical and temperate forests and the export of C by rivers.

Applying the revised estimates to Detwiler and Hall's C budgets (Table 5), reduces almost in half the range of "missing" C, from 3.1 Gt annually in the original Detwiler and Hall paper to a range of 1.7 Gt annually reported in Table 5. In addition, the revised values largely rectify the skewed nature of the Detwiler and Hall results. These revised estimates largely account for the missing C sink.

Table 5. Global C Budget Revised Temperate and Tropical Forest

C	Flux ($\times 10^{15}$ g C yr^{-1})		
	Extreme	Median	Extreme
Released			
Fossil fuel combustion; cement production	4.8	5.3	5.8
Tropical forest clearing	(0.4)	(1.0)	(1.6)
Revised	**1.7**	**1.7**	**1.7**
Carbon exported by rivers	**-0.8**	**-0.8**	**-0.8**
Non tropical forest clearing	(-0.1)	(0.0)	(0.1)
Revised	**-0.7**	**-0.7**	**-0.7**
Accounted for			
Atmospheric increase	-2.9	-2.9	-2.9
Ocean uptake	-2.5	-2.2	-1.8
"Missing"*	(-0.3)	(1.2)	(2.8)
Revised	**-0.4**	**0.4**	**1.3**

 * A minus indicates the need for a source of the size shown; a plus the need for a sink.

Source: Sedjo (1992), and adaptation from Detwiler and Hall (1988)

REFERENCES

Armentano, T.V. and C.W. Ralston.: 1980, Canadian Journal of Forest Research, **10**, pp. 53-60.

Birdsey, Richard A.: 1990, "Inventory of Carbon Storage and Accumulation in U.S. Forest Ecosystems," paper presented at the XIX IUFRO World Congress, Montreal Canada, August.

Brown, Sandra, Andrew J.R. Gillespie, and Ariel E. Lugo.: 1989, Forest Science, **35**, pp. 881-902.

Clawson, Marion.: 1979, Science **204**, 1168-74.

Detwiler, R.P. and Charles A.S. Hall.: 1988, Science, **239**, pp. 43-47.

Economic Commission for Europe/Food and Agricultural Organization of the United Nations (ECE/FAO). 1986. European Timber Trends and Prospects to the Year 2000 and Beyond, (New York, United Nations).

_____. 1989. "Outlook for the forest and forest products sector of the USSR," ECE/TIM/48, p. 75, (New York, United Nations).

_____. 1990. "Timber Trends and Prospects for North America," ECE/TIM/53. p. 68, (New York, United Nations).

Estonian Forest Survey Centre. 1992. "Short View About Estonian Forest and Forest Inventory Works, Tallinn.

Hall, Charles A.S., Marshall R.Taylor and Edwin Evenham.: 1992, Water, Air, and Soil Pollution **64**, 385-404.

Harmon, Mark E., William Ferrell, and Jerry F. Franklin.: 1990, Science, **247**, pp. 699-702.

Holowacz, J.: 1985, Forestry Chronicle, **61**, pp. 366-73.

Honer, T.G., W.R. Clark, and S.L. Gray.: 1990, "Determining Canada's Forest Area and Wood Volume Balance 1977-1986," paper presented at the Conference on Canada's Timber Resources, June 3-6, 1990, Victoria, British Columbia.

Houghton, R.A., Hobbie, J.E., Melillo, J.M., Moore, B., Peterson, B.J, Shaver, G.R. and Woodwell, G.M.: 1983, Ecol. monogr. **53**, 235-262.

Johann, Elisabeth L.: 1990, Forest & Conservation History, vol. **34**, .

Johnson, W.C. and D.M. Sharp.: 1984, Canadian Journal of Forest Research, **13**, pp. 337-383.

Kauppi, Pekka E., Kari Mielikainen, and Kullervo Kuusela.: 1992, Science **256**, pp. 70-74.

Lanly, J.P.: 1982, FAO Forestry Paper 30. Food and Agriculture Organization of the United Nations, Rome.

Marland, Gregg. 1988, "The Prospect of Solving the CO2 Problem through Global Reforestation," Report DOC/NBB-0082 (Washington, D.C. Department of Energy, Office of Energy Research).

Melillo, J., J.R. Fruci, R.A. Houghton, B. Moore, D.L. Stole.: 1988, Tellus **40B**, 116-128.

Olson, J.S., J.A. Watts, and L.J. Allison.: 1983, Carbon in Live Vegetation of Major World Ecosystems. Report ORNL-5862 (Oak Ridge, Tenn.) Oak Ridge National Laboratory.

Persson, R.: 1986, Unpublished report to the Swedish International Development Authority (1985) reported in World Resources 1986, p. 62, Basic Books, New York.

Post, W.M., W.R. Emanuel, P.J. Zinke, and A.G. Strangenberger.: 1982, Nature **298**, pp. 156-159.

Sedjo, Roger A.: 1992, Ambio **21**, pp. 274-77.

_____. 1991, Forest Resources: Resilient and Serviceable, in: Frederick, K. and Sedjo, R.A. (eds), America's Renewable Resources: Historical Trends and Current Challenges, Resources for the Future, Washington, D.C., pp.

Sundquist, Eric T.: 1993, Science **259**, pp. 934-41.

Tans, Peter P., Inez Y. Fung and Taro Takahashi.: 1990, Science **247**, pp. 1431-47.

USDA-Forest Service: 1987, 1926-1967 Forest Fire Statistics, and various annual issues Wildfire Statistics.

_____. 1989. "RPA Assessment of the Forest and Rangeland Situation in   the United States, 1989," Forest Resource Report No. 26, October.

Woodwell, G.M., R.H. Whittaker, W.A. Reiners, G.E. Likens, C.C. Delwiche and D.B. Botkin.: 1978, Science **199**, 141.

FOREST RESPONSES TO CO$_2$ ENRICHMENT AND CLIMATE WARMING

R. J. LUXMOORE, S. D. WULLSCHLEGER, AND P. J. HANSON

Environmental Sciences Division, Oak Ridge National Laboratory
P.O. Box 2008, Oak Ridge, Tennessee, USA 37831-6038

Abstract. Two of the major uncertainties in forecasting future terrestrial sources and sinks of CO$_2$ are the CO$_2$-enhanced growth response of forests and soil warming effects on net CO$_2$ efflux from forests. Carbon dioxide enrichment of tree seedlings over time periods less than 1 yr has generally resulted in enhanced rates of photosynthesis, decreased respiration, and increased growth, with minor increases in leaf area and small changes in C allocation. Exposure of woody species to elevated CO$_2$ over several years has shown that high rates of photosynthesis may be sustained, but net C accumulation may not necessarily increase if CO$_2$ release from soil respiration increases. The impact of the 25% rise in atmospheric CO$_2$ with industrialization has been examined in tree ring chronologies from a range of species and locations. In contrast to the seedling tree results, there is no convincing evidence for CO$_2$-enhanced stem growth of mature trees during the last several decades. However, if mature trees show a preferential root growth response to CO$_2$ enrichment, the gain in root mass for an oak-hickory forest in eastern Tennessee is estimated to be only 9% over the last 40 years. Root data bases are inadequate for detecting such an effect. A very small shift in ecosystem nutrients from soil to vegetation could support CO$_2$-enhanced growth. Climate warming and the accompanying increase in mean soil temperature could have a greater effect than CO$_2$ enrichment on terrestrial sources and sinks of CO$_2$. Soil respiration and N mineralization have been shown to increase with soil temperature. If plant growth increases with increased N availability, and more C is fixed in growth than is released by soil respiration, then a negative feedback on climate warming will occur. If warming results in a net increase in CO$_2$ efflux from forests, then a positive feedback will follow. A 2 to 4°C increase in soil temperature could increase CO$_2$ efflux from soil by 15 to 32% in eastern deciduous forests. Quantifying C budget responses of forests to future global change scenarios will be speculative until mature tree responses to CO$_2$ enrichment and the effects of temperature on terrestrial sources and sinks of CO$_2$ can be determined.

1. Introduction

Expectations that rising atmospheric CO$_2$ concentration will cause global warming are supported by the historical record which shows the co-occurrence of elevated atmospheric CO$_2$ and warmer global temperature (Watson *et al.*, 1990). Although the effects of elevated CO$_2$ and warming on forest carbon budgets are unresolved at the present time, some guidance can be obtained from examination of experimental results and simulation forecasts.

Tree physiological responses to CO$_2$ enrichment and warming effects on soil microbial processes could lead to positive or negative feedbacks between forest ecosystems and global climate depending on whether forests become a net source or sink

Water, Air, and Soil Pollution **70**: 309–323, 1993.
© 1993 *Kluwer Academic Publishers.*

for CO_2. The only known negative feedback between forest ecosystems and global warming is that due to C fixation which may increase with enhanced nutrient mineralization of soil organic matter. Conversely, positive feedback could result from net C releases from such processes as soil respiration, which may increase dramatically with rising soil temperature. These effects could contribute to a net flux of CO_2 to the atmosphere that is supplemental to the rising CO_2 concentration from fossil fuel combustion and deforestation. We evaluate some experimental results of elevated CO_2 and climate warming effects on C and nutrient dynamics of forested ecosystems. We do not address the effects of change in precipitation or in evapotranspiration specifically.

2. Tree Physiological Responses to Atmospheric CO_2 Concentration

2.1. SEEDLING TREE RESPONSES

2.1.1. Photosynthesis

Short-term increases in the CO_2 concentration surrounding healthy C_3 plants such as forest species generally stimulate photosynthetic rates. This enhancing effect of CO_2 on single-leaf photosynthesis is most evident in the relationship of assimilation (A) to substomatal CO_2 concentrations (C_i) as depicted in A/C_i curves. Assimilation rates of seedlings of woody species typically increase 50 to 75% for a doubling of the ambient CO_2 concentration (Wullschleger, 1993). Mousseau and Saugier (1992) point out that short-term CO_2 exposures, such as those used in constructing A/C_i curves, are not necessarily meaningful to understanding elevated CO_2 effects on long-term C gain. We need to understand the capacity of CO_2-enriched plants to maintain enhanced rates of photosynthesis after prolonged periods of exposure, where a diminishing enhancement of photosynthesis by CO_2 with time would be indicative of acclimation (Mousseau and Saugier, 1992).

Most extended-exposure studies indicate that tree seedlings maintain higher photosynthetic rates at elevated CO_2 concentrations than those of seedlings grown at ambient CO_2 concentration, albeit to a lesser extent than indicated in short-term exposure experiments. In a review of 28 controlled-exposure studies, covering 30 boreal, temperate, and tropical tree species, Gunderson and Wullschleger (C.A. Gunderson and S.D. Wullschleger, Environmental Sciences Division, Oak Ridge National Laboratory, Oak Ridge, TN, USA, 37831-6034, unpublished) observed that photosynthesis was increased on average 46% for an approximate doubling of atmospheric CO_2 concentration (Figure 1). For a few species, namely *Alnus rubra* (Arnone and Gordon, 1990), *Quercus prinus* (Bunce, 1992), and *Salix x dasyclados* (Silvola and Ahlholm, 1992) these enhanced photosynthetic rates exceeded twice that of the ambient-grown controls. Studies with *Alnus rubra* showed that whole-plant rates of photosynthesis increased in response to CO_2 enrichment due to greater single-leaf CO_2 uptake and to increased canopy leaf area (Arnone and Gordon, 1990). Growth of seedlings at elevated CO_2 may also be enhanced by a prolonged growing period resulting from a delay in leaf senescence. This was shown by Curtis and Teeri (1992) for *Populus grandidentata*. In contrast, a few tree species display almost complete acclimation to CO_2 exposure, with rates of photosynthesis for plants grown and measured at elevated CO_2 being similar to or lower than those of ambient-grown seedlings. Notable examples of this acclimation response include *Cecropia obtusifolia* (Reekie and Bazzaz, 1989), *Cedrus atlantica* (Kaushal *et al.,* 1989), *Ochroma lagopus* and *Pentaclethra macroloba* (Oberbauer *et al.,* 1985), *Quercus robur* (Bunce, 1992), and *Trichospermum mexicanum* (Reekie and Bazzaz, 1989). Whole-plant studies by Mousseau (1992) with 2-yr-old *Castanea sativa* seedlings indicated that the enhancing effects of CO_2 on canopy photosynthesis were greatest early in the growing season, but declined with time, indicating acclimation. Differences in photosynthetic rates between ambient and elevated CO_2-grown seedlings became negligible after a few months.

Cause-and-effect relationships between photosynthetic acclimation and CO_2 exposure have yet to be established, although several hypotheses about the causal factors have been offered (Bowes, 1991; Stitt, 1991; Mousseau and Saugier, 1992). Likely factors, derived almost exclusively from studies on species other than trees, are feedback inhibition due to carbohydrate build-up, deficiency of inorganic phosphorus, and a decrease in the content, activity, or activation state of the photosynthetic enzyme, ribulose-1,5-bisphosphate carboxylase-oxygenase (Rubisco). In some cases acclimation can be an artifact of plants growing in pots with a restricted rooting volume which limits sink activity (Thomas and Strain, 1991). Alternatively, in two studies, plants grown directly in the soil, where root proliferation may sustain sink activity, elevated photosynthetic rates over multiple growth cycles have been observed. CO_2-enriched plants sustained increased rates of photosynthesis throughout a 28-month exposure period for *Liriodendron tulipifera*

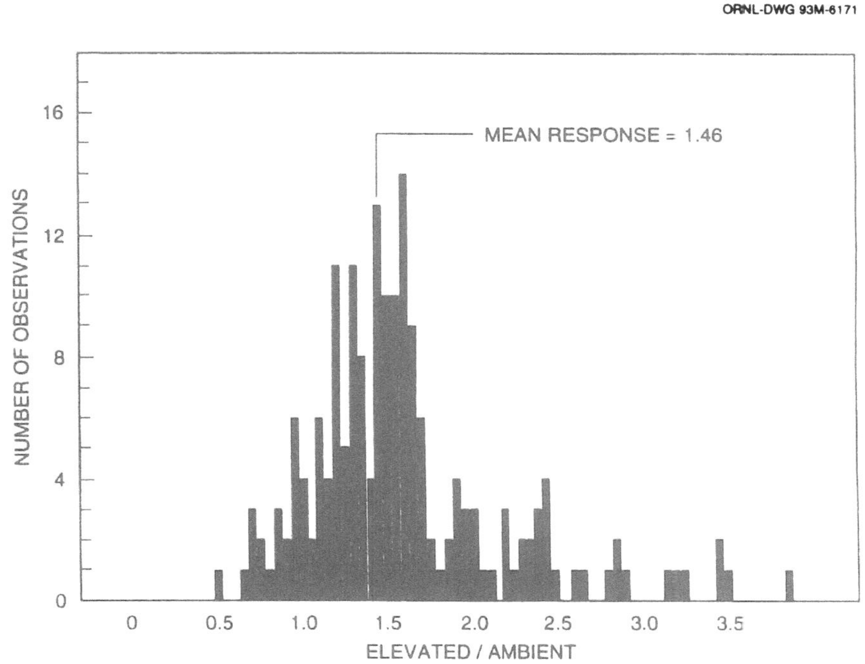

ORNL-DWG 93M-6171

Figure 1. Frequency distribution for the relative increase in photosynthesis as summarized from the results of 28 controlled-exposure studies for 30 tree species grown at ambient and elevated CO_2 concentrations. The mean photosynthetic response shown is the mean of the log-transformed data. (from Gunderson and Wullschleger, unpublished)

(Norby *et al.,* 1992) and throughout a 4-yr exposure period for *Citrus aurantium* (Idso and Kimball, 1992). In the study with *Liriodendron tulipifera,* there was no evidence of photosynthetic acclimation, either in association with leaf age or leaf position within the canopy (Gunderson *et al.,* 1993).

2.1.2. Respiration

There is mounting evidence that trees grown at elevated CO_2 have lower rates of foliar respiration (Bunce, 1992; Mousseau, 1992; Wullschleger *et al.,* 1992; Wullschleger and Norby, 1992). Mousseau (1992) observed that single-leaf and whole-plant rates of respiration in *Castanea sativa* were lower for seedlings grown at elevated CO_2 compared to ambient-grown plants. These differences were greatest early in the year and were no longer apparent during autumn after months of CO_2 exposure. Recent studies by Bunce (1992) indicated that whole-plant respiration rates for CO_2-enriched seedlings of *Acer rubrum, Quercus prinus,* and *Acer saccharinum* were all lower than those of their ambient-grown counterparts. For one species, *Acer saccharinum,* differences in whole-plant respiration due to CO_2 enrichment were greatest early in the season and became less as the exposure period progressed. Evidence obtained in a leaf respiration study of *Liriodendron tulipifera* (Wullschleger *et al.,* 1992) and *Quercus alba* (Wullschleger and Norby, 1992) confirmed lower respiration in leaves grown at elevated CO_2 concentrations. For these latter studies, the effects of long-term CO_2 enrichment were partitioned to the growth and maintenance components, where it was found that maintenance respiration was lower for both *L. tulipifera* and *Q. alba* with CO_2 enrichment, and that growth respiration for *Q. alba* was lower as well. In the case of *L. tulipifera* the decline in maintenance respiration was associated with a decline in leaf N (Wullschleger *et al., 1992),* while this was not the case for white oak (Wullschleger, unpublished).
       A mechanistic basis for the response of leaf and whole-plant respiration to CO_2 enrichment has yet to be resolved, although several explanations have been proposed (Amthor, 1991; Ryan, 1991). For example, Amthor (1991), outlined several potential mechanisms whereby CO_2 could modify respiration with short- and long-term effects. Short-term changes in respiration were defined as the direct, reversible inhibition of CO_2 efflux occurring within minutes of foliar exposure to elevated CO_2, while long-term effects were characterized as the indirect, irreversible inhibition of respiration occurring only after weeks or months of exposure. Possible explanations for lower rates of respiration with short-term CO_2 enrichment include (1) a direct inhibition of respiratory enzymes, (2) end-product inhibition, (3) inhibition of mitochondrial electron transport, and (4) an increase in dark CO_2 fixation by non-photosynthetic carboxylation which reduces net CO_2 release. In contrast to these short-term effects, the long-term effects are mediated in part through changes in leaf nutrient concentration, as shown by Wullschleger *et al.* (1992) for the decrease in maintenance respiration with lower leaf N concentration in CO_2-enriched plants.

2.1.3 Growth and Biomass Allocation

In a recent review, Eamus and Jarvis (1989) noted that for the vast majority of experiments conducted with an approximate doubling of the ambient CO_2 concentration, growth of both broadleaves and conifers increased. A notable exception, however, was the absence of a growth response by *Psuedotsuga menziesii* (Hollinger, 1987). In a survey of 58 controlled CO$_2$-exposure studies, Wullschleger *et al.* (1993) summarized the growth response of over 70 boreal, temperate, and tropical tree species to CO$_2$ enrichment and concluded that the mean relative growth response compiled from 398 observations of seedlings grown at elevated CO$_2$ compared to ambient-grown controls was 1.32 (Figure 2). A relative growth response of 1.32 is equivalent to a 32% increase in total plant dry mass (i.e., above- and below-ground) in response to a doubling of atmospheric CO$_2$ concentration. Boreal species had a mean growth enhancement of 1.38, and the corresponding values for temperate and tropical species were 1.31 and 1.25, respectively. From a subset of observations, this stimulation of whole-plant biomass was found to be more or less equally partitioned to leaves (1.33), stems (1.29), and roots (1.38), but canopy leaf area increased only marginally (1.13) in response to CO$_2$ enrichment.

Limiting supplies of N and water, while known to restrict plant growth, do not appear to constrain the response of tree seedlings to CO$_2$ enrichment. Wullschleger *et al.* (1993) showed that the relative growth response of plants exposed to elevated CO$_2$ was only slightly less with limiting supplies of N and water than when supplies of these resources were adequate. Although increasing photosynthesis, decreasing respiration, shifts in water-use efficiency, and greater canopy leaf area in response to rising CO$_2$ might be interpreted as important determinants of overall plant growth, this need not be the case.

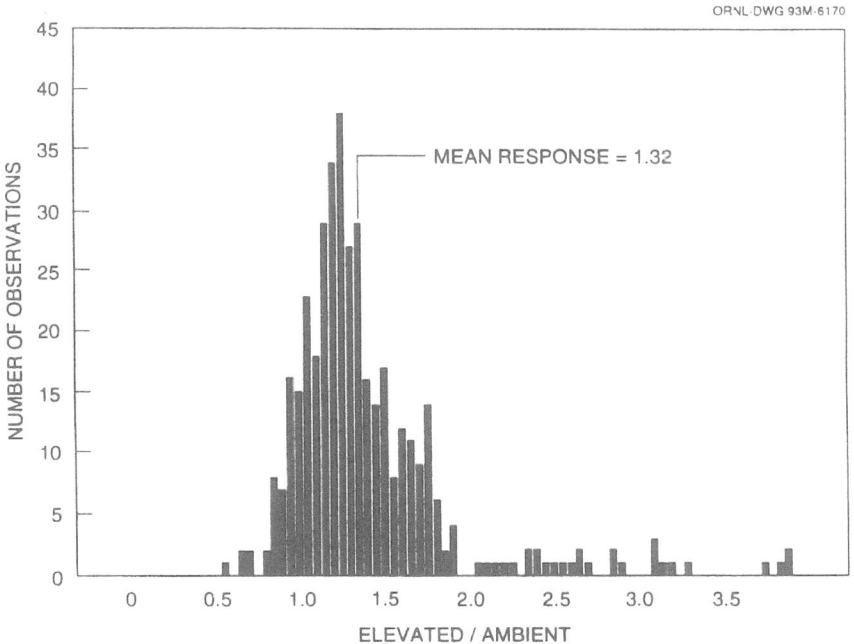

Figure 2. Frequency distribution for the relative growth response (above- and below-ground biomass) as summarized from the results of 58 controlled-exposure studies for 73 tree species grown at ambient and elevated CO$_2$ concentrations. The mean growth response shown is the mean of the log-transformed data. Redrawn with permission from Wullschleger *et al.* (1993).

This surprising result was shown in the review of tropical species response to elevated CO_2 by Hogan *et al.* (1991), who noted that in experiments with *Ficus obtusifolia*, photosynthetic rates increased by 91% and canopy leaf area increased by over 30%, yet total plant biomass displayed only a modest 10% increase with a doubling of CO_2. In contrast, photosynthesis and leaf area of *Trichospermum mexicanum* both decreased by about 20 to 30% in response to a doubling of atmospheric CO_2 concentration, yet total plant biomass increased by 16%. No explanations were given for why short-term measures of gas-exchange and estimates of canopy architecture failed to explain the increase in plant growth. Such reports illustrate the need to interpret results of controlled-exposure studies of photosynthesis and leaf growth in an integrated whole-plant context. Additionally, whole plants need to be viewed as single components of a forest ecosystem, where integration from the physiological scale to ecosystem productivity is not a straightforward task. In view of this need we next examine the historical evidence for a CO_2 enrichment response in field grown trees. It is also recognized that seedlings and mature trees differ physiologically (Donovan and Ehleringer, 1991).

2.2. MATURE TREE RESPONSES

Industrialization has caused many forests to be exposed to rising atmospheric CO_2, some for several decades (regrowth forests) and others for two centuries (old growth forests). These natural exposures may have impacted the growth of mature trees, and a number of investigations have been conducted to quantify mature tree responses to the 25% rise in atmospheric CO_2 from the preindustrial concentration of about 280 µl/l. Examination of tree ring chronologies has been made on a range of tree species as a means for estimating CO_2 enrichment effects on annual stem wood increment. It is well known that temperature and precipitation can influence tree growth and these influences must first be accounted for before drawing inferences about CO_2 effects. An early report of a possible CO_2 fertilization effect was given by LaMarche *et al.* (1984) based on enhanced growth shown in tree ring chronologies obtained from high elevation pine species. Kienast and Luxmoore (1988) reported tree ring chronologies for conifers from 34 sites in four differing climate regions. Eight of the 34 sites showed growth increases in the post-1950 period, and half of these could be explained by favorable climatic conditions. The remaining four sites showed unexplained growth increases. More recently Graumlich (1991) examined tree ring chronologies from conifers at five subalpine sites in the Sierra Nevada, California. Climate variables accounted for growth responses at three of the five sites. Growth enhancement at the two remaining sites could be attributed to CO_2 enrichment, however, current growth enhancements at all five sites were exceeded or equalled during preindustrial periods providing little support for a CO_2 fertilization effect on stem growth of mature trees.

Hari and Arovaara (1988) examined tree ring chronologies from *Pinus sylvestris* from northern Finland. The trees were 300 to 400 yr old. The measured basal area increment exceeded that expected from climatic influences alone by 15 to 43% for the period 1950 to 1983. They cautioned that, although this result correlates with atmospheric CO_2 enrichment, the sample size was too restrictive for a robust conclusion. The authors did not expect to see an increase in volume growth of the old trees investigated.

The biotic growth factor of Bacastow and Keeling {1973, $\beta = [(P_1-P_0)/P_0]/(\ln C_1 - \ln C_0)$, where P_1 and P_0 are plant growth rates at atmospheric CO_2 concentrations of C_1 and C_0, respectively}, can be used as a means for *comparing* various growth responses to

CO$_2$ enrichment. The basal area increment results from Hari and Arovaara (1988) correspond to a ß factor of 1.4 to 4.1 for their low and high responses, respectively. Atmospheric CO$_2$ concentrations of 310 and 344 µl/l were used for 1950 and 1983, respectively, in these ß factor calculations.

West *et al.* (1993) reported a 40% increase in annual ring increment in a mature stand (trees from 100 to 400 yr old) of *Pinus palustris* (longleaf pine) from southern Georgia, USA for the period 1950 to 1987. The growth increase could not be attributed to climatic variables or to stand history. Increased N deposition and rising atmospheric CO$_2$ are factors that could have contributed to the longleaf pine response. However, the growth enhancements from the West *et al.* study, as with the Kienast and Luxmoore (1988) results, exceeded any growth responses that can be expected from CO$_2$ enrichment alone according to seedling response data. Seedlings are considered to give an upper-bound response to elevated CO$_2$. The ß factor value from the West *et al.* study was 3.3 using atmospheric CO$_2$ concentrations of 310 and 350 µl/l for 1950 and 1987 respectively. This ß factor far exceeds the range of 0.1 to 1.2 calculated for short-term growth responses in conifer seedling-CO$_2$ exposure experiments (Kienast and Luxmoore, 1988). Gifford's (1992) recent review of CO$_2$ enrichment effects on vegetation suggests that ß factors in the range of 0.1 to 0.5 seem likely from a "bottom up" physiological approach and also from a "top down" global C pool analysis.

The evidence for elevated atmospheric CO$_2$ causing increased above ground growth in mature trees in natural environments is inconclusive, however, Graybill and Idso (1993) have recently reported a possible CO$_2$ enhancement of growth in high elevation pine species that have a strip-bark growth form. The authors reasoned that in trees with incomplete bark around the trunk, and with weak foliar development, photosynthate would be primarily allocated to cambial growth on the trunk due to the limited sink for photosynthate in foliage or roots. Undisturbed stands of open grown pines in skeletal soil from mountain regions of western USA were selected and individual trees with incomplete bark were cored at breast height for tree ring characterization. Some chronologies from the strip-bark trees showed a growth enhancement of 60% over the last two centuries after accounting for normal growth and climate effects. This growth enhancement was attributed to rising atmospheric CO$_2$ which is probably excessive since it corresponds to a ß factor of 3. A growth enhancement of up to 10% might be expected in a stressed environment, corresponding to a ß factor of 0.5 (Gifford, 1992). The authors did not address the possibility of continuing contraction of the bark strips with aging which could perhaps result in the apparent growth enhancement effect! The carbon allocation patterns of severely stressed old trees is not well enough understood to provide definite detection of a CO$_2$ response in the Graybill and Idso analysis. Nevertheless, their findings are intriguing and worthy of further investigation.

It is possible that mature trees with intact bark may show greater root growth responses than above ground responses to CO$_2$ enrichment, particularly on sites with water deficits or N deficiency (Luxmoore, 1981). Detection of increased C allocation to root systems is very difficult and below ground responses may not be currently measurable since a small change in a large root zone C pool would be difficult to quantify. Norby *et al.* (1992) have shown an increase in fine root mass with CO$_2$ enrichment of field grown *Liriodendron tulipifera* even though above ground growth was not statistically different among CO$_2$ treatments. Photosynthesis was higher with CO$_2$ enrichment as was soil respiration. These results, and those from an exposure of tropical species to elevated CO$_2$ (Körner and Arnone, 1992), suggest that CO$_2$ enrichment may cause CO$_2$ to cycle faster through vegetation without much gain in growth. The effects of elevated CO_2 on vegetation are still unfolding.

3. Forests and Climate Warming

The projected mean annual air temperature rise obtained from general circulation models (GCM) in response to a doubling of atmospheric CO_2 is in the range of 1.9 to 5.2°C with most results being close to 4°C (Houghton et al., 1990), although GCMs that include cloud effects project a lower mean warming of 2°C. Any rise in the *mean annual* air temperature of an area will result in an equivalent rise in the *mean annual* temperature of the litter layer and of the root zone, since the temperatures of these layers are correlated with air temperature (Gupta et al., 1982). Soil warming could have a dramatic effect on organic matter decomposition and mineralization of soil C and nutrient resources.

3.1. SOIL CARBON AND NUTRIENT RESOURCES

Large quantities of C and nutrients are stored in the soil organic matter of temperate forests. For example, on the Walker Branch watershed in eastern Tennessee, about half of the ecosystem C (Edwards et al., 1989) and over 90% of the nitrogen, phosphorus, potassium, and sulfur (Johnson and Henderson, 1989) occurs in the litter or soil. Kimmins et al. (1985) summarized data for carbon and nutrient content from an extensive number of forest study sites. Several of these studies report data for vegetation (including roots), litter, and soil. We compiled the data and found that a very large proportion of N and P are incorporated in the soil and litter components of temperate forests. The summary for N showed that the occurrence of a high proportion of ecosystem N (mean of 90%) in the litter and soil is a general phenomenon for temperate forests. A similar observation, with higher relative variability, was found for the distribution of ecosystem P with a mean of 95% in the soil and litter. Could some of these nutrients become more available with soil warming?

In a review of organic matter retention in soils, Oades (1988) showed that decomposition is the primary process controlling C turnover and nutrient mineralization. While increased soil temperatures are expected to increase organic matter decomposition, it is uncertain how nutrient mineralization, nutrient uptake by vegetation, and leaching processes will be influenced as a whole. For example, wintertime warming may favor mineralization and increased leaching loss of nutrients from the ecosystem. Since N is generally a limiting nutrient for biological activity in many forests, the effect of increased soil temperature on N dynamics will be a critical determinant of forest ecosystem response to warming. Stanford et al. (1973) and Kladivko and Keeney (1987) have shown positive relationships between N mineralization and soil temperature.

3.2. SOIL RESPIRATION

Soil respiration is the combined efflux of CO_2 resulting from the decomposition of litter and soil organic matter by soil microorganisms, and respiration of living roots and mycorrhizal fungi. Of the total CO_2 flux from forest soils, some studies show that approximately half originates from root/mycorrhizae respiration (Billings et al., 1977; Ewel et al., 1987; Behera et al., 1990; Edwards 1991). Soil respiration is strongly influenced by temperature (Schleser, 1982) and a mean Q_{10} of 2.4 was calculated by Raich and Schlesinger (1992) with data from a wide range of environments. In addition, Crill

(1991) has shown that both CO_2 emission and CH_4 oxidation increase with soil temperature. Globally, Jenkinson *et al.* (1991) estimated that about twice as much C exists in the top meter of soil as in the atmosphere, thus, the potential impact of soil warming on increased soil respiration could be a highly significant positive feedback to the greenhouse effect (Schlesinger, 1993).

The efflux rate of CO_2 from the forest floor was measured in an upland oak (*Quercus*) stand on Walker Branch watershed in eastern Tennessee (Hanson *et al.*, 1993) during 1992. Using a Q_{10} of 2.4 (Raich and Schlesinger, 1992), calculations show that a warming of 2 and 4 °C would increase soil respiration above ambient rates by 15-32% (Figure 3). These estimates do not account for possible impacts of reduced precipitation predicted to accompany global warming in some environments (Wigley *et al.*, 1984). Models of soil respiration that include soil water content as an independent variable in addition to temperature (Hanson *et al.*, 1993; Norman *et al.*, 1992; Schlentner and Van Cleve 1985) indicate that limited precipitation and reduced soil water associated with increasing temperatures may inhibit CO_2 efflux from forest soils. Excess water leading to flooded upper soil horizons has also been shown to limit CO_2 efflux at riparian tundra sites (Oberbauer *et al.*, 1992). In the absence of specific projections of future soil water status with increasing soil temperatures, the importance of future changes in soil water content can not be specifically quantified. However, the Raich and Schlesinger (1992) summary of

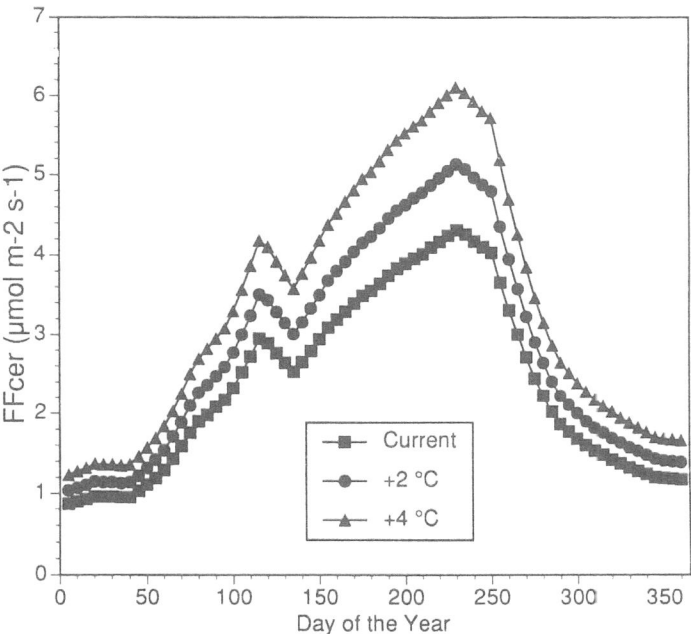

Figure 3. The 1992 annual cycle of forest floor carbon dioxide exchange rates (FF_{cer}) for an upland oak forest in eastern Tennessee (Hanson *et al.*, 1993) plotted against simulations of the FF_{cer} of a similar forest under hypothetical future climates having soil temperatures increased by 2 or 4 °C.

published soil respiration data covered a wide range of ecosystems encompassing mean annual temperatures from -15 to 30°C and mean annual precipitation from 200 to 2800 mm and their analysis suggests a possible interactive effect of temperature rise and precipitation decrease on soil respiration. Raich and Schlesinger (1992) also estimated the turnover time of soil C in temperate forests to be about 29 yr; turnover times are expected to decrease with warming.

A field study of elevated CO_2 effects on the growth of *Liriodendron tulipifera* has shown increased allocation of photosynthate to belowground biomass and increased CO_2 efflux from the soil/root complex (Norby *et al.*, 1992). Soil respiration has the potential to increase under a warmer elevated CO_2 atmosphere not only as a function of rising temperature, but as a result of increased belowground biological activity associated with CO_2 fertilization. Models of soil respiration driven solely by temperature and soil water content may need to explicitly include root responses to elevated atmospheric CO_2. Hanson *et al.* (1993) suggested that their empirical model of forest floor CO_2 efflux rate underestimated total CO_2 efflux from the forest floor because of an inability to represent change in CO_2 flux accompanying seasonal variation in root growth. Other issues that need to be considered in projections of long term changes in soil respiration rates include the quantity and the quality of the organic matter substrate available for decomposition (Flanagan and Van Cleve, 1983; Pastor and Post, 1986). Litter quality could decrease with CO_2 enrichment and result in lower litter decomposition rates; however, there has been very little experimental confirmation of this inference and there is evidence for no effect of CO_2 enrichment on litter decomposition (O'Neill and Norby, 1991).

3.3. FOREST RESPONSE TO SOIL WARMING

A 4°C soil warming is expected to have a significant influence on biological activity such as root growth which is known to increase as temperature increases from 10 to 20°C (Barney, 1951; Kaspar and Bland, 1992; Shepperd, 1981). Van Cleve *et al.* (1990) conducted a soil warming experiment in a *Picea mariana* stand in which heating tape was threaded through the organic soil at 15 cm depth. The tape was spaced at 20 cm intervals and soil temperature was increased by 8 to 10°C above the control plot. Soil heating caused an increase in organic matter decomposition, reducing the mass of the O21 horizon. Extractable N and P were higher in the forest floor as a result of enhanced decomposition, and N, P, and K concentrations were higher in spruce needles. The soil solution had elevated N levels. Similar results would be expected with warming in many other forest soils. For example, Joslin and Wolfe (1993) compared soil solution collected from sunny and shaded areas in a recent clearing within a *Picea rubens* stand and found greater nitrate, Mg, and Al in the samples from the sunny area. The soil temperature was higher by 1.2°C at the 15 cm sampling depth with the sunny exposure. Such responses will be relatively short term (a few years) with global warming since depletion of soil C and nutrient pools will diminish subsequent rates of CO_2 and mineral nutrient release. Predictions of the longer term impact of soil warming will be dependent on the understanding gained from short-term warming experiments combined with simulation modeling.

Small percentage increases in decomposition of the large soil reserves of C and nutrients could have either significant positive or negative feedback effects on global warming. Positive feedback would occur if warming resulted in a net increase in soil CO_2 release. Conversely, the fertilization effect of increased nutrient mineralization may lead to increased forest growth and a net increase in CO_2 fixation in photosynthesis, resulting in a negative feedback to global warming. Soil organic matter has a much lower C/N ratio than forest vegetation (approximately 12 vs 160, respectively) and a small amount of N release by mineralization could support a relatively large increase in woody biomass. Nevertheless, if accelerated levels of nutrient mineralization persist during the dormant winter period, then the potential for nutrient leaching and depletion of the ecosystem nutrient pool could occur.

Using C, N, and P data from a *Quercus-Carya* forest from Walker Branch watershed as an example, calculations are presented for the additional N and P required to support a forest growth rate response to CO_2 fertilization as defined by a β factor of 0.5. The mean annual biomass increment for the 1967-1983 period for the oak-hickory forest was 3.1 Mg ha[-1] yr[-1] (Edwards *et al.*, 1989). Assuming (1) a 50% C content for the biomass (Vogt, 1991), (2) 1975 is the mean year for the 1967-1983 observation period, and (3) atmospheric CO_2 concentrations for 1950, 1975, and 1990 were 310, 332, and 355 μl/l respectively, the CO_2-enhanced growth rates on a C basis are estimated to be 1.5 and 1.6 Mg C ha[-1] yr[-1] for 1950 and 1990 respectively. Taking 1950 as the base year, the cumulative C gain due to CO_2 fertilization for the following 40 years is 2 Mg C ha[-1] (i.e. 0.1 x 40 x 0.5). The C:N and C:P ratios for the forest are 174 and 1783, respectively (Edwards *et al.*, 1989; Johnson and Henderson, 1989). Applying these ratios to the CO_2-enhanced C gain of 2 Mg C ha[-1], the estimated N and P requirements for this C gain over the 40 year period are 11.5 kg N ha[-1] and 1.1 kg P ha[-1]. This represents 0.23 and 0.09% of the combined soil and litter N and P pools, respectively. A very small shift in the N and P from soil to vegetation could support CO_2-enhanced C gain. An alternate source of N is from atmospheric deposition which is currently about 4 to 5 kg N ha[-1] yr[-1] for Walker Branch watershed (Hanson *et al.*, 1989).

The C content of roots in the oak-hickory forest is estimated to be 21 Mg C ha[-1] (Edwards *et al.*, 1989). If all of the CO_2-enhanced C gain of 2 Mg C ha[-1] for the 1950-1990 period was associated with roots, there would have been a 9% increase in root mass. These calculations support the suggestion that mature trees could be responding to CO_2 enrichment, though the response may currently be very difficult to measure by direct root harvesting methods. The cumulative effect of incrementally higher annual root growth rates could eventually become measurable; however, we need reference root data against which to test for an atmospheric CO_2 enrichment effect.

3.4. MODELING FOREST RESPONSES TO GLOBAL CHANGE

McGuire *et al.* (1992) conducted simulations of potential warming effects on ecosystems of North America and showed the importance of coupling between C and N cycles. When the C cycle was simulated alone, warming gave a positive feedback effect due to increased soil respiration and net CO_2 release. When C and N cycles were simulated together, an increase in N mineralization with elevated temperature increased plant growth, resulting in a net CO_2 uptake and a negative feedback to global warming. Bonan and Van Cleve (1992) used simulation to show the importance of interactions between net primary production, decomposition, and N mineralization in the response of a boreal forest ecosystem to climatic change. Although simulation models are being used for prediction of ecosystem responses to global change there remains a need for experimental data to test model predictions so that the understanding of forest ecosystem responses can be applied in regional and global assessments and the development of appropriate abatement strategies. For example, forest planting offers one means for sequestering C, and a modeling study of Dewar and Cannell (1992) showed that long term (100 yr) C storage can be achieved with conifer and hardwood plantations under United Kingdom conditions. This study reinforces the finding of Kauppi *et al.* (1992) that C storage in Europe has increased over the last two decades due to forest planting. Forest planting may prove to have a greater effect on the global C budget over the next few decades than forest responses to global warming and rising atmospheric CO_2 concentration.

5. Acknowledgements

Research sponsored in part by the Carbon Dioxide Research Program and the Ecological Response Program of the Environmental Sciences Division, Office of Health and Environmental Research, U.S. Department of Energy under contract DE-AC05-84-OR21400 with Martin Marietta Energy Systems, Inc., and in part by the Southern Global Change Program of the USDA Forest Service under Interagency Agreement No. 1774-D085-A1 with the U.S. Department of Energy. Publication No. 4084, Environmental Sciences Division, Oak Ridge National Laboratory.

6. References

Amthor, J.S.: 1991, Respiration in a future, higher-CO_2 world. *Plant, Cell Environ. 14*, 13-20.

Arnone, J.A. and Gordon, J.C.: 1990, Effect of nodulation, nitrogen fixation and CO_2 enrichment on the physiology, growth and dry mass allocation of seedlings of *Alnus rubra* Bong. *New Phytol.* 116, 55-66.

Bacastow, R., and Keeling, C.D.: 1973, Atmospheric carbon dioxide and radiocarbon in the natural carbon cycle. II. Changes from AD 1700 to 2070 as deduced from a geochemical model, in Woodwell, G.M., and Pecan, E.V. (eds) *Carbon and the Biosphere*, U.S. Atomic Energy Commission, Washington D.C., pp. 86-135.

Barney, C.W.: 1951, Effects of soil temperature and light intensity on root growth of loblolly seedlings. *Plant Physiol.* 26, 146-163.

Behera, N., Joshi, S.K., and Pati, D.P.: 1990, Root contribution to total soil metabolism in a tropical forest soil from Orissa, India. *For. Ecol. Manag.* 36, 125-134.

Billings, W.D., Peterson, K.M., Shaver, G.R., and Trent, A.W.: 1977, Root growth, respiration, and carbon dioxide evolution in an arctic tundra soil. *Arctic Alpine Res. 9, 129-137.*

Bonan, G.B., and Van Cleve, K.: 1992, Soil temperature, nitrogen mineralization, and carbon source-sink relationships in boreal forests. *Can. J. For. Res.* 22, 629-639.

Bowes, G.: 1991, Growth at elevated CO_2: photosynthetic responses mediated through Rubisco. *Plant, Cell Environ. 14*, 795-806.

Bunce, J.A.: 1992, Stomatal conductance, photosynthesis and respiration of temperate deciduous tree seedlings grown outdoors at an elevated concentration of carbon dioxide. *Plant, Cell Environ. 15*, 541-549.

Crill, P.M.: 1991, Seasonal patterns of methane uptake and carbon dioxide release by a temperate woodland soil. *Global Biogeochem. Cycles* 5, 319-334.

Curtis, P.S. and Teeri, J.A.: 1992, Seasonal responses of leaf gas exchange to elevated carbon dioxide in *Populus grandidentata*. *Can.J. For. Res.* 22, 1320-1325.

Dewar, R.C. and Cannell, M.G.R.: 1992, Carbon sequestration in the trees, products, and soils of forest plantations: an analysis using UK examples. *Tree Physiol.* 11,49-71

Donovan, L.A. and Ehleringer, J.R.: 1991, Ecophysiological differences among juvenile and reproductive plants of several woody species. *Oecologia* 86, 594-597.

Eamus, D. and Jarvis, P.G.: 1989, The direct effects of increase in the global atmospheric CO_2 concentration on natural and commercial temperate trees and forests. *Adv. Ecol. Res.* 19, 1-55.

Edwards, N.T.: 1991, Root and soil respiration responses to ozone in *Pinus taeda* L. seedlings. *New Phytol.* 118, 315-321.

Edwards, N.T., Johnson, D.W., McLaughlin, S.B. and Harris, W.R.: 1989, Carbon dynamics and productivity, in: Johnson D.W., and Van Hook, R.I. (eds), *Analysis of Biogeochemical Cycling Processes in Walker Branch Watershed*, Springer-Verlag, New York. pp. 197-232.

Ewel, K.C., Cropper, W.P., and Gholz, H.L.: 1987, Soil CO_2 evolution in Florida slash pine plantations. II. importance of root respiration.*Can. J. For. Res.* 17, 330-333.

Flanagan, P.W., and Van Cleve, K.: 1983, Nutrient cycling in relation to decomposition and organic matter quality in taiga ecosystems. Can. J. For. Res. 13, 795-817.

Gifford, R.M.: 1992, Interaction of Carbon dioxide with growth-limiting environmental factors in vegetative productivity: Implications for the global C cycle, in: *Advances in Bioclimatology*, Springer-Verlag, New York, pp. 24-58.

Graumlich, L.: 1991, Subalpine tree growth, climate, and increasing CO_2: An assessment of recent growth trends. *Ecology* 72, 1-11.

Graybill, D.A. and Idso, S.B.: 1993, Detecting the aerial fertilization effect of atmospheric CO_2 enrichment in tree-ring chronologies. *Global Biogeochem. Cycles* 7, 81-95.

Gunderson, C.A., Norby, R.J. and Wullschleger, S.D.: 1993, Foliar gas exchange responses of two deciduous hardwoods during three years of growth at elevated CO_2: No loss of photosynthetic enhancement. *Plant, Cell Environ.* (in press).

Gupta, S.C., Radke, J.K., Larson, W.E.,and Shaffer, M.J.: 1982, Predicting temperatures of bare and residue covered soils from daily maximum and minimum air temperatures. *Soil Sci. Soc. Am. J.* 46, 372-376.

Hanson, P.J., Rott, K., Taylor, G.E., Jr., Gunderson, C.A., Lindberg, S.E., and Ross-Todd, B.M.: 1989, NO_2 deposition to elements representative of a forest landscape. *Atmos. Environ.* 23, 1783-1794.

Hanson, P.J., Wullschleger, S.D., Bohlman, S.A., and Todd, D.E.: 1993, Seasonal and topographic patterns of forest floor CO_2 efflux from an upland oak forest. *Tree Physiol.*(in press).

Hari, P. and Arovaara, H.: 1988, Detecting CO_2 induced enhancement in the radial increment of trees. Evidence from northern timber line. *Scand. J. For. Res.* 3, 67-74.

Hogan, K.P., Smith, A.P. and Ziska, L.H.: 1991, Potential effects of elevated CO_2 and changes in temperature on tropical plants. *Plant, Cell Environ.* 14, 763-778.

Hollinger, D.Y.: 1987, Gas exchange and dry matter allocation responses to elevation of atmospheric CO_2 concentration in seedlings of three tree species. *Tree Physiol.* 3, 193-202.

Houghton, J.T., Jenkins, G.J., and Ephraums, J.J.: 1990, Climate Change, the IPCC Scientific Assessment. University Press, Cambridge. 365 pp.

Idso, S.B. and Kimball, B.A.: 1992, Effects of atmospheric CO_2 enrichment on photosynthesis, respiration, and growth of sour orange trees. *Plant Physiol.* 99, 341-343.

Jenkinson, D.S., Adams, D.E., and Wild, A.: 1991, Model estimates of CO_2 emissions from soil in response to global warming. *Nature* 351, 304-306.

Johnson, D.W. and Henderson, G.S.: 1989, Terrestrial nutrient cycling, in: Johnson, D.W., and Van Hook, R.I. (eds), *Analysis of Biogeochemical Cycling Processes in Walker Branch Watershed,* Springer-Verlag, New York, pp. 233-300.

Joslin, J.D. and Wolfe, M.H.: 1993, Temperature increase accelerates nitrate release from high-elevation red spruce soils. *Can. J. For. Res.* 23, (in press).

Kaspar, T.C. and Bland, W.L.: 1992, Soil temperature and root growth. *Soil Sci.* 154, 290-299.

Kauppi, P.E., Mielikainen, K., and Kuusela, K.: 1992, Biomass and C budget of European forests, 1971 to 1990. *Science* 256, 70-74.

Kaushal, P., Guehl, J.M. and Aussenac, G.: 1989, Differential growth response to atmospheric Carbon dioxide enrichment in seedlings of *Cedrus atlantica* and *Pinus nigra* sp. *Laricio* var. *Corsicana. Can. J. For. Res.* 19, 1351-1358.

Kienast, F. and Luxmoore, R.J.: 1988, Tree-ring analysis and conifer growth responses to increased atmospheric CO_2 levels. *Oecologia* 76, 487-495.

Kimmins, J.P., Binkley, D., Chatarpaul, L., and de Catanzaro, J.: 1985, Biogeochemistry of temperate forest ecosystems: Literature on inventories and dynamics of biomass and nutrients. Information Rep. PI-X-47E/F, Petawawa National Forestry Institute, Canadian Forestry Service, Chalk River, Ontario

Kladivko, E.J. and Keeney, D.R.: 1987, Soil N mineralization as affected by water and temperature interactions. *Biol. Fertil. Soils* 5, 248-252.

Körner, Ch. and Arnone, J.A., III: 1992, Responses to elevated carbon dioxide in artificial tropical ecosystems. *Science* 257, 1672-1675.

LaMarche, V.C. Jr., Graybill, D.A., Fritts, H.C., and Rose, M.R.: 1984, Increasing atmospheric carbon dioxide: Tree ring evidence for enhancement in natural vegetation. *Science* 225, 1019-1021.

Luxmoore, R.J.: 1981, CO_2 and phytomass. *BioScience* 31, 626.

McGuire, A.D., Melillo, J.M., Joyce, L.A., Kicklighter, D.W., Grace, A.L., Moore, B., and Vorosmarty, C.J.: 1992, Interactions between C and N dynamics in estimating net primary productivity for potential vegetation in North America. *Global Biogeochem. Cycles* 6, 101-124.

Mousseau, M.: 1992, Effects of elevated CO_2 on growth, photosynthesis and respiration of sweet chestnut (*Castanea sativa* Mill.). *Vegetatio* (in press).

Mousseau, M. and Saugier, B.: 1992, The direct effect of increased CO_2 on gas exchange and growth of forest tree species. *J. Exp. Bot.* 43, 1121-1130.

Norby, R.J., Gunderson, C.A., Wullschleger, S.D., O'Neill, E.G. and McCracken, M.K.: 1992, Productivity and compensatory responses of yellow-poplar trees in elevated CO_2. *Nature* 357, 322-324.

Norman, J.M., Garcia, R., and Verma, S.B.: 1992, Soil CO_2 fluxes and the C budget. *J. Geophys. Res. (in press)*.

Oades, J.M.: 1988, The retention of organic matter in soils. *Biogeochem.* 5, 35-70.

Oberbauer, S.F., Gillespie, C.T., Cheng, W., Gebauer, R., Sala Serra, A., and Tenhunen, J.D.: 1992. Environmental effects on CO_2 efflux from riparian tundra in the northern foothills of the Brooks Range, Alaska, USA. *Oecologia* 92, 568-577.

Oberbauer, S.F., Strain, B.R. and Fetcher, N.: 1985, Effect of CO_2-enrichment on seedling physiology and growth of two tropical tree species. *Physiologia Plant.* 65, 352-356.

O'Neill, E.G. and Norby, R.J.: 1991, First-year decomposition dynamics of yellow-poplar leaves produced under CO_2 enrichment. *Bull. Ecol. Soc. Am.* 72, 208.

Pastor, J. and Post, W.M.: 1986, Influence of climate, soil moisture and succession on forest C and N cycles. Biogeochem. 2, 3-27.

Raich, J.W. and Schlesinger, W.H.: 1992, The global carbon dioxide flux in soil respiration and its relationship to vegetation and climate. *Tellus* 44B, 81-90.

Reekie, E.G. and Bazzaz, F.A.: 1989, Competition and patterns of resource use among seedlings of five tropical trees grown at ambient and elevated CO_2. *Oecologia* 79, 212-222.

Ryan, M.: 1991, Effects of climate change on plant respiration. *Ecol. Applic.* 1, 157-167.

Schlentner, R.E. and Van Cleve, K.: 1985, Relationship between CO_2 evolution from soil, substrate temperature, and substrate moisture in four mature forest types in interior Alaska. *Can. J. For. Res.* 15, 97-106.

Schleser, G.H.: 1982, The response of CO_2 evolution from soils to global temperature changes. *Z. Naturforsch.* 37a, 287-291.

Schlesinger, W.H.: 1993, Soil respiration and changes in soil C stocks, in: Woodwell, G.M. (ed) *Biospheric Feedbacks in the Global Climate System: Will the Warming Speed the Warming ?* Oxford University Press. (in press)

Shepperd, W.D.: 1981, Variation in growth of Engelmann spruce seedlings under selected temperature environments. Research Note RM-404. USDA Rocky Mtn. For. Range Exp. Station, Fort Collins, Colorado.

Silvola, J. and Ahlholm, U.: 1992, Photosynthesis in willows (*Salix x dasyclados*) grown at different CO_2 concentrations and fertilization levels. *Oecologia* 91, 208-213.

Stanford, G., Frere, M.H., and Schwaninger, D.H.: 1973, Temperature coefficient of soil N mineralization. *Soil Sci.* 115, 321-323.

Stitt, M.: 1991, Rising CO_2 levels and their significance for C flow in photosynthetic cells. *Plant, Cell Environ.* 14, 741-762.

Thomas, R.B. and Strain, B.R.: 1991, Root restriction as a factor in the photosynthetic acclimation of cotton seedlings grown in elevated Carbon dioxide. Plant Physiol. 96, 627-634.

Van Cleve, K., Oechel, W.C., and Hom, J.L.: 1990, Response of black spruce (*Picea mariana*) ecosystems to soil temperature modification in interior Alaska. *Can. J. For. Res.* 20, 1530-1535.

Vogt, K.A.: 1991, C budgets of temperate forest ecosystems. *Tree Physiol.* 9, 69-86.

Watson, R.T., Rodhe, H., Oeschger, H., and Siegenthaler, U.: 1990, Greenhouse gases and aerosols, in: Houghton J.T., Jenkins, G.J. and Ephraums, J.J. (eds), *Climate Change: The IPCC Scientific Assessment,* Cambridge University Press, pp. 1-40.

West, D.C., Doyle, T.W., Tharp, M.L., Beauchamp, J.J., Platt, W.J., and Downing, D.J.: 1993, Recent growth increases in old-growth longleaf pine. *Can. J. For. Res.* (in press).

Wigley, T.M.L., Briffa, K.R., and Jones, P.D.: 1984, Predicting plant productivity and water resources. *Nature* 312, 102-103.

Wullschleger, S.D.: 1993, Biochemical limitations to C assimilation in C_3 plants - A retrospective analysis of the A/C_i curves from 109 species. *J. Exp. Bot.* (in press).

Wullschleger, S.D. and Norby, R.J.: 1992, Respiratory cost of leaf growth and maintenance in white oak saplings exposed to atmospheric CO_2 enrichment. *Can. J. For. Res.* 22, 1717-1721.

Wullschleger, S.D., Norby, R.J. and Gunderson, C.A.: 1992, Growth and maintenance respiration in leaves of *Liriodendron tulipifera* L. exposed to long-term Carbon dioxide enrichment in the field. *New Phytol.* 121, 515-523.

Wullschleger, S.D., Post, W.M. and King, A. W.: 1993, On the potential for a CO_2 fertilization effect in forest trees - An assessment of 58 controlled-exposure studies and estimates of the biotic growth factor. in: Woodwell, G.M. (ed) *Biospheric Feedbacks in the Global Climate System: Will the Warming Speed the Warming ?* Oxford University Press. (in press).

PRESENT ROLE OF GERMAN FORESTS AND FORESTRY IN THE NATIONAL CARBON BUDGET AND OPTIONS TO ITS INCREASE

P. BURSCHEL, E. KÜRSTEN, B.C. LARSON, M. WEBER

Chair of Silviculture and Forest Management, University of Munich
D-8050 Freising, Hohenbachernstraße 22, Germany

Abstract. The carbon stock in forests and wooden products is quantified for the Federal Republic of Germany. In addition the sink effects of • the actual forests and • the pool of wooden products are determined. The mitigation obtained by • substitution of high energy raw materials by wood, • substitution of fossil fuels by wood as source of energy were evaluated. Finally the C-ecological effects of the following silvicultural measures were quantified: • lengthening of rotation time • underplanting • change of tree species • afforestation of non forest land • increased use of slash and wood from thinning operations as energy source.

1.   Introduction

One of the great global C-reservoirs is the biosphere, with forests - because of their great storage capacity - playing a paramount role. Quantification of this role with respect to the actual state and management potentials is an important issue, elaborated here for a highly industrialized and densely populated but small country as far as area is concerned.

2.   Methods

To quantify the role of forest and forestry in the national C-budget it was considered to consist of five components:
  - C stored in the standing forest,
  - C stored in wood products originated from this forests,
  - Increase in storage of both the forest and product pool,
  - Avoidance of C-emissions by using wood products in place of more energy demanding materials,
  - Reduction of emissions by use of wood as source of energy substituting fossil fuels.

The simultaneous effects of all these C-ecologically important components of forestry are exemplified for a spruce stand in figure 1.

Water, Air, and Soil Pollution **70**: 325–340, 1993.
© 1993 *Kluwer Academic Publishers.*

Figure 1: CO_2-Mitigation Effect of a Spruce Stand (rotation 80 yrs)

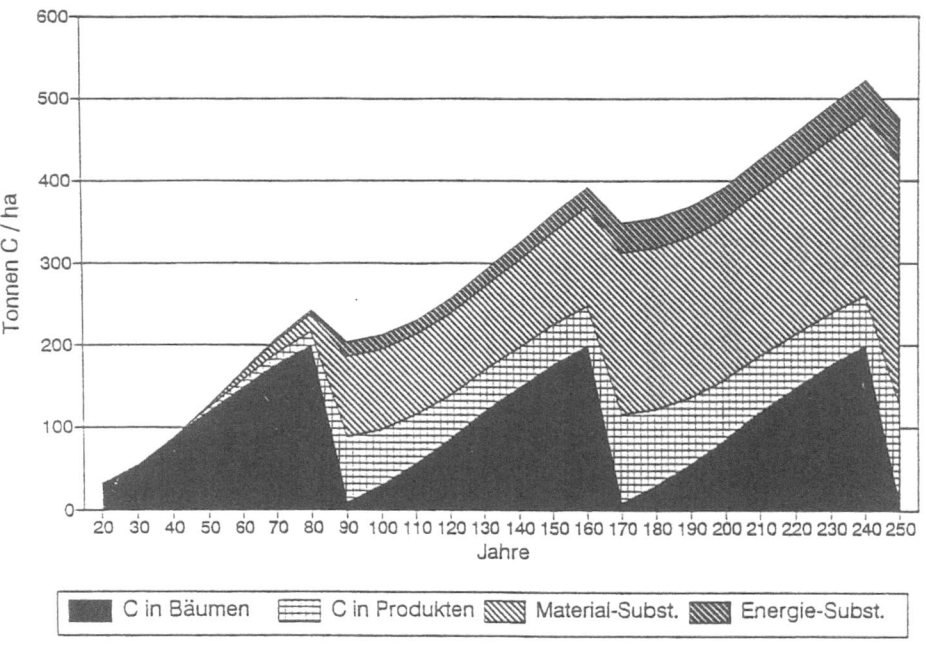

The amount of C stored in the forest and in wood products is determined by an instantaneous measurement, while energy and product substitution effects as well as the increase in forest and product storage is a dynamic measure including a time reference. The equations used for calculation are shown in Table 1.

To assess the validity of the results it is important to know that the inventories of standing volumes, the annual timber cut as well as the calculation of biomass and C-contents are well founded, while all the values pertaining to increment and corresponding increases of stocks are based on estimations made by various authors, referred to in table 1.

C-contents of forest soils are approximations derived from ZIEGLER 1991. Representative studies for the total forest area are not existing. There is an immediate need for additional surveys in this matter. Although the soil is a very large store, no marked and rapid changes are likely. Since many sites in Central Europe are still in a phase of recovery from soil devastation (overcutting, overgrazing, litter raking) in the past, a decrease of C-content is very unlikely. Therefore a low sink-effect has been attributed to it. With respect to possible decreases of soil-C with increasing temperatures however, representative monitoring is advisable.

Table 1: Equations for the calculation of different components of C pool and C mitigation. All data expressed as metric tons.

COMPARTMENT	EQUATION and LEGEND
FOREST Living tree biomass	TOTAL BIOMASS CARBON = ([BOLE]+[BRANCH]+[ROOTS]*[DENSITY])*BIOMASS CARBON BOLE = volume of standing wood > 7 cm diameter (including branches). Data source: Western Germany: 1987 forest survey (BML 1992); Eastern Germany: 1985 complete forest inventory (VEBFP 1986). Volume separately by species and age class BRANCH = volume of branches < 7 cm diameter calculated as % of BOLE based on species, mean diameter and height according to age class (GRUNDNER, SCHWAPPACH 1952). This percentage ranges from 8 % (old larch and pine) to 200 % (all species when very young) ROOTS = volume of living roots based on % of BOLE. Three percentages were used: for all trees under 20 yrs of age 100 % was used, for older spruce 30 %, and for older trees of all other species 25 % (DAUBER, ZENKE, 1978) DENSITY = fresh volume/dry weight factor which was applied separately by species, but kept constant for different tree parts (including bark) and age classes. The values ranged from 0.377 kg m^{-3} (spruce) to 0.561 kg m^{-3} (oak) (KNIGGE, SCHULZ 1966; TRENDELENBURG, MAYER-WEGELIN 1955) BIOMASS CARBON = constant C percentage of 50 % was assumed (KNIGGE, SCHULZ 1966).
Dead biomass	DEAD BIOMASS CARBON = [NON-HARVEST CARBON]+[FOREST FLOOR CARBON]+[STANDING DEAD CARBON] HARVEST RESIDUE = ([UNHARV WOOD]+[BRANCH AND ROOT])*[HARVEST]*[DECAY] UNHARV WOOD = volume of wood over 7 cm left in the woods, including wood contained in stumps and sawdust calculated as a percentage of HARVEST based on species and size of harvested trees (SCHÖPFER, DAUBER 1990; BAYER. STAATSMIN. 1990). This percentage ranged from 3 % (spruce) to 13 % (oak) BRANCH AND ROOT = volume of wood < 7 cm and roots calculated as a percentage of HARVEST (SCHÖPFER, DAUBER 1990). It ranged from 35 % (pine) to 45 % (spruce). HARVEST = annual volume of wood harvested. 40 x 10^6 m^3 was used based on statistics from 1981-1989 (BARTELHEIMER 1990; AGDW 1991) DECAY: time factor based on the decay rate of harvest slash. Assuming that half of the original volume remains in 5 years the factor was 5 (HARMON et al. 1986).
Ground vegetation biomass	GROUND VEGETATION CARBON Based on local studies this was considered as a constant of 1 t ha^{-1} C (BURSCHEL, BINDER 1993; HÖHNE 1962; ELLENBERG et al. 1986). Compared to most other forest systems in the world there is little ground vegetation in German forests, because they are intensively managed and kept at high density
Soil	SOIL CARBON Based on a fairly comprehensive survey the soil carbon was considered as a constant of 157 t ha^{-1} (ZIEGLER 1991)
PRODUCTS	[RES CARBON] = [RES NUM]*[WOOD/RES]*[CARBON/WOOD] [NON-RES CARBON] = [NON-RES NUM]*[WOOD/NON-RES]*[CARBON/WOOD] RES NUM = number of residences in western Germany kept separately by single, double and multiple family dwellings (STAT. BUNDESAMT 1990). WOOD/RES = wood per dwelling based on type of residence (KROTH et al. 1991). CARBON/WOOD = percentage of 20 %, which assumes a dry mass density of 45 % and 50 % C/dry mass NON-RES NUM = number of non-dwelling structures in western Germany kept separately by 3 categories of building (STAT. BUNDESAMT 1990) WOOD/NON-RES = wood per non-dwelling structure based on type of building (KROTH et al. 1991).
MATERIAL **SUBSTITUTION**	[MATERIAL SUBSTITUTION EMMISSION AVOIDANCE] = [CONSTRUCTION WOOD]*[AVOIDANCE FACTOR] CONSTRUCTION WOOD: volume of stemwood harvested and then used as for construction (kept separately by species) per year (BARTELHEIMER 1990; CMA 1987) AVOIDANCE FACTOR: a complex value based on energy used for production of alternative products minus energy used for production of wood corrected for the relative amounts of alternative material and wood needed to build a comparable structured multiplied by the amount of C emitted per production of unit energy in western Germany (EK 1988). This total factor varies from 0.099 to 0.491 per m^3 in different studies. 0.28 was used here after review of studies as typical, but conservative (low) for overall construction types in Germany (BOYD et al. 1976; BURSCHEL, KÜRSTEN 1992; BAIER 1982). This factor could also be low because most substitution for wood would probably be steel which has a higher carbon emission level than energy production as a whole.
ENERGY- **SUBSTITUTION**	[EQUIVALENT OIL] = [ENERGY WOOD]*{[HEATING VALUE$_W$]/[HEATING VALUE$_O$]}*{[HEATING EFFICIENCY$_W$]/[HEATING EFFICIENCY$_O$]} [MJ OIL SAVED] = {[EQUIVALENT OIL]*[HEATING VALUE$_O$]+[PRODUCTION$_O$] - [PRODUCTION$_W$]}* [EMISSION$_O$] [EMISSION AVOIDED] = [MJ OIL SAVED]*[CARBON RELEASED/MJ] ENERGY WOOD: amount of wood used for energy purposes (including mill residues) measured in dry weight HEATING VALUE$_W$: actual energy that can be released by wood combustion which depends on chemical constituents of wood and moisture content. An average value of 25 % moisture was used because some of the wood is mill residues with a low moisture content. We used a value of 13 MJ kg^{-1}. HEATING VALUE$_O$: actual energy that can be released by combustion of oil. A value of 42.7 MJ kg^{-1} was used HEATING EFFICIENCY$_W$: an efficiency value based on piece size, moisture content and type of combustion chamber. A percentage of 72 was used. HEATING EFFICIENY$_O$: an efficiency value based on combustion type. A percentage of 85 was used EMISSION$_O$: amount of C emitted upon combustion of oil. A value of 0.0218 kg C/MJ was used (EK 1988). PRODUCTION$_O$: energy to produce oil. A value of 30.7 % of the oil was used (KÜRSTEN, BURSCHEL 1991) PRODUCTION$_W$: energy used to harvest and transport wood. A value of 0.5 MJ kg^{-1} was used, assuming part of the energy is mill residues with no additional energy input CARBON RELEASED/MJ: the amount of C released for each MJ of energy produced

All values for product storage should be treated as estimations from which only magnitudes can be taken. The same caution applies to the substitution effect which can be attained by utilizing wood instead of other materials or fossil fuels.

In spite of these limitations, the results presented here give an idea of the C-ecological role of forests and forest industries in a densely populated and highly industrialized country. A comprehensive version of this study containing more methodological details is published by BURSCHEL et al. 1993.

3.   Results

3.1   THE FOREST AS STORE OF CARBON

The amount of C in the forests is the total sum of C stored in living biomass, dead biomass and soil. The estimations for each of these compartments as well as for the whole ecosystem are presented in table 2 .

Table 2:   Carbon Pools and Net Carbon Increment in German Forests

Compartments		Stock		Increment
		10^6 t	%	10^6 t yr^{-1}
Living biomass	trees	888	36	5.5
	ground vegetation	10	0	0.0
Dead wood	snags and snails	2	0	0.0
	slash	30	1	0.0
Humus	litter	392	16	(1.0)
	soil organic matter	1174	47	
Forest ecosystems	**total**	**2496**	**100**	**5.5 (6.5)**

Accordingly 900 x 10^6 t C are stored in the living biomass of forests in the FRG. Most forests here are under sustained management, normally as compartmented high forest (i.e., compartment as management unit). These forests are in the aggradation phase of forest ecosystems. Although this phase is not as densely stocked as a mature natural forest, it is characterized by a high increment. This annual increment in German forests amounts to 57.5 x 10^6 m³ (UNITED NATIONS 1986), only about 40 x 10^6 m³ of which are harvested. So, approximately 17.5 x 10^6 m³ of the increment remain in the forests contributing to C-storage and turning the forest into a C-sink. This sink effect may be even greater, since there are indications that the actual increment is higher than the estimation made in this study. But even our conservative estimates show that the forests in the FRG in their present state are an effective sink for C.

The amount of C stored as humus in the forest soil is considerably higher than in the living biomass. However, this reservoir of nearly 1.6×10^9 t of C is most probably in a fairly balanced state where input of fresh litter and decomposition are in equilibrium. Nevertheless, it is likely that there are soils which have not fully recovered from past devastative practices and are still rebuilding to a formerly larger humus body. The magnitude of this process is roughly estimated to be 0.1 t ha^{-1} yr^{-1} corresponding about 1×10^6 t yr^{-1}.

Even if humus is viewed as a fairly static reservoir in this study, it must be given careful consideration for two reasons: First, should temperatures rise markedly, the conditions for the decomposition of organic matter would improve to such an extent that a release of CO_2 could result from this huge reservoir turning it into a potent C-source. Second, if the present amount of humus is not in equilibrium but accumulating C more rapidly than assumed here, the sink-effect would be accordingly higher. A large accumulation however is not possible, since the total annual addition of fresh organic matter (litter) only amounts to between 2 and 4 t ha^{-1} (MOSANDL 1991) of which half is C, and of which almost all eventually gets mineralized (mostly to CO_2) through natural decomposition processes. Therefore it appears reasonable to accept the forest soils as having a more or less neutral role in C-cycle, although they are huge C-pools. However, observation of its behavior through a systematic monitoring is important because it not only may act as moderate sink of C, as is assumed here, but also can turn into a strong source.

The contribution of the dead biomass to the total C-storage of the forests is very small compared with the other two components. Dead wood and harvest residues together make up only 1.3 % of the total amount of 2496×10^6 t.

Finally it can be summarized that the German forests presently act as an annual sink of about 5.5 to 6.5×10^6 t C.

3.2   WOOD PRODUCTS AS STORE OF CARBON

The C remains fixed in the wood until it is released by decomposition or combustion. For their duration-time timber products lengthen the storage effect of the forests, therefore they have to be taken into consideration for a comprehensive estimation of the total effect of forests and forestry for the C-budget.

In Germany 28 % of the harvested wood is used for construction wood, 19 % for furnitures and other products, 28 % for paper and packing (CMA, 1987). Based on mean durations of 65, 15 and 1 yr for these categories, the existing C-stock in timber products was calculated as shown in table 3. With a total of 128×10^6 t C-storage in products represents about 15 % of the C stored in the trees.

The estimated product storage of 1.6 to 1.7 t C per capita is significantly less than the 2.1 t which BRAMRYD 1982 has calculated for highly industrialized countries. The difference is mainly due to his evaluation being based on data from Scandinavia and the USA, where the quota of construction wood as a long-term store is significantly higher than in Central Europe.

Table 3: Carbon Storage in Wood Products in Germany.

Product group	stock	annual increase
	10^6 t C	
Construction wood	89	1.0
Furnitures, other wood products	34	0.1
Paper, packing, fuel wood	5	0.0
Sum	128	1.1

Because of a continuous increase of buildings in the Federal Republic of Germany, this storage represents a C-sink. The magnitude of the sink-effect has been assumed to be 1×10^6 t C yr^{-1} annually. Considering the total emissions in Germany, the magnitude of the sink-effect appears to be low. Nevertheless, there can be no doubt that any utilization of wood for long lasting purposes is advantageous as far as C-ecology is concerned.

3.3   WOOD AS SUBSTITUTE FOR OTHER MATERIALS

3.3.1   Wood as raw material of low energy-input

Considering economic activities not only in commercial terms but also in terms of C-ecology, wood has favorable characteristics, since sustained management of forests in Central Europe, which closely follows natural processes - no soil treatment, no herbicids, no fertilizing, no tree breeding -, requires an extremely low energy-input. And timber processing is relatively little energy-dependent too. Since energy-input is linked with release of CO_2, utilization of wood in place of materials which demand higher energy-inputs results in a considerable mitigation of CO_2 emissions.

While C-storage in wooden products eventually reaches a maximum, mitigation of emission by the use of wood instead of other materials is permanent: Emissions once avoided for a given purpose are permanently avoided (fig. 1). In these calculations, although only the stemwood utilized for construction purposes was considered, the annual mitigation-effect has been estimated at 2.6×10^6 t (table 4).

Table 4: CO_2-mitigation (10^6 t C) by Use of Wood in Germany

Item	Stock	Annual increase
Storage	128	1.1
Material substitution	--	2.6
Energy substitution	--	1.4
Sum	**128**	**5.1**

3.3.2 Wood as energy source

By using wood as energy source, burning of fossil fuels can be avoided. Presently, wood as a source of energy in Germany is mainly burning of industrial waste; heating of residential buildings plays a minor role. However, even then the effect of this substitution is in the range of 1.4 x 10^6 t C (table 4). As in the case of material-substitution, this effect has no maximum (as does storage in forests and products) but is cumulative.

3.4 POTENTIALS FOR STRENGTHENING THE ROLE OF FORESTS FOR CO_2-MITIGATION

The C-budget data presented refer to the actual situation and demonstrate the "business as usual" state. However, there are significant possibilities to further influence the CO_2-mitigation capacity. These possibilities have been compiled in table 5 and are explained below.

3.4.1 Silivicultural measures in existing forests

3.4.1.1 Lengthening of rotation period

Lengthening of rotation periods in managed forests increases the accumulated timber volume per unit area. A similar effect can be achieved in selection forests by increasing the final harvesting diameter (SMITH, 1962). The effect of these changes in German forests has been calculated for three fairly realistic options (table 5) as follows:

Actual situation:

Actual management will be continued and the number of hectares currently in each of the older classes is maintained by harvesting an appropriate number of hectares in these age classes every year.

Medium Option:

The distribution of age classes in private and corporative forests is allowed to reach the distribution of state owned forests in western Germany. This option increases the rotation period for those forests which currently have comparatively short periods of production.

Maximum Option:

An overall lengthening of rotations in all forests was assumed, with the limitation that some stands will be destroyed in catastrophic storms and that some regeneration cutting will have to be made because of general silvicultural requirements (details ref. to BURSCHEL et al. 1993).

As was to be expected, the present timber volume and, accordingly, the C in forests can be considerably increased. By lengthening the rotation period by two decades, an additional 0.7 to 1.8 x 10^6 t C can be fixed annually. However, this effect will be reduced when considered for time periods longer than 20 yrs because of the actual age class distribution. In fact the effect is even reversed for the time period between 40 and 60 yrs.

However, the increase of the average standing volume in German forests by lengthening the rotation periods will only have a limited and transient effect on CO_2-mitigation. The actual not absolutely regular age class distribution together with the reduction of storage in products and substitution effects resulting from a decreased timber harvest partially compensate the increase of standing volume.

3.4.1.2 Spacing, thinning, fertilization

Traditionally German forests are densely stocked, with respect to risks like storm and snow they often can be considered as overdense. Therefore changes in spacing of plantations or decrease of thinning activities cannot be considered as means for an improved C-fixation.

Fertilization never has played an important role in German forestry. Nowadays it is even considered undesirable for ecological reasons (except for liming to compensate acidic immissions). Therefore fertilizing was not included as a means to increase the storage or productive capacity of forests here.

Table 5: Options of CO_2-Mitigation by Forestry in Germany (all values 10^6 t C)

Period (yrs)	20		40		60	
Options in existing forests						
Lengthening of rotation periods						
actual situation	85.1		160.4		226.1	
medium option	99.8	*+14.7*	163.4	*+3.0*	209.5	*-16.6*
maximum option	120.2	*+35.1*	179.2.	*+18.8*	181.4	*-44.7*
Underplanting						
0.5 x 10^6 ha yr^{-1} in 20 yrs	0.2		1.7		6.0	
Substitution of Scots pine by						
Douglas Fir						
10.000 ha yr^{-1}	1.5		5.0		9.2	
5.000 ha yr^{-1}	0.7		2.5		4.6	
Afforestations						
Real model						
5.200 ha yr^{-1}	1.3		5.3		13.2	
Eco model						
1.2 x 10^6 ha in 20 yrs	15.9		63.6		137.9	
0.655 x 10^6 ha shelter belts	4.2	*+18.8*	25.4	*+85.7*	51.9	*+176.6*
Maximum model						
4.5 x 10^6 ha in 20 yrs	92.8		323.4		673.3	
0.5 x 10^6 ha energy plantations	15.5	*+107.0*	46.5	*+364.6*	77.5	*+737.6*
Increased use of wood as fuel						
1. step: 1.4 x 10^6 t wood waste	26.0		52.0		78.0	
3.1 x 10^6 t thinning material						
2. step: wood waste, thinning material	54.0		108.0		162.0	
5.3 x 10^6 t slash						
All options						
medium - sum	**61.2**		**145.4**		**253.2**	
- per yr	**3.1**		**3.6**		**4.2**	
maximum - sum	**197.0**		**495.6**		**865.5**	
- per yr	**9.9**		**12.4**		**14.4**	

Italic and underlined = differences to "actual situation" resp. "real model"

Nevertheless it has to be repeated what was already mentioned in chapter 3.1: there is a profound increase of productivity of forests occuring. It is a consequence of three elements:

- recovery of sites after hundreds of years of degradation by misuse
- immission of considerable amounts of nitrogen, emitted by technical processes and agriculture
- increase of CO_2-content of the air.

These elements are not included in the range of silvicultural measures discussed here, but show up in the annual increment and mitigation estimate stated for the German forests in table 2.

3.4.1.3 Underplanting

The effect of CO_2-mitigation by forests can also be enhanced by underplanting of forests with shade-tolerant tree species. This option was evaluated for large scale underplanting of Scots Pine with European Beech. For this purpose it was assumed that during the course of 20 yrs, 0.5×10^6 ha of pine stands on better sites could be treated in this manner. However, an increased C storage will become effective only on a long-term basis. Over a time-period of 40 yrs the approximate rate of C-accumulation would amount annually to 40 000 t ha^{-1}. By age 60 this value would rise up to 100 000 t. This procedure of underplanting already occurs to a small degree for silvicultural, ecological and economic reasons. The motivating aspects of CO_2-mitigation may further accelerate this practice.

3.4.1.4 Substituting Scots Pine by Douglas Fir

Fast growing exotic species are often considered as alternatives for the regeneration of forests in place of less productive but native ones. Worldwide estimations of C-fixation by forests are often based on such tree species. In Central Europe a higher C-storage could be achieved if western north american Douglas Fir replaces native conifers. Due to its considerably faster growth and higher wood density the effective C-fixation rate by this species would be almost 50 % above that of Norway Spruce stands.

The results presented in table 5 are based on a conversion rate of Scots Pine stands on better sites to Douglas Fir stands such that 6 % of the total forest area would be converted within 60 yrs. A higher proportion of this species is not considered realistic because of stand stability problems and lack of acceptance among people. Based on this conversion rate an additional annual CO_2-mitigation of roughly 1.5×10^5 t C can be achieved, a value which would clearly increase with the increase in age. This mitigation does not represent the total growth of Douglas Fir stands but the additional effective CO_2-mitigation compared to the replaced Scots Pine .

3.4.2 *Afforestation of non-forest areas*

Afforestation of farm lands in Central Europe is gaining importance. The primary, political objective is to reduce surplus agricultural production. However, little is known about the effect on C-ecology which results from afforestation of such land. The area available for this purpose is under debate. In this investigation three options were developed and calculated:

Real model:
   The rate of afforestation remains at its low actual level: 5 200 ha yr^{-1} (Spruce 50 %; Beech 20 %; Douglas Fir 20 %; Oak 10 %)

Eco model:
   a) Afforestation is supposed to be realized on an area of 1.2×10^6 ha within 20 yrs (Beech 40 %; Spruce 30 %; Oak 20 %; Douglas Fir 10 %).

   b) In addition on 0.65×10^6 ha hedges and shelterbelts will be established exclusively with broadleafed species which can be managed as coppice forests for fuel wood production.

Maximum model
   a) 3.85×10^6 ha are afforested (Douglas Fir 40 %; Spruce 40 %; Beech 20 %) within 20 yrs

   b) 0.65×10^6 ha of hedges and shelterbelts are established within the same time.

   c) 0.5×10^6 ha energy plantations of fast growing broadleafed species with a rotation of 10 yrs are to be created within two decades. For these plantations a production of 12 t of dry matter per year was considered realistic.

The first option is to maintain the actual afforestation rate. Even if continued for 60 years and finally covering an area of 3×10^5 ha, afforestations of this type would result an annual sink of only 2.2×10^5 t of C.

If really significant efforts should be made in the shortest possible time - as assumed by second option - a considerable effect can be achieved. This option which is feasible and in many regards ecologically advantageous, results in annual C-storage of 1×10^6 t during the initial 20 yr phase of afforestation. By 60 yrs this value would increase to 3×10^6 t, including substitution effects.

The maximum option requires a great but feasible effort in the first two decades, but an annual 5.5×10^6 t C could be stored or their emission as CO_2 mitigated. For a time-period of 60 yrs this amount would equal 12.3×10^6 t yr^{-1} considering both storage and substitution effect. With an increased duration of this growing stock, this value would increase continuously.

Afforestation in Central Europe represents a major possibility for the mitigation of CO_2-emissions. Depending on the option considered, the annual magnitude of this fixation during the establishment phase of two decades would range between 1 to 5.5×10^6 t, whereas the amounts estimated for 60-yrs would range between 3 and 12×10^6 t. Therefore this possibility is clearly more effective than all measures which can be taken in already existing forests.

3.4.3 Greater use of wood as fuel

The use of wood as fuel is not exploited to the extent that is theoretically possible. This is principaly due to low prices of fossil fuels. There are considerable amounts of wood which could be used as energy sources with a corresponding potential to prevent CO_2 to be emitted from fossil fuels. Two options for greater exploitation of this potential are presented in table 5.

A conservative option - called *first step* in table 5 - considers a more complete use of wood waste, resulting p.e. from demolition of houses, and an increased amount of thinning material to be used as energy source. This would produce an annual CO_2-mitigation effect of approximately 1.3×10^6 t C. The *second step* option assumes that additionally wood currently remaining as slash on the forest floor could be utilized for the same purpose, amounting to approximately 2.7×10^6 t yr^{-1}.

3.5   The role of forests and forestry in the C budget

The results of this investigation about the role of forests and forestry for C-ecology are condensed in table 6. All the data refer to annual storage or substitution values expressing the respective CO_2-mitigation.

The highest estimate corresponds to 10 % of annual emissions, which in 1988 amounted to 272×10^6 t C for the Federal Republic of Germany and the German Democratic Republic (Boden et al., 1990). Since the mining of lignite in eastern Germany has been considerably reduced, the actual magnitude of emissions in Germany may not be over 250×10^6 t yr^{-1}.

Table 6: Forests and Forestry as C-Sinks

Item	Minimum	Maximum
	10^6 t C yr^{-1}	
Forests	5.5	6.5
Wood utilization	5.1	5.1
Intensification of forestry	4.2	14.4
Sum	**14.8**	**27.0**

Two characteristics of the presented data have to be kept in mind:

- The current sink-effects are given for both the forest and wood utilization. The storage in forests continues as a sink of C as long as increment exceeds timber harvesting. Wood utilization however, can clearly be expanded if more timber is harvested. In this way, biological increment would be fully processed and accordingly the sink-effect shifted from forest storage to product storage as well as to material and energy substitution.

- The intensification and extension of forestry instead represent a possibility, and consequently a potential yet to be developed, but could attain a considerable magnitude of CO_2-mitigation even in a densely populated and highly industrialized but small country.

4. Discussion

To evaluate the significance of forests and forestry in the C budget, the total CO_2-emission of anthropogenic origin in the investigated area must be considered. Approximately 250×10^6 t C yr^{-1} produced on a land area of 36×10^6 ha reflects a very high density of 7 t ha^{-1} yr^{-1}. If these emissions are considered only for the forested area of 10×10^6 ha, above density increases to 25 t ha^{-1} yr^{-1}. These data clearly show that on a land area equivalent to that of the Federal Republic of Germany, the mitigation of an emission of this range is absolutely ruled out. Even if the whole of Germany would support a forest cover, and storage as well as the additional substitution effects amount to 3 t ha^{-1} yr^{-1} - a value, reasonably attainable by afforestation on farm lands - a maximum of 100×10^6 t yr^{-1} storage capacity could be attained. On the present forest area it would amount to only 30×10^6 t yr^{-1}. So, forestry and timber industries although contributing towards the mitigation of the CO_2-problem, under no circumstances would represent its national solution. According to estimates presented here the possible CO_2-mitigation rate would range between 15 to 27 x 10^6 t C yr^{-1}, which corresponds to 6 to 10 % of the emissions.

The actual accumulation of additional biomass in existing forests and reduction of emissions through timber utilization are producing mitigation effects which are already occurring. A further effect of up to 14 x 10^6 t C yr^{-1} could be achieved by intensification of forestry. These estimates, however, do not consider the C-ecological potential of an extended use of timber for construction purposes and an increased production of long lasting wood products. At present the magnitude of a generally increased use of timber is difficult to estimate. Considering in addition, that all compilations concerning the mitigation effect of afforestations are made on the basis of low estimates, the figures presented can be taken as conservative.

Table 7 : Most important Potentials of CO_2-Mitigation by German Industry shown on the Basis of Branches with highest Energy Consumption (together 2/3 of Industrial Energy Consumption) (WIEHN, 1990)

Industry, energy consumption 1987 10^6 t coal equivalent/yr	Options	CO_2-mitigation potential 10^6 t C yr^{-1}
Iron and steel 20	Heat recovery from blast furnace and from slag; recovery of converter gas; warm and direct use of rolled steel in steel works	2.7
Chemistry 16	Cl-generation by diaphragm process; H_2-generation by helical coil reactor; NH_3-production by AMV process	1.4
Mining 6	Cement: Extern pre-calcining in furnaces; shortened rotary furnaces Bricks: improved supply air control; flue gas heat exchanger Lime: coke firing in ring shaft kiln; new type of wall up rotary furnaces	0.4
Non-ferrous metals 4	Al-production: Cathodes made of titanium boride; Al-chloride electrolysis after ALCOA	0.8
Pulp and paper 4	Pulp production by continuous boiling processes; new bleaching methods; improved paper dehydrating	< 0.3
	Sum	**≈ 5.6**

The significance of the potential of mitigation becomes obvious if compared with corresponding ones of the industry (table 7). The figures clearly indicate that mitigation potentials of German industry are estimated to be considerably lower than the ones achievable by forestry. They even are below the potential represented by the afforestation of farm lands alone. Therefore, the role of forestry and timber industries can be considered as remarkably great.

One gets to a similar conclusion if the intention of the actual German government is considered to reduce national emissions by 30% until 2005. At the present there is no hint that such an ambituous goal can be reached. Probably it will already be considered a success if emissions will not increase. Here again the forestry options should be looked at as one realistic means to achieve progress in this matter.

5.   References

AGDW (Arbeitsgemeinschaft Deutscher Waldbesitzerverbände e.V.): 1991, Forstwirtschaft und Waldbesitz in der ehemaligen DDR, *DEFO 31*, 40-41.

Baier, B.: 1982, Energetische Bewertung luftgetragener Membranhallen im Vergleich mit Holz-, Stahl- und Stahlbetonhallen, Verlagsgesellschaft Rudolf Müller.

Bartelheimer, P.: 1990, Wirtschaft und Holzmarkt 1989/90. *Allg.ForstZeitschr.* **45**, 1202-1206.

Bayerisches Staatsministerium für Ernährung, Landwirtschaft und Forsten (ed.),: 1990, Hilfstafeln für die Forsteinrichtung. München.

BML (Bundesministerium für Ernährung, Landwirtschaft und Forsten) (ed.).:1992, Bundeswaldinventur 1986-1990. , Bonn, Band II.

Boden, T.A., Kanciruk, P., Farrell, M.P.: 1990, Trends '90, A compendium of data on global change. Carbon Dioxide Information Analysis Center, Oak Ridge, Tenn., USA.

Boyd, C.W., Koch, P., McKean, H.B., Morschauser, C.R., Preston, S.B., Wangaard, F.F.: 1976, Wood for structural and architectural purposes, *Wood and Fiber* **8**, 1-72.

Burschel, P., Binder, F.: 1993, Bodenvegetation - Verjüngung - Waldschäden, *Allg. ForstZeitschr.*, 216-223.

Burschel, P., Kürsten, E.: 1992, Wald und Forstwirtschaft im Kohlenstoffhaushalt der Erde, in: Verbindungsstelle Landwirtschaft-Industrie e.V. (ed.), *Produktionsfaktor Umwelt: Klima - Luft,* etv Landwirtschaftverlag, Düsseldorf, 97-125.

Burschel, P., Kürsten, E., Larson, B.C.: 1993, Die Rolle von Wald und Forstwirtschaft im Kohlenstoffhaushalt - Eine Betrachtung für die Bundesrepublik Deutschland, Forstl. Forschungsber. München, Nr. 126.

CMA (Centrale Marketinggesell. der deutschen Agrarwirtschaft):1987, Distributionsanalyse des Holzes in der Bundesrepublik Deutschland. CMA, Bonn.

Dauber, E., Zenke, B.: 1978, Potential forstlicher Reststoffe (Waldabfälle), Band 1, Fachbereich Forstwissenschaft der Ludwig-Maximilians-Universität München.

EK (Enquete Kommission):1988, Schutz der Erdatmosphäre - Eine internationale Herausforderung. Zwischenbericht der Enquete-Kommission "Vorsorge zum Schutz der Erdatmosphäre" des 11. Deutschen Bundestages, Deutscher Bundestag (ed.), Bonn.

Ellenberg, H., Mayer, R., Schauermann, J.: 1986. Ökosystemforschung - Ergebnisse des Sollingprojekts 1966-1986, Verlag Eugen Ulmer, Stuttgart.

Grundner, F., Schwappach, A.: 1952, Massentafeln zur Bestimmung des Holzgehaltes stehender Waldbäume und Waldbestände, Verlag Paul Parey, Berlin, Hamburg.

Harmon, M.E., Franklin, J.F., Swanson, F.J., Sollins, P., Gregory, S.V., Lattin, J.D., Anderson, N.H., Cline, S.P.,Aumen, N.G., Sedell, J.R., Lienkaemper, G.W., Cromack, K. Jr., Cummins, K.W.:1986, Ecology of coarse woody debris in temperate ecosystems,. *Adv. Ecol. Res.* **15**, 133-299.

Höhne, H.: 1962, Vergleichende Untersuchungen über Mineralstoff- und Stickstoffgehalt sowie Trockensubstanzproduktion von Waldbodenpflanzen, *Arch.f. Forstwes.* **11**, 1085-1145.

Knigge, W., Schulz, H.: 1966, Grundriss der Forstbenutzung. Verlag Paul Parey, Hamburg und Berlin.

Kroth, W., Kollert, W., Filippi, M.: 1991, Analyse und Quantifizierung der Holzverwendung im Bauwesen. IBR-Verlag, Stuttgart.

Mosandl, R.: 1991, Die Steuerung von Waldökosystemen mit waldbaulichen Mitteln - dargestellt am Beispiel des Bergmischwaldes, Mitt. aus der Staatsforstverwaltung Bayerns, Nr. 46.

Moulton, R.J., Richards, K.R.: 1990, Costs of sequestering Carbon through tree planting and forest management in the United States, U.S. Department of Agriculture, Forest Service, General Technical Report WO-58.

Smith, D.M.: 1962, The practice of silviculture, John Wiley & Sons, New York.

Schöpfer, W., Dauber, E.: 1990, Bestandessortentafeln 82/85, in: Bayer. Staatsministerium für Ernährung Landwirtschaft und Forsten (ed.), *Hilfstafeln für die Forsteinrichtung*, München, 241-260.

Statistisches Bundesamt (ed.): 1990, Statistisches Jahrbuch für die Bundesrepublik Deutschland, Metzler-Poeschel Verlag, Stuttgart.

Trendelenburg, R., Mayer-Wegelin, H.: 1955, Das Holz als Rohstoff, Carl Hanser Verlag, München.

United Nations: 1986. European timber trends and prospects to the year 2000 and beyond. Vol. I, United Nations Publication (No. E.86.II.E.19), New York.

VEBFP (VEB Forstprojektierung Potsdam) (ed.): 1986, Der Waldfonds der DDR, Ausgewählte Informationen. Ausgabe 1986, Potsdam.

Wiehn, H.: 1990, Beiträge zur Klimavorsorge am Beispiel industrieller Prozesse, in: Bundesverband der deutschen Industrie (ed.), *Vorsorge zum Schutz der Erdatmosphäre,* Köln, 171-186.

Ziegler, F.: 1991, Die Bedeutung des organischen Kohlenstoffs im Unterboden - Vorratsberechnungen an Waldböden. *Z. Umweltchem. Ökotox.* **3** (5), 276-277.

POTENTIAL FOR CARBON SEQUESTRATION IN THE DRYLANDS

EDWARD GLENN*, VICTOR SQUIRES**, MARY OLSEN* AND ROBERT FRYE*

*Environmental Research Laboratory, 2601 E. Airport Dr., Tucson, Arizona, USA 85706

**Department of Environmental Science and Rangeland Management, University of Adelaide, Roseworthy, South Australia, Australia

Abstract. Non-forested drylands occupy 43% of the world's land surface yet they are not currently regarded as important in sequestering carbon due to overuse and poor management. Seventy percent of drylands have already undergone moderate to severe desertification and an additional 3.5% drops out of economic production each year. Reversing the trend towards desertification through cultivation of halophytes on saline lands, revegetation of degraded rangelands and other innovative conservation measures could result in net C sequestration in dryland soils of 0.5–1.0 Gt yr^{-1} at a cost of $10–18 t^{-1} C, based on a 100 yr scenario. Investment in anti-desertification measures in the world's drylands appears to be an economical method to mitigate CO_2 buildup in the atmosphere while accomplishing a major international objective of restoring dryland productivity.

1. Introduction

The previous conference in this series (Wisniewski and Lugo, 1992) advanced the hypothesis that temperate and tropical forests may be sequestering a large portion of the so-called "missing carbon" that is emitted by fossil fuel burning but does not appear in the atmosphere or oceans. Support for this position has come from measured increases in C storage in European forests since the industrial revolution (Kauppi et al., 1992). Logically, if terrestrial ecosystems are already absorbing some of the excess C, those parts of the landscape under direct human management could be managed to store even more C, suggesting a possible mitigation strategy for global warming should it prove necessary.

Unfortunately, most human activities result in C losses from the landscape rather than gains. Land clearing for agriculture is a particular problem. The world's need for

Water, Air, and Soil Pollution **70**: 341–355, 1993.
© 1993 *Kluwer Academic Publishers.*

food increases every year.   In 1993, food for 5 x 10^9
inhabitants is required; by the year 2000 food for at least
6 x 10^9 will be needed.   Rising standards of living also
require greater land areas for food production; the U.S. diet
requires approximately 3 times as much land to produce as a
typical Asian diet due to greater meat and fat consumption in
the U.S. (World Resources Institute, 1990).   The developed
countries have nearly doubled their agricultural output in the
past 40 yr, but there is serious doubt whether such production
increases can continue.   Even accepting that production
increases will continue, FAO predicts that 200 x 10^6 ha of new
farmland will be needed in the next century.   Forests are
presently being cleared for agriculture at an estimated rate
of 17 x 10^6 ha yr^{-1} with serious implications for the
biospheric C balance (Houghton, 1990).   While attention has
focused on the forests, parallel land use changes have taken
place in the drylands.

2.   Drylands and Desertification

The distribution of drylands is governed by the global
circulation of the atmosphere. A high-pressure zone of sinking
dry air around 30° latitude in each hemisphere is warmed by
compression. The warm dry air provides cloudless skies and
exposes the land to the full effect of the sun.   Brief
seasonal invasions of moist air produce little rain, which is
erratic in both area and time. The result is two series of
deserts surrounded by drylands: in the northern hemisphere lie
the Saharan, North American (Colorado, Gila, Mexican),
Arabian, Thar and Gobi deserts: to the south are the
Patagonian, Kalahari and Australian deserts; all merge
gradually into the habitable drylands which border them
(Figure 1).

Excluding the hyperarid regions which are largely uninhabited
deserts, the drylands total 5.2 x 10^9 ha (Table 1).   These are
areas where there could be some chance of enhancing
productivity either through the use of saline water or better
utilization of rainfall.   This is a very large land base,
exceeding the area of cropland (1.4 x 10^9 ha) or closed forest
(4.4 x 10^9 ha) in the world (World Resources Institute, 1990).
In fact, drylands occupy some 43% of the total world land
area.   Most of the drylands are already under management.   A
significant portion (62%) of the world's irrigated lands are
in the drylands. The remainder are used for dryland farming
(458 x 10^6 ha) or rangelands (4.5 x 10^9 ha).

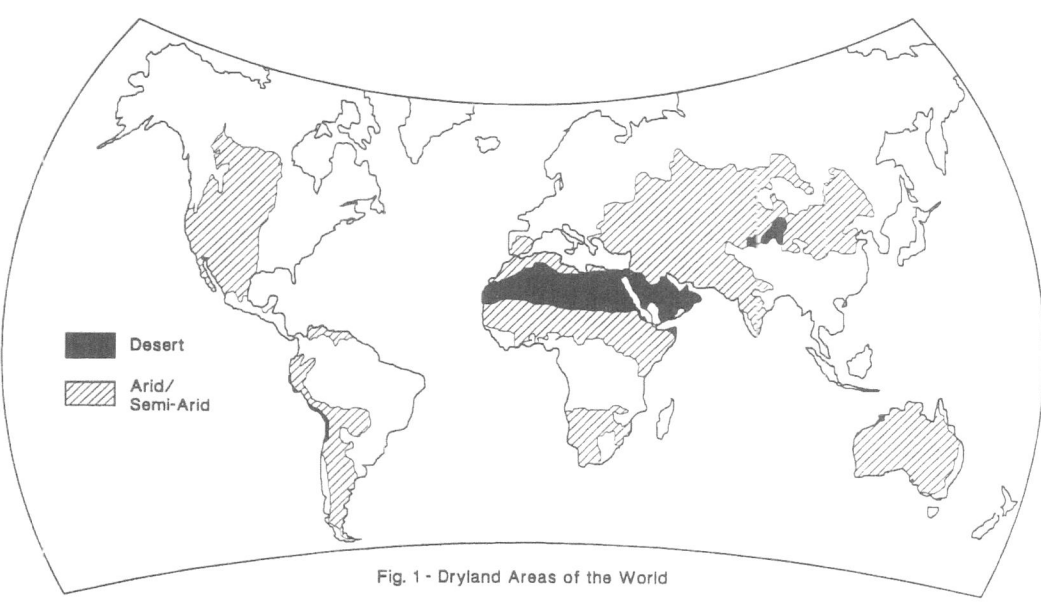

Fig. 1 - Dryland Areas of the World

Table 1. World Drylands (Excluding Hyper-Arid Lands) in Millions of Hectares (Dregne et al., 1991).

	Arid	Semi-Arid	Dry Sub-Humid	Total
Africa	504	514	269	1287
Asia	626	693	353	1672
Australia	303	309	51	663
Europe	11	105	184	300
North America	82	419	232	733
South America	45	265	207	517
World Total	1571	2305	1296	5172

Salt-affected lands form a large subset within the drylands. Coastal deserts and inland salinized soils total 294 x 10^6 and 424 x 10^6 ha, respectively (Glenn et al., 1991a). Most of these are in the hyperarid deserts and do not overlap the areas in Table 1 (see map in Glenn et al., 1991a). Secondary salinization affects irrigated soils. About 43 x 10^6 ha of irrigated lands (or 30% of their total area in the world's drylands) are affected by waterlogging, salinization and alkalinization.

Population increases over the past 30 yr have been rapid in these regions (Table 2), and by 1984 70% of the drylands had undergone moderate to severe desertification due to overuse, resulting in lost primary productivity and CO_2 releases into the atmosphere (Dregne et al., 1991). Current estimates of the amount of drylands lost to economic use due to desertification range from 0.1-10% yr^{-1}, with the highest degradation rates occuring in the driest regions. A weighted mean rate of land lost to desertification is 3.5% yr^{-1} of global drylands (Dregne et al., 1991). It is estimated that annual losses of irrigated land due to abandonment amounts to 1.5 million ha.

Table 2. The Population Increase in the Arid Regions (Eyre, 1985).

	Population (millions)		% Increase	
	1960	1985	1960-1985	Annual
Asia	151.0	270.5	79	3.2
Africa	49.5	96.5	95	3.8
N. Amer.	15.5	26.0	68	2.7
S. Amer.	10.0	17.5	75	3.0
Australia	0.5	0.5	0	0

The UNEP has concluded that whereas at present the drylands contribute little to the global carbon sink, they offer vast areas of land for reforestation, afforestation and other projects to increase carbon storage on the land (Dregne et al., 1991; UNEP, 1991; Kassas et al., 1991).

We will attempt to estimate the potential of these lands for carbon storage and compare the costs with other management options for carbon sequestration. First we will examine the special case for storing carbon in salinized soil using intensive cultivation of halophyte crops then we will consider the case for carbon storage through extensive improvements in management of drylands.

3. Intensive Cultivation of Halophytes

Halophytes are salt-tolerant, native plants which have long been used for grazing; recent developments have shown their value as potential forage (Malcolm, 1986), feed (Glenn et al., 1992a) and oilseed (Glenn et al., 1991b) crops under brackish or even seawater irrigation. The land base for irrigated halophyte production is coastal and inland salt deserts where natural saline water sources are available and in irrigation districts where brackish drain water is available for irrigation of salinized soils. We previously (Glenn et al., 1991a, 1992b) identified 55 desert regions containing an estimated 130 x 10^6 ha of usable land for halophyte plantations. This estimate was approximately 15% of the saline land base and included only flat land that appeared feasible for irrigation from an identified saline water supply.

How much C can be stored on that land depends first upon rates of primary production. We investigated a worst-case scenario for saline water irrigation - the direct use of seawater in a coastal desert environment in the Sonoran Desert (Glenn & O'Leary, 1985; Glenn et al., 1991b). The Electric Power Research Institute and the Salt River Project have funded field trials to measure primary production of halophytes for C storage. Annual dry biomass yields of the best candidate species have ranged from 17-35 t ha^{-1} for a net carbon uptake of 4-8 t ha^{-1} yr^{-1} (Table 3). These yields equal or exceed conventional biomass or forestry yields (e.g. Sedjo, 1989). Projected over the estimated usable world area, intensive halophyte production could absorb approximately 0.6-1.2 Gt C yr^{-1} globally.

Whether significant C sequestration can be achieved depends also upon how much of the primary production enters long-term storage or can be used as replacement for fossil fuels. The particular storage strategy we have investigated is incorporation of biomass into soil for long-term storage in the humic fraction. Dryland soils are typically low in organic carbon, often less than 0.5%. Such soils could conceivably hold greater carbon; in fact, loss of soil organic matter is one of the characteristics of desertification, and

most dryland soils do not contain the amount of carbon they
could conceivably store under restored conditions (UNEP,
1991). Further, residence time of C in dryland soils can be
much longer than forest soils (Gifford et al., 1992). We have
conducted experiments in which two types of halophyte biomass
and wheat straw were incorporated into irrigated or dry desert
soils. Decomposition rates and leaching losses depend upon
the salinity of the irrigation supply, the type of biomass and
the depth of burial (Olsen, Frye and Glenn, in preparation).
Figure 2 shows that decomposition rates of halophyte biomass
are significantly slowed on seawater irrigation compared to
rates on fresh water.

Table 3. Annual Biomass and Carbon Yields of Seawater-
Irrigated Halophytes at Puerto Penasco, 1990-1992. Sample
size (n) refers to number of individual plots of a species
except for Sesuvium, where individual plants within a single
plot were sampled. Carbon content was assumed to be 36% of
ash-free dry matter.

	n	Annual Yield (t^{-1} ha^{-1}) Biomass		Carbon
		mean	SE	mean
Batis maritima	8	33.95	(.99)	8.2
Atriplex linearis	5	24.27	(1.23)	6.7
Salicornia bigelovii				
Year One	22	22.40	(.70)	5.6
Year Two	9	17.72	(1.32)	4.3
Suaeda esteroa	9	17.22	(1.12)	4.3
Sesuvium port-				
ulacastrum	9	16.70	(2.00)	4.2

Using yield data in Table 3 as a production function and the
different decomposition experiments as decay functions, we are
attempting to predict C storage rates in halophyte farms over
time. The decomposition experiments indicate that 30-50% of
C might enter long-term storage, but further monitoring of
soil C levels will be needed to model the decay function
accurately. Nevertheless, all the experiments show that
organic matter accumulates over time in halophyte fields where
the residues are turned under. The experiments support the
hypothesis that even under irrigation, initially C-poor soils

typical of the drylands have the capacity for enhanced carbon storage but the magnitude and duration of the storage are still unresolved.

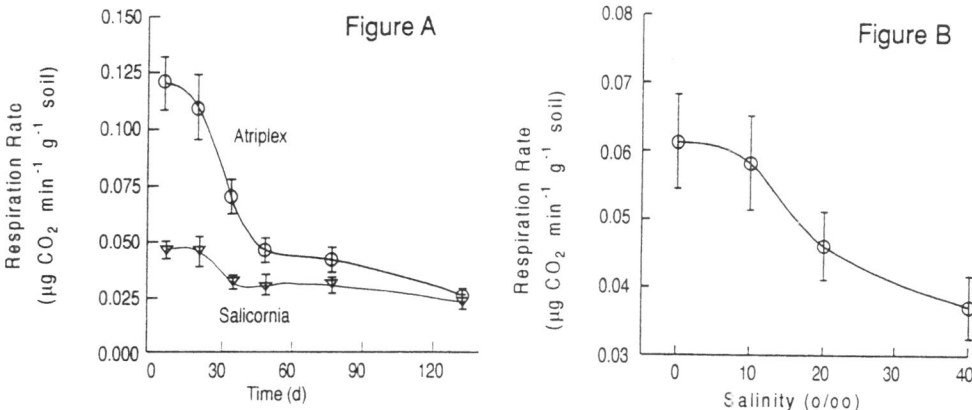

Figure 2. Respiration rate of desert soil amended with 5% w/w _Atriplex_ or _Salicornia_ dry biomass and irrigated with water of 1, 10, 20 or 40 ppt seawater (Olsen, Frye and Glenn, in preparation). Graph A shows differences between biomass types over time across salinities; B shows differences by salinity across biomass types over the whole experiment. Error bars are standard error of estimate.

Water usage and production costs have also been concerns, because early trials utilized much higher quantities of water than could reasonably be used in large-scale practice (Glenn and O'Leary, 1985). However, lysimeter trials with 6 halophyte species have shown that consumptive water use by halophytes is equivalent to conventional crop plants and does not increase with increased salinity of the irrigation supply (Miyamoto et al., in preparation; Glenn et al., in preparation). Even with very salty water (i.e. seawater), high-frequency irrigation with a 0.5 leaching fraction allows high yields with water usage rates within the range of irrigated biomass crops in a similar climate. We have estimated that halophyte biomass can be produced for $44-53 per t compared to $30-45 for conventional biomass crops, and that carbon usage to produce the crop would be 22.5-30.0% of the harvested carbon, assuming all energy inputs were supplied by diesel fuel (Glenn et al., 1992b).

As with tree plantations, the costs of sequestering carbon are high if a halophyte plantation is established solely for the purpose of storing carbon. However, it is more likely that halophytes will be grown as food and animal feed crops with the residues used for carbon storage. In the case of halophyte oilseeds (Glenn et al., 1991b) the straw (containing 90% of total C) is available for carbon sequestration as a by-product; in the case of halophytes grown for forage, the woody stem biomass and roots (ca. 75% of total C) will be available as a by-product. The marginal costs of storing carbon are the costs of plowing under the unused residues - approximately \$20 ha^{-1} using typical (Arizona) farm costs, or \$4 t^{-1} C for a 5 t ha^{-1} yield. If one-third of buried C enters long-term storage, the cost is \$12 t^{-1} C. These costs do not reflect payments to the farmers, however.

The feasibility of halophyte cultivation on a large scale is now under test in the coastal desert near Jubail, Saudi Arabia, where a 320 hectare, seawater farm is growing the oilseed halophyte, Salicornia bigelovii, to supply edible oil and meal for animal feed.

Another role for halophytes is in arresting desertification in drylands. An expert consultation, involving scientists from over 20 countries, met in Nairobi in November 1992 to develop guidelines for returning degraded salt-affected land to productive use via halophytic forage species. The technology (both hardware and software) dealt with in these guidelines is highly relevant to UNEP's plan of action to combat desertification. The guidelines include specific strategies for using extensive (non-irrigated) halophyte plantings to sequester carbon. In these cases, carbon storage is mainly via root biomass, litter accumulation and the standing crop rather than through deliberate harvesting and burial of biomass. They also provide guidelines for planting halophytes in salinized areas of irrigation districts for intensive production.

The largest areas of degraded irrigated lands are situated in the drylands of Asia, followed by North America, Europe, Africa, South America and Australia in descending order. UNEP (1991) estimates that US \$250 ha^{-1} (1990 prices) is the average yearly income foregone due to degradation of irrigated land. Dregne et al. (1991) estimates that it would cost between US \$500 and \$5000 ha^{-1} to restore degraded irrigated land for conventional crops. If we add this to the income foregone (US \$250) there is considerable scope to pay for the cost of establishing halophyte plantations (estimated by Le Houerou 1992 to be US \$ 500-1000 ha^{-1}).

4. Extensive Biomass Production through Management of Dryland Pastures and Crops

As mentioned, desertification is a severe prcblem throughout the drylands. UNEP (1991) concluded that 30 years of past anti-desertification programs failed due in large part to insufficient funding. They estimate (Kassas et al., 1991) that $171-363 \times 10^9$ will be needed to fund a 20 yr program to arrest the process of land degradation and to restore the already-degraded areas of drylands (Table 4). The program includes afforestation, reforestation, planting of shrubs and grasses, control of grazing lands, planting halophytes on salinized land and numerous other remediation methods.

As desertification proceeds, costs rise dramatically as land passes into higher degradation categories which require greater expense per unit of land. UNEP (1991) estimates the average annual income foregone due to degraded rangelands to be $23 billion per year. They calculate the cost:benefit of restoring the rangelands as 1:3.5 on a global basis. The funds required for restoration are available internally in the industrialized regions (N. America and Australia) and the oil-producing regions (the Middle East), but the developing countries in arid zones of Africa and Asia will require external support to conduct anti-desertification on a meaningful scale.

Table 4. Global costs of direct anti-desertification measures (billions $US for a 20 yr program, UNEP, 1991).

Type of Action Needed	Croplands		Rangelands
	Irrigated	Rainfed	
Preventative Measures	10-31	12-36	6-18
Corrective Measures	17-50	18-55	13-38
Rehabilitation Measures	21-41	22-59	80-120
Totals:	48-12	252-150	99-176

In theory, if an economic incentive were offered in the form of a subsidy for each ton of carbon sequestered on such land, dryland farmers and pastoralists could change their management practices to store carbon.   For example, marginal desert rangeland could be allowed to return to perennial shrubs and grasses by restricting clearing and grazing. Stubble could be plowed under where at least a portion would enter long-term storage in the humic fraction.   Such speculations are difficult to quantify. However, if a net sequestration rate of only 100 kg C ha^{-1} yr^{-1} could be achieved on the drylands, a total sequestration of 0.5 Gt yr^{-1} would be achieved on a global basis.   This would be a significant contribution to removing excess carbon from the atmosphere. Several lines of evidence, reviewed briefly below, suggest that the target figure is acheivable.

The rainfall use efficiency (RUE) calculated for perennial shrub drylands is about 2.5-4.0 kg ha^{-1} mm^{-1} annual rainfall (Le Houerou, 1984).   UNEP (cited in Dregne et al.,1991) estimates that 37% of drylands are classified as semi-arid with rainfall in the range of 250-400 mm and a further 17% is classified as subhumid (rainfall 400-550 mm).   Applying the RUE factors we calculate on a global basis that semi-arid regions have the potential to produce 2.76 Gt yr^{-1} of above ground biomass and the subhumid regions 2.33 Gt yr^{-1} (total C = 2 Gt yr^{-1}).

The biomass figures quoted above are for above ground biomass. Root biomass, so critical in the equation, is likely to be at least as high or even higher (Kozlowski, 1972). Furthermore, C sequestered in roots of dryland plants is likely to remain locked away for a long time because of slower decomposition rates belowground. Sims and Singh (1971) found that about 85% of the total standing crop of photosynthesizing plants in North American drylands was belowground.   Gifford et al. (1992) estimated that 30% of total primary productivity entered long-term soil storage in Australian soils.   Hence, dryland soils could theoretically become net sinks for 0.6 Gt yr^{-1} C, based on RUE production rates and soil storage estimates, whereas at present they are probably net sources (Dregne et al., 1991).

Can the theoretical RUE be approached by revegetating degraded drylands? Le Houerou (1992) evaluated standing crops and productivities of Mediterranean arid zones that were revegetated with Atriplex spp. In a 100-400 mm yr^{-1} rainfall area, plant stands were as high as 20 t ha^{-1} with annual productivity rates of 3-5 t ha^{-1}. He estimated that standing crops and productivity rates at least half these values could be realistically acheived over the area of 550 million ha that had been subjected to desertification. Hence, the theoretical RUE were actually exceeded under semi-managed conditions.

Rainfed croplands offer a further opportunity to sequester carbon through burial of the stubble. Worldwide, rainfed croplands occupy about 1.3 x 10^9 ha although an estimated 7-8 x 10^6 ha are lost each year. Improved stubble (crop residue) handling practices could lead to major savings in C sequestering. For example, a 2 t crop of wheat grain leaves about 4 t of stubble. Many regions have yields of grain in excess of 3 t ha^{-1} with concomitant yields of residues. Globally, 1.8 Gt of grain are produced in the drylands, with a yield of stubble of at least 7.5 Gt (containing 2.7 Gt C). If the stubble were plowed under and even 5% entered long term storage in the soil, 0.14 gt C yr^{-1} would be sequestered by this practice alone.

5. Potential for increased production as function of rainfall

It is predicted that the greenhouse effect will cause changes in the distribution of rainfall. Climatic changes will be uneven and some dryland areas will receive less rather than more rainfall but globally the drylands are expected to become more moist (Henderson-Sellars & Blony, 1989). In some cases this will stimulate greater potential productivity over large areas of marginal land (Squires & Tow, 1991). For example, yield increases up to 10-15 per cent might be expected for wheat, rice, and barley and increases from 0-10 per cent for corn, sorghum and sugar cane. Any augmentation of rainfall as a result of climatic change could influence biomass production. Rangeland vegetation responds positively to any increase in precipitation. It has been reported (NRC, 1975) that a direct ratio of increase in herbage biomass to increase in supplementary water occurs in dry areas: yield is increased about 6,700 kg/ha of forage per 2.5 cm of annual precipitation in semidesert, and about 10,700 kg/ha in midgrass areas. If the potential productivity increases are to be realized, however, the lands that have been lost to desertification will have to be managed or restored.

6.   Conclusions

Restoring productivity in the drylands will have a beneficial effect on at least short term carbon uptake in these lands (over a period of approximately 20 years). If the pasture drylands were restored to natural rates of productivity based on rainfall, they would absorb some 2-3 Gt C yr^{-1}. Much of this would be stored in the standing crop of trees and shrubs regrowing on the landscape. Additional carbon amounting to 0.6-1.2 Gt yr^{-1} could be absorbed through intensive halophyte plantations, and plowing under of cereal stubble could store 0.1-0.2 Gt C yr^{-1}. The annual enhanced production of approximately 3 Gt C yr^{-1} would lead to long-term carbon storage in the soil although at reduced efficiency. The pertinent question is: would investment in dryland programs be cost effective in terms of C storage compared to other potential offsets?

During this workshop but independent of our analyses in this paper, the CENTURY model was used to estimate rates of C storage in non-agricultural dryland soils over a 100 yr period, using two management scenarios (Ojima, 1993). In the first scenario, a 80% biomass removal rate was used, to simulate present overuse of drylands; in the second scenario, a 30% removal rate was used, to simulate optimum management practices. The results indicated that under good management and given likely increases in primary production due to climate change in these regions, pasture dryland soils could become a net C sink of 0.5 Gt yr^{-1}, whereas under present management they will continue to represent a net source of C, due to releases from the soil not compensated by inputs on overgrazed or eroded lands.

If anti-desertification programs to restore rangelands were funded at levels recommended by UNEP (1991) and resulted in 0.5 Gt yr^{-1} of C sequestration as predicted, the costs would be $10-18 t^{-1} C. This is a reasonable cost compared to reforestation, tree plantations and other C offsets (Kinsman and Trexler, 1993). A similar amount of C could theoretically be sequestered at similar cost using halophyte crop residues on saline lands. Cereal stubble could contribute an additional 0.1-0.2 Gt of C storage. Total C sequestration could well exceed 1 Gt yr^{-1}. We conclude that C sequestration in dryland soils is feasible and cost effective.

Creating an economic linkage between fossil fuel burning and restoration of drylands to store C would appear to be a worthwhile goal in view of the importance of the drylands in global food production. Since the effects of CO_2 enrichment on climate are uncertain, those actions that make sense even in the absence of global warming have been recommended as present action steps (the "no regrets" policy). Restoration of productivity to the drylands through halophyte crops, rehabilitation of rangelands and other management practices falls into this category of action step.

Acknowledgements

This work is supported by the Electric Power Research Institute and the Salt River Project under RP8011-3. We thank Al Qöyawayma, Jan Miller, Lou Pitelka and Sy Alpert for support and encouragement throughout the study.

References

Dregne, H., Kassas, M. and Rosanov, B.:   1991, "A new assessment of the world status of desertification", Desertification Control Bulletin 20:   6-18.

Eyre, A. L.:   1985, "Population pressure on arid lands: Is it manageable?", in E. E. Whitehead, C. F. Hutchinson, B. N. Timmerman and R. G. Varady (eds.), Arid Lands, Today and Tommorrow, Westview Press, Boulder, Colorado, pp. 989-996.

Gifford, R. M., Cheney, N. P., Noble, J. C., Russell, J. S., Wellington, A. B. and Zammit, C.: 1992, "Australian land use, primary production of vegetation and carbon pools in relation to atmospheric carbon dioxide levels", Bureau of Rural Resources Proceedings 14:   151-187.

Glenn, E. P., Hodges, C., Leith, H., Pielke, R. and Pitelka, L.: 1992a, "Halophytes to remove carbon from the atmosphere", Environment 34:   40-43.

Glenn, E. P., Kent, K. J., Thompson, T. L. and Frye, R. J.: 1991a,   Seaweeds and Halophytes to Remove Carbon from the Atmosphere, EPRI ER/EN-7177, Electric Power Research Institute, Palo Alto, California.

Glenn, E.P. and O'Leary, J. W.:   1985,   "Productivity and irrigation requirements of halophytes grown with seawater in the Sonoran Desert", Journal of Arid Environments 9:   81-91.

Glenn, E.P., O'Leary, J.W., Watson, M. C., Thompson, T. L. and Kuehl, R. O.: 1991b, "Salicornia bigelovii Torr.:   an oilseed halophyte for seawater irrigation", Science 251:   1065-1067.

Glenn, E. P., Coates, W., Riley, J. J., Kuehl, R. and Swingle,

R. S.: 1992b, "Salicornia bigelovii Torr.: a seawater-irrigated forage for goats", Animal Feed Science and Technology 40: 21-30.

Henderson-Sellars, A. and Blony, R.: 1989, The Greenhouse Effect, University of New South Wales Press, Sydney.

Houghton, R. A.: 1990, "The global effects of tropical deforestation", Environmental Science and Technology 24: 414-421.

Kassas, M., Ahmad, Y. and Rosanov, B.: 1991, "Desertification and drought: an ecological and economic analysis", Desertification Control Bulletin 20: 19-29.

Kauppi, P. E., Mielikainen, K. and Kuusela, K.: 1992, "Biomass and carbon budget of European forests, 1971 to 1990", Science 256:70-75.

Kinsman, J. and Trexler, M.: 1993 (in press), "Carbon offsets from the perspective of the electrical utility industry", Water, Air & Soil Pollution 70.

Kozlowski, T.T.: 1972, "Physiology of water stress", in: C.M. McKell, J.P. Blaisdell and J.R. Goodin (eds), Wildland Shrubs - Their Biology and Utilization, USDA Forest Service INY-1.

Le Houerou, H. N.: 1984, "Rain use efficiency: a unifying concept in arid-land ecology", Journal of Arid Environments 7: 213-247.

Le Houerou, H. N.: 1992, "The role of saltbushes (Atriplex spp.) in arid land rehabilitation in the Mediterranean Basin: a review", Agroforestry Systems 18: 107-.

Malcolm, C. V.: 1986, "Production from salt-affected soils", Reclamation and Revegetation Research 5: 343-361.

National Research Council: 1975, Climate and Food -Climatic Fluctuation and US Agricultural Production, The National Academy of Sciences, Washington.

Ojima, D.: 1993 (in press), "Grasslands and deserts working group report, workshop on quantification of terrestrial sinks for carbon dioxide", Water, Air and Soil Pollution 70.

Sedjo, R.A.: 1989, "Forests: a tool to moderate global warming?", _Environment_ 31: 14-21.

Sims, P.L. and Singh, J. S.: 1971, "Herbage dynamics and net primary productivity in certain ungrazed and grazed grasslands in North America", in: N.R. French (ed), _Preliminary Analysis of Structure and Functioning of Grasslands_, Col State Univ., Ser. 10.

Squires, V.R.and Tow, P. (eds): 1991, _Dryland Farming: a Systems Approach_, Oxford University Press, Melbourne.

UNEP: 1991, "UNEP Governing Council decision - desertification", _Desertification Control Bulletin_ 20: 3-5.

Wisniewski, J. and Lugo, A. E.: 1992, _Natural Sinks of CO_2: Workshop Statement_, Kluwer, Boston, pp. 1-6.

World Resources Institute: 1990, _World Resources, 1988-1989_, Basic Books, Inc., N.Y.

ANALYSIS OF AGROECOSYSTEM CARBON POOLS

C. VERNON COLE

USDA Agricultural Research Service
Natural Resource Ecology Laboratory
Colorado State University
Fort Collins, CO 80523 USA

KEITH PAUSTIAN

Natural Resource Ecology Laboratory
Colorado State University
Fort Collins, CO 80523 USA

EDWARD T. ELLIOTT

Natural Resource Ecology Laboratory
Colorado State University
Fort Collins, CO 80523 USA

ALISTER K. METHERELL

Ag Research, c/o Soil Science Department
Lincoln University
Lincoln, New Zealand

DENNIS S. OJIMA

Natural Resource Ecology Laboratory
Colorado State University
Fort Collins, CO 80523 USA

WILLIAM J. PARTON

Natural Resource Ecology Laboratory
Colorado State University
Fort Collins, CO 80523 USA

Abstract. We present analyses of major driving variable controls on soil C in agroecosystems. Historical changes in soil C storage in agricultural soils are characterized by large losses during transition from natural grasslands and forests. A major driver in more recent times is the steadily increasing rate of net primary production of major land areas in agriculture. Simulation and analytical models are used to predict trajectories and potential soil C storage under possible scenarios of changed management and climate. Database and analytical requirements for extrapolation from regional to global scales are outlined.

1. Introduction

Of all the major terrestrial biomes, agroecosystems are the ones most subject to continuous anthropogenic disturbance. Land use changes in forests, grasslands, wetlands and savannas have transformed large areas from relatively stable ecosystems to agroecosystems under extensive and intensive management. The

Water, Air, and Soil Pollution **70**: 357–371, 1993.
© 1993 *Kluwer Academic Publishers.*

introduction of agriculture involving land clearing, drainage or breaking of sod, cultivation, replacement of perennial vegetation by annual crops, and nutrient subsidies in the form of fertilizers, has had major impacts on C pools and fluxes throughout the globe. In the initial phases of these transformations there have been major losses of CO_2 to the atmosphere as soil C levels adjusted to reduced C inputs and increased soil disturbance. In many areas there has been intense pressure for production which has led to serious soil degradation by erosion and nutrient losses. These trends continue in many areas of the world. However, in countries able to provide subsidies of energy and technology, agricultural productivity has shown continuing increases, land degradation has slowed or reversed, and soil C pools have stabilized or increased. In analyzing the amounts and fluxes of C in agroecosystems, the primary focus needs to be on current land use and management as well as future changes in management as agricultural communities respond to climatic change.

Soil organic matter is the major sink/source pool for C in agroecosystems. As in native systems, soil C pools and fluxes in agroecosystems are influenced by a number of factors including C inputs (e.g., amount and type of plant residue) and climatic (e.g., temperature, precipitation) and edaphic (e.g., soil texture, pH, drainage) conditions. In addition, land use and management strongly influence C in agricultural soil which complicates our understanding of its role in the global C cycle. With management, we are usually manipulating ecosystems in one of three ways, (1) adding and/or removing species of plants and animals (2) disturbing the soil and its resident litter and (3) subsidizing water or nutrient inputs with irrigation and fertilization, respectively. A common effect caused by shifts in management is to change the way that macroclimate is experienced as microclimate by the components of the system (e.g., shifts in the balance of evaporation and transpiration or the erosive potential of rainfall). Another important influence of management is changing the productive potential of the plant and animal components of the system. Management may change the components of the system but economic and socio-political conditions determine what land use system and specific management regimes will be applied to a particular ecosystem.

The interactions between climatic, edaphic and management influences on soil C storage can be analyzed using an ecosystem-level model such as the CENTURY model (Parton *et al.*, 1987). CENTURY couples processes determining primary production, soil organic matter and nutrient dynamics and soil water balance. Using the model, process interactions and aggregate ecosystem responses to changes in climate regimes, atmospheric CO_2 concentrations and management practices can be analyzed.

In this paper we discuss the impacts of climate change, increased atmospheric CO_2 concentrations, and management controls on soil C levels in agroecosystems. First we examine the interactions between changes in temperature and precipitation as they affect potential soil C levels for nine field sites distributed globally. This steady-state analysis illustrates the sensitivity of climate change effects to present climate and productivity levels. However, it does not address changes in productivity as a result of climate change, increased CO_2, or management and the subsequent feedbacks of plant residue production on soil organic matter formation. These dynamic feedbacks and the influence of different agricultural management scenarios are detailed in a case study analysis of the Great Plains region of the central U.S. This analysis focuses on historical as well as future impacts of land management impacts on soil C. Finally, we consider the data and modeling resources needed for improving predictions of soil C dynamics as a function of global change.

2. Steady State Analysis of C Sequestration Potentials

A rapid first approximation of potential soil C level, as a function of plant residue inputs rates, residue quality, climate, soil type and management system, can be calculated using a steady-state analysis. The CENTURY model, as with most SOM models, divides soil organic matter into several pools having different turnover rates. Changes in total organic matter are computed by integrating the time-dependent differential equations for the set of SOM pools. These equations are of the form

$$dS_i/dt = I_i - k_i * S_i; \quad k_i = f(\text{climate, soil texture, lignin content, etc.})$$

where I is the rate of input of organic matter to the ith pool (from crop residues or from other organic matter pools) and $k_i * S_i$ is the decomposition rate of the pool. The specific decomposition rate of a given pool may be governed by a variety of climatic and edaphic factors.

At steady-state, organic matter levels are at maximum and remain constant (by definition) and thus the rate equations can be set equal to zero ($dS_i/dt=0$) and solved analytically. One important result of the steady-state analysis is that total C at steady-state (= maximum soil C level) is shown to be directly proportional to the rate of C input (Paustian et al., in press). This has important implications for C sequestration potential since, if the theory holds, then the amount of organic matter in a given soil can be increased in direct proportion to an increase in C inputs, if other factors (soil climate, residue quality, etc.) are unchanged. Several long-term field experiments (e.g., Larson et al., 1972; Rasmussen et al., 1980; Paustian et al., 1992), where different rates of organic matter addition to soil have been maintained over several decades, show linear relationships between C inputs and soil C levels, which supports the theory (Fig. 1).

Whether climate change increases or decreases C in agricultural soils will depend on the balance between climate change effects on productivity (hence C inputs to soil) and effects on decomposition processes. In intensively managed systems, productivity is strongly controlled by management which may act to buffer climate change and CO_2 effects on crop productivity. In contrast, decomposition processes are generally less directly controlled by management practices and are more variable in their responses to changes in temperature and precipitation across different

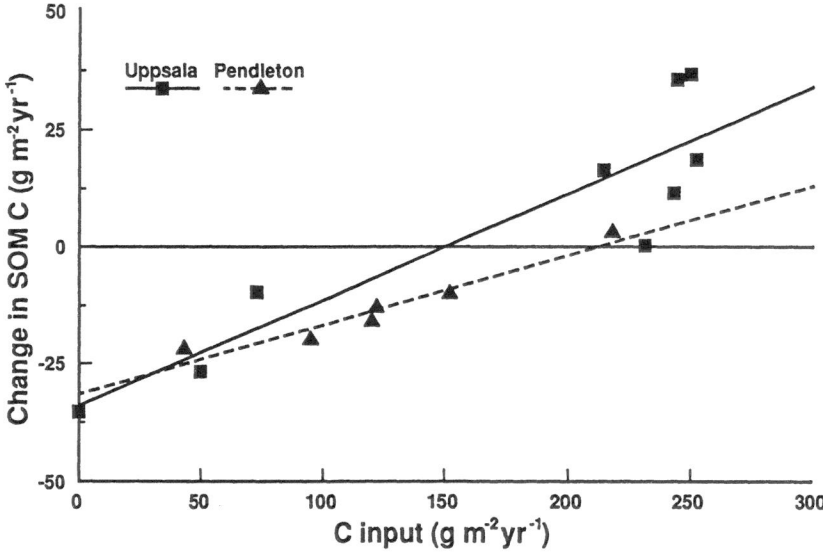

Figure 1. Changes in soil organic C (SOC) as a function of C addition rate, averaged
over the duration of long-term field experiments at Uppsala, Sweden (30 yr)
and Pendleton, Oregon (60 yr). C input rates of ~ 150 g yr^{-1} were required to
maintain pre-experimental SOC levels at Uppsala versus ~220 g yr^{-1} at
Pendleton. Data from Paustian et $al.$ (1992) for Uppsala and Parton and
Rasmussen (in press) for Pendleton. Lines represent least-squares regressions.

climatic regions. By solving the model for steady-state conditions we can analyze
the interactions between temperature and precipitation and soil C levels for different
soils and climatic regions.

Climate change effects on potential soil C levels were analyzed for nine grassland
sites distributed globally, from tropical to sub-boreal regions (Table 1). Grassland
sites were selected since these systems are more likely to approach maximum
potential soil C levels than would annual cropping systems. Both the rate of C
inputs and calculated equilibrium soil C vary substantially across sites. For similar
climatic regimes, SOM levels increase with increasing C addition and are highest for
fine textured soils (Table 1). For sites where climatic factors are most limiting to
decomposition due to low temperatures (e.g., Kjettslinge, Khomutov) or moisture
stress (CPER), SOM levels are highest per unit of C input. Conversely, where
decomposition is rapid, as in the tropics (Nairobi, Hat Yai) steady-state SOM levels
per unit of C input are lower.

Table 1. Calculated steady-state soil C values for surface soils of 9 grassland sites based on climate (mean annual temperature and precipitation), soil texture and estimated mean annual C inputs (data from Ojima *et al.*, 1993 and (for Sweden) Paustian *et al.* 1990).

Site	Lat./Long.	MAT (C)	MAP (cm)	sand	silt	clay	C input (g m^{-2} yr^{-1})	SS SOM C (kg m^{-2})
				-------(fraction)-------				
Warra, Australia	26S,146E	20.5	49.2	.32	.18	.50	110	1.3
Bari, Italy	41N,17E	15.8	57.4	.40	.30	.30	100	1.6
Davis, CA, USA	38N,120W	15.6	42.0	.40	.30	.30	100	2.0
CPER, CO, USA	40N,105W	8.7	31.0	.70	.15	.15	100	2.9
Hat Yai, Thailand	6N,101E	27.6	154.1	.58	.40	.02	750	3.9
Nairobi, Kenya	1S,36E	19.0	67.7	.13	.17	.70	350	5.1
Konza, KS, USA	39N,97W	12.9	83.2	.25	.35	.40	390	6.9
Khomutov, Ukraine	47N,38E	9.5	43.8	.20	.28	.52	240	7.1
Kjettslinge, Sweden	60N,17E	5.3	52.6	.44	.37	.19	270	9.0

Steady-state SOM values were computed assuming increases in mean annual temperature from 0-5°C and a range of annual precipitation changes of ±10% from measured long-term averages. The response surfaces show the predicted changes in equilibrium SOM C levels assuming no change in current C input rate (Fig. 2). Current equilibrium C levels are denoted by a star.

The overall trend is that increasing temperature and precipitation will usually enhance decomposition rates and thereby decrease equilibrium soil C levels if C inputs remain unchanged. However, the response surfaces show patterns which vary sharply according to climatic zone, i.e., Mediterranean and subtropical (Bari, Davis and Warra), North American temperate (CPER and Konza), tropical (Nairobi and Hat Yai, Thailand) and cool temperate (Khomutov and Kjettslinge). Note the shape of the response surface for the two North American prairie sites and the Australian site (Warra). These sites are strongly limited by water availability such that an initial temperature increase causes higher evapotranspiration, greater moisture limitation on

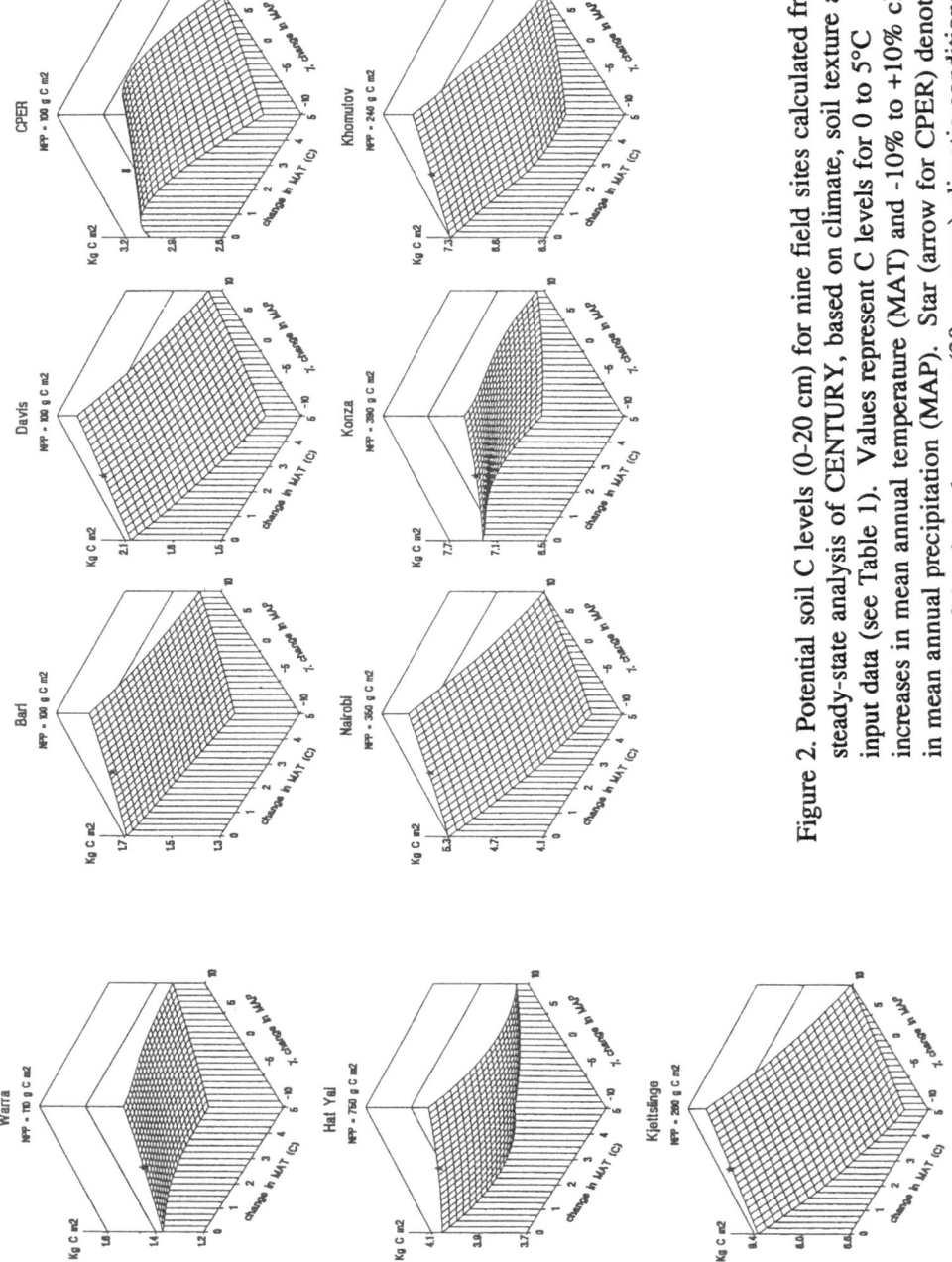

Figure 2. Potential soil C levels (0-20 cm) for nine field sites calculated from steady-state analysis of CENTURY, based on climate, soil texture and C input data (see Table 1). Values represent C levels for 0 to 5°C increases in mean annual temperature (MAT) and -10% to +10% change in mean annual precipitation (MAP). Star (arrow for CPER) denotes equilibrium C levels under current (30 yr mean) climatic conditions.

decomposition, and therefore an initial increase in potential SOM levels. Further warming will, however, eventually override this effect and cause a net increase in decomposition and therefore lower SOM levels (again, assuming no change in inputs). For Hat Yai, temperatures are already near optimum (MAT = 28°C) so that with a 5°C increase decomposition is reduced during the warmest season. The greatest absolute change in soil C is for the cool temperate site (Kjettslinge) which has little moisture stress (hence little response to increasing precipitation) but a strong response to increased temperature.

The steady-state analyses illustrate the sensitivity of climate change effects to the preexistent climatic conditions as well as the variations in temperature and moisture interactions as they influence soil C levels. However, it is important to consider that potential increases in primary productivity in conjunction with global change, e.g., CO_2 fertilization effects, longer growing seasons, increased N mineralization in high-latitude N-limited ecosystems, improved agricultural management, etc., may increase C inputs to soil and thus compensate for or even override increased decomposition rates associated with global warming. The following section addresses these dynamics with respect to climate change, CO_2, and management systems for U.S. Great Plains agroecosystems.

3. U.S. Great Plains Case Study

3.1. HISTORICAL CHANGES IN SOIL C

The North American Great Plains provide an ideal case study of management effects on soil C dynamics. The combination of a north-south gradient in temperature and an east-west gradient in precipitation provides a matrix of climatic driving variables across the region. In addition, the land use history of the region is relatively well documented. Sod breaking for row crop cultivation did not occur until the early 1900's and a network of seventeen USDA dryland research stations measured changes in crop productivity and soil changes under a variety of cropping systems for a forty yr period to 1945 (Haas et al., 1957).

Organic matter in the soils of this semiarid region was seriously depleted under the intense cultivation practices of the pioneer farmers. Haas et al. (1957) reported losses of C and N from surface soils ranging from 35 to 50%. Amounts of C and N lost ranged from 10,000-20,000 kg/ha and 500-1000 kg/ha for C and N, respectively. Soil N content declined 49% and soil productivity (indexed by corn grain yield) declined 71% during the 28 yr time interval after sod breaking. The combination of low yielding varieties, complete straw removal at harvest, and intensive cultivation resulted in serious losses of organic C and available nutrients (Burke et al., 1989). These changes were associated with detrimental physical conditions, poor water conservation and accelerated erosion. The interaction of temperature, precipitation and management on SOC levels across the U.S. Great Plains during the first 40 yr of cultivation are illustrated by model simulations (Fig. 3). After 1940 management

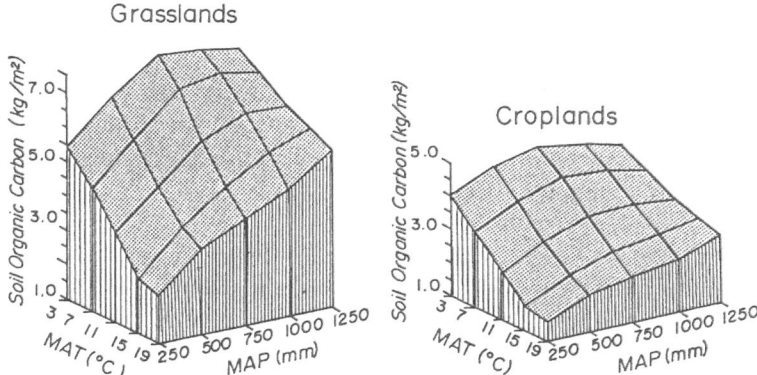

Figure 3. Simulated soil organic C levels in fine textured soils (10-20 cm) of the
Great Plains. C varies as a function of mean annual temperature and
precipitation. Native grasslands and croplands after 40 yr of wheat-fallow
cultivation (redrawn from Cole *et al.*, 1989).

practices were modified to include higher yielding varieties, application of N and P
fertilizers, no straw removal at harvest and stubble mulch tillage. Analyses based on
statistical and mechanistic models (Cole *et al.*, 1990) indicate that these management
changes arrested soil C and N losses and that soil C levels stabilized. At present, C
inputs and outputs are nearly balanced in most agroecosystems in the region.

3.2. PROJECTION OF FUTURE CHANGES IN SOIL C WITH GLOBAL CLIMATE CHANGE

Projection of future soil C changes under various land management alternatives
requires consideration of the dynamic changes in climate and atmospheric CO_2
concentrations and how these factors affect plant growth, nutrient cycling and soil
water and energy balance. Effects of changing climate can be incorporated in
CENTURY model simulations using, for example, output from GCM simulations.
Effects of changing CO_2 concentrations, however, involve a number of different
processes in the model. These effects and their incorporation into CENTURY are
outlined below.

The direct effects of an increase in atmospheric CO_2 concentration on soil processes
would likely be insignificant because CO_2 concentrations in the soil are already

greatly elevated with respect to atmospheric concentrations. However, the feedbacks between CO_2-mediated changes in plant processes and SOM processes could be substantial and must be accounted for in simulations of the effect of global change on SOM (Long, 1991). These plant processes include changes in net primary production, litter quality, and transpiration (Kimball, 1983). The response to CO_2 is not simply due to the removal of a single limiting factor (Sinclair, 1992), but results from a hierarchy of effects (Acock, 1990).

First, increasing CO_2 has a direct effect on C availability by stimulating photosynthesis and reducing photorespiration. There is an important difference between C_3 species, such as wheat, and C_4 species, such as corn, in this response. The growth response to CO_2 is usually lower in C_4 species than in C_3 species (Wong, 1979; Rogers et al., 1983; Morrison and Gifford, 1984b; Cure and Acock, 1986). The main reason for responses to CO_2 in C_4 species is due to improved water use efficiency as discussed below. The second effect is a decrease in stomatal conductance (Moss et al., 1961; Akita and Moss, 1973; Wong, 1979; Rogers et al., 1983; Morrison and Gifford, 1984a; Cure and Acock, 1986) at high CO_2 concentrations, which reduces the transpiration rate per unit leaf area and improves water use efficiency. The effect on stomatal conductance and transpiration is observed in both C_3 and C_4 species. The third major effect of increased CO_2 is a decrease in the plant N concentration in C_3 species (Schmitt and Edwards, 1981; Hocking and Meyer, 1991). With a fixed nutrient supply, an increase in C assimilation is likely to result in lower plant nutrient concentrations due to a dilution effect, but this is not the only effect.

3.3.   IMPLEMENTING CLIMATE, CO_2, AND LAND MANAGEMENT CHANGES IN THE CENTURY SIMULATIONS

The CO_2 effects were taken into account in CENTURY model simulations by linearly increasing CO_2 concentrations from 350 µmol/mol to a final concentration of 700 µmol/mol over a 50 yr period. The effects of CO_2 described above were controlled by a logarithmic function of the CO_2 concentration (Gifford, 1979; Goudriaan, 1992) applied to the calculation of potential production (see Metherell 1992 for a detailed explanation). For C_3 species potential primary production was increased by up to 30% while the minimum plant N concentrations were decreased to maintain the same N uptake. For all species transpiration was decreased by 23%. Native grass and reestablished grasslands were treated as a 50/50 mixture of C_3 and C_4 species.

To evaluate potential effects of management, direct effects of increasing CO_2 and a changing climate, we simulated changes in productivity and soil organic C (SOC) in Weld County Colorado in the Central Great Plains of the U.S. Historical changes in SOC in native grasslands and in a wheat-fallow cropping system were simulated based on weather records from 1917 to the present with stepwise changes in management including improved wheat varieties, reduced tillage and improved residue management over that period. For the period 1990 to 2040 new management practices involving an intensification of cropping in a wheat-corn-fallow system and reestablishment of grass in lands under the Conservation Reserve Program (CRP) were simulated with a repeat of the historical weather record.

Ecosystem sensitivity to the temporal and regional resolution of climate change was evaluated by modifying the current monthly weather record for the past 25 years with output from a high resolution general circulation model (GCM) climate experiment from the Geophysical Fluid Dynamics Laboratory (GFDL) models (IPCC, 1990). We tested the sensitivity of grassland and cropland ecosystems to the following global change effects:

- Climate change effects (i.e., alterations to monthly mean temperature and precipitation);
- Combined effect of climate change and doubling of atmospheric CO_2 (+CC+CO_2)

A current weather file of monthly precipitation and monthly mean maximum and minimum temperatures was created using existing weather station data. We used a 25-year weather record as the base climate to simulate the equilibrium grassland. To generate the $2XCO_2$ climate, we used the representative GCM grid value of projected $2XCO_2$ climate changes of monthly temperature and precipitation for Weld County. Under the business-as-usual scenarios for the GFDL high resolution runs the mean annual temperature averaged ~3°C higher and annual precipitation was increased by 5 mm, with some changes in seasonality of precipitation. We applied these projected monthly values equally to the minimum and the maximum mean monthly temperature values used by CENTURY, with a linear increase during a 50 year ramp.

3.4 IMPACT OF CLIMATE AND CO_2 ON SOIL C IN AGROECOSYSTEMS

Cultivation in the Great Plains resulted in soil C losses with the breaking of sod (Haas *et al.* 1957) which is reflected in the historical simulation of soil C dynamics of Weld County, Colorado during the past 80 yr (Fig. 4). Actual climate patterns and realistic land use practices were used in the simulation. The net result was a 1.8 kg C m^{-2} loss in soil C over the past 70 to 80 yr compared to native grasslands in the area. Two additional management systems, a wheat-corn-fallow (WCF) rotation and a reestablished grass vegetation as part of the Conservation Reserve Program (CRP) were simulated for 50 yr from 1990 (Fig. 4). This period was simulated using an extension of the historical weather and thus represents only the effects of management practices. Recovery of soil C stocks result from revegetation of croplands with perennial grass reseeding (CRP, 0.35 kg C m^{-2} recovery after 50 yr of climate change) and/or implementation of the wheat-corn-fallow rotation (WCF, 0.3 kg C m^{-2} recovery) compared to conventional wheat-fallow system. Plant C inputs

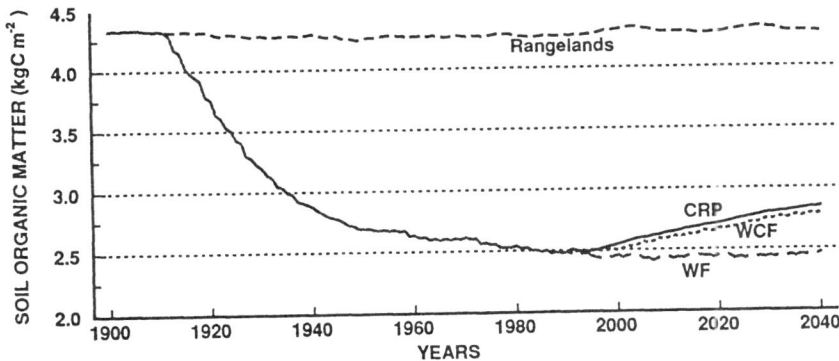

Figure 4. Simulated soil C levels for native grassland and cropland in Weld County,
CO, USA. Cropping system prior to 1990 was wheat-fallow (WF). Included
in crop simulations are historical changes in crop management and
production potentials from 1909 (sod broken) to 1990. Simulation past 1990
assumed unchanged climate (i.e., same as 1910 to 1990) but included two
additional management systems, a wheat-corn-fallow rotation (WCF) and
reseeded grassland in the Conservation Reserve Program (CRP).

are much greater in the CRP and WCF land uses compared to the WF land use and
this is reflected in the increased soil C levels.

Effects of climate change and increased CO_2 were compared to effects of
management in GCM-driven simulations described above (Fig. 5). The simulations
including CO_2 and climate change effects (denoted by plus symbols) result in much
greater soil C levels in the two annual crop systems. The WF+ system had a net
increase of 0.5 kg C m^{-2} compared to 0.35 kg C m^{-2} for the WCF+ system (Fig. 5).
The wheat-corn fallow showed a response to the increased C inputs with cropping
two out of three yr (as compared to one out of two yr in WF), as well as a positive
response to the CO_2 and climate change scenario. The greater increase in the WF+
system is due to the greater response of wheat versus corn to the elevated CO_2 levels
implemented in these CENTURY simulations. It should be noted that the grain
cropping systems assume full fertilization and thus no N limitation. In contrast, the
two CRP simulations (CRP and CRP+) were almost indistinguishable from each
other. CRP lands showed steady increases in SOC after grass was reestablished, but
in the absence of nitrogen fertilization did not show a significant response to CO_2
and climate changes. The minimal increase in production due to N limitation was
essentially neutralized by increased decomposition rates.

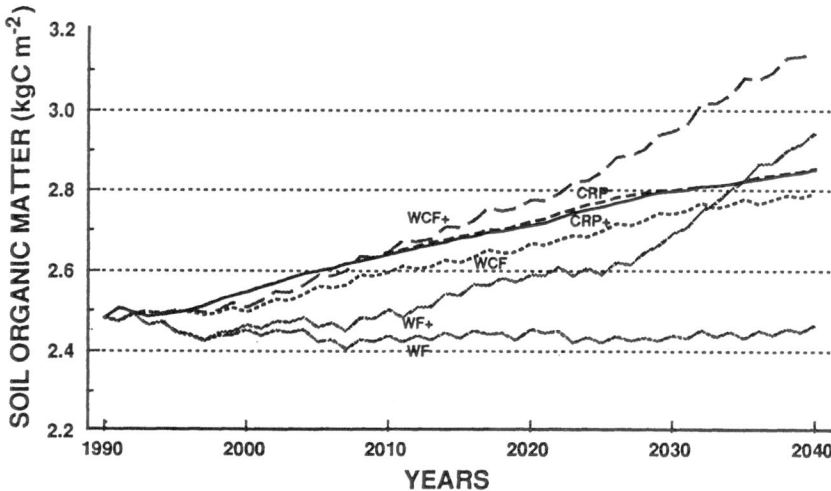

Figure 5. Simulated soil C for wheat-fallow (WF), wheat-corn-fallow (WCF) and
reseeded grassland (CRP) systems, assuming present climate (as in Fig. 4)
and climate change and increased atmospheric CO_2 scenario denoted by (+).
Climate change scenario based on GCM simulations for Northeastern
Colorado, assuming a doubling in CO_2 concentration by 2040, giving an
increase of 3°C and 5 mm in MAT and MAP, respectively.

Although we have considerable confidence in model results for the historical period,
based on comparative data analysis and statistical models (Burke *et al.*, 1989; Cole *et
al.*, 1990) our predictions of potential changes must be considered as preliminary
until our understanding of the many interacting processes improves. In spite of the
limitations, these analyses highlight the complex nature of responses of
agroecosystems to global change. In terms of the global C balance it is significant
that agriculture in the U.S. Great Plains could change from being a net source of
atmospheric CO_2 to being a net sink.

4. Information Requirements for Regional Analysis of C Pools

Assessing present soil C stocks and predicting future changes in soil C at local,
regional and global scales requires the integration of a variety of data sources and
data analysis and organizational tools. Our concept of how these data resources can
be structured is outlined in Fig. 6. We consider the ecosystem as the fundamental
unit at which process-level information and geographically referenced databases (for
soils, climate, vegetation, landuse, etc.) can be combined to drive models.

REGIONAL AND GLOBAL ANALYSIS OF CARBON POOLS

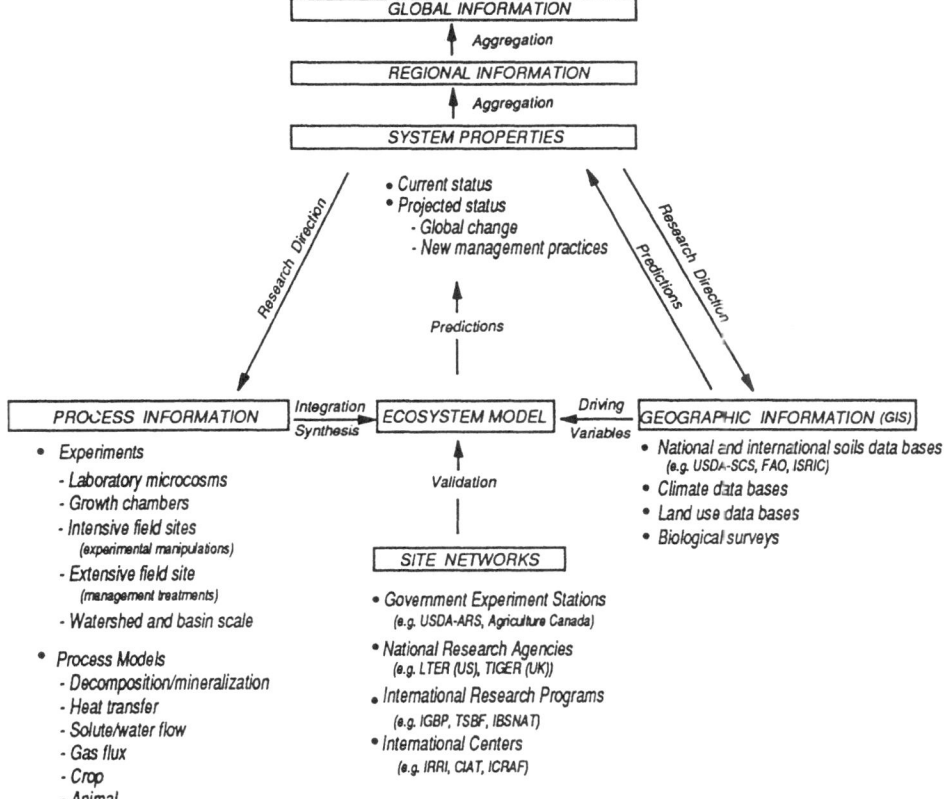

Figure 6. Information requirements and data synthesis tools needed for local, regional and global analysis of soil carbon pools. Process information and spatially-resolved databases drive generalized ecosystem models. Detailed site-specific information from regional site networks are needed for parameterization and evaluation of models. (Adapted from Elliott and Cole, 1989).

Generalized ecosystem models embody our best understanding of the complex set of interactions and feedbacks that determine the character of particular ecosystems. By integrating models and databases organized with geographic information systems we can make projections of the long-term effects of existing and new management practices, changing weather patterns and impending climate change on soil organic matter for analysis of global change effects.

Broad-based regional and global databases of soils, climate, vegetation, topography and landuse will be required for these projections. Some of these resources, particularly topographic and climatic databases are already well advanced. New products such as the international soils database being developed at ISRIC (International Soil Resource Information Center) in Wageningen, Netherlands, in cooperation with IGBP, FAO, and USDA (Scholes and Skole, in prep.) will greatly facilitate soil C assessments and modeling. More comprehensive information on land use and management practices is also a priority requirement.

In addition to large-scale comprehensive databases, more detailed information from research sites and long-term field experiments is needed to parameterize and validate ecosystem models before they can be confidently applied to regional and global analysis. These sites need to encompass a wide range of management practices, climates, vegetation types, soils and time scales. Several efforts have been started to form such site networks, many under the auspices of IGBP projects and/or national global change research initiatives (e.g., Elliott *et al.*, 1993, Molina 1992). Analysis and synthesis of this information will be invaluable in improving our understanding of global C cycling as well as in advancing knowledge within traditional areas of soil and ecosystem science.

5. Acknowledgements

We thank Ms. Laura Harding for assistance with programming simulations. Support from the USDA Global Change Program and from US EPA to "Agroecosystems Carbon Pools and Dynamics" project (AERL 9101) is gratefully acknowledged.

5. References

Acock, B.: 1990, in: *Impact of carbon dioxide, trace gases, and climate changes on global agriculture.* American Society of America Special Publication No. 53.

Akita, S. and Moss, D.N.: 1973, *Crop Science* **13**, 234-237.

Burke, I.C., Yonker, C.M., Parton, W.J., Cole, C.V., Flach, K. and Schimel, D.S.: 1989, *Soil Sci. Soc. Am. J.* **53**, 800-805.

Cole, C.V., Stewart, J.W.B., Ojima, D.S., Parton, W.J. and Schimel, D.S.: 1989, in: *Ecology of Arable Land*, Kluwer Academic Publishers.

Cole, C.V., Burke, I.C., Parton, W.J., Schimel, D.S., Ojima, D.S. and Stewart, J.W.B.: 1990, in: *Challenges in Dryland Agriculture, A Global Perspective.* Proceedings of the International Conference on Dryland Farming, Aug. 15-19, 1988, Amarillo/Bushland, Texas. P.W. Unger, T.V. Sneed, W.R. Jordon and R. Jensen, editors.

Cure, J.D. and Acock, B.: 1986, *Agricultural and Forest Meteorology* **38**, 127-145.

Elliott, E.T. and Cole, C.V.: 1989, *Ecology* **70**, 1597-1602.

Elliott, E.T., Paustian, K., Collins, H.P., Paul, E.A., Cole, C.V., Burke, I.C., Blevins, R.L., Lyon, D.J., Frye, W.W., Halvorson, A.D., Huggins, D.R., Turco, R.F., Hickman, M., Monz, C.A. and Frey, S.D.: 1993, SSSA Spec. Publ. (in press).

Gifford, R.M.: 1979, *Search* **10**, 316-318.

Goudriaan, J.: 1992, *Journal of Experimental Botany* **43**, 1111-1119.

Haas, H.J., Evans, C.E. and Miles, E.F.: 1957, *Nitrogen and carbon changes in Great Plains soils as influenced by cropping and soil treatments*. Tech. Bull. 1164, U.S. Dept. Agric.

Hocking, P.J. and Meyer, C.P.: 1991, *Australian Journal of Plant Physiology* **18**, 339-356.

Intergovernmental Panel on Climate Change: 1990, *Climate change: The intergovernmental panel on climate change (IPCC) scientific assessment*. Univ. Press, Cambridge, England.

Kimball, B.A.: 1983, *Agronomy Journal* **75**, 779-788.

Larson, W.E., Clapp, C.E., Pierre, W.H. and Morachan, Y.B.: 1972, *Agronomy Journal* **64**, 204-208.

Long, S.P.: 1991, *Plant, Cell and Environment* **14**, 729-739.

Metherell, A.K.: 1992, *Simulation of Soil Organic Matter Dynamics and Nutrient Cycling in Agroecosystems*. Ph.D. Dissertation, Colorado State University, Fort Collins.

Molina, L.T.: 1992, *Trace Gas Exchange: Mid-latitude Terrestrial Ecosystems and the Atmosphere (TRAGEX)*. MIT, Cambridge, MA.

Morrison, J.I.L. and Gifford, R.M.: 1984a, *Australian Journal of Plant Physiology* **11**, 361-374.

Morrison, J.I.L. and Gifford, R.M.: 1984b, *Australian Journal of Plant Physiology* **11**, 375-384.

Moss, D.N., Musgrave, R.B. and Lemon, E.R.: 1961, *Crop Science* **1**, 83-87.

Ojima, D.S., Parton, W.J., Schimel, D.S., Scurlock, J.M.O. and Kittel, T.G.F.: 1993, (this volume). *Modeling the effects of climatic and CO_2 changes on grassland storage of soil C.*

Parton, W.J. and Rasmussen, P.E.:(in press) *Soil. Sci. Soc. Am. J.*

Parton, W.J., Schimel, D.S., Cole, C.V. and Ojima, D.S.: 1987, *Soil Sci. Soc. Am. J.* **51**, 1173-1179.

Paustian, K., Andrén, O., Clarholm, M., Hansson, A.C., Johansson, G., Lagerlöf, J., Lindberg, T., Pettersson, R. and Sohlenius, B.: 1990, *J. Appl. Ecol.* **27**, 60-84.

Paustian, K., Parton, W.J. and Persson, J.: 1992, *Soil Sci. Soc. Am. J.* **56**, 476-488.

Paustian, K., Robertson, G.P. and Elliott, E.T.: *Advances in Soil Science* (in press).

Rasmussen, P.E., Allmaras, R.R., Rohde, C.R. and Roager, Jr., N.C.: 1980, *Soil Sci. Soc. Am. J.* **44**, 596-600.

Rogers, H.H., Bingham, G.E., Cure, J.D., Smith, J.M. and Surano, K.A.: 1983, *Journal of Environmental Quality* **12**, 569-574.

Schmitt, M.R. and Edwards. G.E.: 1981, *Journal of Experimental Botany* **32**, 459-466.

Scholes, R.J. and Skole, D: (in prep.) *Global soils data: a proposal for a synthesis task*. Global Change Report No. 27, Stockholm.

Sinclair, T.R.: 1992, *Journal of Experimental Botany* **43**, 1141-1146.

Wong, S.C.: 1979, *Oecologia* **44**, 68-74.

MANAGING CROP RESIDUES FOR THE RETENTION OF CARBON

B.A. Stewart
U.S.D.A. Agricultural Research Service
P.O. Drawer 10, Bushland, TX 79012, U.S.A.

Abstract. Soil organic matter, a major sink for carbon, is controlled by many factors that have complex interactions. The management of crop residues is of primary importance. Reduced tillage and no-tillage practices result in a significant build-up of soil organic matter because they greatly reduce the rates of decomposition of both the native soil organic matter and of the crop residues. The crop residues decompose slower because most remain on the soil surface where there is less biological activity, and the native soil organic matter decomposes slower because there is less tillage for aerating the soil and for breaking the aggregates that expose organic compounds to the soil microorganisms. Crop residues, however, are highly variable. Although most crop residues contain about 40 percent carbon, the nitrogen contents range from very low to more than 3.5 percent. For carbon to be stabilized in the soil as organic matter, there must be adequate nitrogen available in the system and this factor is frequently overlooked. Climate is often the most critical factor determining the sustainability and enhancement of soil organic matter. As temperatures increase, organic matter decomposition, particularly in frequently tilled soils, is greatly accelerated. As precipitation decreases, there is less biomass produced for replenishing decomposed carbon. Consequently, soil organic matter maintenance becomes increasingly difficult in either hot or arid regions, and particularly difficult in areas that are both hot and arid. Semiarid regions comprise almost 40% of the world's land area so management of crop residues in these fragile areas is important in relation to the global C picture.

1. Introduction

Soil organic matter is a major sink for C. Soil organic matter is also significantly correlated with soil productivity and soil quality. It acts as a storehouse for nutrients, increases the cation exchange capacity, and reduces the effects of compaction. It builds soil structure and increases the infiltration and retention of water. It serves as a buffer against rapid changes in pH and serves as an energy source for soil microorganisms; therefore, soil organic matter is of critical importance.

Cultivation increases biological activities in the soil, and this is often due to better aeration. But cultivation also exposes fresh topsoil to rapid drying, and after each drying there occurs a burst of biological activity for a few days following rewetting. This is because the drying process releases organic compounds, probably from the breakdown of soil aggregates that are bound together by humic materials. Considerable organic nitrogen is mineralized as ammonia and later oxidized in large part to nitrates. Other nutrients are also made available from the decomposition of organic matter. This is particularly true for phosphorus since much of the phosphorus in soils is present in organic forms. The nutrients released as a result of tillage are readily available to growing plants and increased yields are generally obtained. Tillage also increases rainfall infiltration, controls weeds, and often helps control insects and diseases. Therefore, intensive and frequent tillage has frequently been considered essential for good crop production. However, unless the organic matter supply is replenished, the soil degrades, and if the degradation continues unchecked, the cropping system cannot be sustained. This has a very negative effect on the C balance because a decline in soil organic matter content adds significant amounts of CO_2 to the atmosphere, and a decrease in crop production uses less CO_2 from the atmosphere.

2. The Issue of Sustainability

Hornick and Parr (1987) reported that, for most agricultural soils, degradative processes and conservation practices occur simultaneously. The relationship of soil productivity to soil degradative processes and soil conservation practices is illustrated in Figure 1. As soil degradative processes proceed and intensify, there is a concomitant decrease in soil productivity. Therefore, the productivity level of an agricultural soil at any time is a result of the interaction of degradative processes and conservation/reclamation practices shown in Figure 1. In natural ecosystems, productivity and sustainability are achieved through the efficient but delicate balance between all necessary inputs and outputs. Recent efforts, for example, alley cropping in the tropics, attempt to duplicate in cropping systems the achievement of the balance that occurs in natural ecosystems.

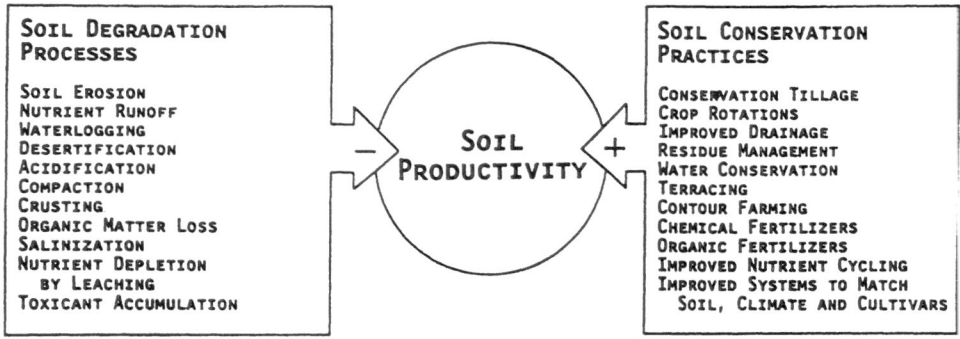

Figure 1.   Relationship of soil productivity to soil degradative processes and soil conservation practices. (Adapted from Hornick and Parr, 1987.)

Using the Hornick and Parr (1987) concept, a sustainable system is any system in which the benefits from soil conservation practices are equal to or greater than the negative effects of the soil degradative processes. Practices that are generally considered positive can, in some instances, result in soil degradation. For example, the use of organic wastes increases organic matter, improves soil structure, enhances soil water storage, and reduces erosion. Under some circumstances, however, use of organic wastes could result in toxic accumulations or nutrient depletion caused by increased leaching. The important point illustrated is that degradative processes and soil improvement processes always occur simultaneously, and the net result can be positive or negative.

The negative impacts of soil degradative processes can become very dominant. In extreme cases, soil productivity can be reduced to zero. On the other hand, soil conservation practices rarely improve the productivity of cropland beyond the level experienced initially after virgin land is cultivated. However, soil restorative practices, to be economically feasible and forward-looking, must, of necessity, be focused on improving the soil productivity beyond the level experienced initially after virgin land is cultivated. Examples of such practices include adding fertilizers in areas where essential plant nutrients are deficient and installing water conservation practices in water deficient regions.

2.1   CLIMATIC EFFECT

Climate is often the most critical factor determining the sustainability of agricultural systems. A generalized view of the effect of varying temperature and water regimes on the difficulty of achieving sustainability in an agricultural system was developed by Stewart et al (1991), and is presented in Figure 2. Since organic matter content is closely correlated with soil productivity and sustainability, the relationships presented in Figure 2 also apply to the difficulty with maintaining the soil organic matter content. As temperatures increase and the amounts of precipitation decrease, the maintenance of soil organic matter becomes more difficult. The reasons for these effects are readily apparent when the processes and practices presented in Figure 1 are analyzed. As temperatures increase, organic matter decline, particularly in frequently tilled soils, is greatly accelerated. Not only is the rate of organic matter decomposition accelerated under these conditions, the production of biomass is decreased so there are smaller amounts of crop residues and roots available that are necessary to replenish the soil organic matter reserve.

The relationships presented in Figure 2 were developed primarily for the drier regions, and do not apply to all climatic regimes. Soil degradation processes also accelerate under very high precipitation regimes, due mainly to erosion, nutrient depletion by leaching, and acidification. Cold conditions severely limit the choice of cropping systems and can also result in water-logging. Therefore, the shape of the lines in Figure 2 would change significantly if extended to very cold or exceedingly high rainfall areas.

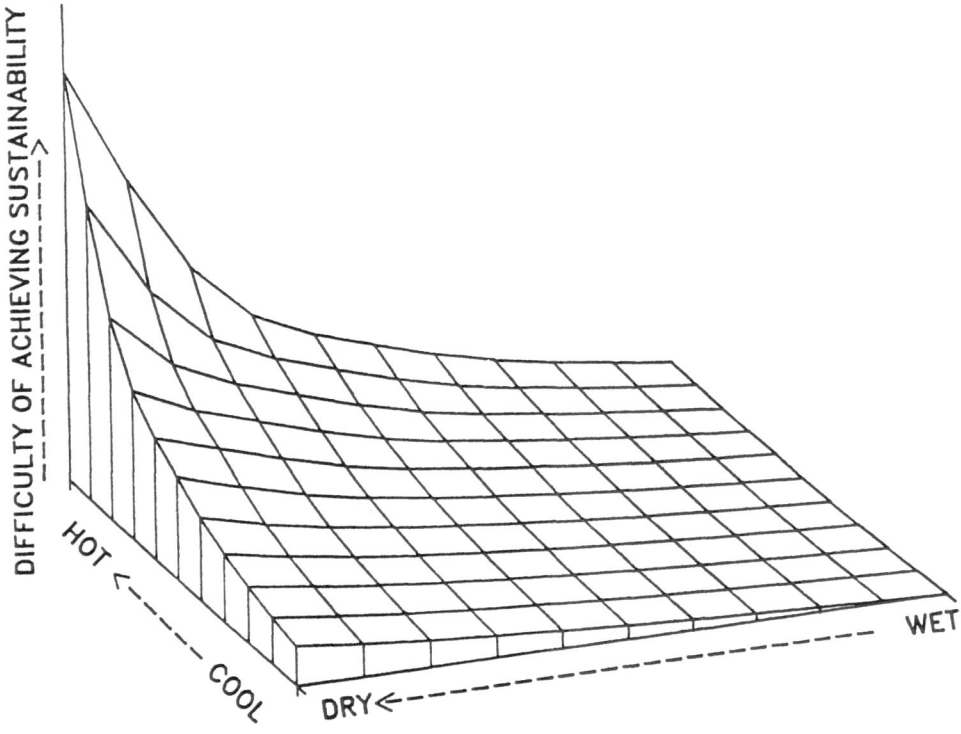

Figure 2. Generalized representation of the effects of temperature and precipitation on the difficulty of developing sustainable agricultural systems. (From Stewart *et al*, 1991.)

2.2   SOIL EFFECT

U.S. Soil Taxonomy (Soil Survey Staff, 1975) classified all soils into 10 orders. The orders are differentiated by the presence or absence of diagnostic horizons or features that mark differences in the degree and kind of the dominant sets of soil-forming processes that have occurred. Stewart *et al* (1990) showed where six of the soil orders are dominant, the worldwide area occupied by these orders, and their major constraints (Figure 3). The Entisols, Histosols, Inceptisols, and Spodosols which are not shown, occupy 11.1, 1.2, 11.7, and 5.6 million km^2, respectively. Referring again to Figure 2, the maintenance and enhancement of soil organic matter in hot and dry regions become extremely difficult not only because of climatic difficulties but also because of soil constraints. For example, the Sahel region of West Africa is dominated by Aridisols, which have low organic matter and major soil constraints. This is in contrast to the Mollisols that dominate the Cornbelt region of the United States and many areas in Europe. Mollisols are located mainly in cool humid regions and contain high amounts of organic matter and have few severe constraints. The data in Figure 3 provide

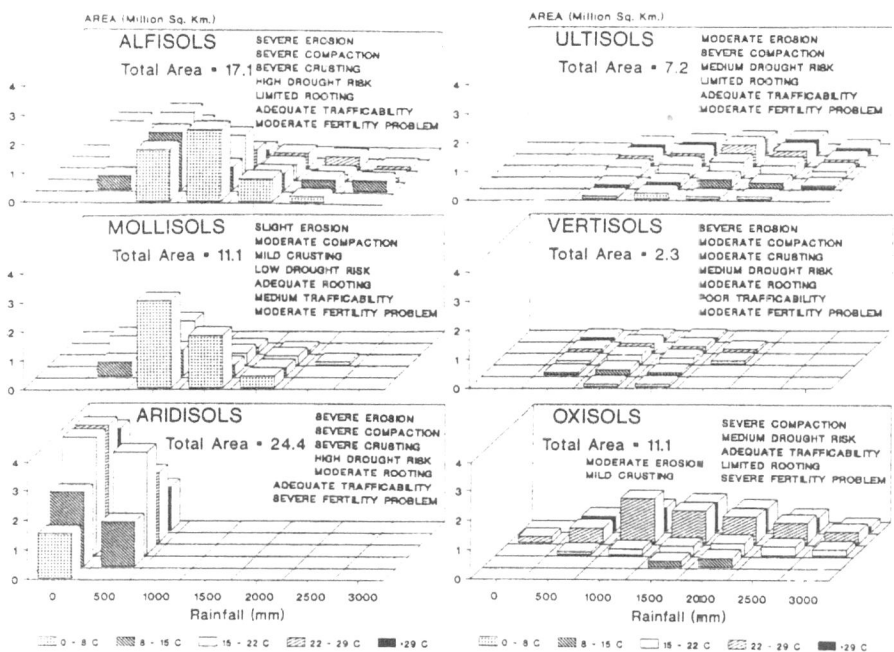

Figure 3. Approximate climatic distribution, area, and chief constraints of selected soil orders. (From Stewart *et al*, 1990.)

important information with regard to development priorities and the potential for enhanced management practices for the sequestration of carbon.

3. Management Practices to Increase Soil Organic Matter Content

Allison (1973) stated that soil organic matter has, since the dawn of history, been the key to soil fertility and productivity. He further stated that historically the best farmers cultivated their soils frequently even when there were few weeds present. The benefits were ascribed in large part to the dust mulch produced, but Allison said we now know that, aside from weed control, the main reasons that cultivation increased yields was because it released nutrients from the soil, and mostly from the organic portion. Cultivation year after year, however, can markedly lower the soil organic matter content, and as this occurs the soil physical properties can markedly deteriorate. The soil then gradually becomes a continually poorer medium for plant growth. This has usually been the picture in the past, but in recent years farming systems have been developed that tend to maintain, and in some situations enhance, the soil organic matter level of soils. These systems are often based on the use of abundant commercial fertilizer together with the return of the crop residues to the soil. More recently, cultivation has been reduced thereby retarding the destruction of soil organic matter.

Manures and legume cover crops have historically been used to increase the soil organic matter content of soil. Farmers often state that they can see benefits of such practices for 20 or more yr. These long lasting benefits are also borne out by research. Jenkinson (1991) reported that the Hoosfield plot at Rothamsted received farmyard manure for 20 yr between 1852 and 1871 and nothing thereafter. More than 100 yr after the last addition, this particular plot still contains more organic matter than the unfertilized control.

Westerman (1992) summarized the data from the Magruder plots in Oklahoma that have been continuously cropped to wheat since 1892. Figure 4 was prepared from data Westerman presented to illustrate how the organic matter content had decreased over the years in spite of the fact that there had been very significant increases in yield of wheat grain. Even where manure has been added, organic matter has continued to decrease although it has been fairly constant for the last 10 yr indicating that a new equilibrium was reached. There were also plots that received N, P, and K fertilizers and the yields from those plots were similar to those shown for the manured plots. Although it is somewhat surprising that manure did not increase the organic matter, it is important to state that only enough manure was added to supply adequate N for wheat production. Beef manure was applied at a rate sufficient to add 134 kg ha^{-1} N and 268 kg ha^{-1} N every fourth year for periods 1930-1967 and 1967–1991, respectively. The important point is that the organic matter content continued to decrease even though increasing amounts of crop residue were being returned to the soil. Intensive tillage, however, was used which strongly indicates that if organic matter levels are going to be maintained or enhanced, cropping systems that involve less intensive tillage practices will be necessary.

The effect of tillage on organic matter decline was further illustrated by Lamb *et al* (1985). They cultivated a native grassland site in western Nebraska for winter wheat production in a crop-fallow rotation under three tillage systems; no-till, stubble mulch, and plow. After 12 yr of cultivation, losses of soil N from the 0 to 30 cm depth were 3% for the no-till, 8% for the stubble mulch, and 19% for the plow tillages. Although not reported, it is assumed that comparable losses of organic C occurred.

Crop residues can be managed in manners that will lead to increased organic matter levels, thereby sequestering C. No-till systems often show increased soil organic matter contents within the first few years of practice. Kemper (1993) stated that one of the gratifying consequences of no-till management is associated increases in organic matter of the soils which has ranged from 100 to over 1,000 kg ha^{-1} yr^{-1}. The higher rates are usually associated with leguminous winter cover crops whose residues were left on the soil. This buildup of organic matter is restoring the "nutrient bank" that can help tide plants over periods of deficiency. Also, the sequestration of N into this accumulating organic matter is probably one of the factors causing the nitrate concentration of water percolating below no-till fields to be less than the nitrate concentration below conventionally tilled soils. Even in semiarid regions where only limited amounts of crop

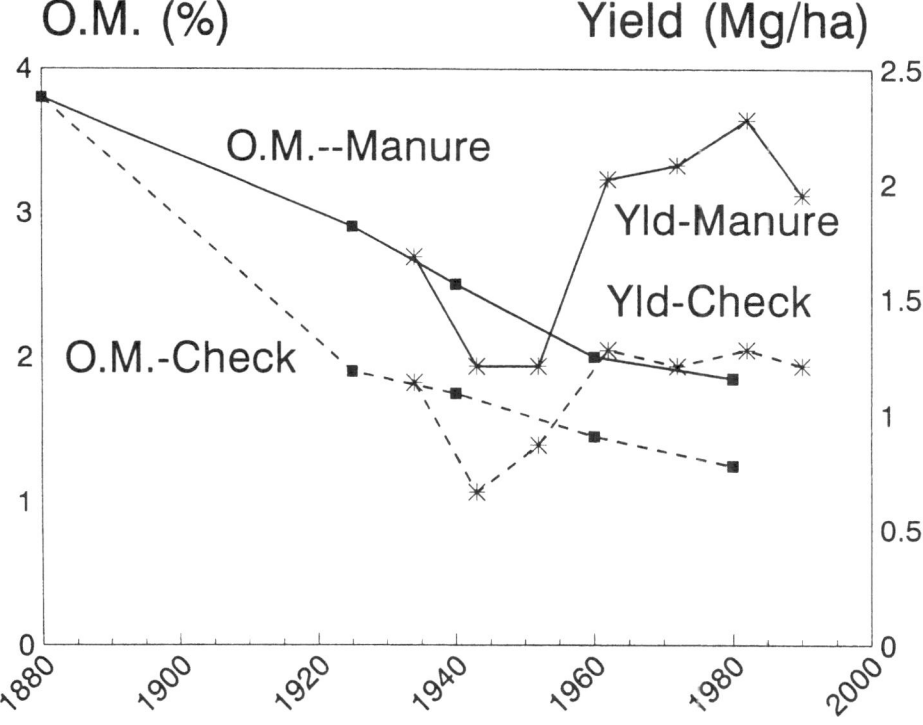

Figure 4. Changes in organic matter and wheat grain yields in a continuous wheat cropping system on manured and control plots in Oklahoma. (Drawn from data of Westerman, 1992.)

residues are produced, significant increases in organic matter occur when no-till cropping systems are adopted. Unger (1991) evaluated the distribution with depth of organic matter in wheat-grain sorghum-fallow plots. The work was done in the Texas High Plains where the annual precipitation averages 465 mm. The no-till fields had higher organic matter levels than the stubble mulch field, although the differences were relatively small and confined mostly to the top 1- and 2-cm soil depths.

4. Conclusions
Cropland offers a huge potential for sequestering C. However, changes must occur in how the crop residues are managed. Historically, soils were regenerated by the use of manures or legume cover crops. These practices added both C and N to the soil so the needs of both the plants and soil were addressed. Following World War II, commercial fertilizers became abundant and were in most cases more economical than hauling

manure or growing cover crops. Also, farm machinery became bigger and tillage was intensified in many cases. In recent years, there has been a concerted effort by scientists and policy makers to convince farmers that residues should be managed in such a way that much of the residue remains on the soil surface. This greatly reduces the potential for wind and water erosion, improves infiltration, and in water deficient areas leads to increased amounts of soil water storage. The evidence is clear that such practices lead to increased soil organic matter levels. In many cases, the increased soil organic matter will lead to increased crop production. Therefore, more CO_2 will be used by the crops and more of the carbon used by the plants will be stored in the soil as organic matter. The amounts, however, will vary greatly from region to region. The potential for sequestering carbon decreases as annual precipitation amounts decrease, and generally as mean temperatures increase.

References

Allison, F.E.: 1973, *Soil Organic Matter and its Role in Crop Production*, Elsevier Scientific Publishing Company, Amsterdam, The Netherlands, 637 pp.

Hornick, S.B. and Parr, J.F.: 1987, Restoring the productivity of marginal soils with organic amendments. *American Journal Alternative Agriculture* 2, 64-68.

Jenkinson, D.S.: 1991, The Rothamsted long-term experiments: Are they still of use?, *Soil Sci. Soc. Am. J.,* **83**, 2-10.

Kemper, W.D.: 1993, Private communication.

Lamb, J.A., Peterson, G.A., and Fenster, C.R.: 1985, Wheat fallow tillage systems' effect on a newly cultivated grassland soils' nitrogen budget, *Soil Sci. Soc. Am. J.* **49**, 352-356.

Soil Survey Staff: 1975, Soil Taxonomy, *A Basic System of Classification for Making and Interpreting Soil Surveys,* Handbook 436. U.S. Department of Agriculture, Soil Conservation Service, Washington, DC.

Stewart, B.A., Lal, R., El-Swaify, S.A., and Eswaran, H.: 1990, Sustaining the soil resource base of an expanding world agriculture, *Transactions 14th International Congress Soil Science*, Kyoto, Japan, Aug. 12-18, 1990. Vol. VII:296-301.

Stewart, B.A., Lal, R., and El-Swaify, S.A.: 1991, Sustaining the resource base of an expanding world agriculture, in: Lal, R. and Pierce, F.J. (eds), *Soil Management for Sustainability*, Soil and Water Conservation Society, Ankeny, IA.

Unger, P.W.: 1991, Organic matter, nutrient, and pH distribution in no- and conventional-tillage semiarid soils, *Agron. J.* **83**, 186-189.

Westerman, Robert L.: 1992, *Efficient Use of Fertilizers*, Agronomy 92-1, Oklahoma State University, Stillwater, OK.

CO_2-EMISSIONS FROM AGRICULTURE:

SOURCES AND MITIGATION POTENTIALS

D. R. SAUERBECK

Institute of Plant Nutrition and Soil Science
German Federal Research Centre of Agriculture
Bundesallee 50, D-3300 Braunschweig, FRG

Abstract. In 1991/92, the subgroup "Agriculture and Forestry" of the IPCC Working Group III developed an update of its earlier statements in the first IPCC-Report, concerning the release of greenhouse gases from agriculture. The present paper is an extract from this updated IPCC Supplement, concentrating on the CO_2 emissions from agriculture and its potentials for mitigation.

1. Introduction and State of the Art

In 1991/92, the Subgroup "Agriculture and Forestry" (AFOS) of the Intergovern-mental Panel on Climate Change - Working Group III on "Response Strategies" up-dated its earlier statements in the first IPCC-Report (IPCC, 1990), concerning the release of greenhouse gases (GHG) from agriculture (IPCC, 1992a, b). The goal was an improved assessment of present and future agricultural greenhouse gas emissi-ons, together with considerations about the potentials of agriculture for mitigation.

In this special volume on "Quantification of Sinks and Sources of CO_2", an extract from the corresponding statements in the updated WG-III report about CO_2 may help set the scene. The primary task given to the author by IPCC was to contribute some specific considerations on "temperate agricultural cropping systems" to the supporting materials, but he was also committed with drafting the summary about agriculture in the IPCC Supplement (IPCC, 1992b).

According to earlier statements in the first IPCC Report, temperate agriculture does not produce any drastic net fluxes of CO_2, because its fixation by photosynthesis and release by biomass breakdown balance each other fairly well. Soil C in long-standing agricultural systems tends to attain an equilibrium level which does not change very much unless cropping and soil management practices are altered.

Water, Air, and Soil Pollution **70**: 381–388, 1993.
© 1993 *Kluwer Academic Publishers.*

Careful organic residue management and reduced soil tillage are ways to increase the C contents of arable soils to a certain extent, which can be beneficial in its own right. However, this increase is limited, both in time and extent, and therefore does not counteract significantly the world-wide anthropogenic CO_2 release.

Fuel and energy requirements of the agricultural sector in the developed countries in the northern hemisphere constitute about 3 % of their overall fossil fuel consumption. Reducing the inputs of mineral fertilizers and the specific soil tillage intensity may help to diminish this energy demand to some extent.

2. Supplementary Findings and Considerations

2.1 IMPROVED ESTIMATES ON GHG EMISSIONS

Several attempts have been made since 1990 to improve what is known about the quantities of GHG produced as a consequence of agriculture (BOUWMAN, 1990a, b; BURKE and LASHOF, 1990; CAST, 1992; ISERMANN, 1993a, b: SAUERBECK and BRUNNERT, 1990; SAUERBECK 1993b; USEPA, 1990). Since most of the statistical data (e.g. arable land, crops, fertilizer consumption) are based on countries and not on agroclimatic zones, a precise separation of relevant calculations and conclusions into temperate and tropical/subtropical agricultural systems has not yet been achieved. Based on the data of BOUWMAN *et al.* (1991), and using their subdivision into geopolitical regions, however, a tentative comparison about GHG emission from mainly temperate versus mainly tropical/subtropical areas is now possible.

As far as the net CO_2 release from agricultural activities is concerned, it may now be safely stated that land use-related net deforestation and biomass burning is negligible in most of the temperate zones. Ongoing and planned reforestation in temperate areas may even sequester some 0.06 Gt C yr^{-1} (IPCC, 1992a), whereas the deforestation and land clearing elsewhere in the tropical and subtropical zones of the world still continues (BURKE and LASHOF, 1990; HOUGHTON 1990, 1991; WISNIEWSKI and LUGO, 1992), resulting in an estimated net loss of about 1.6 to 2.2 Gt C yr^{-1}.

Old mineral field soils have usually attained a site- and cultivation-specific equilibrium level of organic C which no longer contributes significantly to the net release of CO_2 (CAST, 1992; HAIDER, 1993; SAUERBECK and BRUNNERT, 1990). However, apart from the ongoing forest conversion, this is not the case with large

TABLE 1: Projections of human population, arable land requirement, and N-fertilizer consumption for 1990-2025, specified for countries with predominantly temperate and others with mainly tropical and subtropical climate

Region	Hum.popul. 1990 2025 $x\ 10^6$		Arab. land 1990 2025 10^6 ha		N-fertil.cons. 1990 2025 kton N yr^{-1}	
1. Mainly temperate areas						
1.1 Europ. Commun.	341	341	83	72	10448	8876
1.2 Rest-OECD-Eur.	90	123	36	42	1837	1560
1.3 East Europe	124	138	49	47	4654	6446
1.4 USSR	288	351	232	242	11587	16048
1.5 OECD-East	37	49	53	62	1067	1108
1.6 North America	276	333	236	255	10834	11251
Sub-total 1 *)	**1156**	**1335**	**689**	**720**	**40427**	**45289**
% of total 1+2	*22*	*16*	*47*	*36*	*51*	*38*
% change till 2025		*+ 15*		*+ 6*		*+ 12*
2. Mainly tropical and subtropical areas						
2.1 Latin America	448	760	180	285	3809	7942
2.2 Africa	648	1581	187	347	2113	5810
2.3 Middle East	146	356	38	78	1268	2821
2.4 Ctr. Pla. Asia	1236	1670	110	179	19549	28923
2.5 So./Soea. Asia	1606	2675	272	397	12398	27585
Sub-total 2 **)	**4084**	**7042**	**787**	**1286**	**39137**	**73081**
% of total 1+2	*78*	*84*	*53*	*64*	*49*	*62*
% change till 2025		*+ 72*		*+ 63*		*+ 87*
Total 1+2	**5240**	**8407**	**1476**	**2006**	**79564**	**118370**
% change till 2025		*+ 63*		*+ 36*		*+ 49*

2025 (ref.) = Projection for 2025 reference scenario
2025 (pol.) = Projection for 2025 policy scenario
 *) = Quantities are somewhat too low because Argentina,
      Chile and South Africa are not included
**) = Quantities are somewhat too high because Argentina
      Chile and South Africa are included
(Data derived from BOUWMAN et al., 1991)

areas of drained and ploughed peat soils, which may still lose several tons of organic C ha^{-1} yr^{-1}. The same is true for the significant number of grassland soils which are still being converted into arable fields. This is a matter of considerable concern not only in North America, Eastern Asia and the Southern Hemisphere, but even in the otherwise relatively stable agricultural systems of the temperate zone.

2.2 FUTURE DEVELOPMENTS

Estimates of the CO_2 emissions related to land use changes are still rather uncertain. Projections of increases in arable land until 2025 assume little growth (< 5 %) in the temperate areas, contrary to a calculated demand of, at least theoretically, more than 50% in the tropical and subtropical zones (BOUWMAN *et al.*, 1991; TABLE 1). Assuming a maximum drop in the C content of virgin soils from 5 to about 2 % as a consequence of their long-term cultivation, the predicted 30×10^6 ha of additional farmland in the temperate zones might suffer a potential loss of 2.7 Gt C over about 60 to 100 ys.

This certainly would be of great importance for the fertility of these soils. However, in terms of the resulting additional greenhouse effect, it would only correspond to less than 50% of the about 6 Gt C which are presently released world-wide during just 1 yr only from fossil fuel.

There may also be another effect in that global warming will speed up the decomposition of soil organic matter, thereby releasing additional CO_2 to the atmosphere. Based on a temperature rise of 0.03 $^{\circ}$C yr^{-1} as predicted by IPCC, JENKINSON *et al.* (1991) calculated that this CO_2 evolution over the next 60 ys, assuming no increase in organic residue returns, could correspond to some 60 Gt of C world-wide. This would be about one sixth of the CO_2 that might be released from the combustion of fossil fuel during the same period of time.

Since soils in the temperate zones usually contain considerably more C than soils in the tropics, and since warming is predicted to be greater in the middle and higher latitudes, it may well be that temperate agriculture would suffer more from this humus breakdown effect than other agricultural systems. On the other hand, however, there are indications that this could be offset by an enhanced plant residue formation both above and below ground, due to the CO_2 fertilization effect (e. g. BAZZAZ, 1990; GOUDRIAAN *et al.* 1990; KIMBALL *et al.* 1990).

3. Mitigation Potentials

In order to reduce the net CO_2 release from agriculture, the conversion of virgin land into arable fields should be avoided wherever possible, particularly by improving the productivity of the already existing farm land (USEPA, 1990). It may be that in the most developed countries, unless the farm inputs are excessively increased, this potential is gradually becoming difficult to exploit any further (ISERMANN, 1993b; SAUERBECK, 1993b).

However, there are still good possibilities in both the tropical and the temperate zones, to achieve a considerable productivity increase and consequent reductions of forest and rangeland clearing, or of draining peat bogs for agricultural purposes. Improving soil productivity may even allow setting aside considerable areas of marginal arable land for either grassland or forest use (CAST, 1992).

Furthermore, altered soil tillage procedures will help to reduce humus decomposition and net C losses from the more recently cultivated soils, or even help to slightly increase the equilibrium C levels of long-standing arable soils. At the same time, the fuel consumption for soil management operations would decrease accordingly, even though the resulting savings might be relatively small.

The extent to which this reduced C release or increased C fixation by soils would be possible in practice is hard to predict. However, if one assumes a C increase by this way of 0.3 % for the already existing arable soils in the temperate zone (690×10^6 ha), this would correspond to a C sequestration in the order of about 6.2 Gt during a period of, say, 50 ys.

Here again, however, the effect would be limited in both time and extent, and should therefore not be overestimated in its long-term potential of alleviating the future world-wide GHG situation. This is because soils have a finite capacity for accumulating organic matter. Nevertheless, these improved management practices are essential if soil fertility is to be maintained or enhanced, thus constituting an important part of future "no regret" policies.

4. Conclusions

As far as the net CO_2 release from agricultural activities is concerned, forest clearing in the tropical and subtropical areas of the world continues to be the main source. Additionally, the cultivation of virgin land, whether in tropical or temperate agricultural areas, results in gradual C losses. Its actual contribution, however, to the CO_2 enrichment in the atmosphere is still not quite clear.

At the other extreme, long-standing arable soils do not constitute a major net source of CO_2, because their C contents reach equilibria which do not change very much unless cropping and management practices are altered. However, large areas of recently drained and ploughed peat and/or grassland soils may still lose several tons of organic C ha^{-1} yr^{-1}.

It may also be that global warming will speed up the decomposition of soil organic matter. Model calculations predict this C loss to be up to 60 Gt within about 60 ys. This corresponds to a current global CO_2 release from fossil fuel of about 10 ys. On the other hand, there are indications that this could be offset by an enhanced plant residue return due to the CO_2 fertilization effect.

In order to reduce the net CO_2 release from agriculture, improving the productivity of existing arable land should be given priority over the cultivation of virgin soils. Furthermore, in the less populated areas of the world, marginal farm land could be set aside for either range or forest use.

Reduced soil tillage, improved recycling of organic wastes and crop rotations including forages are agricultural practices which reduce soil C losses and/or incorporate additional C in the soils. However, although all this may sequester several Gt of organic soil C in the course of some decades, this gain represents only a small fraction of the CO_2 release from fossil fuel.

The fact to be kept in mind is that most agricultural crops are not durable enough to represent a significant C sink. The same is true in principle for the potential soil C losses or gains, because both lead to new equilibria in just a few decades and are therefore rather limited in their extent as well as in time.

5.   References and Related Literature Since 1990

BAZZAZ, F.A.: 1990. The response of natural ecosystems to the rising global CO_2 levels. - *Ann. Rev. Ecol. Syst. 21, 167-196.*

BOUWMAN, A.F.: 1990a, Land use related sources and sinks of greenhouse gases. Present emissions and possible future trends. - *Land Use Policy*, April 1990: 154-164.

BOUWMAN, A. F. (ed): 1990b, *Soils and the Greenhouse Effect.* - Wiley & Sons, Chichester.

BOUWMAN, A.F., VAN DEN BORN, G.J. and SWART, R.J.: 1991, Land use-related sources of CH_4 and N_2O - Current global emissions and projections for the period 1990 - 2025 and onwards. - Report to the Enquete Commission of the German Bundestag *"Measures to Protect the Earth's Atmosphere"*, Bonn, November 1991.

BURKE, L.M. and LASHOF, D.A.: 1990, Greenhouse gas emissions related to agriculture and land use practices. - in: REICHEL, G., STUBER, C.W. and KISSEL, G.E. (eds): *Impact of carbon dioxide, trace gases and climate change on global agriculture.* - ASA Publication, Anaheim, CA., 27-43.

CAST - Council for Agricultural Science and Technology: 1992, Preparing U. S. Agriculture for global climate change. - Task Force Report No. 119, 96 pp., CAST, Ames, IA, USA

FAO - Food and Agriculture Orgasnisation: 1989, Fertilizer Yearbook, Vol. **38**. - FAO Statistics Series No. 71. Rome, FAO.

GERMAN BUNDESTAG (ed): 1991, *Protecting the Earth: a Status Report with Recommendations for a New Energy Policy*, Vol I+II. - German Bundestag, Publ.. Sect., Bonn, FRG.

GOUDRIAAN, J., van KELLEN, H. and van LAAR (eds): 1990, *The greenhouse effect and primary productivity in European agro-ecosystems.* - PUDOC, Wageningen, NL.

HAIDER, K.: 1993, Auswirkungen zunehmender Temperaturen auf die organische Bodensubstanz mittlerer Breiten. - Mitt. Dtsch. Bodenkundl. Ges. (in press)

HOUGHTON, R.A.: 1990, The future role of tropical forests in affecting the carbon dioxide concentration of the atmosphere. - Ambio 19, 204-209.

HOUGHTON, R.A.: 1991, Tropical deforestation and atmospheric carbon dioxide. - Climatic Change 19, 99-118.

IPCC - Intergovernmental Panel on Climate Change: 1990, Climate Change: The IPCC Scientific Assessment. Houghton, J.T., Jenkins, G.J. u. Ephraums, J.J. (eds.). I: Scientific Assessment of Climate Change; II: Potential Impacts of Climate Change; III: Formulation of Response Strategies. WMO/UNEP, Geneva, June 1990. - Cambridge University Press.

IPCC - Intergovernmental Panel on Climate Change: 1992a, Climate Change 1992: The Supplementary Report to the IPCC Scientific Assessment. Houghton, J. T., Callander, B. A. u. Varney, S. K. (eds) - Cambridge University Press.

IPCC - Intergovernmental Panel on Climate Change: 1992b, Climate Change: The 1990 and 1992 IPCC Assessments. (1) IPCC First Assessment Report - Overview and Policymaker Summaries; (2) 1992 IPCC Supplement. WMO/UNEP, Geneva, June 1992.

ISERMANN, K.: 1993a, Territorial, continental and global aspects of C, N, P and S emissions from agricultural ecosystems. - in: Wollast, R., McKenzie, S. T. and Lei Chu, (eds): *Interactions of C, N, P and S Biochemical Cycles*, 79-121. NATO ASI Series, Vol. 14. Springer, Berlin, Heidelberg.

ISERMANN, K.: 1993b, Agriculture's share in the emission of trace gases affecting the climate and some cause-oriented proposals for sufficiently reducing this share. - Environmental Pollution, Elsevier Science (in press).

JENKINSON, D.S., ADAMS, D.E: and WILD, A.: 1991, Model estimates of CO_2 emissions from soil in response to global warming. - Nature **351**, 304-306.

KIMBALL, B.A., ROSENBERG, N.J. and ALLEN, L.H. (eds): 1990, Impact of carbon dioxide, trace gases and climate change on global agriculture. - ASA Spec. Publ. No. 53, American Society of Agronomy, Madison, USA:

PARRY, M.: 1990, Climate Change and World Agriculture. - Earthscan Publ. Ltd. London, GB.

PHILLIPS, D.E., WILD, A. and JENKINSON, D.S.: 1990, The soil's contribution to global warming. - Geographical Magazine, April 1990, 36-38.

SAUERBECK, D., and BRUNNERT, H. (eds): 1990, Klimaveränderungen und Landbewirtschaftung. - Landbauforschung Voelkenrode, Spec. Vol. 117, 1-75. With contributions from Ahlgrimm, H.-J., Brunnert, H., Gaedeken, D., Haider, K., Heinemeyer, O., Rath, D. and Schoedder, F. German Federal Research Centre of Agriculture, Braunschweig, FRG.

SAUERBECK, D., and HAIDER, K.: 1991, Mögliche Einflüsse von Klimaveränderungen auf die Bodenfruchtbarkeit. - Mitt. Dtsch. Bodenkundl. Ges. 66, 713-718.

SAUERBECK, D. (1992): Landbewirtschaftung und Treibhauseffekt. - in: Verbindungsstelle Landwirtschaft-Industrie (ed): *Produktionsfaktor Umwelt - Klima/Luft*, 73-96. etv Landwirtschaftsverlag, Düsseldorf, FRG.

SAUERBECK, D.: 1993a, Wechselseitige Beeinflussungen von Klima und Böden: Fragen - Bereiche - Prozesse. - Mitt. Dtsch. Bodenkundl. Ges. (in press)

SAUERBECK, D.: 1993b, Herkunft und Strategien zur Verminderung der Emission von klimarelevanten Spurengasen: Bereich Land- und Forstwirtschaft. - in: IWW-Kongreßband *Klimaentwicklung und die Zukunft der Wasserwirtschaft in Europa*, Rhein.-Westfäl. Inst. f. Wasserchemie u. Wassertechnol. Mühlheim/R. (in press)

SOMBROEK, W.G.: 1990, At global change - do soils matter? - Internat. Soil Sci. Soc. Publ., ISRIC, Wageningen, NL.

USEPA: 1990, Greenhouse Gas Emissions from Agricultural Systems, Vol. I+II. Report prepared by IPCC, WG III on Response Strategies, Subgroup on Agriculture, Forestry and Other Human Activities. - US Environmental Protection Agency, Office of Policy Analysis, Washington D. C., USA.

WISNIEWSI, J., LUGO, A.E. (eds): 1992, *Natural Sinks of CO_2*. - Reprinted from Water, Air and Soil Pollution 64, No. 1-2. Kluwer, Dordrecht, Boston, London.

The Effect of Trends in Tillage Practices on Erosion and Carbon Content of Soils in the US Corn Belt

J. J. Lee and D. L. Phillips

US Environmental Protection Agency, Environmental Research Laboratory, Corvallis, OR, 97333, USA

R. Liu

Department of Civil Engineering, Oregon State University, Corvallis, OR, 97331, USA

Abstract. The EPIC model was used to simulate soil erosion and soil C content at 100 randomly selected sites in the US corn belt. Four management scenarios were run for 100 years: (1) current mix of tillage practices maintained; (2) current trend of conversion to mulch-till and no-till maintained; (3) trend to increased no-till; (4) trend to increased no-till with addition of winter wheat cover crop. As expected, the three alternative scenarios resulted in substantial decreases in soil erosion compared to the current mix of tillage practices. C content of the top 15 cm of soil increased for the alternative scenarios, while remaining approximately constant for the current tillage mix. However, total soil C to a depth of 1 m from the original surface decreased for all scenarios except for the no-till plus winter wheat cover crop scenario. Extrapolated to the entire US corn belt, the model results suggest that, under the current mix of tillage practices, soils used for corn and/or soybean production will lose 3.2 x 10[6] tons of C per year for the next 100 years. About 21% of this loss will be C transported off-site by soil erosion; an unknown fraction of this C will be released to the atmosphere. For the base trend and increased no-till trend, these soils are projected to lose 2.2 x 10[6] t-C yr-1 and 1.0 x 10[6] t-C yr-1, respectively. Under the increased no-till plus cover crop scenario, these soils become a small sink of 0.1 x 10[6] t-C yr-1. Thus, a shift from current tillage practices to widespread use of no-till plus winter cover could conserve and sequester a total of 3.3 x 10[6] t-C yr-1 in the soil for the next 100 years.

1. Introduction

Soil is an important reservoir for C, representing perhaps twice the amount of C in the atmosphere and close to three times the amount in vegetation (Post *et al,* 1990). When soils are converted to agricultural production using conventional tillage practices which stir and mix the soil, they typically lose a significant portion of organic C, especially in the first few decades of cultivation (Schlesinger, 1985). Thus, conventionally cultivated soils have acted as net sources for atmospheric CO_2. Increasing atmospheric CO_2 is of concern because of its potential for causing global greenhouse warming (IPCC, 1992).

Water, Air, and Soil Pollution **70**: 389–401, 1993.
© 1993 *Kluwer Academic Publishers.*

Conservation tillage practices, which maintain at least 30% residue cover on the soil surface (CTIC, 1990), are being increasingly used in the United States (US). Some form of conservation tillage is projected to be adopted on 62 to 82% of agricultural land in the US by the year 2010 because it is an effective and economical method for reducing soil erosion (Schertz, 1988). When tillage is reduced or eliminated, soil C may stabilize or even increase (Kern and Johnson, 1993).

Much of the increase in conservation tillage within the US is occurring within the "corn belt", a highly productive agricultural region in the north-central US (Figure 1). Because such a large area of productive cropland is involved, the shift in tillage practices could change net C balances in this area. The purpose of this study was to project the changes in soil erosion and C content for the US corn belt under four tillage scenarios. This was done by selecting a statistically representative sample of 100 sites in the corn belt, and running 100 year simulations at each site using the Erosion Productivity Impact Calculator (EPIC) model (Williams and Renard, 1985; Sharpley and Williams, 1990a,b).

2. Methods

For the purposes of this study, the corn belt was defined as Land Resource Region M - Central Feed Grains and Livestock Region, which includes Major Land Resource Areas (MLRAs) 102 through 115 (Figure 1; SCS, 1981). Sites were selected from National Resource Inventory (NRI; SCS, 1989) sites within these MLRAs. The 1987 NRI is a statistically-based inventory of the status of non-Federal land in the US. Data on land use and condition were obtained at more than 300,000 sample points. The area represented by each sample point (the "expansion factor") is derived from the probability of selection for each point.

Only NRI sites on which corn and/or soybean were grown for the years 1984-1987 were considered. Of the 11,301 potential corn/soybean sites, 70 were rejected as having inadequate soil information. A randomized procedure was used to select 100 sites from the 11,231 eligible sites. The expansion factors appropriately weight data from the sample points to obtain regional averages.

The EPIC model was used to simulate crop growth, soil erosion and soil C content at each site. Four management scenarios were run for 100 years: (1) current mix of tillage practices maintained; (2) base trend to mulch-till and no-till maintained; (3) trend to no-till increased; (4) trend to increased no-till with addition of winter wheat cover crop. In the first scenario ("current mix"), each site used only the management schedule assigned to it for the first year. The second scenario ("base trend") was based on the trend toward conservation tillage projected by Schertz (1988). A trend approximately halfway between Schertz' upper and lower projections was represented by linear increases in the percent of sites using conservation tillage, from 34% in the first year (based on the NRI data) to 65% in the fourteenth year and 75% in the thirty fourth year of simulation, remaining constant thereafter. Sites were randomly chosen for conversion to conservation tillage from among those identified as using conventional tillage. New conservation tillage sites were assigned to mulch-till or no-till so as to keep the ratio of no-till to mulch-till sites approximately equal to its value in the first year (*i.e.* 1:16).

The third scenario ("no-till trend") was identical to the second, except all new conservation tillage sites were assigned no-till schedules, and all mulch-till sites were linearly converted to no-till during the first fourteen years of simulation. The fourth ("no-till plus cover") scenario was identical to the third, except no-till plus winter cover schedules were used instead of no-till schedules.

The schedules of specific management operations (*e.g.* disking, planting, cultivating, fertilizing, harvesting) for continuous corn and corn-soybean rotation developed by Phillips *et al* (1993) were used for conventional tillage and no-till (Table 1). A "mulch-till" schedule was defined by eliminating all but one of the disking operations of the conventional tillage schedule. A "no-till plus cover" schedule was defined by modifying the "no-till" schedule to include planting winter wheat after harvest; the wheat was shredded and killed before planting the spring crop. The three climatic regions defined by Phillips *et al* (*i.e.* north, central, and south; 1993) for Illinois were extended approximately eastward and westward to cover the entire corn belt (Figure 1). Different planting and harvest dates, and different cultivar degree-day requirements, were used to approximate typical practices in each climatic zone. Thus, a total of 24 management schedules were defined as input to EPIC: 2 rotations x 3 regions x 4 tillage practices.

Each site was assigned an initial management schedule based on the practices for that site as shown in the NRI. Sites identified as using conservation tillage were assigned mulch-till schedules unless they were identified in the NRI as using no-till; all others were assigned conventional tillage schedules. Similarly, sites were assigned to continuous corn or corn-soybean rotation according to the crop history given in the NRI record. Following Phillips *et al* (1993), the fertilizer application rates (N, P) at each site were based on the expected corn and soybean yields given for each site's soil in the Soils-5 database (see below). (For corn, 22.3 kg-N ha[-1] and 0.3 kg-P ha[-1] per Mg ha[-1] expected yield, less 44.8 kg-N ha[-1] if follows soybean; for soybean, 0 kg-N ha[-1] and 0.4 kg-P ha[-1] per Mg ha[-1] expected yield.)

EPIC was developed by the US Department of Agriculture to analyze the relationship between soil erosion and agricultural productivity. It uses a daily time-step to simulate crop growth, soil erosion, nutrient cycling (including soil C), and hydrology. A generic model, parameterized for specific crops, is used to simulate crop growth. Soil erosion on an event basis is simulated by the Universal Soil Loss Equation (USLE; Wischmeier and Smith, 1978), Modified Universal Soil Loss Equation (MUSLE; Williams, 1975), or by the Onstad-Foster modification of USLE (Onstad and Foster, 1975). Detailed descriptions of EPIC modules, validation, and input requirements are contained in the model documentation and users' manual (Sharpley and Williams, 1990a,b).

Table 1. Operations for four management schedules (adapted from Phillips *et al*, 1993).

Operation	Conventional	Mulch Till	No-till	No-till + Cover
Continuous Corn				
Tandem Disk	Y			
N Fertilizer	Y	Y	Y	Y
Tandem Disk	Y	Y		
Field Cultivator	Y	Y		
Row Planter (Corn)	Y	Y	Y	Y
Rotary Hoe	Y			
Row Cultivator	Y	Y		
Row Cultivator	Y			
Harvest	Y	Y	Y	Y
Tandem Disk	Y			
P Fertilizer	Y	Y	Y	Y
Twisted Point Chisel	Y	Y		
Plant Winter Wheat				Y
Shred, Kill Wheat				Y
Soybean-Corn Rotation				
Tandem Disk	Y	Y		
Tandem Disk	Y			
Field Cultivator	Y	Y		
Row Planter (Soybean)	Y	Y	Y	Y
Rotary Hoe	Y	Y		
Row Cultivator	Y	Y		
Row Cultivator	Y			
Harvest	Y	Y	Y	Y
P Fertilizer	Y	Y	Y	Y
Plant Winter Wheat				Y
Shred, Kill Wheat				Y
N Fertilizer	Y	Y	Y	Y
Tandem Disk	Y			
Field Cultivator	Y	Y		
Row Planter (Corn)	Y	Y	Y	Y
Rotary Hoe	Y			
Row Cultivator	Y	Y		
Row Cultivator	Y			
Harvest	Y	Y	Y	Y
Tandem Disk	Y			
P Fertilizer	Y	Y	Y	Y
Twisted Point Chisel	Y	Y		
Plant Winter Wheat				Y
Shred, Kill Wheat				Y

EPIC requires site-specific data on climate, soils, and management practices. As part of the EPIC system, a database has been developed that includes climate data from 1041 National Weather Stations, parameterized for input to EPIC. Data from the station closest to the centroid of the county in which each selected NRI site was located was used. (NRI does not include the exact location of each site.) EPIC uses these parameters in a stochastic model to generate daily weather at each site. Daily weather values from the current mix scenario were retained for use in the alternative scenarios. Thus, for a given site, the simulated weather on any specific date was the same for all scenarios.

Soils data were obtained from a version of the USDA Soil Conservation Service Soil Interpretation Record ("Soils-5") database developed for input to EPIC. The Soils-5 identification number (corresponding to a phase of a soil series) was obtained from the NRI record for each site. Soil surface albedo was set to 0.135 for all sites, which is typical for Mollisols and Alfisols, the dominant soils in the region.

All three equations used by EPIC calculate soil erosion rate as the product of factors for rainfall or runoff energy (R), percent slope (S), slope length (L), soil erodibility (K), cover (C), and erosion control practice (P). For USLE, R is determined by rainfall variables, which provides an estimate of erosion that includes soil moved only a short distance within a field. USLE requires use of a sediment delivery ratio to estimate net erosion off-site (Robinson, 1979). For MUSLE, R is based on runoff rather than rainfall, and was designed to more closely represent net erosion (Williams, 1975). Because we wanted to account for C losses associated with soil transported off-site, MUSLE was selected to calculate soil erosion rate. S, L, and P were taken from the NRI data. K was obtained from the Soils-5 data. C and R were calculated by EPIC from the simulated crop and residue cover, and from the simulated daily weather, respectively.

EPIC simulates soil dynamics in each of ten layers, derived from the three to four horizons listed in the Soils-5 database. The first layer is defined as the top 1 cm of the first horizon. Lower layers are defined by halving or quartering each horizon, provided the split will not result in a layer narrower than 10 cm, or in more than 10 layers. Priority is given to splitting layers nearest the soil surface. Organic matter decomposition is driven by temperature, moisture, C:P or C:N ratio, and residue condition in each layer as simulated daily by EPIC. Two major soil C pools, corresponding to fresh residue and soil humus, are defined. Fresh residue is further divided into carbohydrate-like, cellulose-like, and lignin-like pools, depending on the stage of decay. The main mechanisms for introduction of fresh residue into a layer are incorporation of surface residue through tillage, and root growth and death within the layer. Soil humus is sub-divided into stable and active pools, initially based on how long the soil has been cultivated.

Soil C for each site was tracked in three ways. The first method calculated total C for the top 15 cm from the current soil surface, as the surface moved downward due to erosion. This represented the common way of measuring C in agricultural soils. For the second method, total soil C was calculated to a depth 1 m below the original soil surface. This method accounted for changes in soil C due to the combined effects of soil compaction, C dynamics, and losses from erosion. For both of these methods, soil organic C was calculated as:

$$SOC = \sum_{i=1}^{n} ((100\% - \%rock_i)/100\%) \times T_i \times BDD_i \times (\%OC_i/100\%) \times (1000 \text{ kg t}^{-1})$$

where SOC is soil organic C (kg m^{-2}), n is the number of soil layers down to the required depth (15 cm or 1 m), and, for each layer, $\%rock_i$ is the coarse fragment content (%), T_i is the thickness (m), BDD_i is the dry bulk density (t m^{-3}), and $\%OC_i$ is the percent organic C. BDD_i and $\%OC_i$ were simulated daily by EPIC.

The third method explicitly tracked total soil C lost to erosion from the beginning of the simulation. For each year, C in soil eroded from a site was calculated from the C content and bulk density of the top layer of soil, and from the thickness of soil that was removed from the site, as calculated from MUSLE. In this way, changes in SOC due to erosion could be separated from changes due to other processes.

EPIC was run for 100 years at each site for each of the four management scenarios. Average values of erosion rate per ha, soil C per ha to 15 cm, soil C per ha to 1 m, and eroded C per ha for the corn belt for each year were estimated as the mean of all selected sites, weighted by each site's expansion factor from the NRI record. Average values per ha were extrapolated to totals for the corn belt by multiplying by the total area represented by all continuous corn and corn/soybean sites in the NRI sample (25,918,000 ha).

3. Results and Discussion

All trends toward conservation tillage resulted in substantial decreases in soil erosion compared to the current mix of tillage practices (Table 2). The base trend scenario showed a relatively modest decrease in erosion rate from the current mix. The no-till and no-till plus winter cover scenarios produced similar erosion control benefits, with 100-yr average erosion rates approximately half those of the base trend scenario. Thus the greatest erosion benefits were associated with the widespread adoption of no-till, with adoption of winter cover having little additional effect.

All the alternative scenarios displayed year-to-year variability of erosion benefits compared to the current tillage mix (Figure 2), indirectly reflecting the stochastic nature of the EPIC weather generator. Simulated weather was the same for a given site on a specific day, regardless of scenario. The erosion caused by a specific storm varied among scenarios, however, depending on the vulnerability of the soil to erosion as influenced by crop and residue cover. Thus, years with no storms during a period of vulnerability for the current mix would show little benefit from conservation tillage. In

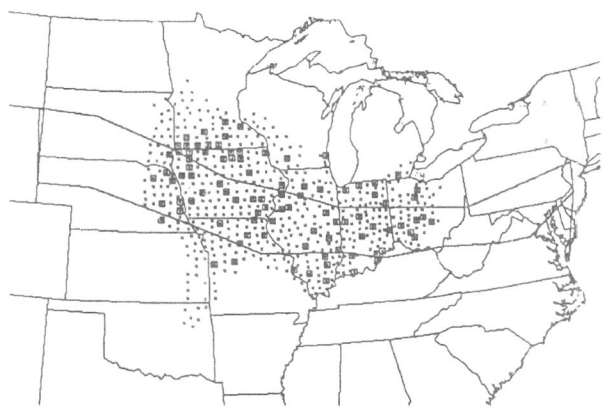

Figure 1. Location of sites. x : centroid of county in US corn belt with at least one potential NRI site; □ : centroid of county with NRI site selected for simulation. Curved lines show boundaries of north, central, and south climatic zones.

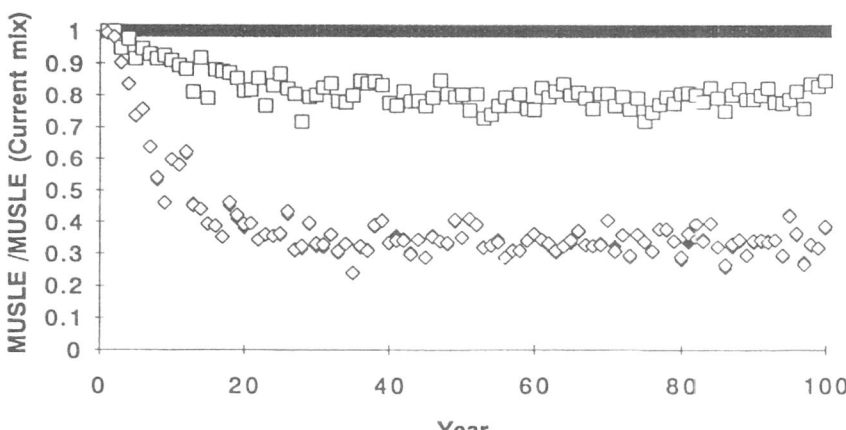

Figure 2. Ratio of annual average erosion rate for alternative scenarios, compared to current tillage mix. Most no-till symbols are obscured by no-till + cover symbols.
■ Current mix □ Base trend ◆ No-till trend ◇ No-till + cover

Table 2. Comparison of soil erosion rate (MUSLE) for the four scenarios. To convert to total annual erosion for continuous corn and corn/soybean soils in the corn belt, multiply by 25,918,000 ha.

Scenario	Average erosion rate t ha^{-1} yr^{-1}	Ratio to current mix
Current tillage mix	1.68	1.00
Base trend	1.37	0.82
Increased no-till	0.64	0.38
Increased no-till plus cover	0.64	0.38

Table 3. Changes in soil carbon content for the four 100-year management scenarios. To convert to t ha^{-1}, multiply by 10.

Scenario	Top 15 cm kg m^{-2}	To 1 m from original surface kg m^{-2}	In eroded soil kg m^{-2}
Current tillage mix	-0.04	-1.23	0.26
Base trend	+0.13	-0.86	0.21
Increased no-till	+0.37	-0.39	0.11
Increased no-till plus cover	+0.73	+0.03	0.11

contrast, major storms occurring at times of low soil cover for the current mix could have substantially reduced erosion rates under conservation tillage because of increased residue cover and, in the case of winter cover, crop cover.

Soil C content of the top 15 cm remained approximately constant for the current tillage mix, decreasing by 0.04 kg m^{-2} from an original value of 4.86 kg m^{-2}. In contrast, C content of the uppermost soil increased for the base trend, increased no-till, and no-till plus cover scenarios (Table 3, Figure 3). For the base trend, at the end of 100 years, there was about 0.2 kg m^{-2} more C in the top 15 cm of soil than for the final year of the current-mix scenario. Adopting no-till instead of mulch-till resulted in double the gain in C, and planting a cover crop doubled the gain again, to four times that from the base trend. Thus, unlike erosion, the addition of a winter cover crop caused a major change in soil C.

Total soil C to a depth 1 m below the original depth decreased for all scenarios except the no-till plus cover scenario (Table 3, Figure 4). This suggests that, after more than a century of agriculture, these soils continue to lose C below plow depth. Mulch-till and no-till reduced the rate at which C was lost. However, it was only with the increased input of C from a winter wheat cover that C accumulated throughout the soil profile.

Data presented by Smith *et al* (1990) support the importance of soil processes below 15 cm for loss of organic C. During the 60 years following cultivation of a native grassland, 40% of the organic C in the soil profile (0-90 cm) was lost. Of this, 60% was from the 15-90 cm soil depth. Smith *et al* compared the measured data with EPIC results, and concluded that "reasonable agreement of the changes in total N, organic P, and organic C following cultivation was obtained."

Extrapolated to the entire US corn belt, model results suggest that, under the current mix of tillage practices, soils used for corn and/or soybean production are losing 3.2 x 10^6 t-C yr^{-1}. For the base trend and increased no-till trend, these soils are projected to lose 2.2 x 10^6 t-C yr^{-1} and 1.0 x 10^6 t-C yr^{-1}, respectively. Under the increased no-till plus cover scenario, these soils are projected to accumulate 0.1 x 10^6 t-C yr^{-1}. Thus, a shift from current tillage practices to widespread use of no-till plus winter cover could conserve and sequester a total of 3.3 x 10^6 t-C yr^{-1} for the next 100 years in these soils.

Not all C lost from the soil will necessarily be released directly to the atmosphere. An unknown fraction of C transported from a site through erosion will accumulate in depositional positions in the landscape, or in freshwater or marine sediments, where it might remain sequestered for long periods. In this study, the total amount of C lost through soil erosion was approximately the same for the current mix and base trend scenarios. Erosion-related C losses were virtually identical for the no-till and no-till plus cover scenarios (Table 3). Soil erosion accounted for 21, 24, and 28 % of total C lost from the top 1 m of soil for the current mix, base trend, and no-till trend scenarios, respectively. The gain in soil C in the no-till plus cover scenario was reduced by 79%

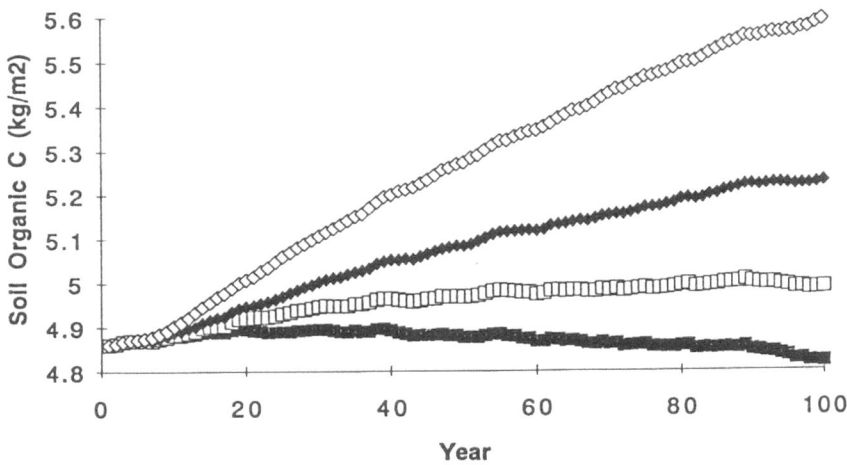

Figure 3. Average total carbon in the top 15 cm of soil for the four scenarios.
■ Current mix ☐ Base trend ◆ No-till trend ◇ No-till + cover

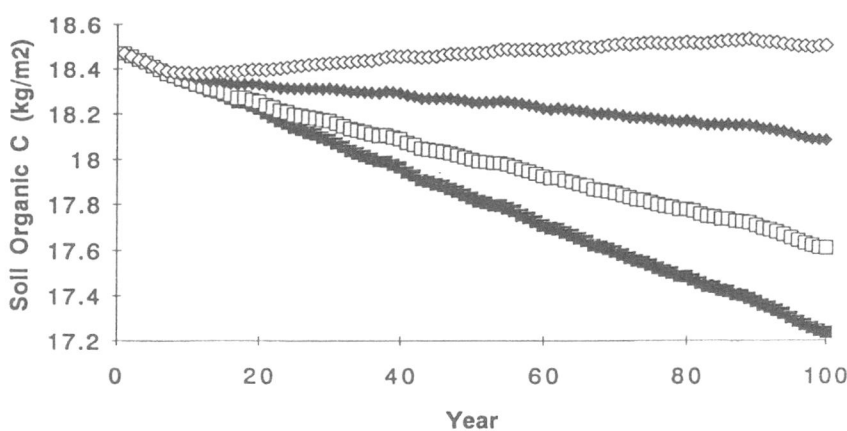

Figure 4. Average total carbon in the top 1 m of soil (measured from original surface) for the four scenarios. ■ Current mix ☐ Base trend ◆ No-till trend ◇ No-till + cover

because of soil erosion. Thus, soil erosion *per se* had a substantial effect on soil C content. Nonetheless, for all scenarios that resulted in C loss throughout the soil profile, the loss was primarily due to C dynamics within the soil. It is not clear, however, how much of the C lost from the profile would be released directly to the atmosphere, versus being transported as dissolved C in groundwater or runoff.

4. Conclusions

The results of this study suggest the following hypotheses:

1. The current trend toward adoption of conservation tillage is projected to reduce soil erosion in the US corn belt. Adoption of no-till would substantially further reduce erosion, while adoption of a winter wheat cover would have little additional effect.

2. The C content of the top 15 cm (*i.e.* approximate plow depth) of soils in the US corn belt is approximately in equilibrium, although erosion may slowly lower the location of the surface. The current trend toward conversion to conservation tillage will increase C content of the top 15 cm by 0.2 kg m^{-2} during the next 100 years, compared to the current mix of tillage practices. Adoption of no-till and no-till plus winter cover would cause increases of 0.4 and 0.8 kg m^{-2}, respectively.

3. Under the current mix of tillage practices, during the next 100 years, soils in the US corn belt will lose 3.2×10^6 t-C yr^{-1} from the top 1 m, measured from the current soil surface. About 20% of this loss will be due to erosion. Adoption of conservation tillage and no-till could reduce loss of C from erosion and soil C dynamics to 2.2 and 1.0 x 10^6 t-C yr^{-1}, respectively. Widespread use of no-till plus winter wheat cover could result in a net accumulation of 0.1×10^6 t-C yr^{-1}. Thus, a shift from current tillage practices to widespread use of no-till plus winter cover could conserve and sequester a total of 3.3×10^6 t-C yr^{-1} for the next 100 years.

6. Acknowledgements

We gratefully acknowledge the assistance of Verel Benson, Dan Taylor, and Jimmy Williams (USDA Agricultural Research Service and Soil Conservation Service, Texas) in providing information on EPIC, and of Richard Dickerson (USDA Soil Conservation Service, Illinois) in developing the management schedules. The information is this document has been funded wholly by the US Environmental Protection Agency. It has been subject to the agency's peer and administrative review, and it has been approved for publication as an EPA document.

5. References

CTIC (Conservation Tillage Information Center): 1990, *National Survey of Conservation Tillage Practices*, Conservation Tillage Information Center, Fort Wayne, IN.

IPCC (Intergovernmental Panel on Climate Change): 1992, *Climate Change 1992: The Supplementary Report to the IPCC Scientific Assessment*, Cambridge University Press, 200 pp.

Kern, J.S., and Johnson, M.G.: 1993, *Soil Sci. Soc. Am. J.* **57**, 200.

Onstad, C.A. and Foster, G.R.: 1975, *Transactions of the ASAE* **18**, 288.

Phillips, D.L., Hardin, P.D., Benson, V.W., and Baglio, J.V.: 1993, *J. of Soil and Water Conservation*, in press.

Post, W.M., Peng T.H. , Emanuel W.R., King A.W., Dale V.H., and DeAngelis, D.L.: 1990, *Amer. Sci.* **78**, 310.

Robinson, A.R.: 1979, Sediment Yield as a Function of Upstream Erosion, in: Peterson, A.E. and Swan, J.B. (eds), *Universal Soil Loss Equation: Past, Present, and Future*. Soil Sci. Soc. Am., Madison, WI, pp. 7-16.

Schertz, D.L.: 1988, *J. of Soil and Water Conservation* **43**, 256.

Schlesinger, W.H.: 1985, Changes in Soil Carbon Storage and Associated Properties with Disturbance and Recovery, in: Trabalka, J.R. and Reichle, D.E. (eds), *The Changing Carbon Cycle: A Global Analysis*, Springer- Verlag, New York, NY, pp. 194-220.

SCS (Soil Conservation Service): 1981, *Land Resource Regions and Major Land Resource Areas of the United States*, US Department of Agriculture Agricultural Handbook 296, Washington, DC, 156 pp.

SCS (Soil Conservation Service): 1989, *Summary Report, 1987 National Resources Inventory*, Iowa State University Statistical Laboratory, Statistical Bulletin Number 790, Ames, IA, 37 pp.

Sharpley, A.N. and Williams, J.R. (eds): 1990a, *EPIC--Erosion/Productivity Impact Calculator: 1. Model Documentation*, US Department of Agriculture Technical Bulletin No. 1768, Washington, DC, 235 pp.

Sharpley, A.N. and Williams, J.R. (eds): 1990b, *EPIC--Erosion/Productivity Impact Calculator: 2. User Manual*, US Department of Agriculture Technical Bulletin No. 1768, Washington, DC, 127 pp.

Smith, S.J., Sharpley, A.N., and Nicks, A.D.: 1990, Evaluation of EPIC nutrient projections using soil profiles for virgin and cultivated lands of the same soil series, in: Sharpley, A.N. and Williams, J.R. (eds), *EPIC--Erosion/Productivity Impact*

Calculator: 2. Model Documentation , US Department of Agriculture Technical Bulletin No. 1768, Washington, DC, pp. 217-219.

Williams, J.R.: 1975, Sediment yield prediction with Universal Soil Loss Equation using runoff energy factor, in: *Present and Prospective Technology for Predicting Sediment Yields and Sources*, US Department of Agriculture Agricultural Research Service Report ARS-S-40, Washington, DC, pp. 244-252.

Williams, J.R. and Renard, K.G.: 1985, Assessment of soil erosion and crop productivity with process models (EPIC), in: Follett R.F. and Stewart B.A. (eds), *Soil Erosion and Crop Productivity*, American Society of Agronomy/Crop Science Society of America/Soil Science Society of America, Madison, WI, pp. 67-103.

Wischmeier, W.H. and Smith, D.D.:1978, *Predicting Rainfall Erosion Losses, a Guide to Conservation Planning*, US Department of Agriculture Agricultural Handbook No. 537, Washington, DC, 58 pp.

THE IMPACT OF CULTIVATION ON CARBON FLUXES IN WOODY SAVANNAS OF SOUTHERN AFRICA

Paul L. WOOMER

Tropical Soil Biology and Fertility Programme, UNESCO-ROSTA, P.O. Box 30592, Nairobi, Kenya

Abstract. The rapid transition from miombo woodland and savanna to maize-based agriculture in Southern Africa results in a near universal loss of total system and biomass carbon. Forests and savannas occupy approximately 3.1 million km^2 in southern Africa. Two natural ecosystems, a miombo woodland (Zimbabwe) and a broadleafed dry savanna (South Africa), contained 48 and 94 Mg C ha^{-1}, respectively. Clearing of the miombo and establishment of maize-based agriculture on a sandy Alfisol resulted in a decline in total soil organic carbon from 28 to as little as 9 Mg ha^{-1}. This decline is not related to the annual aboveground productivities which, in many cases is greater in the cropping system than in the savanna or forest. Severe declines in total soil organic matter resulting from shifting cultivation were also observed in coastal Mozambique. The CENTURY plant/soil simulation model was used to simulate long-term carbon dynamics a miombo woodland and maize-based cropping system in Marondera, Zimbabwe. The miombo woodland continues to accumulate total system C but shifting cultivation and commercial cultivation of maize result in annual carbon losses of 0.15 and 0.14 Mg ha^{-1} yr^{-1}. Increases in temperature (2° C) accompanied by 25% increases in photosynthetic efficiency did not effect the decline in total system carbon, however, improved organic matter management within the agroecosystem reduced the losses in total system carbon within the agroecosystem by 57% under the climate change scenario.

Water, Air, and Soil Pollution **70**: 403–412, 1993.
© 1993 *Kluwer Academic Publishers.*

1. Introduction

The transition from natural to agroecosystems has resulted in a near universal decline in total system and soil organic carbon in temperate (Cole *et al.*, 1987, Post and Mann, 1990) and tropical regions (Nye and Greenland, 1960; Ayodele, 1986; Ayanaba *et al.*, 1987; Detwiler and Hall, 1988). In tropical forests, this decline is related to the initial clearing and burning of aboveground forest biomass (Martins *et al.*, 1991), subsequent tillage (Lal, 1986; Resck and Silva, 1990; Martins *et al.*, 1991) and a decline in litter additions to the soil system. The mechanisms responsible for soil organic carbon decline within agroecosystems include the disruption of aggregates that physically protect organic matter (Stevensen, 1982; Tiessen and Steward, 1983), accelerated cycles of soil wetting and drying (Ladd *et al.*, 1977) and increased microbial activities (Stevenson, 1982).

The conservation of soil organic matter is important at global, farming systems and soil process scales. Land use practices account for approximately 1-2 GT yr^{-1} of the total annual release of CO_2 into the atmosphere (see Detwiler and Hall, 1988; Hall, 1989; Post *et al.*, 1990). At the same time, the mineralisation of decomposing residues is an important source of plant nutrients in farming systems receiving low external inputs (Sanchez *et al.*, 1989). Complexes of soil organic carbon are responsible for improved nutrient (Russell, 1973) and water (Lal, 1986) retention, the promotion of soil aggregation (Oades, 1984) leading to reduced erosion and the complexation of toxic cations in soils (Hue *et al.*, 1986) all of which lead to a more productive rooting environment.

While tropical forests have great potential to accumulate and sequester organic carbon due to rapid growth rates and large biomass (Brown and Lugo, 1982), African forests and savannas are under great pressures for conversion to agriculture, due in large part to high rates of population growth (see FAO, 1992) and the expectation of higher living standards. Furthermore, the sandy soils characteristic of large areas of the Southern Africa sub-region (FAO-UNESCO, 1977) result in relatively low inherent soil fertility that leads to either the need of extended fallow intervals or additions of chemical fertilisers and organic additions in order to maintain crop productivity. This sub-region contains approximately equal land areas of natural forests/savannas (Figure 1) and croplands/fallow (Figure 2).

This paper provides information on the partitioning of total system C of natural woodland and savanna ecosystems in Southern Africa, the loss of soil organic matter following conversion to agriculture and employs the CENTURY plant/soil simulation model to develop scenarios of organic carbon sink-source fluxes as lands are converted to agriculture at the present time, and managed under changed climatic conditions. Also included within these simulations is maize productivity.

2. Materials and Methods.

Site data was obtained from cooperating scientists of the Tropical Soil Biology and Fertility Programme (Woomer and Ingram, 1990). Biomass pools and fluxes of the miombo dry woodland were provided by past and present researchers of the University of Zimbabwe Department of Biological Science (UZBS) including M.J. Swift, P.G.H. Frost, J.C. Hatton and B. Campbell. The broadleafed dry savanna site was described by Mary C. and R.J. Scholes. Biomass measurements and soil analyses were conducted using methodologies described by Anderson and Ingram (1989). The soil organic matter contents of the adjacent miombo woodland and maize soils in Marondera, and for the cultivated communal farmlands in Zimbabwe were contributed by L. Mukurumbira of the Soil Productivity Research Laboratory (SPRL), Marondera, Zimbabwe. Note that different sets of soil data, corresponding to two different miombo woodland sites are presented in Table 1 (from SPRL) and Figure 3 (from UZBS). The data describing the adjacent coastal forest and shifting cultivation in Mozambique were provided by J.C. Hatton of Eduardo Mondale University Department of Biological Sciences. The information on the Guinean savanna and adjacent maize fields was provided by P. Lavelle and A. Martin of L'Ecole Normale Superior, Paris France and is derived from the OSTROM research site in Lamto, Cote d'Ivoire.

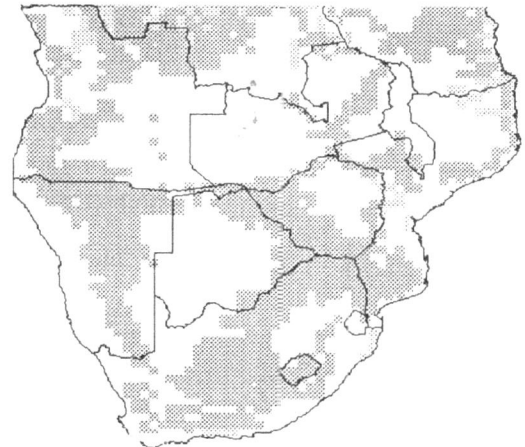

Figure 1. Distribution of savannas (darker shade) and forests (lighter shade) in Southern Africa. Natural ecosystems occupy approximately 3.1 million km^2 (developed using RIS software, Jagtap, 1992).

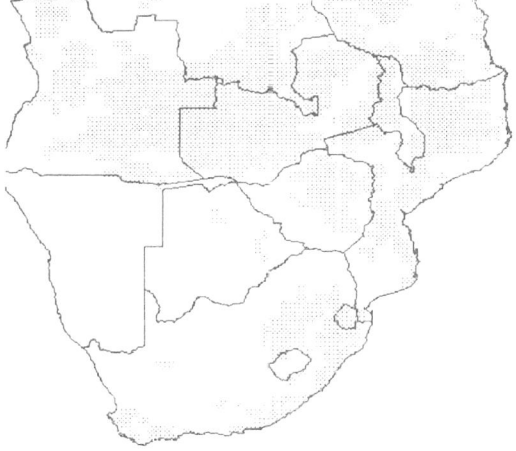

Figure 2. Croplands and fallows occupy approximately 2.7 million km^2 in Southern Africa (developed using RIS software, Jagtap, 1992).

The CENTURY Model (Parton *et al.* 1987, 1992) was initialised using data provided by the scientific cooperators in Zimbabwe. CENTURY is a plant/soil model that simulates soil organic matter and plant nutrient (N, P and S) dynamics and is suitable for the preliminary evaluation of many land use management strategies (Parton *et al.*, 1987). The model runs on a monthly time step and the major input requirements of the model include monthly temperature and precipitation, initial soil C and nutrient levels, plant lignin and C:nutrient ratios, soil pH and texture

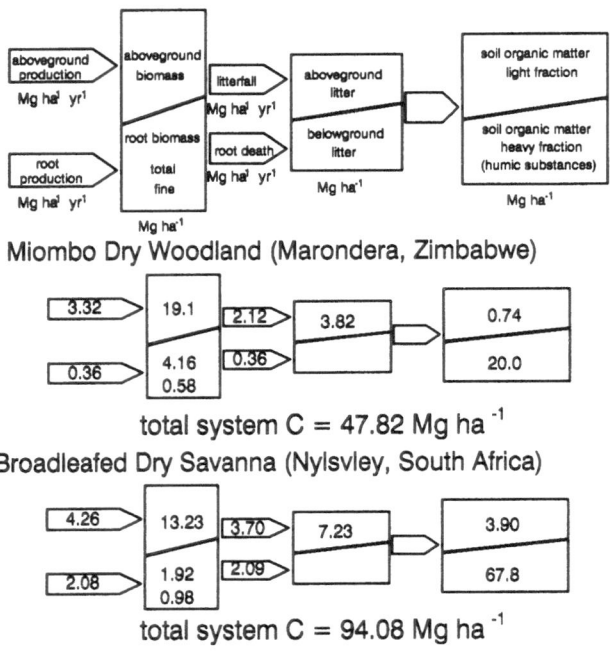

Figure 3. Carbon pools and fluxes within a miombo woodland and broadleaf dry savanna in Southern Africa.

and additional nutrient inputs such as atmospheric deposition and biological N fixation. Although this model was first developed to simulate carbon and nutrient dynamics of the North American Great Plains Grasslands, subsequent routines were added that allow for simulation of croplands, forests and savannas (Parton *et al*, 1992). Long-term monthly precipitation and temperature data were used to initialise the simulation of the effects of forest fallow/maize rotation (Figure 3). The effects of climate change (2° C temperature increase, 25% increase in potential photosynthetic efficiency) were initialised using crop dry matter allocation data provided by L. Mukurumbira and were assumed to remain constant regardless of climatic regime.

3. Organic Carbon Pools and Fluxes in Natural and Managed Lands.

Despite relatively low levels of precipitation, woodlands and broadleaf savannas are able to accumulate significant levels of total system carbon (Figure 3) as in Marondera, Zimbabwe where a 27 year-old miombo dry woodland contained 48 Mg C ha^{-1} and Nylsvley, South Africa where total C sequestration is estimated to be 94 mg C ha^{-1} . In Southern Africa, cropping systems now occupy a minimum of 2,723,000 km^2 (Figure 2).

Table 1. Decline in soil organic matter resulting from cultivation at selected sites in Africa.

land use	site	annual aboveground productivity	total soil organic C
		Mg ha^{-1} yr^{-1}	Mg ha^{-1}
ZIMBABWE			
miombo woodland	Marondera	3.32	27.8
maize field	Marondera	3.14	25.3
maize field	Chihota	6.68	9.2
maize field	Murawa	6.74	11.0
MOZAMBIQUE			
Coastal forest	Inhaca	5.75	51.2
Maize	Inhaca	1.80	19.2
COTE d'IVOIRE			
Guinean savanna	Lamto	8.50	23.2
maize field	Lamto	9.75	18.0

The consequences of conversion to a maize-based cropping system for selected natural vegetation formations are presented in Table 1. On-station research on a sandy Alfisol at Marondera, Zimbabwe indicates that the soil releases 9% of its organic matter over an 8 year period of continuous maize cultivation. Soil organic matter levels in similar soils are considerably lower in nearly communal farming areas (Chihota and Murawa) where maize-based cropping has been practised for extended periods of time. This situation is more severe when compared to a coastal forest formed on tertiary sand dunes. A singe cycle of maize-based shifting cultivation reduced soil organic matter levels decline to 37% of the original forest. This effect was not as great when a Guinean savanna was cleared from a Vertisol for maize cultivation in Côte d'Ivoire.

4. Simulation Modelling of Carbon Dynamics Using the CENTURY Model.

The complexity of the controls on biological soil processes has led the TSBF Programme toward multivariate hypothesis testing. This is facilitated by the use of computer simulation models. The adoption of a model by a research programme is a dynamic process during which site characterisation studies are used to initialise the model and subsequent research results are compared with simulations as a means of model validation.

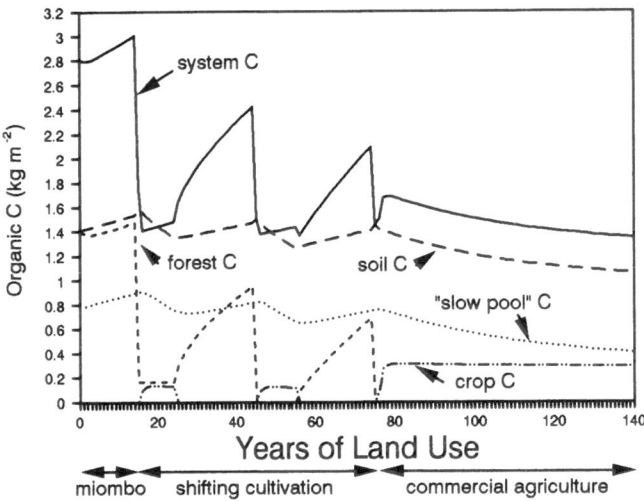

Past and Present Land Use of a Miombo Woodland. The CENTURY Model was used to estimate the carbon pools and sink-source fluxes of a miombo woodland and subsequent land management (Figure 4).

Figure 4. CENTURY Model simulation of the carbon dynamics of a miombo woodland, shifting and permanent maize cultivation at Marondera, Zimbabwe.

The miombo woodland used to initialise the model occurs at the Grasslands Research Station, Marondera, Zimbabwe. The vegetation is dominated by *Brachystegia spiciformis* and the soil is a sandy Alfisol derived from granitic parent material (Woomer and Ingram, 1990). The mean annual precipitation is 910 mm yr^{-1} and distributed unimodally between November and March. A simulation was conducted during which a near-climax miombo woodland was established for 20 years and then subject to two 30 year cycles of shifting maize cultivation. Following this, the land was cleared for 70 years of continuous maize cultivation fertilised with 40 kg N ha^{-1} yr^{-1}. Model outputs included total soil organic carbon, the "slow" pool (Parton *et al.*, 1987, Cambardella and Elliot, 1992), forest biomass carbon and annually accumulated crop carbon.

Model outputs suggest that total system carbon was greatest in the miombo woodland, declined during successive cycles of shifting cultivation and declined even more when the secondary fallow was converted to permanent agriculture. When converted to permanent maize cultivation, soil organic carbon continues to decline. Most of this decline was due to a loss of the "slow" pool, which, in the real world, corresponds to particulate soil organic matter (50 to 250 um) of fairly recent origin (Cambardella and Elliot, 1992).

Carbon Dynamics of Maize Cultivation under a Climate Change Scenario. The CENTURY Model was then used to simulate the impact of climate change on the organic carbon dynamics of continuous maize cultivation (Figure 5). First, the model was initialised with the long-term maximum and minimum monthly temperatures and mean monthly precipitation. No external inputs were applied and 90% of the maize stover was removed from the field following harvest (current practice scenario) The soil organic carbon status resembles that of Figure 4 when the permanent cultivation of maize is initiated (year 75). Next the same simulation was extended after year 10 with a 2° increase in temperature and a 25% increase in the efficiency of dry matter accumulation when nitrogen availability is non-limiting (climate change scenario).

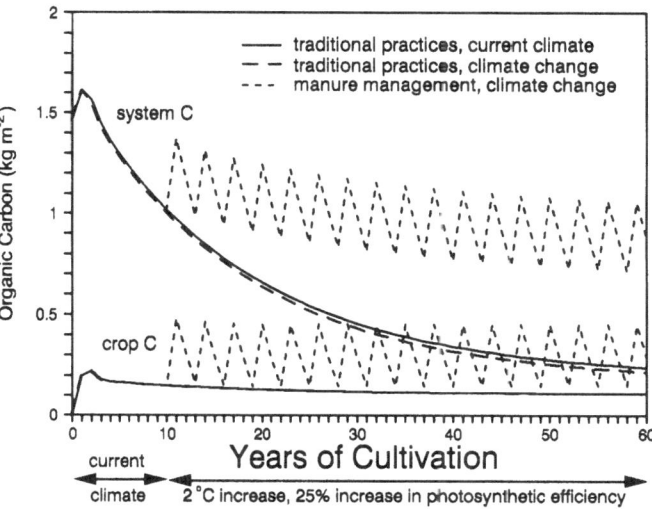

Figure 5. CENTURY Model simulation of the carbon dynamics of maize cultivation under present and expected climate change conditions at Marondera, Zimbabwe.

Finally, a simulation was run where the climate change scenario is combined with improved organic matter management. In this case only 50% of the maize stover was removed and an addition of 5 T ha⁻¹ of manure (150 kg N ha⁻¹) was applied every 3 years (improved management scenario). An important component of this scenario is the partial removal of stover from the farming system and its return every 3 years as manure after passage through livestock. The model outputs include total system and total crop carbon.

After 50 years, crop productivity and total system carbon were greatest in the improved management scenario. Total system carbon includes crop productivity, standing dead, above and below ground litter and total soil organic matter. While this pattern of manure application does not quite establish a steady state of total system carbon, crop productivity is sustained and carbon losses are greatly reduced. The current practice and climate change scenarios differed only slightly (2.4 vs 3.3 Mg total system C ha⁻¹, respectively), presumably due to the offsetting effects of increased temperature on litter and soil organic matter decomposition rates and the increased plant productivity and litter contribution resulting from the increased efficiency of dry matter conversion.

5. Conclusions

Based on the CENTURY Model outputs presented in Figures 4 and 5, the sink-source fluxes for past, present and future land uses may be calculated (Table 2). The removal of natural forests prior to the establishment of annual field crops results in massive and

Table 2. CENTURY simulation of annual sink-source flux of CO_2 (total C loss from ecosystem) within a miombo woodland converted to maize cultivation on a sandy Alfisol in Zimbabwe.

Land use	sink-source flux (Mg C ha^{-1} yr^{-1})
Miombo woodland	-0.12
Shifting cultivation (20 year fallow)	0.15
Land clearing (of 20 year fallow)	6.44
Current traditional maize cropping	0.15
Current commercial maize cropping	0.14
Cropping of degraded land (after 40 years of cropping)	0.03
Traditional maize cropping during climate change[1]	0.15
Manure management of maize during climate change	0.07

[1] Climate change includes a 2° C temperature increase and a 25% increase in dry matter accumulation during non-limited growth conditions.

unavoidable loss of biomass carbon. More manageable is the fate of soil organic matter in agricultural systems. The potential for sequestration of soil organic carbon through the management of organic resources appears to more than offset the threat to existing reserves to soil carbon posed by the anticipated effects of climate change. Furthermore, the improved management of farmer-available resources which leads to maintenance or increases of soil organic carbon need not be applied as a farmers immediate objective but rather as an additional benefit of integrated nutrient management and increased nutrient cycling. While this paper documents the impacts of land disturbance on carbon sequestration for select sites in Southern Africa, additional studies of this sort are necessary. Given the vast differences in scale between the georeferenced description of natural and cultivated land areas (10^3 km^2), the resolution of the CENTURY Model (g C m^{-2}) and the relative paucity of sites described in sufficient detail, it is inappropriate to extrapolate the simulation results presented in this paper across the entire Southern African region.

An inevitable consideration in our attempts to improve carbon sequestration in tropical agroecosystems is that smallhold farmers have little direct interest in conserving soil organic resources unless an immediate benefit in terms of yield or labour efficiency is realised. Recent survey results from Uganda suggests that smallhold farmers practising intensive cultivation become attracted to organic matter management as the fertility of the soil declines (Tukahirwa, 1992). Manuring and mulching were the most frequent organic matter management practices (Table 3). The infrequency of fertiliser use is a reflection

Table 3. Organic resource and inorganic fertiliser use by farmers within individual Parishes of different Districts of Uganda (after Tukahirwa, 1992).

District	n	-------- percentage of farmers practising -------			
		manuring[1]	mulching[2]	composting	fertilisation[3]
Rukungiri	96	83	68	18	0
Bushenyi	100	54	2	5	0
Mpigi	102	21	41	6	9
Iganga	115	8	44	0	4

[1] use of animal manures; [2] does not include mulching of bananas; [3] use of inorganic fertilisers.

of its scarcity within Uganda where the use of nitrogenous fertilisers declined by 64% between 1980 and 1990 (FAO, 1991). In order to have a realistic opportunity of adoption by farmers, recommendations of organic matter management practices that enhance the carbon sequestration of tropical agroecosystems must be carefully weighed against the availability and alternative values of organic resources, labour requirements to exploit those resources and the short-term impact of the organic management strategy on crop performance.

6. Acknowledgements

The preparation of this paper would not have been possible without the contribution of site characterisation data provided by the Tropical Soil Biology and Fertility Programme cooperating scientists in Southern Africa. These scientists include M.J. Swift, P.G.H. Frost, J.C. Hatton, B. Campbell, Mary C. and R.J Scholes, L. Mukurumbira, J.C. Hatton and students, P. Lavelle and A. Martin. The support the author by the Natural Environment Research Council (UK) and the UNESCO Regional Office for Science and Technology for Africa is gratefully acknowledged.

7. References

Anderson, J.M. and J.S.I. Ingram: (1989). *Tropical Soil Biology and Fertility: A Handbook of Methods.* CAB International, Wallingford, UK.
Ayanaba, S.B. Tuckwell and D.S. Jenkinson: 1976, *Soil Biol. and Biochem.* **8**, 519-525.
Ayodele, O.J.: 1986, *Biol. Fertil. Soils* **2**, 519-525.
Bouwman A.F.: 1990, Exchange of Greenhouse Gasses between Terrestrial Ecosystems and the Atmosphere in A.F. Bouwman (ed.) *Soils and the Greenhouse Effect*, J. Wiley and Sons, pp. 59-127.

Brown S. and A.E. Lugo: 1982, *Biotropica* **14**, 161-187.

Cambardella C.A. and E.T. Elliot: 1992, *Soil Sci. Soc. Am. J.* 56, 777-783.

Cole, C.V., J. Williams, M. Shaffer and J. Hanson: 1987, *Soil Sci. Soc. Am. Spec. Publ.* **19**, 147-166.

Detwiler, R.P. and C.A.S. Hall: 1988, *Science*, **239**, 42-47.

FAO: 1991: *FAO Yearbook: Fertilisers 1990, Vol. 40,* Food and Agriculture Organisation of the United Nations.

FAO, 1992: *FAO Yearbook: Production 1991, Vol. 45,* Food and Agriculture Organisation of the United Nations.

FAO-UNESCO: 1977, *Soil Map of the World: Africa*, UNESCO, 297 pp.

Hall, D.O.: 1989, *J. Geographical Soc.* **146**, 175-181.

Hue, N.V., G. Craddock and F. Adams: 1986, *Soil Sci. Soc. Am. J.* **50**, 28-34.

Jagtap, S.: 1992, *The Resource Information System (computer software)*, IITA.

Ladd, J.N., J.W. Parsons and M. Amato: 1977, *Soil Biol. Biochem.* **9**, 309-325.

Lal, R.: 1986, Soil Surface Management for Intensive Land Use and High Productivity, in: B.A. Steward (ed.) *Advances in Soil Science*, Springer-Verlag, pp. 2-97.

Martins, P.F., C.C. Cerri, B. Volkoff, E. Andreux and A. Chauvel: 1991, *Forest Ecol. Management* **38**, 273-282.

Nye, P.H. and D.J. Greenland: 1960, *The Soil Under Shifting Cultivation*, Commonwealth Bureau of Soils, 156 pp.

Oades, J.M.: 1984, *Plant and Soil* **76**, 319-337.

Parton, W.J., D.S. Schimel, C.V. Cole, & D.S. Ojima: 1987, *Soil Sci. Soc. Am. J.* **51**, 1173-1179.

Parton, W.J., B. McKeown, V. Kirchner and D. Ojima: 1992, *CENTURY Users Manual*, Colorado State University.

Post, W.M. and L.K. Mann: 1990, Changes in Soil Organic Carbon as a Result of Cultivation, in: A.F. Bouwman (ed.) *Soils and the Greenhouse Effect*, John Wiley and Sons, pp. 401-406.

Post, W.M., T. Peng, W.R. Emmanuel, A.W King, V.H. Dale and D.L. DeAngeils: 1990, *American Scientist* **78**, 310-326.

Resck D.V.S. and J.E. da Silva: 1990, *Internat. Cong. Soil Sci.* **6**, 325-326.

Russell, E,W.: 1973, *Soil Conditions and Plant Growth*, Longman Group Ltd.

Sanchez P.A., C.A. Palm, L.T. Szott, E. Cuevas and R. Lal: 1989, Organic Input Management in Tropical Agroecosystems, in: D.C. Coleman, J.M. Oades and G. Uehara (eds.) *Dynamics of Soil Organic Matter in Tropical Ecosystems*, NifTAL Project, pp 125-152.

Stevensen, F.J.: 1982, *Humus Chemistry, Genesis, Composition, Reactions.* Wiley-Interscience.

Tiessen H. and J.W.B. Steward: 1983, *Soil Sci. Soc. Am. J.* **47**, 509-514

Tukahirwa, E.M.: 1992, *Uganda - Environmental and Natural Resource Policy and Law: Issues and Options*, Makerere University.

Woomer P. and J.S.I. Ingram,: 1990, *The Biology and Fertility of Tropical Soils*: TSBF Report 1990, TSBF, pp. 44.

Potential Impacts of Elevated CO_2 and Above- and Belowground Litter Quality of a Tallgrass Prairie

CLENTON E. OWENSBY
Department of Agronomy
Throckmorton Hall
Kansas State University
Manhattan, KS 66506-5501 USA

Abstract. Increased atmospheric CO_2 will likely impact the productivity of arid and semiarid ecosystems through increased C, N, and water use efficiencies at the individual plant level. Tallgrass prairie has had increased above- and belowground biomass production under elevated CO_2, primarily due to increased water use efficiency. There is an apparent decreased N requirement to sustain increased productivity in CO_2-enriched tallgrass prairie, and C:N ratios of plant litter above and below ground have increased. The tallgrass prairie ecosystem level response to elevated CO_2 on the C cycle could potentially increase C storage. Reduced litter quality associated with elevated CO_2 in tallgrass prairie has the potential to reduce decomposition rates, and ruminant digestion rate of plant biomass apparently has been lowered. Reduced intake by ruminants would shunt more of the plant biomass directly into the detrital food chain, thereby slowing decomposition further. The potential impact is for increased C to be retained as soil organic matter in the tallgrass prairie.

1. Introduction

The fact that CO_2 is increasing in our atmosphere has been well documented (Boden *et al.*, 1990). The consequences of that increase have been subject to much speculation (Dahlman, 1984), although evidence collected and summarized by Keeling (1989) suggested that the northern biosphere has increased its C fixation by 0.5 % yr^{-1} over the past 20 yr. Recently, much attention has been focused on whether Global Circulation Models can accurately predict future climate without adequate estimates of the major sources and sinks of CO_2. Tans *et al.* (1990) suggested that the majority of the missing C in the global CO_2 balance may be accounted for by a temperate terrestrial sink. In order to assess the source-sink relationships on an ecosystem level, research that allows for the integration of all ecosystem properties must be accomplished. Even though organism-level responses to CO_2 enrichment have been documented, terrestrial ecosystem level C balances remain unresolved (Strain and Cure, 1985). Bazzaz (1990) has extensively reviewed responses of species endemic to natural ecosystems to elevated CO_2 in which he comes to much the same conclusion. Mooney *et al.* (1991) concluded that there was an urgent need for additional research on terrestrial ecosystem response to elevated CO_2. They further state that these experiments must be long-term, a decade or more, to allow a response trajectory to be determined.

Water, Air, and Soil Pollution **70**: 413–424, 1993.
© 1993 *Kluwer Academic Publishers.*

Hall (1989) has reported that global net C fixation by the biosphere to be 20 times that of fossil fuel burning, at approximately 106 Pg yr^{-1} with 60 Pg from terrestrial ecosystems and 46 Pg from aquatic systems. He also estimated that the net atmospheric gain in CO_2 was 3 Pg yr^{-1}. He concluded that many of these estimates were subject to large error and stated that we must attempt to quantify the CO_2 uptake, emissions, and storage mechanisms in soils, vegetation, and oceans. Dyson (1990) supported that view in the Radcliffe Lecture at Oxford University in England where he called for increased research at the ecosystem level that is 'widely distributed in space and extended in time'. The importance of quantifying the ecosystem level response to increased CO_2 in the atmosphere has been illustrated by Houghton *et al.* (1987) who estimated that a 1% change in the C stocks in terrestrial ecosystems could equal 2 Pg, an amount equal one-third of the C released annually to the atmosphere by worldwide combustion of fossil fuels.

Plants with the C_3 pathway generally have increased C fixation rates when CO_2 levels are increased, while C_4 plants do not increase in C fixation to the degree that C_3 plants do (Nijs *et al.*, 1988; Wray and Strain, 1986; Kimball, 1983). In ecosystems with frequent water stress, enhanced water-use efficiency due to partial stomatal closure in CO_2-enriched environments (Gifford *et al.*, 1990) is likely more important than photosynthetic response.

Naturally-occurring ecosystems differ from the managed cropland systems primarily in resource supplies. In natural systems, essentially all nutrient resources are supplied by the system through nutrient cycling. In temperate grasslands, the primary and secondary productivities of organisms may be substantially limited by both nutrients and water (Owensby *et al.*, 1969). Even though numerous authors have hypothesized that due to the differential response of C_3 and C_4 plants to CO_2 enrichment, competitive relationships may be altered (Wray and Strain, 1986), water and nutrient limitations may negate any competitive advantage that C_3 species may have in CO_2-enriched atmospheres due to their photosynthetic pathway. Indeed, the nutrient limitations may increase C that remains fixed in long-term storage in soil organic matter. Tallgrass prairie has been subjected to elevated CO_2 for a 4-yr period, and data on above- and belowground biomass accumulation, litter and forage quality, water relations, and microbial biomass and respiration have been collected (Owensby *et al.*, 1993a,b; Owensby and Auen, 1993c). This paper attempts to use those data to assess the potential impacts of elevated CO_2 on decomposition and soil organic matter. In order for C to be stored in natural grassland ecosystems, two assumptions must be addressed:

Assumption No. 1.   *Natural grassland ecosystems will have sustained, increased C fixation under elevated CO_2.*

Assumption No. 2.   *Changes in litter quality under elevated CO_2 will slow decomposition.*

2. Evaluation of Assumption No. 1.

The mechanisms for increasing C fixation under elevated CO_2 are primarily associated with C, N, and water use efficiencies, alone and with interactions among them. Natural ecosystem productivities are resource-limited, and the extent to which elevated CO_2 affects resource use and availability will largely determine the potential for increased C fixation in a changing atmosphere (Field et al., 1992).

2.1 CARBON USE EFFICIENCY

Carbon use efficiency is associated with the manner by which C is fixed in plants with differing carboxylation mechanisms. Photosynthetic capacity of plants with the C_3 pathway is limited by current atmospheric CO_2 levels due to oxygenase activity of ribulose-1,5-bisphosphate carboxylase (Rubisco). Innumerable studies have shown increased C_3 photosynthesis with elevated CO_2 (summarized by Kimball, 1983; Bazzaz, 1990; Newton, 1991; Woodward et al., 1991). Concern that these increased photosynthetic levels would be sustained arose as a result of a reduced photosynthetic response to elevated CO_2 after a period of time, termed acclimation. Arp (1991) and Thomas and Strain (1991) have suggested that the reduced photosynthetic response to elevated CO_2 reported by many authors was likely due to a reduced belowground sink mediated by the size of the rooting medium. They concluded that plants growing in the field would maintain a high photosynthetic capacity as atmospheric CO_2 level continues to rise. C_4 photosynthesis is not considered C-limited, because C is initially fixed in the mesophyll by phosphoenolpyruvate carboxylase (PEPc) which does not have oxygenase activity (Edwards and Walker, 1983). Carbon fixed by PEPc is moved to the bundle sheath cells of C_4 plants where it is decarboxylated and refixed by Rubisco.

Arp and Drake (1991) confirmed that, over four seasons, a wild C_3 sedge species growing in an estuarine saltmarsh under elevated CO_2 maintained increased photosynthesis for four seasons. However at the same study site, canopy photosynthesis of a saltmarsh ecosystem dominated by two C_4 grasses was not improved by elevated CO_2 (Drake and Leadley, 1991). Therefore, it appears that the tallgrass prairie ecosystem which is dominated by C_4 grasses will likely not increase C acquisition as a result of improved photosynthetic capacity. Knapp et al. (1993a) reported little change in maximum photosynthetic capacity of Andropogon gerardii when exposed to elevated CO_2 in a native tallgrass prairie ecosystem. The tallgrass prairie has a C_3 component, primarily represented by Poa pratensis. Owensby et al. (1993a) reported that P. pratensis did not have increased production under elevated CO_2 and declined in the stand over a 3-yr period. They concluded that under the ungrazed conditions of the study, the shorter P. pratensis was probably light limited in the presence of the taller C_4 grasses, but taller perennial C_3 forbs had increased biomass production under elevated CO_2.

2.2 WATER USE EFFICIENCY

Atmospheric CO_2 concentration affects water use through changes in stomatal conductance (g) (van Bavel, 1974). Eamus and Jarvis (1989) reviewed water use efficiency (WUE) under elevated CO_2 and reported that g was reduced 30 to 40% for both C_3 and C_4 species without reduced photosynthesis. Since water stress is a common occurrence in tallgrass prairie, decreased g under elevated CO_2 should be manifested in increased productivity. Indeed, increased above- and belowground biomass production under elevated CO_2 without input of additional water has been reported for tallgrass prairie (Owensby *et al.*, 1993a) (Figures. 1 and 2). Therefore, by definition, WUE has been increased. Owensby *et al.* (1993a) and Knapp *et al.* (1993a) have both reported improved water relations for *A. gerardii* exposed to elevated CO_2. They concluded that the improved biomass production over three seasons of CO_2 enrichment was almost certainly due to improved WUE. Morrison (1985) reviewed the literature concerning CO_2 enrichment and water relations and reported a range of 60-160% increase in WUE for both C_3 and C_4 plants. Improved root exploration of the soil probably occurs under elevated CO_2, favoring increased water uptake. In natural systems, both Curtis *et al.* (1990) and Owensby *et al.* (1993a) have reported increased root production with CO_2 enrichment. Knapp *et al.* (1993b) reported additional water savings as a result of CO_2 enrichment on dynamic stomatal responses to sunlight of *A. gerardii*. Stomatal conductance of this C_4 grass achieved new steady state levels more rapidly after abrupt changes in sunlight at elevated CO_2. Consequently, there was less water loss (6.5%) for elevated CO_2 plants compared to ambient. Changes in stomatal density may also impart water savings for plants under elevated CO_2. Woodward (1987) indicated that CO_2 enrichment over the past century has likely reduced stomatal density. Potentially, species that respond to elevated CO_2 by increasing water use efficiency most may competitively displace those that respond less, thereby improving community-level water relations and productivity. Gifford *et al.* (1990) concluded that, because of the increased water use efficiency in high CO_2 environments, arid and semi-arid ecosystems would likely respond relatively more to elevated CO_2 than more mesic ones and, therefore, increase in geographical area.

2.3 N USE EFFICIENCY

Productivity in tallgrass prairie is N-limited (Owensby *et al.*, 1969). Any change in ecosystem function that improves productivity without additional N input will increase N use efficiency (NUE). Field *et al.* (1992) have proposed a resource-based approach for response to CO_2 enrichment and have concluded that NUE will increase in most N-poor ecosystems. Numerous authors (Bazzaz, 1990; Newton, 1991) have postulated that increased production under elevated CO_2 will diminish over time as N becomes more limiting due to increased growth. However, in natural systems, increased NUE has been reported by Curtis *et al.* (1989) in an estuarine saltmarsh and Owensby *et al.* (1993b) in tallgrass prairie over 4-yr and 3-yr periods, respectively. NUE may result from a reduced carboxylation enzyme requirement, particularly for C_3 species (Wong ,1979; Long, 1991; van Oosteen *et al.*, 1992). CO_2 enrichment has reduced chlorophyll content of *Trifolium subterraneum* leaves, as well (Cave *et al.*, 1981). These reduced requirements may improve production response by alleviating N deficiencies and, thereby, increase NUE. Increased acquisition, stimulated by increased root growth, increased N fixation, and/or increased mycorrhizal activity, may also improve NUE (Bazzaz, 1990; Newton, 1991).

2.4 SYNERGY OF RESOURCE USE EFFICIENCIES

Alleviation of a resource stress improves the use of other non-limiting resources. For example, Schulze (1991) indicated that any water stress affected almost all plant functions, including the ability of leaves to assimilate C and roots to take in nutrients. With relief from water stress under elevated CO_2, C assimilation and nutrient uptake would increase. Under saturating light intensity, photosynthesis normally increases as leaf N content increases. That has not been true for plants in CO_2 enriched atmospheres. Reductions in Rubisco by as much as 40% for plants under elevated CO_2 has not reduced C assimilation rates (Long, 1991). Apparently, there is a reduced N requirement with CO_2 enrichment, therefore an increased NUE. If resource use efficiency of any one of the most commonly limiting resources, *i.e.* water, N, C, occurs, then all efficiencies are affected. Therefore, a true synergistic relationship among the required resources for plant growth is likely promulgated under CO_2 enrichment.

3. Evaluation of Assumption No. 2.

Plant litter decomposition is controlled by temperature, moisture, and litter quality (Swift *et al.*, 1979). Effects of elevated CO_2 on litter quality in CO_2-enriched environments is therefore an important consideration. The potential impact of CO_2 enrichment on biomass C:N ratios may affect the C storage in temperate terrestrial ecosystems. Hall (1989) reported that there was 1,515 Pg C in soil organic matter, and a relatively small increase could account for sequestration of C from all fossil fuel burning. Since nutrient resources in natural systems are fixed to a great degree, the impact of increased C fixation and reduced N concentration of plant biomass may lead to dietary deficiencies of essential nutrients for ruminant herbivores and soil microbes. Reduced N concentration in above- and belowground biomass results in reduced decomposition rates by microbes. Soil water may be affected through improved water use efficiency. Owensby *et al.* (1993a) indicated that soil water contents were higher under elevated CO_2 than ambient, indicating a potential for improved decomposition rates. However, a decline in litter quality may negate that potential improvement.

3.1 N CONCENTRATION

Reduced N concentration of above- and belowground plant biomass is likely the most constant and prevalent response to elevated CO_2 (Bazzaz, 1990; Newton, 1991). In natural ecosystems, CO_2 enrichment has reduced N concentration for both C_3 and C_4 species, regardless of biomass production response (Curtis *et al.*, 1989b; Owensby *et al.*, 1993b). As previously stated, the cause of the change in N content can be attributed to dilution, reduced carboxylation enzyme amount, and/or reduced chlorophyll content. Owensby *et al.* (1993b) indicated that the reduced N content of biomass under elevated CO_2 was almost immediate, manifesting itself by the first sampling date after only two weeks of exposure to elevated CO_2. The rapid reduction of leaf N concentration is probably not due to dilution alone, and additional N added to the system would not produce biomass of equal N concentration for ambient and elevated CO_2. Unpublished data showed a reduction in N concentration from N-fertilized plots in CO_2-enriched tallgrass prairie compared to ambient. Therefore, N concentration of biomass from natural grassland ecosystems will likely be reduced as atmospheric CO_2 concentration rises. Curtis *et al.* (1990) indicated that there was an increased root biomass on a salt marsh community dominated by the C_3 sedge, *Scirpus olneyi*, under a 680 ppm CO_2 environment, and that the root biomass had a reduced N concentration. A similar response has been reported for tallgrass prairie (Owensby *et al.*, 1993ab)

3.2 FIBER CONCENTRATION

Fiber concentration of plant litter in natural ecosystems under CO_2-enriched atmospheres has been measured only in the tallgrass prairie ecosystem (Owensby and Auen, 1993). They reported that CO_2 enrichment increased fiber concentration compared to ambient throughout the growing season. Melillo (1983) reported increased fiber concentration with CO_2 enrichment for tree litter.

3.3 DECOMPOSITION RATES

Increased fiber concentration and higher C:N ratios associated with plants grown under elevated CO_2 reduce mineralization rates of nutrient resources (Melillo 1983; Berendse *et al.*, 1987; Ågren *et al.*, 1991). The increased fiber concentration and C:N ratios of plant litter and the increased C allocation below ground under elevated CO_2 in tallgrass prairie increased microbial biomass C, microbial biomass N, and soil respiration (Rice *et al.*, 1992). When N was added as a soil amendment, microbial biomass C, microbial biomass N, and soil respiration in the CO_2-enriched plots was increased more compared to ambient than in unfertilized plots. They concluded that there was a definite increase in N limitation to soil microbial activity induced by elevated CO_2. Owensby and Auen (1993) provided further evidence that changes in litter quality would reduce decomposition rates. They showed that rate of *in vitro* dry matter digestibility of forage collected by esophageally fistulated sheep was slower on CO_2-enriched plots than ambient. Since ruminal digestion is microbial, one could infer reduced surface and soil litter decomposition rates as well. Due to reduced intake by ruminants of lower quality forages (Huston and Pinchak, 1991), a greater portion of the biomass will remain as surface litter which will increase the time required for decomposition. Thomas and Harvey (1983) reported that CO_2-enriched plants had increased cuticle and layers of epidermal cells which further reduce the ability of microbes to decompose plant biomass.

4. Potential for Carbon Storage

There is a continuous exchange of C among live biomass, litter, soil organic matter pools, and the atmosphere (Gifford *et al.*, 1990). In natural systems constrained by resource limits, addition of CO_2 to the atmosphere should hypothetically be accompanied by an increase in C storage in live plant tissue, the dead litter, and the soil organic matter. In tallgrass prairie, it has been shown that total C fixed by plants has increased, and that there has been an increase in microbial biomass C. Average peak aboveground biomass from

1989 through 1991 for tallgrass prairie exposed to elevated CO_2 was 27% greater than that for ambient (Owensby *et al.*, 1993a) (Figure 1), with the greatest increases coming during years with the most water stress. Warrick and Gifford (1985) concluded that the greater the water limitation to growth, the greater the relative increase in growth in response to elevated CO_2. Therefore, tallgrass prairie, commonly water stressed, should have sustained increased C fixation as atmospheric CO_2 rises. In the long term, C stored in the system should increase accordingly. There is concern that the increased production with CO_2 enrichment will result in N limitation to production. The discussion earlier indicates that there may be a reduced N requirement to sustain accelerated C fixation due to reduced carboxylation enzyme requirement and reduced chlorophyll content in leaves. Hocking and Meyer (1985) predicted that there would be increased mycorrhizal activity and root growth which would mitigate N limitations in the short run. In the long term, there will likely be adjustments in the total N in ecosystems exposed to elevated CO_2. Gifford *et al.* (1990) stated that total N in the ecosystem is determined by the C cycle, and that increased C in the atmosphere would ultimately lead to increased total N. Jones and Woodmansee (1979) proposed that N-limited systems may foster increased biological N fixation. However, the rate of ecosystem adjustment may not match the accelerated rate of CO_2 increase in the atmosphere and subsequent increased C fixation. Certainly, there have been indications that increased C is being allocated below ground under elevated CO_2 which would stimulate symbiotic and asymbiotic N fixation, and, in support of that contention, Finn and Brun (1982) have shown an increase in symbiotic N fixation with CO_2 enrichment. Nonetheless, if there is a reduced N requirement for growth which will sustain increased C acquisition, then C stored in the various pools should increase. Hunt *et al.* (1991) modeled grassland ecosystem response to elevated CO_2 and concluded that there would be a sustained increase in primary production even though there appeared to be N limitation. They further concluded that soil C would increase in temperate grasslands. Data collected from CO_2-enriched tallgrass prairie support that conclusion. The degree to which C will be stored is yet to be determined, but using logic proposed by Gifford *et al.* (1990), we estimate that C stored in soil organic matter will increase by 12-15% with the incremental doubling of atmospheric CO_2 over the next 50 to 70 yr.

4. Illustrations

Figure 1.   Total aboveground peak biomass for indicated treatments and years for
                 plots of native tallgrass prairie exposed to twice ambient atmospheric CO_2
                 24 hr a day from April 1 to late October over a 3-year period. Bars with a
                 common letter do not differ (P< 0.10).   Adapted from Owensby et al.
                 (1993a).

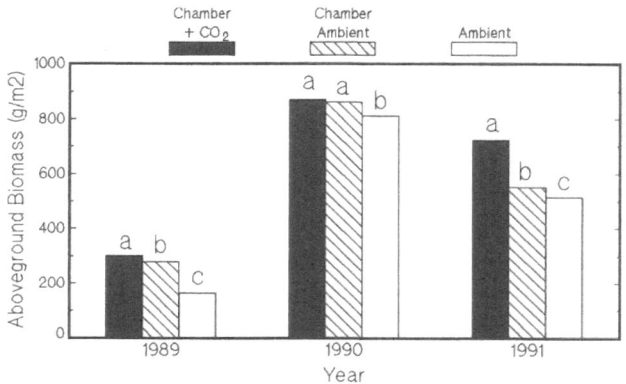

Figure 2.   Root ingrowth biomass for indicated treatments and years for plots of
                 native tallgrass prairie exposed to twice ambient atmospheric CO_2 24 hr a
                 day from April 1 to late October over a 3-year period. Bars with a common
                 letter do not differ (P< 0.10).   Adapted from Owensby et al. (1993a).

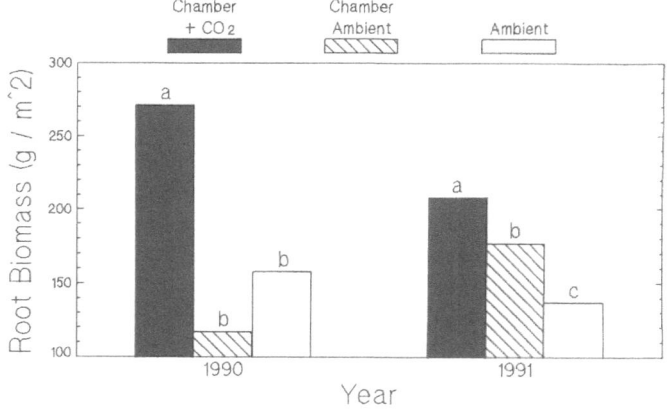

5. Acknowledgements

The research on tailgrass prairie reported here was supported by the U.S. Department of Energy, Carbon Dioxide Research Division. Patrick Coyne, Jay Ham, Alan Knapp, Charles Rice, Lisa Auen, Dale Strickler, Denise Garrett. Dean Larson, Garry Harter, and Neal Adam were valuable members of the tallgrass prairie research team. Terry Bolger helped in the initial setup of the study.

6. References

Ågren, G.I., R.E. McMurtrie, W.J. Parton, J. Pastor, and H.H. Shugart. 1991. State-of-the-art of models of production-decomposition linkages in conifer and grassland ecosystems. *Ecological Applications* **1**(2), 118-138.

Arp, W.J. 1991. Effects of source-sink relations on photosynthetic acclimation to elevated CO$_2$. *Plant, Cell and Environment* **14**, 869-875.

Arp, W.J. and B.G. Drake. 1991. Increased photosynthetic capacity of *Scirpus olneyi* after 4 years of exposure to elevated CO$_2$. *Plant, Cell and Environment* **14**, 1003-1006.

Bazzaz, F.A. 1990. The response of natural ecosystems to the rising global CO$_2$ levels. *Ann. Rev. Ecol. Syst.* **21**, 167-196.

Berendse, F., B. Berg, and E. Bosatta. 1987. The effect of lignin and N on the decomposition of litter in nutrient-poor ecosystems: a theoretical approach. *Can. J. Bot.* **65**, 1116-1120.

Boden, T.A., P. Kanciruk, and M.P. Farrell. 1990. *Trends '90: A Compendium of Data on Global Climate Change*. CO$_2$ Information Analysis Center, Oak Ridge National Laboratory, Oak Ridge TN. 257 pp.

Cave, G., L.C. Tolley, and B.R. Strain. 1981. Effect of carbon dioxide enrichment on chlorophyll content, starch content and starch grain structure in *Trifolium subterraneum* leaves. *Physiol. Plant.* **51**, 171-174.

Curtis, P.S., B.G. Drake, P.W. Leadley, W.J. Arp, and D.F. Whigham. 1989a. Growth and senescence in plant communities exposed to elevated CO$_2$ concentrations on an estuarine marsh. *Oecologia* **78**, 20-26.

Curtis, P.S., B.G. Drake, and D.F. Whigham. 1989b. N and C dynamics in C$_3$ and C$_4$ estuarine marsh plants grown under elevated CO$_2$ in situ. *Oecologia* **78**:297-301.

Curtis, P.S., L.M. Balduman, B.G. Drake, and D.F. Whigham. 1990. Elevated atmospheric CO$_2$ effects on belowground processes in C$_3$ and C$_4$ estuarine marsh communities. *Ecology* **71**, 2001-2006.

Dahlman, R.C. 1984. *Vegetation responses to carbon dioxide research plan*. DOE/ER-0187. Dist. Category UC-11. U.S. Department of Energy, Co2 Res. Div., Office of Basic Sci., Washington, DC. 32 pp.

Drake, B.G. and P.W. Leadley. 1991. Canopy photosynthesis of crops and native plant communities exposed to long-term elevated CO$_2$. *Plant, Cell and Environment* **14**, 853-860.

Dyson, F.J. 1990. *Carbon dioxide in the atmosphere and the biosphere*. Radcliffe Lecture. Green College, Oxford University, England. Oct. 11, 1990.

Eamus, D., and P.G. Jarvis. 1989. The direct effects of increase in the global atmospheric CO_2 concentration on natural and commercial temperate trees and forests. *Advances in Ecological Research* **19**, 1-57.

Edwards, G., and G. Walker. 1983. C_3, C_4: *Mechanisms, and Cellular and Environmental Regulation, of Photosynthesis.* 542 p. Blackwell, Oxford.

Field, C.F., F.S. Chapin III, P.A. Matson, and H.A. Mooney. 1992. Responses of terrestrial ecosystems to the changing atmosphere: A resource-based approach. *Annu. Rev. Ecol. Syst.* **23**, 201-235.

Finn, G.A., and W.A. Brun. 1982. Effect of atmospheric CO_2 enrichment on growth, nonstructural carbohydrate content, and root nodule activity in soybean. *Plant Physiology* **69**, 327-331.

Gifford, R.M., N.P. Cheney, J.C. Noble, J.S. Russell, A.B. Wellington, and C. Zammit. 1990. Australian land use, primary production of vegetation and carbon pools in relation to atmospheric carbon dioxide concentration. In: Australia's Renewable Resources: Sustainability and Global Change, R.M. Gifford and M.M. Barson (eds.) *Bureau of Rural Resources Proceedings No. 14.* Resource Assessment Commission, Queen Victoria Terrace, Parkes ACT 2600. pp.151-187.

Hall, D.O. 1989. Carbon flows in the biosphere: present and future. *J. Geol. Soc.* **46**, 175-181.

Hocking, P.J. and C.P. Meyer. 1985. Responses of noogoora burr (*Xanthium occidentale* Bertol.) to nitrogen supply and carbon dioxide enrichment. *Annals of Botany* **55**, 835-844.

Houghton, R.A., R.D. Boone, J.R. Fruci, J.E. Hobbie, J.M. Melillo, C.A. Palm, B.J. Peterson, G.R. Shaver, G.M. Woodwell, B. Moore, D.L. Skole, and N. Myers. 1987. The flux of carbon dioxide from terrestrial ecosystems in 1980 due to changes in land use: Geographic distribution of the global flux. *Tellus* **39B**, 132-139.

Hunt, H.W., M.J. Trlica, E.F. Redente, J.C. Moore, J.K. Detling, T.G.F. Kittel, D.E. Walter, M.C. Fowler, D.A. Klein, and E.T. Elliott. 1991. Simulation model for the effects of climate change on temperate grassland ecosystems. *Ecological Modeling* **53**, 205-246.

Huston, J.E., and W.E. Pinchak. 1991. Range Animal Nutrition. p. 27-64. *In: Grazing Management: An Ecological Perspective.* R.K. Heitschmidt and J.W. Stuth, eds. Timber Press. Portland OR.

Jones, M.B., and R.G. Woodmansee. 1979. Biogeochemical cycling in annual grassland ecosystems. *Botanical Review* **45**, 111-114.

Keeling, C.D. 1989. *Overview of Scripps Program to observe atmospheric carbon dioxide.* Scripps Institution of Oceanography Preprint. January 2, 1989.

Kimball, B.A. 1983. Carbon dioxide and agricultural yields: An assemblage and analysis of 430 prior observations. *Agron. J.* **75**, 779-788.

Knapp, A.K., E.P. Hamerlynck, and C.E. Owensby. 1993a. Photosynthesis and water relations responses to elevated CO_2 in the C_4 grass *Andropogon gerardii*. *Environmental and Experimental Botany* (submitted)

Knapp, A.K., J.T. Fahnestock, and C.E. Owensby. 1993b. Elevated atmospheric CO$_2$ alters dynamic stomatal responses to sunlight in a C$_4$ grass. *Plant, Cell and Environment* (submitted)

Long, S.P. 1991. Modification of the response of photosynthetic productivity to rising temperature by atmospheric CO$_2$ concentrations: Has its importance been underestimated? *Plant, Cell and Environment* **14**, 729-739.

Melillo, J.M. 1983. Will increases in atmospheric CO$_2$ affect decay processes?. *Ecosys. Center Ann. Rpt.*, Marine Biological Laboratory, Woods Hole, MA, USA. pp. 10-11..

Mooney, H.A., B.G. Drake, R.J. Luxmoore, W.C. Oechel, and L.F. Pitelka. 1991. What has been learned from laboratory experiments on plant physiology and field observations? *BioScience* **41**, 96-104.

Morrison, J.I.L. 1985. Sensitivity of stomata and water use efficiency to high CO$_2$. *Plant, Cell and Environment* **8**, 467-474.

Newton, P.C.D. 1991. Direct effects of increasing carbon dioxide on pasture plants and communities. *New Zealand J. Agr. Res.* **34**, 1-24.

Nijs, I., I. Impens, and T. Behaeghe. 1988. Effects of rising atmospheric carbon dioxide concentration on gas exchange and growth of perennial ryegrass. *Photosynthetica* **22**, 44-50.

Owensby, Clenton E., Robert M. Hyde, and Kling L. Anderson. 1969. Effect of clipping and added moisture and nitrogen on loamy upland bluestem range. *J. Range Manage.* **23**, 341-347.

Owensby, C.E., P.I. Coyne, J.M. Ham, L.M. Auen, and A.K. Knapp. 1993a. Biomass production in a tallgrass prairie ecosystem exposed to ambient and elevated levels of carbon dioxide. *Ecological Applications.* (in press)

Owensby, C.E., P.I. Coyne, and L.M. Auen. 1993b. Nitrogen and phosphorus dynamics of a tallgrass prairie ecosystem exposed to elevated carbon dioxide *Plant, Cell and Environment.* (in press)

Owensby, C.E., and L.M. Auen. 1993. Forage quality for ruminants on a tallgrass prairie ecosystem exposed to elevated carbon dioxide. *J. Range Manage.* (submitted)

Rice, C.W., F.O. Garcia, C.O. Hampton, and C.E. Owensby. 1992. Soil microbial biomass and respiration under increased levels of atmospheric CO$_2$. *Agron. Abstr.*, Am. Soc. Agron. p. 258. 1992 Annual Meeting. Minneapolis, MN.

Schulze, E.D. 1991. Water and nutrient interactions with plant stress. In: *Response of Plants to Multiple Stresses.* H.A. Mooney, W.E. Winner, and E.J. Pell, (eds.). Academic Press, New York. pp. 88-101.

Strain, B.R. and J.D. Cure (eds.). 1985. *Direct effect of increasing carbon dioxide on vegetation.* Carbon Dioxide Research, State of the Art. U.S. Department of Energy, Washington, DC.

Swift, M.J., O.W. Heal, and J.M. Anderson. 1979. Decomposition in terrestrial ecosystems. *Studies in Ecology* **5**, 1-372. Blackwell Scientific Publications, Oxford.

Tans, P.P., I.Y. Fung, and T. Takahashi. 1990. Observational constraints on the global atmospheric CO$_2$ budget. *Science* **247**, 1431-1438.

Thomas, J.F. and C.N. Harvey. 1983. Leaf anatomy of four species grown under continuous CO_2 enrichment. *Bot. Gaz.* **144**, 303-309.

Thomas, R.B. and B.R. Strain. 1991. Root restriction as a factor in photosynthetic acclimation of cotton seedlings grown in elevated carbon dioxide. *Plant Physiology* **96**, 627-634.

van Bavel, C.H.M. 1974. Anti-transpirant action of carbon dioxide on intact sorghum plants. *Crop Sci.* **14**, 208-212.

van Oosten, J.J., D. Afif, and P. Dizengremel. 1992. Long-term effects of a CO_2-enriched atmosphere on enzymes of the primary C metabolism of spruce trees. *Plant Physiol. Biochem.* **30**, 541-547.

Warrick, R.A., and R.M. Gifford. 1985. CO_2, climate change and agriculture: assessing the response of crops to the direct effects of increased CO_2 and climate change. In: *The Greenhouse Effect, Climatic Change, and Ecosystems*, B. Bolin, B. R. Doos, J. Jager, and R.A. Warrick,. (eds.). John Wiley, Chichester, pp. 393-474.

Wong, S.C. 1979. Elevated atmospheric partial pressure of CO_2 and plant growth. I. Interactions of nitrogen nutrition and photosynthetic capacity in C_3 and C_4 plants. *Oecologia* **44**, 68-74.

Woodward, F.I., G.B. Thompson, and I.F. McKee. 1991. The effects of elevated concentrations of carbon dioxide on individual plants, populations, communities and ecosystems. *Annals of Botany* **67**, 23-38.

Woodward, F.I. 1987. Stomatal numbers are sensitive to increases in CO_2 from pre-industrial levels. Nature 327:617-618.

Wray, S.M. and B.R. Strain. 1986. Response of two old field perennials to interactions of CO_2 enrichment and drought stress. *Amer. J. Bot.* **73**, 1486-1491.

DIURNAL AND SEASONAL CARBON DIOXIDE EXCHANGE AND ITS COMPONENTS IN TEMPERATE GRASSLANDS IN THE NETHERLANDS - AN OUTLINE OF THE METHODOLOGY

B.O.M. DIRKS

Department of Theoretical Production Ecology, Wageningen Agricultural University,
P.O. Box 430, 6700 AK Wageningen, The Netherlands

Abstract. A methodological outline is presented of a study into the diurnal and seasonal cycle of carbon fluxes within grassland ecosystems in the Netherlands in relation to their environment. At experimental sites Lelystad and Zegveld - predominantly *Lolium perenne* L. at a clay and peat soil, respectively - measurements will be made on *(1)* net CO_2 assimilation of the grassland vegetation using infrared gas analysis; *(2)* carbon distribution within the plant using ^{14}C pulse labeling; and *(3)* carbon and CO_2 fluxes associated with root respiration and soil organic matter decomposition using ^{14}C pulse labeling. At both sites and at experimental site Cabauw additional measurements will be made on total CO_2 fluxes between the grassland vegetation and the lower part of the atmospheric boundary layer. For the analysis of the experimental results and generalisation of the relationships between carbon fluxes and environmental and plant factors use will be made of dynamic simulation models of grass growth and soil organic matter dynamics.

1. Introduction

Long-term records of atmospheric CO_2 concentrations indicate a continuous increase from the 18th century onwards (Boden *et al*, 1991). Since the human perception of atmospheric change in the 1960s, geosphere and biosphere research associated with increasing CO_2 levels has increased concurrently.

In the discipline of geosphere research, numerous authors have indicated that atmospheric CO_2 could well serve as a climatic factor (Bolin *et al*, 1986; Houghton *et al*, 1990). Indeed, models of the atmospheric general circulation (GCMs) indicate changes in average global temperature and global temperature patterns induced by changing atmospheric CO_2 concentrations (e.g. Gates, 1985; Manabe *et al*, 1991). Projected changes in the global hydrological cycle are less accurate, at least partly as a result of the disparity between the resolutions of simulated processes and processes determining the hydrological cycle (Gates, 1985). Actual climatic changes have not been detected yet (Coops, 1991), and, despite all theoretical research efforts, are difficult to predict (Goudriaan, 1987, 1992; Kahl *et al*, 1993). Attempts to validate the projected climatic changes cannot extend beyond comparisons of current climate and GCM performance for those conditions (Gates, 1985).

Water, Air, and Soil Pollution **70**: 425–430, 1993.
© 1993 *Kluwer Academic Publishers.*

Whereas under controlled conditions CO_2 effects have been well assessed on biospheric subsystems on a small temporal and spatial scale, the effects in field situations have yet to be determined unequivocally (Strain and Thomas, 1992). Detection of effects in the field is difficult because of the large natural heterogeneity (Goudriaan, 1992) and associated complex of feedbacks (Grace, 1991). Conversely, the biosphere displays a major influence on the course of diurnal and annual atmospheric CO_2 cycles, with vertical extensions into the atmospheric boundary layer and the full atmospheric column, respectively (Goudriaan, 1987). Sud *et al* (1990) showed that GCM results with respect to surface fluxes, and therefore the general circulation and hydrological cycle, are modified considerably after accounting for the biosphere.

Carbon sequestering into, and release from, living biomass and soil organic matter constitute long-term biospheric control of atmospheric CO_2. The terrestrial biosphere holds substantial amounts of carbon (approximately 500 Gt in living biomass and 1,400 Gt in soil organic matter) as compared to the atmosphere (700 Gt) (Goudriaan, 1987, 1992). Inasmuch as living biomass will reach steady state conditions with respect to carbon sequestration relatively fast, soil organic matter on the other hand may accumulate for prolonged periods of time. Therefore, soil organic matter constitutes a major carbon reservoir (Goudriaan, 1987, 1992) - a carbon sink or source, in case of a net sequestration or release, respectively.

Goudriaan (1992) distinguished between six major vegetation types for the world. In this subdivision, grasslands in 1980 covered 1,800 Mha of Earth's surface, although a part of the sparsely vegetated surface of 3,000 Mha could equally be denominated as more broadly defined grassland. Along with its large surface cover, grassland is characterised by the highest soil carbon density as compared to the other vegetation types.

Detailed studies on carbon fluxes within the biosphere and between biosphere and atmosphere, either under current or under projected atmospheric and climatic conditions, serve more comprehensive models on the global carbon cycle (Goudriaan, 1987, 1990, 1992). However, to a large extent they seem to have encompassed predominantly forest ecosystems (Wisniewski and Lugo, 1992; this volume). Most efforts in the field of grasslands have been made for the Central Grasslands region in the United States (e.g. Schimel *et al*, 1990). However, as a result of the still relatively low temporal and spatial resolution most of this research implicitly integrates several relevant processes over time and space. Schimel *et al* (1990) pointed out substantial intra-regional differences in grassland productivity, most closely related to precipitation patterns. For a mixed grassland ecosystem, Redmann (1978) related measured daily average grass CO_2 assimilation to environmental factors using multiple regression analysis. Most detailed studies on CO_2 fluxes between grassland and atmosphere were reported for a tallgrass prairie ecosystem (Kim & Verma, 1990; Verma *et al*, 1989), encompassing leaf area measurements and eddy correlation flux measurements with a high temporal and spatial resolution. In a concise theoretical study, Kim *et al* (1992) quantified the components of similar flux measurements on a daily basis.

In the following, the methodological outline is presented of a study aiming at elucidation of the carbon fluxes within a grassland ecosystem in relation to its environment. Year-round measurements will be made of the total CO_2 flux and its components between grassland and the atmospheric boundary layer at a high temporal resolution. Measurements will be made on two grassland ecosystems in the Netherlands, on two different soil types. The experimental data will be analysed and the relationships between carbon fluxes and environmental and plant factors generalised on the same temporal and spatial resolution as the measurements, applying simulation models for grass growth and soil organic matter dynamics. In the first instance, the generalised relationships between CO_2 fluxes and environment and plant will be applied to analyse total CO_2 flux measurements at a third grassland ecosystem.

2. Methodology

2.1. EXPERIMENTAL STUDY

The study encompasses grassland ecosystems at three experimental sites in the Netherlands: Lelystad (52°32'N 5°33'E), Zegveld (52°07'N 4°52'E) and Cabauw (51°58'N 4°55'E). The sites are, depending on the degree of drainage, more or less dominated by perennial ryegrass (*Lolium perenne* L.). The soils consist of clay, peat, and peat with a top clay layer of 1 m depth, respectively. Measurements on carbon fluxes will be done with the involvement of the Netherlands Energy Research Foundation (ECN), KEMA Testing Research and Development and Engineering Consultants in the Electric Power Industry (NV KEMA), DLO Institute for Soil Fertility Research (IB-DLO), Royal Netherlands Meteorological Institute (KNMI) and Department of Theoretical Production Ecology, Wageningen Agricultural University (WAU-TPE). The experimental study will last approximately 2 years, and concerns measurements on *(1)* total CO_2 fluxes between the grassland vegetation and the lower part of the atmospheric boundary layer; *(2)* net CO_2 assimilation of the grassland vegetation and carbon distribution within the plant; and *(3)* carbon and CO_2 fluxes associated with root respiration and soil organic matter decomposition. Additional measurements will be made on meteorological parameters and the soil moisture content and soil temperature profiles.

Measurements of the total CO_2 flux will be made for the lower 4 to 5 m of the atmospheric boundary layer.

Measurements on growth and net CO_2 assimilation of the grassland vegetation will be made at different temporal and spatial resolutions: *(a)* instantaneous net CO_2 assimilation of approximately 1 m^2 of grassland vegetation in two replications, using an infrared gas analyser; *(b)* growth analyses of approximately 1 m^2 of grassland vegetation in several replications at two-week intervals, to determine leaf area, and biomass and dry matter of sheath, and green, yellow and dead leaves; and *(c)* harvest of larger sections of grassland vegetation in several replications at one-week to two-week intervals, to determine harvestable biomass.

Carbon distribution and redistribution within the plant, in dependence of the plant's physiological development stage and management practices, will be measured by a 1 to 2 hour application of ^{14}C pulse labeling under field conditions.

Carbon and CO_2 fluxes associated with the soil subsystem will equally be determined by applying ^{14}C pulse labeling under field conditions. In separate objects ^{14}C-labeled CO_2 will be applied to the plant and ^{14}C-labeled glucose will be injected into the soil. CO_2 fluxes from columns of soil will be measured after sealing the soil surface. Subsequent measurements on *(a)* the distribution of ^{14}C over the different soil organic matter fractions; *(b)* the respiration from decomposition of shoots, roots and soil organic matter; and *(c)* the microbial respiration from root exudates, will be made during a period of approximately 2 years.

Table 1 indicates which of the measurements mentioned in the preceding will be made on the different experimental sites by the different institutions.

Table 1. Matrix of measurements to be made at experimental sites Lelystad, Zegveld and Cabauw, indicating the directly responsible institutions.

measurement	experimental location		
	Lelystad	Zegveld	Cabauw
total CO_2 flux	NV KEMA	ECN	ECN/KNMI
net CO_2 assimilation	WAU-TPE	-	-
growth analysis	WAU-TPE	WAU-TPE	-
harvestable biomass	experimental farm	experimental farm	-
^{14}C pulse labeling	IB-DLO	-	-
soil temperature profile	WAU-TPE	WAU-TPE	-

2.2. THEORETICAL STUDY

For the analysis of the results from the experimental study and generalization of the relationships between CO_2 fluxes and environmental and plant factors use will be made of dynamic simulation models.

A simulation model for grass vegetation growth at a temporal resolution of less than one hour is under development. It is based on the SUCROS crop growth model, originally developed for spring wheat (Van Laar *et al*, 1992; Spitters *et al*, 1989). Adaptations of SUCROS for grass growth have been made before, however have proven to be inadequate. Aspects of a simulation model for single plants of perennial ryegrass (Van Loo and Lantinga, in prep.) will be used for further adaptations of SUCROS. Summarization of the modeled grass growth processes to a coarser temporal resolution is required for simulation of the cycle of seasonal CO_2 exchange.

The principles of a simulation model for the dynamics of soil organic matter are based on a spatial and temporal subdivision, into vertically distributed soil layers and fractions of soil organic matter with different residence times (Parton *et al*, 1987; Parton *et al*, 1988; Verberne *et al*, 1990). To assess the contributions of the different grassland ecosystem components to the diurnal cycle of CO_2 exchange, the temporal resolution of the model should be similar to that of the grass growth model.

Acknowledgment

Drs. J. Goudriaan and C.E. Owensby are acknowledged for their comments on the manuscript.

References

Boden T.A., Sepanski R.J., and Stoss F.W. (eds): 1991, *Trends '91: a compendium of data on global change*, Carbon Dioxide Information Analysis Center.

Bolin B., Döös B.R., Jäger J., and Warrick R.A. (eds): 1986, *The Greenhouse Effect, Climatic Change, and Ecosystems*, Wiley and Sons.

Coops A.J.: 1991, *Norsk Geologisk Tidsskrift* **71**, 179-182.

Gates W.L.: 1985, *Climatic Change* **7**, 267-284.

Goudriaan J.: 1987, *Netherlands Journal of Agricultural Science* **35**, 177-187.

Goudriaan J.: 1990, Atmospheric CO_2, global carbon fluxes and the biosphere, in: Rabbinge R., Goudriaan J., Keulen H. van, Penning de Vries F.W.T., and Laar H.H. van (eds). *Theoretical Production Ecology: reflections and prospects,* Pudoc, pp. 17-40.

Goudriaan J.: 1992, *Journal of Experimental Botany* **43**, 1111-1119.

Grace J.: 1991, *Functional Ecology* **5**, 192-201.

Houghton J.T., Jenkins G.J., and Ephraums J.J. (eds): 1990, *Climate Change - the IPCC Scientific Assessment,* Cambridge University Press.

Kahl J.D., Charlevoix D.J., Zaitseva N.A., Schnell R.C., and Serreze M.C.: 1993, *Nature* **361**, 335-337.

Kim J., and Verma S.B.: 1990, *Boundary-Layer Meteorology* **52**, 135-149.

Kim J., Verma S.B., and Clement R.J.: 1992, *Journal of Geophysical Research* **97**, 6057-6063.

Laar H.H. van, Goudriaan J., and Keulen H. van: 1992, *Simulation of Crop Growth for Potential and Water-Limited Production Situations (as Applied to Spring Wheat).* Centre for Agrobiological Research and Department of Theoretical Production

Ecology, Wageningen Agricultural University.

Loo E.N. van, and Lantinga E.A.: in prep.

Manabe S., Stouffer R.F., Spelman M.J., and Bryan K.: 1991, *Journal of Climate* **4**, 785-818.

Parton W.J., Schimel D.S., Cole C.V., and Ojima D.S.: 1987, *Soil Science Society of America Journal* **51**, 1173-1179.

Parton W.J., Stewart J.W.B., and Cole C.V.: 1988, *Biogeochemistry* **5**, 109-131.

Redmann R.E.: 1978, *Canadian Journal of Botany* **56**, 1999-2005.

Schimel D.S., Parton W.J., Kittel T.G.F., Ojima D.S., and Cole C.V.: 1990, *Climatic Change* **17**, 13-25.

Spitters C.J.T., Keulen H. van, and Kraalingen D.W.G. van: 1989, A simple and universal crop growth simulator: SUCROS87, in: Rabbinge R., Ward S.A., and Laar H.H. van (eds). *Simulation and Systems Management in Crop Protection*, Pudoc, pp. 147-181.

Strain B.R., and Thomas R.B.: 1992, *Water, Air, and Soil Pollution* **64**, 45-60.

Sud Y.C., Sellers P.J., Mintz Y., Chou M.D., Walker G.K., and Smith W.E.: 1990, *Agricultural and Forest Meteorology* **52**, 133-180.

Verberne E.L.J., Hassink J., Willigen P. de, Groot J.J.R., and Veen J.A. van: 1990, *Netherlands Journal of Agricultural Science* **38**, 221-238.

Verma S.B., Kim J., and Clement R.J.: 1989, *Boundary-Layer Meteorology* **46**, 53-67.

Wisniewski J., and Lugo A.E. (eds): 1992, *Natural sinks of CO_2 - Palmas del Mar, Puerto Rico, 24-27 February 1992*, Kluwer.

MAJOR CARBON RESERVOIRS OF THE PEDOSPHERE ; SOURCE - SINK RELATIONS; POTENTIAL OF $D^{14}C$ AND $\delta^{13}C$ AS SUPPORTING METHODOLOGIES

H.W.SCHARPENSEEL

Institute of Soil Science, Hamburg University, Allendeplatz 2, D-2000-HAMBURG 13 , GERMANY

Abstract. I tried to identify and assess the C reservoirs in the pedosphere with C-source-sink-relations whereever it seemed possible and pointing to the poorly known rather enigmatic remainder. A special endeavor focussed on the chances and limitations of applying radiocarbon dating and stable isotope $\delta^{13}C$ measuring techniques.

1. Introduction

The atmospheric CO_2 content, presently rising about 0.5 % y^{-1}, showed a long term inverse trend of decrease throughout the Archean, Proterozoic and Phanerozoic (Budyko et al,1987, Berner,1990, Lovelock,1978, 1979, 1988, 1991). After sustained anthropogenic CO_2 inputs from forest clearing and fossil fuel consumption will cease the trend may be inverted. It seems, there is a geochemical superiority of the C sink - by photosynthesis, chemical weathering, dissolution or precipitation in the ocean - compared to the C source - by respiration, volcanism, eduction from subduction zones in conjunction with carbonate metamorphism -. CO_2 may in accordance with its atmospheric residence time of ca 100 years become the minimum factor of plant growth and food production for the overboarding world population.
A likewise uncertainty is the size of the economically accessable fossil C pool. While Moore et al. (1989) opt for about 10,000 Gt of C, Grassl and Klingholz (1990) cite the International Energy Agency in Paris, whose estimate is but 1000 Gt C from a total fossil C pool of 5 to 13,000 Gt. If the latter estimate is correct, with 250 Gt of C being so far consumed, the remaining 750 Gt C could at best produce a further rise of CO_2 in the atmosphere of 50%, considering, that the atmospheric C reservoir is about of the same magnitude. Otherwise,from the anthropogenic CO_2 input of 7 to 7.5 Gt C yr^{-1} only about 3.5 Gt go in the atmosphere, another 2.0 Gt in the ocean and 1.5 to 2.0 Gt in the still disputed "missing C fraction". According to Esser (1990) the CO_2-source from wood clearing / slash & burn is already exceeded by the CO_2 sink due to CO_2 fertilization (Table 1).
Another important uncertainty is the pool size of C in the SOM (soil organic matter) - C reservoir, being estimated by various authors to comprise between 1000 and 3000 Gt C (Batjes and Bridges , 1992).
Table 2 indicates the different C reservoirs in the pedosphere. It appears, that about 1200 Gt C for the terrestrial pedosphere is with an average of 9.4 kg C m^{-2} alraedy a plush estimate, considering, that 12 kg C m^{-2} and 1 m depth are the threshold of the "Hum - form" in the Soil Taxonomy.
For the about 570 x 10^6 ha of natural wetlands (Aselmann and Crutzen, 1990) about 300 Gt C, equal to 53 kg C m^{-2} seemed adequate. This workshops wetland discussion group expanded the area to 1024 x 10^6 ha and 631 Gt pool size of C.

COMPARTMENTS IN C CYCLE	ESTIMATED POOL SIZE OF C
SEDIMENTS	65,000,000 Gt
OCEANS BELOW 100 m	38,000 Gt
OCEANS 0 to 100 m DEPTH	700 Gt
ATMOSPHERE IN CO_2	740 Gt
LIVING BIOMASS	650 Gt
FOSSIL ENERGY SOURCES	5 - 13,000 Gt
ECONOMICALLY USEFUL (INT.EN.AG.PARIS)	1,000 Gt
MEANWHILE CONSUMED	250 Gt
USEFUL POOL STILL LEFT	750 Gt
ANTHROPOGENIC C ADDED PER YEAR	7.5 Gt
- FROM COAL,OIL,GAS	5.5 Gt
- FROM FOREST CLEARING, SLASH & BURN	1.5 to 2.0 Gt
RELEASED INTO ATMOSPHERE PER YEAR	3.5 Gt
RELEASED INTO OCEANS PER YEAR	2.0 Gt
DIFFERENCE "MISSING CARBON"	1.5 to 2.0 Gt
(CO_2-fertilization, C-sequestration in SOM)	

Table 1   Estimated compartments in C-cycle

The 1.5×10^9 ha of cropland out of the 12.8×10^9 ha terrestrial soils contribute some 150 to 300 Gt C, which stands for 10 to 20 kg C m^{-2}. The reality could be closer or even below the lower threshold.

2. Carbon reservoirs of the pedosphere ; source - sink relations

The grassland has in its humus compartment C sequestered of a magnitude as in the living biomass of tropical rain forests (Whittaker and Likens , 1973). Grassland shows especially long C residence time in the SOM, not the least in comparison with woodland of the same soil environment. This applies also to soils with preferably grass vegetation like Mollisols and Alfisols in comparison with other soil orders. Figure 1 shows age versus depth curves of australian Vertisols and Krasnozems under forest and grassland. The C residence time at different soil depth levels is higher in the grassland profiles. Figure 2 reflects age versus depth regression curves of Inceptisols, Spodosols, Vertisols, Alfisols, Mollisols. Alfi- and Mollisols, typic grassland ecosystems, show the highest C residence time versus depth. Calcretes contribute ca 1000 Gt of C, an amount close to the terrestrial SOM pool size, but only ca 50% of the C originates from organic C due to biomass decomposition. Since the amount of C depends on the Calcium of the preceeding primary carbonates calcretes represent in total no additional C sequestration.

CARBON IN SOIL EPIPEDON ESTIMATED COMPARTMENTS	10^6 ha	~Gt of C	kg C m^{-2}
ICEFREE CONTINENTAL SURFACE	13,400	2000	
TERRESTRIAL SOILS	12,800	1500	11.7
		1200	9.4
		1000	7.8
WETLANDS	570	500	88
		300	53
Discussion group,this workshop	1,024	646	63
CROPLANDS	1,500	300	20.0
		150	10.0 *
GRASSLAND	> 3,000		
WOODLAND	> 4,000		
in Tropics	1,700		
Rain Forest	655		
Moist Woods	626		
Dry Woods	252		
Mountain Woods	178		
PEDOLITH´ CALCRETES	~1,000	1000	100 **

Table 2   Carbon reservoirs in the pedosphere
     * (> 12 kg C m^{-2} begin of "Hum-" in Soil Taxonomy)
    ** (if Ca concentration same in primary and secondary pedo-
       genic carbonates no additional C sequestration)

Additional C sinks in the pedosphere - mostly outside the epipedon - are
the illuvial horizons, argillic and spodic horizons, as well as the
fA-horizons of paleosols (Table 3). Taking the areas with argillic horizons, spodic horizons and an estimate of 50% of the area of steppe and
alluvial soils for paleosols with distinctive fA-SOM containing horizons
and assessing the C m^{-2} with 1 kg, all together ca 40 Gt of C would have
to be added, which remains within the disputed range of uncertainty
with regard to the size of the SOM-C reservoir. 1 kg of C m^{-2} is no arbitrary estimate, but results from summation of C in 2 cm thin layers of
corresponding soil profile scan (Becker-Heidmann, 1989, 1990). Although
being a rough estimate from a few thin layer wise investigated soil profiles, it may in a first approximation indicate the magnitude also of
the argillic and spodic horizon plus paleosol - C sink.
Table 4 identifies major soils and land use forms as belonging to C sources or sinks. Backed-up with the knowledge on soil formations the items
of the table are self explaining.
The disputed "missing C fraction´s" input into terrestrial ecosystems in
the light of some fluxes is reflected in table 5.

3. Potential contribution of D^{14}C and δ^{13}C measurements

Radiocarbon dating (D ^{14}C) and δ^{13}C measurements in soils are principally
excellent methodologies to substanciate or alleviate working hypothesis´s
in relation to C residence time, source or sink problems.

Especially the slow apparent aging effect due to sustained input of anthropogenic mostly fossil "dead C", becoming measurable and interpretable
as "Suess Effect", could be helpful in soil C cycle studies. In case of
materials of known age the Suess Effect would precipitate as shift of
^{14}C age, which could be interpreted as C source or sink properties. Houtermans et al (1967) identified the Suess Effect till 1950 to amount to
an aging in D^{14}C of 250 ys.

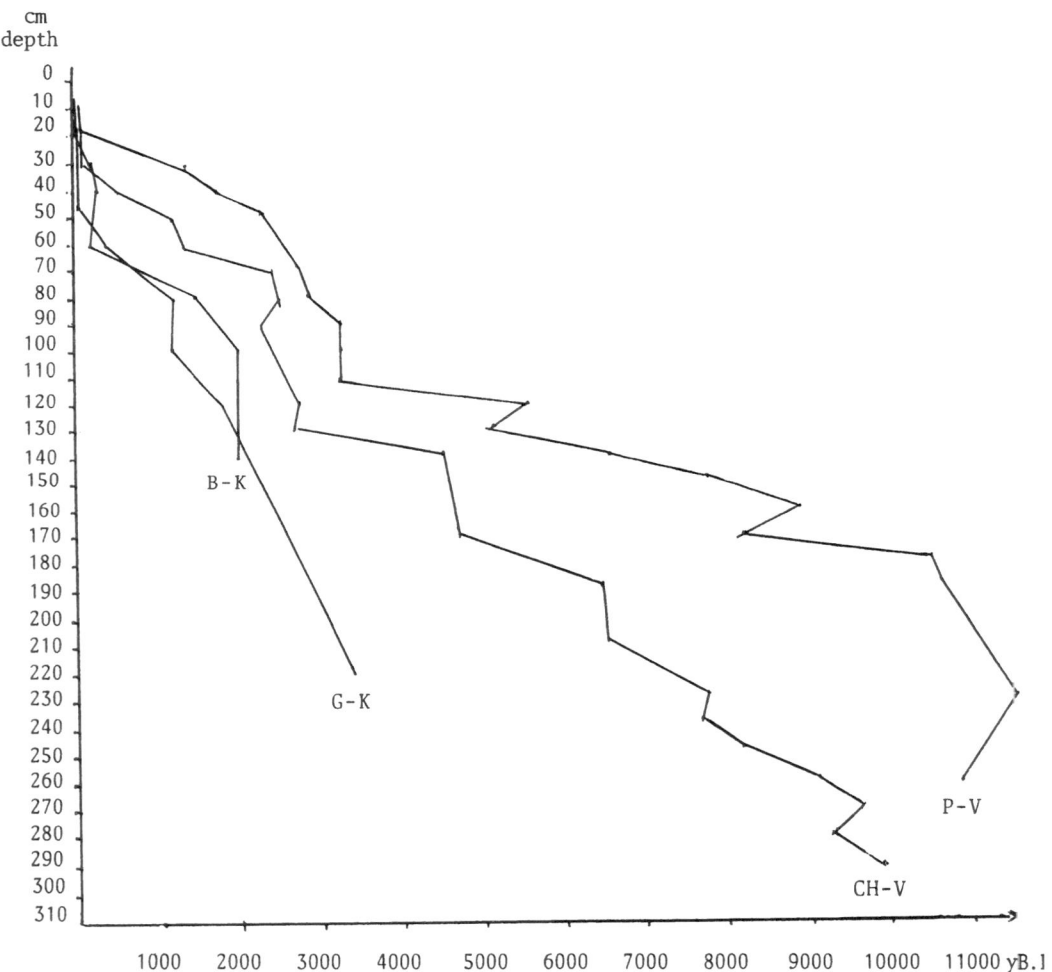

Fig. 1 Australian Vertisols and Krasnozems under forest or grassland

    (P-V= Paget-Vertisol under grassland,sampled by G.D.Hubble,CSIRO,Queensland
    CH-V=Chinchilla-Vertisolunder Acacia harpophylla,forest,sampled by G.D.Hubbl
    G-K=Gabbinbar-Krasnozem,wooded grassland,sampled by G.D.Hubble
    B-K=Beechmont-Krasnozem,under subtropical rainforest,sampled by G.D.Hubble)

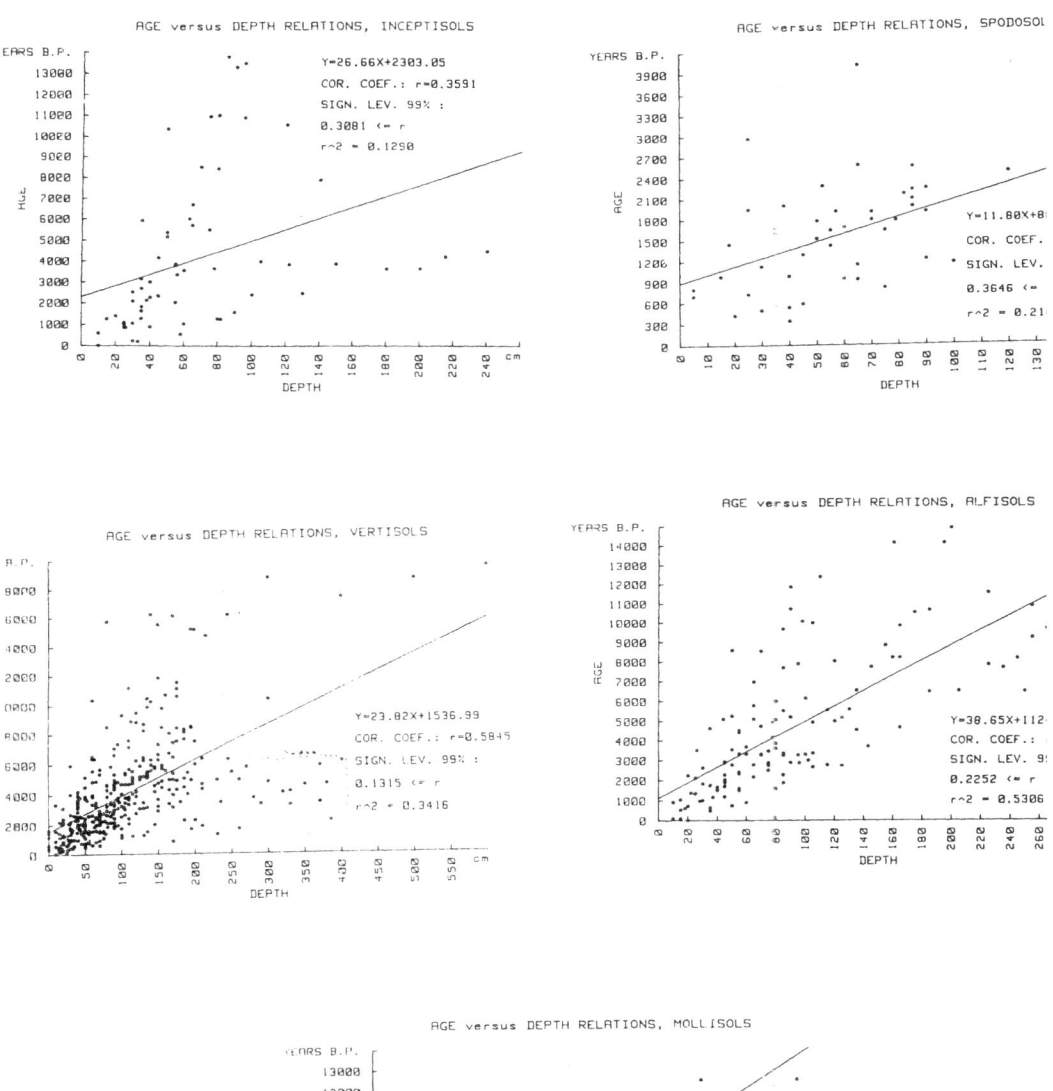

Fig. 2 Age versus depth measurements (from all over) Inceptisols, Spodosol
Vertisols, Alfisols, Mollisols

SOIL UNIT	In 1000 ha World Soil Resources FAO 66	Comparison: Driessen & Dudal (1989)		
		Argillic Horizon (mil ha) *	Spodic Horizon (mil ha)**	Gleysols+Fluvisols +Cherno-,Kastano-, Phaeo-, Greyzem **
LEPTOSOLS	1,655,318			
CAMBISOLS	1,573,402			
ACRISOLS	996,600 *	800 *		
ALISOLS		100 *		
ARENOSOLS	901,885			
CALCISOLS	796,169			
FERRALSOLS	742,600			
GLEYSOLS	718,800 ***			620 ***
LUVISOLS	648,505 *	600 *		
REGOSOLS	578,971			
PODZOLS	487,513 **		480 **	
KASTANOZEMS	467,757 ***			400 ***
LIXISOLS	436,520 *	200 *		
FLUVISOLS	356,238 ***			320 ***
VERTISOLS	337,322			
PODZOLUVISOLS	321,068 *	260 *		
HISTOSOLS	273,248			
CHERNOZEMS	229,218 ***			300 ***
(CH+KA+PH+GR-zem)	(885,097 ***)			(828 ***)
NITISOLS	205,518 *	250 *		
SOLONTCHAKS	187,325			
PHAEOZEMS	154,239 ***			100 ***
SOLONETZ´	135,267			
PLANOSOLS	129,896 *	150 *		
ANDOSOLS	107,045			
GYPSISOLS	90,017			
PLINTHOSOLS	61,135			
GREYZEMS	33,883 ***			28 ***

	2,360 mil ha	480 mil ha	1,770 mil ha
12,625 mil ha	(2,738 mil ha	487 mil ha	1,960 mil ha)

If 1 kg C per 1 m^2
of spodic, argillic
horizon, paleosol,and
1/2 of Gley-,Fluvisol
plus CH-KA-PH-GR-zems
have Paleosols, seque-
stered Gt of C are:

	23.6 to 27.4	ca 5	8.8 to 9.8

Table 3   Additional C sequestration in argillic-,spodic horizons,paleosols

SOILS AND LAND USE FORMS	C - SOURCE	C - SINK
Terrestrial soils with enough clay to sustain clay-organic complexes as C-sink,unmanaged	mild source (varying)	mild sink (varying)
Cropland in ploughing zone, Ap	X	
Soil dynamic processes, forming argillic horizon, spodic horizon, paleosols		X
Fe-toxicity and Acid Sulfate soils SOM-C fuel for Fe and S reduction	mild source	
Bio-, Pelo-, Cryoturbation	<————————>	
Plaggepts		X
Andisols (ca 1000 to 10,000 ys) Andisols after change of Allophane - Humus - Complexes to Clay formation with humus - C loss	X	X
(Submerged) rice soils till 28°C (Submerged) rice soils beyond 28° C	mild source	mild sink
Biological conquest of subsoil after deep ploughing or subsoiling		X
Calcisols with caliche/calcrete only if gaining Ca by weathering		X
Annual application of ca 100 mil t of N, P, K nutrients to cropland by corresponding increase of biomass production		X (ca 1.25 $Pg C.yr^{-1}$)
In correlation with CO_2-fertilization , by anthropogenic C-input (Missing C fraction ?)		X ($\geq$ 1.5 $Pg C . yr^{-1}$)

Table 4   Soils and land use forms as C sources or C sinks

MISSING CARBON FLUXES	MIL T C x Y^{-1}
Missing C total	1,500 to 2,000
C-flux in organic soils (Armentano, 1980)	140
C-flux into soil carbonates (Schlesinger,1985)	10
Riverine transport of C (Schlesinger & Melach,1981)	450
Biomass-C gain from CO_2-fertilization (Esser,1990) (claim, that higher than C-source of forest clearing,savanna-burning)	> 1,500

Table 5   Carbon fluxes against the background of "Missing C" fraction

Unfortunately, the thermonuclear bomb-^{14}C has since 1956 intercepted the
chances via the Suess Effect due to builtup of a 190% modern level of radi
ocarbon at the peak in 1963 (Figure 3). At present the annual input of ca
7.5 Gt C, ca 75% "dead" fuel-C and ca 25% almost "modern" biomass - C remains in the 0.5 % standard error range of conventional C-14 dating.
In near future, with bomb-C content stabilizing on a low level, sustained
measurements at selected sites of biomass and SOM-fractions in conjunction
with the precision of AMS (accelerator mass spectrometry) will open up new
possibilities.
Table 6 shows the climate related depth penetration of bomb ^{14}C in the
soil profiles.
δ^{13}C, indicative for changing land use in the past, accompanied with chan
ges of C^3, C^4, CAM photosynthetic mechanism gives great hope especially by
repeated and longer term measurements. If the change of land use is historically datable, the shift in δ^{13}C can be quantitised in regard to carbon
residence time and mixing(Martin et al,1990; Scharpenseel et al,1992,1993
Roeloffzen et al (1990) could show, that δ^{13}C in the atmosphere decreased
in the period from 1978 till 1989 from about -7.44 to -8.01 ‰. Just one
single only or a one year measurement of δ^{13}C in SOM will not be enough
to substanciate changes due to input of anthropogenic C. This comprises
about 98% of biomass- and fossil C derived CO_2 of -25 to -30‰ and only
ca 2% of CH_4 with a δ^{13}C of about - 60 ‰. Individual measurements therefore remain below the conventional standard error range of ca 0.05 ‰ or
very close to it in their sample specific deviation. Repetitive, longer,
sustained measurements bear good chances for successful trend analysis.

4. Conclusions

- Treatment of C reservoirs in the pedosphere clashes with ambiguity in the size of economically manageable fossil C pool and a reliable size estimate of the SOM - C compartment.
- The size, nature and principle itself of entrance of the "missing C" fraction into the soil related C cycle is still obscured.
- C - compartments of terrestrial soils, wetland, cropland, grassland, woodland and shares of annual net photosynthesis (60 Gt C) are poorly resolved.
- Additional C from pedogenesis and paleosols remains below the present uncertainty limit of the gross SOM C pool.
- Source - sink attribution to soils and land uses are fairly clear. Quantitization is still bleak.
- $D^{14}C$ and $\delta^{13}C$ are promising methods. Utilization of the Suess Effect due to sustained anthropogenic fossil C input is unfortunately jeopardised by the inverse principle of bomb - C since 1956. Thin layer wise soil profile scan by $D^{14}C$ and $\delta^{13}C$ reveals the dynamics and C^3, C^4, CAM - switch related land use changes with impact on C dynamics.

LOCATION OF SOIL PROFILES	CLIMATE	MEAN TEMPERATURE °C	ANNUAL RAINFALL MM	DEPTH PENETRATION OF BOMB - C in cm
Wohldorf, Ohlendorf,Timmendorf, all of Germany	udic-temperate	8 to 9	650	10 (till 45 in sandy soil)
Akko and similar but drier Qedma, both Israel (sampling with D.H.Yaalon)	xeric	~18	620	10 to 15
ICRISAT,Patencheru,India	ustic - semiarid	25.8	760	12
Khon Kaen,NE Thailand,similar Pangil,Luzon, San Dionisio, Panay both Philippines	ustic	27	1300	20
Tiaong, similar Los Banos, both Laguna,Philippines in old volcanic ashes on top of tuffaceous bedrock	udic - tropical	26	2150	60 to 80
Bugallon, Pangasinan,Philippines and Klong Luang, Thailand, both almost constantly reduced and wet	udic - tropical	27	~2000	~5

Table 6   Bomb - ^{14}C / SOM - ^{14}C migration in soil profiles of different climate regions, based on thin layer wise soil profile scanning by natural ^{14}C and $\delta^{13}C$ measurements

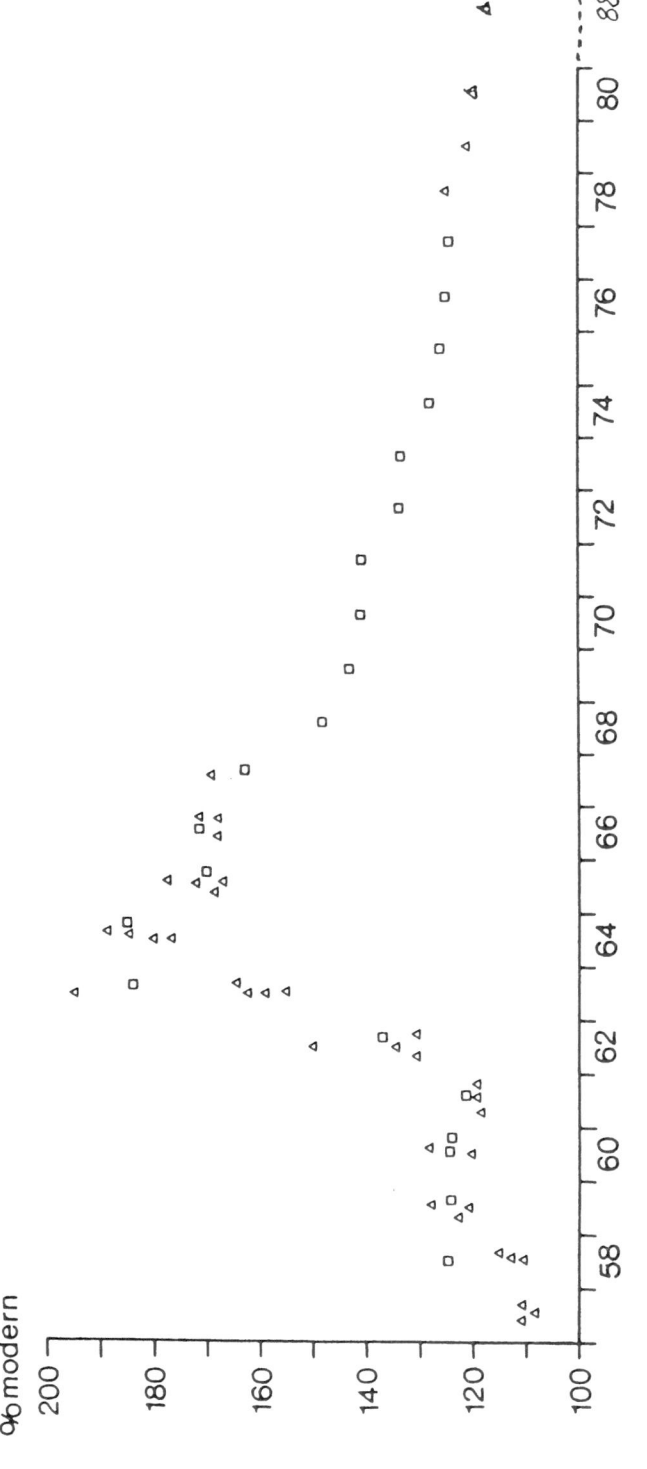

Fig . 3 : Radiocarboncurve of distribution of natural ^{14}C in the atmosphere, measured in the plant substance; □ = wine △ = gras (till 1967 by Tamers et al. 1968)

References

Aselmann,I.,Crutzen,P.J.:1990, A Global inventory of wetland distribution and seasonality, net primary productivity and estimated methane emissions. in:Soils and the Greenhouse Effect,(Ed.A.F.Bouwman), Wiley,pp.441-449.

Armentano,T.V. (Ed.): 1980, The role of organic soils in the world carbon cycle, CONF.7905135, United States Department of Energy, Washington D.C.

Batjes,N.H., Bridges,E.M.:1992, World Inventory of Soil Emissions. ISRIC, WISE Report 1, p.132, Wageningen.

Becker-Heidmann,P.:1989, Die Tiefenfunktionen der natürlichen Kohlenstoff-Isotopengehalte von dünnschichtweise beprobten Parabraunerden und ihre Relation zur Dynamik der organischen Substanz in diesen Böden.(Dissertation), Hamburger Bodenkundliche Arbeiten,Vol 13,1-228.

Becker-Heidmann,P.:1990, Carbon fluxes in important soil classes, with emphasis on lessivé soils and on soil of the terrestrial, of the hydromorphic and temporarily submerged environment. Terminal Report to GTZ (contract 72.7866.6-01.400/1420), pp.1-177.

Berner,R.A.: Atmospheric carbon dioxide levels over Phanerozoic time, Science 249, pp. 1382-1386.

Budyko,M.J., Ronov,A.B., Yanshin,A.L.:1987, History of the Earth Atmosphere,p.80, Springer,Berlin.

Driessen,P.M.,Dudal,R.:1989 Lecture notes on the geography, formation,properties and use of the major soils of the world. Agricultural University Wageningen & Katholieke Universiteit Leuven, p.1-296.

Esser,G.:1990, Modelling global terrestrial sources and sinks of CO_2 with special reference to soil organic matter. In: Soils and the Greenhouse Effect, (Ed. A.F.Bouwman),Wiley,pp.247-263.

FAO,World Resources Reports : 1991, Vol.66, Food and Agricultural Organization of the United Nations ,Rome, p.10.

Graßl,H.,Klingholz,R.:1990, Wir Klimamacher, p.70, S.Fischer Publ.Frankfurt.

Lovelock,J.E.;1979,1978, Gaia, p.39, Oxford University Press.

Lovelock,J.E.:1988, The Ages of Gaia, p. 84, W.W.Norton & Co,London,N.Y.

Lovelock, J.E.: 1991, Healing Gaia, p.23, Harmony Books, N.Y.

Houtermans,J.,Suess,H.E.,Munk,W.:1967,Effect of industrial fuel combustion on the carbon-14 level of atmospheric CO_2. In: Radioactive Dating and Methods of Low Level Counting, Vienna, IAEA, pp. 57-68.

Martin,A.,Mariotti,A.,Balesdent,B., Lavelle,P.,Vuattoux,R.:1990, Estimate of organic matter turnover rate in a savanna soil by ^{13}C natural abundance measurements. Soil Biol.Biochem. Vol. 22, No.4, pp.517-523.

Moore III,B.M.P., Gildea,L.J.,Vorosmarty plus five other contributors: 1989, Biogeochemical cycles. In: Global Ecology Towards a Science of the Biosphere (Eds. M.B.Rambler, L.Margulis, R.Fester). Academic Press, Boston, pp.113-141.

Roeloffzen,J.E., Mook,W.G., Keeling,C.D.: 1990,Trends and variations in stable carbon isotopes of atmospheric carbon dioxide. Proc. IAEA/FAO-Intern. Sympos. on the Use of Stable Isotopes in Plant Nutrition, Soil Fertility and Environmental Studies. IAEA-SM-313, pp.1-24.

Scharpenseel,H.W.,Schiffmann,H.,Hintze,B.: 1984, Hamburg University Radiocarbon Dates III, Radiocarbon. (listed C-14 dates of Chinchilla-, Paget-, Gabbinbar-, Beechmont- Profiles, samples by Dr. Hubble, Australia.

Scharpenseel,H.W.,Becker-Heidmann,P.: 1992, The dilemma of conflicting interests between CO_2's and CH_4's IR trapping capacity and role, in case of CO_2 even as limiting factor for plant growth. Proceedings of "Global Warming, A Call for International Coordination" , in World Resource Review   ,Chicago, April 1992.

Scharpenseel,H.W.,Becker-Heidmann,P.:1993, Carbon storage by grassland soils in different climate zones as revealed by carbon-14 dating. Paper 81+1, Proceedings of International Grassland Congress, February 1993, Palmerston North, N.Z. and Rockhampton, Australia.

Schlesinger,W.H.: 1985, The formation of caliche is soils of the Mojave desert, California. Geochim. Cosmochim. Acta,Vol 49, pp 57-66.

Schlesinger,W.H., Melack,J.M.:1981, Transport of organic carbon in the world rivers. Tellus, Vol 33, pp.172-187.

Tamers,M.A.,Balke,K.D.,Scharpenseel,H.W.: 1968, Untersuchungen zur Fließgeschwindigkeit des Grundwassers durch Bestimmung der Radiokohlenstoff- und Tritiumaktivität. Zeitschr. Kulturtechnik,Flurbereinigung, Vol 9, p.364. (Curve in later years extended).

Whittaker,R.H., Likens,G.E.: 1973, The primary production of the biosphere. Human Ecology Vol 1, pp.299-369.

RIVERINE TRANSPORT OF ATMOSPHERIC CARBON : SOURCES, GLOBAL TYPOLOGY AND BUDGET

Michel MEYBECK

Laboratoire de Géologie Appliquée, C.N.R.S.

Place Jussieu, 75257 Paris Cedex 05

France

Abstract. Atmospheric C (TAC) is continuously transported by rivers at the continents' surface as soil dissolved and particulate organic C (DOC, POC) and dissolved inorganic C (DIC) used in rock weathering reactions. Global typology of the C export rates ($g.m^{-2}.yr^{-1}$) for 14 river classes from tundra rivers to monsoon rivers is used to calculate global TAC flux to oceans estimated to 542 $Tg.yr^{-1}$, of which 37 % is as DOC, 18 % as soil POC and 45 % as DIC. TAC originates mostly from humid tropics (46 %) and temperate forest and grassland (31 %), compared to boreal forest (14 %), savannah and sub-arid regions (5 %), and tundra (4 %). Rivers also carry to oceans 80 Tg. yr^{-1} of POC and 137 $TG.yr^{-1}$ of DIC originating from rock erosion. Permanent TAC storage on land is estimated to 52 $Tg.yr^{-1}$ in lakes and 17 $Tg.yr^{-1}$ in internal regions of the continents.

1. Introduction

Delivery of inorganic C budget to the oceans by rivers was first estimated by Clarke (1924) but reliable budgets of total organic carbon (TOC) have become available only recently. There is little mention of TOC in Livingstone's master review on world's rivers (1963). The first budgets were published in 1981 by Schlesinger and Melack, and by Meybeck. Since then, our understanding of the C content in world's rivers has improved substantially largely by means of the SCOPE-CARBON

Water, Air, and Soil Pollution **70**: 443–463, 1993.
© 1993 *Kluwer Academic Publishers.*

program (1982) headed by E. Degens and his Hamburg team which lead to numerous published workshops and the book by Degens *et al.* (1991), as well as numerous reviews of the global budget of riverine TOC (Milliman *et al.,* 1984 ; Kempe, 1985 ; Meybeck, 1988 ; Ittekkot, 1988). Related topics, such as lake retention of TOC and anthropic influences on the river C budget, have also been considered by Mulholland and Elwood (1982), Kempe (1984) and Kempe *et al.* (1991).

Numerous methods have been employed to evaluate global riverine budget (Meybeck, 1988). Perhaps the most interesting is that which relies on a global typology of specific export rates of C (in t $C.km^{-2}.yr^{-1}$ or g $C.m^{-2}.yr^{-1}$) attributing one mean value for each environmentally classified category (e.g., the taiga, the tropical rain forest). This method is particularly well suited for the estimation of global elemental budgets for each major biome, such as the boreal regions or the humid tropics, and has already been used for major ions - including dissolved inorganic carbon (DIC) by Meybeck (1979), and TOC by Schlesinger and Melack (1981), as well as Meybeck (1981, 1982, 1988). The budget of particulate organic carbon (POC) derives from another approach based on the global distribution of total suspended matter in rivers (TSS) and on the TSS *vs* POC relationship (Meybeck, 1982 ; Ittekkot, 1988).

Considering the more extensive data base on river C now available, a new C budget is possible through the combination of several budget development methodologies (Meybeck, in preparation), of which only the global results have been published (Meybeck, 1993). In this paper, I focus on the global classification of river C export of all C species, i.e., on major river C sinks on land. Anthropic influences in terms of the global river C budget are treated by Downing *et al.* (this volume).

2. Natural Sources of Atmospheric Carbon and Rock Carbon in Rivers

Table 1 lists riverine C in its multiple forms, its characteristics, residence time, and its different sources and origins (Figure 1). Riverine C may be classed as dissolved (DIC, DOC) or particulate (PIC, POC), organic (DOC, POC, TOC) or inorganic (DIC, PIC).

Dissolved inorganic carbon exists mostly in the form of bicarbonate (HCO_3^-) in rivers where the pH range is commonly between 6 and 8.4. In non-carbonate environments, such as plutonic, metamorphic and volcanic regions, and most of shale and sandstone regions, river DIC results entirely from soil and atmospheric CO_2 according to the general weathering equation :

$$\text{Non-carbonate mineral} + CO_2 + H_2O \rightarrow HCO_3^- + \text{clay mineral} + \text{cation} + SiO_2$$

In carbonate environments (e.g., with limestone, dolomite or carbonated shales, etc.), half of the DIC originates from atmospheric and soil CO_2, the other half from weathered rock :

$$\text{Carbonate mineral} + CO_2 + H_2O \rightarrow 2HCO_3^- + Ca^{++} \text{ (or } Mg^{++})$$

In mixed river basins where both carbonate and non-carbonate rocks are found, the proportion of atmospheric CO_2 in riverine DIC is variable, generally between 60 and 80 %.

Table 1. Specific forms of carbon in river-borne material.

Major-Specific form			Natural Origins	Approximate Age (1) (years)
Total carbon	Dissolved carbon	DIC diss. inorganic carbon	. Carbonate mineral weathering . Atm. and Soil CO_2	10^8 $0 - 10^2$
		DOC* diss. organic carbon	Soil leaching	$10^1 - 10^2$
	Particulate carbon	POC* part. organic carbon	. Soil erosion . Sedimentary rocks . Autochtonous	10^2 10^8 10^{-2}
		PIC part. inorganic carbon	. Autochtonous . Sedimentary rocks	10^{-2} 10^8

* Specific forms somewhat affected by human activities.

(1) Age of river carbon from the beginning of atmospheric CO2 uptake by biochemical reactions.

TAC = total atmospheric carbon = DOC + soil POC + riverine atm. DIC.

PIC in rivers originates mostly from the mechanical erosion of sedimentary rocks, so it does not participate in the global CO_2 cycle. That is PIC represents only a transfer of carbonate minerals from highlands to lowlands, oceans or internal basins (e.g., Caspian Sea). However, there are two exceptions. First, in eutrophied rivers where pH can exceed 9.0 during the day, $CaCO_3$ is likely to precipitate in the river, thus reducing the DIC level ; this process occurs mainly during the low water stage of rivers in summer time and does not affect the global balance. Second, in few oversaturated rivers such as the Huang He (Yellow River), $CaCO_3$ may form without any influence of planktonic activity (Feng Jian-Xiang and Kempe, 1987).

DOC in non-polluted rivers originates from soil leaching. In small streams still flowing under forest canopy, throughfall leaching of DOC is likely to occur, but such DOC is so reactive that it is not likely to be carried more than 100 to 1,000 km downstream and thus does not reach the ocean. DOC is much linked to the drainage pattern of river basins (Brinson, 1976 ; Mulholland *et al.*, 1990). Moore (1987) has compared two stream basins in Westland, New Zealand. The well-drained forested Mamaï basin exported 6.8 g C.m^{-2}.yr^{-1} as DOC, while the poorly drained Larry river exported 78.1 g C.m^{-2}.yr^{-1}, which is probably a world maximum.

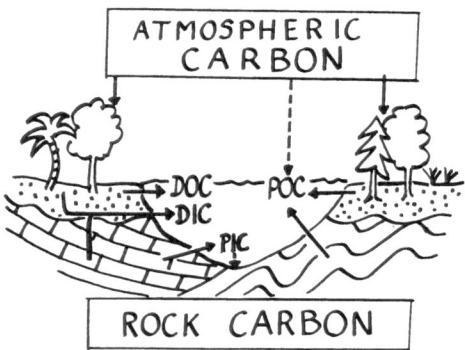

Figure 1. Sources and sinks of riverine C. Direct CO_2 uptake through algal production and calcite precipitation are negligible on global scale. Acronyms, see Table 1.

POC is still not regularly measured in rivers. Most of environmental agencies prefer to survey TOC, i.e., a more rapid measurement of organic C on unfiltered sample. POC can either be expressed in content of suspended matter (TSS) generally as %, abbreviated here as POC %, or as a concentration in mg.l^{-1} : POC (mg.l^{-1}) = POC % x TSS (mg.l^{-1}). In most rivers, the POC content ranges from 1 to 5 % but it can exceptionally reach 0.8 % for the Huang He and 20 % for lowland rivers draining swamps such as the Sopchoppy river (Malcolm and Durum, 1976). POC % decreases with TSS within a given river and in the global variation of average POC % and average TSS in terms of world's rivers (Meybeck, 1982 ; Ittekkot, 1988 ; Martins and Probst, 1991). In most rivers, POC is allochtonous, i.e., not formed within the aquatic environment.

In poorly turbid waters (TSS < 100 mg.l^{-1}), riverine POC originates mostly from soil erosion. In the Amazon, POC represents 45 % of TOC and 90 % of POC is found as fine material. Coarse POC, about 5 % of TOC, is made of various debris (70 % leaves, 20 % wood, and only 10 % of grasses from the varzea flood plain)(Hedges *et al.*, 1986 ; Ertel *et al.*, 1986). In streams (stream order 1 to 3) and small rivers (stream order 4 to 7), the direct fallout of leaves and wood debris from the canopy is a major POC source which decreases rapidly with stream order. This allochtonous C is quickly recycled along the river system : most of it is stored in small streams and the majority of this organic C is eventually metabolized in the 7th-9th order according to Naiman *et al.* (1987) who worked on the Moisie river (9th order, 20,000 km^2 in Quebec).

In highly turbid rivers such as the Huang He, or during major river floods in semi-arid environment such as the Missouri or Indus, river POC % is found to be between 0.5 and 1 % (Zhang *et al.,* 1992 ; Malcolm and Durum, 1976 ; Ittekkot and Arain, 1986). These values are very close to those found in eroded sedimentary rocks, about 0.5 % for the Huang He loess and for average shales (Ronov, 1976). Therefore, most of this river POC probably originates from rock erosion, not from soil erosion, which must be taken into account when constructing the global budget of POC.

In eutrophied rivers, phytoplanktonic production can produce high chlorophyll concentrations (exceeding 100 μg.l^{-1}, which would put them in the hypertrophic category based on any lake classification. If a POC/chlorophyll ratio of 30 g.g^{-1} is applied, maximum plankton biomass peaks may correspond to 6 mg algal POC.l^{-1} (Meybeck, 1993). This ratio is based on direct observations during riverine algal blooms (Dessery *et al.,* 1983). In streams and medium sized rivers, emergent aquatic plants may also contribute to the autochtonous C pool. This autochtonous C certainly represents a substantial fraction of the riverborne C that derives from the atmosphere. However, as for autochtonous DIC, uptake of atmospheric CO_2 by primary production in rivers must be negligible because, on a global scale, pristine rivers have average chlorophyll contents less than 5 μg.l^{-1} and eutrophied rivers are only found in some regions (e.g., Western Europe, USA)(see Downing *et al.,* this volume). Attached and floating vegetation are a secondary POC source in river systems. In the Zaire river, the numerous water hyacinth mats *(Eichhornia crassipes)* have been counted (Leprun, quoted by Davies, 1986). When converting this biomass into organic C, this much coarse autochtonous POC corresponds to about 1 mg.l^{-1}, or about 1/3 of fine and coarse POC.

3. Spatial Distribution of Riverine Carbon Export

3.1.   DATA BASE

For the C cycle, all forms of atmospheric C (here abbreviated as TAC) in rivers must be considered and summed up : TAC = atm DIC + DOC + atm POC.

For a given river, needed measurements are typically HCO_3^-, DOC and POC (generally available as TOC), TSS (for the estimate of rock POC in total POC), river discharge and area, all taken at the same stations. I have compiled this data base from more than 40 rivers (Meybeck, in preparation) including most of the world's largest rivers : Amazon, Zaire, Orenoco, Chang Jiang (Yang Tse Kiang), Brahmaputra, Mississippi, Yenissei, Lena, Parana, Saint-Lawrence, Ob, Ganges, Amur, Mackenzie, Columbia, Danube, Yukon, Niger, Uruguay, Ogooue, Fraser and Rhine (in decreasing order or river discharge). Some rivers are not used in all C export typologies because their present conditions have been dramatically affected by human activities. I have chosen not to use some data from the Columbia and Indus rivers because of extensive damming ; the Rhine and Seine rivers, because of eutrophication ; and other rivers, because of organic pollution.

Out of all the world's rivers, this data base covers from 38 % in terms of DOC to 45 % in terms of DIC based on total river discharge to oceans (exorheic runoff) as estimated by Baumgartner and Reichel (1976) to be 37,400 $km^3.yr^{-1}$. Their water budget gives precise spatial distribution of water runoff on a global scale and has been used to set up the typology of river runoff. Few rivers draining into the internal part of the continents (endorheic runoff), such as the Volga and the Terek, have been used to set up the typology. Also some smaller rivers (Alpine Rhone and Dranse that drain into lake Geneva) and a few tributaries of bigger rivers, previously in the data base, have also been used when no major rivers were available to appropriately describe the exportation from a given environment (Great Bear, Athabaska, Peace, for the Mackenzie ; Apure, Caroni, Caura, for the Orenoco). Furthermore, due to its enormous size and inherent heterogeneity, the Amazon has been split into its main tributaries (Madeira, Solimoes, Negro, Tapajos and Xingu) wherever possible. Part of this data base for organic C originates from the SCOPE-CARBON program launched by the Geology Institute of Hamburg (Degens *et al.,* 1991) and regularly published in the Mitteilungen aus dem Geologisch-Paläontologischen Institut der Universität Hamburg from 1982 to 1989. Important references for individual rivers include Lewis and Saunders (1989) for the Orenoco, Zhang *et al.* (1992) for the Huang He, Lesack *et al.* (1984) for the Gambia, Cadée (1984) for the Sanaga and Ogooue and Milliman *et al.* (1984) for the Chiang Jiang. Most of the DIC data employed here are taken from Meybeck (1979).

3.2.   RIVER TYPOLOGY AND MATERIAL TRANSPORT

For lack of measurements, it is extremely difficult to gather together data from more than 50 % of the world's rivers regardless whatever they are ranked according to water discharged to the ocean or river basin area. For these ranking methods, the first 200 rivers correspond to 70 % of the river basin area and 60 % of the river discharge (Meybeck, 1988) ; yet, even most of these rivers remain unstudied.

Extrapolation of the results from documented rivers sample is a necessary step for any global budget. When the Amazon is included in the river set, the representativity of the documented river set is biased, due to its enormous discharge : latest estimates by M. Molinier (personal communication) give 6,200 $km^3.yr^1$ or 16 % of total river discharge for only 6.4 % of the exorheic area. In order to avoid this bias, extrapolation by typology from known river concentrations, or export rates, was chosen as early as 1953 by Pardé for the TSS budget ; also it is employed by Milliman and Meade (1983) for TSS (with 30 types), as well as by Meybeck (1979, 1988) for major ions and TSS (with 12 types). Also Schlesinger and Melack (1981) and Meybeck (1981, 1982 updated in 1988) considered a river typology to establish TOC budgets (respectively with 10 and 6 types).

The river typology used here was first set up in 1979 on the basis of average annual river basin temperature and average annual river runoff (q in $mm.yr^1$ or in $l.s^{-1}.km^{-2}$). Four temperature classes determine four latitudinal zones : cold (< 4°C) - temperate (4-15°C) - arid (15-25°C) - warm (> 20°C). Five runoff classes separate the arid ($q < 30$ $mm.yr^1$), semi-arid (30 to 120 $mm.yr^1$), average (120-280 mm. yr^1), humid (280-630 $mm.yr^1$) and wet regions (> 630 $mm.yr^1$ or > 20 $l.s^{-1}$. km^{-2}). The savannah region combines two runoff classes and ranges from 30 to 280 $mm.yr^1$. The wet tropics (> 20°C and > 630 $mm.yr^1$) have been split into two components : the lowland wet tropics (lower and central Amazon, most of the Orenoco, Zaire) and the highland wet tropic (upper Amazon Papua New Guinea, Burma, parts of Indonesia, Viet-Nam and South China), a distinction necessary for two reasons : (i) the bedrock lithology is different between lowlands (mostly shields and non-carbonated continental sediments) and highlands (folded or uplifted regions rich in carbonate rocks), and (ii) mechanical erosion rates are likewise different, sometimes differing by two orders of magnitude (e.g., between the Negro and the upper Madeira in the Amazon basin), which can affect both the nature of POC and its export rate. Meybeck (1979, 1988) provides complete details on the boundaries of these river types as well as a global map of their distribution in both the endorheic and the exorheic parts of the continents. Names are more convenient than numbers, and types were chosen that reflect their environments as much as possible (e.g., taiga, humid temperate, and savannah). However, it must be emphasized that the naming convention used here may not fully reflect traditional definitions for the boundaries of these environments.

Runoff is the corner stone of this typology because of its key influence on river export rates. River runoff strictly controls the export rates of most elements, including DIC and DOC. When the global boundaries of river types were set up on the basis of Baumgartner and Reichel's map (1976), they were adjusted so that global river budgets (i.e., sum of the area x runoff for each river type) were equal to

the Baumgartner and Reichel's sum of exorheic runoff (37,400 $km^3.yr^1$) and
endorheic runoff (2,000 $km^3.yr^{-1}$). Additionally, there also exists the non-perennial
desert river type that corresponds to runoff less than 7.5 $mm.yr^1$. It represents only
1.7 Mkm^2 of the exorheic regions but 21.4 Mkm^2 of the endorheic regions and thus,
it is neglected here.

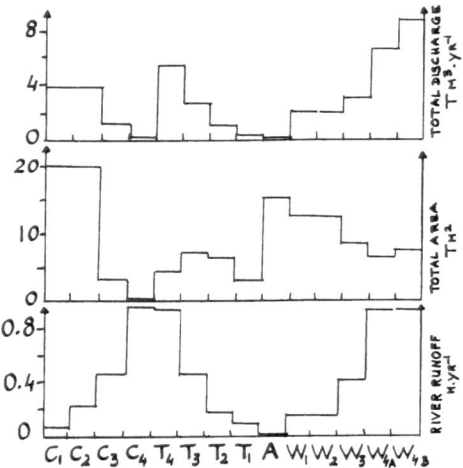

Figure 2. Schematic north-south distribution of river types used in global budgets to
oceans (Meybeck, 1979, 1988). 14 types and subtypes are defined based on average
river runoff and basin temperature. For each type, the total area and water discharge
have been mapped or computed. C_1 tundra ; C_2 taiga ; C_3 humid taiga ; C_4 wet
taiga ; T_4 wet temperate ; T_3 humid temperate ; T_2 temperate ; T_1 semi-arid
temperate ; A = arid ; W_1 and W_2 savannah ; W_3 humid tropics ; W_{4A} lowland wet
tropics ; W_{4B} highland wet tropics.

Figure 2 presents, for each river type, the global distribution of total basin area, of
discharge rate and of corresponding average runoff. Any typology must properly
consider the problem of transition between neighbouring classes. Generally, there is
a north-south distribution of the main types from tundra to wet tropics with the arid
zone and savannah being near the 40°-10° latitudes. Figure 2 places the most humid
types of the temperate and boreal regions next to one another, as found in the northwest coast of North America. However, due to its schematic nature, Figure 2 cannot
show the wet temperate regions adjacent to the wet warm regions as found in South-
East Asia. Figure 2 clearly demonstrates the inverse relationship between area and
discharge : arid and savannah types represent 28.7 % of total exorheic area but only
6.5 % of its runoff, while the total area of the most humid regions (defined here as
> 630 $mm.yr^1$, i.e., the sum of the cold [C_4], temperate [T_4] and warm [W_{4A},
W_{4B}] humid zones) represent 19.6 % of total area but 58.4 % of total runoff. This
discrepancy is the primary factor controlling global scale of any river budget.

3.3.   TYPOLOGY OF RIVER CARBON EXPORT

Each major river documented in this analysis has been assigned to a river type on the basis of its average runoff and its average basin temperature, as determined from the Soviet Physical Geography Atlas of the World (1964). In each class, a set of representative rivers, from 1 (tundra) to 7 (temperate), were used to determine the average export rates of the various C species. Then average values for each type were extrapolated to the total area of each type. As previously mentioned, the rivers in the data base cover between 38 and 45 % of the exorheic basin area ; their average runoff is within 10 % of the average global river runoff for exorheic area (374 mm. yr^{-1}).

Two important corrections have been made to determine export rates of atmospheric DIC and soil POC. First, on the basis of DIC concentrations in rivers and on our knowledge of their basic basin lithology, I have attributed a percentage of atmospheric DIC ranging from 100 % for non-carbonate basins (where DIC concentrations are typically less than 10 mg.l^{-1}) to 50 % for pure carbonated basins (where DIC levels can reach 30 mg.l^{-1}).

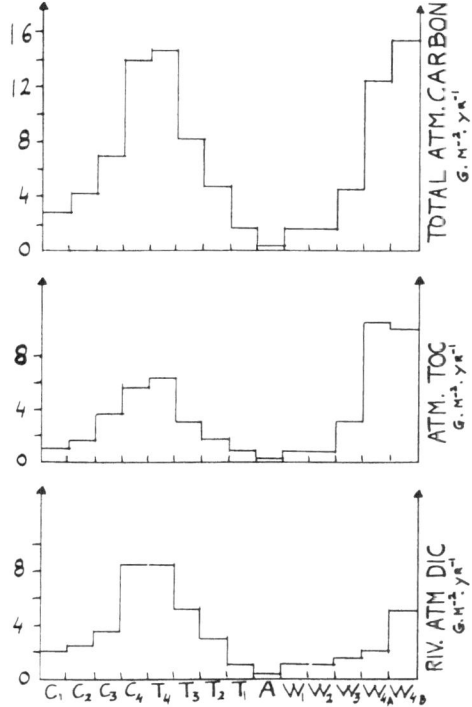

Figure 3. Schematic north-south distribution of river export rates of total atmospheric C (g C.m^{-2}.yr^{-1}).C$_1$ to W$_{4B}$ river types of the cold to warm zone, see Figure 2. TAC : DOC + soil POC + riverine atm DIC.

Second, a correction for rock POC has been made for each river on the basis of (i) its POC % and TSS levels, and (ii) its POC % *vs* TSS inverse relationship (Meybeck, 1982 ; Ittekkot, 1988) with the assumption that a POC content of 0.5 % was entirely of rock origin. This correction is important ; the most turbid rivers (average TSS > 1 500 mg.l^{-1}) carry to oceans as much as 30 % of the world's suspended sediments ; 77 % of the world's suspended sediments is carried by these rivers and those with TSS between 500 and 1 500 mg.l^{-1} (Meybeck, 1982 and in preparation ; Ittekkot, 1988). For these two groupings, cumulative river runoff are only 2 % and 25 % of total river discharge. On other words, few turbid rivers are responsible for the main input of suspended matter, and thus of total POC, and most of this POC is thought to be rock derived ; conversely, most rivers waters are characterized by low and medium turbidity, high POC, and they carry suspended matter, rich in POC and essentially of soil origin.

Figure 3 shows the global distribution of all atmospherically derived C exported to the ocean in rivers, both in organic and inorganic forms. As mentioned above, the highest export rates are found in the wettest types (C_4, T_4, W_{4A} and W_{4B}), but the proportion of TOC and atm DIC are not constant. In the temperate regions, DIC is the dominate form of C export, due to the prevalence of sedimentary carbonate rocks, as well as relatively steeper relief, which favors efficient water drainage, and thus reduces the levels of DOC. In the humid tropics, conditions are opposite, so TOC dominates. There are few exceptions to this global trend. For example, some wet tropical karsts exist, as in Papua New Guinea and its archipelago where the 5 m.yr^{-1} rainfall produces record riverine atmospheric DIC export rates of about 100 g C.m^{-2}.yr^{-1}.

3.4.   INFLUENCE OF CARBONATE WEATHERING

Riverine atmospheric inorganic C resulting from the weathering of carbonate rocks may play an important role in riverine TAC export. The Figure 4 presents a schematic evolution of the relative proportions of atmospheric DIC, POC and DOC when TAC increases : the proportion of DIC jumps from 20 % to 70 % when TAC goes from 5 mg.l^{-1} (its recorded minimum in the Sahel region, Gambia), to 30 mg.l^{-1} (its noted maximum in major rivers; e.g., Danube and Indus). The TAC budget is therefore much influenced by surficial lithology. In highly turbid rivers such as the Missouri and the Huang He, the TAC export may exceed 30 g C.m^{-2}.yr^{-1} ; however, in these cases, it derives mostly from enormous amounts of "soil POC" which actually may still partly originate from the erosion of shales or loess, even after correction made for rock C.

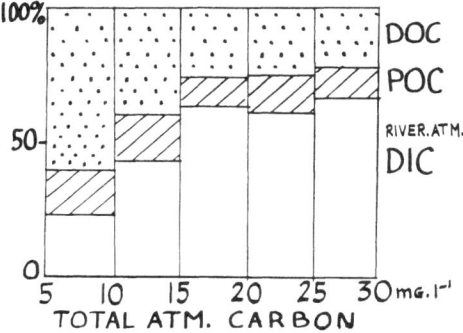

Figure 4. Average proportions of sources of atmospheric C in 5 groups of major rivers arranged according to their concentrations in total atmospheric C. DOC = dissolved organic carbon ; POC = particulate organic carbon of soil origin ; DIC = dissolved inorganic carbon originating from soil and atmospheric CO_2 during weathering reactions (riverine atm.DIC).

As for the global distribution of carbonate rock, a much greater TAC proportion (as limestone) is found in the mid-northern latitudes (Balazs, 1977) ; substantially, less is found in the tropics and in the northern latitudes (> 60° N). These differences are responsible for some of the latitudinal variations in DIC carried by rivers, as noted before by Meybeck (1979), even when averaged over extended area (most area for which river types are defined here exceed 5 Mkm²).

3.5.   GLOBAL BUDGET OF RIVERINE CARBON

Any global budget of riverine material is based on a framework of the global water balance. The data base relies on the detailed work by Baumgartner and Reichel (1976). They estimate that the total water discharge from the non-glaciated parts of the continents (exorheic runoff) to be 37,400 km³.yr⁻¹ over 99.9 Mkm², while the total surface water drained towards the internal parts of the continents (endorheic runoff) represent only 2,000 km³.yr⁻¹ over 33.2 Mkm². Glaciated parts of the continents, Greenland, Canadian Archipelago and Antarctica, account for 2,300 km³.yr⁻¹ over 15.8 Mkm². I assume this latter contribution to oceans to be negligible for the balance of atmospheric C with comparison to the river inputs. If the external drainage has been the focus of most river budgets, the internal river drainage which eventually leads to the formation of continental sediments, i.e., atmospheric C sinks, has not been properly addressed. Finally, it must be pointed out that two ways of global extrapolation are possible : (i) consideration of concentration and discharges *versus* (ii) consideration of export rates and areas. The latter is much preferable (Alekin and Brazhnikova, 1960 ; Meybeck, 1979) since

Table 2. Global budget of atmospheric carbon carried to ocean.

	DOC	atm DIC	atm DC	soil POC	TAC	A Mkm2	Q km^3.yr^1	q m.yr^1
Total Input to Oceans								
Cexp (g.m^{-2}.yr^1)	1,99	2,44	4,43	0,99	5,42	99,9	37400	0,375
MatmC (Tg.yr^1)	199	244	443	99	542			
%	37	45	82	18	100			
Total Cold								
Cexp (g.m^{-2}.yr^1)	1,31	2,5	3,8	0,42	4,25	23,35	5500	0,235
MatmC (Tg.yr^1)	30,5	59,2	89,7	9,9	99,6			
% carbon species (1)	30,6	59,4	90,0	10,0	100			
% climatic zones (2)	15,3	24,2	20,3	10,0	18,4	23,3		14,7
Total Temperate								
Cexp (g.m^{-2}.yr^1)	1,5	4,5	6,0	1,5	7,5	22,0	10250	0,465
MatmC (Tg.yr^1)	32,2	100	132,2	33,7	165,9			
% carbon species (1)	19,4	60,3	79,7	20,3	100			
% climatic zones (2)	16,2	41	29,9	33,8	30,7	22,0		27,4
Total Arid+Savannah								
Cexp (g.m^{-2}.yr^1)	0,15	0,6	0,75	0,23	0,98	28,7	2430	0,085
MatmC (Tg.yr^1)	4,45	17,1	21,5	6,75	28,3			
% carbon species (1)	15,7	60,4	76,1	23,9	100			
% climatic zones (2)	2,2	7,0	4,9	6,8	5,2	28,7		6,5
Total Humid Tropics								
Cexp (g.m^{-2}.yr^1)	5,1	2,6	7,7	1,9	9,6	25,8	19210	0,745
MatmC (Tg.yr^1)	131,5	67,6	199,1	48,9	248			
% carbon species (1)	53	27,3	80,3	19,7	100			
% climatic zones (2)	66,2	27,7	45,0	49,4	45,8	25,8		51,4

Cexp transport rate of carbon
MatmC Total mass of riverine carbon carried to ocean
A : Exorheic drainage area of climatic zone in Mkm^2 (Meybeck, 1979)
Q : Total river discharge (km^3.yr^1) ; q : Average river runoff (m.yr^1)
DOC Dissolved organic carbon ; DC = DOC + DIC ; POC Particulate organic carbon
(1) Percentage of carbon species in total carbon budget
(2) Percentage of climatic zone participation in carbon budgets.

area can be determined with great precision while river discharge estimates vary considerably. For instance, our understanding of the total Amazon discharge went from 100,000 $m^3.s^{-1}$ to 200,000 $m^3.s^{-1}$ in just the last 30 years.

The sum of river inputs for all boreal (C_1 to C_4), temperate (T_1 to T_4), dry (A and W_{1-2}), humid and wet tropical (W_3, W_{4A} and W_{4B}) regions leads to the global budget presented on Table 2. It results from summing over all types the product of each type's total area (Figure 2) times its export rate of C species (Figure 3). Thus, the global river budget is split into four climatic zones, similar to classification by biomes ; areas of which vary from 22 to 28.7 Mkm^2.

Globally, of the C in rivers originating from the atmosphere, 45 % are carried by DIC, 37 % by DOC and only 18 % by soil POC. This small fraction of POC arises because of the correction of rock POC which would otherwise represent about 40 % of total riverine POC. In any case, the relative proportions of DOC, DIC and POC vary dramatically between the different climatic zones.

Cold areas are characterized by the dominance of DIC (DIC > DOC > POC). Temperate areas, where mountainous regions are abundant, exhibit nearly equal proportions of POC and DOC export, reflecting their higher mechanical erosion rates (DIC > POC ≥ DOC). This same trend is accentuated in the arid and savannah zones (DIC > POC > DOC). Finally, rivers in the humid tropics are remarkable because of their low proportion of DIC and much higher DOC (DOC > DIC > POC), a result of less carbonate rock and of more lowland areas which favor the formation of high DOC contents.

4. River Carbon Storage in Internal Areas

The global budget of river transport to internal areas (e.g., the Caspian and Aral "seas"', lakes Titicaca, Tchad, Eyre, Okawango), has been extrapolated in the same fashion as for exorheic runoff. For transport to internal areas, the occurence of river types (Meybeck, 1979) shows that the dominate class are those draining desert land (known as arheic areas in hydrographic terms) which are defined here as surface average runoff < 7.5 $mm.yr^{-1}$ and cover 21.4 Mkm^2 of desert out of a total of 33.2 Mkm^2. Arid and savannah areas are estimated to contribute another 9 Mkm^2 while more humid regions represent only 2.8 Mkm^2, as the Volga river basin (entirely found in the "temperate" regions). Recall that the temperate class here is based only on average temperature (4 < T < 15°C) and not latitudinal position ; for instance, the Titicaca drainage basin is considered here as part of the temperate areas. Assuming average export rates of C, determined globally for each river class in exorheic areas, is equally valid for endorheic areas, I determined a budget for the global endorheic areas, again summing the products of area times rate for each river class. This global budget for endorheic areas determines riverine C transport as : atmospheric POC, rock POC, DOC, rock DIC, riverine atmospheric DIC to be 4, 9, 4, 12.5 and 17.5 Tg C.yr[1], respectively. When endorheic river waters ultimately evaporate, half of the riverine DIC returns to the atmosphere as CO_2 according to the reaction : $Ca^{++} + 2HCO_3^- \rightarrow CaCO_3 + CO_2 + H_2O$, so only 8.75 Tg.yr[1] of atmospheric DIC is stored.

Finally, the long-term storage of atmospheric C in internal area is 4 + 4 + 8.75 or 16.75 Tg C.yr[1]. This flux is only a few percent of the total mass of atmospheric C discharged directly to oceans by rivers.

5. Riverine Carbon Storage in Lakes

Lakes participating in external drainage *via* rivers are another permanent sink of atmospheric C through various processes including settling of suspended organic matter brought to them by tributaries, precipitation of carbonate minerals (mostly calcite) and permanent deposition of part of their algal production. The data base employs the total lake area of around 2.35 Mkm[2] for lakes exceeding 10 ha, and 1.57 Mkm[2] for lakes exceeding 100 km[2], and an average ratio drainage basin area/lake area is around 15 (Meybeck, in preparation). Therefore, total area upstream of major lakes is around 23.5 Mkm[2]. After correction for the few but important lakes located in the internal areas which have been previously considered for the global storage of C, an estimated 20 Mkm[2] of area upstream major lakes within the external drainage is likely. Smaller lakes are considered to be within the drainage area of larger ones as in glaciated shields. If the global average soil POC export rates in rivers draining to oceans (around 1g C.m[-2].yr[1]) estimated previously is used here, the corresponding soil POC storage in lakes is 20 Tg C.yr[1].

Lake budgets of organic C are not numerous, but lake outlets have generally lower DOC levels than tributaries. Reduction may be attributed to bacterial activities or, more likely, to the precipitation of colloïdal organic C. Kortelainen and Mannio (1988) have reported that, in Finland, river TOC, of which 90 % is as DOC, may be decreased by 60 % when the percentage of lake area within the river watershed exceeds 20 %. Assuming 35 % of the DOC in tributaries of lakes to be precipitated and estimating the river waters input to lakes to 7,500 km[3].yr[1], 15 Tg C.yr[1] would be stored as DOC in lakes.

Calcium carbonate precipitation is important in carbonated environments such as in the French, Swiss, Italian and Austrian lakes of the Prealps. In pure karstic regions, the sedimentation rate of DIC can be over 600 g C.m[-2].yr[-1], as in the Plitvice lakes of Yugoslavia (Kempe, 1988). The total lake area found within carbonated regions is estimated here to be around 0.18 Mkm^2 or 10 % of the total lake area of the exorheic regions. This proportion is slightly less than that for the global distribution of carbonate rocks (16 %) since these environments are much less favorable to lake formation. Based on a review of average calcite precipitation in lakes (Meybeck, in preparation), I allocated an average precipitation rate of 100 g DIC.m^{-2}.yr^{-1} for carbonate-rich lakes (mentioned above) and 5 g DIC.m^{-2}.yr^{-1} for non-carbonate lakes (such as Baikal, Titicaca and Taupo) over a total area of 1.6 Mkm^2. Thus, the total DIC retention in lakes is estimated to be around 26 Tg C.yr^{-1}. In addition to this DIC storage in lakes, another 26 Tg C.yr^{-1} go back to the atmosphere as CO_2 during this $CaCO_3$ precipitation process. Part of these 52 Tg C.yr^{-1} stored in lakes and released to the atmosphere originates from the dissolution of carbonate rocks : estimates range from 27 % (Alekin and Brazhnikova, 1960) to 44 % (Meybeck, 1979) ; this budget found around one-third. In summary, the world's lakes store about 17 Tg C.yr^{-1} from riverine atmospheric DIC out of TAC stored in lakes estimated to be around 50 Tg C.yr^{-1}. Although this latter figure is equal to the one estimated by Mulholland and Elwood (1982), it is still much less accurately known than river inputs to oceans.

6. Discussion

Global budgets of riverine C are based here on one method only : river typology that permits the regionalization of C budgets ; also the comparison to net terrestrial primary production is possible because, for such, budgets are made in the same fashion. Other methods (Meybeck, 1982 and in preparation) further employ the TSS distribution and the POC-TSS relationship for the POC budget, and thus estimate global riverine POC input to oceans as 110 Tg C.yr^{-1}, as well as the use of global statistics of C export rates. With this latter method, the budgets for riverine C inputs to oceans of atm DIC, DOC, POC and soil POC are 244, 199, 180 and 99 Tg C.yr^{-1}, respectively. These figures are very close to the budgets previously presented and are also similar to the most recent estimates obtained by various authors : 330 Tg.yr^{-1} for TOC (Degens et al., 1991$_B$), 220 and 77 Tg.yr^{-1} for DOC and total POC (Spitzy and Leenheer, 1991), 206 and 169 Tg.yr^{-1} for DOC and total POC (Meybeck, 1988), 231 Tg.yr^{-1} for POC (Ittekkot, 1988). Most variable are estimates for the POC budget because turbid rivers that contribute most to that budget are very seldom documented and thus are not considered by the conventional budget (i.e., extrapolation of documented rivers) ; therefore, specific POC budgets (Meybeck, 1982 ; Ittekkot, 1988) must be set up. When compared with the first estimates of

TOC budget by Schlesinger and Melack (1981) and by Meybeck (1981, 1982), the figures presented here have not changed much although this data base is considerably larger. Schlesinger and Melack use two methods : simple extrapolation (finding 376 Tg C.yr^{-1} for TOC) and typology (410 Tg C.yr^{-1}), quite similar to the 378 Tg C.yr^{-1} found by Meybeck, also with the river typology. TOC budgets seem now to converge to 380 ± 20 Tg C.yr^{-1}.

Total budgets of atmospheric DIC have converged long since. The average total DIC export rate by world rivers has been estimated as 3.7 g C.m^{-2}.yr^{-1} (Clarke, 1924), 3.6 (Livingstone, 1963) and 3.8 (Meybeck , 1979). The global inputs to oceans of riverine atmospheric DIC are estimated as 234 Tg C.yr^{-1} (Holland, 1978, based on Livingstone's analyses), 213 (Meybeck, 1979, and Wollast and Mackenzie, 1983, based on Meybeck's analyses), and 255 (Meybeck, 1987). Alekin and Brazhnikova (1960) did not give a global input of atmospheric DIC to oceans but their proportion of atm DIC/total DIC is quite similar to other authors : they give 73 % for all Soviet rivers *versus* other estimates at 56 % (Wollast and Mackenzie, 1983), 67 % (Meybeck, 1987), 65 % (Holland, 1978) and 70 % for the Amazon basin alone (Stallard and Edmond, 1983). Wollast and Mackenzie (1983) also attribute 9 % of river DIC to the oxydation of fossil organic matter.

Anthropogenic influences on the river C budget at a global scale are little studied, although some scenarii are developed by Downing *et al.* (this volume). As for organic wastes discharged to rivers, most are re-mineralized by bacterial activity, and in North America and in Europe, abatment of organic pollution, through primary and secondary treatment, is now effective.

Kempe *et al.* (1991) present a documented trend of the partial CO_2 pressure (pCO_2) at various stations of the Rhine river system. The annual average pCO_2 ranges from about 600 ppmv for the outlet of lake constance to 65,000 ppmv for the Braubach station near mouth. This over-saturation results from bacterial activity ; it increased between 1963 and 1978. According to Kempe *et al.*, the long-term average proportions of C in the Rhine near mouth are the following (on the basis of 2.3 Tg C.yr^{-1}) : DIC 33.3 mg.l^{-1} ; TOC 7.6 mg.l^{-1} ; C-CO_2 3 mg.l^{-1}.

The CO_2 measured in the Rhine river is probably much higher than in most other rivers. If it is assumed that the CO_2/DIC ratio in unpolluted world's rivers is about 5 to 10 %, the global riverine flux of dissolved CO_2 would be between 20 and 40 Tg C.yr^{-1}.

7. Conclusions

The global budget of riverine C is presented in Figure 5. Inland long-term storage of atmospheric C, i.e., DOC + soil POC + atm DIC, is estimated to 17 Tg C.yr[1] for the internal part of the continents not drained to oceans, compared to around 50 Tg C.yr[-1] for lakes within the external drainage to oceans. About 542 Tg C.yr[-1] of atmospheric C is discharged by rivers to oceans. Other estimates made with the same data base (Meybeck, in preparation) lead to about 600 Tg C.yr[1].

The breakdown of the global river C budget is presented in Table 3.

Table 3. Global budget of river C, sources and sinks (Tg C.yr[1])(mostly based on river typology method)

Sources / Pathways	Atmosphere			Surficial rocks		
River transfer to oceans	DOC atm POC atm DIC atm PIC		199 99 244 -	DOC rock POC rock DIC rock PIC[2]		- 80 137 170
	Total (TAC)		542	Total		380 - 400
		Lakes	Int.[1]		Lakes	Int.
Sinks on continents	DOC atm POC atm DIC [3] atm PIC	15 20 17 -	4 4 9 -	DOC rock POC rock DIC [3] rock PIC[2]	- 18 9 35	- 9 6 20
	Total (TAC)	52	17	Total	62	35

(1) Internal drainage of continents ; (2) This estimate does not take into consideration the enormous amount of eroded particles from highlands which are barely carried by rivers, and which settle in talus slope, river beds, etc. ; (3) Carbonate precipitation ; same amount of C-CO$_2$ is released to the atmosphere.

The riverine atmospheric dissolved inorganic C derived from the weathering reaction accounts represents about 45 % of the total atmospheric C carried by rivers to the ocean ; however, it has often been neglected. Proportions of DIC, POC and DOC in world's rivers are variable, largely due to differences in lithology which controls DIC levels, and differences in relief which controls the DOC/POC ratio. In the wet tropical rivers, DOC is by far the dominating form of atmospheric C that is exported, in contrast with rivers from all other biomes. Lakes and internal drainage area are responsible for long-term storage of only 10 % of riverine C. Soil POC has been distinguished from rock POC because of relationships between POC *vs* total suspended solids in world's rivers.

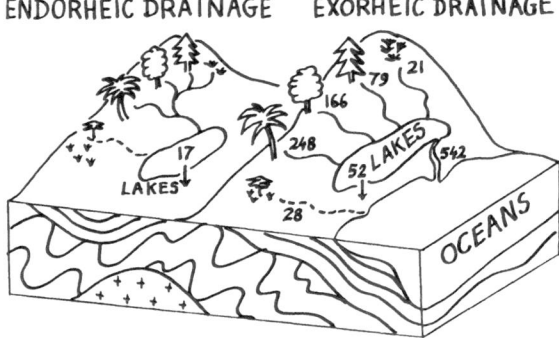

Figure 5. Schematic global budget of atmospheric C carried by rivers to oceans and trapped in lakes and in internal regions of continents (Tg C.yr^{-1}). River transport within the exorheic regions grossly correspond to the tundra (21), boreal forest (79), temperate regions (166), humid tropics (248) and arid regions plus savannah (28).

Finally, river transfer of C to the ocean is related to surface runoff. High runoff regions (defined here as $q > 20$ l.s^{-1}.km^{-2}, i.e., twice the average exorheic runoff) represent only 19.6 % of the total non-glaciated exorheic runoff, or 13 % of the continental area, but they may contribute to 51 % of the total atmospheric C carried by surface waters to oceans of which 38 % derive only from the wet tropics (15 % of exorheic drainage area). Much of this enhancement is attributable to the extensiveness of glaciated area, endorheic regions, arid and dry regions that contribute none or very little to the transfer of atmospheric C to oceans.

Riverine C budgets now present a remarkable convergence ($\pm$ 10 %) and most of world's major rivers are now documented : additional work on rivers will only augment the data base by few %. Still the origins, pathways and residence time of riverine atmospheric C have to be precised :

- the breakdown of soil and rock particulate organic carbon proposed here must be confirmed using C isotopes and organic tracers.

- global land storage in flood plains, lakes and endorheic regions must be more studied.

- anthropogenic influences on riverine C must now be taken into account, particularly eutrophication and organic matter pollution that alter the production/ respiration balance in rivers.

Finally, one must keep in mind that river fluxes to oceans presented here are valid for the upstream boundary of the land-ocean interface (estuaries, deltas, coastal wetlands) which may considerably modify the net TAC budget to oceans. Our understanding of their global influence on C budget is still to be improved.

Aknowledgements

The author expresses his warm thanks to J. Orr for his careful and most welcome corrections.

References

Alekin, O.A. and Brazhnikova, L.V. : 1960, Contribution to the Knowledge of Dissolved Matter Runoff at the Earth Surface (in Russian), *Gidrochim. Mater.,* 32, pp. 12-34.

Balazs, D. : 1977, The Geographical Distribution of Karst Area, *Proc. 7th Int. Congress Speleology,* Sheffield, pp. 13-15.

Baumgartner, A. and Reichel, E. : 1976, *The World Water Balance,* Elsevier, 179 pp.

Brinson, M.M. : 1976, *Limnol. Oceano.,* 22, pp. 572-582.

Cadée, G.C.: 1984, *Neth. J. Sea Res.,* 17, pp. 426-440.

Clarke, F.W. : 1924, *Data of Geochemistry,* Vth Edition, US Geol. Survey Bull., 770, 841 pp.

Davies, B.R. : 1986, The Zambezi River, in : Davies B.R. and Walker K.F. (eds), *The Ecology of River Systems,* pp. 225-267.

Degens, E.T., Kempe, S. and Richey, J.E. (eds) : 1991$_A$, *Biogeochemistry of Major World Rivers,* John Wiley and Sons, 356 pp.

Degens, E.T., Kempe, S. and Richey J.E. : 1991$_B$, Summary : Biogeochemistry of Major World Rivers, in : Degens E.T., Kempe S. and Richey J.E. (eds), *Biogeochemistry of Major World Rivers,* John Wiley and Sons, pp. 323-347.

Dessery, S., Dulac, C., Laurenceau, J.M. and Meybeck, M. : 1984, *Archiv. Hydrobiol.,* 100, pp. 235-260.

Downing, J.P. *et al* : 1993, this volume.

Feng, Jian-Xiang and Kempe, S. : 1987, *Mitt. Geol. Paläont. Inst. Univ. Hamburg,* 64, pp. 161-170.

Ertel, J.R., Hedges, J.I., Devol, A.H., Richey, J.E. and Ribeiro, N. : 1986, *Limnol. Oceanogr.,* 31, pp. 739-754.

Hedges, J.I., Clark, W.A., Quay, P.D., Richey, J.E., Devol, A.H. and Santos, U. de M. : 1986, *Limnol. Oceano,* 31, pp. 717-738.

Holland, H.D. : 1978, *The Chemistry of the Atmosphere and Oceans,* Wiley-Interscience, 351 pp.

Ittekkot, V. : 1988, *Nature,* 332, pp. 436-438.

Ittekkot, V. and Arain, R. : 1986, *Geochim. Cosmochim. Acta,* 50, pp. 1643-1656.

Kempe, S. : 1984, *J. Geophys. Res.,* 89, pp. 4657-4676.

Kempe, S. : 1985, *Mitt. Geol. Paläont. Inst. Univ. Hamburg,* 52, pp. 91-332.

Kempe, S. : 1988, Freshwater Carbon and the Weathering Cycling, in : Lerman A. and Meybeck M. (eds), *Physical and Chemical Weathering in Geochemical Cycles,* Kluwer Acad. Publ., pp. 197-224.

Kempe, S., Pettine, M. and Cauwet, G. : 1991, Biogeochemistry of European Rivers, in Degens E.T., Kempe S. and Richey J.E. (eds), *Biogeochemistry of World Rivers,* John Wiley and Sons, pp. 169-212.

Kortelainen, P. and Mannio, J. : 1988, *Water, Air and Soil Poll.,* 42, pp. 341-352.

Lesack, L.F.W., Hecky, R.e. and Melack, J.H. : 1984, *Limnol. Ocean,* 29, pp. 816-830.

Lewis, W.M. and Saunders, J.F. : 1989, *Biogeochemistry,* 7 , pp. 203-240.

Livingstone, D.A. : 1963, *Chemical Composition of Rivers and Lakes,* in Data of Geochemistry, *US Geol. Survey Prof. Paper,* 440 G, 64 pp.

Malcolm, R.L. and Durum, W.H. : 1976, *Organic Carbon and Nitrogen Concentration and Annual Organic Carbon Load in Six Selected Rivers of the USA, US Geol. Survey Water Supply Paper,* 1817, 21 pp.

Martins, O. and Probst, J.L. : 1991, Biogeochemistry of Major African Rivers : Carbon and Mineral Transport, in Degens E.T., Kempe S. and Richey J.E. (eds), *Biogeochemistry of Major World Rivers,* John Wiley and Sons, pp. 127-155.

Meybeck, M. : 1979, *Rev. Géographie Physique Géologie Dynamique,* 21, pp. 215-246.

Meybeck, M. : 1981, River Transport of Organic Carbon to the Ocean, in *Flux of Organic Carbon by Rivers to the Ocean,* CONF 8009140, US, Dept. of Energy, Office Energy Res., Washington DC, pp. 219-269.

Meybeck, M. : 1982, *Amer. J. Sci.,* 282, pp. 401-450.

Meybeck, M. : 1987, *Amer. J. Sci.,* 287, pp. 401-428.

Meybeck, M. : 1988, How to Establish and Use World Budgets of River Material, in Lerman A. and Meybeck M. (eds), *Physical and Chemical Weathering in Geochemical Cycles,* Kluwer Acad. Publ., pp. 247-272.

Meybeck, M. : 1993, C, N, P and S in Rivers : from Sources to Global Inputs, in Wollast R., Mackenzie F.T. and Chou L. (eds), *Interaction of C, N, P and S Biogeochemical Cycles and Global Change,* Springer-Verlag, pp. 163-193.

Milliman, J.D. and Meade, R.H. : 1983, *J. Geol.,* 91, pp. 1-21.

Milliman, J.D., Quinchun, X. and Zuosheng, Y. : 1984, *Am. J. Sci.,* 284, pp. 824-834.

Moore, T.R. : 1987, *Int. Ass. Hydrol. Sci. Publ.,* 167, pp. 481-487.

Mulholland, P.J. and Elwood, J.W. : 1982, *Tellus,* **34**, pp. 490-499.

Mulholland, P.J., Dahm, C.N., David, M.B., Ditoro, D.M., Fisher, T.R., Hemond, H.F., Kögel-Knabner, I., Meybeck, M., Meyer, J.L. and Sedell J.R. : 1990, What are the Temporal and Spatial Variations of the Organic Acids at the Ecosystem Level ?, in : Perdue M. and Gjessiky E.T. (eds), *Organic Acids in Aquatic Ecosystems,* Dahlem Workshop Rpt, John Wiley, pp. 315-329.

Naiman, R.J., Melillo, J.M., Lock, M.A., Ford, T.E. and Reice S.R. : 1987, *Ecology,* **68**, pp. 1139-1156.

Pardé, M. : 1953, *Revue Géographie Alpine,* **41**, pp. 399-421.

Ronov, A.B. : 1976, *Geochem. Int.,* **13**, pp. 172-195.

Schlesinger, W.H. and Melack, J.M. : 1981, *Tellus,* **33**, pp. 172-187.

Soviet Physical Geography Atlas of the World : 1964, Nauka, Moscow.

Spitzy, A. and Leenheer, J. : 1991, Dissolved Organic Carbon in Rivers, in Degens E.T., Kempe S. and Richey J.E. (eds), *Biogeochemistry of World Major Rivers,* John Wiley and Sons, pp. 213-232.

Stallard, R.F. and Edmond, J.M. : 1983, *J. Geoph. Res.,* **88**, pp. 9671-9688.

Wollast, R. and Mackenzie, F.T. : 1983, The Global Cycle of Silica, in : Aston S. (ed), *Silicon Geochemistry and Biogeochemistry,* Academic Press London, pp. 39-76.

Zhang, S., Gan, W.B. and Ittekkot, V. : 1992 , *Mar. Chem.,* **38**, pp. 53-68.

ACCORD BETWEEN OCEAN MODELS
PREDICTING UPTAKE OF ANTHROPOGENIC CO_2

James C. Orr
Laboratoire de Modélisation du Climat et de l'Environnement
DSM / CEN Saclay / CEA
L'Orme des Merisiers, Bât. 709
F-91191 Gif-sur-Yvette, FRANCE

ABSTRACT.   Models of the ocean provide the best estimate of how much anthropogenic CO_2 the ocean can and will absorb. Yet their agreement is only within 40% as characterized by the range of 2.0 ± 0.8 Gt C yr^{-1} computed by the Intergovernmental Panel on Climate Change (IPCC) in 1990 from four model estimates. Since then, one of the former results has been updated and two new model estimates have become available. In a reassessment, now with six ocean models and concern for individual model uncertainties, this study found a narrower range of 2.0 ± 0.5 Gt C yr^{-1} (38% less than the former uncertainty). Less uncertainty for oceanic uptake of anthropogenic CO_2, means greater certainty for two combined terms in the budget for the global carbon cycle. First the uncertainty of the combined atmosphere plus ocean sink is also nearly halved (now at ± 0.5 Gt C yr^{-1} for 1980–1989). Second, the uncertainty of the imbalance term (or missing sink) is reduced, but only slightly because most of its large uncertainty remains associated with the difficulty in precisely quantifying deforestation and land use change.

1. Introduction

Our understanding of the global carbon cycle can to some extent be measured by our ability to balance its budget. Such is not only of academic interest if we are to make correct predictions of future levels (and corresponding side effects) of rapidly rising CO_2 concentrations in the atmosphere. Unfortunately, our accounting over man's recent history is plagued by uncertainties, and the global C budget is far from being balanced. This paper refines the budget of the global C cycle by showing that the uncertainty associated with the anthropogenic CO_2 flux into one of its three major reservoirs, the ocean, is nearly half of what it was reported to be just three years ago.

Atmospheric CO_2 has increased noticeably since the onset of the industrial revolution. Now about 1/4 higher than the preindustrial concentration ($\sim$280 ppm), it is predicted to double that by the middle of the next century. All anthropogenic CO_2 is emitted to the atmosphere, but only a little more than half has stayed there. The rest has been partitioned between the other two reservoirs: the ocean and the terrestrial biosphere.

In 1990, the IPCC provided a budget for recent changes in the global C cycle (Watson *et al.*, 1990) that later received some small modifications (see Sarmiento and Sundquist, 1992 as well as Watson *et al.*, 1992). This global budget (Table 1) separates the three major reservoirs into their net quantifiable behavior as sources or sinks, with the terrestrial biosphere distinguished as both a source and a sink (the imbalance term). The uncertainty of the imbalance term results from our lack of ability to quantify the responsible factor, and not having that, from the uncertainty attributable to all other components.

Water, Air, and Soil Pollution **70**: 465–481, 1993.
© 1993 *Kluwer Academic Publishers.*

Table 1.   The global budget for anthropogenic CO_2 during 1980-1989 with recent modifications (in bold).

	Average change [GT C yr⁻¹]		
	IPCC (1990)[a]	IPCC (1992)[b]	This Study
Sources			
Fossil C emissions	5.4 ± 0.5	5.4 ± 0.5	5.4 ± 0.5
Deforestation + Land Use Change	1.6 ± 1.0	1.6 ± 1.0	1.6 ± 1.0
Total	7.0 ± 1.2	7.0 ± 1.2	7.0 ± 1.2
Sinks			
Atmosphere	3.4 ± 0.2	$\mathbf{3.2 \pm 0.1}$	3.2 ± 0.1
Oceans (model estimates)	2.0 ± 0.8	2.0 ± 0.8	$\mathbf{2.0 \pm 0.5}$
Total	5.4 ± 0.8	$\mathbf{5.2 \pm 0.8}$	$\mathbf{5.2 \pm 0.5}$
Imbalance (Sources − Sinks)	1.6 ± 1.4	$\mathbf{1.8 \pm 1.4}$	$\mathbf{1.8 \pm 1.3}$

[a]Original IPCC budget by Watson *et al.* (1990)

[b]Changes implemented by Sarmiento and Sundquist (1992) from results reported by Watson *et al.* (1992)

One can rate each component in Table 1 as to how well it is quantified. Most certain is the recent atmospheric record, a tribute to the precise measurements taken by D. Keeling at Mauna Loa since 1958, and before that time, to measurements of CO_2 gleaned from air bubbles trapped in ice cores that have records extending back to long before the industrial revolution. Our next most precise information comes from those who have been able to track man's use of his fossil C reserves (and cement production) from 1860 to present with a precision of about $\pm 10\%$ (Rotty and Masters, 1985; Marland, 1989). Third most certain are the ocean estimates that derive not from measurements of recent changes in oceanic CO_2, a difficult task indeed (see discussion in Sarmiento *et al.*, 1992), but from a summary of model results. Least measurable are the complexities of the much more heterogeneous biosphere: a relatively large uncertainty is associated with the term for deforestation and land use change, while no current method exists to determine the historical behavior of the terrestrial sink, except of course, by calculating a residual of all other terms.

Ocean modelers determine the response of the terrestrial biosphere over the industrial era through a method known as deconvolution (Siegenthaler and Oeschger, 1987; Keeling *et al.*, 1989; Sarmiento et al., 1992; Siegenthaler and Joos, 1992), shown graphically in Figure 1a. In this approach, an ocean model is forced with observed CO_2 concentrations in the atmosphere, the ocean uptake being determined as a function of gas exchange, CO_2 chemistry, and the physical ocean model. The deconvolution procedure sums curves for the oceanic uptake and the observed change in the atmosphere, then subtracts the curve for fossil C emissions. The result, the deconvolved terrestrial response or the "non-fossil CO_2 emissions" (Siegenthaler and Oeschger, 1987), shows the biosphere as a net source until about 1940 when it gradually becomes negative and thus a net sink until the 1980's. In 1990, the biosphere is neutral, based on this deconvolution from the Princeton ocean model (Sarmiento *et al.*, 1992) whose CO_2 uptake is slightly lower than the middle of the range given by the IPCC; thus, a curve more representative of all ocean models would show the 1990 biosphere as a slight source of anthropogenic CO_2. The deconvolved curve accurately represents the biosphere if (1) the ocean model is representative of the real ocean and (2) the C cycle in the ocean can be accurately approximated as being in steady state, as discussed below.

(a)

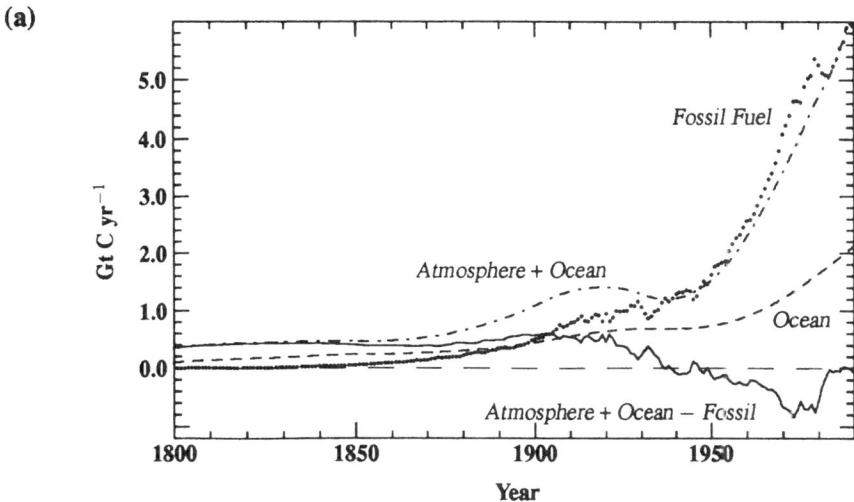

(b)

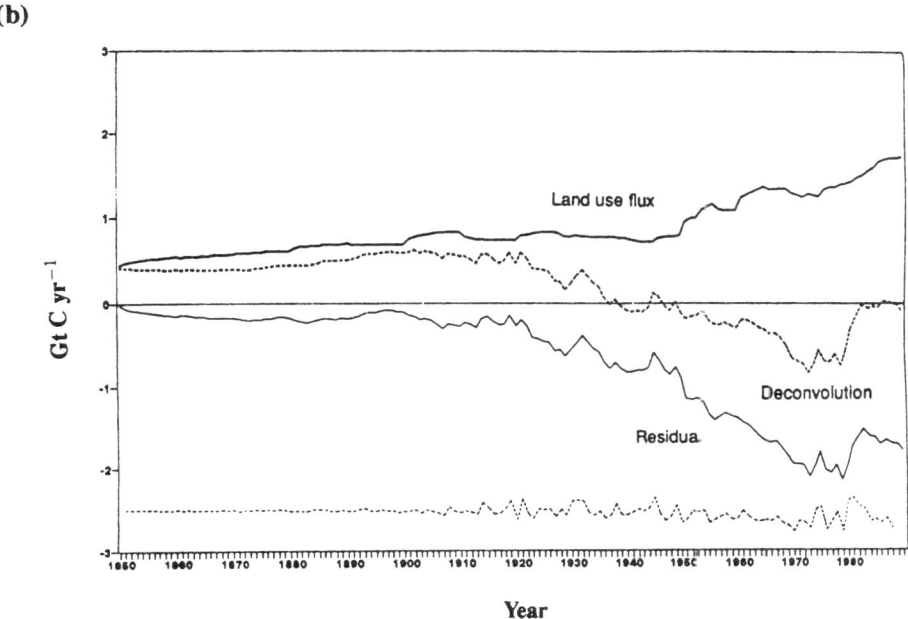

Figure 1.   Panel a: Deconvolution of the behavior of the terrestrial biosphere as determined by subtracting the emissions of fossil C from the sum of ocean model uptake (Sarmiento *et al.*, 1992) and observed atmospheric change (as characterized by a spline fit to pCO_2 measurements from the Siple ice core [Neftel *et al.*, 1985; Friedli *et al.*, 1986] combined with those taken at Mauna Loa [Keeling *et al.*, 1989a]). Panel b: Houghton (1992) subtracts his land use flux from the deconvolved curve (the same as in *a*) to determine the residual or missing carbon, i.e., that which man has yet to quantify directly. Also represented are the interannual fluctuations in fossil C emissions (thin dashed line at the bottom).

With the terrestrial response curve from Sarmiento *et al.* (1992), Houghton (1992) specifies a curve for land use change (including deforestation), then computes the difference (Figure 1b). This residual is termed the "missing sink" and is a measure of our lack of understanding of how the terrestrial system operates. It should not be confused with the deconvolved terrestrial response curve, which is derived without any information concerning the terrestrial system, but from records of CO_2 in the atmosphere, CO_2 from fossil C emissions, and CO_2 absorbed by an ocean model; the missing sink derives from our inability to balance the terrestrial budget.

Of course, a deconvolution is only as good as the ocean model used to derive it, so it is important to understand how reliable ocean uptake estimates might be. Unfortunately, we cannot yet adequately compare absorption of anthropogenic CO_2 in ocean models to that in the real ocean, but we can do the next best thing: compare models to one another. In the first IPCC report (working group I), Watson *et al.* (1990) showed that for uptake of anthropogenic CO_2, ocean models agree reasonably well with one another; however, their range $(2.0 \pm 0.8$ Gt C yr$^{-1})$, determined with results from two box models and two general circulation models (GCM's), implies that we understand oceanic uptake to no better than $\pm 40\%$, even though all models begin with the same basic assumption.

All ocean models that estimate uptake of anthropogenic CO_2 have been grounded on the steady-state assumption, i.e., that the operation of the ocean's C cycle has not changed since about a millennium before the onset of the industrial revolution because ocean physics and biology have not changed dramatically. It may be that the steady-state assumption is not completely justifiable; however, the marked lack of variability $(\pm 10$ ppm) in the CO_2 record from the 10th to the 18th centuries in the Siple ice core (Siegenthaler *et al.*, 1988), as well as the little change evident during interglacials in longer records suggest otherwise. In any case, given that ocean models start with the same assumptions, it seems surprising that the range of their results is not narrower.

Before one can speculate further about the cause for such variability, it is appropriate to consider the new and improved ocean models that have become available during the three years since the IPCC first published their report. This study reassesses the agreement between estimates for uptake of anthropogenic CO_2 from ocean models. It features one model estimate that has been updated since the IPCC assessment, two new model estimates, and attention to the uncertainty attached to individual model estimates.

2. Description of Models

One will not find here an analysis that includes all models that have ever estimated the ocean's absorption of anthropogenic CO_2. Instead this work is only an update of the IPCC analysis: it uses their four model estimates plus new results. The IPCC neglects results from many box models; I did the same. Those not included range from a simple two-box ocean model, through modified versions of the box-diffusion model (see section 2.1), to more complicated 2-D models, such as that of Peng and Broecker (1985), which includes effects of isopycnal mixing but finds total uptake of anthropogenic CO_2 similar to the box-diffusion model. Baes *et al.* (1985) and Emanuel *et al.* (1985) offer reviews of box models used to estimate oceanic uptake of anthropogenic CO_2.

If all box models were to be included here, each result would have to be adjusted so that it would be consistent with (1) uptake during 1980–1989, and (2) an atmosphere forced by observed pCO$_2$ (many older models are forced by emissions of fossil C). Adjustment for the latter is highly dependent upon the model and how it is calibrated (see section 4.2). Furthermore, if every box model that has ever estimated uptake of anthropogenic CO_2 were to be incorporated here, resulting statistics would be biased towards box models, particularly older ones to which one would have to apply the imprecise second adjustment. The IPCC ascribes equal weight to box models and GCM's, with two estimates from each category; this re-analysis maintained the same weighting but added the recently available estimates: from one box model and one GCM.

The focus of this paper is on comparison of the different model estimates for total oceanic uptake of anthropogenic CO$_2$, not on the models themselves, although limited detail and references are provided for the interested reader. If possible, when model estimates are first presented or re-presented after modification, they are given here with a precision of 0.01 Gt C yr^{-1}. The same precision is employed during all calculations. However, for final presentation, model results are given with a precision of 0.1 Gt C yr^{-1}, consistent with estimates summarized by IPCC.

2.1. BOX MODELS

Nearly 20 years ago, Oeschger *et al.* (1975) used their box-diffusion model to evaluate the ocean's uptake of anthropogenic CO$_2$. The box-diffusion model simplifies ocean circulation and mixing by averaging horizontally over the whole ocean and considering transport to occur only by vertical eddy diffusion. Calibration of the box-diffusion model employed a bomb-^{14}C inventory that is now known to be too large (Siegenthaler and Joos, 1992); hence, vertical mixing is somewhat too rapid (see section 4.3).

The outcrop-diffusion model (Siegenthaler, 1983; Siegenthaler and Oeschger, 1987) also considers vertical mixing in the ocean to occur by eddy diffusion, but further, it explicitly considers the connection between the deep sea and the high latitudes (where deep water formation processes dominate). This connection in the real ocean is considered to occur along surfaces of constant density and thus relatively rapidly; however, the outcrop-diffusion model's direct link between high latitude surface waters and the deep sea is allowed to act instantaneously, unlike the real ocean. It seems not so surprising then, that for anthropogenic CO$_2$, it absorbs 3.6 Gt C yr^{-1} for the 1980's, at least 50% higher than any other model discussed here. Thus the outcrop-diffusion model overestimates uptake of anthropogenic CO$_2$. This model is not used in the IPCC assessment; likewise it was not used here.

The multi-box model put forth by Goudriaan and Ketner (1984) and later modified for the studies by Goudriaan (1989) details the oceans by latitude and depth. Goudriaan (1989) separates the Atlantic from the rest of the ocean and connects it to the Pacific and Indian Oceans via a surface flowing circumpolar current. He runs simulations to determine the effect of changes in ocean physics and marine photosynthesis, thereby offering some test of the steady-state assumption for the ocean's C cycle. Goudriaan finds relatively little effect due to changes in biology and the depth of penetration of North Atlantic Deep Water (NADW). Yet he did show that large changes in ventilation rate would alter CO$_2$ uptake dramatically, as might be expected.

The most recent box model to report estimates for oceanic uptake of anthropogenic CO$_2$ is the High-Latitude Exchange/Interior Diffusion-Advection (HILDA) model that was formulated analytically by Shaffer and Sarmiento (1993). Siegenthaler and Joos (1992) converted HILDA to a numerical model and redetermined its transport parameters for their transient tracer studies. They were able to model simultaneously both bomb and natural ^{14}C by introducing an eddy diffusivity that decreases with depth. With this formulation, they also model anthropogenic CO$_2$. Their assumption of a decreasing eddy diffusivity with depth is justified based on analysis of the transport of both anthropogenic CO$_2$ and natural ^{14}C in the Princeton GCM (Joos, Orr, Siegenthaler, in prep.). With the HILDA model and the same depth-dependent eddy diffusivity, Joos (1992) models ^{39}Ar, CFC-11, and CFC-12 and finds good agreement with observations.

2.2. GENERAL CIRCULATION MODELS

GCM's differ from box models, which average processes over large areas of the ocean and parameterize ocean circulation and mixing by rates of diffusion and transfer between boxes. As a calibration procedure in box models, such rates can be adjusted so that simulated tracer distributions best match measurements made in the real ocean. GCM's are more realistic because their circulation fields, temperature, and salinity derive from basic principles of geophysical fluid dynamics. Yet reality comes with a price. GCM's are necessarily much more complicated and divide the ocean into many thousands of boxes; therefore, adjusting a GCM's transport (e.g., with coefficients of explicit eddy diffusion) is not only difficult but would demand too much computer time, i.e., at least with present day resources. Typically then, GCM's are only validated (not calibrated) against well-measured tracers (such as bomb and natural ^{14}C for C cycle studies, e.g., Toggweiler et al., 1989a,b) and later used with some understanding of their weaknesses. Oceanic uptake of anthropogenic CO_2 has been characterized by three GCM's: in models from Hamburg, Princeton, and Paris.

2.2.1. *The Hamburg Model*. Maier-Reimer and Hasselman (1987) study uptake of anthropogenic CO_2 with the so-called "off-line" approach, coupling the equilibrium-derived advection fields from the Hamburg ocean GCM (Maier-Reimer et al., 1982) to an inorganic C cycle model that computes the chemical distribution of CO_2 in seawater from equilibrium equations for hydrolysis of CO_2, borate, and water as well as conservation equations for alkalinity, total borate and total carbon. Alternatively, one could determine uptake of anthropogenic CO_2 with an "on-line" approach, i.e., where tracer transport is computed simultaneously along with fields for temperature, salinity, velocity, and convection; however, this is much more computer intensive. To date, all reported GCM simulations that determine oceanic uptake of anthropogenic CO_2 employ the off-line method. Bacastow and Maier Reimer (1990) employ a slightly different version of the Hamburg model and include effects for ocean biota, which dramatically affect the distribution of natural CO_2 but play little role in the ocean's absorption of anthropogenic CO_2.

The Hamburg model employs the quasi-geostrophic approximation, filtering out gravity waves, and dividing the circulation into baroclinic and barotropic components. Tracers are transported by the advection field and numerical diffusivity attributable to the finite grid. Bacastow and Maier-Reimer (1990) introduce a small explicit horizontal diffusivity and perform the convective adjustment directly; whereas Maier-Reimer and Hasselman (1987) parameterize these factors as part of their numerical diffusion. The Hamburg ocean model has passed through several versions but is now referred to as the Hamburg Large Scale Geostrophic (LSG) Ocean GCM (Maier Reimer and Mikolajewicz, 1992). With its one-month time step, possible because of its quasi-geostrophic approximation, the Hamburg LSG ocean model runs much faster than other primitive-equation GCM's of the same resolution. Also, the LSG model resolution ($3.5° \times 3.5°$, 15 layers) has been improved from that of Maier-Reimer and Hasselmann ($5° \times 5°$, 10 layers).

2.2.2. *The Princeton Model*. The Princeton model (Sarmiento et al., 1992) employs a primitive-equation, world-ocean GCM based on the model developed by Bryan (1969) at the Geophysical Fluid Dynamics Laboratory (GFDL) in Princeton. With a grid divided into 12 layers vertically, $4.5°$ meridionally, and $3.75°$ zonally, its resolution is comparable to that of the Hamburg model. The maximum depth is 5000 m, and as with the Hamburg model, bathymetry is as reasonable as allowed by the grid resolution. The surface layer is forced by the annual mean of observed temperature and salinity (Levitus, 1982) and yearly averaged wind stress (Hellerman and Rosenstein, 1983).

Unlike with the Hamburg model, the Princeton effort uses a perturbation approach (Siegenthaler and Joos, 1992; Sarmiento *et al.* 1992) that relates changes in oceanic pCO_2 and ΣCO_2 as a function of temperature through the thermodynamic equilibrium constants describing the hydrolysis of CO_2. The perturbation approach tracks changes to the natural C cycle after assuming one can approximate the ocean as having (1) a preindustrial surface pCO_2 in equilibrium with the preindustrial atmosphere of 280 ppm, (2) constant salinity (35 ppt), and (3) constant alkalinity (2300 μmol kg^{-1}). Changes in oceanic pCO_2 determined with the perturbation approach match values calculated explicitly to better than 1% (Sarmiento *et al.*, 1992).

The perturbation formulation for the ocean's CO_2 chemistry allows more rapid simulations than if they were performed in the same model but with the full carbonate system. Furthermore, the perturbation approach does not require first running the CO_2 chemistry model to equilibrium (i.e., simulations begin in the year 1750 with no perturbation CO_2 in either the ocean or atmosphere); without the perturbation approach, the model must first be run for several thousand model years so that the natural (or preindustrial) distribution of ΣCO_2 in the ocean reaches equilibrium. Other details concerning the perturbation approach and its use in related simulations in the Princeton model can be found in Sarmiento and Orr (1991) as well as Orr and Sarmiento (1992).

2.2.3. The Paris Model. The most recent GCM to estimate the ocean's uptake of anthropogenic CO_2 employs the OPA (Océan Parallelisé) model (Chartier, 1985; Andrich, 1988; Madec and Crépon, 1991; Madec *et al.*, 1991a,b; Blanke and Delecluse, 1993) developed at the Laboratoire d'Océanographie Dynamique et de Climatologie (LODYC, Paris) and converted to a global version by Marti (1992). Anthropogenic CO_2 simulations were recently made at the Laboratoire de Modélisation du Climat et de l'Environnement or LMCE (Orr and Monfray, in prep.). Marti's global model carries 19 vertical levels and a grid size that varies between 1° and 2°; hence, its resolution is substantially greater than GCM's used for estimating CO_2 uptake at either Princeton or Hamburg.

Yet to afford such high resolution, Marti's model is run in the so-called robust-diagnostic mode, i.e., where potential temperature (θ) and salinity (S) are restored to observations (Levitus, 1982) throughout the water column. In this manner, the model approaches equilibrium relatively rapidly. Such an approach goes counter to conventional wisdom: Toggweiler *et al.* (1989a) showed in their simulations with the on-line version of the lower-resolution Princeton model that the robust-diagnostic procedure yields a circulation field inferior to that from the prognostic approach (where θ and S are restored only at the surface). Nevertheless, the circulation field with Marti's model appears quite reasonable, and in fact, Marti (1992) argues it is (1) better than that of Toggweiler *et al.*'s prognostic model in several respects, and (2) rather similar to that from Semtner and Chervin (1988), a model with much higher resolution ($\frac{1}{2}° \times \frac{1}{2}°$), but also run in robust-diagnostic fashion.

Other differences in Marti's model are particularly noteworthy. Marti's robust-diagnostic method differs from that of Toggweiler *et al.* (1989a) in that restoration of θ and S is relaxed when closer to the equator (similar to Fujio and Imasato, 1991), nearer the coasts (as consideration for more vigorous dynamics particularly near western boundaries), as well as in regions of deep convection. Also, unlike other GCM's, Marti's global version of OPA features a curvilinear grid in all directions, thereby allowing numerical accuracy to the second order (Marti *et al.*, 1992); otherwise, GCM's are accurate only to the first order (Yin and Fung, 1991). Furthermore, the grid is contorted so that its North pole singularity lies over North America, thereby avoiding filtering or other numerical fixes (and their inherent problems) that are otherwise required to satisfy the Courant-Friedrichs-Lewy (CFL) stability criterion.

For tracer studies, Marti has developed an off-line version of his model that runs eight times faster than the on-line version by using a method of Flux Corrected Transport (Smolarkiewicz, 1982; Smolarkiewicz, 1983; Smolarkiewicz and Clark, 1986). Still though, such high resolution means a one-year simulation with one tracer requires about eight times longer than the same simulation in the Princeton off-line model; an equivalent simulation in both these models is much slower than in the Hamburg model.

3. The IPCC Range

The IPCC (Watson *et al.*, 1990) uses results from two box models and two GCM's to derive its range for the ocean's anthropogenic CO_2 uptake of 2.0 ± 0.8 Gt C yr^{-1} during the period 1980–1989. Of these four estimates, the IPCC specifies three: 1.2 (from Hamburg), 1.9 (from Princeton), and 2.4 Gt C yr^{-1} (from the box-diffusion model). As for the fourth estimate from the Goudriaan (1989) model, Watson *et al.* (1990) do not list its value, although they use it for the IPCC range calculation. Furthermore, neither does Goudriaan (1989) provide uptake estimates that are directly comparable to the those from the other three models. Therefore, in an attempt to reproduce IPCC range calculation, which serves as a base for this new analysis, I modify Goudriaan's results to make them comparable to other models, as detailed below.

Goudriaan provides results from two simulations in the form of airborne fractions: (1) the marginal airborne fraction α_m (the atmospheric increase a over the total fossil C emissions E for a given time), and (2) the total airborne fraction α_t (the atmospheric increase a over the sum of all emissions). When only fossil fuel emissions are considered, $\alpha_t = \alpha_m$, but when a biosphere component is included or when a model is forced by observed pCO_2 in the atmosphere, $\alpha_t \neq \alpha_m$. Goudriaan (1989) forces his one-box atmosphere with fossil C emissions in two simulations: one with a biosphere and the other without.

For Goudriaan's model with the biosphere, one can determine oceanic uptake as o_t (in Gt C) or ϑ_t (in Gt C yr^{-1}), given the two airborne fractions and E during a given time Δt,

$$\vartheta_t = \frac{o_t}{\Delta t} = \frac{E\alpha_m(1 - \alpha_t)}{\alpha_t \Delta t} \tag{1}$$

where Goudriaan (1989) provides α_m (0.564) and α_t (0.595), E is 83.379 Gt C for 1960–1980 (Rotty and Masters, 1985), and $\Delta t = 21$ yr. Thus Goudriaan's oceanic uptake for the simulations with a biosphere is 1.52 Gt C yr^{-1} for 1960–1980.

Determining oceanic uptake for the simulation with no biosphere is straightforward. One simply employs the definition of the marginal airborne fraction

$$\alpha_m = \frac{a_e}{E} = \frac{E - o_e}{E} \tag{2}$$

(which in this case is the same as the total airborne fraction) where the subscript e denotes simulations with only fossil C emissions. Upon rearranging terms

$$\vartheta_e = \frac{o_e}{\Delta t} = \frac{E(1 - \alpha_m)}{\Delta t} \tag{3}$$

With E and Δt as before, and $\alpha_m = 0.634$ (from Goudriaan), oceanic uptake for Goudriaan's simulation with only fossil C emissions is 1.45 Gt C yr^{-1} for 1960–1980.

Goudriaan's model estimates for 1960–1980 must be adjusted to 1980–1989. This time adjustment was made by employing the Princeton model (Sarmiento *et al.*, 1992) as a reference, i.e., by constructing a ratio from Princeton's uptake in the latter period over its uptake in the former, then multiplying that ratio by Goudriaan's uptake in 1960–1980 (see section 4.1). Thus for 1980–1989, Goudriaan's model ocean should absorb approximately 2.27 and 2.16 Gt C yr^{-1} for simulations with and without the biosphere, respectively. During 1960–1980, Goudriaan's biosphere behaves as a net sink; if such behavior differs substantially from that if it were to be modeled in 1980–1989, the time change adjustment will be affected.

Although it makes no difference for the final range calculation, in my attempt to reconstruct the IPCC range, I used Goudriaan's result from his simulation with a biosphere (2.3 Gt C yr^{-1}) because this simulation should act more like those from other models forced by the observations, assuming the biosphere in his model resembles that in reality. Later for my updated range calculation, I adjusted Goudriaan's result from his model without a biosphere (see section 4.1).

With these four results (1.2, 1.9, 2.3, and 2.4 Gt C yr^{-1}), the range spans from 1.2 to 2.4 Gt C yr^{-1}; however, in a conservative approach, the IPCC gives a range that is larger (2.0 ± 0.8 Gt C yr^{-1}) by simply doubling the difference between their mean and the most distant result (U. Siegenthaler, pers. comm.). Similarly, my re-analysis calculates a range from six model estimates, but I widen the range based on some understanding of individual model uncertainties.

4. Modified Range

Of the tracer model estimates given by the IPCC in 1990, that from the Princeton model remains unchanged, the Hamburg model has been updated, the box-diffusion model has been shifted (due to its calibration procedure), and the Goudriaan model was adjusted further here to make it more compatible with the other models. Additionally, two new model estimates have become available since 1990. Uncertainties associated with individual models also affect the range.

4.1. ADJUSTMENTS

For proper comparison, models should be compared during the same time period and with the same atmospheric forcing. If a model result is provided for a period other than the decade of the 1980's, it was adjusted here by multiplying it by the ratio formed from two results of a reference model, i.e., uptake for 1980–1989 over uptake for the period from which it must be corrected. Fortunately, it appears that the time adjustment depends little upon which model is used as a reference. For instance, to adjust from the 1960–1980 period (as done earlier for Goudriaan's results), the Sarmiento et al. (1992) model results yielded a ratio of 1.49; whereas, those from the Siegenthaler and Joos (1992) HILDA model (F. Joos, pers. comm.) gave a ratio of 1.48. I made all time adjustments with results from Sarmiento et al. (1992).

Another adjustment is necessary for results from the model of Goudriaan (1989) because he forces his atmosphere with fossil C emissions, while all other models discussed here run simulations with an atmosphere held to the observed record of pCO$_2$. Alternatively, one could employ Goudriaan's simulation that includes a biosphere (as done earlier), but that approach is not as rigorous because Goudriaan's biosphere is arbitrarily defined; other models in this comparison consider the effect of the true biosphere that is implied (through deconvolution) by using observed atmospheric CO$_2$. For this atmospheric forcing adjustment, a ratio could be constructed from oceanic uptakes for both simulations run in another model, just as with the time correction. Yet most papers do not provide results from both simulations in the form of oceanic uptakes. Fortunately though, some papers do report both results in the form of airborne fractions, which can be used to derive the appropriate ratio.

Employing the definition of α_m for the fossil C emissions scenario in equation (2) and the total airborne fraction for the simulation forced by observations

$$\alpha_t = \frac{a_t}{a_t + o_t} \tag{4}$$

one can solve for oceanic uptake in each, then form the following ratio

$$\frac{\vartheta_t}{\vartheta_e} = \frac{o_t}{o_e} = \frac{a_t(1 - \alpha_t)}{E\alpha_t(1 - \alpha_m)} \tag{5}$$

where E is specified from the fossil C emission data (Rotty and Masters, 1985), while a_t, α_t, and α_m must be from the model that serves as a base for the correction. Unfortunately, the reference model must be well chosen, because each model provides a different ratio. For instance, with airborne fractions for 1959–1983 (nearly the same as the 1960–1980 period given by Goudriaan) from Sarmiento et al. (1992), $\vartheta_t/\vartheta_e = 0.978$, while results from HILDA depend upon its calibration: $\vartheta_t/\vartheta_e = 1.012$ when calibrated with both natural and bomb ^{14}C (version K(z)) and $\vartheta_t/\vartheta_e = 1.087$ when calibrated only by bomb ^{14}C. To correct Goudriaan's fossil-emissions-only estimate, the most comparable model appears to be the K(z) simulation of HILDA: α_m's differ by less than 0.2%. Thus Goudriaan's model with no biosphere, if forced by observed atmospheric pCO_2, should absorb approximately 2.19 Gt C yr^{-1} during 1980–1989.

4.2. NEW RESULTS

The latest results from the Hamburg LSG model show an oceanic uptake of 1.33 Gt C yr^{-1} for 1973–1988 (E. Maier-Reimer, pers. comm.). With the time adjustment (section 4.1), the Hamburg model should absorb 1.47 Gt C yr^{-1} during 1980–1989.

Results from two new models have become available since the 1990 IPCC report. The first is that of Siegenthaler and Joos (1992) with the HILDA box model, which estimates oceanic uptake for 1980-1989 as 2.14 Gt C yr^{-1} (Joos, pers. comm.). The second new result is that from the Paris GCM (Marti, 1992), to which modelers at LMCE have added the perturbation approach. They predict the ocean absorbs 2.12 Gt C yr^{-1} during 1980–1989.

4.3. LIMITS, SHIFTS, AND UNCERTAINTIES DERIVED FROM ^{14}C

With GCM's, one can gain some understanding as to how well model estimates represent the real ocean's absorption of CO_2 by comparing simulated versus measured ^{14}C. The Hamburg model's thermocline is too shallow, so its uptake of anthropogenic CO_2 is considered as a lower limit (E. Maier-Reimer, pers. comm.). A version of the Hamburg model is now under development that incorporates increased vertical resolution and more explicit diffusion (as found necessary to produce younger deep Pacific waters based on comparison of simulated and measured natural ^{14}C); this new version is expected to exhibit higher uptake of anthropogenic CO_2. As for the Princeton model, it underestimates oceanic uptake of bomb ^{14}C by 16% (Toggweiler et al., 1989b); hence its uptake of anthropogenic CO_2 is taken as a lower limit (Sarmiento et al., 1992; Sarmiento and Sundquist, 1992). Validation of the Paris GCM with bomb ^{14}C is in progress.

Box models are typically calibrated with the ocean's surface concentration and inventory of bomb ^{14}C. Broecker et al. (1985) estimate that the bomb-^{14}C inventory is known to between 5% and 20%. Similarly, Siegenthaler and Joos (1992) find a precision of 15% for their bomb-^{14}C

inventory computed from regional inventories of Broecker *et al.* (1985). Yet the uncertainty associated with the bomb ^{14}C inventory does not translate directly (through a model's calibration) to a model's uncertainty for its CO$_2$ uptake. Oceanic absorption of bomb ^{14}C is limited largely by gas exchange; uptake of anthropogenic CO$_2$ is relatively insensitive to gas exchange and is controlled more by exchange between the surface and deep ocean. If the bomb-^{14}C inventory were to prove to be 20% higher than that reported by Broecker *et al.* (1985), the calibration procedure would likely augment both the model's gas exchange rate and its rate of mixing by eddy diffusion. If much of the higher inventory were to be explained by more intense gas exchange, the change in the CO$_2$ uptake would be much less than 20%. For the extreme case, studies in the Princeton GCM reveal that when only gas exchange is increased by 20%, the total inventory of bomb ^{14}C changes by 14.4% (Toggweiler *et al.*, 1989b), whereas total absorption of anthropogenic CO$_2$ changes by just 2.5% (Sarmiento *et al.*, 1992).

More appropriate is the work by Monfray *et al.* (1989) who study regional uncertainties in the bomb-^{14}C inventory, and find the related total oceanic uptake of anthropogenic CO$_2$ to vary by $\pm 9\%$: they employ a modified version of the Peng and Broecker (1985) model and a Monte Carlo procedure that randomizes this models circulation about a mean determined by the ^{14}C data. For the range calculation in this re-analysis, I conservatively employed the maximum $\pm 20\%$ uncertainty in the bomb-^{14}C inventory and assume that variability of CO$_2$ uptake is about half that. Of course, the relationship is more complicated (e.g., calibrations also employ the surface bomb-^{14}C concentration), and further study is clearly warranted (see section 6). The box-model estimates for oceanic CO$_2$ uptake lie at or near the upper limit of the range determined from all six models; thus, variability of up to 10% would expand the range only in the upward direction. Considering just this uncertainty, CO$_2$ uptake by the box-diffusion model (the upper limit) would be at most 10% higher than 2.4 Gt C yr^{-1}; however, the upper limit must be further adjusted because of two other shifts.

The first shift concerns the calibration of the box-diffusion model by an older bomb-^{14}C inventory that was 10% too high, based on the more recent analysis of Broecker *et al.* (1985), which takes advantage of a more extensive data set for ocean ^{14}C measurements (Siegenthaler and Joos, 1992). With this shift alone (halved because of the aforementioned relationship with CO$_2$), maximum CO$_2$ uptake by the box-diffusion model could only be 5% higher than 2.4 Gt C yr^{-1}. The second shift arises because of simplifications inherent in the calibration of the box-diffusion model. To simulate uptake of anthropogenic CO$_2$, the box-diffusion model is calibrated with the bomb-^{14}C transient; however, calibration of the same model but with only the natural distribution of ^{14}C produces much smaller diffusivities (Siegenthaler, 1983). Unlike natural ^{14}C, the nearly pulse-like input of bomb ^{14}C has had only enough time to penetrate into surface regions (over most of the ocean) where more rapid physical processes dominate. Oceanic uptake of anthropogenic CO$_2$ resembles more closely the absorption of bomb ^{14}C, because both remain mostly in the surface ocean; however, anthropogenic CO$_2$, with its longer transient (with an exponential growth rate of about 30 yr) has penetrated deeper than has bomb ^{14}C. Thus, box models calibrated only with bomb ^{14}C must overestimate uptake of anthropogenic CO$_2$. Taking this into consideration, Siegenthaler and Joos (1992) were able to calibrate simultaneously their HILDA model with both bomb and natural ^{14}C by introducing a vertical eddy diffusivity that decreases with depth (see section 2.1). The HILDA box model absorbs 12% less anthropogenic CO$_2$ when calibrated with both natural and bomb ^{14}C than it does when calibrated only with bomb ^{14}C (Siegenthaler and Joos, 1992, Table 5). Therefore, assuming about the same offset for other box models, combining both shifts, and including the uncertainty, the box-diffusion model should absorb 2.0 ± 0.2 Gt C yr^{-1}.

As for the other two box models, neither of the shifts are applicable. The Gourdriaan (1989) model is not calibrated in the conventional manner that targets bomb-^{14}C inventory and surface concentration. Rather, Goudriaan adjusts his model to match the observed decline of bomb ^{14}C in the atmosphere. Goudriaan's estimate (2.2 Gt C yr^{-1}) is too high, based upon his resulting oceanic distribution of bomb ^{14}C, which reveals surface concentrations that are too low and a whole-ocean inventory that is too high, relative to observations. As for the HILDA model, neither shift is appropriate, so I apply only the uncertainty, which at 10% means the HILDA model absorbs 2.14 ± 0.21 Gt C yr^{-1}. Thus after considering the two shifts for the box-diffusion model and the bomb-^{14}C results from Goudriaan, the HILDA model with its uncertainty sets the upper limit of the range. Finally, I increased the upper limit by another 5% (to 2.5 Gt C yr^{-1}) to be more conservative (the bomb ^{14}C vs. CO_2 uptake relationship remains poorly defined) and to make the final range (1.5–2.5 Gt C yr^{-1}) symmetric about the mean, a convenience for budget calculations.

5. Discussion and Conclusions

Considering new and updated results as well as shifts and uncertainties associated with individual model estimates (Table 2), this study finds better agreement between ocean models estimating anthropogenic CO_2 uptake, relative to that reported in 1990 by the IPCC (Figure 2). For the period 1980–1989 and an atmosphere forced by observed pCO_2, all six model estimates including their uncertainties lie in the range of 2.0 ± 0.5 Gt C yr^{-1}.

Table 2.   Modeled uptake of anthropogenic CO_2 for 1980–1989.

	Oceanic Uptake [GT C yr^{-1}]	
Model	IPCC (1990)	This Study
Hamburg GCM	1.2	≥1.5[a,b]
Princeton GCM	1.9	≥1.9
Goudriaan	2.3[c,d,e]	≤2.2[c,f]
Box-diffusion	2.4	2.0 ± 0.2[g,h,i]
HILDA	–	2.1 ± 0.2[i,j]
Paris GCM	–	2.1[j]

[a] Updated result
[b] Adjusted result from 1973–1988 to 1980–1989
[c] Adjusted result from 1960–1980 to 1980–1989
[d] Simulation with a biosphere + fossil C emissions
[e] The IPCC did not specify this result
[f] Adjusted the simulation with only fossil C emissions
[g] Shifted from calibration with excess bomb-^{14}C inventory
[h] Shifted from calibration with only bomb ^{14}C
[i] Uncertainty associated with bomb-^{14}C calibration
[j] New result since Watson et al. (1990)

Will the range of oceanic uptake estimates improve or shift substantially? For this we must look at the source of the variability. Modeling oceanic uptake of anthropogenic CO_2 requires three ingredients: (1) an atmospheric boundary condition defined by gas exchange and the record of observed atmospheric CO_2 (or fossil C emissions); (2) a formulation for the ocean's nonlinear CO_2 chemistry, and (3) a model describing the ocean's circulation. Of these, uncertainties in formulations for gas exchange and the carbonate chemistry may lead to errors of several percent (e.g., see Sarmiento et al., 1992; Siegenthaler and Joos, 1992), but it is the simulated ocean circulation that is responsible for most of the variability between models. The lower limit (1.5

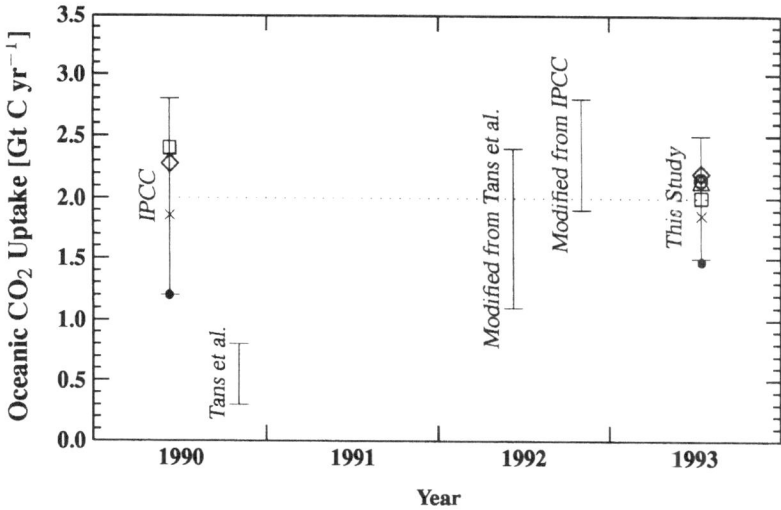

Figure 2. Range bars for estimates of oceanic uptake of anthropogenic CO_2 during the period 1980–1989. The IPCC (Watson *et al.*, 1990) finds a range of 2.0 ± 0.8 Gt C yr^{-1} using estimates from the Hamburg (•), Princeton (×), Goudriaan (◇) , and box-diffusion (□) models. Tans *et al.* (1990) infer an oceanic sink of 0.3–0.8 Gt C yr^{-1} (in their scenarios 5–8) from an analysis that compares the latitudinal distribution of observed pCO_2 with that simulated in an atmospheric GCM, as forced by spatially varying fossil fuel emissions; they partition the sink between the biosphere and oceans based on oceanic measurements of pCO_2. Sarmiento and Sundquist (1992) modify the Tans *et al.* estimate to 1.1–2.4 Gt C yr^{-1} because of corrections due to (i) the ocean's surface skin temperature, (ii) CO effects (i.e., its production during combustion of fossil C, subsequent transport in the atmosphere, and finally oxidation to CO$_2$), and most importantly (iii) river transport of dissolved carbon. Sarmiento and Sundquist also modify the IPCC's range to 1.7–2.8 Gt C yr^{-1} using the IPCC's upper limit and and a lower limit from the Sarmiento *et al.* (1992) model estimate of 1.9 Gt C yr^{-1} (for 1980-1989) or 1.7 Gt C yr^{-1} (for 1972-1989 as compatible with the sampling period of ocean measurements reported by Tans *et al.*), based upon the validation of the Princeton model against observed bomb ^{14}C (Toggweiler *et al.*, 1989b). This study finds a range of 2.0 ± 0.5 Gt C yr^{-1} after adding recent estimates from the HILDA box model (○), and the Paris GCM (△) to the updated and adjusted set of four model estimates from the IPCC; it also considers the uncertainty and offsets associated with the calibration of box models by bomb ^{14}C.

Gt C yr^{-1}) from the Hamburg GCM derives in part from its simulated thermocline that is too shallow. The latest version of the Hamburg model now under development, attempts to resolve this problem by increasing vertical resolution and including more explicit diffusion. If the Hamburg estimate were to increase to within the realm defined by the other models, the mean calculated from all model estimates would shift upward only 0.1–0.2 Gt C yr^{-1}, but the range would become 40% narrower. The conservative upper limit (2.5 Gt C yr^{-1}) is not as likely to change, considering the relative position of the other model estimates; however, it may move 0.1–0.2 Gt C yr^{-1} closer to the mean if the bomb-^{14}C inventory and its precision can be better defined.

Box model estimates rely on calibration with the bomb-^{14}C inventory that may not be known to better than $\pm 20\%$. Assuming half that uncertainty is associated with CO_2 uptake (see section 4.3), the computed range of estimates could be no narrower than $\pm 10\%$, even if all model results were identical. GCM's also rely on comparing simulated versus measured bomb ^{14}C. For example, although the Princeton model underpredicts the bomb-^{14}C inventory (Toggweiler *et al.*, 1989b) by 16%, using this result to place limits on the same model's anthropogenic CO_2 uptake is tentative in light of greater uncertainty for the measured bomb-^{14}C inventory.

Therefore, an improved understanding of the precision of the ocean's bomb-^{14}C inventory is vital if we are to better bracket oceanic uptake of anthropogenic CO_2. Besides more and better ^{14}C measurements, ocean modelers must develop more precise methods to estimate how far off bomb- and natural-^{14}C inventories might be (e.g., via objective mapping and modeling efforts). Furthermore, we must clearly demonstrate how uptake of anthropogenic CO_2 changes as a function of the bomb-^{14}C inventory (and surface concentration). Furthermore, the role of the uncertainty in the natural ^{14}C distribution must be better characterized, particularly for box models studies that will follow the lead of Siegenthaler and Joos (1992) with the HILDA model, and perform simultaneous calibration with bomb and natural ^{14}C.

GCM's provide the most realistic descriptions of ocean circulation and mixing. Yet GCM's contain many simplifications. More realistic circulation fields are or will likely be found in on-line GCM's that incorporate such effects as seasonality, higher resolution, better parameterization of sub-grid scale convection events, and a more realistic mixed layer. The effect of these simplifications upon uptake of anthropogenic CO_2 is uncertain. Preliminary results from bomb-^{14}C simulations in the Princeton model (J. R. Toggweiler, pers. comm.) seems to show that neither adding seasonality nor increasing GCM model resolution (from about $4°$ to $2°$) will have much affect upon ocean CO_2 uptake. Surprisingly, however, the on-line Princeton model absorbs 10% less bomb ^{14}C than does the off-line version (Sarmiento *et al.*, 1992), probably because convection is handled differently. Coupling a GCM to a model of the mixed layer and adding more realistic convection may substantially impact simulated CO_2 absorption.

As for the global C budget, the new estimate for the uncertainty of the ocean sink has about the same magnitude of that for fossil C emissions (Table 1). Propagating the improved uncertainty of the oceanic sink through the budget, the uncertainty of the combined atmosphere and ocean sink is reduced by 40%. Despite these improvements, however, the uncertainty associated with the computed imbalance term becomes only slightly smaller, because most of its uncertainty still derives from that associated with deforestation and land use change. Unfortunately, we remain unable to quantify directly the missing sink. Furthermore, all improvements in the budget of the global C cycle that result from this re-analysis rely on ocean models as representative of the real ocean, and all ocean models employ the steady-state assumption.

A necessary simplification, the steady-state assumption appears justified based on the ice core record of atmospheric pCO_2 extending back to long before the industrial revolution as well as little evidence for recent dramatic changes in ocean circulation or productivity (see Sarmiento, 1991; Sarmiento and Siegenthaler, 1991). Yet prognostic ocean models concerned with the rapid changes that are likely to occur in the near future probably will have to go beyond the steady-state assumption. They will have to use coupled atmosphere-ocean models, whose physical processes change over time. The ocean's biology may also change but it would only affect anthropogenic CO_2 if it becomes more or less efficient at removing nutrients from surface waters, changes its carbon to nutrient ratio, or there are compositional shifts between species of siliceous versus calcareous forming plankton.

This paper compares total CO_2 uptake amongst ocean models as a means to gauge our understanding of CO_2 uptake in the real ocean; however, it is only a beginning. For instance, models might yield the same total uptake but for different reasons. Thus more intricate evaluations are necessary, particularly for GCM's, which should compare the nature of CO_2 uptake in three dimensions. Furthermore, because our understanding of the ocean's carbon cycle relies so heavily on ocean models (as does much of our understanding about ocean circulation and mixing), models must be benchmarked with more than just anthropogenic CO$_2$. GCM simulations with other tracers (e.g., natural ^{14}C, bomb ^{14}C, δ^{13}C, CFC's, ^{3}H, and ^{39}Ar) are now becoming more commonplace; therefore, outlines should be set forth so that models might better be compared not only against observations but amongst one another. With similar information from more than one model, one could better pinpoint model inconsistencies and direct ocean sampling campaigns for tracers such as ^{14}C and ^{39}Ar whose tremendous potential to characterize oceanic mixing is limited because of the large effort required for each measurement.

ACKNOWLEDGEMENTS. I sincerely thank E. Maier-Reimer who generously provided recent unpublished data from the Hamburg model and F. Joos who kindly made available detailed output for the HILDA model. Much appreciation goes to J. Downing, M. Heimann, F. Joos, K. Kurz, C. Lambert, S. Mateu, P. Monfray, and two anonymous reviewers whose critical comments have improved this manuscript. I gratefully acknowledge C. Le Quere, J. Sarmiento, and U. Siegenthaler for open discussions about model differences. I also thank the French CEA for supporting this effort, and the German Umweltbundesamt who provided travel funds for the Bad Harzburg workshop that is detailed further in this volume.

References

Andrich, P., 1988. OPA—A multitasked Ocean General Circulation Model, *Tech Rep.*, LODYC, Université Paris VI, France, 60 pp.

Bacastow, R., and E. Maier-Reimer, 1990. Ocean-circulation model of the carbon cycle, *Clim. Dyn., 4*, 95-125.

Baes, C. F., Jr. , A. Björkström, and P. J. Mullholland, 1985. Uptake of carbon dioxide by the oceans, in J. Trabalka (ed.), *Atmospheric Carbon Dioxide and the Global Carbon Cycle.* DOE/ER-0239, U.S. Dept. of Energy, Washington, DC, 20545, 81-111.

Blanke, B., and P. Delecluse, 1993. Low frequency variability of the tropical ocean simulated by an OGCM with mixed layer physics, *J. Phys. Oceanogr.*, in press.

Broecker, W. S., T.-H. Peng, G. Ostlund, and M. Stuiver, 1985. The distribution of bomb radiocarbon in the ocean. *J. Geophys. Res., 90*, 6953-6970.

Bryan, K. 1969. A numerical method for the study of the circulation of the world ocean, *J. Comput. Phys., 4*, 347-376.

Chartier, M., 1985. Un modèle numérique tridimensional aux équations primitives de la circulation générale de l'océan. CEA Report R-5372, France.

Emanuel, W. R., I. Y.-S. Fung, G. G. Killough, Jr., B. Moore III, and T.-H. Peng, 1985. Modeling the global carbon cycle and changes in atmospheric carbon dioxide levels, in J. Trabalka (ed.), *Atmospheric Carbon Dioxide and the Global Carbon Cycle*, DOE/ER-0239, U.S. Dept. of Energy, Washington, DC, 20545, 141-173.

Friedli, H., H. Loetscher, H. Oeschger, U. Siegenthaler, and B. Stauffer, 1986. Ice core record of the ^{13}C/^{12}C ratio of atmospheric carbon dioxide in the past two centuries. *Nature, 324*, 237-238.

Fujio, s., and N. N. Imasato, 1991. Diagnostic calculation for circulation and water mass movement in the deep pacific, *J. Geophys. Res., 96*, 759-774.

Goudriaan, J., 1989. Modelling biospheric control of carbon fluxes between atmosphere, ocean, and land in view of climatic change, in A. Berger, S. Schneider, and J. Cl. Duplessy (eds.), *Climate and Geo-sciences*, NATO-ASI Series C285, 481-499, Kluwer.

Goudriaan, J., and P. Ketner, 1984. A simulation study for the global carbon cycle, including man's impact on the biosphere, *Climatic Change, 6*, 167–192.

Hasselmann, K., 1982. An ocean model for climate variability studies, *Prog. Oceanogr., 11*, 69–92.

Hellerman, S. and M. Rosenstein, 1983. Normal monthly wind stress over the world ocean with error estimates, *J. Phys. Oceanogr., 13*, 1093–1104.

Houghton, R. A., 1992. Paper presented at the workshop on Biotic Feedbacks in the Global Climatic System, Woods Hole, MA, October 25–29.

Joos, F., 1992. Modellierung der Verteilung von spurenstoffen im ozean und des globalen kohlenstoffkreislaufes, Ph.D. Thesis, Universität Bern, Switzerland.

Keeling, C. D., R. B. Bacastow, A. F. Carter, S. C. Piper, T. P. Whorf, M. Heimann, W. G. Mook, and H. Roeloffzen, 1989. A three-dimensional model of atmospheric CO_2 transport based on observed winds: 1. Analysis of observational data, *Geophysical Monograph, 55*, 165–236.

Levitus, S., 1982. Climatological atlas of the World Ocean, *NOAA Prof. Pap. 13*, U.S. GPO., Washington, D.C., 173 pp.

Madec, G., and M. Crépon, 1991. Thermohaline-driven deep water formation in the Northwestern Mediterranean Sea, in P. C. Chu and J. C. Gascard (eds.), *Deep Convection and Deep Water Formation in the Oceans*, Elsevier, pp. 241–265.

Madec, G., M. Chartier, and M. Crépon, 1991a. The effect of thermohaline forcing variability on deep water formation in the western Mediterranean Sea: a high resolution three-dimensional study, *Dyn. Atmos. Oceans, 15*, 301–332.

Madec, G., M. Chartier, P. Delecluse, and M. Crépon, 1991b. A three-dimensional numerical study of deep water formation in the Northwestern Mediterranean Sea, *J. Phys. Oceanogr., 21*, 1349–1371.

Maier-Reimer, E., and K. Hasselmann, 1987. Transport and storage of CO_2 in the ocean – an inorganic ocean-circulation carbon cycle model, *Clim. Dyn., 2*, 63–90.

Maier-Reimer, E., and U. Mikolajewicz, 1992. The Hamburg Large Scale Geostrophic Ocean General Circulation Model (Cycle 1), *Tech. Report, No. 2*, German Climate Computer Center, Institut für Meereskunde der Universität Hamburg, Hamburg, Germany.

Maier-Reimer, E., K. Hasselmann, D. Olbers, and J. Willebrand, 1982. An ocean circulation model for climate studies, *Tech Rep.*, Max Planck Institut für Meteorologie, Hamburg, Germany.

Marland, G., 1989. Fossil fuels CO_2 emissions, *CDIAC Communications, Winter 1989*, Carbon Dioxide Information Analysis Center, Oak Ridge National Laboratory, Oak Ridge, TN, 1–3.

Marti, O., 1992. Etude de l'ocean mondial: Modélisation de la circulation et du transport des traceurs anthropiques, Thèse, Docteur de l'Université Paris VI, LODYC, France, 201 pp.

Marti, O., G. Madec, and P. Delecluse, 1992. Comment on "Net diffusivity in ocean general circulation models with nonuniform grids" by F. L. Yin and I. Y. Fung, *J. Geophys. Res. 97*, 12763–12766.

Monfray, P. , M. Heimann, and E. Maier-Reimer, 1989. Uptake of anthropogenic CO_2 by ocean models calibrated or tested by bomb radiocarbon: sensitivity to uncertainties in gas exchange formulation and ^{14}C data, in *Extended Abstracts of Papers Presented at the Third International conference on Analysis and Evaluation of Atmospheric CO_2 Data Present and Past*, Environmental Pollution Monitoring and Research Programme, WMO, Geneva, 215–216.

Neftel, A., E. Moor, H. Oeschger, and B. Stauffer, 1985. Evidence from polar ice cores for the increase in atmospheric CO_2 in the past two centuries, *Nature, 315*, 45–47.

Oeschger, H., U. Siegenthaler, U. Schotterer, and A. Gugelmann, 1975. A box diffusion model to study the carbon dioxide exchange in nature, *Tellus, 27*, 168–192.

Orr, J. C., and J. L. Sarmiento, 1992. Potential of marine macroalgae as a sink for CO_2: constraints from a 3-D general circulation model of the global ocean, *Water, Air, and Soil Pollution, 64*, 405–421.

Peng,T.-H., and W. S. Broecker, 1985. The utility of multiple tracer distributions in calibrating models for uptake of anthropogenic CO_2 by the ocean thermocline, *J. Geophys. Res., 90*, 7023–7035.

Rotty, R. M., and C. D. Masters, 1985. Carbon dioxide from fossil fuel combustion, in J. Trabalka (ed.), *Atmospheric Carbon Dioxide and the Global Carbon Cycle*, DOE/ER-0239, U.S. Dept. of Energy, Washington, DC, 20545, 63–80.

Sarmiento, J. L., 1991. Oceanic uptake of anthropogenic CO_2: the major uncertainties, *Global Biogeochem.*

Cycles, 5, 309–313.

Sarmiento, J. L., and J. C. Orr, 1991. Three dimensional ocean model simulations of the impact of Southern Ocean nutrient depletion on atmospheric CO$_2$ and ocean chemistry, *Limnol. Oceanogr., 36*, 1928–1950.

Sarmiento, J. L., and U. Siegenthaler, 1991. New production and the global carbon cycle, in P. Falkowski and A. Woodhead (eds.), *New Production and the Global Carbon Cycle*, Plenum, New York.

Sarmiento, J. L., and E. Sundquist, 1992. Oceanic uptake of anthropogenic CO$_2$: a new budget, *Nature 356*, 589–593.

Sarmiento, J. L., J. C. Orr, and U. Siegenthaler, 1992. A perturbation simulation of CO$_2$ uptake in an ocean general circulation model, *J. Geophys. Res., 97*, 3621–3645.

Semtner, A. J. J., and R. Chervin, 1988. A simulation of the global circulation with resolved eddies, *J. Geophys. Res. 93*, 15502–15552.

Shaffer, G., and J. L. Sarmiento, 1993. Biogeochemical cycling in the global ocean 1: A new analytical model with continuous vertical resolution and high latitude dynamics, *J. Geophys. Res.* , in press.

Siegenthaler, U., 1983. Uptake of excess CO$_2$ by an outcrop-diffusion model of the ocean, *J. Geophys. Res., 88*, 3599–3608.

Siegenthaler, U., and H. Oeschger, 1987. Biospheric CO$_2$ emissions during the past 200 years reconstructed be deconvolution of ice core data, *Tellus, 39B*, 140–154.

Siegenthaler, U., and F. Joos, 1992. Use of a simple model for studying oceanic tracer distributions and the global carbon cycle, *Tellus, 44B*, 186–207.

Siegenthaler U., H. Friedli, H. Loetscher, E. Moor, A. Neftel, H. Oeschger, and B. Stauffer, 1988. Stable isotope ratios and concentration of CO$_2$ in air from polar ice cores, *Annals of glaciology, 10*, 1–6.

Smolarkiewicz, K. P., 1982. The multidimensional Crowley advection scheme. *Month. Weath. Rev., 110*, 1968–1983.

Smolarkiewicz, K. P., 1983. A simple positive definite advection with small implicit diffusion, *Month. Weath. Rev., 111*, 479–486.

Smolarkiewicz, K. P., and T. L. Clark, 1986. The multidimensional positive definite advection transport algorithm: further development and applications, *J. Comput. Phys., 67*, 396–438.

Tans, P. P., I. Y. Fung, and T. Takahashi, 1990. Observational constraints on the global atmospheric CO$_2$ budget, *Science, 247*, 1431–1438.

Toggweiler, J. R., K. Dixon, and K. Bryan, 1989a. Simulations of radiocarbon in a course-resolution world ocean model. 1. Steady state pre-bomb distributions, *J. Geophys. Res., 94*, 8217–8242.

Toggweiler, J. R., K. Dixon, and K. Bryan. 1989b. Simulations of radiocarbon in a course-resolution world ocean model. 2. Distributions of bomb-produced ^{14}C, *J. Geophys. Res., 94*, 8243–8264.

Watson R. T., H. Rodhe, H. Oeschger, and U. Siegenthaler, 1990. Greenhouse gases and aerosols, in J. T. Houghton, G. J. Jenkins, J. J. Ephraums (eds.), *Climate Change, The IPCC Scientific Assessment*, WMO/UNEP, Cambridge U. Press, Cambridge, England, pp. 1–40.

Watson, R. T., L. G. Meira Filho, E. Sanhueza, and A. Janetos, 1992. Greenhouse gases: Sources and sinks, in J. T. Houghton, B. A. Callander, S. K. Varney (eds.), *Climate Change 1992: The Supplementary Report to the IPCC Scientific Assessment*, WMO/UNEP, Cambridge U. Press, Cambridge, England, pp. 24–46.

Yin, F. L., and I. Y. Fung, 1991. Net diffusivity in ocean general circulation models with nonuniform grids, *J. Geophys. Res., 96*, 10773–10776.

U.S. CARBON OFFSET POTENTIAL USING BIOMASS ENERGY SYSTEMS

L. L. Wright
Environmental Sciences Division, Oak Ridge National Laboratory
Oak Ridge, TN 37831 USA

and

E. E. Hughes
Hydroelectric Generation and Renewable Fuels Program
Electric Power Research Institute, Palo Alto, CA 94303 USA

Abstract. A previous analysis had assumed that about 20% of 1990 U.S. C emissions could be avoided by the substitution of biomass energy technologies for fossil energy technologies at some point in the future. Short-rotation woody crop (SRWC) plantations were found to be the dedicated feedstock supply system (DFSS) offering the greatest C emission reduction potential. High efficiency biomass to electricity systems were found to be the conversion technology offering the greatest C emission reduction potential. This paper evaluates what would be required in terms of rate of technology implementation and time period to reach the 20% reduction goal. On the feedstock supply side, new plantings would have to installed at an average a rate of 1×10^6 ha yr^{-1} while average yields would have to increase by 1.5% annually over the 35-year period. Such yield increases have been observed for high value agricultural crops with large government research support. On the generation side, it requires immediate adoption of available technologies with a net efficiency of 33% or higher (such as the Whole Tree Energy™ technology), installation of approximately 5000 MWe of new capacity each year, and rapid development and deployment of much higher efficiency technologies to result in an average of 42% efficiency by 2030. If these technology changes could be achieved at a linear rate, U.S. C emission reduction could progress at a rate of about 0.6 % yr^{-1} over the next 35 years.

1. Introduction

Biomass energy systems are comprised of a broad range of technologies that vary greatly in the feedstocks and conversion systems used, the efficiency with which biomass is converted to usable energy, and the cost of implementation and their C benefits. Of the two major options for large-scale use of biomass in the energy sector, electric power generation and liquid fuels for transportation, substitution of electric

Water, Air, and Soil Pollution **70**: 483–497, 1993.

power generation from woody DFSS for coal-fired electric power was found to be about twice as effective in offsetting C emissions as the conversion of woody crops to ethanol for substitution of gasoline (Graham *et al.*, 1992; Wright *et al.*, 1991). To meet the goals for C emission reductions suggested by the Framework Convention on Climate Change held in June 1992 at the United Nations Conference on Environment and Developments, optimal strategies for C offset must be developed. Since biomass energy systems do require the use of land resources, it is extremely important to select and develop systems which obtain the maximum energy output and C emission reduction with the minimum use of land.

Conflicts between land uses are sure to occur as more and more people search for places to live, work, grow their crops, produce livestock, produce cellulosic resources, conserve wild areas for recreation and biological diversity, and designate areas for the sequestration of C. Finding ways to meet all of these needs will be an ever increasing challenge. An analysis by Marland and Marland (1992) has suggested that within a 50-year time horizon, establishment of woody DFSS on cropland results in a larger C offset than would reforestation when the land has the potential of producing greater than 4.0 MgC ha^{-1} yr^{-1} in a biomass energy system with an energy conversion efficiency of 33% or greater.

Our previous analysis of the C emission reduction potential of biomass energy systems suggested that a 20% reduction below 1990 levels could be achieved "in the future" by using woody DFSS to fuel efficient biomass electric power systems. This analysis establishes the target time as 2030 and evaluates the rate of change that is required in SRWC and biomass power systems to meet a 20% reduction goal by 2030. Factors considered include the probability of developing and building new, high-efficiency electric generation capacity, the possibility of achieving rapid SRWC yield increases while simultaneously expanding the feedstock production land base, and the contribution that soil C sequestration in SRWC systems could make to meeting the goal.

2. Estimation of C Offset Potential

Estimates of C offset potential in the United States require several assumptions about feedstock yields, land areas available, conversion efficiency, fuel substitution factors, and C inputs to producing the feedstock (Graham *et al.* 1992). The information requirements for calculating C offset potential can be broken down into four linked categories as follows:

> Amount of feedstock as function of:
> > (biomass yield/hectare, harvest and storage losses, number of hectares)
> Amount of energy produced as a function of:
> > (amount of feedstock, conversion efficiency)
> Fossil fuel displaced as a function of:
> > (energy produced, fuel substitution factors, fossil fuel C level)

Net C offset as a function of:
(fossil fuel displaced, C input to feedstock, C sequestration)

All of these assumptions should be tempered by environmental considerations. While the environmental benefits for soils and water quality offered by DFSS and the air quality benefits offered by advanced conversion system may be very positive, those attributes will have to be confirmed. Current and future regulations and societal concern about pollution potential could initially slow the implementation of biomass energy systems. Delaying implementation of advanced biomass energy systems could also have negative environmental risks. Developers of the technologies must be sensitive and responsive to local environmental issues, while pursuing implementation of the technology to assure that global environmental goals are met.

2.1 FEEDSTOCK POTENTIAL

The amount of land that might be economically available in the United States for production of energy crops is estimated to range between 14 and 28×10^6 ha (Wright et al., 1991). The higher amount of 28×10^6 ha equals about 17.5% of the cropland in the United States that is considered capable of growing energy crops (Graham, 1993).

The land currently in the Conservation Reserve Program (CRP) has been hypothesized to be the land most likely to be first used for energy crop production. However, most of that land lies in the Great Plains, which is not very productive nor is it an area of high electricity demand. Furthermore even in regions of the country quite suitable for energy crops production, the CRP land tends to be the least productive and profitable cropland. The best cases where there may be a match between CRP land and wood energy crop requirements will be on croplands with moderate wetness. Tree crops such as poplars, sycamores, and silver maple, which have high moisture requirements, may perform very well on such lands if weed control problems can be resolved.

Fifteen years of research by U.S. DOE on the development of economically-viable, woody DFSS has demonstrated the need for relatively good quality land to attain high yields. Cost-supply analyses, constrained by current agricultural policies, indicate that the land most likely to be converted to dedicated feedstock production will be the better quality cropland, especially if it is located close to an energy conversion facility (Graham et al., 1993).

The amount of feedstock which can be grown depends on the location, type and amount of land which can be converted to feedstock production as well as the plant material selected. The range of yields indicated in Figure 1 is primarily a function of site quality and genetic makeup of the plants. The high experimental and commercial yields in the Pacific Northwest can be attributed to the availability of selected clonal plant materials, favorable climate, and the establishment of both experimental and commercial plantings on fertile soils. The rest of the country does not have the

ORNL-DWG-93M1256R

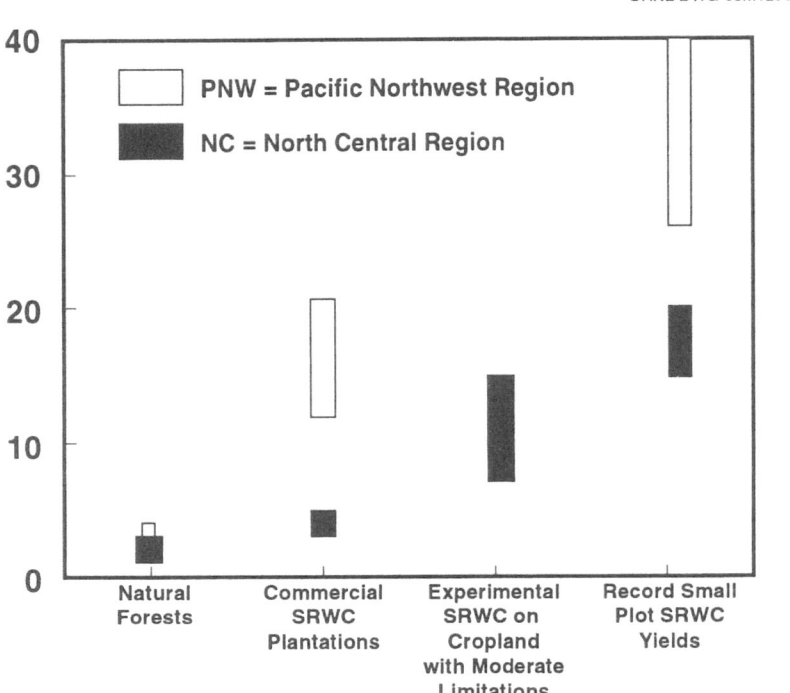

Figure 1. Mean biomass yield at harvest (Mg dry biomass/ha/year).

favorable climatic conditions of the Pacific Northwest, nor the commercial availability of select clonal plant material. Commercial yields are unlikely to ever equal the record yields shown in Figure 1 due to climatic variations, less than perfect management techniques, and environmental considerations. Achievement of larger plot yields on a commercial basis should be entirely possible if improved plant materials and technology transfer programs are widely available.

The net yield level of 11.2 dry Mg ha^{-1} yr^{-1} selected as representative of current DFSS technology assumes careful matching between land types and species or varieties, successful use of best available technology for weed control, and no unexpected disease or pest problems. Achieving such yields with most woody crops appears to require that water be available to the trees most of the year. Previously cropped bottomlands subject to flooding once or twice per year could be very suitable for trees but limiting for annual crops. There are over 40×10^6 ha of relatively good cropland in this category in the United States. Nearly half of that land is in the north central region of the United States, but all parts of the United States contain such lands. There would seem to be a reasonably good potential that suitable lands for woody DFSS will also be economically available.

Figure 2 shows the relationship which exists between average net yield assumptions and available land base in the calculation of total exajoules (EJ or 1×10^{18} J) of feedstocks which can be produced. Calculation of EJ is obtained by multiplying total net Mg of feedstock produced by the average energy content of the feedstock. Total net Mg equals (annual yield ha^{-1} minus harvest and handling losses) × number of hectares). Harvest and handling losses can vary considerably among feedstocks and handling systems. They may be as small as 5% or less for trees harvested and hauled in whole form and dried under cover, but as large as 17 to 20% for trees chipped in the field and stored under open conditions for 6 months or more. The curves in Figure 2 (originally developed for another paper) assumed an average of energy content of 18.5 Gj Mg^{-1} derived from averaging the higher heating energy values of woody and herbaceous crops. The set of curves allows an approximate reference to the hectares and average net yields required to produce a given level of primary energy. These curves clearly show the importance of achieving high average yields of DFSS, under any assumptions about land availability.

ORNL-DWG 91Z-13913

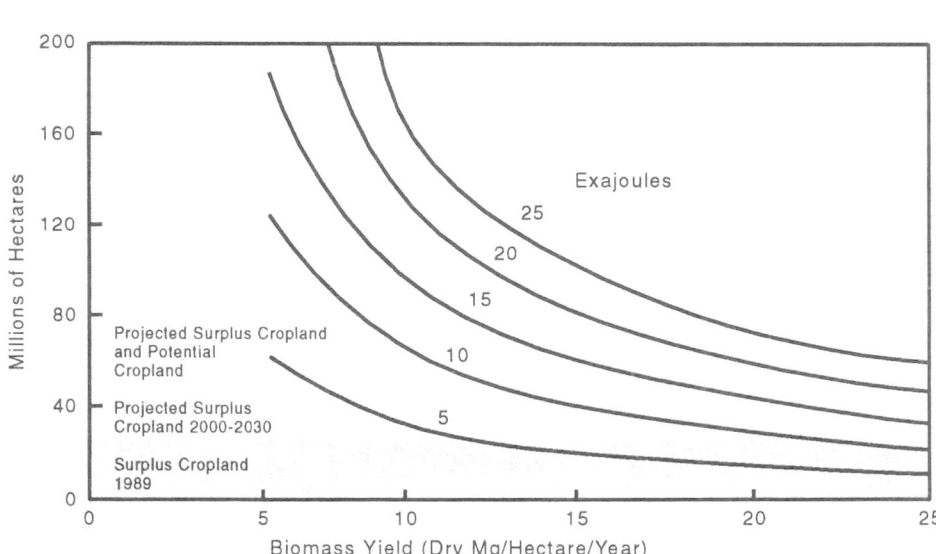

Figure 2. Potential primary energy production as a function of biomass yield and number of hectares.

2.2 ENERGY PRODUCTION POTENTIAL

Electric power generation from biomass is a conversion technology which is already in commercial use. These technologies use solid fuel, not liquid or gaseous fuel. The feedstock for nearly all of this existing power generation is wood, mostly wood wastes and residues already in the hands of wood products industries, especially the pulp and paper industry. (A liquid intermediate product, namely "black liquor," is a significant fraction of the fuel actually fed to boilers in the pulp and paper industry. Even in this case the original feedstock in the pulp/paper process is wood in the solid form, usually chips of 2.5 or 5.0 cm, top size.) The technologies now used commercially for electric power generation from wood or other biomass fuels are traveling grate stoker boilers, Hydrograte™ boilers and fluidized-bed combustion (FBC) boilers. All three accept fuel in the form of chipped wood (or other solid biomass residue) and burn the fuel mostly on or in a bed at the bottom of a boiler, which generates steam that is converted to electric power in a steam turbine/generator unit. All three of these boiler technologies continuously move the unburned fuel (i.e., ash) out of the boiler: the traveling grate via slow continuous moving of the floor, the Hydrograte™ via vibration of water-cooled tubes that constitute the grate, and the FBC via separation of ash from the sand that forms most of the mass in the bed, a bed fluidized by air blown in from underneath.

The existing power generation from biomass fuel is generally characterized by relatively small unit size (in the 10- to 50-MWe range, mostly less than 30 MWe) and low efficiency (on the order of 20% net efficiency on a higher heating value basis, or a heat rate of 17,000 Btu or 17,918 J per kWh). The low efficiency is the result of economic decisions reflecting the small unit size and the low cost of fuel that is often readily available waste material. The largest (about 50 or 60 MWe) and most efficient have efficiencies closer to 25%. These plants fire wood fuel that is often about 50% moisture, by weight on a wet basis. The high moisture content contributes to a low boiler efficiency, in the 65 to 70% range, compared to the 84 to 89% boiler efficiency expected of a 200 MWe coal-fired boiler. (Johnston *et al.*, 1991).

Future wood-fired, and other biomass-fired, power plants are expected to be higher in efficiency. The higher efficiencies will result from (1) market forces, assuming that biomass wastes and residues will be less readily available and cost more in the future; (2) successful research and development efforts; and (3) selection of larger unit sizes (on the order of 100 MWe or larger, perhaps even to 400 MWe). With very advanced technologies, such as fuel cell processes, net thermal efficiencies may reach 50% or higher (higher heating value basis) within the next two to four decades. High temperature, modified Rankine cycles (steam turbines with direct combustion) may provide an opportunity to achieve relatively high efficiencies now with biomass fuels. Steps are being taken to optimize efficiency of biomass conversion by (1) using the existing higher efficiency steam cycles to cofire coal with wood and (2) planning for a demonstration plant using Whole Tree Energy™ technology.

The Electric Power Research Institute (EPRI) is cosponsoring case studies of the co-firing option and has sponsored an assessment (Johnston *et al.*, 1991) and field tests (Ostlie *et al.*, 1993) of the Whole Tree Energy™ option. Both of these options are capable of net thermal efficiencies, on a higher heating value basis, above 30%. [In fact, both are anticipated to convert wood to electric power with a heat rate of about 10,000 Btu (10,540 GJ)/kWh, i.e., about 34% net thermal efficiency.]

Co-firing of wood with coal has been done in the United States in a few commercial operations for several decades (Ostlie, personal communication) and more recently in some experimental conditions. Under these operational or test conditions only a small fraction of the heat input to the boiler, usually 5% or less, is from wood. Because the combustion environment is determined by the primary fuel (coal), which is not a high moisture fuel, the combustion occurs with a relatively small loss of heat rate (about 10%). The larger size of the coal-fired units (i.e., 200 to 500 MWe) allows the steam cycle to achieve economies of scale and, therefore, higher performance at acceptable cost, giving net efficiencies in the 34% range typical of large, efficient coal-fired units. The case studies being cosponsored by EPRI (with the Tennessee Valley Authority and the U.S. Department of Energy) are investigating co-firing at higher percentages where the heat contributed by the wood is in the 10% to 15% range. By moving from the conventional pulverized-coal boilers (tiny particles burning in suspension) or cyclone-fired boilers (larger coal particles burning in a molten slag of coal ash in rotation flow on the wall of a cylindrical barrel) to the new fluidized-bed combustion (FBC) boilers, utilities could cofire wood in higher percentages, say up to 50%. However, there are as yet very few FBC boilers in utility service.

Co-firing offers a low-capital-cost option for introducing wood firing into utility power systems. The boiler, turbine/generator and balance of plant already exist. From the perspective of the public looking for a low-cost way to achieve the benefits of being CO_2-neutral by using a renewable fuel rather than a fossil fuel, co-firing may seem to be an obvious choice. However, from the perspective of a utility already operating existing capacity on coal, the wood co-firing does not add any new capacity (i.e., no additional MWe). Furthermore, the only economic benefits accompanying the additional cost of adding wood handling, wood drying (required in the case of co-firing above about 4 to 7% heat input in a pulverized-coal unit), and wood firing are the benefits of avoided pollution control (most likely SO_2 control), avoided CO_2 emissions, and the unlikely benefits of finding adequate wood supplies costing less than the coal. In the absence of taxes or limits on CO_2 emissions the benefits may be small, and, therefore, the economic incentive to co-fire wood may also be small or even negative.

Whole-Tree-Energy™ is a patented technology for dedicated wood firing (100% wood fuel) in new or converted boilers. The EPRI-sponsored evaluation (Johnston *et al.*, 1991) indicated this technology could be a low-cost way to produce electricity from wood, with an estimated cost of electricity at $0.046/kWh compared to $0.062/kWh for new state-of-the-art combustion of 50% moisture wood chips and $0.057/kWh for new coal-fired plants with scrubbers for SO_2 removal from the flue gas. (The plant sizes used in that evaluation were 100 MWe for the whole-tree and woodchip cases and 200

MWe for the coal.) The cost advantage of the Whole-Tree Energy™ power plant is expected from several features of the technology: (1) no chipping cost and harvesting operations that are free of any delays or timing constraints possibly imposed by the chipping operation; (2) handling and stacking wood in its natural form for drying; (3) improved boiler efficiency, combustion rate, and combustion completeness due to use of dried wood; (4) combustion similar to a gas-fired boiler, above a bed of whole trees that produce the fuel gas as they are heated and volatilized; (5) an efficient steam cycle using high pressure, high temperature superheat and reheated steam; (6) stack gas cleaned and cooled by a condensing heat exchanger that transfers heat to the air used to dry the trees and also scrubs particulates from the flue gas; and (7) elimination of many items that add to the cost of coal-fired plants, such as the SO_2 scrubber, coal bunkers, coal pulverizers and related starters, controls and electrical wiring, structural steel and foundations.

As suggested above, several features of Whole Tree Energy™ are different from currently existing biomass power systems. Wood is handled its natural whole tree form until just before combustion. After harvest and transport to the power plant site, the whole trees are piled in a tall stack (such as a circular pattern some 120 m in diameter and 30 to 50 m high) inside a tent-like structure. The trees are dried from their typical 40 to 50% moisture down to less than 25% moisture by warm air (waste heat that would otherwise go up the smoke stack). After being dried, trees are removed from the stack and randomly dropped onto an open-topped conveyer channel which moves in intermittent steps to the boiler. Near the boiler, the trees are batch-sawed into load sections of a length close to that of the boiler wall (perhaps about 8 to 10 m). The wood is pushed into a 2-stage sealed door and ram system, dropped into the boiler, and burned in a 3-stage combustion process. The combustion process takes advantage of the relatively low moisture content of the dried wood to perform complete combustion with relatively low overall excess air (10 to 15%). This brings about high-temperature combustion in air with low levels of NO emissions. Above the deep bed of whole trees, the boiler is very much like one built for firing natural gas. Much of the release of the energy in the fuel emerges as heat above the bed, as the gaseous fuel released by volatilization of the wood burns in the tall (25 to 35 m) space above the fuel bed (bed depth of about 4 or 5 m). Fully implemented, the Whole-Tree-Energy™ concept also encompasses innovations in the growing, harvesting, and transportation of the feedstock. While the system can use wood from overage, declining forest stands and residue from existing logging operations, the optimum environmental benefits of the system will be derived from the use of DFSS.

2.3   FOSSIL C DISPLACEMENT

The amount of raw energy levels of biomass and coal required to produce a kilowatt hour of electricity from coal or biomass is the same if the conversion efficiencies are the same. A simplifying assumption made in our analysis is that the conversion efficiencies of new biomass energy systems will be similar to those of the coal systems being displaced. The conversion efficiency for most modern coal-burning facilities averages about 33% but can be much higher.   Process conversion efficiencies for converting coal to energy are anticipated to greatly improve in the future (Starr *et al.*, 1992). Improvements in coal conversion technologies will be applicable to biomass.

Another comparison that is often made when wood replaces coal as a feedstock is the relative tonnage of wood and coal that is required. Since wood is delivered with a much higher moisture content than coal and has a slightly lower energy content per unit of dry weight (compared to sub-bituminous coal), much greater tonnage of delivered feedstock will be necessary to produce the same amount of electricity. Delivery and storage of the feedstocks will have sociological and environmental impacts that may contribute to limiting the amount of biomass energy that can be used for a given facility, especially those located near urban centers.   On the positive side, however, a greater number of smaller energy production facilities scattered throughout the rural sections of the United States could have very positive economic impacts with positive environmental attributes as well.

2.4   NET C OFFSET

The greatest C benefit of biomass energy systems can be attributed to the effect gained by leaving fossil C fuels in the ground. This benefit is cumulative with time. However, DFSS also provides a significant amount of C sequestration both in the soil and in the average standing stock of biomass materials. As average yields increase, the amount of standing C ha^{-1} was estimated to increase from 16.8 to 27.7 Mg. This represents the maximum C sink offered by the standing trees which is attained by the end of the first rotation on any given hectare.   The value of that C sink must be divided by the number of years over which the analysis is conducted in order to add the value to the cumulative C offset gained by fossil energy substitution.   If divided by a value of 35 years (1996 to 2030), the annualized value of carbon sequestration in the trees varies from 0.48 to 0.80 Mg C ha^{-1} yr^{-1}.

The C increment that might be occurring in the soil also offers an additional C sink. It cannot be assumed, however, that soil C will always be incremented on DFSS sites. Conversion of pastureland, moist bottomlands, and peat soils to DFSS may in fact result in an initial loss of soil C. Both above and belowground C losses would occur if forested land were converted to woody crop plantations. It is not recommended that forests be cleared for energy plantations.   After a recent evaluation of published literature, Ranney *et al.*, (1991) concluded that changes in soil C levels may range between a loss of 5 Mg to a gain of 10 Mg C ha^{-1} before reaching equilibrium.

Some recent experimental information (Hansen, 1993) suggests that soil C values increase at an average rate of about 1.24 Mg C yr^{-1} over the first 18 to 20 years of an unharvested hybrid poplar stand established on land previously managed for row crops. This was determined by comparisons between soil C contents of soil in the stand and the adjacent land in row crops. The comparisons showed that soil C increments were occurring below 30 cm depth and that soil surface levels of C were similar to that on row crop land.

The experimental data are inadequate to predict the period of time over which these increases will continue, and the extent to which harvesting at 6- to 12-year intervals would affect the rate of C turnover. Given the unknowns, our calculations are based on a conservative assumption that the soil C is likely to increase at about 1.0 Mg C ha^{-1} yr^{-1} over the first 12 years of the plantation after which an equilibrium is assumed to occur. To simplify the calculations, an average soil C increment of 0.3 Mg C ha^{-1} yr^{-1} was used over the 35-year period of our calculations (1995–2030).

If the C sequestration assumptions made above are added to the C offsets achieved by fossil fuel substitution, then the total C offset per hectare increases from 6.0 Mg C ha^{-1} yr^{-1} to 9.7 Mg C ha^{-1} yr^{-1} by 2030. It should be noted that the C offset values have already been reduced by the C emissions which result from woody crop plantation management activities (Graham *et al.*, 1992).

2.5   CONVERTING POTENTIALS TO REALITY

Taking advantage of the global C benefits offered by biomass energy systems will be complex because it requires changing attitudes, habits, and institutions in the agricultural, energy, and environmental sectors of society. However the interest is becoming stronger, new policies are being initiated, and the utilities are looking seriously at both co-firing and the opportunities offered by new technologies. Discussion on the pros and cons of C taxes and incentives can be found almost daily in the news. Analyses by utilities are indicating that biomass energy and reforestation will be among the lower cost options available for meeting reduced emission requirements while meeting energy needs.

One set of assumptions which achieves the theoretical goal of reducing U.S. fossil fuel emissions by 20% includes; (1) a land base of 28 × 10^6 ha, (2) average delivered biomass yields of 18.5 Mg ha^{-1} yr^{-1}, and (3) average conversion efficiencies of 42%. Table 1 shows that to bring those assumptions to reality by 2030, the United States would need to be planting about 1 × 10^6 ha per year and building or retrofitting 5000 MWe of biomass electric capacity per year. Additionally DFSS commercial yields would have to improve by 1.5% per year, and conversion facilities would have to improve efficiencies by 0.7% per year. Since installed capacity and established plantations will be difficult to improve, the implication is that by the time 2030 approaches, some new capacity will have to be capable of 50% or better conversion efficiency and some plantations will have to achieve yields higher than 18.5 Mg ha^{-1} yr^{-1}. The important question is, are these rates of change conceivable?

There are several coal and wood conversion technologies under development and in the demonstration phase which have the potential of achieving net efficiencies of 34 to 41% (Starr *et al.*, 1992). Modern commercial coal stations have efficiencies of 34%

Table 1. Annual rate of technology change required for meeting C offset goals

	1996	2010	2030	Annual rate of change
C offset Goals[1]	<1%	~10%	~20%	~0.6%
MWe	4,500	72,000	170,000	~5000MWe
Capacity factor	80%	80%	80%	
Conversion efficiency	33%	37%	42%	~0.7%
DFSS Mg ha^{-1} yr^{-1}	11	14	18.5	~1.5%
DFSS ha	1.5×10^6	18×10^6	28×10^6	~1×10^6

[1]Carbon offset as a percent of 1990 total U.S. fossil fuel C emissions of 1310×10^6 Mg (Marland, personal communication); utility C emissions from electric power production in 1989 equaled 478×10^6 Mg.

or higher, and they offer the opportunity for wood co-firing at the same efficiency. The Whole Tree Energy™ technology offers a new, wood-specific, steam-cycle technology that has 33 to 40% efficiency potential and which is now ready for a commercial-scale trial with little, if any, additional research. The developer of this technology believes that high temperature steam cycles could improve in efficiency up to 50% (Ostlie, personal communication). Gasification systems are also expected to achieve efficiencies in the range of 50% (DOE, 1992). Fossil-fuel efficiency improvements beyond 50% are anticipated to come from developments of the fuel cell. The molten carbonate fuel cell, which is the current focus of development, can directly replace the combustion turbine in an integrated cycle. All of these advanced systems have somewhat higher capital costs, but continuing development and environmental externalities are anticipated to make them competitive with use of coal as the feedstock (Starr et al., 1992). They will likely be even more competitive if wood is used as part or all of the feedstock.

While the Whole Tree Energy™ technology could improve biomass energy efficiencies from 25 to ~40% now, it is expected that several decades will be required for a significant transition from today's conventional electricity generation systems to those of >50% efficiency. The history of energy fuel transitions (wood-coal-oil) shows that in a peacetime commercial environment almost a half century is required to significantly shift fuel patterns (Starr et al., 1992). It has taken 30 years to get 50,000 MWe of gas turbine equipment installed by the U.S. electric industry. Catalytic cracking, commercially introduced in 1942, took about 20 years before it was generally used in refineries. Molten carbonate fuel cell technology is just now being tested with 200 kW size units. Starr et al., (1992) suggest that it would take 35 to 50 years to get 75,000 to 125,000 MWe of fuel cell equipment installed. These types of projections suggest that installment of up to 170,000 MWe of new capacity with efficiencies ranging from 35 to 55% would be pushing the limits of feasibility, but it might be possible. It will also be very expensive. The fuel cell development and deployment is anticipated to require \$80 to \$150 billion and the IGCC is projected to require \$110 to \$175 billion (Starr *et al.*, 1992). However advanced steam cycle technologies may be available first and at costs of less than \$80 billion.

To complete the transition to high efficiency biomass energy systems, supplies of dedicated biomass feedstocks must be assured. Thus, in addition to an industrial transition, the United States must solve the challenges of introducing a new crop on large amounts of land. Soybeans provide the closest analogy to what would be required for energy crops. Prior to the 1920s soybeans were essentially an unknown crop. Between 1924 and 1979, the planted area increased from 0.18 to 28.58×10^6 ha (Specht and Williams, 1984). The increase in planted area and average yields was rather steady over that period of time. Yield improvements increased at an average annual rate of 1.9% from 1924 to 1980, with about 50 to 85% of that attributed to genetic improvements and the rest to agronomic practices. While soybeans represent a major "new crop" success, several other crop introductions have resulted in failure (Jaycor, 1985). The reasons for these successes and failures should be studied carefully.

The expectation of an average yield improvement change on the order of 1.5% yr^{-1} is possible based on experience with major agricultural crops (Table 2). Over a 30-year period, sorghum has been observed to increase yields at an average of 7% yr^{-1} though current increases are in the range of 1.5 to 2.0% yr^{-1}. Both corn hybrids and soybeans have shown commercial yield increases of close to 2% yr^{-1} over a 50+ year time frame. In all three of these cases, major yield increases were seen all at once with the introduction of greatly improved genetic materials. Adoption by farmers was very quick and average yields were able to rise quickly. Cotton and wheat yields have risen more slowly and have likely reached a plateau where further yield increases are anticipated to be very slow. Evans (1980) has suggested that average increases for yield potential have been in the range of 0.5 to 1.0% for many different crops.

Table 2. Rate of average yield improvements of major agricultural crops

Crop	No. of yr	Annual yield income	No. of Ha increase	Expected annual increase
Sorghum[1]	30	7.0%	0.5×10^6	1.5–2%
Corn hybrids[2]	50	1.9%	—	1.4%
Soybeans[3]	56	1.9%	28×10^6	<1.9%
Cotton[4]	30	0.7%	—	<0.7%
Wheat[5]	20	0.7%	—	<0.7%

[1]Miller and Kebede, 1984.
[2]Duvick, 1984.
[3]Specht and Williams, 1984.
[4]Meredith, Jr. and Robert Bridge, 1984.
[5]Schmidt, 1984.

With the availability of clonal propagation techniques it is entirely possible that large initial advances in yield potential of woody crops can be achieved. Gains achieved through breeding and genetic transformation can be quickly captured by the propagation of large numbers of copies of the genotypes. The potential of such advances has been demonstrated by work with hybrid poplars (Heilman and Stettler, 1990). It will be very important from an environmental and risk reduction standpoint that woody crops not be limited to a few genotypes. Thus, genetic advances will need to be made in several species simultaneously. Once initial advances in genetic improvement are realized through clonal technology, it may be very difficult to maintain the rate of yield improvements seen in annual agricultural crops. The breeding cycles of trees are much slower and yield improvement per breeding cycle is not generally very high. Average commercial yield increases of 1.5% yr^{-1} represent an optimistic, but not impossible, view of what could be achieved.

If both the woody crop yield improvements and the conversion technology improvements do occur as speculated, then approximately 10×10^{18} J of biomass energy could be produced by 2030 without emitting additional C into the atmosphere (Table 3). If substituted for coal-based electric power generation, about 272×10^6 Mg of C would be offset. If utility emissions of C increase by 2% yr^{-1} over the next 35 years then current levels would be doubled. Thus 1989 emissions 478×10^6 Mg C would increase to 956×10^6 Mg C . The level of emission reduction achievable by biomass energy would thus only offset about one-third of utility C emissions in 2030.

Table 3. Electricity production with DFSS and C offsets

	1996	2010	2030
Energy (J $\times10^{18}$)	0.3	5	10
Electricity (TWh)	32	500	1200
C offset per hectare (MgC ha^{-1} yr^{-1})	5.2	6.6	8.6
C sequestered per hectare (MgC ha^{-1} yr^{-1})	0.78	0.89	1.1
Total C offset (Mg)	10.9×10^6	136×10^6	272×10^6
Percent reduction[1]	—	10%	21%

[1]Artificially assumes C sequestion is evenly spread over the 35-year period.

The rates of technology advances required are very optimistic but potentially achievable. However, the rates of genetic improvement required in the woody crops have only been achieved in agricultural crops receiving substantial research and technology transfer support. Similarly, the technology advances required to produce conversion systems with greater than 40% conversion efficiency will require a significant research effort. It may require the levels of support now devoted to developing "clean coal" technologies. A very strong commitment by government and industry working together will clearly be required.

3. Acknowledgment

The preparation of this paper and the research on energy crop production reported in this paper has been funded by the Biofuels Systems Division, U.S. Department of Energy, under contract DE-AC05-84OR21400 with Martin Marietta Energy Systems, Inc.

4. References

Duvick, D.N.: 1984, Genetic Contributions to Yield Gains of U.S. Hybrid Maize, 1930 to 1980. in: Fehr, W. R. (ed). Genetic Contributions to Yield Gains of Five Major Crop Plants. Crop Science Society of America Special Publication Number 7. American Society of Agronomy. Madison, Wisconsin 53711.

Evans, L. T.: 1980, The Natural History of crop yield. *Am. Sci.* 68:388-397.

Graham, R.L.: 1993, An analysis of the potential land base for energy crops in the conterminous United States. *Biomass and Bioenergy* (in press).

Graham. R. L., English, B. C., Alexander, R. R. and M. G. Bhat: 1993, Biomass fuel costs predicted for east Tennessee power plant. *Biologue* 10(1):23-26.

Graham, R. L., Wright, L. L. and A. F.Turhollow: 1992, The Potential for Short-Rotation Woody Crops to Reduce CO_2 emissions. *Climatic Change* 22:223-238.

Hansen, E.: 1993, Unpublished Annual Report submitted to Oak Ridge National Laboratory's Biofuels Feedstock Development Program. Forest Experiment Station. Grand Rapids, MN.

Heilman, P. W. and R. F. Stettler: 1990, Genetic variation and productivity of *Populus trichocarpa* and its hybrids. IV. Performance in short-rotation coppice. *Can. J. For. Res.* 20:1257-1264.

Jaycor: 1985, The successful Introduction of New Crops. An unpublished final draft report prepared for Oak Ridge National Laboratory's Biofuels Feedstock Development Program. Oak Ridge, TN 37831.

Johnston, S. A., J. G. Cleland, *et al.*: 1991, Whole-Tree-Energy™: Engineering and Economic Evaluation. Draft final report to EPRI by the Research Triangle Institute, to be published in April 1993 as EPRI Report TR-101564.

Johansson, T.B., Kelly, H., Reddy, A.K.N and R. Williams: 1993, Renewable Energy, Sources for Fuels and Electricity. Island Press, Washington D.C. 1160 pp.

Marland, G. and S. Marland: 1992, Should we store C in trees? *Water, Air and Soil Pollution* 64: 181-195.

Marland, G. and A. Pippin: 1990, United States Emissions of VCarbon Dioside to the Earth's Atmosphere by Economic Activity. Energy Systems and Policy 14:319-336.

Meredith, W. R. Jr. and R. R. Bridge: 1984, Genetic Contributions to Yield Changes in Upland Cotton. IN: Fehr, W. R. (ed). *Genetic Contributions to Yield Gains of Five Major Crop Plants*. Crop Science Society of America Special Publication Number 7. American Society of Agronomy. Madison, Wisconsin 53711.

Miller, F. R. and Y. Kebede: 1984, Genetic Contribution to Yield Gains in Sorghum, 1950 to 1980. in: Fehr, W. R. (ed). *Genetic Contributions to Yield Gains of Five Major Crop Plants*. Crop Science Society of America Special Publication Number 7. American Society of Agronomy. Madison, Wisconsin 53711.

Ostlie, D. and M. Gilbertson: 1992, *Watts Bar Conversion of Whole-Tree-Energy™ Final Report, Volumes 1 and 2*. Report to Electric Power Research Institute, Tennessee Valley Authority, and U.S. DOE Southeast Regional Biomass Energy Program by Energy Performance Systems, Inc.

Ostlie, D., B. Schaller, and R. Sundberg: 1993, *Program to Test Key Elements in the Design and Operation of a Whole-Tree-Energy™ System*. Draft final report to EPRI to be published in June 1993.

Ranney, J. W., L. L. Wright, and C. P. Mitchell: 1991, Carbon Storage and recycling in short-rotation energy crops. pp. 39-60 in: Mitchell, C.P. (ed) *Bioenergy and the Greenhouse Effect, Proc. of a Seminar Organized by International Energy Agency/Bioenergy Agreement and National Energy Administration of Sweden*. NUTUK B 1991:1. Stockholm, Sweden. 141 p.

Schmidt, J. W.: 1984, Genetic Contributions to Yield Gains in Wheat. in: Fehr, W. R. (ed). *Genetic Contributions to Yield Gains of Five Major Crop Plants. Crop Science Society of America Special Publication Number 7*. American Society of Agronomy. Madison, Wisconsin 53711.

Scurlock, J.M.O. and D. O. Hall: 1990, The Contribution of Biomass to Global Energy Use (1987). Short Communication. *Biomass* 21:75-81.

Specht J.E. and J. H. Williams: 1984, Contribution of Genetic Technology to Soybean Productivity - Retrospect and Prospect. in: Fehr, W. R. (ed). *Genetic Contributions to Yield Gains of Five Major Crop Plants*. Crop Science Society of America Special Publication Number 7. American Society of Agronomy. Madison, Wisconsin 53711.

Starr, C., Searl, M.F. and S. Alpert: 1992, Energy Sources: A Realistic Outlook. *Science* 256:981-987.

U.S. Department of Energy: 1992, Electricity from Biomass: A Development Strategy. DOE/CH10093-152. Available from NTIS, Springfield, VA 22161.

U.S. Environmental Protection Agency: 1990. *Policy Options for Stabilizing Global Climate, Report to Congress, Main Report*. Report No. 21P-2003.1, EPA Office of Policy, Planning and Evaluation. Washington, D.C.

Williams, R. H. and E. D. Larson: 1993, Advanced Gasification-Based Biomass Power Generation. pp.729-785 in: Johansson T.B., Kelly, H., Reddy, A.K.N. and R.H. Williams (ed). *Renewable Energy, Sources for Fuels and Electricity*. Island Press. Washington D.C. 1160 pp.

Wright, L.L., Graham, R. L., A. F. Turhollow, and B. C. English: 1991, The Potential Impacts of Short-Rotation Woody Crops on Carbon Conservation. in: Sampson, R. N. and D. Hair (ed), *Forests and Global Change Volume One: Opportunities for Increasing Forest Cover*. American Forests. Washington D.C. 285 p.

UTILISING BIOMASS CROPS AS AN ENERGY SOURCE:
A EUROPEAN PERSPECTIVE

J.M.O. SCURLOCK, D.O. HALL, J.I. HOUSE
Division of Life Sciences, King's College London W8 7AH, UK
AND R. HOWES
IIED, 3 Endsleigh St., London WC1H 0DD, UK

Abstract. Biomass can be grown to act as a carbon (C) store, or as a direct substitute for fossil fuels (with no net contribution to atmospheric CO_2 if produced and used sustainably). There is great potential for the modernisation of biomass fuels to produce convenient energy carriers such as electricity and liquid fuels. Bioenergy accounts for about 15% of primary energy used throughout the world, and 4% of energy used in Western Europe. Several European countries plan to significantly increase their use of bioenergy and some already obtain over 10% of their energy from biomass fuels. The European Community (EC) is planning to implement policies which will more than double the use of biomass by 2005, with biofuels taking 5% of the motor vehicle fuel market, and a resultant reduction in CO_2 emissions of about 180 million tonnes (Mt), equivalent to 50 Mt C/yr. The potential contribution of biofuels is even greater, especially with all the 'set-aside' land being taken out of production. Use of 15–20 million hectares (Mha) of agricultural land for biomass crops could represent an annual sink of some 90–120 Mt C or else offset between 50 Mt C and 120 Mt C from fossil fuel emissions, depending on the fuel displaced (7–17% of total EC carbon emissions). Policies are needed that will encourage the penetration of biofuels into the market such as increased support for research, development and demonstration, subsidies for biofuels, and carbon taxes on fossil fuels.

1. Biomass Energy Today

Biomass is considered one of the key renewable energy resources of the future due to its large potential, economic viability and various social and environmental benefits (Johansson *et al.*, 1993). Although it is rarely considered in official energy statistics, it is already the fourth largest source of energy in the world, supplying about 15% of primary energy (1985) ie. 55 exajoules (EJ) per year or the equivalent of 25 million barrels of oil per day. Developing countries as a whole derive 38% of their

Water, Air, and Soil Pollution **70**: 499–518, 1993.
© 1993 *Kluwer Academic Publishers.*

energy from biomass, and in many it provides over 90% of
total energy used in the form of traditional fuels, e.g.
fuelwood, straw and dung. Since 90% of the world's
population may reside in developing countries by 2050
biomass energy is likely to remain a substantial energy
source (Hall et al., 1993).

Industrialised countries obtain about 3% of their
energy from biomass. In the EC countries, biofuels provide
about 1.7 EJ or 40 million tonnes of oil equivalent (Mtoe);
3.3% of primary energy consumption. With non-EC countries
such as Finland obtaining 18% of their energy from biomass,
Sweden 14% and Austria 10%, the contribution of bioenergy
to Western Europe as a whole is nearer 4% (Hall *et al.*,
1993). A number of European countries plan to
significantly increase bioenergy production; it has been
has estimated that 100 Mtoe could be derived from biomass
in the next ten years, with a long-term potential of 300
Mtoe (Grassi, 1992; Grassi and Bridgewater, 1993).

When considering a large-scale bioenergy programme,
whether at global or local scale, the following questions
need to be addressed:

1. Land availability (short and long-term)
2. Productivities, species and mixtures.
3. Environmental sustainability.
4. Social factors.
5. Economic feasibility.
6. Ancillary benefits.
7. Disadvantages and perceived problems.

Perceived problems concern nutrient cycling, fertiliser and
pesticide requirement, energy input/output ratios, effects
on biodiversity and the landscape, possible contributions
to erosion, possible conflicts with food production on
high-productivity land, and the level of subsidies required
(Beyea *et al.*, 1991).

Many of these problems diminish if biomass energy is
seen as a long term entrepreneurial opportunity for
improved land management, based on optimal productivities
using minimum inputs of fertilisers and pesticides
(Turnbull, 1993).

1.1 GROWING FORESTS AS A CARBON STORE VERSUS GROWING BIOMASS FOR FOSSIL FUEL SUBSTITUTION

Various strategies have been proposed for using trees to help stabilise atmospheric CO_2 levels and ameliorate global warming:

1. Stopping deforestation
2. Afforestation and rehabilitation of degraded lands
3. Agroforestry
4. Reforestation/plantations

While sequestration and storage of carbon by afforestation is a relatively cheap, simple solution, it has to be long-term and successful and it must be recognised that its impact on atmospheric carbon levels ceases once the trees have grown to maturity. To continue carbon sequestration beyond forest maturity, the woody biomass can be put into permanent storage ("pickled") or used as long-lived forest products which could also raise revenue. Unfortunately, the storage option is more costly, while the demand for long-lived wood products is relatively small in comparison to the amount of carbon that needs to be sequestered. Alternatively, the biomass could be used as an energy source, displacing fossil fuels and therefore offsetting carbon emissions.

Table 1 examines the theoretical potential for offsetting CO_2 emissions by planting trees to sequester and store CO_2 in Western Europe, and includes the USA and Japan for comparison. Assuming a dry weight productivity of 12t/ha/yr (about 6tC/ha/yr) which is the target yield in the EC Biomass for Energy programme, the USA could theoretically sequester all its CO_2 emissions by afforesting 25% of its land, Western Europe would require 42% of its land, and Japan 129% (Hall et al., 1992). However, if Western Europe had the same efficiency as Japan in terms of tC/GNP, land requirements would fall to 27%.

Table 2 considers the theoretical potential for biomass energy supply in Western Europe, and also includes the USA and Japan. Assuming a productivity of 12 t/ha/yr and an energy content of oven-dry biomass of 20 GJ/t, this shows how much land would theoretically be needed to meet all of present energy consumption with biomass (a very unlikely outcome). If the use of residues is included, we can see that the USA would need 34% of its land and Western Europe 63%, whilst Japan simply does not have enough land area. However, if we subtract the energy provided by non-fossil sources (nuclear and hydro), which account for 18% of primary energy in Western Europe, the land requirement falls to 56% (Hall et al., 1992). With improved efficiencies of production, conversion and end-use, this

TABLE 1. OFFSETTING CO2 EMISSIONS WITH AFFORESTATION TO SEQUESTER CARBON

	I	II	III	IV	V	VI	VII
	POPULATION (1990)	TOTAL LAND AREA (Mha)	1987/88 CARBON EMISSIONS		% LAND AREA REQUIRED TO SEQUESTER PRESENT CARBON EMISSIONS		
	(millions)		(tonnes carbon per capita)	(tonnes C per million US$ GNP)	Productivity 4t/ha (= 2tC/ha)	Productivity 12t/ha (= 6tC/ha)	If Japanese tC/GNP and 6tC/ha
W. EUROPE	372	384	2.4	219	119	40	27
AUSTRIA	8	8	2.0	163	89	30	28
BELGIUM-LUX	10	3	3.3	287	525	175	92
DENMARK	5	4	3.1	206	186	62	45
FINLAND	5	30	2.8	190	22	7	6
FRANCE	56	55	1.8	146	94	31	33
GERMANY(W+E)	77	35	3.5	251	385	128	77
GREECE	10	13	2.0	507	78	26	8
IRELAND	4	7	2.0	346	55	18	8
ITALY	57	29	2.0	192	194	65	51
NETHERLANDS	15	3	3.7*	314	804	268	129
NORWAY	4	31	1.9	114	13	4	6
PORTUGAL	10	9	1.0	353	55	18	8
SPAIN	39	50	1.4	246	57	19	12
SWEDEN	8	40	2.0	126	20	7	8
SWITZERLAND	7	4	1.8	83	145	48	88
UNITED KINGDOM	57	24	2.8	269	330	110	62
UNITED STATES	249	917	5.6	308	76	25	12
JAPAN	124	38	2.4	151	386	129	129

Table based on Hall et al., 1992b, Table 1 - an analysis of individual countries in W. Europe

Column I: (WRI, 1990)

Column II: (FAO, 1990)

Columns III and IV: Carbon emissions calculated using BP Statistics for fossil fuel consumption (BP, 1990) with oil having an energy content of 42 GJ/t. Emission coefficients for each fuel are from Edmonds & Reilly (1983), i.e. Coal = 0.0238 tC/GJ, Oil = 0.0193 tC/GJ, Gas = 0.0138 tC/GJ. Data for population, GNP & carbon emissions from cement manufacture are from WRI (1990). Land use data is from FAO (1989).

Columns V, VI, VII: Estimates of land area required to grow enough biomass to sequester all 1987/88 carbon emissions.

Col. V: - based on present average productivity of 4t/ha

Col. VI: - based on the target productivity for the EC Biomass for Energy programme (12 t/ha)

Col. VII: - based on 12t/ha, plus the assumption that each country can achieve Japanese levels of energy efficiency per $GNP.

The proportion of land required in many cases is significantly lower.

*The Netherlands uses about 20% of its oil and gas as feedstocks for the chemical industry and for bunkers; thus the indigenous carbon emissions would be about 3.1 tC/capita/yr according to IEA statistics.

TABLE 2. POTENTIAL ENERGY PRODUCTION FROM BIOMASS AND ESTIMATES OF LAND REQUIRED IN WESTERN EUROPE

	I PRESENT ENERGY CONSUMPTION (EJ)	II TOTAL LAND AREA (Mha)	III POTENTIALLY HARVESTABLE RESIDUES (EJ)	IV % LAND NEEDED TO PRODUCE ALL ENERGY FROM ENERGY CROPS (12 t/ha) No residues	V + 25% residues	VI % OF PRESENT ENERGY USING 10% OF ALL LAND FOR ENERGY CROPS (var. prod.) + 25% RESIDUES	VII % OF PRESENT ENERGY USING 20% OF ALL ARABLE, PERMANENT CROPS AND FOREST LAND (12 t/ha)
W. EUROPE	57	384	4.40	68	63	14	18
AUSTRIA	1.1	8	0.20	55	46	18	21
BELGIUM-LUX	2.2	3	0.08	279	269	4	3
DENMARK	0.8	4	0.11	79	68	13	18
FINLAND	1	30	0.46	14	7	66	123
FRANCE	8.4	55	0.95	64	56	15	19
GERMANY	14.7	35	0.40	175	171	5	7
GREECE	0.9	13	0.12	29	25	29	35
IRELAND	0.4	7	0.06	24	21	35	16
ITALY	6.4	29	0.38	91	85	10	14
NETHERLANDS	3.1	3	0.07	381	372	3	2
NORWAY	1.4	31	0.12	19	17	42	32
PORTUGAL	0.5	9	0.12	23	17	39	65
SPAIN	3.5	50	0.39	29	26	28	49
SWEDEN	2.4	41	0.52	24	19	36	62
SWITZERLAND	1.2	4	0.06	126	119	7	6
UK	8.7	24	0.37	150	144	6	5
USA	82.6	917	8.14	38	34	22	28
JAPAN	16.9	38	0.08	187	186	4	8

Table based on Hall et al., 1992b, Table 2 an analysis of individual countries in W. Europe

Column I: (BP, 1990) Commercially traded fuels only.

Column II: (FAO, 1990)

Columns III: (Hall et al., 1992a)

Potentially harvestable residues - the total biomass produced from forest, crop and dung resources.
Recoverable residues - 25% of potentially harvestable residues (all that is likely to be used)

Columns IV and V: Land required to provide all present energy consumption from biomass by growing energy crops with an assumed productivity of 12t/ha, in absence and presence of supplementary 25% of harvestable residues.

Column VI: Proportion of present energy consumption which could be produced by growing energy crops on 10% of all land at variable productivities (depending on land type), and using 25% of residues

Column VII: Proportion of present energy consumption which could be produced using 20% of all arable, permanent crops and forest land (FAO, 1989) at an assumed productivity of 12 t/ha.

12 t/ha is the projected yield in the EC Biomass for Energy programme.

figure could decrease considerably further.

It would appear from the estimates in Tables 1 and 2 that afforestation to absorb the carbon released by fossil fuels requires less land than to produce the biomass to replace these fuels. This, however, is not necessarily true in the long term since periodic harvesting of biomass for energy allows land to be used indefinitely. Furthermore, producers of bioenergy crops will choose short-rotation woody or herbaceous crops, for which annual yields are 2-3 times as large as for long rotation species (Cannell, 1989; Hummel et al., 1988; Kulp, 1990). There are also many other considerations to be taken into account when considering the most effective strategy/policy for using biomass to ameliorate global warming.

For example, we can contrast the two strategies by comparing the net effect on atmospheric carbon levels of growing one tonne of biomass. Assuming that it contains 50% carbon, one oven-dry tonne of biomass will sequester and store 0.5 tonnes of carbon. Alternatively, based on a heating value of 20 GJ/t and equivalent conversion efficiencies for biofuels and fossil fuels, it could be used to displace 20 GJ of fossil fuel. This represents between 0.28 and 0.5 tC displaced, since the amount of carbon per GJ varies for each fossil fuel; about 0.014 tC/GJ for natural gas, 0.20 tC/GJ for oil, and 0.025 tC/GJ for coal. One GJ of biomass contains roughly the same amount of carbon as one GJ of coal, so it makes little difference whether biomass is used to substitute for coal or to absorb the carbon released by coal combustion. However, both oil and gas release less carbon per GJ, so substitution of biomass for oil or gas is theoretically not as effective as sequestering and storing the carbon in trees (Hall et al., 1990).

Nevertheless, reducing atmospheric carbon emissions is not the only environmental advantage of bioenergy substitution. Others include negligible sulphur and lower NO_x emissions, a means of disposing of and recycling organic wastes, and the decreased production of surplus agricultural commodities. Social benefits include greater job creation in rural areas since production of bioenergy crops is more labour intensive than afforestation for storage.

What of the economic aspects? The cost of offsetting CO_2 emissions by sequestering carbon in trees is directly related to the costs of growing biomass, e.g. average and marginal unit costs for a tree-growing programme offsetting 56% of USA fossil CO_2 are estimated to be $27/tC and $48/tC, respectively, with an annual cost of about $19.5 billion (Moulton and Richards, 1990). This cost could be absorbed by a carbon tax of $15/tC on all fossil fuels consumed, the effect of which would be to increase the cost of coal-based electricity generation by US$0.004/kWh and the cost of gasoline by $0.01/litre (Hall et al., 1990). By comparison, recovering 90% of the CO_2 from the flue-gases of coal power plants and piping it into abandoned natural gas wells has been estimated to cost around $120/tC for the Netherlands (Hendriks et al., 1989).

The production of biomass for energy purposes (or long-lived products) is more costly than simply growing trees to sequester carbon because of the added costs of harvesting, processing, drying and storage. However, revenues from the sale of the energy produced can be taken as a credit against the cost of providing it. Since biomass-derived electricity and liquid fuels can be produced competitively in many circumstances, the net cost of offsetting CO_2 emissions by substitution by could be near zero or negative and therefore lower than the cost of sequestration (Hall et al., 1990). Economics would become even more favourable if the costs of environmental degradation from fossil fuels were taken into account.

In short, growing biomass to substitute for fossil fuels has a number of advantages over simply growing trees as a carbon store, although this will not always be the case. Carbon-sequestration strategies will be important where the biomass cannot be practically harvested for energy, or where the creation of new forest reserves is deemed desirable for environmental or economic reasons. Furthermore, old forests should be left intact. Marland and Marland (1992) have modelled carbon flows and found that, depending on the assumptions made, sequestration may be more appropriate on low productivity land. However, "where high productivity can be expected, the most effective strategy is to manage the forest for a harvestable crop and to use the harvest for maximum efficiency either for long-lived products or to substitute for fossil fuels."

2. Biomass as a Carbon Sink in Europe and Elsewhere

2.1 EUROPE

Biomass yields for European forests presently average about 4 t(oven-dry biomass)/ha/year, compared with 10-20 t/ha/year for tropical regions. The target productivity under the EC Biomass for Energy Programme is 12 t/ha/yr, and has been used as the basis for the estimates presented here. According to Nilsson (1993) and other studies, most of the boreal forest trees of Northern Europe are severely nutrient-limited and show a strong response to application of nitrogen. Optimisation of nutrients for this biome type

could therefore create a significant carbon sink in the
short-term and, in the longer term, would provide biomass
for fossil fuel substitution.

Large areas of surplus agricultural land in North
America and Europe (possibly as much as 150 million
hectares (Mha) in the next century) are potentially
significant biomass producing areas. For the European
Community, at least 15-20 Mha of good agricultural land (an
area the size of England and Wales) is expected to be taken
out of production by the year 2000/2010 as a result of set-
aside policies. If all this land were used for biomass
crops in the form of trees, it would represent an annual
sink of some 90-120 Mt C for the near future (assuming that
oven-dry biomass is about 50% carbon and that the trees
will therefore absorb 6t C/ha/yr, if the productivity
target of 12 t/ha is met). Alternatively, this area of
land could provide 3.6-4.8 EJ/yr of biomass energy,
displacing 90-120 Mt of carbon emissions from coal, 72-96
Mt from oil, or 50-67 Mt of carbon emissions from natural
gas. Together, these figures represent a range of 7-17% of
total EC carbon emissions (709 Mt C in 1991).

There may be even more land available according to
some estimates. One study carried out for the Dutch
Government concluded that whatever future strategy was
followed (eg. free market and free trade, environmental
protection, etc.) at least 40 Mha of land presently in
agricultural production (present total about 130 Mha) will
be surplus to European requirements (Netherlands Scientific
Council for Government Policy, 1992).

2.2 USA

The above figure for European carbon storage may be
compared with that presented at the 1992 CO_2 Workshop in
Puerto Rico by Sampson (1992), who estimated that 59-189 Mt
C per annum could be taken up in the USA if marginal
agricultural lands were converted to forest (21-104 Mha),
and if short-rotation coppice, shelterbelts and urban
planting were encouraged (16-33 Mha). This is equivalent
to about 5-15% of US fossil fuel carbon emissions.

In the USA, 30 Mha of cropland has already been set
aside to reduce production or conserve land. However,
another 43 Mha of croplands have erosion rates exceeding
the maximum rate consistent with sustainable production and
a further 43 Mha have "wetness" problems, both of which
could be eased by shifting this land from annual food crops
to various perennial energy crops (Hall et al., 1993). The
amount of cropland idled to reduce production in the USA
may increase by 60 Mha over the next 25 years (USDA, 1987).

TABLE 3. COMMERCIAL ENERGY USE AND POTENTIAL SUPPLIES OF BIOMASS FOR ENERGY (Hall et al., 1992a)

(EJ/year) REGION	I COMMERCIAL ENERGY USE	II RECOVERABLE RESIDUES CROP	III FOREST	IV DUNG	V TOTAL	VI BIOMASS PLANTATIONS
Industrialised						
USA/CANADA	87.9	1.7	3.8	0.4	5.9	34.8
EUROPE	79.8	1.3	2.0	0.5	3.8	11.4
JAPAN	16.6	0.1	0.2	-	0.3	0.9
AUSTRALIA/NZ	3.6	0.3	0.2	0.2	0.6	17.9
FORMER USSR	56.9	0.9	2.0	0.4	3.3	46.5
Sub-total	244.8	4.3	8.2	1.5	13.9	111.5
Developing						
LATIN AMERICA	17.4	2.4	1.2	0.9	4.5	51.4
AFRICA	9.2	0.7	1.2	0.7	2.6	52.9
CHINA	23.0	1.9	0.9	0.6	3.4	16.3
OTHER ASIA	27.7	3.2	2.2	1.4	6.8	33.4
OCEANIA	-	-	-	-	-	1.4
Sub-total	77.3	8.2	5.5	3.6	17.3	155.4
World	322.1	12.5	13.7	5.1	31.2	266.9

N.B. All energy values are expressed on a higher heating value basis. Regional residue estimates are aggregates of country-by-country inventories compiled at the Information and Skills Centre of the Biomass Users' Network, King's College London.

Column I: Commercial energy use for 1985, as estimated by the US Department of Energy, excluding biomass.

Column II: Included are residues from cereals, vegetables and melons, roots and tubers, sugar beet and sugar cane. Not included are residues from pulses, fruits and berries, oilcrops, tree nuts, coffee, cocoa and tea, tobacco, or fibre crops. Crop production data are from FAO (1989). Except for sugar cane, it is assumed that 1/4 of all residues generated are recoverable. For sugar cane, it is assumed that all bagasse is recoverable, together with 1/4 of the tops and leaves. For crops other than sugar cane, the following are the assumed residue generation rates per tonne of crop, and the assumed heating values of these residues:

	Generation rate tonnes per tonne	Heat content GJ/t, air dry basis
Cereals	1.3	13.0
Vegetables and melons	1.0	6.0
Roots and tubers	0.4	6.0
Sugar beet	0.3	6.0

Column III: It is assumed that 3/4 of the milling and manufacturing wood wastes and 1/4 of the forest residues are recoverable.

Column IV: It is assumed that 1/8 of the dung generated is recoverable.

Column VI: It is assumed that plantations having an average yield of 15 t/ha/year (dry weight) with a heating value of 20 GJ/t are established on 10% of the total land area now under forests, woodlands, cropland or permanent pasture - some 372 Mha in industrialised countries and 518 Mha in developing countries

2.3 REST OF WORLD

Worldwide, there is even more land potentially available
for biomass energy crops. Bekkering (1992) estimates that
385 Mha is available from only 11 tropical countries.
Similarly, Nakicenovic et al. (1993) estimate that 265 Mha
is "suitable", with an additional 84.5 Mha considered
available for agroforestry.

Table 3 summarises the theoretical potential of
biomass energy production worldwide, assuming that
plantations were established on 10% of all the land now in
forests/woodlands, croplands and permanent pasture, that an
average yield of 15t/ha were attained and that
"recoverable" agro/industrial residues were being fully
utilised. It shows that the world as a whole could
theoretically produce 83% of its present commercial energy
use from biomass plantations and 10% from residues.
Industrialised countries could produce 46% from plantations
and 6% from residues, over half their present energy. In
developing countries 201% of the energy could come from
plantations and 22% from residues, with Africa capable of
providing 575% from plantations and 28% from residues.
Many developing countries have such a great potential for
substitution of fossil fuels with modernised bioenergy that
they could become totally self-sufficient in energy,
perhaps even exporting it in the future (Hall and House,
1992).

3. Biodiesel and Bioethanol in Europe

There has been a recent revival of interest in the
production of liquid biofuels from crops in a number of
European countries as well as in the USA. An EC Committee
has set a target that up to 5% of the liquid fuel market
could be supplied by biodiesel and bioethanol by 2010 and
that an agricultural area of 7 Mha (5.5% of the utilised
agricultural area of the EC) will be necessary to produce 7
Mtoe (European Communities Economic and Social Committee,
1992). Biodiesel is more likely to be used than bioethanol
in the short term because it involves less disruption to
the energy supply and distribution systems, and it fits in
with current EC practices and policies (POST, 1993).

3.1 BIODIESEL

Currently, commercial quantities of biodiesel are produced
in Italy, France and Austria with production plants planned
in Germany, Denmark, Belgium, Spain and Czechoslovakia.
Italy plans to double its production to 100,000 tonnes this

year (1993). Biodiesel comprises ethyl or methyl esters of edible oils. Rape methyl ester (RME) produced from oilseed rape is the main source in Europe, while soya oil is used in the US and canola oil in Canada. One hectare of oilseed rape yields 2-3 t rape seed and about 1 t useable RME. Biodiesel can be used pure or blended with mineral diesel in existing engines with only minor modifications and a small reduction in engine performance. Its main market so far has been buses and taxis in capital cities (POST, 1993).

One reason for the rising interest is the set-aside restrictions on the production of oilseeds and other crops for food. The oilseed area is supposed (according to GATT negotiations) to be reduced from 5.5 Mha to 5 Mha by 1995, but this does not apply to non-food uses such as the production of biofuels. In the UK, for example, if the 0.6 Mha set-aside in 1992 were used for biodiesel, this could provide 6.4% of the UK's total annual consumption of diesel and would reduce CO_2 emissions by 0.2% (Culshaw and Butler, 1992). The CO_2 benefits of biodiesel depend on the inputs used and are therefore variable. Other environmental benefits include negligible sulphur, reduced particulates and a product that biodegrades within a few days (making it particularly suitable for use on inland waterways). Furthermore, rape-seed-based lubricants have superior lubrication properties than mineral oils. Finland produces 1,500 tons of rape-seed-based lubricants annually as agroindustrial byproducts (VTT, 1992).

The current cost per litre of biodiesel is UK£0.25-0.30 (US$0.39-0.46), which compares to £0.15-0.20 (US$0.23-0.31) per litre of mineral diesel. If biofuels are to become commercially attractive they need preferential tax concessions and subsidies. Present EC regulations apply the same tax to diesel substitutes as to diesel itself. A draft directive has been proposed that would cut excise duty on biofuels to 10% or less of that charged on petrol and diesel. The EC Commissioner for taxes feels this is justified on environmental grounds and as a means of increasing energy security and protecting farmers incomes (POST, 1993).

3.2 BIOETHANOL

Trials of bioethanol have been carried out in Germany, Italy and Sweden. A small amount is blended with petrol in France. More experience has been gained in Brazil which produces about 12 billion litres of ethanol per annum; a third of its cars run on pure ethanol, and the rest use an ethanol/petrol blend. The USA also has a major bioethanol programme producing about 3.4 billion litres per annum. Bioethanol is produced from crops with high sugar or starch

contents.   In the EC, promising sources include cereals,
sweet sorghum, Jerusalem artichoke, maize and surplus wine.
       There is considerable discussion on the net energy and
CO_2 benefits from producing and using alcohol from different
feedstocks – see studies by ORNL (Marland & Turhollow,
1991) and the OECD (1993).   Calculations show that energy
ratios greater than one are feasible especially when
coproduct credits are used, and that a CO_2 benefit is
obtained compared to gasoline.   Bioethanol has similar
environmental advantages to biodiesel but is also likely to
require a subsidy to encourage market penetration.   The
environmental sustainabiltiy of using row-crop arable
agriculture for producing energy crops compared to woody
biomass for energy is the subject of considerable debate.

4. Examples from European Countries

4.1   FINLAND

Finland obtains 18% of its primary energy from biomass,
with a total biomass consumption of 5 Mtoe in 1990.   45% of
their bioenergy is produced from waste liquors from
pulping, 19% from peat (peatlands cover one third of
Finland's land area), 18% from firewood, 18% from wood
waste, and 1% from municipal refuse.   The pulp and paper
industries use waste wood and liquor for 60% of their fuel
requirements, with modern pulp mills capable of providing
all their own energy requirements as well as significant
amounts of excess electricity and liquid fuels.   District
heating has been used since 1952 and supplies more than 40%
of the country's heat demand.   Over half of the large
district heating plants use biofuels, as do many smaller
stations, and peat-fired combined heat and power plants are
being introduced in some larger towns (VTT, 1992).
       One of the reasons for the success of the bioenergy
industry in Finland is government support.   Currently, 10%
of the total financing of energy research (2 million ECU
(European Currency Units) or US$2.5 million) goes to
bioenergy.   According to the Finish Ministry of Trade and
Industry, there are enough biomass resources available to
allow for a doubling of their use.   Finland's considerable
forest resources would allow for a large increase in this
amount, as will land released as set-aside.   Peatlands
cover a third of Finland's land area and, at present rates
of consumption of 1 Mtoe, can provide 200-300 years of
industrially utilisable peat.   This consumption could be
doubled or even tripled with the land returned to forest
after extraction (VTT, 1992).

4.2   SWEDEN

Sweden obtains 16% of its energy (69 TWh or 250 PJ per
annum) from biofuels. The use of these fuels can be split
into three different sectors:
1. The forest products industry has traditionally converted
its by-products into heat and electricity for its own use.
In 1991, it obtained 29 TWh from pulp digester liquors, 8.4
TWh from cellulose waste and 7 TWh from sawmill wastes.
2. Individual households used about 11 TWh of woodfuels in
1991, this was mostly in the form of logs for heating.
3. The use of biofuels for district heating is growing fast
and accounted for 12.5 TWh in 1991. Of this, 4.8 TWh were
obtained from wood fuels (mostly unprocessed), 4.3 TWh were
obtained from refuse and 3.4 TWh came from peat. Energy
crops such as trees, straw and grass were also used but as
yet contribute only a very small amount (NUTEK, 1992).
        The Swedish Government has made a commitment to close
all 12 nuclear power plants by 2010, that the four
presently unexploited rivers will remain free of hydropower
plants, and that atmospheric CO_2 will be stabilised at 1988
levels. Therefore they will need to develop new non-fossil
energy sources. There is much more potential to produce
energy from indigenous biomass fuels particularly from
agroindustrial wastes and energy crops grown on marginal
land. Sweden also imported a small amount of biomass fuels
indicating the potential for the future development of an
international trade in biofuels (NUTEK, 1992).

4.3   AUSTRIA

Austria obtains about 10% of its primary energy/total
energy/ electricity from wood (Howes, 1992). Bioenergy use
has increased by 2% in total over the last 5 years and is
expected to rise further, may be accounting for 20% of
total energy consumption. This is important for a country
like Austria which is dependent on imported fuel for 90% of
their oil, and 80% of their gas and coal (Unteregger,
1992).
        Biomass is mostly used for space heating. Over 50% of
farmhouses use fuel wood for this purpose, and district
heating is being introduced as an efficient and convenient
way of heating houses concentrated in a small area
(Unteregger, 1992). Austria now has 11,000 biomass heating
systems with a combined capacity in excess of 1200 MW
(equivalent to two very large coal fired power stations).
80 district heating schemes have recently been installed
(1-2 MW capacity), and these may be developed to
incorporate combined heat and power (CHP) systems. This
has been achieved by political commitment creating a
encouraging institutional framework that includes

favourable legislation, capital grants, cheap finance and education. Farmers groups developing woodfuel-fired district heating schemes can receive grants covering up to 50% of the capital costs (Howes, 1992).

4.4 UNITED KINGDOM

The UK obtains only 0.3% of its energy from biofuels (25 PJ or 0.6 Mtoe per annum) (ETSU, 1993; 1992b), but biomass has repeatedly been recognized by the UK Energy Division of the Department of Trade and Industry as a major renewable energy source which is already economically viable in Britain. Its economic potential has been estimated at 9 Mtoe per annum or 0.25 EJ - 3% of present UK primary energy use. Two-thirds of this would come from wastes, and one-third from energy crops. In the longer term, the estimated annual contribution from biomass in the UK may be as large as 40 million tonnes of coal equivalent (1.1 EJ), 12% of primary energy use in Britain today (UK Department of Energy, 1988).

A likely scenario for the UK is based on farmers producing woodfuel for cooperatives which supply heat, or heat and power for local industry. Space and process heating is the biofuel market which is most likely to expand in the immediate future (Howes, 1992).

4.5 OTHER EUROPEAN COUNTRIES

In Denmark, considerable efforts are made to use the annual straw surplus of 3 Mt (Nielsen, 1992). The Danish Government programme for CHP plant utilising biomass or natural gas has a target of 450 MWe. More than 50 straw-fired district heating plants totalling 170 MW (thermal) produce energy for district heating schemes at a cost of $12/GJ, competitive with coal and oil-fired heating plant. Four cogeneration (CHP) plants fired by straw are in operation, ranging in size from 7 MW to 28 MW (thermal), and two more rated at 60 MW (thermal) are under construction (Ravn-Jensen, 1992). Since the mid-1980s, 10 centralised and 10 single-farm biogas plants have been established, with an output of 14 million m^3 per annum (0.5 PJ). However, this is only a fraction of the biogas resource potentially available (25-30 PJ), and production could increase by as much as 5-10 times by the year 2005 (DEA, 1992).

France obtains approximately 9 Mtoe or 5% of its primary energy consumption from wood, with half of all single-family dwellings using fuelwood as a major or secondary energy source (Savanne et al., 1992).

The contribution of biomass to primary energy is about 5% also in Greece, although in many rural areas and islands

more than 10% of energy is supplied by biomass. During the
1980s, growth in biomass energy matched the continued
growth in fossil fuel use, so this proportion has remained
constant (Koukios and Umealu, 1992).

5. European Policy

Most biomass energy technologies have not yet reached a
stage where market forces alone can make the adoption of
these technologies possible. One of the principal barriers
to the commercialisation of all renewable energy
technologies is that current energy markets mostly ignore
the social and environmental costs and risks associated
with fossil fuel use (Johansson et al., 1993).
Conventional energy technologies are able to impose upon
society various external costs (such as environmental
degradation and health care expenditures) which are
difficult to quantify. Meanwhile, renewable energy
sources, which produce few or no external costs and may
even cause positive external effects (such as decreased SO_2
& CO_2 emissions, job creation, rural regeneration and
savings in foreign exchange), are systematically put at a
disadvantage. Furthermore, conventional energy sources
tend to receive large subsidies. Internalising external
costs and benefits and re-allocating subsidies in a more
equitable manner must become a priority for all renewables
to be in a better ("level playing field") position to
compete with fossil fuels
        Pursuit of all policy options needs stimulus from
appropriate policy instruments, such as carbon taxes. In
Sweden carbon taxes were introduced in 1991 - and were
roughly equivalent to US$150 /tC, although they have now
been reduced. Norway's carbon tax is about $120 /tC. Cost
estimates for reducing CO_2 emissions vary quite considerably
- often by a factor of two or three. An OECD working paper
estimated that to cut the output of CO_2 by 20% between 1990
and 2010, and to stabilize it thereafter, would need a tax
averaging $210/tC for the world as a whole which is
equivalent to $36 per barrel of oil (petrol taxes in some
European countries are presently equivalent to a carbon tax
of $200/tC) (The Economist, 1991). The EC has proposed a
carbon tax of US$3/barrel of oil in 1993 rising to
$10/barrel of oil in 2000 (which will be equivalent to
US$0.05/litre of petrol or $0.015/kWh for electricity)
(Greenpeace Energy Policy Project, 1993)
        The Economic and Social Committee of the EC have
stated their commitment to continue encouraging research
and development of renewable energies, to introduce a
Community Energy/CO_2 Tax, and to create an environment

favourable to the market penetration of renewable energies.
They believe these measures will make it possible for
renewables energies to increase their contribution to total
energy demand from 4% in 1991 to 5-6% in 2000 and 8% in
2005; that electricity production from renewables will
treble; and that biofuels share of the transport fuel
market will increase to 5%. As a result of this, they
think that there will be a 180 Mt reduction of CO_2 emissions
by 2005 (European Communities Economic and Social
Committee, 1992).

The European Commission expects that half the
predicted increase in renewable's contribution will come
from greater exploitation of biomass which will more than
double. It considers that biomass "is the only renewable
energy source which will be able to make a substantial
contribution to the replacement of conventional fuels". It
also stresses that "priority will be given to the
commercial penetration of biofuels and fuels of
agricultural origin" (European Communities Economic and
Social Committee, 1992).

6. Conclusions

Of the major alternatives to reduce atmospheric CO_2 levels,
revegetation (reforestation) to sequester CO_2 and the
substitution of fossil fuels with biomass are among the
most promising (Alpert et al., 1992; Rubin et al., 1992).
However, biomass is only one component of an overall
strategy of energy efficiency and use of renewable energy.

Bioenergy industries have already been launched in
several countries. Nevertheless the techniques and
technologies for growing biomass and converting it into
modern energy carriers must be more fully developed. If
bioenergy research and development is given high priority,
and if policies are adopted to nurture the development of
bioenergy industries, these industries will be able to
innovate and diversify as they grow and mature. Suitable
policy options include a review of energy subsidies to
create a "more level playing field", internalising external
negative costs of fossil fuels by methods such as carbon
taxes, and the use of these taxes to support bioenergy as
well as other carbon mitigation options. Direct combustion
of wastes for electricity production, substitution of
liquid fuels by bio-ethanol and bio-diesel, and combined
heat and power production from energy crops are all likely
to make a major contribution to European biomass energy
supply in the near future.

References

Alpert, S.B., Spencer, D.F. & Hidy, G.: 1992, Biospheric options for mitigating atmospheric carbon dioxide levels, *Energy Conservation and Management*. **33(5-8)**,729-736.

Bekkering, T.D.: 1992, Using Tropical Forests to Fix Atmospheric Carbon: The Potential in Theory and in Practice. *Ambio*, **21(6)**,414-419.

Beyea, J., Cook, J., Hall, D.O., Socolow, R. & Williams, R.: 1991, Toward ecological guidelines for large-scale biomass energy development. National Audobon Society, New York/CEES, Princeton University, New Jersey.

BP (British Petroleum): 1990, *BP Statistical Review of World Energy*, British Petroleum Company PLC. London, UK.

Cannell, M.G.R.: 1989, Physiological Basis of Wood Production: A Review. *Scand. J. Forest Res.*, **4**,459-490.

Culshaw F. & Butler, C.: 1992, *A Review of the Potential of Biodiesel as a Transport Fuel*, (ETSU-R-71), HMSO, London.

DEA (Danish Energy Agency): 1992, *Update on centralised biogas plants*, Danish Energy Agency, Copenhagen.

Economist (The): 1991, Cool it: a survey of energy and the environment. 31st August, 1991. The Economist Publ. Co., London.

ETSU (UK Energy Technology Support Unit): 1993, *RESTATS Newsheet*, **No.1**, Dec, 1992. ETSU, Harwell, Oxfordshire OX11 ORA.

ETSU: 1992b, *1992 Digest of UK Energy Statistics*, HMSO, London.

European Communities Economic and Social Committee: 1992, *Opinion on the Proposal for a Council Decision Concerning the Promotion of Renewable Energy Sources in the Community*, (Doc.COM(92) 180 final). Document CES(92) 1314, November 1992, Brussels.

FAO (Food and Agricultural Organisation): 1989, *Agrostat Database*, FAO, Rome.

Grassi, G.: 1992, Address to the 6th European Conference on Biomass, in: G. Grassi, A. Collina & H. Zibetta (eds.), *Biomass for Energy, Industry and Environment - 6th EC Conference*, Elsevier Applied Science, London, pp. 27-29.

Grassi, G. & A.V. Bridgewater: 1993, The opportunities for elelctricity productiion from biomass by advanced thermal conversion technologies. *Biomass and Bioenergy* (in press).

Greenpeace Energy Policy Project: 1993, *Powerhouse* Issue 11, 23 March 1993. Greenpeace UK, London N1 2PN.

Hall, D.O., Mynick, H.E. & Williams, R.H.: 1990, Alternative Roles for Biomass in Coping with Greenhouse Warming, *Science and Global Security*, **2**, 113-151; *Nature* (1991) **353**, 11-12.

Hall, D.O. & House, J.I.: 1992, Conversion of Biomass into Ethanol and Electricity, in *Application of Biomass Energy Projects in Developing Countries*. United Nations Centre for Human Settlements (HABITAT), Nairobi.

Hall, D.O., Rosillo-Calle, F., Williams, R.H. & Woods, J.: 1993, Biomass for Energy:Supply Prospects. In B.J.Johansson, H. Kelly, A.K.N. Reddy & R.H. Williams (eds) *Renewables for Fuels and Electricity*, Island Press, Washington D.C. pp.583-652.

Hall, D.O., Woods, J. & House, J.I.: 1992, Biological systems for uptake of carbon dioxide. *Energy Conservation and Management*, **33(5-8)**:721-728.

Hendriks, C.A, Blok, K. & Turkenburg, W.C.: 1989, The Recovery of Carbon Dioxide From Power Plants, in P.A. Okken, R. Swart & S. Zwerver (eds), *Climate and Energy, the Feasibility of Controlling CO_2 Emissions*, Kluwer Academic Publishers, Netherlands, pp.107-124.

Howes, R.: 1992, Set Aside and Energy Policy: The Potential of Short Rotation Energy Forestry. M.Sc. Thesis, Center for Environmental Technology, Imperial College, London University.

Hummel, F.C., Palz, W. and Grassi, G. (ed.): 1988, *Biomass Forestry in Europe: A Strategy for the Future*. Elsevier Applied Science, London, pp.1-7.

Johansson, T.B., Kelly, H., Reddy, A.K.N. & Williams, R.H.: 1993, Renewable Fuels and Electricity for a Growing World Economy: Defining and Achieving the Potential, in T.B. Johansson et. al. (eds) *Renewables for Fuel and Electricity*, Island Press, Washington D.C., pp.1-72.

Koukios, E.G. and Umealu, O.S.: 1992, Present contribution of bioresources to satisfy the energy needs of Greece, in: G.Grassi, A.Collina & H.Zibetta (eds.), *Biomass for Energy, Industry and Environment - 6th EC Conference*, Elsevier Applied Science, London, pp. 41-47.

Kulp, J.L.: 1990, *The Phytosystem as a Sink for Carbon Dioxide*, EPRI Report EN-6786, Electric Power Research Institute, Palo Alto, CA, USA.

Marland, G. & Marland, S.: 1992, Should we Store Carbon In Trees?, *Water, Air and Soil Pollution*, **64**,181-195.

Marland & Turhollow: 1991, CO_2 emissions from production and combustion of fuel ethanol from corn. *Energy*, **16**,

1307-1316.

Moulton R.J. & Richards, K.R.: 1990, *Costs of Sequestering Carbon Through Tree Planting and Forest Management in the United States*, USDA Forest Service General Technical Report. USDA, Washington, DC.

Nakicenovic, N. *et al.*: 1993, Long Term Strategies for Mitigating Global Warming. *Energy International J.*, **18(5)**,97-106

Neilsen, C.: 1992, Straw-fired decentralised combined heat and power plants in Denmark, in: G.Grassi, A.Collina & H.Zibetta (eds), *Biomass for Energy, Industry and Environment - 6th EC Conference*, Elsevier Applied Science, London, pp. 1143-1145.

Netherlands Scientific Council for Government Policy: 1992, *Ground for Choices - Four Perspectives for the Rural Areas in the European Community*, Netherlands Scientific Council for Government Policy, P.O. Box 20004, 2500 EA The Hague, The Netherlands.

Nilsson, L.O.: 1993, Carbon fixation in Norway Spruce in Southern Sweden as a function of air pollution, water availability and optimal fertilisation with irrigation, in: J. Wisniewski and A.M. Solomon (eds), *Quantification of Sinks and Sources of CO_2* Kluwer Academic Publishers, Dordrecht.

NUTEK (Swedish National Board for Industrial and Technical Development): 1992, *Energy in Sweden 1992*, NUTEK, S-117 86 Stockholm, Sweden.

OECD (1993) Biofuels for Transport. Organisation for Economic Cooperation and Development, Paris. In press.

POST (UK Parliamentary Office for Science and Technology): 1993, *Biofuels for Transport* Briefing Note 41, March 1993. POST, Houses of Parliament, London.

Ravn-Jensen, L.: 1992, Danish experience with combustino of straw, in: G.Grassi, A.Collina & H.Zibetta (eds) *Biomass for Energy, Industry and Environment - 6th EC Conference*, Elsevier Applied Science, London, pp. 884-888.

Rubin, E.S, Cooper, R.N., Frosch, R.A., Lee, T.H., Marland, G., Rosenfield, A.H. & Stine, D.D.: 1992, Realistic Mitigation Options for Global Warming. *Science*, **257**,148-266.

Sampson, R.N.: 1992, Forestry opportunities in the United States to mitigate the effects of global warming, in: Wisniewski, J. and Lugo, A.E. (eds.) *Natural Sinks of CO_2*. Kluwer Academic Publishers, Dordrecht, pp. 157-180.

Savanne, D., Dufour, N., & Degand, C.: 1992, A Fresh Appraisal of Fuelwood Consumption in France.

Poceedings: 7th European Conference on Biomass for Energy and Environment, Agriculture and Industry, Florence, Italy, October 1992. Ponte Press, Bochum (in press).

Turnbull, J.: 1993, *Strategies for achieving a sustainable, clean and cost-effective biomass resource.* Electric Power Research Institute, Palo Alto, California.

UK Department of Energy: 1988, *Renewable Energy in the UK: The Way Forward*, Energy Paper 55. HMSO, London.

Unteregger, E.: 1992,. Country Report for Short Rotation Forestry: Austria, in: S. Ledin & A. Alriksson (eds.), *Handbook on How to Grow Short Rotation Forests*, Swedish University of Agricultural Sciences, Uppsala, pp. 10.1.1–10.1.4.

USDA: 1987, *A Report From the New Forest Products Task Force*, United States Department of Agriculture, Washington D.C.

VTT (Technical Research Centre of Finland): *1992, New Options in Bioenergy From Finland*, VTT, Laboratory of Fuel and Process Technology, P.O.Box 205, SF-02151 Espoo, Finland.

WRI (World Resources Institute): 1990, *World Resources 1990– 91*, Oxford University Press, UK.

FOREST MANAGEMENT AND BIOMASS IN THE U.S.A.

R. NEIL SAMPSON

American Forests
1516 P St NW
Washington, DC 20005 USA

Abstract. There are a variety of opportunities to change land and forest management and, at the same time, create a positive impact on the current use of fossil energy. To the extent that these opportunities can be captured, they address the root cause of greenhouse warming--fossil fuel emissions--while, at the same time, improving economic opportunities, ecosystem productivity, and environmental conditions over broad areas. The need for better markets to absorb biomass energy, plus research to make biomass conversion more efficient, is probably the most important deterrent to achieving these possibilities.

1. Introduction

For much of human history, there was not much question about the connection between forest biomass and energy. Wood was the energy source, not just of choice but of necessity. One defining difference between the Americas and Europe for the first three Centuries after European settlement was the enormous wood supply found in the New World.

As important as that wood was for building everything from war ships to wagon roads, it has been estimated that, in the late 1700's, close to two-thirds of the volume of wood removed from the American forest was used for energy (MacCleery, 1992). Single households consumed 70 to 145 m³ of wood annually for heating and cooking in inefficient fireplaces. Charcoal smelters formed the basis of the iron industry, which was such a critical component of the early years of the industrial revolution.

Wood usage for fuelwood grew rapidly and steadily in the United States throughout the 19th Century, with wood providing over 90% of the energy needs of the United States. That percentage began to drop as industry turned to coal and petroleum, so that by 1920 wood was providing only 10% of energy needs and today, that percentage is about 3%, two-thirds of which is produced in industrial processes (MacCleery, 1992).

In America's forests, fuelwood remained the primary product until the 1880's, when the volume of lumber produced began to exceed it (Sedjo, 1990). The fuelwood share of forest product output dropped to almost nothing by 1970, but has seen a rebound since that time (Sedjo, 1990).

That rebound has been driven by several factors. The oil price shocks of 1973 and 1979 encouraged substitution. Air and water pollution regulations began to impact significantly upon the forest products industry. Wood wastes which had been burned in open burners or released in effluent streams were no longer tolerable, and needed different treatment. Combustion in co-generating plants became the technology of choice, until today almost all such wastes are treated as a fuel source (Row and Phelps, 1992).

We are now faced, however, with a new set of issues and pressures. Fossil fuels, while they remain plentiful in supply for the near future, are not inexhaustible. A shift toward more renewable sources of energy, particularly for some energy needs, has long been advocated (Lovins, 1976). Biomass fuels, while they create pollution problems of their own which cannot be overlooked, create far less of some of the more damaging compounds than fossil or nuclear competitors.

As the world grows more concerned with reducing the emission of greenhouse gas, biomass that can be produced in a rather brief, renewable cycle becomes an attractive alternative to fossil sources that represent long-term net transfers from the terrestrial biosphere to the atmosphere (Chandler, 1990). It also becomes a potentially important feedstock for energy production itself, particularly if a suite of technical problems in both production and conversion technology can be resolved (Wright *et al.* 1992).

The issues involved today involve both science and public policy. In general, they can be grouped in three categories: (1) Management of forest biomass for energy production; (2) Increasing the area of forests, and using other forest management methods to protect or increase terrestrial C storage, as mitigating measures to offset the C production in energy generating plants; and (3) Using trees and forests in ways that reduce energy consumption, thereby cutting the need for energy generation and its associated C emissions (Sampson, 1992).

2. Biomass Energy Production

2.1 USE OF EXISTING FOREST BIOMASS

Most forest biomass in the late 20th Century is utilized for building materials and pulp production (Sedjo, 1990). That may be more of a reflection, however, of the market demand for products than of the material available within the forest systems. Many forests in the United States, because of their condition, need to have wood removed. In fire-adapted forest ecosystems, for example, a century of fire control has resulted in unnatural buildups of highly flammable fuels (Steele *et al.* 1986). The continuation of this type of management will result in continued forest health problems and an increasing risk of high-intensity wildfires (Mutch *et al.* 1993).

In the Blue Mountains of eastern Oregon, an appraisal of 4 National Forests has concluded that, of 2.6 Mha of total forest, half are considered to be "out of ecological balance" and in need of immediate management attention (Schmidt *et al.* 1993). Much of the management attention involves removing fuels that have built up, thinning existing stands to prevent additional stress and tree death, and restoration of riparian zones to halt streambank erosion and provide wildlife habitat. Where appropriate, the resulting plan proposes the harvest of of dead and dying trees as merchantable logs. Trees that need to be removed for fuels reduction, but which do not meet merchantable sawlog standards will be sold as feedstock for power generation, paper production, fiberboard, and otehr products (Schmidt *et al.* 1993).

A major problem encountered in many of these forest improvement proposals, however, is the lack of a market for small and defective logs, low quality species, or partially rotted material. Without such a market, the costs of thinning and fuel reduction are too high for most landowners, including the public agencies.

One solution is to utilize the low-grade material as feedstock for biomass energy generation. This is being done in some areas of the United States, particularly where energy prices are high. Three sawmills operated by Sequoia Forest Industries in the San Joaquin Valley of California, for example, now operate on a split-shift basis during the summer, shutting down the sawmill operations in mid-day while the co-generation plant runs full speed to produce and sell electrical power into the regional power grid during the hours of peak power demand (Shultz, 1992). That is seen as a major opportunity by local forest managers, who can market timber sales that are a combination of thinning and fuel removal, designed to remove the small trees, excess trees, and reduce the dead fuels in stands where, without such treatment, the risk of stand-replacing wildfires was exceptionally high. That same opportunity has not been available to forest managers in other parts of the Intermountain West, however, where the lack of co-generation facilities nearby or low prices for electricity due to abundant hydroelectric power have failed to generate a market.

In addition to the need for increased forest management of over-crowded and stressed stands, there are millions of hectares of U.S. timberland where management changes to increase production have been identified. If any significant portion of these opportunities are instituted to the degree that they seem feasible, it seems likely that total wood output of forest products will be more than traditional wood and paper markets will absorb.

2.2 NEW FOREST PLANTATIONS

Large-scale increases in afforestation have been proposed as one of the ways in which greenhouse gas emissions could be mitigated (Moulton and Richards, 1990; Helme *et al.* 1993; Parks and Hardie, 1993; Dixon, *et al.* 1993; Kinsman and Trexler, 1993). This builds on recent experience in the United States in recent years. A national effort to improve the use of marginal croplands in the U.S., the Conservation Reserve Program (CRP), helped spur the four highest tree-planting years in the Nation's history during the period 1987-1991, with over 1 Mha of new forests planted in each of those years (Mangold *et al.* 1992). In spite of its accomplishments, the CRP fell well short of its announced tree planting goal, since many farmers with land suitable for trees planted grass as a means of retaining maximum flexibility to re-convert the land to cultivated crops at the end of the 10-year contract period (Esseks, *et al.* 1992).

The United States government launched the "America the Beautiful" program in

1990 with an avowed goal of increasing tree planting in the U.S. by 1×10^9 trees yr^{-1} for a decade. While that effort has not met the goal, due largely to a turndown in the forest products industry that dampened their large-scale plantings, along with a reluctance by private landowners to commit to tree planting costs and risks in the face of market and weather pressures, tree planting numbers have remained high through 1991 (Mangold *et al.* 1992).

A new international tree planting and forest management effort called "Forests for the Future" has been launched as a cooperative effort of the U.S.D.A. Forest Service, U.S. Environmental Protection Agency, and the U.S. Agency for International Development. Working partnerships are being sought with several nations to test the idea of large-scale afforestation and forest care projects that can help offset C emissions in the industrial world, while expanding forests, improving environments, and creating employment and economic opportunities in the countries where the new forests will be located (U.S. EPA, 1993).

Those plantations, no matter where they are grown, will need pre-commercial thinning within two decades, followed by at least one commercial thinning after 20 to 35 years, depending on the climate and growth rate. Growth and yield estimates have been prepared for most of the commonly-used species and management regimes in the U.S. (Birdsey, 1992). Those data can be used to estimate the net C sequestration that will be produced over the life of the forest, although their extension to individual sites or international situations must be done with caution to assure that the soil C status at the time of planting is known and that growing conditions are similar (Sampson, 1993).

While there are commercially-profitable products possible on the second thinning or final harvest in most of these situations, the advantages of having a biomass energy market for all the thinned materials are evident. Whether or not that will be a possibility is a major question, however, depending far more on public policy regarding incentives for biomass energy and the structure of the electric power industry in the vicinity of the forest than upon forest management or productivity.

2.3 SHORT-ROTATION WOODY CROP PRODUCTION

There are significant opportunities in the United States to expand the use of short-rotation woody crop production methods. These plantations rely on fast-growing woody species that can be harvested on a 4 to 8 yr rotation, then re-grow from sprouts without the need for re-planting. Production methods are more nearly allied with agriculture than with forestry, including fertilization, cultivation and weed control, and mechanical harvesting. Because of the intensive production methods used, the land needed is good cropland with adequate natural moisture. Thus, this production competes with crops for land, but would not be a major competitor with the goal of converting marginal crop and pasture land to forests, since so much of that marginal land is on slopes of greater than 8%, leading to serious erosion and mechanical harvesting problems in short-rotation woody crops.

Current estimates indicate that, with 14 to 28 Mha devoted to this type of biomass production, the potential impact on greenhouse emissions could range from 51 to 339 x 10^6 t C yr^{-1} (Sampson, 1992). The wide range in variation is due to the current gap in conversion efficiencies between using the wood for ethanol production or directly burning it to produce electricity, as well as the differences believed to be possible if research priorities are focused on bringing conversion technologies for commercial production up to levels of current laboratory achievements (Wright *et al.* 1992).

2.4 ECONOMIC ISSUES

The indications are that there are both great needs and potentials to expand the utilization of forest-produced biomass for energy production in the United States. It seems clear, however, that such an expansion depends more upon the creation of the necessary markets and infrastructure than upon changes or expansion of forests or forest productivity.

The most likely place to expand electrical production from forest biomass would appear to be in association with the existing forest products industry. With a co-generation system that can burn the wood wastes to run the mill, plus sell electricity back to the regional power company, the potential for using wood as an energy feedstock increases. Harvesting and transporting the full range of wood products available from the forest at one time, then sorting them for the best industrial application (fuel-pulp-lumber) at the mill, can be a highly cost-effective way to handle this heavy, low-value material.

Expansion of co-generation capacity will depend on the perceived profitability inherent in selling surplus electricity, and that will be a reflection of the price structure of competing sources of fuel. In areas where cheap or subsidized power is available, this opportunity will not be significant. In those areas where fossil fuels are used, and prices rise, biomass will become a more attractive option.

Finally, there are opportunities for public policies to affect the economics of biomass energy, either through incentives or barriers. These include, but are not limited to, tax incentives, cost-sharing for land conversion, forest practice regulations, environmental regulations, and energy policy.

2.5   ENERGY ISSUES

Energy policy will be an important determinant of energy prices, which will, in turn, be critical in whether or not biomass-produced energy becomes competitive. That will almost certainly be controlled by several factors, including world oil prices, increased prices for meeting environmental restrictions with coal or oil, and the success or failure of nuclear energy technology to build a bridge around the current stalemate in expanding nuclear generation.

But high energy prices will not, in themselves, be totally sufficient. There will need to be willingness on the part of large regional energy producers to purchase co-generated power from relatively small, widely-dispersed sources. That will be a function both of the energy marketplace, the national energy policy, and the policies that emerge in state regulatory agencies.

Research priorities, particularly as they affect support for more efficient biomass conversion technologies, will also be critical in helping tilt the economic balance toward biomass.

3. Carbon Offset Plantings

A relatively new idea being intensively studied in both the United States and Europe is that of subsidizing tree planting as a means of offsetting fossil C emissions. Under the idea, a government could impose greenhouse gas restrictions, or taxes, then provide "C credits" to emitters willing to plant trees and assure their maintenance and growth (Hasenkamp, 1993). The first such attempt in the United States was initiated in 1988 by Applied Energy Systems, who worked with World Resources Institute and CARE to plant trees in Guatemala as a means of offsetting the C emissions from a new coal-fired power plant in Connecticut (Trexler et al. 1989). Although the U.S. has no C regulation or tax under which such a scheme would get official credit, the program was undertaken by AES largely for altruistic reasons and public relations (Trexler et al. 1992).

The problems are more complex than they may appear on the surface, however. Land must be available that is biologically suited to trees, at reasonable rents, under the care of an individual or agency committed to its protection and proper management. The planning must include reliable estimates of the amount of C likely to be sequestered in the new plantings, and costs need to be documented. Once the planting is made, some form of accountability must be established in order to prevent "C credit fraud" (Trexler, 1993).

The problems have not kept additional projects from being studied and initiated, however. In a study prepared by the Center for Clean Air Policy, it is noted that forest offsets offer great transitional benefits. Forest offsets, it was determined, "would buy time for a smoother transition toward use of less CO_2 intensive fuels, further improvements in energy efficiency, or other alternatives" (Helme et al. 1993).

One idea that has been explored informally by the author, but not fully analyzed, is to locate forest offset plantings in the near vicinity to the power plant being served. This could be done in the U.S. by creation of a new private-sector cost-sharing program that could provide landowners with the additional cash needed to overcome their reluctance to commit their land to a long-term crop such as trees that might or might not be profitable at the time of maturity (Esseks et al. 1992). If the power plant owner wanted to gain the option of conversion from fossil to biomass fuels within the next decade or so, the forest plantings could begin reaching the size needed to produce biomass and provide both feedstock for the plant and a market for the forest thinnings. Properly done, it seems possible to meet the goals of both the power plant operator and the cooperating landowner--C offsets, a reliable feedstock source, and economically productive land, while furthering society's goal for a 21st Century conversion from fossil to renewable biomass fuels.

4. Energy Conservation

Another area of opportunity lies in the use and management of trees and other plants in ways that reduce the need for energy and thus cut power plant emissions. The United States, with its energy-intensive economy, can realize as much or more C benefit from using trees, forests, and wood products to reduce fossil fuel usage as is possible through the sequestration of additional C in soils and woody plants (Table 1).

4.1 MATERIALS SUBSTITUTION

Using wood for building materials results in far less energy use and C emissions that if steel or aluminum is substituted. For example, 10 m^2 of interior framing would result in the following C emissions (CORRIM, 1976):

Wood frame	11.8 kg. C
Steel frame	29.0 kg. C
Aluminum frame	35.5 kg. C

Using wood instead of concrete results in reducing the significant CO_2 emissions that come from the cement-making process. Calcining, or heating, of limestone creates about 1 t CO_2 for each tonne of limestone. The reaction is basically:

$$CaCO_3 + heat = CaO + CO_2.$$

Thus, the use of wood in construction of buildings and other long-lived products such as furniture results not only in the storage of C for an additional period of years, it is also a more CO_2-friendly industrial product than its more energy-intensive competitors.

4.2 CONSERVATION TREES

Windbreaks and shelterbelts are associated mainly with farm and ranchlands, and their effect on C emissions lies both in the biomass that they create (within woody plants and soils), and the energy savings they incur. It has been estimated that a windbreak and shelterbelt program in the U.S. would require about 1 to 2.3 x 10^9 trees and shrubs, which would occupy about 2 to 3.4 Mha, and sequester from 20 to 45 x 10^6 t C in biomass (not including new soil C) by the time they reached 20 to 25 yr of age (Brandle et al. 1992).

Indirect savings of around 1 to 3 x 10^6 t C yr^{-1} would result from such a program, due to the reductions in fuel and fertilizer caused by removing acres from cultivation (both because the windbreaks take up land and because they raise yields on the remaining area enough to compensate for the area devoted to trees) and the energy value of the topsoil saved, as well as the reduction in heating and cooling costs in farm homes and livestock shelters (Brandle et al. 1992).

There are over 1 x 10^6 km of rural roads in the northern half of the United States that would benefit from the planting of a living snow fence (Brandle et al. 1992). Living snow fences had only been installed on about 200 km of U.S. highways by 1991, but the potential benefits are significant. Where road departments erect temporary wood-slatted snow fences, prices run as high as $115 per unit of snow storage per km per year, and the structures last an average of 7 yr. A 3-row living snow fence drops unit costs for snow storage to about $2.00 per km per year, with a service life of 75 yr. In Wyoming, snow fence protection can result in a 70% reduction in winter weather-related accidents, while reducing the cost of snow and ice removal by one-third (Brandle et al. 1992). The resulting fuel savings alone add up to emissions reductions of about 0.5 t C km^{-1} of snow fence per year.

A goal of protecting 1% of the rural roads would add up to 6.4 to 12.7 x 10^6 trees on 7,500 to 13,000 ha. Direct C sequestration would be on the order of 27,000 t yr^{-1}, and indirect C reductions could run 4,500 to 9,000 t yr^{-1}.

4.3 URBAN AND COMMUNITY TREES

In addition to sequestering C directly, properly placed urban trees can have a significant impact on atmospheric C buildup through energy conservation. Studies in the U.S. indicate that the daily electrical usage for air-conditioning could be reduced by 10 to 50% by properly located trees and shrubs (Akbari et al. 1988). Savings of 1,351 to 1,665 kWh per year for a 137 m^2 house have been recorded (McPherson et al. 1990). On the other side of the calendar for energy conservation, properly placed trees can also reduce winter heating costs by 4 to 22% (DeWalle, 1978).

From a C perspective, an urban energy-saving tree can be 4 to 15 times as effective at managing atmospheric CO_2 as a rural tree, which primarily is only involved in C sequestration (Akbari et al. 1988). A detailed guidebook has been published on low-cost, low-technology methods to improve energy conservation, describing methods for scientific landscaping methods that maximize energy savings and surface albedo modification through painting dark-colored surfaces in a lighter color (U.S. EPA, 1992).

The practices in the guidebook are the basis for an action and demonstration program called Cool Communities, currently being implemented in 6 U.S. cities under the guidance of American Forests, a non-profit U.S. conservation group. With participation and support from the U.S. Environmental Protection Agency, U.S. Department of Energy, and the USDA Forest Service, as well as local government agencies, electric utilities, and other businesses in the demonstration cities, the program seeks to make a major change in the effectiveness of the community's energy conservation program. An intensive monitoring program will establish improved data on the actual impact of such community efforts. These data will, it is hoped, be useful in assisting other communities to plan and implement effective community-based energy conservation programs using trees and light surfaces.

Monitoring and research in the Cool Communities demonstration cities is focused on three areas of intended impact: (1) citizen familiarity and skill with energy-saving practices; (2) measurable changes in the urban environment (tree canopy cover, runoff, air quality, albedo, etc.); and, (3) energy usage per capita and per dwelling unit (Semrau, 1992).

Despite the many advantages of urban and community trees, several limitations to urban tree planting for energy conservation on the global scale should be noted. First, little energy is used for cooling in many parts of the world, and where there is no air conditioning, the potential energy savings are few.

Second, proper maintenance of urban trees is expensive, difficult, and essential. Urban trees live notoriously short lives, succumbing to many different stresses such as heat, drought, vandalism and urban air pollution (Moll *et al.* 1992). An effective urban and community forestry program must have a significant tree care component, as well as a tree planting and replacement effort.

Third, the U.S. electric utility industry spends approximately $1.5 billion annually to clear trees away from power lines (Kinsman and Trexler, 1993). These trees can be a major cause of power outages where they are placed wrong or have not been properly trimmed. Urban trees need to be carefully chosen and located to prevent them from becoming a problem when they grow to a large size.

Clearly, however, urban tree planting programs that plant proper trees in appropriate places make environmental and economic sense in most communities. In the U.S., such programs have been most successful when they are supported by an active citizen's organization, rather than being the responsibility of local government alone. The reason for such added success, it has been shown, is that when local citizens are involved in the planning, work and expense of establishing trees in their neighborhood, the trees receive better care and maintenance, suffer less vandalism, and live significantly longer and more beneficial lives (Moll *et al.* 1992).

With energy conservation as a major goal, tree planting programs should focus first upon those trees (including species selection and location) that provide direct shading and energy benefits to individual homes and buildings. The next priority for planting would be trees that would provide maximum shade for parking lots, streets, and other dark-surfaced areas. The lowest priority would be to plant those open areas where trees could "fill in" the open spaces that, while not directly shading or protecting buildings, would help reduce the urban heat island effect by modifying albedo and wind patterns as part of the total urban forest.

Sampson *et al* (1992b) proposed a goal for a 10-year program aimed at increasing the canopy cover by 10% on residential lands, and 5 to 20% on other urban lands in the U.S. They estimate the effect of such a program on U.S. urban forests could result in sequestration of 2 to 5 x 10^6 t C yr^{-1} in trees and soils, and a 7-29 x 10^6 t C yr^{-1} reduction in C emissions due to energy conservation from improved shading, increased evapo-transpiration, and reduction of the urban heat island, along with wintertime heat savings.

Expanding urban and community forests on an international scale would enlarge the amount of C sequestered in trees and soils, and notably improve the living quality in many cities. The energy savings would not rise in proportion to the amount of C sequestered, however, due to the reduced dependance on air conditioning in much of the world compared to the U.S.

5. Conclusions

There are many reasons to improve the management of trees and forests, as well as other biomass systems such as grasslands and croplands. Obvious among them are the benefits derived by humans from the products and services that flow from those systems. It is now also generally agreed that management of those systems in a way that is ecologically sustainable is also likely to be positive from a C emissions point of view. The dimension added in this paper is the likelihood that improved management of forest ecosystems can also be linked directly to the goal of reducing fossil energy burning, which is the major forcing factor in C emissions (Chandler, 1990).

If the development of a robust biomass energy industry can be the basis for an economic market for the currently-unmarketable biomass that foresters must remove from many overcrowded forests as a means of assuring fast growth, plant health, and a reduced risk of ecosystem stress or collapse, the environmental benefits would flow from a variety of aspects in addition to reduced C emissions.

There is always concern about competition for land in order to support schemes to increase forests, or convert croplands to biomass production. For the time being, in the United States, land availability seems to be a concern only in the event that the upper-most estimates of afforestation and land use change are implemented.

In 1982, the United States had a total of 170 Mha of available cropland. Projections done by the Department of Agriculture estimated that conversion to other land uses and removal of highly erodible land from cultivation would reduce that supply to 140 Mha by 2030. Projections of land needed for food production in 2030 range from a low estimate of around 60 Mha to a high estimate of 140 Mha. The high estimate assumed the effects of smaller increases in yield due to technology than are currently being experienced, combined with high export demand for U.S. commodities (U.S. Department of Agriculture, 1989).

In their "intermediate" scenario, USDA analysts estimated that the U.S. could meet its domestic and export need for crops in 2030 on about 90 Mha of cropland. If so, that would mean a surplus land capacity of around 50 Mha -- after 30 Mha of marginal cropland were already removed. Even if conversions of 32 to 90 Mha were realized as Table 1 indicates, much of that conversion would occur on the marginal cropland or the land headed for urbanization because of other reasons. At the upper end of the conversion estimate, land competition would become a factor, it appears.

Table 1. Average annual C impacts estimated to occur with various types of land use change and afforestation in the U.S.

Type of Action	Area	C in Soil and Trees[1]	Fossil C Offset	C Impact	Total C Impact
	$(10^6$ ha)	\multicolumn{3}{c}{(t C ha^{-1} yr^{-1})}		$(10^6$ t)
Converting Marginal Land	9.5 to 47	2.5 to 3	1	3.5 to 4	40 to 160
SRWC for Fuel	14 to 28	1.85	4.6 to 8[2]	6.5 to 10	50 to 340
Shelterbelts, Windbreaks, Snow Fences	2 to 4.6	0.5 to 0.6	0.6 to 0.9	1.1 to 1.5	2 to 7
Urban and Community Trees	6 to 10[3]	0.4	1.1 to 3	1.5 to 3.4	9 to 34
Total U.S. Opportunity	32 to 90				101 to 541

Source: Tables 13.2 through 13.5, Sampson et al. 1992a.

[1] The amount of C stored in the soil depends on the degree to which soil organic matter was depleted prior to afforestation, and the climate regime of the site. Generally, cooler and moister soils build up and maintain higher C levels than warmer or drier soils. The estimates shown are calculated as an average of U.S. conditions.

[2] Variation due to: (1) whether the biomass is used to produce ethanol or directly burned to produce electricity; and (2) whether we continue to use existing technology or are successful in commercially adapting technologies that have been laboratory-proven (Wright et al. 1992).

[3] Estimated available growing space for trees within existing and newly-developing urban areas in the United States.

In spite of the fact that much of the "surplus" cropland that may come out of the U.S. inventory is arid or semi-arid and, therefore, not suited for trees, the conclusion is that, in the U.S., conversion of significant areas to forest or biomass production will not be limited by competition for crops, nor will it reduce food supplies. In fact, the creation of a new and profitable crop on those lands might result in a reduction of the need for the federal government to pay farmers to keep such lands out of surplus crop production.

For many of the opportunities to improve the inter-face between forest management and the energy sector, the impact on C emissions from energy conservation is as high or higher than can be achieved from the biological activity of the plants in capturing and storing C in biomass and soils (Table 1). That ratio may not hold for other nations, where the energy costs and/or type of energy usage (e.g. air conditioning) is not similar.

References

Akbari, H.; Huang, J.; Martien, P.; Rainer, L.; Rosenfeld, A.; Taha, H.: 1988, The impact of summer heat islands on cooling energy consumption and global CO_2 concentration. In: *Proceedings of ACEEE 1988 summer study on energy efficiency in buildings*; August 1988; Asilomar, CA. Washington, DC: American Council For An Energy Efficient Economy: 11–23.

Birdsey, Richard A.; 1992, Changes in Forest Carbon Storage from Increasing Forest Area and Timber Growth, in: Sampson, R. Neil and Dwight Hair (eds), *Forests and Global Change, Volume 1: Opportunities for Increasing Forest Cover*, Washington DC: American Forests, pp. 23-40.

Brandle, James R., Thomas D. Wardle, and Gerald F. Bratton: 1992, Opportunities to Increase Tree Planting in Shelterbelts and the Potential Impacts on Carbon Storage and Conservation, In R. Neil Sampson and Dwight Hair, eds., *Forests and Global Change, Volume 1: Opportunities for Increasing Forest Cover,* Washington, DC: AMERICAN FORESTS, pp. 157-176.

Chandler, William U.: 1990, *Carbon Emissions Control Strategies: Executive Summary*. Washington, DC:World Wildlife Fund, 42 p.

CORRIM: 1976, *Renewable resources for industrial materials: Report of the Committee on Renewable Resources for Industrial Materials*, Board of Agriculture and Renewable Resources, Commission on Natural Resources, National Research Council, Washington, DC: National Academy of Sciences, 266 p.

DeWalle, D. R.: 1978, Manipulating urban vegetation for residential energy conservation. In: *Proceedings of the 1st national urban forestry conference*; November 13–16, 1978; Washington, DC. Washington, DC: U.S.D.A. Forest Service: 267-283.

Dixon, R.K., Andrasko, K.J., Sussman, F.G., Lavinson, M.A., Trexler, M.C. and Vinson, T.S.: 1993, Forest Sector Carbon Offset Projects: Near-Term Opportunities to Mitigate Greenhouse Gas Emissions, this volume.

Esseks, Dixon, Steven E. Kraft, and Robert J. Moulton: 1992, Land Owner Responses to Tree Planting Options in the Conservation Reserve Program, in: Sampson, R. Neil and Dwight Hair (eds), *Forests and Global Change, Volume 1: Opportunities for Increasing*

Forest Cover, Washington DC: American Forests, pp. 23-40.

Hasenkamp, Karl Peter: 1993, Large-Scale Afforestation as a Contribution to Warding off a Climate Catastrophe, Testimony before a non-public hearing held by the Enquete Commission on "Protecting the Earth's Atmosphere," May 15, 1993, Bonn: German Bundestag.

Helme, Ned, Mark G. Popovich, and Janet Gille: 1993, *Cooling the Greenhouse Effect: Options and Costs for Reducing CO$_2$ Emissions from the American Electric Power Company*, Washington, DC: Center for Clean Air Policy, 55 p.

Kinsman, J.D. and Trexler, M.C.: 1993, Terrestrial Carbon Management and Electric Utilities, this volume.

Lovins, Amory: 1976, Energy Strategy: The Road Not Taken? *Foreign Affairs*, **55**:1, pp. 65-96.

Mangold, Robert D., Robert J. Moulton, and Jeralyn D. Snellgrove: 1992, *Tree Planting in the United States*, 1991, U.S. Department of Agriculture, Cooperative Forestry, 14 p.

MacCleery, Douglas W.: 1993, *American Forests: A History of Resiliency and Recovery*, FS-540, U.S. Department of Agriculture Forest Service, 59 p.

McPherson, E. Gregory; Woodward, Gary C.: 1990, Cooling the urban heat island with water- and energy-efficient landscapes. *Arizona Review* Spring 1990: pp. 1–8.

Moll, Gary A. and Stanley Young: 1992, *Growing Greener Cities*, Los Angeles, CA: Living Planet Press, 128 pp.

Moulton, Robert J. and Kenneth R. Richards: 1990, *Costs of Sequestering Carbon through Tree Planting and Forest Management in the United States*, Gen Tech Rep WO-58, U.S. Department of Agriculture Forest Service.

Mutch, R.W., Arno, S.F., Brown, J.K., Carlson, C.E., Ottmar, R.D. and Peterson, J.L.: 1993, *Forest Health in the Blue Mountains: A Management Strategy for Fire-Adapted Ecosystems*, Gen. Tech. Rep. PNW-GTR-310, Portland, OR: USDA Forest Service, Pacific Northwest Research Station, 14 pp.

Parks, Peter J. and Ian W. Hardie: 1993, (In Review) Least-Cost Forest Carbon Reserves: Cost-Effective Subsidies to Convert Marginal Agricultural Land to Forests.

Parks, Peter J., Susan R. Brame, and James E. Mitchell: 1992, Opportunities to Increase Forest Area and Timber Growth on Marginal Crop and Pasture Land, in: Sampson, R. Neil and Dwight Hair (eds), *Forests and Global Change, Volume 1: Opportunities for Increasing Forest Cover*, Washington DC: American Forests, pp. 97-122.

Row, Clark and Robert B. Phelps: 1992, Carbon Cycle Impacts of Improving Forest Products Utilization and Recycling, in: Ata Qureshi (ed), *Forests in a Changing Climate*, Washington, DC: Climate Institute, pp. 208-219.

Sampson, R. Neil: 1992, Forestry Opportunities in the United States to Mitigate the Effects of Global Warming, in: Wisniewski, J. and Lugo, A.E. (eds), *Natural Sinks of CO$_2$*, Kluwer Academic Publishers, pp. 157-180.

Sampson, R. Neil: 1993, Increasing Forest Areas as a Carbon-Fixing Strategy, Testimony before a non-public hearing held by the Enquete Commission on "Protecting the Earth's Atmosphere," May 15, 1993, Bonn: German Bundestag.

Sampson, R. Neil and Thomas E. Hamilton: 1992a, Forestry Opportunities in the United States to Mitigate the Effects of Global Warming, in: Sampson, R. Neil and Dwight Hair (eds), *Forests and Global Change, Volume 1: Opportunities for Increasing Forest Cover*, Washington DC: American Forests, pp. 231-245.

Sampson, R. Neil, Gary A. Moll, and J. James Kielbaso: 1992b, Opportunities to Increase Urban Forests and the Potential Impacts on Carbon Storage and Conservation, in: Sampson, R. Neil and Dwight Hair (eds), *Forests and Global Change, Volume 1: Opportunities for Increasing Forest Cover*, Washington DC: American Forests, pp. 51-72.

Schmidt, T., Boche, M., Blackwood, J., and Richmond, B.: 1993, *Blue Mountains Ecosystem Restoration Strategy: A Report to the Regional Forester*, Portland, OR: USDA Forest Service, Pacific Northwest Region, 12 pp.

Sedjo, Roger A.: 1990, *The Nation's Forest Resources*, Discussion Paper ENR90-07, Washington, DC:Resources for the Future.

Semrau, Anne: 1992, Introducing Cool Communities, *American Forests*, **98**:7-8, pp. 49-52.

Shultz, Sheldon: 1993, personal communication.

Steele, R., S.F. Arno and K. Grier-Hayes: 1986, Wildlife Patterns in Central Idaho's Ponderosa Pine-Douglas-Fir Forest. *Western Journal of Applied Forestry* 1, pp. 16-18.

Trexler, Mark C., Paul Faeth, and J.M. Kramer: 1989, *Forestry as a Response to Global Warming: An Analysis of the Guatemala Agroforestry and Carbon Sequestration Project*, Washington, DC: World Resources Institute, 66 p.

Trexler, Mark C., Christine A. Haugen, and Lisa A. Loewen: 1992, Global Warming Mitigation through Forestry Options in the Tropics, in: Sampson, R. Neil and Dwight Hair (eds), *Forests and Global Change, Volume 1: Opportunities for Increasing Forest Cover*, Washington DC: American Forests, pp. 73-96.

Trexler, M.C.: 1993, Manipulating Biotic Carbon Sources and Sinks for Climate Change Mitigation: Can Science Keep Up with Practice?, this volume.

U.S. Department of Agriculture: 1989, *The Second RCA Appraisal: Soil, Water, and Related Resources on Nonfederal Land in the United States, Analysis of Condition and Trends*, Washington, DC: U.S. Government Printing Office (1989 242-141/03004), 280 pp.

US Environmental Protection Agency: 1993, *Forests for the Future: Launching Initial Partnerships*, Report of an Interagency Task Force, January 15, 1993.

Wright, Lynn L., Robin L. Graham, Anthony F. Turhollow, and Burton C. English.: 1992, Growing Short-Rotation Woody Crops for Energy Production, in: Sampson, R. Neil and Dwight Hair (eds), *Forests and Global Change, Volume 1: Opportunities for Increasing Forest Cover*, Washington DC: American Forests, pp. 123-156.

CO2-MITIGATION BY AGROFORESTRY

E. KÜRSTEN and P. BURSCHEL

Chair of Silviculture and Forest Management, University of Munich, Hohenbachernstr. 22, D 8050 Freising, Germany

Abstract. Agroforestry is a very important option for C-fixation in the tropics because of the large area suitable for this practice. The amount of C directly sequestered by trees within the different agroforestry systems normally ranges from 3 to 25 t·ha-1 and in some cases up to 60 t·ha-1. Additional CO2-mitigating effects can achieve more than 20 times this quantity. Such effects are:
- the protection of existing forests,
- the conservation of soil productivity,
- the reduction of fossil energy consumption by use of timber instead of more energy-intensive raw materials and
- the substitution of fossil fuel by wood as an energy source.

1. Introduction and Methology

Planting trees is an important possibility for mitigating the increase of CO2 in the atmosphere. In addition to measures reducing the emission of this greenhouse gas from fossil fuels, and preserving existing forests as C stores, afforestation as an important options for climate protection (IPCC, 1990; EK, 1991).

The largest potential for C-sequestering afforestations in terms of area and growth rates appears to be in the tropics. But looking at the vast areas of unused land there (Grainger, 1988), one has to take into account the degradation of many sites and the needs of the local population to produce food. To consider these aspects, Houghton et al. (1991) have evaluated satellite imagery of the entire tropics. For each area-unit investigated they determined the actual land-use and the most and the less effective potential of the site for CO2-sequestering (Table 1). On potential forest sites for example, agroforestry is usually the pessimistic option (C), whereas plantations represent the optimistic one (A). Similarly, when soils are less degraded, protection of regrowth is the favorable option (B). The feasibility of each variant, however, is dependent on economic needs and possibilities at the given site.

Table 1. Land available for carbon sequestration in the tropics (Houghton et al., 1991)

Option	Area (106 ha)
A. Degraded lands with potential for plantations	0-579
B. Secondary and/or fallow forest suitable for protection	15-858
C. Agricultural areas with potential for agroforestry	356-499

Water, Air, and Soil Pollution **70**: 533–544, 1993.
© 1993 *Kluwer Academic Publishers.*

Table 1 clearly shows that planting trees within the context of agroforestry systems in every case should play a significant role as far as afforestation in the tropics is considered. In spite of this fact there is very little knowledge about the C-mitigation potential of agroforestry systems. For global estimations of the average C-sequestration capacity Houghton et al. (1991) used a value of 60 t·ha⁻¹ for America and Asia and 30 t for Africa. Winjum et al. (1992) consider 95 tC·ha⁻¹ as an average C-pool for agroforestry.

Swisher (1991) developed a distinct estimation procedure for determination of the C stock using the quotient of average temperature and precipitation as driving forces and C density of respective natural forests and rotation period of the trees in agroforestry systems as a basis. In the case of a rotation period of 35 yrs, in the premontane tropical wet forest life zone an average C stock in the trees is calculated as 38 t·ha⁻¹. Thereby the author assumes that an agroforestry plantation attains an equivalent to 75% of a fully stocked forest plantation.

This paper analyzes results of biomass studies conducted in Central American agroforestry plantations and compares them with overall estimates mentioned above. Furthermore, CO_2-mitigation effects attainable by agroforestry in addition to C storage in trees will be investigated.

2. C-Storage in Different Types of Agroforestry Plantations

Agroforestry comprises a wide range of combinations of trees and agricultural crops and /or livestock on the same land. This however does not only imply simultaneous agricultural or forestry production. Pure forest stands, resulting either from afforestation or from natural succession are also included if they alternate with periods of agricultural use. Various forms of agroforestry can be distinguished. From the C storage point of view the average number and size of trees are decisive. The spectrum ranges from loosely scattered trees to closed stands, and from living tomato-stakes to timber stems. Table 2 presents the results of various cultivation types. Biomass data given by various publications were multiplied by 0.5 to get the content of C (see Brown and Lugo, 1982).

All C-stocks in the agroforestry systems in Table 2, except for the first one, are clearly lower than the values mentioned in the foregoing section. Even the addition of root C (not shown in Table 2) does not alter this. For cases 1c and 1d root carbon amounts to 4.9 and 2.8 tC·ha⁻¹, respectively (including root biomass of Cacao plants). In no other example was the root biomass determined. However, approximate values can be taken from other biomass studies. Accordingly, in tropical wet forest life zone, values from 2 tC·ha⁻¹ (1 to 4 yr old plantations) to 6 to 12 tC·ha⁻¹ in old-growth and mature stands can be assumed (Golley et al., 1969; Ewel, 1971; Berish, 1982; Young, 1989; Lugo, 1992). In 10 to 14 yr old plantations of Eucalyptus camaldulensis, values from 4.5 to 5 tC·ha⁻¹ were measured, which clearly is within the above range (Ranasinghe and Mayhead, 1991). In general, it was observed that secondary forests develop a higher root biomass than plantations (Lugo, 1992).

Table 2. Above-ground C storage in trees of some agroforestry practices in
        Central America

Agroforestry System Tree Species	Life Zone (1)	Trees (N·ha^{-1})	Age (yrs)	C-Storage (t·ha^{-1})	Source (2)
1 *Shade Trees in Coffee and Cacao*					
a Gliricidia sepium	TPW	330	30	51.6	1
b Inga densiflora	TPW	400	20	24.3	2
c Cordia alliodora	TPW	278	10	24.9	3
d Erythrina poeppigiana	TPW	278	10	19.0	3
e Mimosa scarabella	TPW	650	2	14.2	4
2 *Fuelwood Plantations*	3				
a Leuceana leucocephala	TLD	3800	5	28.9	5
b Eucalyptus saligna	TPW	1378	2.5	27.0	6
3 *Secondary Forests*					
a Miconia lonchophylla/Vismia guiansis	TPW	3400	8	31.0	7
b Lonchocarpus spp. and others	TLM	7300	3	17.9	8
c Lonchocarpus spp. and others	TLD	3400	3	7.6	8
d Cassia grandis	TLD	1700	3	12.3	8
e Guazuma ulmifolia	TLD	28250	4	5.8	9
4 *Trees in Corrals and Annual Crops*					
a Alnus acuminata	TLM	35	30	25.0	10

(1) TPW = tropical premontane wet forest, TLD = tropical lowland dry forest;
    TLW = tropical lowland wet forest, TLM = tropical lowland moist forest
(2) Sources: 1 = Salazar, 1984; 2 = Salazar, 1985; 3 = Fassbender et al., 1990; 4 = Picado, 1986;
    5 = Martinez et al., 1989; 6 = Salazar, 1986a; 7 = Salazar 1986b; 8 = Queseda, 1986;
    9 = Escobar and Salazar, 1990; 10 = Canet, 1986a

In addition to the missing root biomass the data given in table 2 are problematic for another reason: They mostly are taken from stands at the end of rotation time. For the estimation of permanent C-storage achievable through agroforestry however, the average stock during the whole rotation period (normal stock) has to be taken into consideration, which generally amounts to little over 50% of the final stock. Considering this viewpoint, and adding a reasonable supplement of C in the roots, a relatively dense tree crop in agroforestry plantations of Central America would permanently store only 3 to 25 tC·ha^{-1}.

The C stock in trees, strip-wise mixed with agricultural crops in the form of shelterbelts, boundary plantings and living fences, can be placed in the lower part of above range. In case of alley-cropping, aboveground stock eventually is harvested twice annually and used as mulch. Then an average C-stock of only 1 to 2.3 tC·ha^{-1} results (after Kass, 1987). About the same quantity may be stored in the roots.

Considerably higher C-stocks can be attained if the Taungya system is applied, which after a short time of agricultural use, leads to closed plantations (fully stocked) with longer rotation. According to Schroeder (1992) a normal stock between 20 to 60 tC·ha^{-1} can be attained if the rotation period is 15 yrs. Such values become only slightly higher with the rotation periods extended to 30 yrs since intensive thinnings maintain the stock up to the final cutting close to that of a 10 to 15 yr old stand (Hughell and Chaves, 1990; Hughell, 1991). A substantial C stock of over 50 t·ha-1 may also be accumulated in tropical mixed household orchards, which are characterized by a multistoreyed permanent stocking with a wide range of trees and shrubs producing fruits, timber and fuelwood (Budowski, 1991).

3. Carbon Storage in Wood Products

Harvesting of trees does not end the C storage in wood in every case. If the stems are processed into long lasting products, they act as C store until their decomposition. For the evaluation of this effect, a stand of 100 trees is considered (e.g. Tectona grandis, Bombacopsis quintanum, Cupressus lusitanica) planted in rows or irregularly in fields or pastures. According to Martinez (1989) during a rotation period of 25 yrs they produce 92 m3 of round wood in addition to small sized fuelwood. Assuming that - reduced by processing losses - 50% of such wood ends up as building material, an amount of 11.5 t C can be fixed on a long term basis. The duration of this storage depends on the average lifetime of wood products. For temperate regions, a decay rate of 1% is assumed (Burschel et al., 1993), a rate which Swisher (1992) also uses for the tropics. However, because natural durability of wood under tropical conditions is half of that given in temperate regions (Findlay, 1985), it appears reasonable to assume a decomposition rate of 2% annualy.

Figure 1. Longterm C-storage in timber products resulting from the sustainable production of 100 trees·ha-1 in agroforestry

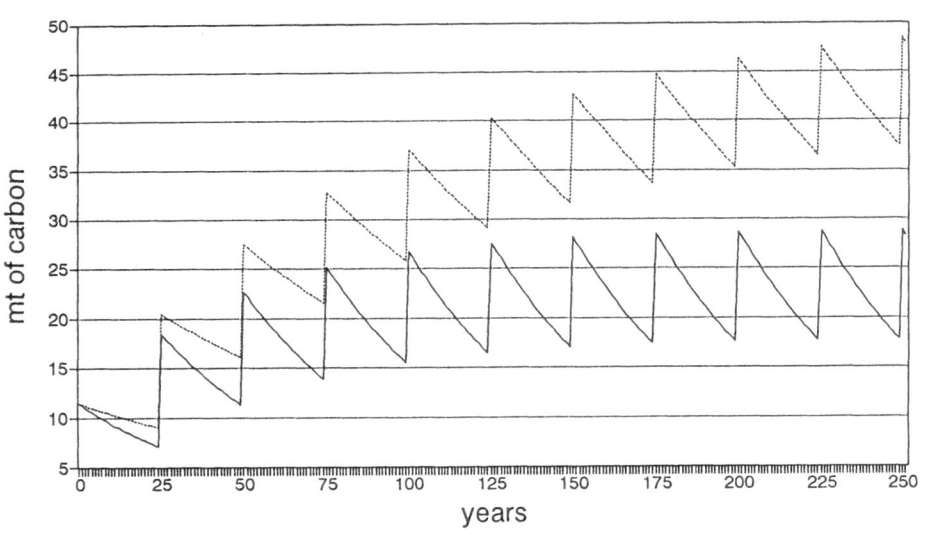

From figure 1 the development of C-storage in the wood products for both the cases can be taken. When decay and production of wood products become equal on a long run, a C stock is reached, which oscillates around a mean value which can be added to the average growing stock of the trees (Burschel et al., 1993). Assuming an annual decay rate of 2%, according to this calculation an amount of 23 tC·ha^{-1} of cultivated area is stored in the wood products. For a fully stocked plantation, the Taungya-system being an example, this stock is up to four times higher. On the other hand stands with a very short rotation period only produce fuelwood or poles and posts with a lifetime of a few years. Summarizing, the C storage in wood products may range between 1 t (stored fuelwood) and 100 t·ha^{-1} (a fully stocked timber plantation with about 400 trees·ha^{-1} in case of the Taungya system).

4. Conservation of carbon stores

Agroforestry sytems not only create carbon sinks in the form of trees and wood products but also may help to prevent existing natural C stores from being depleted. Primarily this concerns the humus C at a given site, the prerequisite for soil productivity. If agriculture by this means can be practiced on a sustained basis and possibly with increasing yields, the pressure on the remaining forests as land reserves for farming and cattle grazing would be reduced. In this regard, agroforestry systems may provide an important contribution to the preservation of natural forests. In addition, by producing fuelwood in areas of fuelwood scarcity the need for destructive exploitation may be minimized or totally avoided.

4.1   SOIL ORGANIC CARBON (SOC)

Atmospheric C assimilated by trees is partly fixed in organic matter. All of it eventually reaches the soil as litter and dead roots, compensating the decomposition of SOC caused by microbial and fungal oxidation. Annually this amounts to 3% of the soil C stock in tropical forests. On farm lands, these C losses are even 4% (Young, 1989). In the wet tropics, addition of agricultural crop residues is often not sufficient to conserve the SOC. Agroforestry trees can compensate the steady loss of C by providing additional organic material. According to estimates made by Young (1989), a C stock of 35 t·ha^{-1} (in SOC) under a common crop like maize gets reduced by 9 t within 50 yrs in the zone of Tropical Lowland Wet Forests. In the case of an initial stock of 80 tC·ha^{-1} which is close to pristine rain forest values (Brown and Lugo, 1982), a loss of 48 tC·ha^{-1} results under such conditions. By alley-cropping and other agroforestry systems this loss of SOC can be avoided and partly reversed where it has occured already.

These findings were confirmed by studies mentioned in Table 2 (numbers 1c and 1d) with Cordia alliodora and Erythrina poeppigiana cultivated as shade trees for cacao, where a substantial increase of C content in the soil occured. Within 9 yrs C stock grew by 8.9 and 24.1 t·ha^{-1}. Moreover layers of litter were produced, containing another 7 to 8 t of C (Fassbender et al., 1990; C-content in humus = 58% (Young, 1989)). This increase is similar to the loss of SOC mentioned earlier, and leads to a complete compensation of decomposition by the addition of new organic material. The difference between the C stock resulting from long-term purely agricultural use, and the one which is established through enhanced litter addition in agroforestry systems, is the quantity of C that is permanently removed from the atmosphere. The data cited here indicate that it amounts to between 10 and 50 $tC \cdot ha^{-1}$.

4.2   FORESTS

The extent to which the introduction of agroforestry systems may facilitate the conservation of forests is strongly related to local conditions and difficult to estimate. Obviously no forest conservation can take place in an area where no forest cover has been left. The opposite extreme is represented by a region where shifting cultivation affects primary forests with a relatively high C stock of 200 $t \cdot ha^{-1}$ (Brown and Lugo, 1982) turning it into degraded wasteland within few rotations. So, if an area of 1 ha of pristine forest is cut and cultivated for 2 yrs, and thereafter left untouched 8 yrs for the development of a secondary forest, such depletion may be attained after 4 more of such rotations. During the period of 8 yrs without cultivation, another 4 ha would be subjected to same treatment. After 50 yrs, 5 ha of pristine forests will be lost and about 1000 t of C emitted from the vegetation into the atmosphere. The conservative assumption thereby is that C released from humus is compensated by the resettlement of the waste lands.

Theoretically, this destruction of 5 ha of forests can be avoided by introducing sustained agroforestry on 1 ha of existing farm land (Dixon and Andrasko, 1992). According to the above calculations, a 17 to >300 times larger effect than by sequestering C through agroforestry trees can be achieved by avoidance of CO_2 emissions from the conserved forest. If by this means instead of a pristine forest, secondary regrowth with a lower growing stock but high increment can be protected from further cuttings, it will accumulate C until reaching maturity, or permanently produce timber and other types of wood if managed sustainably.

5. Reduction of CO_2-Emissions Resulting from Fossil Fuels

To this point a discussion has been presented as to what extent trees, soils and wood products can be developed as C sinks by means of agroforestry or can be preserved as C stores. In addition, planting trees and their sustainable use possibly may reduce the burning of fossil fuels, and thereby the emissions of CO_2 from the lithosphere.

5.1   FUELWOOD AS SUBSTITUTE FOR FOSSIL FUELS

Carbon, which is sequestered by growing trees, gets re-emitted into the atmosphere when they are used as a source of energy. If increment and burning are in balance, - as is the case with sustained production - the burning of fuelwood is neutral in terms of C-emissions. To quantify the CO_2-reduction by substituting wood for fossil fuels, the heating values of alternative energy sources - for wood with a realistic water content-, the energy-inputs required for their processing, and differences in heating efficiencies have to be considered. The comparison made in Table 3 for tropical countries assumes, that well air-dried wood in the form of billets, substitutes fuel oil. This type of energy substitution is realistic in Central America, especially for small industries and workshops such as coffee roasting, salt works, lime kilns, bakeries and sugar industries which use fuelwood as an energy source if it is cheaper than fuel oil (Canet, 1986b).

Table 3. Comparison of carbon-emissions from heating with wood (1000 kg d.m.) and fuel oil (194.2 kg) (after Kürsten and Burschel, 1991)

Typ of Fuel	Wood	Fuel Oil
Quality	20% Water Content	Light
Heating Value	14.1 MJ·kg-1	35.5 MJ·liter-1
Efficiency	50%	85%
Procecessing	Growing, Harvesting, Transport,Cleaving	Extraction, Transport, Refining
Energy-input	0.8 MJ·kg-1	10.9 MJ·liter-1
C-Emissions	0.0 kg from Wood 17.5 kg from Preparation	180.8 kg from Oil 55.5 kg from Preparation
Total Emissions	17.5 kg C	236.3 kg C

From Table 3 it can be concluded that by utilization of 1 t of air-dry fuelwood in place of fuel oil, 218.8 kg of C emissions can be avoided. Table 4 shows the mitigation effects resulting from the use of fuelwood produced by different agroforestry systems as presented in Table 2. They range from 0.1 to 3.6 tC·ha-1·yr-1.

Table 4. CO₂-mitigation by fuelwood production in agroforestry systems

Agroforestry System Tree Species	Trees· ha-1	Rotation Period (yrs)	Fuelwood Production (t)	CO₂-mitigation (tC·ha-1·yr-1)
Shade Trees in Coffee and Cacao				
a Gliricidia sepium	330	30	101.4	0.7
b Inga densiflora	400	20	42.8	0.5*
c Mimosa scarabella	650	2	18.3	2.0
2 *Fuelwood Plantations*				
a Leuceana leucocephala	3800	5	46.2	2.0
b Eucalyptus saligna	1378	2.5	41.3	3.6
3 *Secondary Forests*				
a Miconia lonchphylla/Vismia guiansis	3400	8	54 0	1.4
b Lonchocarpus spp. and others	7300	3	27.8	2.0
c Lonchocarpus spp. and others	3400	3	10.6	0.8
d Cassia grandis	1700	3	21.2	1.6
e Guazuma ulmifolia	28250	4	5.2	0.3
4 *Trees in Corrals and Annual Crops*				
a Alnus acuminata	35	30	18.3	0.1
5 *Living Fences*				
a Gliricidia sepium	per km and year		6.3	1.4**

* = without prunings; ** = Picado and Salazar, 1984; other quotations see Table 1

5.2 UTILIZATION OF TIMBER IN PLACE OF MORE ENERGY-INTENSIVE RAW MATERIALS

Production, transport and processing are less energy demanding in the case of timber than in the cases of other raw materials used for construction purposes, such as aluminum, steel, concrete or plastics. If timber is used instead of those substances a considerable reduction of energy expenditure can be achieved, that represents a CO_2 mitigation effect. The reduction of C-emissions by this type of material substitution is estimated at 0.28 t C·m-3 of timber for highly industrialized countries (Burschel et al., 1993). Lacking better ones this value was used here too, although in less developed countries the mitigation effect might be even higher because of more labour and less energy intensive lumber processing on the one hand and of a less energy efficient production of alternative materials on the other hand. Not considered were possible energy savings by the utilization of wood products in place of plastic boxes, concrete or steel masts etc.. They are much lower as compared with all types of construction and less investigated.

Within 25 yrs the stand of 100 trees mentioned in section 3 delivers a timber volume (for construction) of 46 m3, which corresponds to 13 t of C-emission avoided, which means a CO_2-mitigation of 0.5 tC·ha-1·yr-1. For the agroforestry systems mentioned in Table 2 this effect ranges from 0 t (purely fuelwood production) and 2 tC·ha-1·yr-1 (e.g. Taungya system, producing about 200 m3·ha-1 of construction timber).

5.3 SAVINGS OF FERTILIZERS AND PESTICIDES

Growing agricultural crops together with trees allows a reduction of the input of fertilizer for different reasons:
- Tree roots are able to capture nutrients which are being transported out of the crop root zone. By litter fall or mulching they are recycled back to the upper soil horizons.
- The protection of the soil by tree crowns and litter reduces SOC and nutrient losses by erosion.
- Litter fall and mulching lead to an elevated SOC content of the soil and thereby to an increased cation retention capacity, which makes fertilization more effective if not unnecessary.
- If N-fixing tree species are used, the need for mineral N-fertilization becomes reduced, in some cases down to zero (Rippin, 1991).

Indirectly a reduction of CO_2-emissions is the consequence of all these effects since the production of fertilizers is an energy-intensive process. According to Mittal et al. (1992) by alley cropping 20 kg N·ha-1·yr-1 can be saved. Assuming an energy-input of 60 MJ·kg-1 for N-fertilizers (Zweier, 1985) and C-emissions of 0.02 kg C·MJ-1 (see Burschel et al. 1993), the annual reduction in emissions would be 0.024 tC·ha-1. If it is possible to save all 80 kg of N-fertilizers normally applied, this value increases to 0.1 tC·ha-1·yr-1.

By mulching, an additional reduction in herbicide needs can be achieved (Rippin, 1991). Since the production of such chemicals requires an energy-input of 185 MJ·kg-1 and the application rate is 2 kg·ha-1 (Zweier, 1985), the CO_2-mitigation effect of reduced chemical weed control would amount to 0.007 tC·ha^{-1}·yr^{-1} and be much smaller than the one obtained by diminished use of fertilizers.

6. Conclusions

The C-ecological role of trees within agroforestry systems is diverse and complex. Two groups of effects can be distinguished:

1. Accumulation as well as conservation of C stores in trees, soils and wood products.

2. Reduction of C-emissions resulting from energy and material substitution.

The first group of effects can only be realized to an upper limit, resulting from site conditions and the type of system and management. Effects achieved through the conservation of natural forests and SOC may be great, but are restricted by the type and size of existing C-stores.

Reductions of C-emissions, on the other hand, can be achieved on a perpetual basis by a sustained production and harvest. Each unit of fuel oil saved by energy conservation through substitution of timber for more energy demanding materials, or by the use of equivalent amounts of fuelwood, is never converted into an additional quantity of CO_2 emitted to the atmosphere. Agroforestry systems under sustained yield management act continuously and practically unlimited in this manner.

Table 5. Estimated CO2-mitigation effects of agroforestry systems
          (in tC·ha^{-1} agroforestry land)

Accumulation and Conservation of Carbon Stores	
Trees in Agroforestry Systems	3... 60
Wood Products	1... 100
Soil Organic Matter	10... 50
Protection of Existing Forests	0...1000
Sum	*(14...1210)*

Reduction of CO2-Emissions within 50 Yrs	
Energy-Substitution	5... 360
Material-Substitution	0... 100
Reduction of Fertilizer-Input	1... 5
Sum	*(6... 465)*

Total	(20...1675)

Effects of accumulation as well as permanent management effects are presented in Table 5. The latter ones are calculated for a period of 50 yrs, since it may take quite a while until C stocks in stands, soils and wood products reach their equilibrium levels (see Fig. 1). The data presented are a gross simplification of a complex situation, but facilitate an analysis.

These considerations accepted, it can be seen from Table 5, that agroforestry practices have CO_2-mitigation effects of very different kinds and magnitudes. The calculated minimum of 20 t C within 50 yrs will become surpassed in almost every real case, because each type of agroforestry system provides more than the lowest output in at least one type of effect. Pure fuelwood plantations, for instance, have a fairly low C storage capacity and no effects as far as material substitution is concerned. Instead they have a considerable C mitigation effect mainly based on energy substitution. Similarly there will be no case, where all values are at maximum.

C storage in standing trees generally is of less significance compared to other C-mitigation effects of agroforestry. Among them the potentially largest is the conservation of existing forests as carbon stores. Unfortunately it is by far the most difficult to quantify and the most dependent on local conditions. Therefore it is very hard to estimate the extent to which agroforestry may contribute to C-mitigation efforts on a global scale. Ranges of 30 to 95 tC·ha^{-1} used so far are too high if they only relate to C storage in agroforestry trees. But they might become low if all aspects of carbon-mitigation are included. With respect to the great importance of agroforestry within the context of afforestation strategies for the tropics, additional research on a sound empirical basis is urgently needed to provide better data.

7. References

Beer, J.W., Fassbender, H.W. and Heuveldop, J. (eds.):1987, Advances in agroforestry research. Proceedings of a seminar, Sept. 1 to 11th, 1985, Turrialba, Costa Rica, CATIE and GTZ, Serie Tecnica, Informe tecnico, no. 117.

Berish, C.W.:1982, *Can.J.For.Res.* **12**, 699-704.

Brown, S. and Lugo, A.E.: 1982, *Biotropica* **14**, 161-187.

Budowski, G.: 1991, An Increase in Agroforestry Practices in Central America; The Justification for Carbon Dioxide Sequestering, in: Faeth, P., Trexler, M.C. and Page, D., Sustainable Forestry as a Response to Global Warming: A Central American Perspective, A Workshop Held in Guatemala City, May 23-25, 1990, Workshop Report and Supporting Prepared Papers, World Resources Institute, Washington, DC, 15-24.

Burschel, P., Kürsten, E., and Larson, B.C.: 1993, Die Rolle von Wald und Forstwirtschaft im Kohlenstoffhaushalt - Eine Betrachtung für die Bundesrepublik Deutschland, Forstl. Forschungsberichte München, Nr. 126, 135 p.(short version in english)

Canet. B.G.: 1986a, Carateristicas del sistema silvipastoral jaul (Alnus acuminata) con lecheria de altura en Costa Rica, in: Salazar, R. (ed.) 1986c, 241-249.

Canet, B.G.: 1986b, Consumo y abastecimiento de leña en Costa Rica, Silvoenergia No. 14, CATIE, Turrialba, Costa Rica.

Dixon, R.K. and Andrasko, K.: 1992, Integrated systems: Assessment of promising alternative land-use practices to enhance carbon conservation and sequestration. in: IPCC Response Strategies Working Group, Proceedings of a Workshop on Assessing technologies and management systems for Agriculture and Forestry in relation to Global Climate Change, Canberra, Australia, 29-23 January 1992, 48-54

EK (Enquete-Kommission "Schutz der Erdatmosphäre" des Deutschen Bundestages) (ed.): 1991, Dritter Bericht "Schutz der Erde" der Enquete-Kommission "Vorsorge zum Schutz der Erdatmosphäre", Teilband I, Deutscher Bundestag (Ed.), Economica Verlag, Bonn, 686 p.

Escobar, F. and Salazar, R.: 1990, Manejo de un rodal natural de Guazuma ulmifolia Lam. en Azuero, Panama. Silvoenergia No. 36, CATIE, Turrialba, Costa Rica, 4 p.

Ewel, J.: 1971, *Turrialba* **21** (1), 110-112.

Fassbender, H.W.: 1987, Modelos edafologicos de sistemas agroforestales, CATIE, Serie de Materiales de Enseñenza No. 29, Turrialba, Costa Rica, 475 p.

Findlay, W.P.K.: 1985, Preservation of timber in the tropics, Martinus/Junk Publishers, Dordrecht, Boston, Lancaster, 273 p.

Golley, F.B., McGinnis, J.T., Clements, R.G., Chield, G.I. and Duever, M.J.: 1969, *Bioscience* **19**, 693-696.

Grainger, A.: 1988, *Intern. Tree Crops Journal* **5** (1-2), 31-61

Houghton, R.A., Unruh, J.D. and Lefebre, P.A.: 1991, Current land use in the tropics and its potential for sequestering carbon, The Woods Hole Research Center, P.O. Box 296, Woods Hole, MA 02543 USA, 25 p.

Hughell, D. and Chaves, E: 1990, Modelo preliminar de crecimiento y rendimiento de cipres (Cupressus lusitanica Miller) en America Central, Silvoenergia No. 38, CATIE, Turrialba, Costa Rica, 4 p.

Hughell, D.A.: 1991, Modelo preliminar para la prediccion del rendimiento de Gmelina arborea Roxb. en America Central., Silvoenergia No. 44, CATIE, Turrialba, Costa Rica, 4 p.

IPCC (Intergovernmental Panel on Climate Change): 1990, Policymakers summary of the scientific assessment of climatic change. Report prepared for IPCC by Working Group 1.

Kass, D.:1987, Alley cropping of annual food crops with woody legumes in Costa Rica, in: Beer et al., 1987, 197-214

Kürsten, E. and Burschel, P: 1991, *Holz-Zentralblatt* **117**, 1953-1954, 2010-2012

Lugo, A.E.: 1992, *Ecological Monographs* **62**, 1-41

Martinez, H.H.A.: 1989, El componente forestal en los sistemas de finca de pequeños agricultores, Boletin Tecnico No. 19, Serie Tecnica, CATIE, Turrialba, Costa Rica, 78 p.

Martinez, H.H.A., Sandoval C. and Calderon, N.: 1989, Efecto des espaciamiento en el crecimiento y produccion de Leuceana leucocephala, en San Pedro Sula, Honduras. Silvoenergia No. 31, CATIE, Turrialba, Costa Rica, 4 p.

Mittal, S.P., Greval, S.S., Agnihotri, Y. and Sud, A.D.: 1992, *Agroforestry Systems* **19** (3), 207-216

Picado, V.W.: 1986, Mimosa scabrella, especie con potencial para sombra y produccion de leña en cafetales de Costa Rica, in: Salazar, R. 1986c, 227-286

Picado, W. and Salazar, R.: 1984, Produccion de biomasa y leña en cercas vivas de Gliricidia sepium (Jacq.) Steud de dos años de edad en Costa Rica, Silvoenergia No. 1, CATIE, Turrialba, Costa Rica, 4 p.

Queseda, M.J.: 1986, Potencial de areas en barbecho para produir leña en la Peninsula de Nicoya, Costa Rica, in: Salazar, R. 1986c, 315-325.

Ranasinghe, D.M.S.H.K. and Mayhead, G.J.: 1991, *Forest Ecology and Management* **41**, 137-142

Rippin, M.: 1991, Alley-cropping and mulching with Erythrina peppigiana (Walp.) O.F. Cook and Gliricidia sepium (Jacq.) Walp.: Effects on maize/weed competition and nutrient uptake, Diploma Thesis, Agricultural Fac., University of Bonn, 82 p.

Salazar, R.: 1984, Produccion de leña en arboles de Gliricidia sepium usado como
    sombra en cafetales en Costa Rica, Silvoenergia No. 2, CATIE, Turrialba, Costa
    Rica, 4 p.
Salazar, R.: 1985, Produccion de leña y biomasa de Inga densiflora Benth en San
    Ramon, Costa Rica. Silvoenergia No. 3, CATIE, Turrialba, Costa Rica, 4 p.
Salazar, R.: 1986a, Produccion de leña de Eucalyptus saligna en San Ramon, Costa
    Rica, Silvoenergia No. 15, CATIE, Turrialba, Costa Rica, 4 p.
Salazar, R.: 1986b, Estudio de caso de abastecimiento de leña con Eucalytus saligna
    Smith en una industria rural en San Ramon, Costa Rica, in: Salazar, R. 1986c, 181-
    189.
Salazar, R. 1986c (Ed.), Simposios sobre Tecnicas de Produccion de Leña en Fincas
    Pequeñas y Recuperacion de Sitios Degradados por Medio de la Silvicultura
    Intensiva, Turrialba, Costa Rica, 24-28 de junio de 1985, IUFRO, Wien, Grupo de
    trabajo S1-07-09, CATIE, Turrialba, Costa Rica
Schroeder, P.: 1992, *Forest Ecology and Management* **50**, 31-41.
Swisher, J.N.: 1991, *Biomass and Bioenergy* **1**, 317-328.
Winjum, J.K., Dixon, R.K. and Schroeder, P.E.: 1992, *Water, Air, and Soil Pollution*
    64, 213-227
Young, A.: 1989, Agroforestry for soil conservation, ICRAF Science and Practice in
    Agroforestry 4, CAB International, Wallingford, 276 p.
Zweier, K.: 1985, Energetische Beurteilung von Verfahren und Systemen in der
    Landwirtschaft der Tropen und Subtropen - Grundlagen und Anwendungsgebiete,
    Thesis, Agricultural Faculty, University of Göttingen

TERRESTRIAL CARBON MANAGEMENT AND ELECTRIC UTILITIES

JOHN D. KINSMAN
Edison Electric Institute, 701 Pennsylvania Avenue, N.W.
Washington, D.C. 20004, USA

MARK C. TREXLER
Trexler & Associates, 1131 SE River Forest Road,
Oak Grove, OR 97267, USA

Abstract. In the near future regulations could be imposed affecting emissions of CO_2 and other greenhouse gases. Carbon offsets should be a component of any such regime. This paper addresses: 1) international and domestic policy actions related to C offset forestry, including the United Nations' Framework Convention on Climate Change and the U.S. Energy Policy Act of 1992; 2) forestry-related efforts sponsored by U.S. electric utilities to sequester and store C, protect and manage forests, and conserve energy in urban environments; 3) considerations for implementing such efforts; and 4) electric utility industry research and development on advanced methods to use biomass as fuel.

1. Introduction

The "greenhouse effect" is a natural and beneficial phenomenon, without which the planet's average temperature would be close to -18 degrees C. Because human activities are increasing the concentrations of the greenhouse gases in the atmosphere, the natural greenhouse effect may be enhanced. Projecting long-term climatic impacts of man's greenhouse gas emissions remains contentious and fraught with uncertainties, but public policy development to address global climate change is proceeding, especially for CO_2 emissions. Recommendations to mitigate the potential consequences of climate change that could result from an enhanced greenhouse effect include increased energy supply and use efficiencies, increased renewable energy systems, fuel switching, and adaptation (e.g., planning for sea-level rise). Another option is to sequester CO_2 in "sinks" such as plant biomass. For example, C might be sequestered in growing trees, in harvested trees (forest products), in halophytes (salt-tolerant plants), as organic matter in soil, in oceanic seaweed farms, or in microalgae in the ocean. Alternatively, biomass can be used as a substitute for fossil fuel. This paper will review current information regarding terrestrial C sinks and their management in C offset programs.

Water, Air, and Soil Pollution **70**: 545–560, 1993.
© 1993 *Kluwer Academic Publishers.*

2. Offsets as Part Of a CO_2 Control Program

Although the U.S. has extensive experience regulating many air pollutants ranging from particulate matter to ozone, CO_2 is different from these pollutants. Carbon dioxide production is neither the result of impure fuels nor inefficient combustion; rather, it is the necessary byproduct of any C-based production of energy. Carbon dioxide is produced by almost every facet of human activity, ranging from a subsistence-based peasant existence to the commercial energy consumption associated with industrialized societies. Additionally, while most pollutants principally have local or regional effects, CO_2 creates no direct localized impacts and CO_2 concentrations are relatively uniform around the globe.

The use of emissions offsets in environmental management is not new, having been applied in some cases for as long as 15 yr in the U.S., with new emitters of conventional pollutants such as SO_2 required to reduce emissions somewhere else before starting up a new facility, thus holding net emissions steady. Carbon dioxide offsets, however, would operate within a different scientific and regulatory context. First, stack-based CO_2 controls are both extremely expensive and present severe practicality problems (Fluor Daniel, 1991). Second, CO_2 is long-lived in the atmosphere, mixes globally, and thus can be offset anywhere in the world. Third, CO_2 is different from emissions such as SO_2, because it can be practically removed from the atmosphere after being emitted.

Through C offsets the impact of emitting a tonne of C can be negated or offset by avoiding the release of a tonne somewhere else, or by absorbing a tonne of C from the air that otherwise would have remained in the atmosphere. Several types of C offsets can be identified. Demand-side management, for example, can reduce CO_2 emissions by reducing energy demand. Supply-side efficiencies can reduce CO_2 production by getting more energy out of a unit of fuel. Fuel switching can reduce CO_2 production because fossil fuels differ in the proportion of energy content that is attributable to C as opposed to hydrogen. Several types of forestry-based C offsets are also available; e.g.: 1) protecting or managing existing forests that would otherwise be lost through deforestation in coming decades; 2) natural regeneration or reforestation efforts intended to increase global C reservoirs; or 3) using biomass energy to displace fossil fuels use.

3. Biotic Measures to Offset Carbon Emissions

The fundamental basis for forestry C offsets is that photosynthesizing plants use CO_2 as a raw material. Although much of the C absorbed by growing vegetation is released in the short term through night time respiration, some of the C is stored in woody biomass or soils. At the individual plant level, photosynthesis is dependent on ambient CO_2 concentration and other factors. Strain and Thomas (1992) reviewed the literature on plant response to elevated CO_2 and concluded that 1) if other environmental resources are present at required levels, then CO_2 enrichment will increase photosynthesis and plant growth; 2) plants stringently limited by resource deficiencies (e.g., N or P) will respond slightly or not at all to CO_2 enrichment; 3) CO_2 reduces transpiration and improves plant water status, due to increasing photosynthesis and decreased water loss; and 4) CO_2 and other environmental factors of global change affect species in different ways and will generate ecological changes in ecosystem flora and fauna. The potential of small trees to act as a C sink may be more dependent upon tree response to nutrient resources and environmental stresses than to atmospheric CO_2 concentration (Norby et al, 1992). Because of nutrient deficiencies, inadequate water, weather variations and changes due to climate, one should not expect that CO_2 "fertilization" of natural forests will increase plant productivity and C storage enough to compensate for increasing atmospheric CO_2 (Mooney et al, 1991). However, this does not rule out either a more modest CO_2 stimulation of natural vegetation growth or forest management activities to maintain or increase C sequestration.

3.1   FOREST PROTECTION

Protecting and/or managing standing forests can be an attractive means of implementing a C offset program. Tropical forests are being cleared for timber export, fuel wood, shifting cultivation, permanent agriculture, pasture, and urbanization and infrastructure (Detwiler and Hall, 1988; Postel and Heise, 1988). The most recent estimate of annual deforestation is 17 Mha during the 1980s (Houghton et al, 1992). Deforestation in tropical latitudes is responsible for 10 to 30% of anthropogenic CO_2 emissions.

Secondary forests generally do not accumulate as much C as primary forests for a long time; thus, saving a hectare of existing forest that would otherwise be lost can effectively offset as much C as several hectares of new growth for quite a few years. Forest conservation can take a variety of forms. Direct methods include land purchase for parks or wildlife refuges. Indirect methods include reducing deforestation pressure by increasing agricultural productivity or fuel supplies, and improved harvesting practices.

A C offset benefit due to forest protection can often be realized immediately in contrast to projects that entail new forest growth. Forest conservation can also provide ancillary environmental benefits related to biodiversity, watershed and water quality maintenance, and erosion prevention. Early indications indicate that forest protection can supply C offsets at a relatively low cost. At the same time, forest protection projects must be critically evaluated from a C offset standpoint. For example, although a project may physically protect a specific parcel of forest, the deforestation might simply be shifted somewhere else, resulting in no net C savings.

3.2   REGENERATION OF DEGRADED LANDS

There are hundreds of millions of hectares around the world that previously supported tree cover and could support tree cover again. In some cases formerly sustainable slash and burn agricultural systems have become unsustainable under the pressures of growing populations. Salinization, soil compaction and erosion have rendered agricultural land and pasture unproductive over many millions of hectares. Many millions of hectares have been degraded through logging, fuel wood collection, grazing, and fire. In each of these cases C storage on the land declines, often dramatically. While some degraded forests recover on their own, others do not. Fuel wood pressures might be too intense, logging interventions too frequent, or heavy undergrowth and vines might impede tree regeneration, etc. In many areas the control of wildfire would result in the regeneration of forest cover over large areas.

Regeneration has several advantages when employed for C offset purposes, offering a balance between verifiability and cost. It will often be easier to claim C benefits from new trees on previously bare land than to attempt to prove that standing forest will be effectively protected that would have otherwise been lost. At the same time, the focus on removing barriers to natural regeneration rather than on intensive tree planting and maintenance can result in reduced costs. If regeneration projects are designed so that local people benefit, the chances of success are good.

3.3   AGROFORESTRY

Agroforestry -- the incorporation of tree growing with agricultural and other practices -- can play a significant role in C offset projects, particularly when combined with forest regeneration or protection. A primary disadvantage is that only a portion of each hectare is planted in trees, and since the trees are commonly harvested on a relatively short rotation, large areas could be required to achieve significant long-term C offsets. Social forestry projects also tend to be spread out and difficult to monitor or track. The advantage of these types of projects is that they can conform to the economic development goals and can help reduce pressures on surrounding forested areas. Social forestry projects require intensive involvement of local communities, and site-specific tailoring and education.

3.4   COMMERCIAL FOREST PLANTATIONS

Tree plantations can offer rapid growth rates over large areas of land, uniform management, and quantifiable costs and benefits (Sedjo and Lyon, 1990). The C accumulation associated with plantations is more readily quantifiable than unrestricted natural forest growth. Plantations also offer the possibility of an economic return on harvested wood, thereby decreasing the overall cost of the sequestration project. Additionally, C accumulation rates in properly managed plantations may be more than double the rates found on naturally regenerating areas.

From a C sequestration perspective, several concerns can be raised. First, the history of plantation forestry in the tropics is checkered at best, with many efforts having failed for a wide variety of reasons (Trexler and Haugen, 1993). Furthermore, natural forest has often been removed to make room for plantations, which makes little sense in C offset terms. Additionally, the economics of wood supply and demand are often ignored in estimating the potential for additional commercial plantations. Over reliance on a single species, or overly ambitious planting rates given local demand, can lead to market saturation and project failure.

3.5   WOOD PRODUCTS

Eventually a tree will approach maturity and its rate of net C uptake will decline. Averaged over time, a mature forest is in near C equilibrium as the uptake via photosynthesis is balanced by releases via respiration and decay. One way to prolong the active role of forestry in sequestering C is engage in periodic harvesting, thus setting the stage for a new growth cycle. To maintain the C benefit, however, the harvested wood must be used in such a way as to prevent its oxidation. One choice is to use wood for long-lived products such as lumber in construction. Another option is to increase recycling of building material as well as paper products.

3.6   SOILS, AGRICULTURAL CROPS AND ARID LANDS

Globally, soils contain about 150 to 300% as much C as does above-ground biomass (Dixon and Turner, 1991). Some management practices (e.g., cultivation and intense prescribed fire) lead to soil C loss, while crop fertilization and reduced tillage can increase C storage (Johnson, 1992). In general, increased C accumulation in soil is associated with practices that promote cooler soils (e.g., mulch, shade), wetter soils (irrigation), more fertile soils and soils with reduced aeration (limited tillage). Reversion of agricultural land to forest leads to significant soil C storage (Brown *et al*, 1992; Sedjo, 1992).

Agriculture can provide opportunities to sequester C or reduce CO_2 emissions via storing additional C in soils and utilizing energy crops for fuels (see next section). Strategies such as increasing biomass yield and energy crop usage could be attractive options whether or not the intent is to store C or offset CO_2 emissions (Abelson, 1992; Council for Agricultural Sciences and Technology, 1992).

Approximately one-half of the Earth's land surface is arid or semi-arid. In these systems C may be sequestered in halophytes (salt-tolerant desert plants). About 130 Mha of the world's 700 Mha of salt desert habitat could support halophytes, with C sequestration rates comparable to those of tree plantations (Glenn *et al*, 1992 a,b). Halophytes could potentially sequester up to 0.7 Pg C yr^{-1} (Glenn *et al*, 1992a). The harvested biomass could be stored in desert soils if its decomposition is slow, or it could be burned to produce energy. The key advantage of using halophytes is that they can grow in saline soils that are useless for conventional agriculture and thus can be irrigated with salt or sea water, avoiding the use of fresh water.

4. Using Trees for Energy Conservation and Production

4.1   ENERGY CONSERVATION

Because of the replacement of soil and vegetation with concrete, asphalt and metal, many urban areas have experienced a heat island effect characterized by a several degree higher temperature than in nearby rural areas. The heat island effect has caused the need for an additional 1500 MW of electric power plants in Los Angeles (U.S. EPA, 1992). Trees can counteract this heat island effect through the process of evapotranspiration -- a tree can transpire up to 100 gallons of water per day, equivalent to the cooling effect of 100 hours of air conditioners in a hot, dry climate (U.S. EPA, 1992). In addition, shade tree planting in urban areas of the U.S. can reduce the requirement for cooling residences and buildings, sometimes offsetting the need for fossil fuel use and reducing CO_2 emissions. Shade tree planting on the south and west sides of a home can reduce air conditioning needs by 10 to 50% (U.S. EPA, 1992). These trees also sequester C from the atmosphere. During winter, vegetation can slow wind speeds and help insulate structures from cold winds, reducing winter heating needs by from 4 to 22% (DeWalle, 1978), and trees planted in rural areas as windbreaks and shelterbelts can reduce the energy expended in snow control and removal (Brandle *et al*, 1992).

An urban energy-saving tree can be 4 to 15 times as effective at "managing" atmospheric CO_2 as a rural tree, which primarily is involved only in C sequestration (U.S. EPA, 1992). A guidebook has been published on strategic landscaping for energy conservation, describing tree planting and surface albedo modification (U.S. EPA, 1992). However, there are limitations to urban tree planting for energy conservation purposes. Little energy is used for cooling in many parts of the world and even in many parts of the U.S. Proper maintenance of urban trees is difficult and urban trees live notoriously short lives, succumbing to many different stresses such as drought, vandalism and urban air pollution. The U.S. electric utility industry spends approximately 1.5×10^9 annually and considerable energy clearing trees, which are the number one cause of electricity outages, away from power lines. Still, in many situations, urban tree planting makes environmental and economic sense.

4.2   BIOMASS ENERGY PRODUCTION

Wood or other biomass can be turned into a C offset through its conversion to energy if it is used in lieu of fossil fuel. Net CO_2 emissions are practically zero for a system where CO_2 released during biomass combustion is simultaneously sequestered by the next energy crop. Fossil fuel inputs to facilitate such an essentially closed system are very low.

Biomass power plants (not including municipal solid waste) in the U.S. have a total capacity of about 6000 MW, with utilities owning 300 MW and industrial cogenerators 5600 MW, mostly in the wood products industry (U.S. DOE, 1992). At present, the use of tree biomass as fuel by utilities is economical only for small power plants near the biomass source, with a general rule of thumb that the feedstock needs to be located within 80 to 120 km. A 100 MW power plant, operating at 35% efficiency, will require slightly more that 40,000 ha of land, or about 2% of the area within 80 km (assuming a feedstock yield of 24 green tC ha^{-1} yr^{-1}) (Turnbull, 1993). The Electric Power Research Institute states that biomass could realistically be used to supply 50,000 MW of electric capacity in 2010 and twice that in 2030 (Turnbull, 1993). Research and development activities are focusing on promising energy conversion technologies (such as whole tree burning, biomass gasification, and co-firing wood chips with coal), plus improving feedstock yield.

With advances in energy conversion and crop yield, short-rotation trees grown on a 6 to 12 yr rotations have been estimated to have the potential to reduce U.S. fossil fuel CO_2 emissions by 20% (Graham et al, 1992). Opportunity exists to almost double the energy produced per ha of woody crops due to both yield increases and improved conversion efficiency (Graham et al, 1992). Trexler (1991) estimates a maximum theoretical C emissions reduction benefit of 0.4 to 1.0 Gt C yr^{-1}. Modeling emphasizes that the net C offset can be small or nonexistent unless the wood harvest is used very efficiently to substitute for fossil fuels (Marland and Marland, 1992). One current obstacle is that boilers used to convert wood to electricity do so with lower efficiency than coal-fired boilers. This is largely because of the high water content of the wood and because much of the wood is used in small and inefficient boilers. A scheme for burning 5 to 18 m sections of whole trees, using waste heat for pre-combustion drying, is being evaluated and preliminary tests indicate that such a system may be competitive with medium-sized baseload coal-fired plants (Hughes, 1992). Biomass gasification could also generate considerable efficiency and produces pollution control advantages (Hall, 1991). Preliminary information indicates that co-firing of wood chips with coal is technically and economically feasible in many cases.

In developing nations, biomass energy development leading to fossil fuel displacement can yield C offset opportunities, while at the same time providing needed energy. Nations such as the U.S. may benefit from exporting such technology. Agricultural policy, specifically food crop subsidies and land requirements, will help determine the prospects for increased biomass energy production, both domestically and abroad. Biomass energy projects provide among the most quantifiable and verifiable C offsets available (Trexler and Meganck, 1993).

Turnbull (1993) describes many factors that could affect the future of biomass energy in the U.S., some of which are noted below. Stimuli which may spur development of biomass energy include consideration of environmental externalities and tax incentives such as under the Energy Policy Act of 1992. Secondary benefits of biomass energy include low SO_2 emissions; a smaller quantity of ash, which can be used as soil supplement; and improvement of some degraded lands. Barriers include lack of a current market for biomass feedstocks, agricultural subsidies and limited federal research funding.

5. Estimates of Potential for Carbon Management via Forestry

Dyson (1977) estimated that it would take 700 Mha of new tree plantations in the tropics to offset the approximately 5 Gt C emitted annually by fossil fuels (for reference, the U.S. and Australia are each approximately 800 Mha in size). Marland (1988) calculated that, on a national scale, 164 Mha of American sycamore (Platanus occidentalis) equivalents (a fast-growing species that can take up 7.5 tC ha^{-1} yr^{-1}) would be required to offset U.S. emissions from fossil fuel burning, an area only 16% less than that currently forested in the U.S. To offset direct emissions from a 1000 MW coal-fired electric plant operating at 38% thermal efficiency and 70% availability requires 190,000 ha of American sycamore, equivalent to a plantation with a radius of 24.7 km (Marland, 1989).

In the tropical regions, estimates of land availability range from 350 to 2000 Mha (Dixon *et al*, 1991; Trexler and Haugen, 1993). The upper value reflects an estimate based solely on physical land availability, while the lower value includes evaluation of demography, institutional and policy infrastructures, and past forestry efforts. The most detailed evaluations of C-forestry potential have been for the U.S. (e.g., Trexler, 1991; Moulton and Richards, 1990; Sampson, 1992). In the U.S. there are nearly 125 Mha of non-forested private land capable of supporting tree growth (USDA, 1990). The commitment of large amounts of land to new forest growth would ultimately affect food and other commodity prices. At issue is economic accessibility to land (Trexler, 1991).

6. Carbon Offset Policy Development

Carbon offsets are routinely discussed as a national and international environmental policy objective as part of a global warming mitigation strategy.

6.1   INTERNATIONAL LEVEL

The United Nations Conference on Environment and Development (UNCED), held in Rio de Janeiro in June 1992, was the forum for the signing of the Framework Convention on Climate Change by 154 countries. The Convention commits all parties to develop and make available national inventories of anthropogenic emissions by sources and removals by sinks of all greenhouse gases. The Convention also requires developed country parties, including Eastern European countries and the newly independent states, to adopt policies and implement measures to mitigate climate change by limiting emissions of greenhouse gases and enhancing sinks and reservoirs. The U.S. was the first industrialized nation to ratify the Climate Convention, in October 1992. Implementation and elaboration of the Convention's provisions will occur during the next several years through the work of the Intergovernmental Panel on Climate Change (IPCC) and the Convention's Intergovernmental Negotiating Committee (INC). An interesting provision of the Convention is its authorization of "joint implementation," which allows industrial countries to receive credit towards their own emissions reduction and stabilization objectives by reducing emissions in other countries. The Convention's joint implementation provisions are motivated by the cost-effectiveness associated with increased flexibility.

6.2 UNITED STATES

Policy measures to manage CO_2 emissions that are currently attracting attention include the addressing environmental externalities in electricity planning, market mechanisms such as C taxes, and regulatory intervention such as efficiency standards. In almost every case C offsets would be a likely component of any final strategy.

Section 1605 of the Energy Policy Act of 1992 ((Public Law 102-486), signed into law on October 24, 1992, directs the Secretary of the U.S. Department of Energy (DOE), through the Energy Information Administration (EIA) and in consultation with the EPA Administrator, to issue guidelines within 18 mo for the voluntary collection and reporting of information on greenhouse gas emissions and reductions by entities for 1987 to 1990 and subsequent years. The list of "annual reductions of greenhouse gas emissions and C fixation achieved through any measures" includes fuel switching, forest management practices, tree planting, use of renewable energy and other actions. Section 1605 is silent on the issue of whether actions taken overseas may be counted as reductions. Since the Framework Convention provides for joint implementation and reporting in pursuing commitments, logically, international reductions should be permissible.

Offsets are not the focus of as much dialogue and debate at the state level, although efforts to integrate global warming concerns into electricity planning and pricing are becoming increasingly common. In one case, a proposed C tax in the state of Montana specifically gives CO_2 emitters the option of offsetting up to 25% of total emissions in order to reduce the amount of tax that would be due. In the Pacific Northwest, Bonneville Power Administration is beginning to require CO_2 mitigation as part of its power purchase contracts. The first such contract, in which the power supplier agreed to spend $\$1 \times 10^6$ on C offsets, was negotiated with Tenaska, Inc. as part of an agreement for the construction of a 250 MW natural gas fired facility in Washington state.

7. Carbon Offsets Experience to Date

Formal policy development on C offsets is in some ways lagging behind practical steps being taken by both private and governmental organizations. In 1988, Applied Energy Services Corporation of Arlington, VA announced the first C offset project to offset the CO_2 C emissions from a power plant. Since then a number of offset efforts have been initiated and more are being considered.

7.1   GOVERNMENTAL LEVEL

The U.S. "America the Beautiful" program announced in early 1990 was one of the first governmental programs publicized partially on the basis of its global warming mitigation effects. The U.S. Forest Service was charged with coordinating and funding the planting of 1×10^9 trees yr^{-1} on private lands for at least 10 yr. According to various estimates, the program would offset the emission of from 16 to 65×10^6 tC annually after 10 yr of planting -- roughly equivalent to 1 to 5% of current U.S. CO_2 emissions from fossil fuel combustion.

Announced in January of 1993, the Forests for the Future Initiative (FFI) will result in the negotiation and implementation of C offset projects in a number of countries including Mexico, Russia, Guatemala, Indonesia and Papua New Guinea (The White House, 1993). The FFI is intended to establish the infrastructures through which such offsets can be pursued without U.S. government assistance.

The Global Environmental Facility (GEF), a 3 yr experiment that provides grants for investment projects, technical assistance and research, is being implemented by the United Nations Development Programme, the United Nations Environment Programme and the World Bank (United Nations Development Program, 1993). The GEF has proposed through the International Finance Corporation to fund plantation establishment and management in part to sequester C by a large wood products company in Ecuador (Dixon *et al*, 1993). Areas of pasture, mixed agriculture and secondary forest would be purchased from landowners. Carbon sequestration is expected at a cost of $3 to $4 tC^{-1} (Dixon *et al*, 1993).

7.2   PRIVATE SECTOR

American Forests' (formerly the American Forestry Association) Global ReLeaf program focuses on education and tree planting in urban environments. A key goal of this program is the reduction of energy use by reducing the urban heat island effect. Specific programs have been tailored to individual sponsor's objectives and include educational and informational materials, plus programs focusing on tree planting, care and history. Funds are provided for the best planting and care proposals, which can be specified to be within the sponsor's service area. Hundreds of local urban forestry programs are now underway around the country. Most projects are not being funded for C offsets, though. In 1992, the Edison Electric Institute (EEI) formed a partnership with American Forests to promote the program.

A joint endeavor between American Forests and the EPA called Cool Communities also has the primary purpose of shading dwellings and reducing energy use for air conditioning, which in turn reduces CO_2 emissions. A detailed guidebook has been published on strategic landscaping for energy conservation (U.S. EPA, 1992). In addition, there is a component related to "lightening" the surface colors of buildings and pavement. There are currently seven model communities in the Cool Communities program and four more being considered. In the first year of the program opportunities for planting trees and lightening surface colors are identified. Years 2 through 4 are for planting and resurfacing. Year 5 is to measure energy savings and other benefits. Investor-owned utilities are involved with these programs, including Potomac Edison, West Penn Power, Tucson Electric, Northeast Utilities, Boston Edison, and Georgia Power, and three public power companies are partners.

In southwest Oregon, PacifiCorp, an electric utility company, is experimenting with the concept of reforestation on rural pasture land. This pilot project covers about 140 ha of private, non-industrial land that will be planted with Douglas fir, a high C storage species. PacifiCorp has developed a cost-share program that is premised on an agreement with landowners to maintain trees for 45 to 65 yr.

Most C-offset activity undertaken by U.S. companies is being conducted overseas due to perceived cost advantages. In 1988, Applied Energy Services Corporation (AES) of Arlington, VA announced the first C offset project, a social forestry project in Guatemala intended to offset the C emissions of a 180 MW coal-fired circulating fluidized bed boiler in Connecticut over its 40 yr lifetime (Trexler et al, 1989). The forestry components of the effort consist of tree planting in woodlots and agroforestry applications, increasing biomass yields through soil conservation techniques, conservation of forest biomass through non-planting activities (such as fire prevention efforts), and coordination of technical educational programs to assure that the implemented techniques will be applied beyond the life of the project. Farmers on small holdings are being trained and supported in planting trees to halt erosion, increase agricultural productivity, and increase wood supplies. The project involves nearly 40,000 farm families and the planting of approximately 52×10^6 trees over a 10 yr period. Additionally, many more trees will likely be planted as the project activities become self-sustaining. Some 12,140 ha of new woodlots will be created while agroforestry techniques will affect 60,700 ha of agricultural land. The AES funding level of 2×10^6 has been supplemented with CARE fund-raising efforts, grants from the government of Guatemala and in-kind services through the Peace Corps and the U.S. Agency for International Development (AID).

AES is involved in funding establishment of a nature reserve for C storage in Paraguay, in which a forest parcel targeted for sale to timber interests and containing endangered species has been purchased and is being managed for long-term conservation by the Nature Conservancy. The project was designed to offset the CO_2 emissions of the AES Barbers Point 180 MW coal-fired facility on Oahu, Hawaii. The International Finance Corporation of the World Bank agreed to sell the 58,000 ha area, one of the last large areas of dense humid subtropical forest in Latin America, significantly below its market value of $5 to $7 $x10^6$ to assure its preservation. Funding for the $2 $x10^6$ purchase was provided by U.S. AID and AES. AES has also just announced the funding of several forest management and preservation projects in South America to offset CO_2 emissions from a 320-MW power plant in Oklahoma.

In Malaysia, the New England Electric System (NEES) and Innoprise, a leading Malaysian forests products corporation, are in the early stages of an offset project designed to reduce CO_2 released during the logging process on 1,400 ha. Elements of the project include improved siting of logging trails, directional felling of trees, and vine removal prior to harvest, all intended to decrease damage to undergrowth and unharvested trees during the logging process, thereby decreasing CO_2 released to the atmosphere, as well as facilitating regeneration of the forest. In year one (1992-3), Innoprise personnel will receive specialized training, map the areas to be harvested and acquire the necessary equipment. Year two will implement the improved harvesting practices while year three will be reserved for project evaluation. NEES companies will spend $450,000 on the 3 yr effort. Additionally, NEES will retain the services of the Rainforest Alliance, an environmental and conservation group, as part of an oversight team.

Tenaska Inc., an independent power producer located in Omaha, NE, plans to fund afforestation of up to 20,000 ha in Russia as part of a cooperative effort with Russian government and private organizations, and perhaps the U.S. EPA. The project is proposed for a region in which the soils are sandy and the targeted species are not unusually high in C storage capacity, but it should be effective due to the low cost of implementation and oversight. Initial efforts by the International Forestry Institute (Moscow) and the Russia Forest Service have demonstrated that the targeted natural grasslands are capable of supporting long-term tree growth. The cost estimate is $1 to 2 tC^{-1} (Dixon et al, 1993). The current state of the Russian economy creates a scenario in which international private investment is enthusiastically welcomed.

Internationally, the FACE (Forests Absorbing Carbon Dioxide Emissions) Foundation in the Netherlands was set up by the Dutch Electricity Generating Board to encourage C offset forestry (Dixon et al, 1993; FACE Foundation, 1991). The Foundation is planning to reforest 2,000 ha in Malaysia, offsetting the equivalent emissions of a 600 MW coal-fired power plant. The Foundation is considering additional efforts in Malaysia and other regions.

8. Carbon Offset Economics

The actual cost of forestry offsets depends on 1) the amount of land required for the offset and 2) the cost per unit of land area. Both of these can vary dramatically by region, species and forestry type. Costs per hectare can vary by orders of magnitude. Regeneration and social forestry might cost just dollars per hectare, while intensive plantations can cost as much as $1,200 ha^{-1} just to establish. Using current projects and cost estimates as a basis for analysis, high-quality C offsets can probably be obtained for under $5 t^{-1} on a non-discounted basis (Trexler, 1993). There is a tradeoff between cost and risk, with the cheaper but riskier projects often being found in the tropics, and the more secure but expensive projects often being found in temperate areas. If biotic mitigation mechanisms were to be used to offset 1 Gt C emissions yr^{-1} (10 to 15% of the current global total), the annual cost would likely range from $5 to 20 x$10^9$ (Trexler, 1993).

The potential impacts of a C tax or other emissions control strategy would be significant for C offset economics. With the implementation of a C regulatory regime and the designation of carbon's "value" as determined by a C tax, for example, the "cost" of forestry projects will rise. There is little doubt that C offsets will never be any less expensive than they are today. If a market begins to develop for C offsets, the cost of offsets is likely to increase.

9. Designing Forestry-Based Carbon Offset Projects

Given the tremendous range of potential project types and settings, a comprehensive review of each project must be undertaken, including evaluation of not only the details of the proposed project but its biological, institutional, and social setting. The following questions should be considered.

o **How reliable is the proposed project?** Reliability can be judged based on global experience with similar projects and the background and expertise of project proponents, as well as the prevailing social, economic, and political context in the country where the project is implemented. There has been a history of high failure rates for projects aimed at natural resource management in the tropics.

o **How measurable and verifiable is the proposed project?** Many projects that are desirable in light of other public policy criteria (e.g., forest protection for biodiversity conservation) may prove difficult to measure and verify for C sequestration purposes.

o **How cost-effective is the proposed project?** The "real" cost of C forestry projects will depend significantly upon their credibility, reliability, and verifiability. "Cheap" projects that sacrifice any of these attributes could prove to be a false economy.

o **Is the project expandable and reproducible?** These qualities are desirable in that proven C sequestration concepts can be capitalized upon in the most cost-effective and timely manner feasible.

o **How do forestry offset options compare to non-forestry options?** A utility might choose to stay in the energy sector rather than straying into the forestry sector, usually a relative unknown.

o **How will offsets be treated in the prevailing regulatory environment?** In the absence of a federal CO_2 control regime, the treatment of CO_2 offsets may vary from state to state with respect to cost recovery or ratebasing. Regulated industries would be well advised to consult with their counsel and rate regulation staff and, if possible, obtain the approval of regulatory commissions before embarking on significant voluntary programs, particularly if these programs are implemented overseas.

o **What tradeoff between cost and risk is the organization willing to make?** The price range for C offset opportunities varies widely, due partly to an equally wide range of risks.

o **What value is given to ancillary benefits?** Each offset strategy offers secondary benefits beyond direct C storage (e.g., tropical forestry usually provides resource benefits such as protection of habitat and biodiversity, reduction in soil erosion, flood control and water quality improvement). On the other hand, domestic forestry activities might accrue greater political and public relations benefits.

10. Conclusions

In the near future major sources could face the imposition of some form of regulatory regime affecting emissions of CO_2 and other greenhouse gases. Carbon offsets, properly documented and monitored, should be a component of any such regime. Understanding the issues involved in C offset forestry is important as we prepare ourselves for a possible shift in pollution control priorities and methods in the next century.

Acknowledgments

The authors thank Robert Beck, William Fang, Edward Glenn, C.V. Mathai and Billy McCormac for supplying comments. This paper will also be presented at a September 1993 conference in Hamburg, Germany sponsored by the International Union of Producers and Distributors of Electrical Energy (UNIPEDE) and the International Energy Agency.

References

Abelson, P.H.: 1992, *Science* **257**, 9.

Brandle, J.R., Wardle, T.D. and Bratton, G.F.: 1992, Increasing Tree Planting in Shelterbelts, in: Sampson, R.N. and Hair, D. (eds), *Forests and Global Change, Volume 1: Opportunities for Increasing Forest Cover*, American Forests, Washington, D.C., pp. 157-176.

Brown, S., Lugo, A. and Iverson, L.R.: 1992, *Water, Air, and Soil Poll.* **64**, 139.

Council for Agricultural Sciences and Technology: 1992, Preparing U.S. Agriculture for Global Climate Change, Ames, Iowa.

Detwiler, R.P. and Hall, C.A.S.: 1988, *Science* **239**, 42.

DeWalle, D.R.: 1978, Manipulating Urban Vegetation for Residential Energy Conservation, Proceedings of First National Urban Forestry Conference, U.S. Department of Agriculture, Forest Service, Washington, D.C., pp. 267-83.

Dixon, R.K. and Turner, D.P.: 1991, *Environmental Pollution* **72**, 245.

Dixon, R.K., Andrasko, K.J., Sussman, F.A., Trexler, M.C., Vinson, T.S.: 1993, this volume.

Dixon, R.K., Schroeder, P.E., Winjum, J.K.(eds): 1991, Assessment of Promising Forest Management Practices and Technologies for Enhancing the Conservation and Sequestration of Atmospheric Carbon and Their Costs at the Site Level, EPA/600/3-91/067, U.S. Environmental Protection Agency, Environmental Research Laboratory, Corvallis, OR.

Dyson, F.J.: 1977, *Energy* **2**, 287.

Face Foundation: 1991, Forests Absorbing Carbon Dioxide Emission (FACE), Arnhem, Netherlands, informational brochure.

Fluor Daniel: 1991, Energy and economic evaluation of CO_2 removal from fossil-fuel fired power plants, IE-7365, Electric Power Research Institute, Palo Alto, CA.

Glenn, E.P., Pitelka, L.F. and Olsen, M.W.: 1992a, *Water, Air, and Soil Poll.* **64**, 251.

Glenn, E.P., Hodges, C.N., Leith, H., Pielke, R. and Pitelka, L.: 1992b, *Environment* **34**, 40.

Graham, R.L., Wright, L.L. and Turhollow, A.F.: 1992, *Climatic Change* **22**, 223.

Hall, D.O.: 1991, *Energy Policy*, October, 711.

Houghton, J.T., Callander, B.A. and Varney, S.K. (eds): 1992, *Climatic Change 1992 - - the Supplementary Report to the IPCC Scientific Assessment, Intergovernmental Panel on Climate Change*, Report prepared for IPCC by Working Group I, Cambridge University Press, New York.

Hughes, E.: 1992, personal communication.

Johnson, D.W.: 1992, *Water, Air, and Soil Poll.* **64**, 83.

Marland, G.: 1988, The Prospect of Solving the CO_2 Problem Through Global Reforestation, DE-AC05-76OR00033, U.S. Department of Energy, Office of Energy Research.

Marland, G.: 1989, The Role of Forests in Addressing the CO_2 Greenhouse, in: White, J.C. (ed), *Global Climate Change Linkages -- Acid Rain, Air Quality and Stratospheric Ozone*, Elsevier, New York, pp. 199-212.

Marland, G. and Marland, S.: 1992, *Water, Air, and Soil Poll.* **64**, 181.

Mooney, H.A., Drake, B.G., Luxmoore, R.J., Oechel, W.C. and Pitelka, L.F.: 1991, *BioScience* **41**, 96.

Moulton, R.J. and Richards, K.R.: 1990, Costs of Sequestering Carbon Through Tree Planting and Forest Management in the United States, GTR WO-58, United States Department of Agriculture, Forest Service, Washington, D.C.

Norby, R.J., Gunderson, C.A., Wullschleger, S.D., O'Neill, E.G. and McCracken, M.K.: 1992, *Nature* **357**, 322.

Postel, S. and Heise, L.: 1988, Reforesting the Earth, Worldwatch Paper 83, Worldwatch Institute, Washington, DC.

Sampson, R.N.: 1992, *Water, Air, and Soil Poll.* **64**: 157.

Sedjo, R.A.: 1992, *Ambio* **21**, 274.

Sedjo, R.A. and Lyon, K.S.: 1990, *The Long-Term Adequacy of World Timber Supply*, Resources for the Future, John Hopkins University Press, Washington, D.C., pp. 30-31.

Strain, B.R. and Thomas, R.B.: 1992, *Water, Air, and Soil Poll.* **64**, 45.

Trexler, M.C.: 1991, Minding the Carbon Store: Weighing U.S. Forestry Strategies to Slow Global Warming, World Resources Institute, Washington, D.C.

Trexler, M.C. and Haugen, C.A.: 1993 forthcoming, Keeping it Green: Global Warming Mitigation Through Tropical Forestry, World Resources Institute, Washington, D.C.

Trexler, M.C. and Meganck, R.: 1993, *Climate Research*, in press.

Trexler, M.C., Faeth, P.E. and Kramer, J.M.: 1989, Forestry as a Response to Global Warming: An Analysis of the Guatemala Agroforestry and Carbon Sequestration Project, World Resources Institute, Washington, D.C.

Turnbull, J.: 1993, Strategies for Achieving a Sustainable, Clean and Cost-Effective Biomass Resource, Electric Power Research Institute, Palo Alto, CA.

USDA. (United States Department of Agriculture): 1990, America the Beautiful: National Tree Planting Initiative.

U.S. DOE (United States Department of Energy): 1992, Conservation and Renewable Energy Technologies for Utilities, DOE/CH10093-86, Washington, D.C.

U.S. EPA (United States Environmental Protection Agency): 1992, Cooling our Communities: A Guidebook on Tree Planting and Light-Colored Surfacing, 22P-2001, Washington, D.C.

United Nations Development Programme: 1993, Global Environment Facility, GEF Technical Advisory Division, New York.

The White House: 1993, Office of the Press Secretary, Forests for the Future Launched, press release, Washington, D.C.

FOREST SECTOR CARBON OFFSET PROJECTS: NEAR-TERM OPPORTUNITIES TO MITIGATE GREENHOUSE GAS EMISSIONS

R.K. DIXON
U.S. Environmental Protection Agency,
Environmental Research Laboratory
200 SW 35th Street
Corvallis, OR 97333 USA

K.J. ANDRASKO
U.S. Environmental Protection Agency,
Climate Change Division, Office of Policy Analysis
Washington, D.C. 20460 USA

F.G. SUSSMAN
ICF, Incorporated
850 K Street
Washington, D.C. 20006 USA

M.A. LAVINSON
ICF, Incorporated
850 K Street
Washington, D.C. 20006 USA

M.C. TREXLER
Trexler and Associates
1131 SE River Forest Road
Oak Grove, OR 97267 USA

T.S. VINSON
Oregon State University
Department of Civil Engineering
Corvallis, OR 97331 USA

Abstract. The Framework Convention on Climate Change separately recognizes sources and sinks of greenhouse gases and provides incentives to establish C offset projects to help meet the goal of stabilizing emissions. Forest systems provide multiple opportunities to offset or stabilize greenhouse emissions through a reduction in deforestation (C sources), expansion of existing forests (CO_2 sinks) or production of biofuels (offset fossil fuel combustion). Attributes and dimensions of eight forest-sector C offset projects, established over the past three years, were examined. The projects, mostly established or sponsored by US or European electric utilities, propose to conserve/sequester over 30×10^6 Mg C in forest systems at an initial cost of $1 to 30 Mg C. Given the relative novelty and complexity of forest sector C offset projects, a number of biogeochemical, institutional, socio-economic, monitoring, and regulatory issues merit analysis before the long-term potential and cost effectiveness of this greenhouse gas stabilization approach can be determined.

Water, Air, and Soil Pollution **70**: 561–577, 1993.
© 1993 *Kluwer Academic Publishers.*

1. Introduction

The June 1992 United Nations Conference on Environment and Development (UNCED) in Brazil focused attention on global forests and their role in the global C cycle. Establishment of the UNCED Framework Convention on Climate Change, and the separate non-binding Forest Principles also promulgated during UNCED, emphasize the potential role of C sinks to mitigate the accumulation of greenhouse gases (GHG) in the atmosphere (UNCED, 1992a, b).

The Framework Convention on Climate Change includes provisions to protect and enhance terrestrial sinks that reduce the accumulation of CO_2 and other GHG in the atmosphere (UNCED, 1992b). Carbon dioxide emissions result primarily from combustion of fossil fuels, but forest degradation and deforestation are also sources (Houghton et al., 1992). Carbon dioxide is responsible for more than one-half the radiative forcing associated with the greenhouse effect (Schneider, 1989). Stabilization of atmospheric CO_2 concentrations could be partially achieved by limiting forest cutting and burning (reduce C source), and replacing millions of hectares of forest that have been destroyed (expand CO_2 sink). Up to 2×10^9 ha of deforested or degraded land are technically suitable worldwide for forestation (expansion of forest area) or other improved land management techniques using tree cover (Dixon et al., 1993a,b; Trexler and Haugen, 1993; Winjum et al., 1992).

Under the terms of the Framework Convention on Climate Change, developed countries will strive to reduce their GHG emissions to 1990 levels (UNCED, 1992b). The terms of the agreement allow participating countries to jointly implement measures that help meet the goals of the Convention. Joint implementation potentially allows developed countries to reduce GHG emissions using C offsets within other countries and receive credit toward their GHG emission obligations under the Convention (UNCED, 1992b). Moreover, the Framework Convention provides the international basis for developing a market-based approach to meet GHG reduction obligations. The benefits of this approach include generating resources for developing countries to implement environmentally beneficial projects, providing cost-effective alternatives for industrialized nations to meet obligations, and creating a vested interest on the part of the party providing the funding, in achieving the desired environmental results (Swisher and Masters, 1992). At this time, however, an operational definition of joint implementation has not been developed and an array of issues must be resolved before countries undertake joint implementation projects (Trexler, 1993).

Many activities reduce or offset GHG emissions. These include forest sector projects that stimulate sequestration of CO_2 or conserve terrestrial C, improvements in technology related to fossil fuel combustion, or switching to alternative fuels such as biomass (Sampson et al., 1993; NAS, 1991). The estimated financial costs of GHG emission reduction or offsets range from a net positive benefit to costs of up to $100 Mg C (Rubin et al., 1992). Forest systems provide multiple GHG offset opportunities since they can provide relatively low-cost C offset credits through reduction in deforestation, expansion of C sinks, or production of biofuels (Dixon et al., 1993a,b; Winjum et al., 1992; Graham et al., 1992; Schroeder and Ladd, 1991). There is a large potential for forestation, reforestation and improvement of degraded land to remove CO_2 from the atmosphere and the socio-economic benefits for developing country forest C offset programs may be great (Dixon et al., 1993a,b). The demand for forest sector CO_2 offsets will depend on the ultimate commitments of developed nations to reduce emissions and on the pressures created by economic growth (Bekkering, 1992). In the US, promulgation of Title 16 of the Energy Policy Act of 1992 provides the basis for voluntary C sequestration and conservation activities by producers of GHG, especially CO_2 emitters. The Act requires the Energy Information Administration (EIA) to develop guidelines for, and implement, a database of GHG offsets voluntarily undertaken and submitted to the EIA by participants.

This report is intended to: 1) review national and global opportunities to manage forest systems as C sinks, 2) examine the status of representative forest-sector C offset projects in six nations, and 3) consider future opportunities for, and constraints on, forest-sector C offset activities under the terms of the Framework Convention for Climate Change.

2. Forest systems as carbon sinks

Forest ecosystems can be managed to assimilate CO_2 via photosynthesis and temporarily store C in biomass and soils (Winjum et al., 1992; Brown et al., 1992; Trexler et al., 1992; Harmon et al., 1990). Forest and agroforest establishment and management practices can be grouped by two major functions: 1) maintain or improve existing sinks and stores of GHG; and , 2) expand forest areas that can serve as sinks of CO_2. Several recent studies provide the basis for an overview of the potential for C conservation and sequestration (e.g., Sampson et al., 1993: Trexler and Haugen, 1993; Dixon et al., 1993b; Brown et al., 1992; Sedjo, 1992; Schroeder and Ladd, 1991).

2.1 MAINTENANCE OF EXISTING FOREST CARBON SINKS

Stopping deforestation in tropical latitudes could, in principle, conserve up to 1.2 to 2.2 Pg C annually, but actual implementation of this goal remains elusive (Houghton et al., 1992; Sharma, 1992). Although efforts to slow deforestation and degradation in tropical latitudes have met with mixed success (Brown et al., 1992), forest destruction in some regions such as Brazil's Amazon basin dropped 20% in 1991, relative to deforestation rates of the 1980s. Replacement of shifting agriculture by 1 ha of sustainable agroforestry could potentially offset 5 to 20 ha of deforestation and conserve existing C reservoirs (Sanchez and Benites, 1987). Based on the direct cost of providing economic incentives to practice sustainable forest management in tropical latitudes, C conservation can be achieved for $3 to 60 Mg, with a median cost of $10 Mg (Table 1). For example the Forest Village Program in Thailand, which provides incentives for shifting cultivators to establish agroforest systems and reforest degraded lands has conserved or sequestered approximately 1 Pg C over the past 20 yr (Boonkird et al., 1991). The direct and indirect benefits of sustainable forest or agroforest management in human welfare and protection of biological resources and soil and water systems should also be considered (Sharma, 1992).

Globally, temperate and boreal forest fires contribute 0.5 to 1.0 Pg C to the atmosphere annually. Within Russia, 4 to 8 x 10^6 ha of boreal forest burn annually, contributing 0.2 Pg of direct and indirect C emissions to the atmosphere (Krankina and Dixon, 1993). Over 40% of the boreal forests in the former Soviet Union have no fire monitoring or protection system. In Russia, it is estimated that fire management and silvicultural practices could be employed to temporarily conserve C at $1 to 3 Mg by reducing the occurrence or magnitude of wildfire (Table 1).

Forest soils are significant reservoirs of C and conservation practices could be employed to reduce GHG emissions. Tillage and management can result in significant changes in soil C (Gaston et al., 1993; Johnson, 1992). Management practices to conserve forest soil C include: 1) forestation to reduce erosion, 2) maintaining or improving soil fertility, 3) concentrating tropical agriculture and reducing shifting agriculture, 4) removing marginal lands from agricultural production, and 5) retaining forest litter and debris after silvicultural or logging activities (Dixon et al., 1993a,b).

Table 1   Initial costs of C sequestration and conservation in boreal, temperate and tropical forest systems and estimates of land technically suitable and socio-politically available to expand forest area.

Source	Geographic Scope	Land[1] Technically Suitable (x 10⁶ ha)	Land[2] Socio-Politically Available (x 10⁶ ha)	Carbon Sequestration Option[3]				
				Tropical ($Mg C)		Temperate Plantation ($Mg C)	Boreal	
				Agroforestry	Plantations		Plantation	Protection
Andrasko, 1991	Tropical	865	300	3 - 5	3 - 6	0 - 20	- - -	- - -
Dixon et al., 1993a,b	Global	1,600	570	4 - 16	6 - 60	2 - 50	3 - 27	1 - 4
Krankina and Dixon, 1993	Russia	- - -	- - -	- - -	- - -	1-7	1-8	1 - 3
Houghton et al., 1991	Tropical latitudes	3,125	- - -	3 - 12	4 - 37	- - -	- - -	- - -
Trexler et al., 1992	Tropical latitudes	- - -	462	- - -	- - -	- - -	- - -	- - -

1   Land which when considering edaphic and climatic factors will support forest systems.
2   Land, which is technically suitable, and available for establishment of agroforest and forest systems given prevailing social, economic and political conditions.
3   Estimates consider initial (3 to 5 yrs after establishment) direct costs of seedling production, site preparation, planting and early tending. Long-term direct costs of forest management indirect costs nor the full range of benefits (products, goods, services) were considered.

Table 2. Attributes of forest sector C offset projects at eight locations worldwide.

Project Location	U.S.A Utah	U.S.A. Oregon	Russia	Paraguay	Malaysia	Malaysia	Guatemala	Ecuador
Principal Sponsor	PacifiCorp	PacifiCorp	Tenaska	AES Barbers Point	FACE	NEES	AES Thames	GEF
Project	Urban Tree Planting	Sustainable Forestry	Forestation	Preservation/ Sustainable Agroforestry	Reforestation	Sustainable Forestry	Sustainable Agroforestry	Forestation/ Forest Protection
Total Costs	$ 100,000 /yr	$ 100,000 /yr	$500,000 - ?	$ 2-5 million	$ 1.3 million	$ 450,000	$14 million	--
Utility Contribution	$ 100,000 /yr	$ 100,000 /yr	$500,000-?	$ 2 million	$ 1.3 million	$ 450,000	$ 2 million	--
Other Participants	TreeUtah	Oregon Department of Forestry	--	Nature Conservancy, FMB Foundation	Innoprise	Rain Forest Alliance, COPEC	CARE, Guatemala Government	Durini Group
C Sequestered (Mg)	N/A	64,750 Mg/yr	500,000-?	13.1 million	--	300,000 - 600,000	15.5 - 58 million	375,000
Project Size (ha)	1,000 trees	140	20,000-?	56,800	--	1,400	52 x 10⁶ trees	6,000
$/Mg C Sequestered	$15 - $30	$5	$1 - $2	$1.5	--	less than $2	$9 overall $1 AES	$ 3 - $4
Project Duration (yr)	50[1]	65[1]	25	30	--	10	10	--
Local Benefits	Reduce cooling and heating needs of the community, contribute to the aesthetics of the area.	Assist landowners in productive land management activities.	Habitate improvement; Soil and water conservation and transition to regional market economy.	Create a watershed, promote biodiversity, create sustainable agroforestry opportunities for inhabitants, promote recreation and eco-tourism opportunities.	Promote biomass and soil conservation, landscape protection.	Train local inhabitants in sustainable logging activities, preserve non-harvested trees, improve water quality, maintain biodiversity, reduce soil erosion.	Promote soil and biomass conservation, develop sustainable forestry groups to protect, plant and manage trees, establish a fund to promote continuing agroforestry activities	Foster local forest management, preserve forests

[1] The duration of the project is unknown. However, trees planted cannot be harvested for 65 years.
The duration of the project is unknown. However, PacifiCorp estimated that strategically positioned trees could substantially reduce cooling load needs over the course of 50 years, thereby offsetting fossil fuel emissions.

2.2 EXPANSION OF FOREST AREAS TO SERVE AS CARBON SINKS

Several promising forest and agroforest management practices could be employed to expand forest area and stimulate atmospheric CO_2 sequestration (Table 1). Afforestation of non-forest lands and reforestation of cut-over or abandoned forest lands are promising practices across boreal, temperate and tropical latitudes. These practices can result in establishment of plantations, agroforest systems, or other forest cultures. Although C sequestration can be achieved via forestation for $1 to 60 Mg with a median cost of less than $10 Mg, these estimates vary and regional opportunities and constraints should be carefully considered. Applying intermediate silvicultural practices can stimulate C sequestration for $1 to 15 Mg (data not shown). Because of lower labor, land-rent, transportation and supply costs the near-term costs of C sequestration in forest systems are less than $5 Mg in developing nations and those countries with an economy in transition (Krankina and Dixon, 1993).

Estimates of technically suitable land available to expand forest area vary, but up to 2×10^9 ha may, in principle, be available (Table 1). Within tropical latitudes, 1500 to 1800 $\times 10^6$ ha may be suitable for establishing agroforest or forest systems, including 217, 830, 503 and 339 $\times 10^6$ ha of deforested watersheds, degraded drylands, forest fallow, and logged rainforests, respectively (Houghton *et al.*, 1991). Other reports suggest that the total land available for forest expansion is less than 500 million ha, based on demographic, socio-economic and political factors (Trexler *et al.*, 1992). A recent analysis of European land-use trends reveals that forest area and net C sequestration are increasing (Kauppi *et al.*, 1992). Within Russia, it is estimated 100×10^6 ha may be available for forestation (eg, plantations, shelterbelts) and biomass productivity of another 200 to 300×10^6 ha could be improved via thinning of stands and soil drainage (Krankina and Dixon, 1993). In the US, the Conservation Reserve Program revealed that environmental goals, such as soil conservation and C sequestration, can be achieved by subsidizing forestation and other practices on marginal and degraded crop lands. Over 1.5×10^6 ha of former U.S. crop land were converted to forests at a total land rent cost of 1.7×10^9 over approximately 10 yr (Parks, 1992).

3. Current status of forest-sector carbon offset projects

Six international and two US GHG offset projects are considered in this assessment (Table 2). Of the international projects, the AES Corporation is sponsoring two, one in Guatemala and one in Paraguay, and the New England Electric System (NEES) and the Dutch Electricity Generation Board are sponsoring projects in Malaysia. PacifiCorp has established pilot projects in Oregon and Utah, USA. Tenaska, in cooperation with several government and non-government organizations, is planning a project in Russia. The Global Environmental Facility (GEF) is considering one or more projects in Ecuador. These projects may include both forest and agroforest establishment and management activities, depending on the opportunities and constraints of participants. Despite differences in overall design, project objectives tend to be similar and include: (1) C conservation and sequestration potential, (2) cost-effectiveness, (3) potential for C credits under joint implementation or the 1992 US Energy Policy Act process, (4) sustainability, (5) local participation, and (6) ancillary benefits not associated with C storage.

3.1 AES THAMES GUATEMALA PROJECT

AES Thames, a subsidiary of AES Corporation, is investing $2 x 10^6 to establish agroforest and woodlot systems in Guatemala to help offset CO_2 emissions from its 180 MW coal-fired power plant (Trexler *et al.,* 1989). In consultation with the World Resources Institute (WRI), AES developed a program linking GHG mitigation with the goal of sustainable economic development in the agroforest sector, pioneering the forest-sector C offset concept. Through a grant to CARE, an international relief and development organization, the AES project is intended to assist 40,000 farmers in Guatemala to plant more than 52 x 10^6 trees over a 10 yr period. By managing the new trees for farm and community use, the project is expected to help reduce deforestation pressures in both the highlands and the eastern lowlands of Guatemala. Sequestering atmospheric CO_2 will be achieved by establishing woodlots, plantations, and agroforestry systems, and conserving C will be fostered by soil management, forest fire protection, and displacement of forest explotation pressures. The project will be implemented at the farm-level through technical extension services, by encouraging farmer cooperation and the formation of self-sustaining community organizations to establish, manage and protect agroforest and forest systems beyond the 10 yr investment period. The AES Thames project was originally expected to offset more than the 15 x 10^6 Mg C to be emitted during the expected 40 yr life of the Connecticut, US, power plant. Efforts are underway to model the C benefit of the project and to refine projected C sequestration estimates (Faeth *et al.,* 1993).

Analysis of options to implement a forest sector C offset project revealed the advantages of pursuing such a program within tropical latitudes rather than in temperate systems. These advantages included: (1) greater annual forest C accretion rates; (2) grant leveraging opportunities; and (3) ancillary forest and non-forest socio-economic benefits.

The estimated value of goods and services (including in-kind) to be provided in implementing the AES project is $14 x 10^6. The $2 x 10^6 contributed by AES will be used to establish an endowment that will fund project activities over the initial 10 yr. The government of Guatemala is committed to contributing at least $1.2 x 10^6 to the project, and CARE is contributing approximately $1.8 x 10^6 in matching funds. In-kind services from U.S. AID and the Peace Corps will make up the balance of the $14 x 10^6. No land purchases occur in the project, nor is it necessary to pay for the labor of planting and maintaining the trees, which will be located on farmer or community owned land.

The cost of the AES project is estimated to range from $1 to 9 Mg C depending on implementation costs considered. These cost estimates do not include land or labor costs nor flow of multiple benefits. Project selection was based on four criteria: 1) C sequestration potential up to 15 x 10^6 Mg C over 40 yr; 2) local participation and socio-economic benefits; 3) experience and capability of implementaing organization; and, 4) grant leveraging including debt swaps, block or match grants, food aid programs, and support of existing projects to reduce initial costs.

3.2 AES BARBERS POINT PARAGUAY PROJECT

To help offset CO_2 emissions from a planned coal-fired cogeneration facility, AES Barbers Point, a subsidiary of the AES Corporation, has committed $2 x 10^6 to The Nature Conservancy for land purchase and forest conservation. A Nature Conservancy partner, the Moises Bertoni Foundation (MBF), will invest the AES contribution plus another $3 x 10^6 in 56,800 ha of endangered tropical forest in Paraguay formerly owned by the World Bank. The forest land was originally used for commercial activities by a Paraguayan forest products company. The company defaulted on their loans and the land was forfeited to the World Bank as payment.

The long-term project goals are promoting sustainable agroforestry activities, preserving existing tropical forest, and creating a sustainable watershed management system on the Brazilian/Paraguayan border. Over the course of the next 30 yr, AES expects to offset approximately 13.1 x 10⁶ Mg C at an estimated cost of $1.5 Mg (Faeth et al., 1993). The land will be managed by MBF and preserved for recreational purposes, eco-tourism, and scientific research. Over the course of the project, MBF will also work with local communities to reintroduce native flora that were lost to logging activities.

3.3 DUTCH ELECTRICITY GENERATING BOARD FACE PROJECT

The Dutch Electricity Generating Board established the FACE (Forest Absorbing Carbon Dioxide Emissions) Foundation in 1991, with the aim of offsetting CO_2 emissions equivalent to those from a large coal-fired power station with a life span of 25 yr. The FACE Foundation has established a contract with Innoprise to conduct a pilot tree planting program on 2,000 ha in Danum Valley, Sabah, Malaysia. The project, which will be implemented over the next 3 yr, focuses on the regeneration of degraded forest stands. Initial project cost is $1.3 x 10⁶. If the pilot project is successful, FACE intends to rehabilitate 26,000 ha in Malaysia over the next 25 yr This venture with Innoprise is the first of several FACE forest-sector C offset projects planned worldwide. Other projects under consideration include rehabilitation of forests in the Czech Republic and Costa Rican national parks.

3.4 GLOBAL ENVIRONMENTAL FACILITY (GEF) ECUADOR PROJECT

The Global Environmental Facility (GEF) has four priority areas, one of which is global warming mitigation. In northwestern Ecuador, the GEF through the International Finance Corporation (IFC) proposes to fund plantation establishment and management by the Durini Group as a C sequestration project. The Durini Group, the largest wood products company in Ecuador, controls ownership of sawmill and wood veneer production capacity. At present, Durini companies harvest indigenous hardwood species on between 3,000 and 5,000 ha of primary forest purchased under contract from local landowners. The proposed project is intended to satisfy Durini timber needs using plantation lands. Durini already manages plantations in the area. Although existing plantations are primarily monocultures, Durini intends to research and implement multi-species, multi-age systems.

The GEF project would involve purchasing from local landowners approximately 6,000 ha of land that are currently in pasture, mixed agriculture, secondary forest, and small patches of remnant primary forest. Approximately 5,000 ha would be put under plantation cover (the balance would be preserved in its existing forest cover). An estimated 66 to 82 plots of 40 to 50 ha would be purchased. Colonists farms are usually valued at between $100 and $500 ha-1. Preliminary C sequestration resulting from the project is estimated at approximately 375,000 Mg. Costs of C sequestration are estimated at $3 to 4 Mg.

3.5 NEW ENGLAND ELECTRIC SYSTEM MALAYSIAN PROJECT

The New England Electric System has agreed to assist a state-run forest products enterprise in Sabah, Malaysia (Innoprise) improve rain forest logging techniques to preserve terrestrial C and, in turn, help offset CO_2 emissions from NEES power production facilities in the US. New England Electric estimates that the pilot project will offset between 3 to 6 x 10^5 Mg C at a cost of less than $2 Mg, while also preserving Malaysian forests.

For this pilot program, NEES provided Innoprise with $450,000 to implement improved forest harvesting techniques on 1,400 ha to accomplish the following goals: (1) preserve non-harvested trees and associated vegetation through better-planned logging trails, directional felling of trees, and removing vines before cutting; (2) improve watershed management while maintaining biodiversity and reducing soil erosion; and, (3) prevent release of GHG and preserve forest ecosystems.

Encompassed in the initial $450,000 investment are a series of fixed costs, including (1) contracting the Queensland Forest Service (an agency within the Queensland Department of Primary Industries) to implement forest management practices, (2) training Innoprise personnel to deal with forest management techniques, (3) cutting climbing vines, (4) assembling a team of Innoprise personnel to implement the project, and (5) improving harvesting practices (i.e., the incremental cost for loggers to do more careful cutting). Land costs were not a consideration in this project because the land is owned by the Sabah Malaysia government. Innoprise holds a 100 year concession. The Innoprise project will be monitored for NEES by an independent audit committee, which includes the Rain Forest Alliance, Forest Research Institute of Malaysia and the University of Florida. COPEC, a Los Angeles-based firm, served as a broker for NEES in arranging the financing and logistics of the project.

The Innoprise project is the first in a series of pilot programs planned by NEES to demonstrate the cost-effectiveness and feasibility of reducing or offsetting GHG. The forest sector C offset strategies are one component of the utility's long-term resource plan which seeks to reduce the utility's atmospheric CO_2 emissions up to 20% by the year 2000. The project involves three phases: (1) 1992-3: Innoprise will receive specialized training, map areas to be harvested, and acquire necessary equipment; (2) 1993-4: Innoprise will implement the new harvesting practices on 3,500 ha; (3) 1993-4: NEES, and the environmental audit committee will monitor compliance with forest management and harvest guidelines; and (4) 1993-95: the University of Florida will conduct an assessment of project C conservation and sequestration potential.

3.6 PACIFICORP PROJECTS

PacifiCorp, with assistance from Trexler and Associates, is considering three types of forest sector C offset projects: (1) domestic reforestation programs that assist landowners to establish trees on private non-industrial lands currently in pasture or other non-forest uses. (By providing funds through a cost-share program, the project takes advantage of existing infrastructure to identify property and assist landowners with forest land management); (2) domestic urban tree planting programs that provide C offsets by both conservating energy and directly sequestering C; and, (3) international reforestation projects that may provide offsets at low cost because of lower land costs, high productivity in selected regions, and significant ancillary benefits.

PacifiCorp evaluated project proposals based on the cost of C offsets, including opportunities to leverage internal and external resources. For pilot activities selected in 1992, projects within the area serviced by PacifiCorp were given preference because they allow a more direct link to company goals. Two pilot projects have been announced. The first is a: (1) rural reforestation project in Southern Oregon, USA, where PacifiCorp is working with the Oregon Department of Forestry to assist nonindustrial landowners in planting trees on private lands currently covered with non-forest vegetation. PacifiCorp will provide 75 to 100% of site preparation and tree planting costs and landowners will retain ownership of the forest. Under a PacifiCorp C offset agreement, landowners may not harvest trees for 45 to 65 yr depending on the cost-share selected. The initial project involves planting Douglas fir (*Pseudotsuga menziessi*) on approximately 140 ha in Southern Oregon. The Douglas fir stands will eventually sequester up to 450 Mg C ha. This estimate by PacifiCorp is based on empirical stand yield tables for the region (e.g., Harmon *et al.*, 1990). Using a 4% discount rate applied over a 65 yr period the cost of C sequestered is approximately $5 Mg C.

The second project is an urban forest program. PacifiCorp, in partnership with TreeUtah, a nonprofit organization, is planting trees in residential areas in Salt Lake City, Utah, US, to conserve energy and displace fossil fuel combustion. The pilot project will involve planting trees in at least three different neighborhoods around the Salt Lake Valley. These trees (approximately 1,000) will be sited around houses and multi-family dwellings to achieve maximum windbreak/shade effect. TreeUtah is taking the lead in selecting trees, organizing neighborhood volunteers and planting trees. PacifiCorp, in cooperation with TreeUtah, will monitor the success of the program and determine the C offset benefit. The cost of offsetting fossil fuel emissions via energy conservation in Salt Lake City is estimated to range from $15 to $30 Mg and is highly site specific (based on a 4% discount rate). Further biologic and economic modeling will be used to refine these estimates and take into account additional variables such as reduced winter heating load. Although it appears that urban trees may offset C at a higher cost than trees planted in a rural setting, these cost estimates are not based on site specific data and are likely to be revised (OTA, 1991). Urban shade trees also contribute to the company's expanding energy conservation effort.

3.7 TENASKA SARATOV RUSSIA PROJECT

Russian boreal forests represent one-fifth of the global forest land area and constitute approximately 50% of the world's conifer forests. An analysis of the Russian forest-sector C budget, based on an ecoregion approach, reveals that C sequestration is currently 0.5 Pg annually. The potential to manage forest (Krankina and Dixon, 1993) and agricultural systems (Gaston et al., 1993) to sequester and conserve additional C is highly significant. Thus, Tenaska, working in cooperation with Trexler and Associates, is developing plans to initiate one or more forest sector C offset projects at multiple locations.

The biological, operational and institutional opportunities to manage a Russian boreal forest as a C sink will be evaluated in the Saratov Oblast (region), which is located approximately 700 km southeast of Moscow. The total area of Saratov is 10 x 10^6 ha of which 570,000 ha or 5.7%, is forested. The proposed forest cover will also protect soils and watersheds. Initially it is planned to create 1400 ha of forest plantations on three sites in the Saratov region, as follows:

1st site: Lysogorskyi and Shirokokamyshenskyi Forest Management Units - reforestation of 900 ha of burned area. This is a special project to create a highly productive, ecologically valuable pine (*Pinus*) plantation.

2nd site: Ershovskyi Forest Management Units - 150 ha of water and soil erosion protecting forest plantations on the lands of Yeruslanskyi and Putk-Kommunizmu collective farms. The primary species will be oak (*Quercus*) and elm (*Ulmus*).

3rd site: Yershiovskyi Forest Management Units - Dergachevskyi Forest Melioration station - 350 ha of water erosion protecting forest plantations of the lands of Suvorov collective farm. Species to be established include elm and maple (*Acer*).

Ultimately, 20,000 ha of forests may be created in the Saratov and neighboring Volgogard region. The cost of the Tenaska Russian Project is estimated to range from $1 to 2 Mg C (Krankina and Dixon, 1993).

The project will be implemented in a partnership of US and Russian government and non-government organizations. Russian Federation participants include the Russian Forest Service, Ministry of Ecology, Institute of Market Problems, International Forestry Institute and the Dokuchyaev Soil Institute. U.S. participants will include Tenaska and Trexler and Associates. In addition, the U.S. Environmental Protection Agency in cooperation with Russian government agencies may assist forest sector C offset research and policy analysis activities. Project tasks include: (1) field demonstration and monitoring of forest establishment and management options for large-scale implementation, (2) analysis of project costs and benefits including development of policy options for large scale implementation, and (3) evaluation of institutional, legal and financial framework for C offset project in a transition economy.

4. Future opportunities and limitations of forest sector carbon offsets

The concept of sequestering and conserving C in boreal, temperate and tropical forest systems to mitigate the accumulation of greenhouse gasses in the atmosphere is relatively new (Rubin et al., 1992; Swisher and Masters, 1992; Andrasko, 1991). Moreover, national and international efforts to establish the role of forestry in CO_2 offsets and define a framework for project incentives, finance, guidelines, monitoring and compliance, are in their infancy (UNCED, 1992b). Given the relative novelty and complexity of forest sector C offsets and joint implementation, a number of biophysical, institutional, social and economic issues merit consideration (OTA, 1991). These uncertainties, which are discussed below, include: (1) defining the appropriate role for forest-sector CO_2 offsets in a national and international GHG mitigation offset program; (2) project selection and eligibility criteria to insure credibility; (3) appropriate guidelines for project implementation that are consistent and defensible; (4) quantifying and monitoring C sequestration potential; (5) evaluating the financial mechanisms and incentives that would be required to encourage utility participation in a voluntary or regulatory program; and, (6) compliance monitoring and enforcement of GHG stabilization efforts.

4.1 ROLE OF FOREST SECTOR IN CO_2 SEQUESTRATION

The transitional nature of forest-sector C sequestration and conservation is often stressed, for several reasons (Dixon et al., 1993a,b). First, the C that can be sequestered by any individual forest project is limited. Over time, trees cease to sequester large incremental amounts of C annually as forest stands approach equilibrium or are harvested and replanted. Further, in tropical latitudes, where tree C accretion is relatively large, it is difficult to guarantee the long-term success of forestry projects because of socio-economic factors (i.e., realizing timber, economic and C benefits for which the project was designed) (Bekkering, 1992; Trexler et al., 1992). Long term success depends both on the extent to which the project is integrated into the local cultural, economic, and physical environment, and on future socio-economic developments, such as population growth (Winjum et al., 1993; Gregerson et al., 1989). These factors are difficult to predict or fully account for in C offset project design.

There are considerable uncertainties associated with C sequestration, particularly in developing countries, including the risk of project failure (Swisher and Masters, 1992). These considerations suggest that the purpose of forest sector C offsets must be carefully defined in pursuing a national offset program. For example, it may be desirable to evaluate the merits of forest sector C offsets in comparison to direct energy-related emissions reductions (Sampson et al., 1993; Rubin et al., 1992). Alternatively, because not all projects conserve or sequester C with the same degree of certainty, it may be desirable to promote some projects rather than others, using selection and eligibility criteria. For example, bioenergy derived from agricultural or forest systems sequesters C as well as provides the opportunity to displace fossil fuels (Graham *et al.,* 1992).

4.2 PROJECT SELECTION AND ELIGIBILITY CRITERIA TO INSURE CREDIBILITY

Until more experience with forest sector C offset projects has accumulated and methods of C accounting and program implementation are generally recognized by both utilities and regulators, careful consideration must be given to what constitutes a C sequestration project. Given the relative novelty of C-based forestry and the tremendous range of potential project types and settings, the evaluation process cannot be reduced to a checklist. A more comprehensive approach is justified, including evaluation of not only the details of the proposed project, but its biological, institutional, and social setting. The following six criteria may be valuable for assessing the promise of a proposed C offset project.

4.2.1 *Credibility*

Credibility of a project's conceptual ability to modify otherwise prevailing C flows over a given period is a primary consideration. The link between a forestry practice and its impact on C flows is not always direct. In agroforestry, for example, the primary C benefit can be reflected in decreasing pressure on existing forest systems (which may be difficult to measure), thus contributing to preserving C stores, rather than significantly increasing long-term C storage on agricultural land itself (Dixon *et al.,* 1993b). The more complex the linkage the less credible a project might appear. Afforestation projects on degraded forest or non-forest lands (e.g., PacifiCorp project in Oregon, USA or Tenaska shelterbelt planting) offer conceptually simple C offset models, with few persistent analytic issues confounding efforts to quantify C benefits. Forest protection projects (e.g., AES in Paraguay) require demonstration of assumed deforestation pressures that would otherwise degrade or harvest the stands.

4.2.2 *Reliability*

Reliability should be based on global experience with similar projects and the background and expertise of project proponents, as well as, the prevailing social, economic, and political context in the country where the project is implemented. Given the high failure rates of projects aimed at economic development and natural resource management in developing countries, for example, project reliability in some regions may be a stumbling block.

4.2.3 *Measurability*

Measurability and verifiability of the claimed C sequestration credit is a key factor. Different project types will vary dramatically in the project implementor's ability to demonstrate land-use change and C loading. Many projects that are desirable in light of other public policy criteria, such as forest protection for biodiversity conservation, may prove difficult to measure and verify for C sequestration purposes. The complexity and multiple-component design of the AES Guatemala project, such as fire protection, establishment of agroforestry systems, and soil management required development of a systemtic tracking system. Regional (Harmon *et al.,* 1990) or national C budgets

(Kolchugina and Vinson 1993; Kauppi *et al.*, 1992) or dynamic landscape scale models (Faeth *et al.*, 1993) may offer promise to measure and verify C benefits. For example, dynamic analysis with the WRI LUCS model refined C benefits of AES Guatemala project from 16.3 to 58 x 10^6 Mg C (Faeth *et al.*, 1993).

4.2.4 *Cost-effectiveness*

Cost-effectiveness compared to alternative mechanisms for controlling net CO_2 emissions merits consideration. The "real" cost of C forestry projects will depend significantly upon their credibility, reliability, and verifiability (Winjum and Lewis 1993). "Cheap" projects that sacrifice any of these attributes could prove to be a false economy. Costs and benefits analysis of projects in this paper are preliminary and somewhat inconsistent because of data limitations and methodological issues (e.g., establishment *vs* full project costs, discount rates, labor and land rent assumptions, C *vs* non-C benefits).

4.2.5 *Expandability and reproducibility*

Expandability and reproducibility so that proven C sequestration concepts can be capitalized upon in the most cost-effective and timely manner feasible is a consideration in some projects. For example, the Tenaska project in Russian has a large potential for expansion in the Saratov and Volvograd regions. Forest conservation activities which maintain C in the terrestrial biosphere, such as those being developed by AES and NEES, have potential for implementation across the tropics.

4.2.6 *Opportunity*

Availability of window-of-opportunity and/or windfall benefits has played a prominent role in early project design. Many of the projects reviewed here were undertaken in part because of windfall project opportunities or benefits available within a short time period. These windows of opportunity often were key factors in site selection. Project examples include: (a) default on a World Bank loan in Paraguay, bringing a large forest tract suddenly on the market; (b) the devaluation of the Russian currency and reorganization of the Russian economy, creating forestation opportunities at a fraction of their cost 1 to 2 yr earlier; and (c) refinancing of the CARE Guatemala project, infrastructure and support commitments by Guatemala and US government agencies, and NGO contributions, all of which improved the cost-effectiveness and potential reliability of the AES/Guatemala project. In the future, C offset projects will likely be designed opportunistically until adequate private sector incentives are in place through national and international crediting institutions and processes.

4.3 GUIDELINES FOR PROJECT IMPLEMENTATION AND MONITORING

Although national and international programs or goals to sequester and conserve C have been announced, development of consistent and defensible project implementation and monitoring guidelines has lagged behind. National and international institutional and legal guidelines to assess validity and feasibility of projects are needed to certify C credits. For example, mechanism(s) for joint implementation of projects under the Framework Convention for Climate Change have not been specified or tested.

While measuring and monitoring energy-related emissions reduction credits is not always straightforward, measurement issues involved in forest sector C offsets are more complex (Birdsey, 1992; Eriksson, 1992). The manner in which credits are measured, (e.g., the type of C sequestration that can be credited) affects the type of offset projects that will be undertaken and the perceived cost-effectiveness of alternative projects. Measurement criteria should, therefore, be structured to minimize the risk associated with offset projects, and to ensure that incentives to participate and develop appropriate projects are provided to GHG producers. Four primary measurement design questions have been identified: (1) considering C credits for alternative projects; (2) establishing the baseline against which offsets are measured; (3) defining C offset ratios; and, (4) addressing ultimate fate questions.

At this stage, guidelines for estimating the full range of C sequestration benefits associated with alternative projects have not been developed. The issues are both conceptual and empirical. The AES Guatemala forest-sector CO_2 offset project paced the issue and emphasized several C benefits including: (1) net addition to standing C; (2) demand displacement; (3) utility of C harvested; (4) protection of standing C; and (5) soil C accretion. The treatment of C credits that accrue in later years and the complexity of "C accounting" could discourage the use of forest sector C offset projections, shift project design, and restrict selection opportunity.

At the same time, the long term nature of forestry projects may result in offset credits being issued for C sequestration that never actually occurs. Projects that seek credits for a current operation (rather than averaging future benefits) will tend to be readily verifiable. For example, the NEES Innoprise project in Malaysia seeks annual credit for C conservation resulting from alternative harvest practices without waiting for future C sequestration. Ultimate forest C fate is also important in determining sequestration potential. For example, if the wood grown is used for construction or for durable wood products, the amount of C sequestered over the long-term (e.g., 50 yr) would be greater than if it is burned or if the woody tissue lasts only a few years before decompostion (Graham *et al.*, 1992; Apps and Kurz, 1991).

4.4 FINANCIAL MECHANISMS AND INCENTIVES

Offset activities require investment by GHG producers to cover fixed and operating costs associated with the project as well as unexpected costs (Bekkering, 1992). These costs may include rental or purchase of land, purchase of trees and/or seedlings, planting costs, consultants' or brokers' fees, salaries for site-management personnel, third-party monitoring and evaluation and other costs that might arise during the project (Dixon *et al.*, 1993a,b; Krankina and Dixon, 1993). The projects considered in this paper were developed with a combination of external and internal resources including some in-kind contributions. Direct national or international finance in the form of cost-share arrangements, direct grants, or direct and guaranteed loans, may permit governments to retain more control over projects than if financing occurs solely through a GHG producer or with the assistance of outside entities. Examples of international financial institutions include: multilateral agencies and international foundations. Multilateral institutions have programs that are designed to identify and develop sound projects and foster financial arrangements. The World Bank's Global Environmental Facility (GEF) is the largest multilateral funding organization for environmental purposes. The GEF is capitalized by $1.3 x 10^9 over the next 3 yr and was formed to address large-scale environmental needs. Resources from GEF generally take the form of grants rather than loans. Leveraging opportunities exist to attract local, national and international resources to match initial project funds. Most of the current forest sector C offset projects considered in this paper have a significant resource leveraging component. Finally, international foundations and targeted financial institutions (e.g., FACE Foundation) can provide resources for project implementation.

4.5 COMPLIANCE MONITORING AND VERIFICATION:

Two primary issues merit consideration: monitoring and reporting mechanisms, and legal frameworks and recourse in the event of noncompliance. Several potential monitoring mechanisms have been suggested: (1) self-reporting, with public access to findings (results of offset programs must be quantified and reported to the general public); (2) consensus reporting by GHG producer, regulatory bodies, and/or third party (under this mechanism, the GHG producer, regulator, and an objective party assess the status of a program and report their findings publicly); (3) self-reporting, combined with on-site spot-checks (a regulator would be given the authority to randomly visit a site, unannounced, and assess whether or not the site is in compliance with contractual arrangements and is adhering to C offset agreements); (4) satellite monitoring (satellite data can be used to determine whether or not land is actually being forested, reforested, or preserved, and to what extent); and (5) private third party reporting (a disinterested party monitors and assesses the effectiveness of a program in meeting its C sequestration goals, subject to review by the GHG producer and regulatory body).

These mechanisms vary in reliability, in the level of control that a regulatory entity retains, and in transaction costs to GHS producers, the government, and third parties. Self reporting requires the least involvement by government in actual monitoring, but may require establishing centralized data collection services to make information available to the public. Spot-checking requires more involvement by government or other monitoring entities, but provides additional checks on the accuracy of reports by GHG producer. Consensus reporting requires significant resources but also provides the most certainty and accuracy. Satellite monitoring also requires the use of a sophisticated of resources, some of which are very costly, particularly at a global scale (Houghton *et al.*, 1991).

5.   Conclusion

The opportunities to manage agroforest and forest systems to sequester and conserve terrestrial C are significant (Sampson *et al.*, 1993). In the absence of specific national and international legal authority to require C offsets, voluntary forest sector programs are being developed under existing legislation or agreements. In the US, the Energy Policy Act, specifically Title 16, provides an additional stimulus for forest sector C offsets. US agencies (e.g., DOE/EIA and EPA) are beginning to promulgate a framework for C offsets in cooperation with non-government organizations and GHG producers. Given the relative novelty and complexity of forest sector C offset projects, a number of biogeochemical, institutional, socio-economic, monitoring and regulatory issues merit further analysis before the long-term potential and cost-effectiveness of this GHG stabilization approach can be determined.

6.   Acknowledgement

The information in this document has been funded in part by the U.S. Environmental Protection Agency. It has been subjected to the Agency's peer and administrative review and it has been approved for publication as in EPA document. Mention of trade names or commercial products does not constitute endorsement or recommendation for use.

7. References

Andrasko, K.J. 1991. UN Food and Agriculture Organization, Rome, Italy, 78p.
Apps, M.J. and A.W. Kurz. 1991. World Resource Review 3:333-343.
Bekkering, T.D. 1992. Ambio 21:414-419.
Birdsey, R.A. 1992. USDA FS General Technical Report W059, Washington, DC, 51 p.
Boonkird, S.A., E.C.M. Fernandes and P.K.R. Nair. 1991. In: Agroforestry Systems in the Tropics. (P.K.R. Nair, ed.) Kluwer Academic Publishers, Boston, MA p. 211-228.
Brown, S., A.E. Lugo, and L.R. Iverson. 1992. Water, Air and Soil Pollution 64:139-155.
Dixon, R.K., J.K. Winjum and P.E. Schroeder. 1993a. Global Environmental Change. in press.
Dixon, R.K., J.K. Winjum K.J. Andrasko, J.J. Lee and P.E. Schroeder. 1993b. Climatic   Change. in press.
Eriksson, H. 1992. Ambio 20:146-150.
Faeth, P., R. Livernash and C. Cort. 1993. World Resources Institute, Washington, DC, 141p.
Gaston, G.G., T.P. Kolchugina, and T.S. Vinson. 1993. Agriculture, Ecosystems and Environment. in press.
Graham, R.L., L. Wright and A.F. Turhollow. 1992. Climatic Change 22:223-238.
Gregerson, H. S. Draper and D. Elz (eds.) 1989. EDI, The World Bank, Washington D C
Harmon, M.E., W.K. Ferrell and J.F. Franklin. 1990. Science 247:699-702.
Houghton, J.T., B.A. Callander and S.K. Varney (eds.) 1992. The 1992 Supplementary Report to the IPCC Assessment. University Press, Cambridge, UK, 24 p.
Houghton, R.A., J. Unruh and P.A. LeFebure. 1991. In: Proceedings for Technical Workshop to Explore Options for Global Forest Management (Howlett, D. and C. Sargent, eds.) IIED, London, UK, p. 279-310.
Johnson, D.W. 1992. Water, Air and Soil Pollution 64:83-100.
Kauppi, P., K. Mielikainen and K. Kuusela. 1992. Science 256:70-74.
Kolchugina, T.P. and T.S. Vinson. 1993. Canadian Journal of Forest Research 23:81-88.
Krankina, O.N. and R.K. Dixon. 1993. World Resource Review. in press
NAS. 1991. U.S. National Academy of Science, Washington, DC, 127 p.
OTA. 1991. OTA-0-482, Washington, DC, 354 p.
Parks, P.J.. 1992. American Forests, Washington, DC, p. 97-121
Rubin, E.S., R.N. Cooper, R.A. Frosch, T.H. Lee, G. Marland, A.H. Rosenfeld and D.D. Stine.   1992. Science 257:148-149, 261-266.
Sampson, R.N., L.L. Wright, J.K, Winjum, J. Benneman, J.D. Kinsman, E. Kursten and J. Scurlock. 1993. Water, Air, and Poil Pollution. this volume
Sanchez, P.A. and J.R. Benites. 1987. Science 238:1521-1527.
Schneider, S.H. 1989. Science 243:771-781.
Schroeder, P.S. and L.B. Ladd. 1991. Climatic Change 19:283-290.
Sedjo, R. 1992. Ambio. 21:274-277.
Sharma, N.P. (ed.) 1992. Kendall/Hunt Publishers, Dubuque, Iowa, 605p.
Swisher, J. and G. Masters. 1992. Ambio 21:154-159.
Trexler, M.C. 1993. Water, Air and Soil Pollution. this volume.
Trexler, M.C., P.E. Faeth and J.M. Kramer. 1989. World Resource Institute, Washington, DC,   65p.
Trexler, M.C. and C.A. Haugen. 1993. World Resources Institute, Washington, DC. in press.

Trexler, M.C., L.A. Loewen and C.A. Haugen. 1992. In: Forests and Global Change
(R.N. Sampson and D. Hair, eds.) American Forests, Washington, DC p 73-96.
UNCED. 1992a. Forest Principles. United Nations Conference on Environment and
    Development. Rio de Janeiro, Brazil.
UNCED. 1992b. Framework Convention on Climate Change. United Nations
Conference   on Environment and Development. Rio de Janeiro, Brazil.
Winjum, J.K., R.K. Dixon and P.E. Schoreder. 1992. Water, Air and Soil Pollution
64:213-   227.
Winjum, J.K., R.A. Meganck and R.K. Dixon. 1993. Journal of Forestry 91:38-42.
Winjum, J.K. and D.K. Lewis. 1993. Climate Research. in press.

MANIPULATING BIOTIC CARBON SOURCES AND SINKS FOR CLIMATE CHANGE MITIGATION: CAN SCIENCE KEEP UP WITH PRACTICE?

Mark C. Trexler

Trexler and Associates, 1131 S.E. River Forest Rd., Milwaukie, OR 97267
USA

Abstract. The potential for natural C sinks to be manipulated by human means to mitigate climate change has been discussed in the environmental literature for more than a decade. There now appears to be little doubt that changes in global land-use and land management practices could significantly slow the accumulation of CO_2 in the atmosphere. As a result, some forward-thinking companies and governmental bodies are acting now upon the biotic mitigation literature by developing actual mitigation projects. It is now national policy in the United States to encourage such activities. The future of C offsets, however, is unclear, due in large measure to lagging scientific knowledge. Large-scale private action likely will await regulatory signals that action will be accepted as a legitimate mitigation measure, perhaps providing retroactive regulatory credit, a source of tradeable emission entitlements, or credit against yet-to-be-established C taxes. The practical potential of most biotic mitigation approaches is unknown, and the entire concept remains subject to political challenge domestically and abroad. The ability to predict C benefits of individual mitigation projects is often tenuous and subject to debate. To allow expansion of C offset practices as quickly as possible, and hopefully to fund projects with many ancillary environmental and economic benefits, policymakers and project developers desperately need physical and social science data to be provided in a useable form.

1. Introduction

Identifying and quantifying natural and anthropogenic sources and sinks of CO_2 have been the subject of a rapidly growing literature during the last 25 yr. In the last 10 yr a new element has been added to the picture: discussion of our ability to modify sinks and flows for purposes of mitigating the role of CO_2 in potential future climate change. Many questions remain, however (Wisniewski and Lugo, 1992; Trexler and Haugen, 1993).

Today, public policy intended to manipulate C sources and sinks for purposes of climate change mitigation threatens to outpace the available scientific knowledge. Internationally, the United Nations Framework Convention on Climate Change (Climate Change Convention) establishes a "joint implementation" framework for the pursuit of C offset projects in foreign countries. Domestic legislation in the United States encourages companies to undertake voluntary climate change mitigation measures in the form of greenhouse gas offsets. As a result, private companies as well

Water, Air, and Soil Pollution **70**: 579–593, 1993.

as intergovernmental agencies such as the Global Environmental Facility (GEF) are beginning to initiate offset projects in areas such as forest protection and management, reforestation, and biomass energy (Kinsman and Trexler, this volume). These projects could simultaneously further other important policy goals such as biodiversity conservation.

Notwithstanding a rapidly expanding literature, little information is available directly assisting offset proponents in quantifying the C benefit associated with a particular project or in convincing funders and regulators that pursuing C offset projects is a prudent action to initiate in advance of C control regulations. The global potential of biotic options remains largely unclear, a source of concern to regulators who must allocate limited resources across the whole range of potential climate change mitigation strategies. The uncertainties involved in the long-term nature of most biotic mitigation options will automatically be suspect to bureaucracies accustomed to highly engineered approaches to pollution control. Comparative C offset cost data are so inadequate that regulators may feel themselves politically unable to accept international forestry as an offset approach when pitted against domestic energy and forestry interest groups.

This paper briefly reviews the information needed to advance the development of public and private policy in this area. The information needed spans disciplines ranging from global systems dynamics to silviculture to sociology and economics. By flagging the needs of private sector actors and governmental regulators, perhaps the necessary data gathering and presentation can be accelerated. The paper does not seek to review the relevant literature; the literature is surveyed elsewhere in this volume or in Wisniewski and Lugo (1992).

2. Biotic Carbon Flow Modification in the Context of Overall Climate Change Mitigation Measures

Scientists and policymakers accept that the potential for detrimental climate change resulting from human actions is a significant policy problem. Since the "Noordwijk Declaration" of 1988, in which 68 environmental ministers from around the world established global targets for reversing the decline in global forest cover (Noordwijk, 1988), linking forestry and climate change at the political and policy levels has become common. Existing energy production and consumption patterns in industrialized countries must be modified dramatically if CO_2 emissions are to be significantly reduced. Projected energy consumption trends in developing nations also must be significantly reduced if global emissions are to be capped at or below current levels (IPCC, 1990).

Actions outside the energy sector also hold promise to affect both CO_2 sources and sinks and other greenhouse gases, including modification of land-use practices and land-use conversion rates. The theoretical basis for biotic options has been discussed extensively in the literature (Grainger, 1990; Houghton, 1990; Marland, 1988; Trexler, 1991; Trexler and Haugen, 1993).

3. Practical Issues in Developing and Implementing Public and Private Policy Regarding Carbon Offsets

As discussed by Kinsman and Trexler (this volume), C offset policy is developing rapidly. Several governments are studying how to pursue the joint implementation provisions of the Climate Change Convention. The United States Department of Energy is researching how to implement Section 1605 of the Energy Policy Act of 1992, which provides for a voluntary register of greenhouse gas offset projects. Several utilities in the United States and abroad have pursued C offset projects ranging in cost from 10^5 to 10^6, and the Global Environmental Facility is financing its own projects. Several companies are entering the field as potential C brokers in anticipation of a future regulatory regime.

As governments plan C offset policy and private companies and governmental agencies undertake C offset projects, several issues have appeared regularly: the political legitimacy and scientific soundness of the offset concept, the total climate change mitigation potential of offset measures, and the validity of individual projects as C offsets.

3.1. LEGITIMACY AND PRACTICALITY OF THE OFFSET CONCEPT

From a political standpoint, it is not uncommon to hear CO_2 offset projects in the tropics compared to extra-national hazardous waste dumping. The analogy, although flawed, reflects a skepticism among many developing countries that will be difficult to overcome unless offset projects can explicitly be tied to their own social and economic goals. Even many environmentalists express considerable skepticism over the role tropical forestry should play in climate change mitigation, irrespective of the physical potentials involved. These environmentalists are concerned that large-scale tree planting not undertaken on the basis of meeting local needs could lead to severe impacts on local agricultural production, foster land conflicts between local residents and outside interests, increase rural poverty and landlessness by denying local access to land, promote migration to urban centers, and possibly lead to compensatory conversion of remaining areas of natural forest to agricultural uses (Barnett, 1992).

Domestically, skepticism exists regarding the pursuit of international offset projects. For many regulators the concept that planting trees in Russia or protecting forests in Costa Rica can offset domestic CO_2 emissions is difficult to grasp. However, the domestic potential of biotic mitigation measures is quite limited, and regulators should be convinced to work with international offsets as well.

Other political issues will arise as interest in biotic offsets evolves. Although planting trees to mitigate climate change is perceived as politically painless, widespread implementation of biotic policy options could prove contentious. Withdrawing timber from harvesting will have economic and political repercussions whether done in the United States or the Congo. Increased timberland management on remaining public and private lands would intensify debate over forest management practices and the loss of wildlife habitat. Removal of a significant fraction of agricultural lands in the United States or elsewhere from crop production would cause political opposition from well-entrenched interest groups; in the tropics, this option is out of the question. A large-scale switch to biofuels would significantly alter the economics of the energy production sector. In each case, important political and economic interest groups will be affected by the change.

Scientists also have several technical questions regarding the legitimacy of the C offset concept. Scientists' inability to close the C cycle and explain several Gts of "missing" C every year inevitably causes many people to question the ability to successfully manipulate the C cycle to climate change mitigation ends. After all, the concept of tradeable C emissions rights and offsets generally will depend on successful construction of C baselines for countries funding and hosting C offset projects. Yet estimates of global C emissions from land-use change continue to range from under one to almost three Gts yr^{-1}. These estimates often fail to account for forest degradation and include only actual deforestation. Clearly, a C baseline that credits a forest recovery project for C offset purposes but fails to incorporate forest degradation within the country's C baseline will lead eventually to anomalous results.

The potential relationships between climate change, CO_2 fertilization, and the C offset concept are also of technical concern. To what extent will the impacts of climate change negate or impede particular offset types or projects? Conversely, to what extent should enhanced biomass growth rates attributable to CO_2 fertilization be counted for C offset quantification purposes?

To address these various information needs, several areas of investigation should be given priority:

- Data on biotic C baselines must be gathered, both for the globe and for individual nations. A variety of proposals have been made to implement a GIS-based global forestry monitoring system (IPCC, 1992). Ideally, monitoring should include C fluxes associated not only with deforestation, but also forest degradation and recovery.

- Clear scientific discussion is needed of the current understanding of the global C cycle and of how it affects biotic climate change mitigation and its political credibility. Is the "missing C" relevant to offset policy? If vegetation is a sink for the missing C, does this have positive or negative ramifications for C offsets; for example, will other limiting constraints on biomass growth be activated sooner?

- Scientific guidance is needed regarding the relationships between offset policy and potential impacts of both climate change and CO_2 fertilization. Should biotic mitigation projects be avoided in certain areas given foreseeable impacts of climatic change in those areas? Similarly, are some mitigation projects more likely to benefit from CO_2 fertilization than others? Is it realistic to incorporate

CO_2 fertilization impacts into the quantification of CO_2 offset projects? Although the science in these areas remains ambiguous, the scientific community should provide some input into the policy process regarding how to address these questions.

3.2. POTENTIAL FOR AND COSTS OF CARBON OFFSETS

A question commonly heard in discussions of biotic climate change mitigation involves the project potentials and costs. Biotic options are quite distinct from energy-based options (with the exception of biomass energy development) and will require their own policy guidelines, implementation approaches, and monitoring and verification systems. If the potential of offset projects is modest, development of a policy system specifically tailored to biotic options may not be justified. If some categories of biotic options are not significantly cheaper than energy-based mitigation measures private sector interest may diminish.

The potential and economics of biotic mitigation measures are premised on several sets of physical, social, and economic variables. Among the issues needing to be addressed are:

- Land availability for offset projects. This availability is based on physical variables as well as social, environmental, and demographic constraints;

- Growth rates achievable in offset projects utilizing various management, silvicultural, and protection techniques;

- Total biomass accumulation if harvesting is not anticipated;

- Long-term soil capabilities, if periodic harvesting is foreseen, particularly in intensive management designed for biomass energy production;

- Land and labor costs as well as the economic rent that likely will be extracted by C offset hosts once the value of the commodity is recognized;

- Indirect implications of the large-scale implementation of offset projects, including potential swamping of world timber markets, undercutting of existing forest management incentives, and contributions to or detractions from national economic development goals and sustainable resource management.

Policymakers and many scientists are concerned about whether significant climate change mitigation through offsets is practical given current demographic and other trends, particularly in the tropics. As argued by Trabalka *et al* (1985), "large-scale reforestation is not a realistic option for mitigation if one considers the most probable estimates of further CO_2 increase and concepts of a large net biospheric C sink over the next century fly in the face of realities that include a projected minimum doubling of human population in tropical regions." Proving the legitimacy of the C offset concept will require demonstrating to policymakers that these kinds of variables have been taken into account.

3.2.1. *Land Availability*

Since the initial estimates of Freeman Dyson in 1977 that it would take 700×10^6 ha of new tree plantations in the tropics to absorb the approximately 5×10^9 t C yr^{-1} to the atmosphere from the combustion of fossil fuels, a great deal of attention has focused on how much land could be made available for global reforestation measures as a means to mitigate the effects of climate change. Land availability can be assessed for tropical, temperate, and boreal zones. Although estimates for all areas exist, the discussion below focuses on the methodology applied to the tropics.

Most estimates of land availability and climate change mitigation potentials take little account of the demographic, social, environmental, economic, and political pressures and barriers that will impede efforts to rapidly end deforestation or undertake massive reforestation. Consequently, much of the information results in estimates of land availability that consider only physical/biological criteria, i.e. land that is capable of sustaining tree growth and is "theoretically" available for C forestry.

Since approximately 1985, significant speculation has been made on the availability of land for reforestation, but the reliability of the resulting estimates, particularly with respect to the tropics, has not changed dramatically during that time. This is largely because data available to assess past and current land use trends and to project these trends into the future are severely limited.

As a result, estimates of land availability range widely (Houghton, 1990; Grainger, 1988; Moulton and Richards, 1990). Dixon *et al* (1991) surveyed available information and concluded that land technically available in the tropics for expanded management and agroforestry is somewhere between 620×10^6 and 2×10^9 ha. The study acknowledged that this estimate encompassed physical land availability and did not include economic or social criteria. The World Resources Institute assessed tropical land availabilities in 50 countries, incorporating economic and social variables normally not addressed in land availability estimates (Trexler and Haugen, 1993). The application of this methodology led to the conclusion that hundreds of millions of hectares are available. The study concluded, however, that the rate at which biotic mitigation measures can be realistically implemented in countries that often possess little if any infrastructure to administer them is a considerably greater constraint than basic land availability.

Thus, land availability estimates currently in use probably provide only modest insight into how much land could really be protected or returned to tree cover, how alternative forestry strategies compare to each other in terms of total C storage potential, and how forestry strategies compare in their overall potential to alternative options for the mitigation of climate change. Yet international forestry policy formation is likely to occur soon as the debate over climate change and biological diversity conservation advances. Country-specific modeling of land availabilities, taking into account a wide range of variables, is a high priority.

3.2.2. *Biomass Growth Rates*

As with land availability, many estimates of C accumulation rates and total C accumulation potentials are used in estimating biomass growth rates (See Table 1). The ranges are so wide that it is difficult to discuss reliably how much C ha^{-1} yr^{-1} would be absorbed in the tropics or in other zones through individual offset projects. Experience with reforestation efforts suggests that estimates of high productivities in the tropics or elsewhere, while technically feasible, probably do not mean much in the context of what is achievable across large-scale forestry efforts.

3.2.3. *Carbon Accumulation*

In considering climate change mitigation, it is as important to consider total biomass accumulations as yearly biomass accumulation rates. Should large areas of reforestable land be planted in short-term high yield crops that ultimately store less C than a slower-growing, longer rotation species, the overall effect could be counter-productive if the faster-growing trees are not effectively harvested and used. Once again, the state of scientific knowledge of C accumulation across ecosystem types is in flux (Botkin *et al*, this volume), making it difficult to assess overall global potentials of mitigation options as well as the C benefits of specific offset projects.

3.2.4. *Carbon Economics*

Given uncertainties regarding land availability, growth rates, and C accumulation totals, it is not yet possible to develop a forestry C supply curve for any of the three major geographic regions. In many cases, it is not even possible to even clearly rank the alternative forestry approaches by cost. Some land-use changes being proposed are likely to be cost-effective in their own right, requiring no "C subsidy." Others may not be cost- effective and will require a subsidy. It is also difficult to prepare a proper economic analysis since C storage is only one implication of many forestry programs, particularly in the tropics; many other costs and benefits (e.g. watershed protection, biological diversity conservation, income generation, and income foregone) are also hard to quantify.

　　An additional complication is the potential market feedbacks associated with the future large-scale implementation of biotic policy options. Large-scale forestry undertaken for climate change mitigation, for example, could easily pose a potentially serious threat to the long-term management of forests already being managed for economic return. Although energy policies pose some of the same problems -- that

Table 1: Estimated Rates of Carbon Accumulation

Source:	Tropical (t C ha⁻¹ yr⁻¹)	Temperate (t C ha⁻¹ yr⁻¹)	Boreal (t C ha⁻¹ yr⁻¹)	Notes
Brown et al., 1992	1.4-4.8 in plantations; 1-3.5 in currently forested areas			
Lashof and Tirpak, 1990	18; Potential of 26	5-7 Potential of up to 15		Numbers based on research in intensive plantation management. Tropical plantations used eucalyptus hybrids. Temperate figures based on growth rates of Washington Douglas fir and loblolly pine in North Carolina.
Marland, 1988*	2-5	0.5-1.5		Based on data from current forest areas in Brown and Lugo (1982) and U.S. Forest Service.
Hall et al., 1990**	5-6 (long rotation); 10-15 (short rotation)	2-4 (long rotation); 3-6 (short rotation)		They compare these numbers to 17-35 tC/ha/yr for herbaceous crops to prove that herbaceous crops are preferable when growing plants for biomass energy, rather than for sequestering carbon.
Dixon et al., 1991			0.4-0.6	Derived from data from the former USSR.
Kanowski and Savill, 1992*	0.5-5 in natural forests; 6-17 on plantations	0.5-2.0 in natural forests; 2-10 on plantations		Range of numbers based on data from specific tree types, including natural tropical forests, eucalyptus hybrid plantations, conifers in the United States and Europe, and pine and eucalyptus plantations in Portugal and New Zealand.

* converted from m³ ha⁻¹ yr⁻¹ at a factor of 0.33

** converted from t ha⁻¹ of biomass a factor of 0.5

is, any significant policy-inspired decrease in global energy demand would probably be partially offset by falling prices and resulting upward pressure on consumption -- these undesired market feedback effects are much easier to control in the energy sector.

A number of major uncertainties have been identified that interfere with our ability to effectively quantify the potential of biotic climate change mitigation mechanisms for C sequestration and C offsets. These uncertainties make it difficult to convince regulators that biotic offsets are a legitimate climate change mitigation measure and to lay out for potential offset funders the relative merit of biotic as opposed to non-biotic measures. As a result, both regulators and funders must act on faith. Although this "faith" is sufficient to generate small-scale interest among both groups, it will not sustain large-scale regulatory or investment activity. Data and research necessary to build a proper foundation include:

- Realistic country-by-country assessments of land availability that account for the many relevant variables. This will require a highly interdisciplinary approach to the country assessments.

- Derivation of a formula allowing simple calculation of reasonable C accumulation and total C loading estimates for an individual offset site. Although rough, some method of simplified C accounting must be available to offset proponents unable or unwilling to undertake a more intensive assessment of their own.

- Derivation of a standardized economic assessment methodology so as to avoid the current situation, where each project is assessed differently. Scientific input into the time-value of C sequestration is needed. For example, is C sequestration more valuable now or in the future? Should C accumulation be discounted as monetary flows are discounted? These are questions vitally important to agreement on C sequestration economics.

4. Legitimacy of Individual Project Types and Proposals

A number of land use-based approaches can be used to slow the buildup of CO_2 in the atmosphere. Some land use-based C offset projects are already being pursued (See Dixon *et al*, this volume). The four categories of potential biotic C offset projects are:

1) Slowing or stopping the loss of existing forests, thus preserving current C reservoirs. Project opportunities include forest protection, forest management, harvest management, and off-site agroforestry.

2) Adding to the planet's vegetative cover, thus enlarging living terrestrial C reservoirs. C offset opportunities include reforestation, afforestation, natural

and assisted regeneration, fertilization, farm forestry, and soil management practices.

3)   Increasing the C stored in C reservoirs such as agricultural soils and wood products. Offset opportunities could include paper recycling, substitution of wood for steel and concrete, and changes in tillage and fertilization practices.

4)   Substituting sustainable biomass energy sources for fossil fuel consumption. Offset opportunities include transfer of biomass technologies, efficiency improvements in existing biomass utilization, and subsidization of biomass energy development.

The atmosphere has no particular preference among these options. Each approach, however, can have distinct advantages and disadvantages given land availability, environmental, social, and political criteria. More important for current purposes is their validity and potential for C offset purposes. As discussed in Kinsman and Trexler (this volume), several important regulatory and funding criteria can be identified for each offset concept: credibility, reliability, verifiability, measurability, and cost-effectiveness. Discussion in this paper is limited to illustrating the kinds of technical questions for which answers would be helpful to both C regulators and C offset funders.

4.1.   OFFSET CREDIBILITY

It is easy to identify both credible and tenuous potential offset projects. The use of reforestation to fuel a biomass energy facility that replaces a planned coal plant is clearly a "high-quality" biotic offset. Conversely, proposing to purchase a remote parcel of forest that faces no conceivable threat is a "low-quality" offset, regardless how cost-effective per ton of C it might appear. There is a large range of potential projects between these two extremes. Several examples of the credibility problems that may be encountered can be identified:

4.1.1. *Social Forestry*

Social forestry often involves projects in which community-based reforestation is assumed, through the production of additional fuel and fodder, to result in the protection of remaining natural forest (Sanchez and Benites, 1991). But this assumption may be challengeable: is the causal link between social forestry and forest protection truly sufficient to scientifically validate a project for C offset purposes? Are there particular circumstances and project designs that make a subset of projects more acceptable than others?

4.1.2. *Forest Protection*

Forests not facing threats cannot be protected for C offset purposes. But forests that are under threat, whether through logging concession or imminent invasion, do not necessarily present a simple offset situation. To what extent will the exploitative pressure simply be displaced to another area? Are there variables that allow this assessment or are there project structures that would be considered to "solve" this problem, e.g. forest management in combination with protection?

4.1.3. *Reforestation*

Even reforestation can present a complex context for verification of an offset project. If the alternative to reforestation is natural forest management, and in the absence of an economic incentive to pursue management once the reforestation is in place, reforestation itself could contribute to the loss of adjacent forest. Are there indicators that could be used to predict this problem in advance?

4.1.4. *Forest Management*

Forest management is commonly mentioned as a means of C sequestration. In one existing project, forest harvesting practices are being modified to reduce the damage to non-commercial biomass. At least superficially, the C benefit of this project would appear short-lived if the forest will regenerate to its former C density within several decades. What are the circumstances under which this is not true? Are there specific types of management and harvesting practices that should qualify as credible offsets more than others?

4.1.5. *Reservoir Enhancement*

Paper recycling is often mentioned as a means of increasing C storage, as is forest fertilization and modification of agricultural practices such as soil tillage. It is currently not at all clear whether credible C offset projects can be premised on these approaches; extensive C modeling is required.

4.2.   OFFSET RELIABILITY

It is quite possible for a credible project to not be particularly reliable; thus, the two variables require independent assessment. Assessment of offset reliability will ultimately require agreement on a risk assessment methodology that can commonly be applied across projects and project types. For example, should varying offset multipliers be required to account for medium to high risk projects? Also crucial to offset reliability is the timeline over which projects should be assessed and whether all projects should be subject to a common timeline analysis. If C sequestration is seen primarily as a transitional strategy during which time a global switch is made from fossil to other fuel sources, the determination of an appropriate timeline may primarily be a policy decision. It would seem possible, however, that there are scientific time value of C issues involved in determining the appropriate timeframe for evaluating biotic mitigation projects.

4.3.   OFFSET VERIFIABILITY

Offset projects will involve certain implementing steps, ranging from extension activities in a social forestry project to boundary protection in a forest preservation project. Projects will differ considerably in a sponsor's ability to verify that these steps result in the intended C sequestration. In some cases, such as forest protection or reforestation, offset verification must be maintained on an ongoing basis since the project could "fail" at any time. At this time, however, verification technologies remain poorly developed. It must be determined under what circumstances satellite verification can be used, for example, and what are the realistic alternatives to this approach. Offset verification is important because an overly rigorous approach could add substantially to the costs of the basic offset.

4.4.   OFFSET MEASURABILITY

Beyond verifying that a reforested area is still growing or that a protected forest remains intact, the C sequestration benefit must be assessed. This presents difficult methodological questions that have only begun to be evaluated. Several methods of assessing C benefit are in principle available:

- Reliance on observable C accumulation, e.g. through empirical measurement. This is most appropriate in cases such as commercial reforestation or fossil fuel displacement where C accumulation can be readily observed.

- Reliance on indirect and assumed benefits, e.g. the use of conversion factors to convert from successful sustainable agriculture implementation to an estimate of forest area protected. This is most appropriate where it is implausible to observe the sequestered C; however, this approach can be highly uncertain.

- C flow modeling, e.g. the use of simple or dynamic computer models to attempt to predict C sequestration on the basis of a set of quantifiable variables. Although modeling could prove valuable, it can be a difficult task. The best-known model today, developed by the World Resources Institute, operates largely on the basis of the local population growth rate and ignores many other potential variables.

These approaches can be used independently or in conjunction with each other. In all cases it is necessary to construct a C baseline "with" and "without" the offset project. Specific issues to be addressed are:

- A new plantation could add new biomass, and could result in preserving existing biomass if the drop in demand for forest logs keeps logging roads from being built and agricultural invasion from occurring. Both of these impacts could increase total C from the baseline of no plantation and agricultural invasion of the logged forest.

- A new plantation could affect natural forests negatively, e.g. through relocation of farmers selling their land.

- If natural forest in the area is already being rapidly depleted, it could be concluded that new plantations simply accelerated plantation establishment in the region, rather than adding to the number of plantations than would otherwise have been established.

4.5. OFFSET COST-EFFECTIVENESS

Drawing the bounds around an offset project is not always easy. In funding an agroforestry project, does one count just the C stored in new vegetation, or also the C in forests allegedly protected by the project? In looking at long-term rotations on reforested land, does one include the economic return on the trees, even if the return is not to the project developer? A decision as to whether to discount C accumulation must also be made. No consistency yet exists on how these questions are to be addressed, but they are vital. Additionally, the costs to a project funder may be different than the societal costs of a project if a limited investment can leverage other funds. This is often possible in social forestry projects, for example, where other public funding is likely to be available.

There are clearly many questions relating to the credibility, reliability, verifiability, measurability, and cost-effectiveness of alternative biotic mitigation techniques. Many of these questions affect public policy or social science, but would benefit from physical science input. Ideally, C offset approaches will soon be ranked across these five criteria, and accepted measurement and verification protocols will be available.

5. Conclusions

Private sector C offsets offer an excellent opportunity to fund climate change mitigation activities that can also serve other social and policy goals. From a regulatory perspective, however, the science and practice of biotic mitigation will appear questionable for some time. The practical potential of most biotic mitigation approaches remains unknown, and the entire concept remains subject to political challenge domestically and abroad. Notwithstanding the convincing literature pertaining to potential biotic mitigation measures, there is a fundamental difference between accepting that C flows can in fact be modified to help mitigate climate change and accepting that this modification can take the form of individual projects that can be evaluated and verified as part of a C emissions control system.

If projects undertaken in the near future cannot be scientifically validated, their C benefit could ultimately be challenged or disproven. Sufficient cases of this sort could undercut the credibility of the concept of biotic C offsets. It would be unfortunate if the legitimate concept of using forestry and other biotic means to help mitigate climate change prematurely falls prey to political or technical credibility problems associated with the absence of key scientific knowledge.

REFERENCES

Barnett, A.: 1992, *Desert of Trees: The Environmental and Social Impacts of Large-Scale Tropical Reforestation in Response to Global Climate Change*, Friends of the Earth.

Botkin, D.B., and Simpson, L.G.: 1993, *Water, Air and Soil Poll.* This volume.

Brown, S., and Lugo, A.E.: 1982, *Biotropica.* 14, 161-87.

Brown, S., Lugo, A.E., and Iverson, L.R.: 1992, *Water, Air and Soil Poll.* 64, 139-55.

Dixon, R. K.: 1993, *Water, Air and Soil Poll.* This volume.

Dixon, R.K., Schroeder, P.E., and Winjum, J.K., et al (eds): 1991, *Assessment of Promising Forest Management Practices and Technologies for Enhancing the Conservation and Sequestration of Atmospheric Carbon and Their Costs at the Site Level*, U.S. Environmental Protection Agency.

Grainger, A.: 1988, *Int'l Tree Crops J.* 5, 31-61.

Grainger, A.: 1990, Modeling the Impact of Alternative Afforestation Strategies to Reduce Carbon Dioxide Emissions, in: U.S. Environmental Protection Agency, *Proceedings of the Conference on Tropical Forestry Response Options to Global Climate Change*, Intergovernmental Panel on Climate Change, Response Strategies Work Group, Univ. of São Paolo, pp. 93-104.

Hall, D.O., Mynick, H.E., and Williams, R.H.: 1990, *Carbon Sequestration versus Fossil Fuel Substitution: Alternative Roles for Biomass in Coping with Greenhouse Warming*, Center for Energy and Environmental Studies, Princeton University.

Houghton, R.A.: 1990, *Ambio* 19(4), 204-09.

Intergovernmental Panel on Climate Change: 1990, *Scientific Assessment of Climate Change*, World Meteorological Organization.

Intergovernmental Panel on Climate Change: 1992, *Climate Change: The IPCC Response Strategies*, Island Press.

Kanowski, P.J. and Savill, P.S.: 1992, Plantation Forestry, in: Sharma, N.P. (ed), *Managing the World's Forests: Looking for Balance Between Conservation and Development*, Kendall/Hunt Publishing Company, pp. 375-401.

Kinsman, J.D., and Trexler, M.C.: 1993, *Water, Air, and Soil Poll* This volume.

Lashof, D.A. and Tirpak, D.A. (eds): 1990, *Policy Options for Stabilizing Global Climate*, U.S. Environmental Protection Agency.

Leach, G.: 1990, Woodfuels and Smallholder Afforestation, in: U.S. Environmental Protection Agency, *Proceedings of the Conference on Tropical Forestry Response Options to Global Climate Change*, Intergovernmental Panel on Climate Change, Response Strategies Work Group, Univ. of São Paolo, p. 143.

Marland, G.: 1988, *The Prospect of Solving the CO_2 Problem Through Global Reforestation*, Office of Energy Research, U.S. Department of Energy.

Moulton, R.J., and Richards, K.R.: 1990, *Costs of Sequestering Carbon Through Tree Planting and Forest Management in the United States*, GTR WO-58, United States Department of Agriculture, U.S. Forest Service.

Noordwijk Declaration on Atmospheric Pollution and Climatic Change, 1988: Ministerial Conference on Atmospheric Pollution and Climatic Change.

Sanchez, P.A., and Benites, J.R.: 1991, *Science*. **238**, 1521.

Sedjo, R.A.: 1989, *J. Forestry*. **July 1989**, 12-15.

Trabalka, J.R., Edmonds, J.A., Reilly, R.M., *et al*: 1985, Human Alterations of the Global Carbon Cycle and the Projected Future, in: Trabalka, J.R. (ed), *Atmospheric Carbon Dioxide and the Global Carbon Cycle*, U.S. Department of Energy, p. 247-88.

Trexler, M.C.: 1991, *Minding the Carbon Store: Weighing U.S. Forestry Strategies to Slow Global Warming*, World Resources Institute.

Trexler, M.C. and Haugen, C.H.: 1993, *Keeping it Green: Global Warming Mitigation Through Tropical Forestry*, World Resources Institute.

Wisniewski, J. and Lugo, A.E. (eds): 1992, *Natural Sinks of CO_2*, Kluwer Academic Publishers.

THE INTERACTION OF CLIMATE AND LAND USE IN FUTURE TERRESTRIAL CARBON STORAGE AND RELEASE

ALLEN M. SOLOMON

U.S. Environmental Protection Agency (USEPA), 200 S. W. 35th St.,
Corvallis Oregon 97333 USA

I. COLIN PRENTICE

Department of Plant Ecology, Lund University, Östra Vallgatan 14,
S-223 61 Lund, Sweden

RIK LEEMANS

National Institute for Public Health and Environmental Protection (RIVM),
3720 BA Bilthoven, The Netherlands

WOLFGANG P. CRAMER

Potsdam Institute for Climate Impact Research (PIK),
Telegrafenburg, O-1500, Potsdam, Germany

Abstract. The processes controlling total carbon (C) storage and release from the terrestrial biosphere are still poorly quantified. We conclude from analysis of paleodata and climate-biome model output that terrestrial C exchanges since the last glacial maximum (LGM) were dominated by slow processes of C sequestration in soils, possibly modified by C starvation and reduced water use efficiency of trees during the LGM. Human intrusion into the C cycle was immeasurably small. These processes produced an averaged C sink in the terrestrial biosphere on the order of 0.05 Pg yr^{-1} during the past 10,000 years.

In contrast, future C cycling will be dominated by human activities, not only from increasing C release with burning of fossil fuels, and but also from indirect effects which increase C storage in the terrestrial biosphere (CO_2 fertilization; management of C by technology and afforestation; synchronous early forest succession from widespread cropland abandonment) and decrease C storage in the biosphere (synchronous forest dieback from climatic stress; warming-induced oxidation of soil C; slowed forest succession; unfinished tree life cycles; delayed immigration of trees; increasing agricultural land use). Comparison of the positive and negative C flux processes involved suggests that if the C sequestration processes are important, they likely will be so during the next few decades, gradually being counteracted by the C release processes.

Based only on tabulating known or predicted C flux effects of these processes, we could not determine if the earth will act as a significant C source from dominance by natural C cycle processes, or as a C sink made possible only by excellent earth stewardship in the next 50 to 100 yrs. Our subsequent analysis concentrated on recent estimates of C release from forest replacement by increased agriculture. Those results suggest that future agriculture may produce an additional 0.6 to 1.2 Pg yr^{-1} loss during the 50 to 100 years to CO_2 doubling if the current ratio of farmed to potentially-farmed land is maintained; or a greater loss, up to a maximum of 1.4 to 2.8 Pg yr^{-1} if all potential agricultural land is farmed.

Water, Air, and Soil Pollution **70**: 595–614, 1993.
© 1993 *Kluwer Academic Publishers.*

1. INTRODUCTION

The atmosphere annually exchanges about as much carbon (C) with the terrestrial biosphere as with the oceans. Ocean-atmosphere CO_2 exchanges have been calculated consistently, if not necessarily accurately, within a narrow range of values for the past 20 yrs (Keeling, 1973; Tans et al., 1990; Orr, 1993). The problem of ocean-atmosphere interaction in the global C cycle has been both tractable and attractive for analysis as a three-dimensional problem using the principles of physical chemistry and fluid dynamics.

The terrestrial biosphere requires a different kind of analysis. Vegetation stands and communities are largely independent of one another so the problem becomes one of modelling one-dimensional exchange processes at a very large number of points. These processes are not yet documented accurately enough to allow comparably consistent estimates of net C uptake or release.

More precise measurements of C storage and release under current conditions are therefore a necessity. Improvements in our information on present C stocks are a real possibility and are being undertaken in many places in the world (e.g., Botkin and Simpson, 1992; Simpson and Botkin, 1993; Scharpenseel, 1993). However, with the exception of ice core data reflecting overall atmospheric composition, direct measurements of C dynamics 1000 or 10000 yrs ago are nearly impossible, being replaced at best by proxy data of some kind. Similarly, estimates of the magnitude of future C exchanges can only be undertaken via mathematical models, incorporating the processes which we believe will be most important in defining future C dynamics. The more accurate is our understanding of these processes, the more valid will be the modeled results of future biospheric response to changing atmospheric chemistry and climate. This emphasis on processes may seem self-evident, but it is often disregarded. Projections of future C dynamics are unlikely to be correct if we use a model that merely reproduces the patterns of the present or past world without explicitly representing the processes involved.

The analysis which follows focusses on forests because they possess about 80% of the global terrestrial C stocks. We use the biome as our fundamental ecological unit. We work at the scale of 1/2 X 1/2 º of latitude and longitude, aggregated to the global scale in a static model of the terrestrial biosphere. This scale is apparently a valid one at which to describe terrestrial biosphere C storage changes during the past several millennia although it is still too coarse to directly represent many of the processes determining land-atmosphere C exchange. We first outline the processes causing change in C stocks since the Last Glacial Maximum (LGM). Then, estimated LGM C stocks are compared to C stocks today and to potential future C stocks corresponding to the equilibrium state of a "2 X CO_2 " world. We then assess the so far least evaluated process involved in modifying the trend toward this equilibrium state; namely, the population- and climate-dictated increase in the intensity and cover of agricultural land use. The major simulation data sets used for this study are the result of equilibrium runs of general circulation models with conditions of the LGM and of doubled CO_2; the Biome model of Prentice et al. (1992), which predicts equilibrium biome distribution as a function of climate and soils; and a new scheme to predict the potential distribution of agriculture based on climatic constraints.

2. Carbon Exchanges since the Last Glacial Maximum

The geography of the terrestrial biosphere was considerably different 18,000 yrs ago. The greatest differences from today were above 30o latitude. Continental ice sheets and open tundra covered most areas above 45o (Figure 1a) and many of today's high-latitude biomes then were found only in what now are temperate regions (Figure 1b). The area of "boreal" forest, forming the wide, continuous circumpolar band so characteristic of the modern world (Figure 1b), was fragmented and occupied only about one-fifth of its current area (Figure 1a). This counter-intuitive feature of the colder LGM world is borne out by recent syntheses of paleoecological data describing vegetation distribution at the LGM (e.g., Huntley and Webb, 1989; COHMAP Members, 1989; Prentice *et al.*, 1993). In low latitudes, tropical forest was reduced within its current area, but tropical forests may also have covered a more-or-less compensating area that was later inundated during the melting of the ice sheets (Figure 1a, b).

The total terrestrial C stocks of this colder world were reduced compared to those of today despite lower sea-level. If today's C density values, measured in each biome in biomass (Olson *et al.*, 1983) and in soil C (Zinke *et al.*, 1984), are assigned to the area covered by biomes projected during the LGM, then C storage of the LGM was approximately 300 Pg ($\pm$ 100 Pg) less than that of the pre-industrial modern world (Table 1). This value excludes the great peat blankets of high latitudes which originated after 6,000 yrs ago. Assuming that these peats contain an additional 200 to 400 Pg of C (Gorham, 1991), the difference between modern and LGM terrestrial C stocks was probably between 400 and 700 Pg, i.e., 1/5 to 1/3 less C was stored during the LGM than today (Prentice *et al.*, 1993; Prentice and Sykes, 1993). This range of values is consistent with independent estimates based on mean $\partial^{13}C$ changes in the global ocean, within the rather broad uncertainty limits of those estimates.

In addition to C gains since the LGM suggested from direct projections of present C storage into the past, reconstructions of some LGM forests suggest that less C was stored in biomass per unit area than in the most similar present-day forests (e.g., Webb, 1987a). Forest stand simulations of LGM forests (Solomon and Tharp, 1985) also suggest that the temperate and boreal forest biomes in the eastern U.S. contained lower biomass than do their modern counterparts but this effect has not been quantified at the global scale.

Much of the net C sequestration that appears to have occurred in the terrestrial biosphere since the LGM probably occurred after about 12,000 to 10,000 yrs ago, the time of most rapid deglaciation. Hence, 0.03 to 0.07 Pg C yr^{-1} is a coarse estimate of average C sequestration rates during the Holocene. Recent model analyses (Harden *et al.*, 1993) suggest that soil C sequestration in boreal regions varied during this period, steadily increasing since the early Holocene, increasing most rapidly between 8,000 and 4,000 yrs ago, and maintaining the greatest C sequestration rates since about 4000 yrs ago. But as yet there has been no serious attempt to reconstruct global soil C storage dynamically through the period since the LGM.

The forces responsible for carbon content of the terrestrial biosphere during the past 20,000 yr are discussed below. It is notable that effects of human activity on C fluxes and stocks was probably immeasurably small until the 19th century (Kates *et al.*, 1990).

2.1.   CARBON STARVATION

Enhanced atmospheric CO_2 concentrations are known to increase the rate of photosynthesis, and to increase the amount of C which can be fixed from the atmosphere per

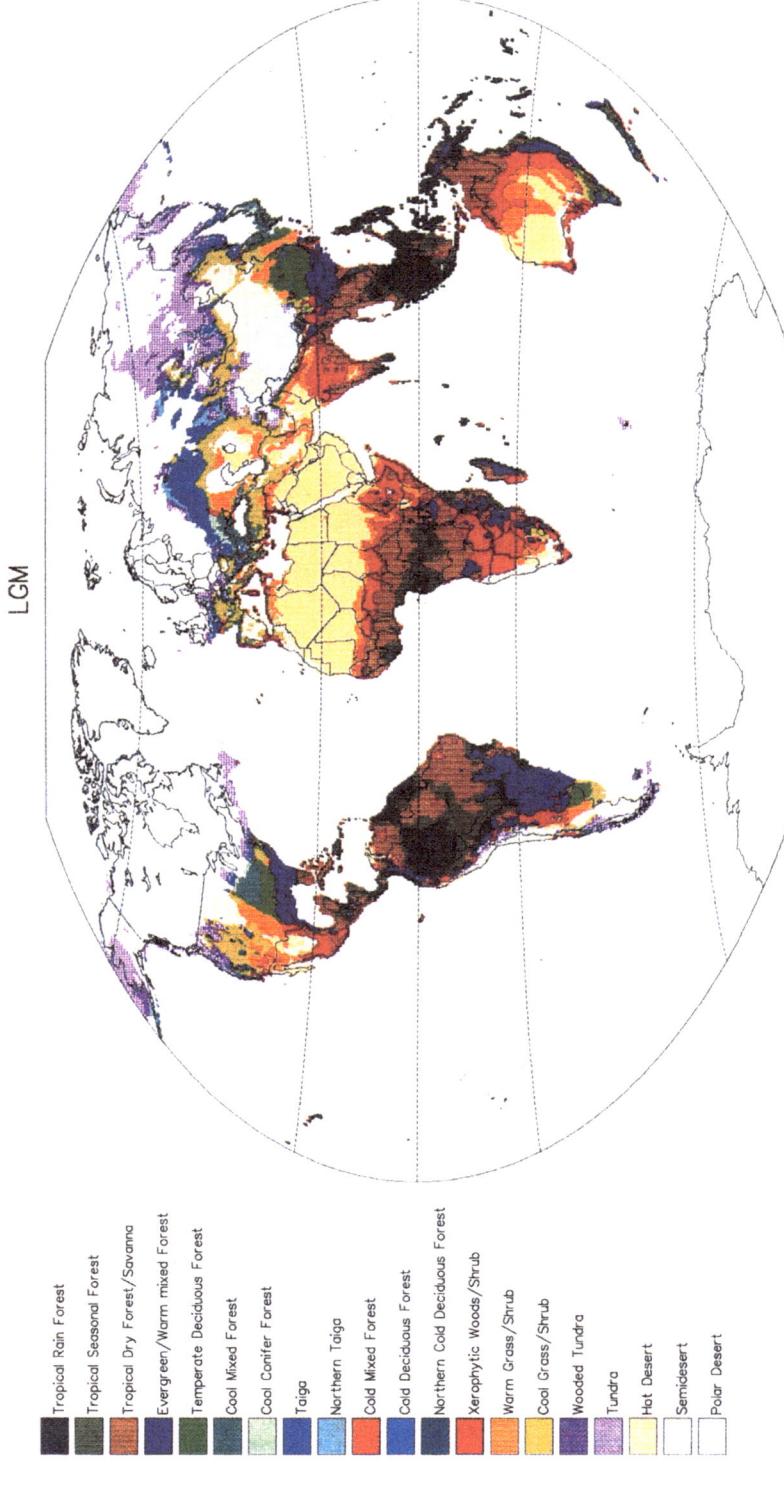

LGM

Tropical Rain Forest
Tropical Seasonal Forest
Tropical Dry Forest/Savanna
Evergreen/Warm mixed Forest
Temperate Deciduous Forest
Cool Mixed Forest
Cool Conifer Forest
Taiga
Northern Taiga
Cold Mixed Forest
Cold Deciduous Forest
Northern Cold Deciduous Forest
Xerophytic Woods/Shrub
Warm Grass/Shrub
Cool Grass/Shrub
Wooded Tundra
Tundra
Hot Desert
Semidesert
Polar Desert

Figure 1a.

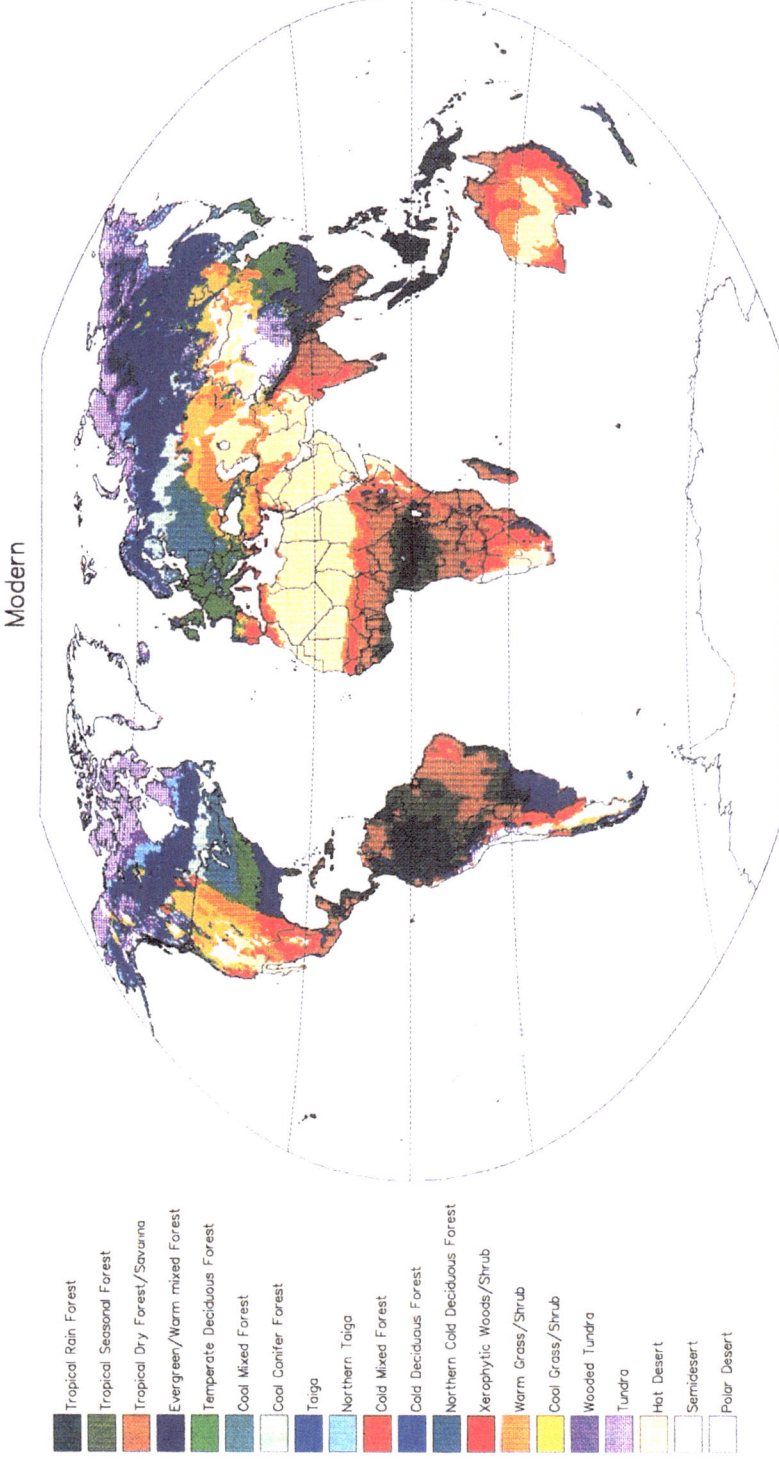

Modern

Tropical Rain Forest
Tropical Seasonal Forest
Tropical Dry Forest/Savanna
Evergreen/Warm mixed Forest
Temperate Deciduous Forest
Cool Mixed Forest
Cool Conifer Forest
Taiga
Northern Taiga
Cold Mixed Forest
Cold Deciduous Forest
Northern Cold Deciduous Forest
Xerophytic Woods/Shrub
Warm Grass/Shrub
Cool Grass/Shrub
Wooded Tundra
Tundra
Hot Desert
Semidesert
Polar Desert

Figure 1b.

unit of water used. In glasshouse environments these effects translate into increased production and biomass especially if water availability is reduced. There is still no direct evidence for such effects in forested wildland vegetation. But if they occur today, then they also must have operated with reversed effects during the LGM; the 190 to 200 ppmv of CO_2 measured in ice cores from the LGM (e.g., Raynaud *et al.* 1993) approaches the minimum concentration (50-100 ppmv) at which C_3 plants can grow under natural conditions. Solomon (1984) suggested that low CO_2 concentrations during the LGM could explain the dearth of late-successional tree species at that time. Combined with a weaker hydrological cycle associated with colder global climate, and greater requirements for water per unit of carbon fixed, low CO_2 concentrations could be responsible for the open park-like woodlands suggested by fossil pollen evidence from unglaciated eastern North America (Webb, 1987).

2.2.   DELAYED TREE IMMIGRATION

A second process of potential importance in regulating rates of C sequestration in the terrestrial biosphere is delayed immigration of trees. Climate may have changed much more rapidly than tree populations can spread by continual and sequential transport of seed, growth to maturity and subsequent transport of seed. This process could be responsible for the absence of tree species that can take maximum advantage of local conditions to accumulate greater C density than the less suitable species that they replace.   M. B. Davis (1976, 1981, 1983) hypothesized that tree species arrival in regions to which they may have been adapted since early in the Holocene could have been delayed by up to several thousand yrs.   However, no definitive evidence has emerged that any barriers other than major water bodies or large mountain ranges have kept species from migrating as rapidly as climate allowed (e.g., Davis *et al.*, 1986; Webb, 1986).   Indeed, recent analyses indicate few anomalies in taxonomic composition during the Holocene (Overpeck *et al.*, 1992) suggesting that continuous climate changes during the whole period since the LGM have paced species migrations on a millennial time scale (Davis, 1990; Prentice *et al.*, 1991). Delayed immigration under rapid future warming could exert a stronger effect because climate is expected to change at a rate about an order of magnitude faster than measured species migration rates (Solomon *et al.*, 1984; Davis, 1990).

2.3.   SOIL CARBON ACCUMULATION

C starvation and delayed immigration are factors that could influence the amount of C stored in terrestrial systems at any specified time.   However, the behavior of the terrestrial biosphere as a continuous source or sink for atmospheric C over long time periods is ultimately controlled by soil processes. Areas too cold or too dry to support decomposition rates which match or exceed net primary production rates accumulate organic C. The process is illustrated by today's organic C content of glaciated soils that were originally free of organic C (e.g., Harden et al, 1993), a process which has continued throughout the Holocene. Net accumulation of soil C in a warmed future will decrease through increased net soil respiration if cold soils warm in the presence of moisture, or dry soils become moist in the presence of warm temperatures.

3. Carbon Cycling in the Modern and Future Terrestrial Biosphere.

The C content of today's terrestrial biosphere has been calculated from isolated samples of above and below-ground vegetation. For the pre-industrial modern world without agriculture, the biomass estimate by Whittaker and Likens (1975) of 827 Pg, and by Matthews (1983) of 840 Pg compares well with the Biome 1.0 model projection of 754 Pg (Cramer and Solomon, 1993) and Biome 1.1 model projection of 811 Pg (Prentice and Sykes, 1993; Table 1). The

Table 1. C in Petagrams (1 X 10^{15} g) stored above and below ground under differing climates (Prentice and Sykes, 1993)

	Biomass	Soil	Total C
LGM Climate (18 Kyr)	682	1171	1853
Today's Climate	811	1356	2167
GFDL Future Climate	891	1292	2183

Biome model versions classify static vegetation geography, defining vegetation classes by climate thresholds determined from plant physiology, moisture capacity of soils and dominance hierarchies of plant functional types. Biome 1.1 differs only in recognizing subdivisions of taiga and tundra with characteristic carbon storage values. Soil C for undisturbed soils has been calculated by Zinke et al. (1984) at 1400 Pg, and by Schlesinger (1977) at 1456 Pg, compared to the soil C estimate from Biome 1.0 of 1367 (Cramer and Solomon, 1993) and from Biome 1.1 of 1356 Pg (Table 1). The calculated and modeled estimates are similar enough to allow consideration of modeled estimates driven by future climate projections.

The Biome 1.1 model estimates of C stocks (Prentice and Sykes, 1993 and Table 1) under the GFDL future climate change scenario (Manabe and Wetherald, 1987) increase above-ground (80 Pg), decrease below-ground (64 Pg), and increase slightly in total C (16 Pg in a budget of 2,200 Pg). Other future climate scenarios (e.g., OSU, Schlesinger and Zhao, 1989; GISS, Hansen et al., 1983; UKMO, Mitchell, 1983) generate the same trends in sources and sinks although C stock values differ somewhat among scenarios (41-103 Pg increases, Cramer and Solomon, 1993). Other analyses of terrestrial C stocks under future changed climate using the Holdridge Life Zone Classification, reach qualitatively similar conclusions (e.g., Sedjo and Solomon, 1989; Leemans, 1989; Prentice and Fung, 1990; Smith et al., 1992; King and Neilson, 1992), i.e., the terrestrial biosphere stores more C under the stable climate and vegetation of a doubled atmospheric CO_2 concentration than it does under present conditions. The results of Cramer and Solomon (1993) indicate that the increase in above-ground C stocks is derived from replacement of low-C density tundra by high-C density boreal forest, while other low and high C density biomes essentially replace one another without significant C gain or loss. Meanwhile, below-ground C losses reflect increased area of low-C soils in tropical regions, and loss of high-C soils in tundra regions again with other replacements essentially canceling out C losses with gains.

We expect that the history of the next 100 to 200 yr will document that the foregoing estimate of global C cycle response to future warming is very inaccurate, probably much less accurate than even the values which result from our reconstructed patterns of postglacial C storage and release. Several difficulties are apparent. Even if the effects of processes involving climate change and C fertilization, which probably dominated C stocks during postglacial time, were the only ones of importance in the future, the foregoing assessment of future C fluxes would be improbable because it assumes significant conditions which are patently false. The foregoing estimate of terrestrial C stocks assumes that no manipulation of C stocks will be induced by new governmental policies, that the successional state of forests will be the same in the future as it is now, that the productivity and C storage capacity of biomes in the future will be as great as it is now, that species adapted to new climates will be available to grow there, that land-use in the form of agriculture will be unchanged, and that biomes with their intact soils and C content can be reformed in new areas instantaneously.

3.1   PROCESSES ENHANCING TERRESTRIAL CARBON STORAGE

3.1.1 Fertilization of forest growth by atmospheric CO_2 .

This topic is the opposite side of the coin from that discussed as "C Starvation" above. Assumed to be a major unmeasured terrestrial C sink (if not "the missing sink") for many years, C fertilization is reviewed in detail by Vloedbeld and Leemans (1993). In essence, it is known in glasshouses to simultaneously increase C fixation, decrease water use, and increase allocation of C to storage organs, particularly to the roots. Because the long lived forests both store the majority of the global terrestrial C, and because their long turnover times identify forests as a future means for C storage management practices, the fertilization question is largely a forest biology question. Absence of certain critical biological information (Körner, 1993) makes current modeling attempts provisional, but the process must be quantified.

3.1.2. Forest management for C sequestration.

This topic is considered in detail both in papers in this volume (e.g., Sampson, 1993; Trexler, 1993; Wright and Hughes, 1993; Winjum *et al., 1993*; Kursten and Burschel, 1993; Dixon *et al., 1993*; Sampson *et al., 1993*). Clearly, mitigation efforts could have a measurable impact on atmospheric CO_2 concentrations, if the efforts enhance C storage in vegetation over variable times (10s to 100s of yrs), if enough young forests can be continuously cropped for forest products, or if C released in one sector is offset by its short-term replacement elsewhere. Models must be developed which allow the simulation of effects by management approaches if we are to calculate their potential value.

3.1.3. Synchronized early forest succession.

Evidence is accumulating from several forest zones (e.g., Boreal forests, Apps and Kurz, 1993; temperate zones, Kauppi *et al.*, 1992; tropical zones, Brown *et al.*, 1992; Brown, *et al.*, 1993) that recent disturbance (by burgeoning human populations?) has produced much more young forest than expected, in essence "synchronizing" the majority of forests into an early successional state. These forests may be significant net C sinks, at least until succession proceeds to more slowly-growing species. If a large proportion of global forest stocks during the next 50 yr or so is replaced by agriculture, or is preserved from recutting, this sink would essentially disappear. This suggests that we must include forest successional processes in C cycle models if they are to project changes over several decades or longer.

Other processes which can produce significant enhancement of terrestrial C storage are also possible but the foregoing seem most important at present. Processes likely to result in net C releases to the atmosphere in the future also need to be incorporated into models. The most important of these are described briefly below.

3.2. PROCESSES ENHANCING TERRESTRIAL CARBON RELEASE

3.2.1. Forest dieback.

The climate shift over the next 100 yrs is projected at a rate so fast that many forest species will be subject over much of their geographic ranges to climates which they do not tolerate today (e.g., Solomon et al., 1984; Davis and Zabinski, 1991). The increasing climate stress is expected to cause widespread tree mortality, a process taking a few years, with subsequent forest regrowth requiring many decades to centuries even in the absence of continuing climate stress. Forest succession model experiments simulated a release 11 Mg C ha^{-1} during 100 yrs (to CO_2 doubling) in eastern North America despite concurrent afforestation in the highest latitudes (Solomon, 1986). In addition, the current presence of synchronized early succession implies a synchronized late-succession in 50 to 100 yrs unless the agents which caused the forest disturbance continue to operate. As forests age, they become increasingly sensitive to stress and to resulting mortality; if synchronized, such dieback would generate a "pulse" of CO_2 release to the atmosphere (Solomon, 1986; King and Neilson, 1992; Smith and Shugart, 1993).

3.2.2. Delayed immigration.

Unlike the apparent similarity of warming rates and migration rates during postglacial time, the migration northward by the projected July isotherm of 400 to 600 km century^{-1} (Solomon et al., 1984) under General Circulation Model (GCM) warming scenarios is about an order of magnitude faster than the fastest tree migration rates calculated from pollen data deposited during the past 20,000 yrs (e.g., Davis, 1983; Gear and Huntley, 1989). The great global latitudinal belts of intense agriculture as barriers to south-north migration are likely to reduce migration speed compared with the slow Holocene rates in many regions. If climate induced by doubled CO_2 requires a century to appear, Solomon and Bartlein (1992) calculated that 50 to 100 yrs later appropriate tree species for a given region would disappear, while new ones would be too far away to immigrate. The result is that the residence time of CO_2 in the atmosphere could increase as species-depauperate forests decline in C capacity over large areas.

3.2.3. Tree life cycles.

A related set of considerations involves life cycles of trees; temperate zone tree species reach reproductive maturity in closed stands some 45 to 50 yrs after establishment, and boreal conifers, after 25-30 yr. The conditions needed by current seedlings may not be present if trees are planted which can survive as adults some 25-50 yr later. Conversely, seedlings appropriate to today's conditions could possess adult stages which cannot survive the future climate (Solomon and West, 1985; Solomon and Leemans, 1990).

In addition, the conditions needed by each species to establish, to grow to maturity and to produce viable seeds differ. Of course, by definition, these differing conditions occur in the same locality within each species' geographic range. The rapid change of certain climate variables, e.g., winter temperature, suggests the disintegration of covarying conditions required for life cycle completion during less than a life cycle of time (ibid.). Populations of trees unable to complete their life cycles become extinct. The processes decoupling tree life cycles from climate and delaying migration have similar effects on C cycling; they act to lower the C carrying capacity of forests by limiting species to those which can grow quickly and can migrate quickly. These are early successional, sun-loving species which grow rapidly but without the large C carrying capacities exhibited by late successional forests in most places.

3.2.4. Soil Carbon losses.

Considerable debate exists about the potential losses of C sequestered in soils (e.g., Scharpenseel, 1993). The concern involves a C pool which is probably twice the size of the above-ground C stocks (Zinke et al., 1984). Predictions of global warming suggest increased soil respiration, a process which may release C at a greater rate than it can be stored by photosynthesis. Raich (this conference) documented process-level considerations to indicate that warm, moist areas of the tropics, subtropics and warm temperate zones are most likely to generate such a pulse of soil C to the atmosphere. Other workers have concentrated on the much larger stocks of C in high latitude soils and peats, visualizing their release both as CO_2 and as CH_4 when temperature-sensitive permafrost soils warm. The soil C cycle eventually could be the most important process in defining transient C pools of the terrestrial biosphere.

3.2.5. Increasing agricultural land use.

A global population which is doubling every 35 yr will require increased food supplies from dwindling stocks of farmland. Warming of winter temperatures should greatly increase the length of high latitude growing seasons, increasing the amount of land which is climatically appropriate for agriculture (Cramer and Solomon, 1993; Leemans and Solomon, 1993). The vegetation of that land is almost entirely forested at present (Cramer and Solomon, 1993). Conversion of forests to agriculture involves destruction and release to the atmosphere of above-ground biomass and subsequent decline of soil C.

The foregoing transient processes will not be as easily modeled as have been the constraints by climate. The time required for each process to operate varies, although it may be critical to policy considerations to note that the forces increasing net C sequestration require short time periods to implement (yrs to decades) while those increasing net C release do so over longer time periods (decades and centuries to millennia).

4. Potential Interaction of Climate Change and Land Use

What might be the effect on the global C cycle of these positive and negative feedbacks? There are few estimates of direct effects on total global terrestrial C storage of C fertilization and water use efficiency although oceanographers (e.g. Tans *et al.*, 1990) and atmospheric physicists (e.g., Wigley, 1992) have attributed 1 to 2 Pg yr-1 to this sink, beyond all other terrestrial sources and sinks of C. Effects of climate change and continued deforestation of only the tropics suggests a C source of 4.2 to a sink of 1.3 Pg yr-1 (Sampson *et al.*, 1993). Efficient vegetation management and management of vegetation specifically for C sequestration are thought to be capable of adding a C sink of 0.4 to 6.8 Pg yr-1 (Ibid.). Smith and Shugart (1993) calculated effects of transient vegetation response as constraints on the global terrestrial C cycle, following an instantaneous climate change, for the GFDL and GISS scenarios, respectively, by parameterizing and aggregating C flux from regrowth after delayed migration (+0.22 to +0.26 Pg yr-1), forest succession (+0.41 to +0.66 yr-1), forest dieback (-2.48 to-2.39 yr-1), and soil C decomposition (-3.94 to -4.27 yr-1). It is notable that most procedures expected to sequester C by vegetation management, and most of the transient and climate change effects which result in C release, are mutually exclusive processes, i.e, they cannot both occur in the same place because the latter assumes stresses which would eliminate the former.

Summing these fluxes (tropical land use @ -4.2 to +1.3 Pg; efficient management @ +0.4 to +6.8 Pg; transient responses depending on climate scenario, @ -5.79 or -5.74 Pg) engenders a terrestrial C flux of -9.5 or 9.6 to +2.3 or 2.4 Pg yr-1 during the next 50 to 100 yr, disregarding the possible countervailing effect of CO_2 fertilization. The C fluxes summed here also do not include estimates of effects of shifting land uses permitted by climate change. These are assessed below.

4.1   PROJECTING CLIMATE-DICTATED AGRICULTURAL LAND USE

The approach we used for estimating agriculture effects is simple but practical; we defined an additional climate-dictated agriculture "biome" in the Biome Model of Prentice *et al.* (1992). If agriculture is possible in a given location, it replaces any other biome (Cramer and Solomon, 1993). Mapped arable land (e.g., Hummel and Reck, 1979; Olson *et al.*, 1983) is shown in Figure 2a. It is located within an "envelope" bounded by 2000 growing degree days (base 0 oC) and a ratio of actual evapotranspiration (AET) to potential evapotranspiration (PET) greater than 0.45 as calculated from the IIASA climate data base (Leemans and Cramer, 1991). If mean coldest-month temperatures greater than 15.5 oC (the effective frost limit) occur and are accompanied by AET:PET exceeding 0.7 (Cramer and Solomon, 1993), the land also was excluded from the envelope, an area corresponding to the wettest of tropical regions where leaching of soil nutrients may make intensive and continuous cropping impossible.

The resulting geographic pattern of agricultural land (Figure 2b) provides a good match for the actual pattern (Figure 3). The cold and dry borders of agriculture in temperate and southern boreal regions are reproduced especially well. The most serious deficiency is the amount of land within the envelope; by definition, the envelope covers all land, while measured patterns demonstrate a more irregular distribution (Figure 2a). This increased the total global agriculture land base defined by Olson, et al. (1983) by 39.5% and is most clearly problematic in tropical regions. We therefore reduced the agricultural land within the envelope. Part of the land which Olson et al. (1983) classed as agricultural is irrigated (ca. 16%), primarily in areas which the envelope defines as too dry to support crops. Hence, we assumed only 50.5% of the land within the agricultural envelope (in all biomes) was devoted to agriculture and the C pool effects of agriculture were calculated accordingly (Cramer and Solomon, 1993).

Use of the Biome Model with an agriculture biome (Table 2) assumed the removal of all

Table 2. C in Petagrams (1×10^{15} g) stored above and below ground under differing climates and land use scenarios (Cramer and Solomon, 1993).

	Biomass	Soil	Total C
Modern Climate			
Pre-industrial C	754	1367	2121
Agricultural C	602	1312	1914
GFDL Climate			
C without Agriculture	834	1303	2137
C with Agriculture	622	1237	1859

above-ground biomass and 20% of soil C (Mann, 1986) from agricultural land. Olson et al. (1983) estimated above-ground low, medium and high biomass values of 360, 558, and 807 Pg, compared to our values of 383, 602, and 829 Pg, respectively. Our soil C estimates compare well with those mentioned earlier; we project a value of 1367 Pg, Zinke et al. (1984) tabulated 1400 Pg and Schlesinger (1977) estimated 1456 Pg. [Note that the modern terrestrial biosphere C estimates from our modified Biome 1.0 without agriculture (Table 2) are not precisely the same as those generated by Biome 1.1 in Table 1].

4.2.   CLIMATE-DICTATED AGRICULTURE UNDER FUTURE CLIMATES

The shift in geographic area capable of supporting agriculture is illustrated in Figures 3a and 3b. We chose the GFDL climate scenario because the GFDL model was the most accurate GCM at reproducing the variables used for the agriculture projections (Cramer and Solomon, 1993). Because most of the change occurs in the northern hemisphere, we have redrawn the map projection used in Figure 2 as a polar projection of today's (Figure 3a) and GFDL future (Figure 3b) potential agricultural land superimposed on natural biomes.

The shift in agricultural land is extraordinary. All areas today covered by undisturbed circumpolar boreal forest virtually disappear under a cover of potential agriculture. Arable land fills the Scandinavian peninsula and all of European Russia along with the majority of Siberia, while the Northwest Territories of Canada and much of Alaska supports crops under the GFDL climate scenario. The agricultural invasion is almost entirely of now forested land (cf. Figure 1b vs. 3a, b), covering much of the area now apparently a northern hemisphere C sink (e.g., Kauppi et al, 1992; Kolchugina and Vinson, 1993; Apps and Kurz, 1993) and being considered for C mitigation through tree planting activity (Dixon et al., 1993).

The difference between C stocks under modern climate, and those under a doubled

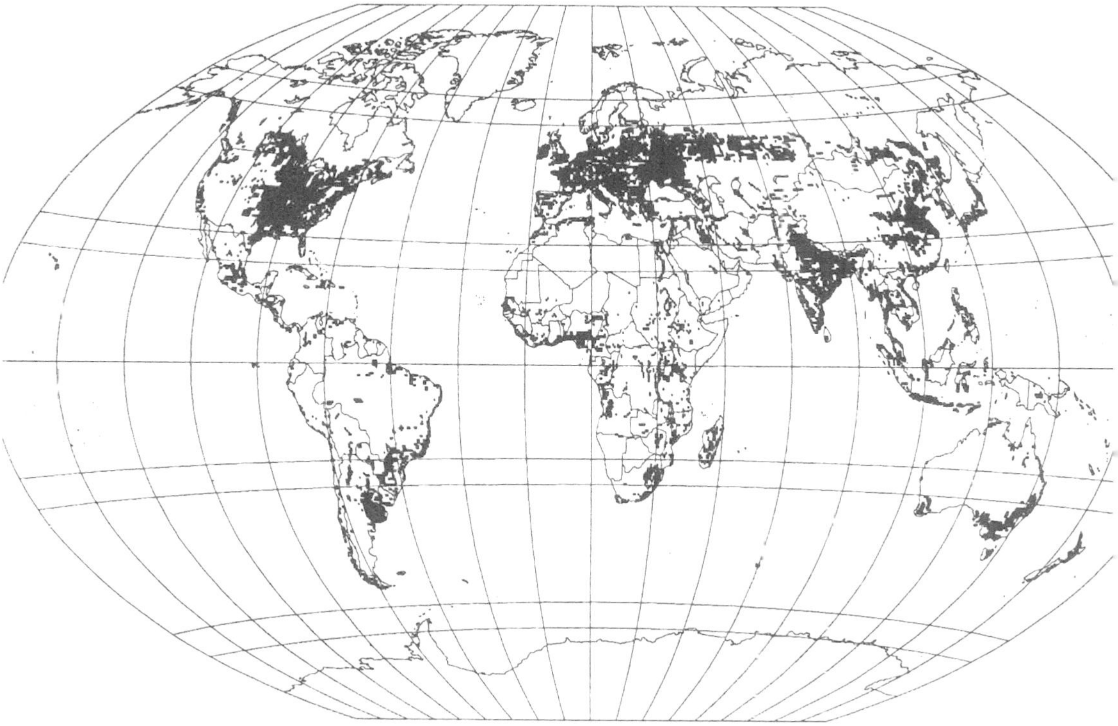

Figure 2a.

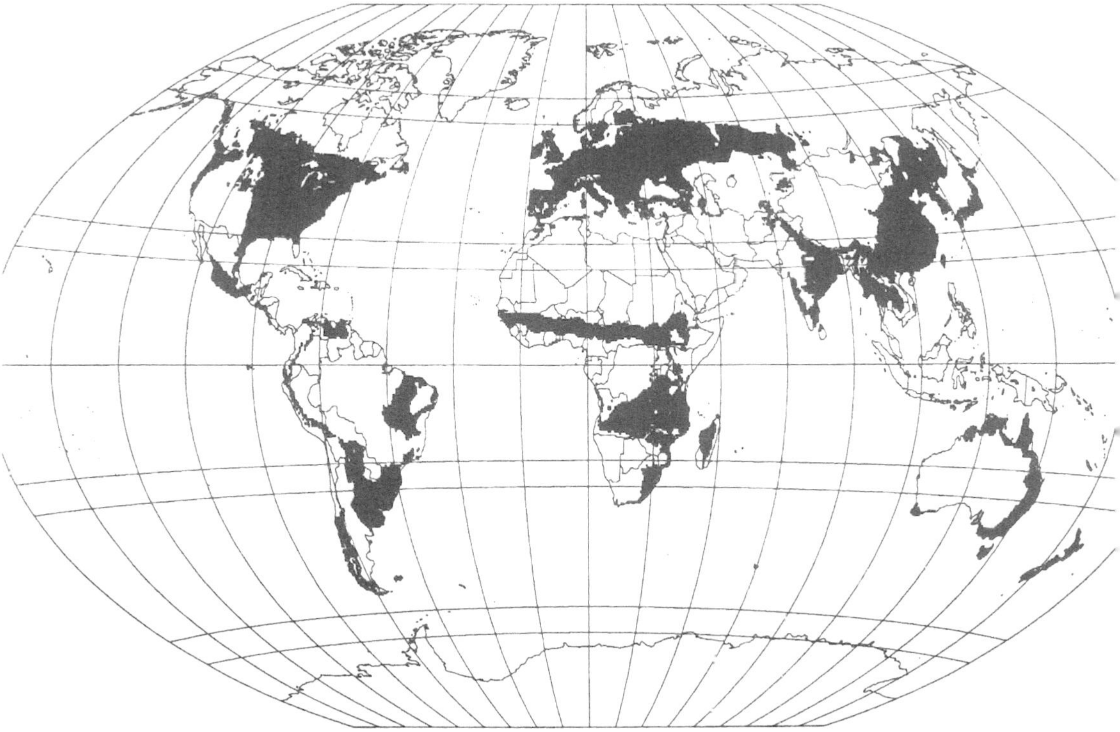

Figure 2b.

Potential Vegetation and Agricultural Land Use (CUR)

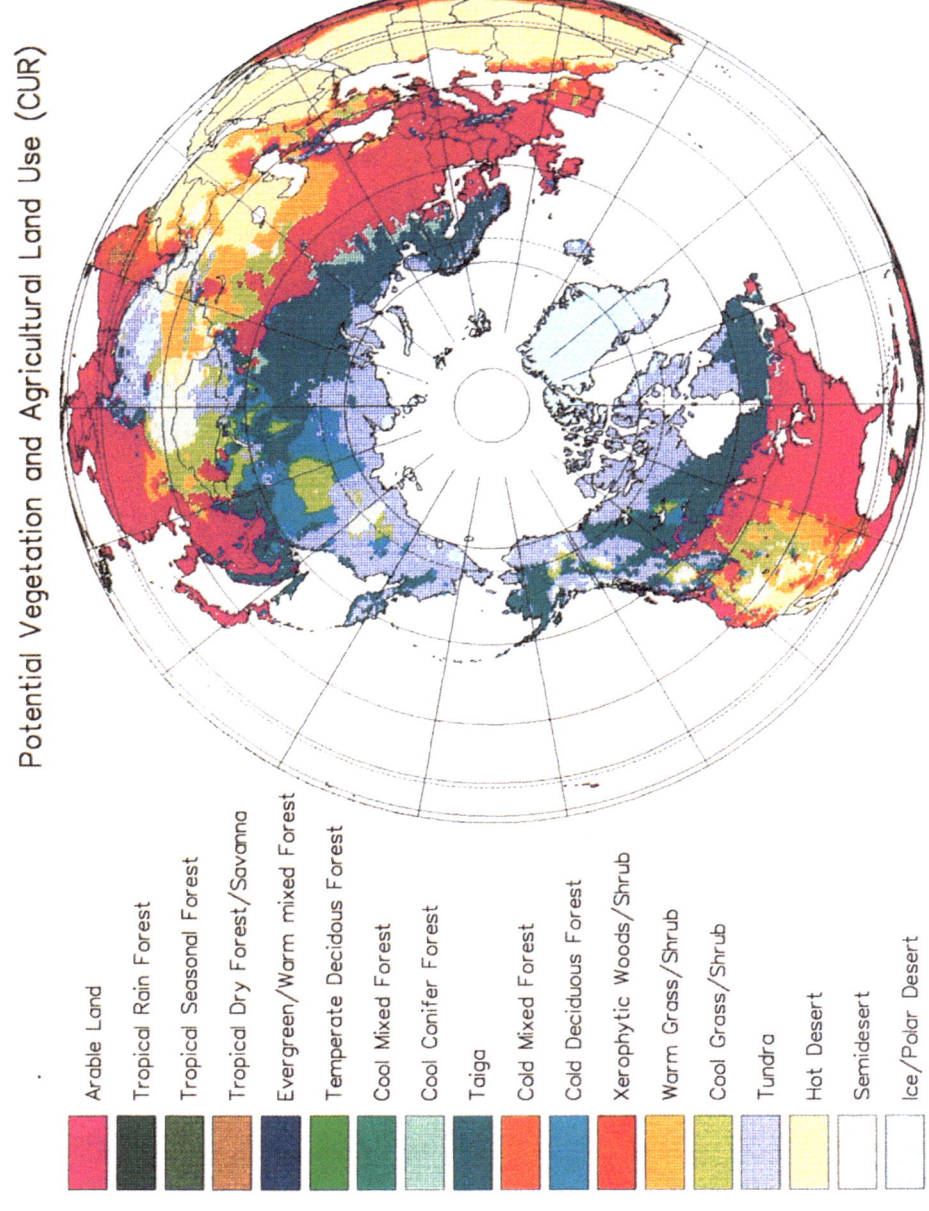

Arable Land
Tropical Rain Forest
Tropical Seasonal Forest
Tropical Dry Forest/Savanna
Evergreen/Warm mixed Forest
Temperate Deciduous Forest
Cool Mixed Forest
Cool Conifer Forest
Taiga
Cold Mixed Forest
Cold Deciduous Forest
Xerophytic Woods/Shrub
Warm Grass/Shrub
Cool Grass/Shrub
Tundra
Hot Desert
Semidesert
Ice/Polar Desert

Figure 3a.

Potential Vegetation and Agricultural Land Use (GFDL)

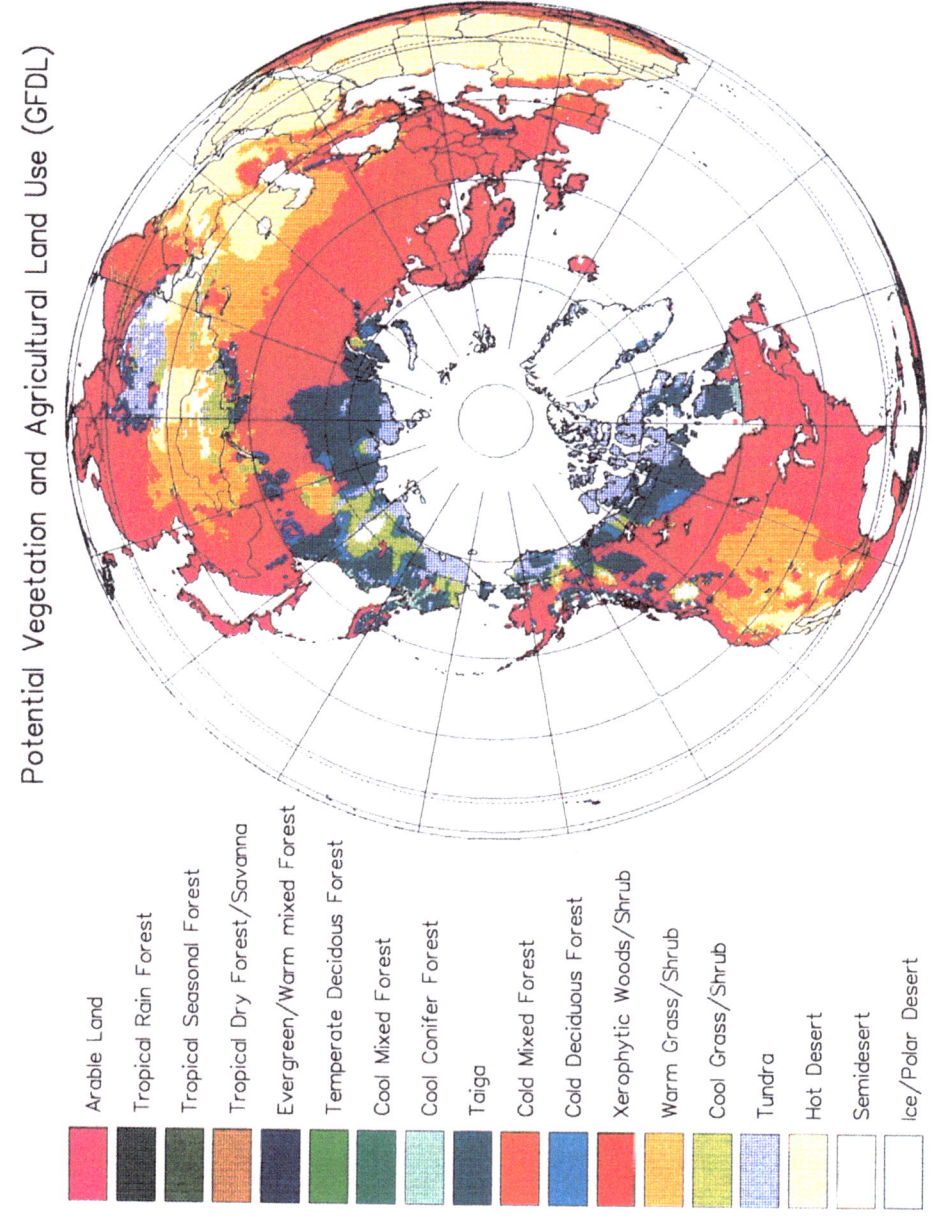

Arable Land
Tropical Rain Forest
Tropical Seasonal Forest
Tropical Dry Forest/Savanna
Evergreen/Warm mixed Forest
Temperate Decidous Forest
Cool Mixed Forest
Cool Conifer Forest
Taiga
Cold Mixed Forest
Cold Deciduous Forest
Xerophytic Woods/Shrub
Warm Grass/Shrub
Cool Grass/Shrub
Tundra
Hot Desert
Semidesert
Ice/Polar Desert

Figure 3b.

concentration of atmospheric CO_2, is significant. With agriculture included in the calculation, C stocks increase above ground (20 Pg) decline in soils (-75 Pg) and thus, decline in the terrestrial biosphere as a whole (-55 Pg; Table 2). Geographic patterns apparent in Figures 3a and b suggest that the loss of C in soils occurs primarily because of the decline in area covered by high-C density boreal and tundra soils, and secondarily, in broadleaved evergreen forests. These suggestions are borne out by the C storage estimates by biome type provided in Table 3.

Table 3: Estimates of total above and below-ground C content in Petagrams, stored in major biomes today and under greenhouse climate (derived from GFDL model). The values were generated by Cramer and Solomon (1993) using the BIOME model described by Prentice *et al.* (1992).

BIOME NAME	Current biomass	Current soil C	Current Total C	GFDL biomass	GFDL soil C	GFDL Total C
TROPICAL FORESTS -						
Tropical Rain Forest	118	88	206	127	94	221
Tropical Seasonal Forest	111	83	194	118	88	206
Tropical Dry Forest	50	116	166	58	132	190
Xerophytic Woods/Shrubs	39	77	116	43	84	127
TEMPERATE FORESTS -						
Broadleaved Evergreen Forest	29	70	99	42	100	142
Temperate Deciduous Forest	30	81	111	37	101	138
Cool Mixed Forest	26	60	86	50	116	166
Cool Evergreen Forest	35	44	79	46	60	106
BOREAL VEGETATION -						
Cold Mixed Forest	1	7	8	1	5	6
Taiga	112	223	335	64	149	213
Cold Deciduous Forest	17	47	64	4	10	14
Tundra	7	160	167	2	46	48
GRASSLANDS/DESERTS -						
Hot Desert	6	58	64	6	60	66
Warm Grasslands	14	101	115	20	144	164
Semidesert	3	31	34	2	18	20
Cool Grasslands	4	66	70	2	30	32
Ice/Polar Desert	0	0	0	0	0	0
TOTALS	602	1312	1914	622	1237	1859

The disparity between biome estimates of warming impacts without agriculture (+16 Pg) and those with sparse agriculture (-55 Pg) is 71 Pg. Warming impacts with dense agriculture (-168 Pg) generates a 152 Pg difference. Cramer and Solomon used three other future climate scenarios to calculate the differences between no agriculture and sparse agriculture as 33 Pg (OSU scenario, Schlesinger and Zhao, 1989), 47 Pg (GISS scenario, Hansen *et al.*, 1983) and 83 Pg (UKMO scenario, Mitchell, 1983). In the perspective of reaching a climate associated with doubling of greenhouse gases in 50-100 yrs (e.g., IPCC, 1992), the averaged differences imply a terrestrial C source of about 0.6 to 1.2 Pg yr-1 which is not accounted for in model runs that exclude impacts of climate-mediated change in agriculture on the global C cycle. The same logic applied to a world in which all potential agricultural land was farmed produces a range of 1.4 to 2.8 Pg yr-1. These values exclude current estimates of non-climatically induced tropical forest destruction, which result in that biome functioning as an average C source of 1.2-2.2 Pg yr-1 (Brown, *et al.*, 1993; Sampson, *et al.*, 1993).

The foregoing model experiment provides an initial estimate of the impact agriculture could play in a future warmed world. History teaches us that the land use-related loss of 60 to 140 Pg of C in a century is not extraordinary. Houghton (1985) calculated that about 158 Pg of terrestrial C had been released to the atmosphere between 1860 and 1980. Solomon *et al.* (1985) surveyed literature to tabulate a range of 90 to 180 Pg C flux from the land since 1800. These losses occurred in support of a much smaller global population, working with much less mechanized equipment than one would expect to be operating in the next 50 to 100 yr.

5. Discussion

The preceding analysis defined roles in the global terrestrial C cycle played by the natural processes controlling C flux, that is, the processes which will function as an indirect consequence of human activities on earth, rather than as a direct response to planned interventions. We first examined the processes controlling carbon fluxes of the past 20,000 yr, concluding that dominance by soil carbon dynamics generated a long-term C sink in the terrestrial biosphere, with effects of C starvation and decreased water use efficiency of probable importance during the LGM. We then examined the processes controlling present and future C fluxes, deciding that these will be very different from those dominating prehistoric C dynamics. The small carbon sink generated by soil C dynamics during the past 10,000 yr would be unlikely to produce much effect in a rapidly changing climate of the next century or so, while the processes which do respond to rapid climate change are predominantly associated with C release (e.g., forest dieback, incomplete tree life cycles, delayed tree immigration, soil C oxidation, and climate-mediated increases in potential agricultural areas).

We most closely examined the role in future C flux to be played by new agricultural land uses which would be permitted by a rapidly warming climate. Our analysis projected a source of C of 0.6 to 1.2 Pg yr-1 based on averaging four future climate scenarios and on imposition of climate from a doubled CO_2 within 50 to 100 yr. The analysis assumed that future agriculture would occupy the same proportion of the land on which climate is adequate for agriculture today. That assumption may be conservative because climatically suitable but marginally productive farmland not used at present may be farmed if global population continues doubling every 35 years. Cramer and Solomon (1993) found 1.4 to 2.8 Pg C yr-1 was released from the terrestrial biosphere under the four climate scenarios if all land climatically suitable for agriculture was counted, than with a constant proportion of farmed to potentially-farmed land. The most probable contribution to terrestrial C release of future agriculture lies between these extremes.

The processes which reduce the amount of potentially-farmed land to the amount of land farmed depend upon the size of the human population to be supported, the level of mechanization of the farms, the nature of the soils, and the topographic relief. Of these, the first two vary over time and may be expected to increase the future proportion of land farmed.

The last two are temporally invariant, and can be predicted now with the proper data sets. For example, the thin soils of the Laurentian Shield in boreal Ontario and Quebec and the steep slopes of the Rocky Mountains are likely to be unproductive agriculturally under any future climate, while the rich peatlands of northcentral Siberia or of the Northwest Territories of Canada could become exceedingly productive if they could be successfully drained.

Some workers believe that the future C releases projected here could be reduced or even eliminated if the natural processes of C dynamics are modified by direct human intervention. A biomass management working group (Sampson *et al.*, 1993) estimated the effects of such intervention techniques if fossil fuels were replaced by fuels from biomass and tree farms optimized carbon sequestration. They calculated 3.8 Pg yr^{-1} would be sequestered if the international community was unanimously agreeable and successful at implementing these measures above local needs, and if the conflicting processes we discuss, which cause slow growth and tree dieback, do not operate.

6. References

Apps, M. J., Kurz, W. A., Luxmoore, R., Nilsson, L.-O., Sedjo, R. A., Schmidt, R., Simpson, L. G., and Vinson, T. S.: 1993, This volume.

Apps, M. J. and Kurz, W. A.: 1993, This volume.

Botkin, D. B. and Simpson, L. G.: 1990, *Biogeochemistry* 9, 161.

Botkin, D. B. and Simpson, L. G.: 1993, This volume.

Brown, S., Lugo, A. E., and Iverson, L. R.: 1992, *Water, Air, and Soil Poll.* 64, 139.

Brown, S.: 1993, This volume.

COHMAP Project Members: 1988, *Science* 241, 1043.

Cramer, W. P. and Solomon, A. M.: 1993, *Climate Research*, in press.

Davis, M. B.: 1976, *Geosci. Man* 13, 13.

Davis, M. B.: 1981, Quaternary history and the stability of forest communities, West, D. C., Shugart, H. H. and Botkin, D. B. (eds), *Forest Succession: Concepts and Application*, Springer-Verlag, pp. 132-154.

Davis, M. B.: 1983, *Ann. Miss. Bot. Gardens* 70, 550.

Davis, M. B.: 1990, Research questions posed by the paleoecological record of global change, Bradley, R.S. (ed), *Global Changes of the Past,* OIES, UCAR, Boulder, pp. 385-395.

Davis, M. B., Woods, K. D., Webb, S. L. and Futyma, R. P.: 1986, *Vegetatio* 67, 93

Davis, M. B. and Zabinski, C.: 1992. Changes in geographical range resulting from greenhouse warming: Effects on biodiversity in forests, Peters, R. L. and Lovejoy, T. E. (eds), *Global Warming and Biological Diversity*, Yale Univ. Press, pp. 297-308.

Dixon, R. K.: 1993, This volume.

Gear, A. and Huntley, B.: 1991, *Science* 251, 544.

Harden, J. W., Sundquist, E. T., Stallard, R. F., and Mark, R. K.: 1992, *Science* 258, 1921

Gorham, E.: 1991, *Ecol. Applications* 1, 182.

Hansen, J., Russell, G., Rind, D., Stone, P., Lacis, A., Lebedeff, S., Ruedy, R. and Travis, L.: 1983, *Monthly Weather Rev.* 111, 609.

Houghton, R. A.: 1985, Estimating changes in the C content of terrestrial ecosystems from historical data, Trabalka, J.R. and Reichle, D. E. (eds.), *The Changing C Cycle: A Global Analysis*, Springer-Verlag, pp. 175-193.

Hummel, J. R. and Reck, R. A.: 1979, *J. Appl. Meteorol.* 18, 239.

Huntley, B. and Webb, T., III: 1988, *Vegetation History*, Kluwer.

Kates, R. W., Turner, B. L., and Clark, W. C.: 1990, The great transformation., Turner, B. L., Clark, W. C., Kates, R. W., Richards, J. F., Matheews, J. T., and Meyer, W. B. (eds.), *The Earth as Transformed by Human Action*, Cambridge Univ. Press, pp. 1-19.

Kauppi, P. E. and others: 1993,This volume.

Kauppi, P. E., Mielikäinen, K. and Kuusela, K.: 1992, *Science* 256, 70.

King, G. A., and Neilson, R. P.: 1992, *Water, Air and Soil Poll.* 64, 365.

Körner, C.: 1993, CO_2 fertilization: The great uncertainty in future vegetation development. Solomon, A. M. and Shugart, H. H., (eds), *Vegetation Dynamics and Global Change*, Chapman and Hall, pp. 53-70.

Kolchugina, T. P. and Vinson, T. S.: 1993, This volume.

Kursten, E.: 1993, This volume.

Leemans, R.: 1989, Possible changes in natural vegetation patterns due to a global warming, Hackl, A. (ed), *Der Treibhauseffekt: das Problem - Mögliche Folgen - Erforderliche Maßnahmen*, Akademie für Umwelt und Energie, 105.

Leemans, R. and Cramer, W. P.: 1991, *The IIASA Climate Database for Mean Monthly Values of Temperature, Precipitation, and Cloudiness on a Terrestrial Grid*, WP-90-18, International Institute for Applied Systems Analysis.

Leemans, R. and Solomon, A. M.: 1993, *J. Clim. Res.* (in press).

Manabe, S. and Wetherald, R. T.: 1987, *J. Atmosph. Sci.* 44, 12:1.

Mann, L. K.: 1986, *Soil Sci.* 142, 279.

Matthews, E.: 1983, *J. Clim. App. Meteorol.* 22, 474.

Mitchell, J. F. B.: 1983, *Quart. J. R. Met. Soc.* 109, 113.

Neilson, R. P.: 1993, This volume.

Olson, J., Watts, J. A. and Allison, L. J.: 1983, *Carbon in Live Vegetation of Major World Ecosystems, ORNL-5862*, Oak Ridge National Laboratory.

Orr, J.: 1993, This volume.

Overpeck, J. T., Webb, R. S., and Webb, T. III.: 1992, *Geology* 20, 1071.

Prentice, I.C., Bartlein, P.J., and Webb, T. III. 1991. *Ecology* 72, 2038.

Prentice, I. C., Cramer, W., Harrison, S. P., Leemans, R., Monserud, R. A. and Solomon, A. M.: 1992, *J. Biogeogr.* 19, 117.

Prentice, I. C., and Sykes, M. T.: 1993, Vegetation geography and global carbon storage changes, Woodwell, G.M. (ed),*Woods Hole Workshop on Biotic Feedbacks in the Global Climate System*, Oxford University Press, In press.

Prentice, I. C., Sykes, M. T., Lautenschlager, M., Harrison. S. P., Denissenko, O. and Bartlein, P. J.: 1993, *Global Biogeogr. Letters,* In press.

Prentice, K. C. and Fung, I. Y.: 1990, *Nature* 346, 48.

Raynaud, D., Jouzel, J, Barnola, J. M., Chappellaz, J., Delmas, R. J., and Lorius, C.: 1993, *Science* 259, 926.

Sampson, R. N.: 1993, This volume.

Sampson, R. N., Wright, L. L., Winjum, J. K., Kinsman, J D., Benneman, J., Kursten, E., and Scurlock, J. M. O.: 1993, This volume.

Scharpenseel, H. W.: 1993, This volume.

Schlesinger, M. E. and Zhao, Z.-C.: 1989, *J. Clim.* 2, 459.

Schlesinger, W. H.: 1977, *Ann. Rev. Ecol. Syst.* 8, 51.

Schneider, S. H. and Thompson, S. L.: 1981, *J. Geophys. Res.* 86, 3135.

Sedjo, R. A. and Solomon, A. M.: 1989, Climate and forests, Rosenburg, N. J., Easterling, W. E., Crosson, P.R., and Darmstadter, J. (eds.), *Greenhouse Warming: Abatement and Adaptation*, Resources for the Future, Washington, D.C.pp. 105-119.

Simpson, L. and Botkin, D.B.: 1993, This volume.

Smith, T. M. and Shugart, H. H.: 1993, *Nature* 361, 523.

Smith, T. M., Shugart, H. H., Bonan, G. B. and Smith, J. B.: 1992, *Adv. Ecol. Res.* 22, 93.

Smith, T. M.: 1993, This volume.

Solomon, A. M.: 1984, *Amer. Quaternary Assoc. Meeting Abstracts*, 120.

Solomon, A. M.: 1986, *Oecologia (Berlin)* 68, 567.

Solomon, A. M. and Bartlein, P. J.: 1992, *Can. J. For. Res.* 22, 1727.

Solomon, A. M. and Cramer, W. P.: 1993, Biospheric implications of global environmental change, Solomon, A. M. and Shugart, H. H. (eds), *Vegetation Dynamics and Global Change*, Chapman and Hall, Publishers, pp. 25-52.

Solomon, A. M. and Leemans, R.: 1990, Climatic change and landscape-ecological response: Issues and analyses, Boer, M. M. and de Groot, R. S. (eds), *Landscape Ecological Impact of Climatic Change*, IOS Press, pp. 293-317.

Solomon, A. M. and Tharp, M. L.: 1985, Simulation experiments with late-Quaternary C storage in mid-latitude forest communities. E. T. Sundquist and W. S. Broecker (eds), *The C Cycle and Atmospheric CO_2: Natural Variations, Archean to Present*. American Geophysical Union, pp.. 235-250.

Solomon, A. M., Tharp, M. L., West, D. C., Taylor, G. E., Webb, J. W. and Trimble, J. L.: 1984, *Response of Unmanaged Forests to C Dioxide-Induced Climate Change: Available Information, Initial Tests, and Data Requirements. TR-009,* United States Department of Energy.

Solomon, A. M., Trabalka, J. R., Reichle, D. E. and Voorhees, L. D.: 1985, The global cycle of C, J. R. Trabalka (ed), *Atmospheric C Dioxide and the Global C Cycle."* *DOE/ER-0239,* United States Department of Energy, pp. 1-13.

Solomon, A. M. and West, D. C.: 1985, Potential responses of forests to CO_2-induced climate change, White, M. R. (ed), *Characterization of information requirements for studies of CO_2 effects: water resources, agriculture, fisheries, forests and human health, DOE/ER-0236,* U.S. Department of Energy, pp. 145-169.

Tans, P. P., Fung, I. Y. and Takahashi, T.: 1990, *Science* 247, 1431.

Trexler, M. C.: 1993, This volume.

Vloedbeld, M. and Leemans, R.: 1993, This Volume.

Webb, T., III: 1986, *Vegetatio* 67, 119.

Webb, T., III: 1987, *Vegetatio* 69, 177.

Whittaker, R. H. and Likens, G. E.:1975. The biosphere and man, Lieth, H. and Whittaker, R. H. (eds.), *Primary Productivity of the Biosphere*, Springer-Verlag, pp. 305-328

Wigley, T. M. L. and Raper, S. C. B.: 1992, *Nature* 357, 293.

Winjum, J. K., Dixon, R. K., and Schroeder, P. E.: 1993, This volume.

Wright, L. L and Hughes, E. E.: 1993, This volume.

Zinke, P. J., Stangenberger, A. G., Post, W. M., Emanuel, W. R. and Olson, J. S.: 1984, *Worldwide Organic Soil C and Nitrogen Data, ORNL TM-8857,* Oak Ridge National Laboratory.

QUANTIFYING FEEDBACK PROCESSES IN THE RESPONSE OF THE TERRESTRIAL CARBON CYCLE TO GLOBAL CHANGE: THE MODELING APPROACH OF IMAGE-2.

M. VLOEDBELD and R. LEEMANS

Global Change Department, National Institute of Public Health and Environmental Protection, RIVM, POBox 1, 3720 BA Bilthoven, the Netherlands.

Abstract. The terrestrial biosphere component of the Integrated Model to Assess the Greenhouse Effect (IMAGE 2) uses changes in land cover to compute dynamically the C fluxes between the terrestrial biosphere and the atmosphere. The model explores the potential impact of feedback processes incorporated in the model, which are the enhancement of plant growth (CO_2 fertilization) and a more efficient use of water under increased CO_2 concentrations in the atmosphere; the temperature response of photosynthesis and respiration of plants; the temperature and soil water response of decomposition processes; and the climate–induced changes in vegetation and agricultural patterns with the consequent changes in land cover. In this paper we discuss the implementation and operation of the different feedback processes in the IMAGE 2 model. Results are shown for each process separately as well as the combined processes. The aim of this paper is to quantify the importance of these feedback processes geographically. The main results are that vegetation shifts due to climatic change and increased water use efficiency, CO_2 fertilization decreases net C emissions, while changed decomposition rates strongly increase C emissions to the atmosphere. Changes in the global balance between photosynthesis and respiration make little net difference. With the IPPC business-as-usual scenario the terrestrial biosphere continues to emit C into the atmosphere. This behavior is governed by changes in land-use, caused by a multitude of anthropogenic processes.

1. Introduction

The accuracy of greenhouse–gas emission estimates and of climate–change predictions are still met with scepticism. This is largely due to the degree of ignorance about feedback processes in many assessments of such global change. Feedback processes can enhance (positive feedback) or slow down (negative feedback) the buildup of atmospheric greenhouse gases. Several geophysical feedback mechanisms have been taken into account already in atmospheric general circulation models (AGCMs), but most biogeochemical feedback processes are not yet incorporated in these models (Houghton et al., 1992).

A proper assessment of global change can be made only if all important processes and their interactions are included. One of the first attempts to do so was the IMAGE model (=Integrated Model to Assess the Greenhouse Effect; Rotmans , 1990; Rotmans et al., 1990). This model simulated global greenhouse–gas emissions, biogeochemical cycling, atmospheric processes and climatic forcing, and sea–level rise. Sea–level rise only led to a projection of possible climatic-change impacts. No socio–economic or ecological impacts were included in this earlier version. The modeling approach of IMAGE led to the ESCAPE model (CRU and ERL, 1992). In this model a more advanced module for the socio-economic driving factors of greenhouse–gas emissions is included. Further, there is an elaborate set of climate–change impacts (ecosystems, agriculture, heating/cooling needs and tourism). However, the feedback mechanisms that are modeled in IMAGE and ESCAPE are highly aggregated and parameterized, and the impacts do not influence the driving factors of greenhouse gas emissions. Recent analyses have shown that the impacts of global change can have significant feedbacks on future emissions and C storage of the terrestrial biosphere (e.g. Prentice and Fung, 1990; Smith et al., 1992).

We use the improved version 2 of the IMAGE model (Figure 1) to quantify biogeochemical feedbacks. This version includes modules for the Energy/Industry system (technology, energy production and energy efficiency), the terrestrial biosphere (ecosystems, agrosystems and land–use/cover changes with their related emissions) and the atmosphere–ocean system (atmospheric chemistry, C uptake by the oceans, atmosphere–ocean interactions and climatic change). The outcomes of the subsystems are fed directly in to the other subsystems. In this way we calculate important aspects of global change.

Water, Air, and Soil Pollution **70**: 615–628, 1993.

Socio–economic parameters, like population density and gross national products, are all based on regional assessments and statistics, while the environmental parameters, like climate, vegetation and soil characteristics, are based on geographically explicit databases (cf. Leemans, 1992) The calculations are performed for a 150 yr time horizon and can be used to evaluate, for example, different options of energy efficiency improvements, regional C sequestration, and impacts of climatic change. Most objectives of the model are policy orientated, and state–of–the–art scientific understanding must therefore at least be included. This enables IMAGE 2 to serve as a research tool for exploring linkages and feedbacks in the various physical/chemical/biological components of the global system.

The feedback processes that we have incorporated into the terrestrial biosphere model are CO_2 fertilization, the impacts of climate change on plant growth and soil respiration, the impacts of climate change on vegetation and agricultural patterns, and human–induced changes in land–cover. All these feedback processes in the terrestrial biosphere model will be explained in terms of their theoretical and experimental conditions, their relations with other environmental factors and the implementation in the terrestrial biosphere component. We did not implement feedbacks related to the global N and P cycle. Their effects on the C cycle are still poorly understood and therefore difficult to implement unambiguously. Finally, we use the model to quantify the importance of the different feedbacks, both separately and in combination.

2. Modeling Approach of Feedbacks within IMAGE 2

2.1 CARBON FERTILIZATION

Plant growth is supported by photosynthesis, which is the process where CO_2 and water are transformed to carbohydrates. In this process oxygen is released. However, particularly in areas with water and nutrient stress, growth controls photosynthesis (Luxmoore and Baldocchi, 1993).Atmospheric CO_2 is thus an important factor for plant growth. Elevated atmospheric CO_2 concentrations could enhance plant growth and this is shown by many studies (e.g. Gifford, 1988; Lemon, 1983; Strain and Cure, 1985). In particular the short time response of plants under controlled conditions is well-known for many species, but long term vegetation response at regional and global levels is still much less clear (Körner, 1993).

Carbon dioxide diffuses from the atmosphere into the leaves through the stomata. At high atmospheric CO_2 levels, more CO_2 diffuses in to the leaf and the photochemical uptake of CO_2 increases, while photorespiration simultaneously decreases. Consequently gross photosynthesis increases, but this increase may decline with time (e.g. Körner, 1993). Enhanced photosynthesis occurs mainly in C_3 plants. The photosynthetic pathway of C_4 plants is seldom CO_2 limited and photorespiration does not occur (e.g. Bazzaz, 1990; Mooney et al., 1991).

Stomatal conductance controls the rate of gas diffusion through stomata and is directly proportional to stomatal pore width. In general, high ambient CO_2–concentrations reduce stomatal conductance of all plants (Woodward, 1987) and less water may be lost through evapotranspiration. Lower stomatal conductance thus improves water use efficiency (WUE) of plants. WUE is the ratio between dry-matter production to water consumption. This process could enhance plant growth especially under water limited conditions (e.g. Gifford, 1979; Morison, 1985; Körner, 1993). Lower stomatal conductance also decreases the diffusion of CO_2 through stomata, but this has little effect on photosynthesis because CO_2 diffusion is faster than the photochemical uptake of CO_2 (Mooney et al., 1991). Both C_3 and C_4 plants show an enhanced WUE. Increased photosynthesis and increased WUE could thus result in an enhancement of plant growth and net primary production (NPP).

Increased atmospheric CO_2–concentrations could affect plants in various other ways as well. Besides reducing water-stress through increased WUE, elevated atmospheric CO_2–levels can increase plant growth at low light intensity, under nutrient, SO_2–pollution or sodium excess stress (Melillo et al., 1990; Mooney et al., 1991). The response could be different for C_3 or C_4 species. The C allocation in plants changes as well. In trees, particularly more C is sequestered in roots (Melillo et al., 1990; Norby et al., 1992). Moreover, elevated CO_2 can affect tissue density, tissue quality (increasing C:N ratio), phenology and senescence (Melillo et al., 1990). Many of these aspects of plant response to elevated atmospheric CO_2–levels

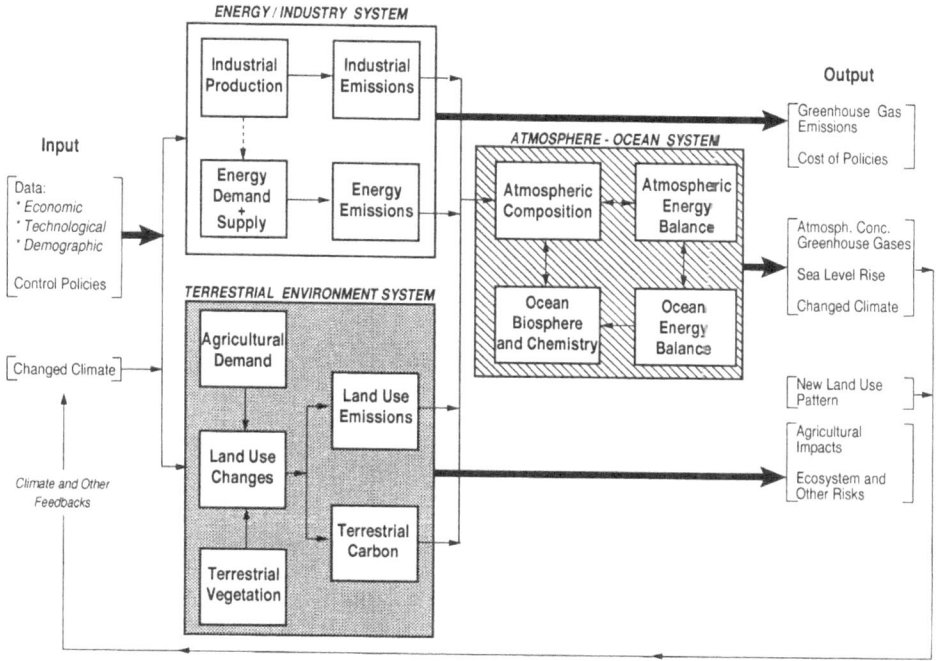

Figure 1. Flowdiagram of the linkages between the different components of the IMAGE 2 model.

determine competitive abilities of plants and could therefore have important feedback influences. We have not (yet) included them, because most of these aspects are not yet generalized enough to be quantified and parameterized towards the PFTs of the BIOME model.

2.1.1. Implementation of the Carbon Fertilization Feedback

The C–cycle model of IMAGE 2 is based on the approach of Goudriaan and Ketner (1984). The main difference is that we have adapted their model to run on a global terrestrial grid and have expanded the number of modeled ecosystems to seventeen (Table 1). This expanded number of ecosystems is used throughout the terrestrial biosphere module of IMAGE 2 and is compatible with the other models (c.f. BIOME; Prentice et al., 1992) and data bases (e.g. Leemans and Cramer, 1991; Olson et al., 1983) used. The net uptake or release of C by the terrestrial biosphere is given for each ecosystem as Net Ecosystem Productivity, NEP (Pg C/yr). Each ecosystem is divided into different compartments (Table 1). NEP is defined as the Net Primary Productivity, NPP, minus the C release through decay of roots, litter, humus, and charcoal (dROOT, dLIT, dHUM, and dCHAR respectively):

$$\textbf{NEP = NPP - (dROOT + dLIT + dHUM + dCHAR)} \qquad (1)$$

The relation between elevated CO_2 and NPP is defined by the fertilization factor ß; ß is the fractional increase in NPP due to a fractional increase in atmospheric CO_2 concentration:

$$\textbf{NPP}_{i,t} = \textbf{NPP}_{i,c}(1+\beta_{i,t}\ln(\frac{pCO_{2_t}}{pCO_{2_c}})) \qquad (2)$$

Table 1 Area (Mha), Net Primary Productivity (NPP, t C ha-1 yr-1) and Initial Biomass (t C ha-1) for each compartment of the different landcover types in IMAGE 2 for 1970.

nr	Landcover type	Area	NPP	Leaf	Branch	Stem	Root	Litter	Humus	Charcoal
1	Agricultural Lands	2509.6	4.0	3.20	0.0	0.0	0.80	3.20	16.0	20.0
2	Ice	278.5	0.0	0.00	0.0	0.0	0.00	0.00	0.0	0.0
3	Cool Semidesert	200.8	0.5	0.25	0.5	2.5	0.45	0.35	12.0	5.0
4	Hot Desert	1622.8	0.5	0.25	0.5	2.5	0.45	0.35	10.0	5.0
5	Tundra	1111.9	1.0	0.50	1.0	5.0	3.00	3.50	24.0	10.0
6	Cool Grass and Shrubs	481.9	3.0	1.80	0.0	0.0	3.60	1.80	72.0	30.0
7	Warm Grass and Shrubs	1736.0	4.0	2.40	0.0	0.0	4.80	2.40	32.0	40.0
8	Xerophytic Woods	850.9	4.0	1.20	8.0	60.0	8.00	3.20	80.0	40.0
9	Taiga	1146.1	4.5	4.05	27.0	67.5	9.00	18.00	108.0	18.0
10	Cool Conifer Forest	309.4	5.0	4.50	20.0	75.0	10.00	12.00	100.0	50.0
11	Cool Mixed Forest	153.5	6.0	1.80	24.0	90.0	12.00	4.80	96.0	60.0
12	Temperate Deciduous For.	78.3	6.0	1.80	24.0	90.0	12.00	4.80	96.0	60.0
13	Warm Mixed Forest	431.5	6.0	1.80	24.0	90.0	12.00	4.80	96.0	60.0
14	Tropical Dry Savanna	1151.1	3.5	2.10	7.0	31.5	3.50	2.80	28.0	35.0
15	Tropical Seasonal Forest	621.0	8.0	4.80	16.0	72.0	12.80	6.40	64.0	80.0
16	Tropical Rain Forest	427.1	9.0	5.40	18.0	81.0	14.40	7.20	72.0	90.0
17	Wetlands	370.7	9.0	5.40	18.0	81.0	14.40	7.20	72.0	90.0
	Totals	13481.1		41.25	188.0	748.0	121.20	82.80	978.0	693.0

where $NPP_{i,t}$ is Net Primary Production of cell i at time t, $NPP_{i,c}$ the initial primary production of cell i in 1970, $ß_{i,t}$ is the fertilization factor for cell i at time t; $pCO_{2,t}$ the atmospheric CO_2 concentration at time t; and $pCO_{2,c}$ the current (i.e. 1970) atmospheric CO_2 concentration. The main factors controlling the response of plants to elevated CO_2 through $ß_{i,t}$, are temperature, soil water and nutrient availability, species characteristics, and altitude. We assumed that the decline of $CO2$ fertilization with time is neglectable. The main factors are modeled using a multiplicative approach:

$$ß_{i,t} = (\gamma_{i,t}^T \ \gamma_{i,t}^W \ \gamma_i^N \ \gamma_{i,t}^C \ \gamma_i^A) \ ß_{i,c} \tag{3}$$

where $ß_{i,c}$ is the initial value of the fertilization factor (i.e. a ß of 0.7 for fertilized, mesic, herbaceous agricultural C_3 crops at 25°C at doubling of CO_2). The different multiplication factors, γ, adjust the ß factor with respect to the initial value. The γ's represent multiplication factors respectively for temperature (T), soil water (W), nutrient availability (N), species characteristics (C) and altitude (A) (Figure 2) and their exact implementation is discussed below. The values of these γ's lay between 0 and 3 and such range reduces the disadvantages of the multiplicative approach (c.f. Leemans, 1992).

2.1.2.1 Temperature. Enhancement of CO_2-fertilization occurs with increasing temperatures. However, this temperature response can change with atmospheric CO_2-concentrations and is different for each species (Coleman and Bazzaz, 1992). This interacting effect is demonstrated both by theoretical analyses (Long, 1991), single species experiments (e.g. Idso and Kimball, 1987; Sage and Sharkey, 1987) and long term experiments on ecosystems (Drake and Leadley, 1991). The temperature response applies only within a well-defined temperature domain, below which enhanced atmospheric CO_2 levels even could reduce plant growth (Idso et al., 1987; Figure 2 and Table 2).

High atmospheric CO_2-concentrations generally also increase the optimum temperature for photosynthesis (T_{opt}). Long (1991) indicates that a doubling of atmospheric CO_2 causes T_{opt} to increase by 5°C. A similar increase applies to maximum temperature (T_{max}), but not to minimum temperature (T_{min}). We implemented those temperature shifts assuming a linear relationship with atmospheric CO_2 concentrations.

2.1.2.2 Soil water availability. Adequate soil water supply does not influence plant growth through increased WUE. CO_2 fertilization will be most pronounced in situations of water-stress. The effect of water stress on ß is modeled through the Priestly-Taylor coefficient α, which is estimated by the ratio of actual annual evapotranspiration to annual equilibrium evapotranspiration, and which determines the soil water limits of the BIOME model. α can be considered a measure of the annual amount of growth-limiting drought stress on plants (Prentice et al., 1992, 1993).

Our reference value of ß ($ß_{i,c}$) which originates from agricultural crops applies for mesic conditions, because these crops are supposed to have a adequate soil water supply. These conditions coincide with a high Priestly-Taylor coefficients (α larger than 0.8). ß is constant under those conditions. Under water-stress or xeric conditions (α less than 0.3), NPP increases about 30% with doubling atmospheric CO_2-concentrations (consistent with data from e.g. Körner (1993) and Morison (1985)) Interpolating between these two extreme conditions and assuming that no other γ-factor is limiting, leads to the function given in Figure 2.

2.1.2.3 Nutrient availability. Carbon fertilization will be neutralized if nutrient supply is not sufficient. Although nutrient use efficiency increases at elevated CO_2, this cannot prevent a lower response of plants under severe nutrient-stress conditions. The availability of nutrients depends on several factors such as soil characteristics, vegetation, atmospheric inputs and climate. We use only a nutrient level that is characterised for each soil type. The influence of deforestation on soil nutrients, and changing C-N ratios is not taken into account.

Goudriaan and de Ruiter (1983) found that the ß factor decreases to about half its original optimal value when plants suffer from N deficiency, and drops to zero when P is strongly limiting. We have defined a generic soil fertility, which is based on the FAO/UNESCO Soil Map of the World (FAO/UNESCO, 1974). We have classified all soils into 5 fertility classes. ß will be decreased with 70% for the lowest fertility class, 30% for moderate fertility class and no decrease for fertile soils.

2.1.2.4 Species characteristics. Photosynthesis of C_3 plants is more responsive to elevated CO_2 levels than that of C_4 plants, but WUE increases in a similar way for both plant types. Annual herbaceous plants in general benefit more from high CO_2 than perennials and woody plants (Körner, 1993). Other plant characteristics which could influence the ß response are the occurrence of special storage organs, fruits, and leaf characteristics. The age and the developmental stage of plant individuals is also of major importance. Seedlings appear to respond more strongly than older individuals; this is especially observed for trees.

Körner (1993) ranks response hierarchies of the most relevant species to elevated CO_2: C_4 plant are much less responsive than C_3 plants; evergreens are less responsive than deciduous trees; and woody plants are less responsive than herbaceous plants. We have assumed that agricultural C_3 crops have the highest ß values, because most crops are herbaceous, annual or perennial species, that are grown under optimal conditions, often with the addition of fertilizers. The division in C_3 and C_4 plants will be made only for herbaceous plants, such as agricultural crops and grasses. Other species are supposed to be of the C_3 type. The vegetation types hot desert, xerophytic woods/shrub, and tropical dry forest/savanna are assumed to incorporate 50% C_4 species. The actual values are given in Table 2 for all PFTs, and the aggregated values for biomes in Table 3. We have thus expanded the original PFTs in the BIOME approach with some additional divisions for some types. C_3 and C_4 are already modeled explicitly in the FAO crop suitability approach.

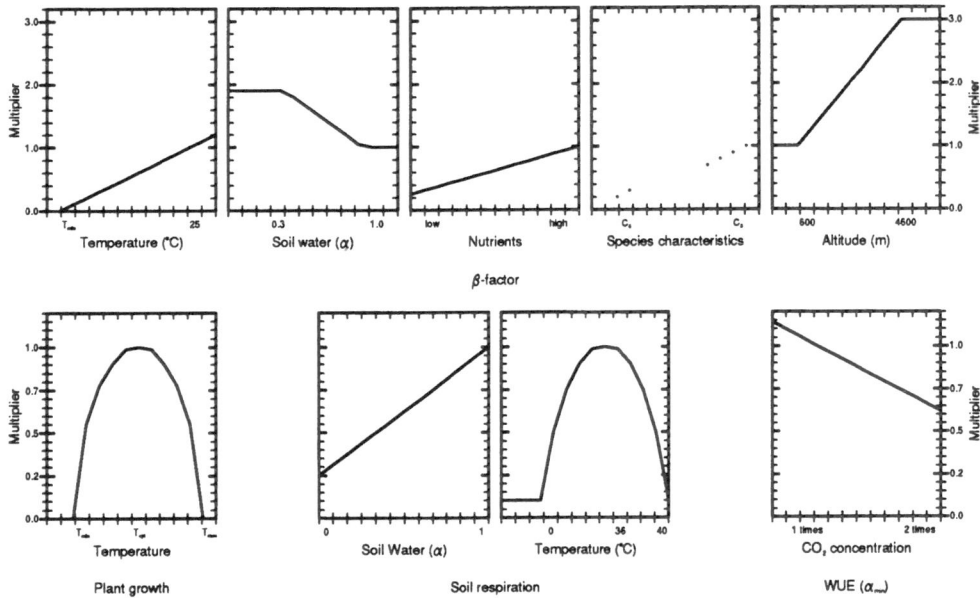

Figure 2. Factors influencing the strength of different feedback processes. The above 5
diagrams define the specific multipliers on the ß factor, below respectively
temperature effect on plant growth, temperature and soil water effect on soil
respiration and WUE by adjusting the minimum moisture requirements for plant
distribution.

2.1.2.5 *Altitude*. Since atmospheric CO_2 pressure decreases with altitude, plants at high
altitudes are strongly C limited. Therefore, these plants should profit more from CO_2
fertilization than plants at low altitudes. Körner and Diemer (1987) found that plants growing at
2600 m show a twice as large photosynthetic response to doubled CO_2 compared with plants
growing at 600 m. Below altitudes of 600 m no significant effect can be observed and no
correction factor is applied for these low altitudes. At high altitudes (> 4600 m) we assume that
the altitudinal correction factor has obtained a maximum value of 3.0, mainly because higher
values are physiologically unlikely. The altitudinal correction factor is based above 600 m on a
lapse rate approach with an increase of ß of 0.05 per 100 m (Figure 2). The topography fields
for each grid cell were taken from Leemans and Cramer's (1991) global data base.

2.2 RESPONSE OF PLANT GROWTH TO CHANGING TEMPERATURES

Like atmospheric CO_2 levels, temperature influences the photosynthetic and respiration rates of
plants directly. Photosynthetic rates increase rapidly from a minimum temperature T_{min}, until
the optimum temperature range is reached, after which it decreases (Fitter and Hay, 1981). The
value of T_{min} is often assumed to be around 0°C for temperate and boreal plants, but could be
considerably higher for tropical plants. Respiration starts slowly at low temperatures and
increases as a power function. At a certain temperature, T_{max}, gross photosynthesis balances
photorespiration. At T_{max} net photosynthesis is zero (Figure 2).

Table 2 Plant Functional Types and their related Gamma values (= relative response of plant type to CO_2 fertilization), and minimum, optimum, and maximum temperatures for photosynthesis (Larcher, 1980; Anonymous, 1978; Leemans and Solomon, 1993).

PFT nr	Plant Functional Type (PFT) (Prentice et al., 1992)	Gamma	T_{min} °C	T_{opt} °C	T_{max} °C
	Natural Vegetation				
1	Polar desert	0.90	-5	10	30
2	Cool desert/shrub	0.90	-5	22	45
3	Hot desert/shrub	0.90	5	32	50
4	Cold grass/shrub	0.90	-3	15	35
5	Cool grass/shrub	0.90	-1	25	45
6	Warm grass/shrub C_3	0.90	1	30	50
7	Warm grass/shrub C_4	0.20	6	35	55
8	Sclerophyll/succulent shrub	0.70	-3	25	50
9	Boreal summergreen	0.80	-3	18	40
10	Boreal evergreen	0.70	-5	15	35
11	Cool-temp. evergreen	0.70	-3	20	40
12	Temperate summergreen	0.80	-1	22	45
13	Warm-temperate evergreen	0.70	0	25	45
14	Tropical raingreen	0.70	5	30	50
15	Tropical evergreen	0.70	5	30	50
	Agricultural Crops (Anonymous, 1978)				
1	Temperate C_3 Crop	1.00	4	18	33
2	Temperate C_4 Crop	0.30	9	25	50
3	Tropical C_3 Crop	1.00	9	28	48
4	Tropical C_4 Crop	0.30	14	33	55

Table 3 Conversion of Plant Functional Types into IMAGE 2 Land Cover Types together with the related Gamma values (= relative response of plant type to CO_2 fertilization), and minimum, optimum, and maximum temperatures for photosynthesis.

Combination of PFT's	LCT nr	Land Cover Type (LCT) (IMAGE 2)	Gamma °C	T_{min} °C	T_{opt} °C	T_{max} °C
	1	Agricultural Lands	1.00	-5	22	55
1	2	Ice	0.90	-5	10	30
2	3	Cool Semidesert	0.90	-5	24	45
3	4	Hot Desert (50% C_4)	0.55	5	32	50
4	5	Tundra	0.90	-3	18	40
4,5	6	Cool Grass and Shrubs	0.90	-2	20	30
6,7	7	Warm Grass, Shrubs (100% C_4)	0.20	6	40	55
8	8	Xerophytic Woods (50% C_4)	0.45	-3	25	50
9,10	9	Taiga	0.75	-4	18	38
9,11	10	Cool Conifer Forest	0.75	-4	18	38
9,10,11,12	11	Cool Mixed Forest	0.75	-2	21	43
10,11,12	12	Temperate Deciduous Forest	0.70	-1	22	44
13	13	Warm Mixed Forest	0.70	0	25	45
14	14	Tropical Dry Savanna	0.45	5	30	50
14,15	15	Tropical Seasonal Forest	0.70	5	30	50
15	16	Tropical Rain Forest	0.70	5	30	50
	17	Wetlands	0.90	-2	20	35

The temperature dependence of NPP (photosynthesis minus respiration) varies among different habitats, species and altitudes (Larcher, 1980). Plants in cold regions have lower minimum, optimum and maximum temperatures than plants in warmer regions. C_4 species, which often grow in the warmest regions, have the highest temperature optima (30 to 40°C), while those of most C_3 species are in the range of 10 to 25°C (Fitter and Hay, 1981; Table 2 and 3).

Relative NPP rates at T_{min} and T_{max} are zero, while at T_{opt} it equals 1 (figure 2). To determine the temperature response of photosynthetic and respiration rates mean monthly temperature values are used from a climate data base (Leemans and Solomon, 1993).

Another important temperature effect of climate change on plant growth is the change in length of growing and dry seasons. Plants can accumulate C for a longer period during an extended growing season and thus NPP increases. The influence of the changing length of the growing season is modeled proportional with a square root function of growing season extension.

In the model we have defined some constraints on the length of the growing season. In severe cold and/or dry climates, the length of the growing season could show an enormous relative change. In these cases no NPP correction factor is used and NPP is re-initialized with the values of the potential biome types (Table 1). We have also done this for situations where the land–cover type changes through climatic or anthropogenic forcing.

2.3 RESPONSE OF SOIL RESPIRATION TO CLIMATE CHANGE

Decomposition rates increase nearly linearly with soil water content up to saturation. If soil water saturation is reached water is no longer a limiting factor for decomposition, at least in well-drained sites. In poorly drained sites excess of precipitation can limit soil respiration.

To determine the effect of water availability on soil respiration we use again α. Parton et al. (1987) specifies a relation between soil respiration and soil water using a similar index, the ratio of precipitation to potential evapotranspiration (P/PET). He assumes soils to be saturated at ratios larger than 1.25. For values up to 1.0 both indices are similar, but α actually has a maximum value of 1. This value specifies per definition no depletion of soil water by evaporation. We assume a linear correlation exists with decomposition rates and α (Figure 2).

Decomposition rates increase with rising soil temperatures (T_s) until $T_s = 36°C$, after which decomposition rates decline (Parton et al., 1987). Decomposition rates are low below a temperature of about 5°C. We assume that soil respiration occurs mainly when soil temperatures are positive. This coincides closely with the average day-temperatures during the growing season (T_{gs}, see above). We therefore use T_{gs} to describe the temperature effect on soil respiration (Figure 2). We assumed that soil respiration always occurs at a slow rate, even outside the T_{gs} domain. The minimum value of the temperature multiplier was therefore set to 0.1 (Figure 2).

Although other factors, such as nutrient availability and soil texture can also affect soil respiration, temperature and soil water are the dominant factors (Carlyle and Than, 1988). The correction factors of both soil water content and temperature are combined in a multiplicative way (cf. Parton et al., 1987).

2.4 SHIFTS IN VEGETATION DUE TO CLIMATIC CHANGE

The potential vegetation in the terrestrial biosphere component of IMAGE 2 is based on BIOME (Prentice et al., 1992). This model simulates global vegetation patterns by defining climatic limits to plant functional types. These plant functional types (Table 2) are defined in order to characterize major physiognomic features of vegetation. The climatic envelopes are defined with extreme temperatures, growing season characteristics and soil water requirements through the α index. Combinations of PFTs make up the different biomes of the world (Table 3). The simulated patterns of BIOME for current climate are compatible with the global data base on landcover of Olson et al. (1983), which is used to initialize global vegetation and agricultural patterns (Table 1).

The climate variables can be computed for different climate-change scenarios and corresponding vegetation patterns are determined. This methodology has been recently reviewed by Leemans (1992). Each biome is assigned a NPP and C values for the different compartments (Table 1). With shifting vegetation zones new C dynamics are computed based on the C values of the previous land cover type. This approach is a dynamic extension of those of Prentice and Fung (1990) and Smith et al. (1992).

2.5 SHIFTS IN VEGETATION DUE TO INCREASED WUE

Gifford (1979) found that at extreme xeric conditions (α in the range from 0.21 to 0.27) plants could grow at high atmospheric CO_2-conditions were no plants grew at lower CO_2 levels. This has far-reaching consequences for setting the soil water requirements of the PFTs that are used to determine global vegetation patterns. If water use efficiency increases under elevated atmospheric CO_2 concentrations, plants can grow in more xeric areas. Plants tolerate a larger drought stress, thus the lower α-limit that determines the distribution of each plant functional type should decrease. We have implemented this by assuming that WUE increases 30% with a doubling of atmospheric CO_2 concentrations (Körner, 1993). We assumed a linear response between current and doubled CO_2 conditions and adjusted a accordingly for transient simulations (Figure 2).

2.6 SHIFTS IN LAND-USE/COVER

A unique feature of the IMAGE model is the possibility to simulate changes in land cover by implicitly taking into account the dynamics of socio-economic factors. The drivers for the demand for agricultural land are regional population and gross national products (GNPs). These set the demand for agricultural land to produce human food, and animal fodder and grasslands used for grazing. The agricultural demands are combined with the already determined crop productivity and distribution. If more agricultural land is needed, we use simple transparent rules to determine the expansion patterns of agricultural land. Such rules could be 'Use the nearest areas with the highest potential productivity first'. This system allows for a deterministic, geographically explicit simulation of land use change patterns and not a simple probabilistic simulation like in earlier C cycle models (e.g. Goudriaan and Ketner, 1984; Goudriaan, 1992). Another important advantage is that any landcover type (cf. Table 1) can be converted into agricultural land, if productivity allows. This gives much more realistic simulations of land cover patterns. The systems also allows for abandoning of agricultural land when socio-economic pressures decrease or the climate becomes less appropriate.

The transitions between the original vegetation type and grassland or agricultural lands is used to determine the immediate fluxes into the atmosphere. The emission rate is depending on environmental factors (see above) and the C contents of the different compartments.

3 Simulations with different feedback mechanisms

IMAGE 2 is still under development, in particular the Energy/Industry and Atmosphere/Ocean component are not yet completed. We therefore used a standard socio-economic and climatic scenario to show the different feedbacks. Changes in land use are only simulated for South America, because the input-data for some other regions is not yet available. Land cover changes in South America are well documented and therefore used as an illustration of its influence on the global C cycle.

The socio-economic scenario was based on the IPCC business-as-usual scenario (Houghton et al., 1992), with an increase of world population of 93% at 2050 (82% for South America). The accompanying change in GNP is 134% (242% for South America). The climate scenario is based on the GFDL-AGCM (Manabe and Wetherald, 1987) which shows a strong temperature increase in high latitude regions during winter. The creation of climate scenarios with AGCMs is described by Leemans (1992).

We initialized the model with the global landcover data base of Olson et al. (1983), which combines natural vegetation and agriculture. We assumed that this data base is a relatively good description of the landcover situation at the beginning of the seventies. The simulation with the Terrestrial Biosphere component several times starts at 1970 and continues up to 2050, which resulted in more then a doubling of CO_2-equivalents in greenhouse gases (686 ppm) and almost a doubling of the CO_2 concentration (570 ppm). To quantify the impacts of the different feedback mechanisms, we ran the model in different modes.

Table 4 Preliminary effects on Net Primary Productivity, Net Ecosystems Productivity and C
flux from the biosphere with different feedbacks for Latin America. The simulation
ran until 2050 using the IPCC business as usual scenario (Houghton et al., 1992) for
the atmospheric composition and the GFDL-AGCM climate scenario (cf. Leemans,
1992).

		NPP	NEP	C flux in 2050
		(Pg C yr-1)	(Pg C yr-1)	(Pg C)
	Current Situation (1970)	10.8	0.0	–
Individual Effects:				
1.	CO_2 fertilization	13.7	0.93	-0.90
2.	Climate Change and Productivity	11.9	0.24	-0.23
3.	Climate Change and Soil respiration	10.8	0.13	-0.24
4.	Socio–Economic Landuse Change	9.2	-0.30	0.39
Combined Effects:				
5.	4+1	11.7	0.20	0.16
6.	4+2	10.9	0.17	0.54
7.	4+3	9.1	-0.56	0.83
8.	4+1+2	13.9	0.52	-0.08
9.	4+1+3	11.7	0.04	0.35
10.	4+2+3	10.9	-0.47	0.78
11.	4+1+2+3	13.9	-0.13	0.25
12.	1+2+3	15.4	1.01	-0.76

Table 5. Preliminary effects on Net Primary Productivity and Net Ecosystems Productivity of
the different feedbacks on the C cycle of the terrestrial biosphere. The simulation ran
until 2050 using the IPCC business as usual scenario (Houghton et al., 1992) for the
atmospheric composition and the GFDL-AGCM climate scenario (cf. Leemans,
1992).

		NPP (Pg C yr-1)	NEP (Pg C yr-1)
	Current Situation (1970)	50.2	0.0
Individual Effects:			
1.	CO_2 fertilization	63.5	3.4
2	Climate Change and Productivity	73.8	5.8
3	Climate Change and Soil respiration	50.2	-1.4
4	Climate Change and Vegetation shifts	55.3	–
5	Climate change and WUE shift	63.2	–
Combined Effects:			
6.	1+2	59.9	4.4
7.	1+3	63.5	1.8
8.	2+3	73.8	3.4
9.	1+2+3	63.4	5.0

4 Results and discussion

We have first analyzed the geographic patterns of CO_2 fertilization with respect to the standard ß factor of 0.7 for fertilized agricultural plants on soils with a good soil water status. The geographic patterns are very apparent. Most tropical savanna woodland and grasslands in the tropics show a large downward deviation from the standard ß factor. This is mainly caused by the abundance of C_4 plants. This reduction of ß is somewhat offset by the improved growth through increased WUE. Globally, the ß factor is reduced somewhat almost everywhere. The responsible processes are low temperatures, soil fertility and plant characteristics. The only regions that show an enhanced ß factor are some dry temperate and tropical grasslands. The enhancement is caused mainly by growth enhancement of WUE. Due to climatic limitations the productivity, however, remains low in these regions.

The different feedback mechanisms all have a significant effect on the emission rates, sources and sinks in the C cycle (Tables 4 and 5). The negative CO_2 fertilization feedback is significant. However it is the most important negative feedback (Goudriaan, 1992; Idso, 1990): Changes in productivity under a changed climate is as significant. The feedbacks resulting from changes in soil decomposition rates are less obvious. Large regional differences exists and the process sometimes even results in a negative feedback. This is especially true for areas that become drier under climatic change. Global changes in soil respiration do not compensate the other negative feedbacks. The combined effect of CO_2 fertilization, climate change and soil decomposition consists still of a negative feedback. Under those conditions the sink–function of the terrestrial biosphere increases when compared with the current situation.

The increase in C storage capacity of the terrestrial biosphere due to vegetation shifts resulting from an improved WUE is large, not only for the arid regions and desert, but especially for the different forest vegetation zones. These forests could, under future atmospheric conditions, expand into more arid regions. This expansion of forest is even larger in the tropics than the vegetation shifts due to climatic change only. However, the magnitude of this feedback could be exaggerated, due to our assumptions in the changes in soil water requirements of PFTs.

Land use changes have the most pronounced effect on the C characteristics of the biosphere. With increasing, more developed populations, the demand for new agricultural lands is enormous. This analysis for South America clearly shows the importance of this issue. All other feedback processes become insignificant with respect to changes in landuse and cover during the next century.

5 Conclusions

The results of the included feedbacks into the Terrestrial Biosphere Component of the IMAGE model show that it is important to be able to understand and quantify the fluxes from the terrestrial biosphere. It illustrates the importance of combining different processes quantitatively. For example, several authors emphasize the importance of the negative feedback 'CO_2 fertilization' for the functioning of the terrestrial biosphere. We confirm this importance, but also found that in combination with changes in soil decomposition, the net result is much less. Another feedback that shows a large response is the shift in vegetation zone due to improved WUE. All these negative feedbacks cannot, however, neutralize the large impacts of changes in land cover. In the coming century, human-induced land use changes are the most important determinant of the terrestrial C cycle fluxes.

The most important advantage of this approach is the determination of the site–specific characteristics and magnitudes of each feedback process. Environmental constraints are efficient in altering those. Highly aggregated models are per definition not capable to incorporate such spatial variability. These models are therefore less applicable to be interpolated in to the future, the main objective for which IMAGE 2 was designed.

Finally, the most eminent factor in determining the role of the biosphere are changes in land use and land cover. If policies to mitigate the negative impacts of global change (and the science to support them) do not consider these changes, they are due to fail.

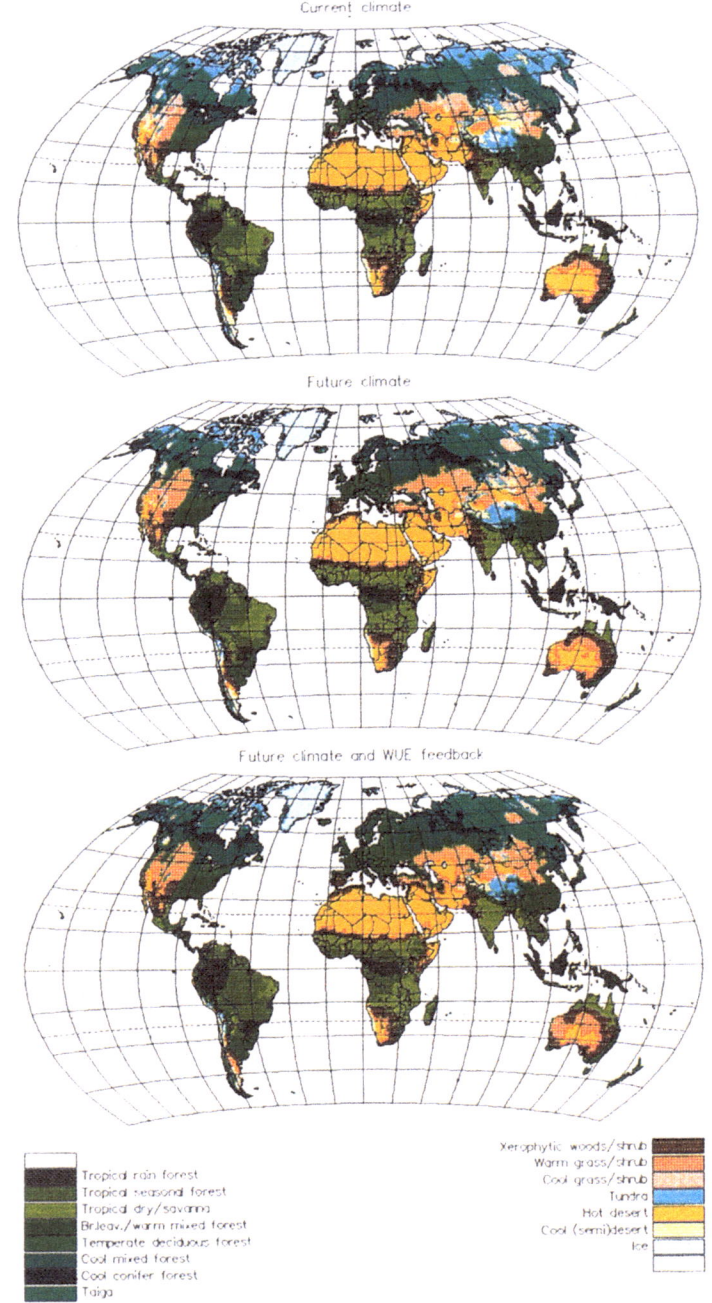

Figure 3. Spatial differences in the impact of climate change on vegetation pattern, using climate change only or combined with a changed Water Use Efficiency. The top maps presents from top to bottom respectively vegetation pattern based on current climate, future climate and future climate with increased water use efficiency.

6 Acknowledgments

The research was funded by the Dutch National Research Programme "Global Air Pollution and Climate Change" under agreement NOP 851045 and MAP 482509 to RIVM and a grant to Marieke Vloedbeld of the University of Nijmegen. Gé Zuidema, Kees Klein Goldewijk en Eric Kreileman assisted to implement all feedbacks into a consistent programme. This paper could not have been written without their help. We would further like to thank Joe Alcamo, Bob Luxmoore and Charles Hall for critical, but useful comments on earlier drafts of the manuscript.

7 References

Anonymous: 1978, *Report on the agro-ecological zones project. Vol 3. Methodology and results for South and Central America*, World Soil Resources Report 48/3, Food and Agricultural Organization, Rome.

Bazzaz, F.A.: 1990, *Annu. Rev. Ecol. Syst.*, 21: 167-196.

Carlyle, J.C. and U.B. Than: 1988, *J. Ecol.*, 76: 654-662.

Coleman, J.S. and F.A. Bazzaz: 1992, *Ecology*, 73: 1244-1259.

CRU and ERL: 1992, *Development of a framework for the evaluation of policy options top deal with the Greenhouse Effect. A scientific description of the ESCAPE model version 1.1*, Climatic Research Unit and Environmental Resources Limited, London.

Drake, B.G. and P.W. Leadley: 1991, *Plant, Cell Environm.*, 14: 853-860.

FAO/UNESCO: 1974, *Soil Map of the World, 1:5,000,000*, Food and Agriculture Organisation, Rome.

Fitter, A.H. and R.K.M. Hay: 1981, *Environmental physiology of plants*, Academic Press.

Gifford, R.M.: 1979, *Austr. J. Plant. Physiol.*, 6: 367-378.

Gifford, R.M.: 1988, Direct effects of higher carbon dioxide concentration on vegetation, in: G.I. Pearman (eds), *Greenhouse. Planning for climate change*, E.J. Brill, pp. 506-519.

Goudriaan, J.: 1992, *J. Exp. Bot.*, 32: 1111-1119.

Goudriaan, J. and H.E. de Ruiter: 1983, *Netherl. J. Agricult. Sci.*, 31: 157-169.

Goudriaan, J. and P. Ketner: 1984, *Clim. Change*, 6: 167-192.

Houghton, J.T., B.A. Callander and S.K. Varney (eds): 1992, *Climate Change 1992. The Supplementary Report to the IPCC Scientific Assessment*, Cambridge University Press.

Idso, S.B.: 1990, Interactive effects of carbon dioxide and climate variables on plant growth, in: B.A. Kimball, N.J. Rosenberg and L.H. Alle Jr. (eds), *Impact of Carbon Dioxide, Trace Gases, and Climate Change on Global Agriculture*, American Society of Agronomy, pp. 61-69.

Idso, S.B. and B.A. Kimball: 1987, *Environm. Experim. Bot.*, 29: 135-139.

Idso, S.B., B.A. Kimball, M.G. Anderson and J.R. Mauney: 1987, *Agricult. Ecosyst. Environm.*, 20: 1-10.

Körner, C.: 1993, CO_2 fertilization: The great uncertainty in future vegetation development, in: A.M. Solomon and H.H. Shugart (eds), *Vegetation Dynamics and Global Change*, Chapman and Hall, pp. 53-70.

Körner, C. and M. Diemer: 1987, *Funct.l Ecol.*, 1: 179-194.

Larcher, W.: 1980, *Physiological Plant Ecology*, Springer-Verlag.

Leemans, R.: 1992, The biological component of the simulation model for boreal forest dynamics, in: H.H. Shugart, R. Leemans and G.B. Bonan (eds), *A Systems Analysis of the Global Boreal Forest*, Cambridge University Press, pp. 428-445.

Leemans, R.: 1992, *J. Sci. Ind. Res.*, 51: 709-724.

Leemans, R. and W.P. Cramer: 1991, *The IIASA database for mean monthly values of temperature, precipitation and cloudiness on a global terrestrial grid*, Research Report RR-91-18, International Institute of Applied Systems Analyses, Laxenburg.

Leemans, R. and A.M. Solomon: 1993, *Clim. Res.*, (in press):

Lemon, E.R.: 1983, CO_2 *and Plants: The Response of Plants to Raising Levels of Carbon Dioxide*, Westview Press.

Long, S.P.: 1991, *Plant Cell Environm.*, 14: 729-739.

Luxmoore, R.J. and D.D. Baldocchi: 1993, Modelling interactions of carbon dioxide, forests, and climate, in: G.M. Woodwell (ed), *Biotic Feedbacks in the Global Climate System: Will the Warming Speed the Warming*, Oxford University Press, (in press).

Manabe, S. and R.T. Wetherald: 1987, *J. Atmosph. Sci.*, 44: 1211-1235.

Melillo, J.M., T.V. Callaghan, F.I. Woodward and E. Salati: 1990, Effects on ecosystems, in: J.T. Houghton, G.J. Jenkins and J.J. Ephraums (eds), *Climate Change: The IPCC Scientific Assessment*, Cambridge University Press, pp. 283-310.

Mooney, H.A., B.G. Drake, R.J. Luxmoore, W.C. Oechel and L.F. Pitelka: 1991, *Biosci.*, 41: 96-104.

Morison, J.I.L.: 1985, *Plant Cell Environm.*, 8: 467-474.

Norby, R.J., C.A. Gunderson, S.D. Wullschleger, E.G. O'Neill and M.K. McCracken: 1992, *Nature*, 357: 322-324.

Olson, J., J.A. Watts and L.J. Allison: 1983, *Carbon in Live Vegetation of Major World Ecosystems*, ORNL-5862, Oak Ridge National Laboratory, Oak Ridge, Tennessee.

Parton, W.J., D.S. Schimel, C.V. Cole and D.S. Ojima: 1987, *Soil Sci. Soc. Am. J.*, 51: 1173-1179.

Prentice, I.C., W. Cramer, S.P. Harrison, R. Leemans, R.A. Monserud and A.M. Solomon: 1992, *J. Biogeogr.*, 19: 117-134.

Prentice, I.C., M.T. Sykes and W. Cramer: 1993, *Ecol. Mod.*, 65: 51-70.

Prentice, K.C. and I.Y. Fung: 1990, *Nature*, 346: 48-51.

Rotmans, J.: 1990, *IMAGE: An Integrated Model to Assess the Greenhouse Effect*, Kluwer Academic Publishers.

Rotmans, J., H. de Boois and R.J. Swart: 1990, *Clim. Change*, 16: 331-356.

Sage, R.F. and T.D. Sharkey: 1987, *Plant Physiol.*, 84: 658-664.

Smith, T.M., R. Leemans and H.H. Shugart: 1992, *Clim. Change*, 21: 367-384.

Strain, B.R. and J.D. Cure: 1985, *Direct Effects of Increasing Carbon Dioxide on Vegetation*, DOE/ER-0238, Washinghton D.C..

Woodward, F.I.: 1987, *Nature*, 327: 617-618.

The Potential Response of Global Terrestrial Carbon Storage To A Climate Change

T. M. Smith and H. H. Shugart

Department of Environmental Sciences
University of Virginia, Charlottesville, VA 22903 USA

Abstract. An analysis is undertaken to examine the potential impacts of a global climate change on patterns of potential terrestrial C storage and resulting fluxes between terrestrial and atmospheric pools. A bioclimatic model relating the current distribution of vegetation to global climate patterns is used to examine the potential impacts of a global climate change on the global distribution of vegetation. Climate change scenarios are based on the predictions of two general circulation model equilibrium simulations for a $2XCO_2$ atmosphere. Current estimates of C reserves in the vegetation types and associated soils are then used to calculate changes in potential terrestrial C storage under the two climate change scenarios. Results suggest a potential negative feedback to increasing atmospheric concentrations of CO_2, with the potential for terrestrial C storage increasing under both scenarios. These results represent an equilibrium analysis, assuming the vegetation and soils have tracked the spatial changes in climate patterns. An approach for providing an estimate of the transient response between the two equilibria (i.e., current and $2XCO_2$ climates) is presented. The spatial transitions in vegetation predicted by the equilibrium analyses are classified as to the processes controlling the transition (eg., succession, dieback, species immigration). Estimates of the transfer rates related to these processes are then used to estimate the temporal dynamics of the vegetation/soils change and the associated C pools. Results suggest that although the equilibrium analyses show an increased potential for C storage under the climate change, in the transient case the terrestrial surface acts as a source of CO_2 over the first 50 to 100 yrs following climate change.

1. Introduction

The potential impacts of the rising atmospheric concentration of CO_2 (and other greenhouse gasses) on the global climate system have been the focus of much research over the past decade. The response of the global climate system to a doubling of atmospheric CO_2 concentration as simulated by general circulation models of the earth's atmosphere would have major impacts on the distribution and abundance of terrestrial vegetation (Emanuel *et al.*, 1985; Prentice and Fung, 1990; Smith *et al.*, 1992a,b). An important question relating to the response of terrestrial vegetation to a climate change is "How will the changes in vegetation and associated soils

Water, Air, and Soil Pollution **70**: 629–642, 1993.
© 1993 *Kluwer Academic Publishers*.

influence the net transfer of C between the atmosphere and terrestrial biosphere?" In the case of increased uptake of CO_2 by terrestrial ecosystems, the terrestrial surface may act as a negative feedback to rising atmospheric concentrations of CO_2. Conversely, in the case of a net positive flux from ecosystems, the terrestrial surface could act as a positive feedback on increasing atmospheric CO_2 concentration and associated patterns of global climate change.

A number of studies have examined the potential impacts of a climate change on the global distribution of terrestrial vegetation and the subsequent affect on the potential for terrestrial C storage (Lashoff, 1987; Prentice and Fung, 1990; Smith *et al.*, 1992a). These studies use simple bioclimatic models which relate the current distribution of vegetation/ecosystem types with biologically important features of the climate (eg., temperature and precipitation). Estimates of C storage (per unit area) in these vegetation types and associated soils are combined with estimates of their areal coverage to provide an estimate of terrestrial C storage under current climate patterns and as well as predictions of future stores under climate change scenarios based on GCM simulations. Such calculations presumably estimate a long-term equilibrium response to a climate change, assuming that vegetation distribution is able to track changes in climate in both space and time. The results of these "equilibrium" analyses suggest that the terrestrial surface will have the potential to store more C, and will thereby provide a negative feedback to rising atmospheric CO_2 concentrations. However, the terrestrial surface may well act as a net C source to the atmosphere during the transition period required to achieve the new "equilibrium" patterns of vegetation and soils under the changed climate conditions (Smith and Shugart, 1993). In this paper we present a methodology for estimating the transient dynamics of terrestrial C storage by 1) using a bioclimatic model to provide estimates of the transitions in vegetative cover under a climate change and then 2) examining the timescales associated with those processes.

2. Equilibrium Estimates of Changes in Global Terrestrial Vegetation and Associated Carbon Storage

2.1. METHODS

2.1.1 Vegetation Distribution

The potential distribution of terrestrial vegetation under current climate and climate change scenarios was modelled using the Holdridge Life Zone Classification (Holdridge, 1967). The Holdridge Life Zone Classification (Figure 1) is a bioclimatic model relating the distribution of major ecosystem complexes to the climatic variables of biotemperature, mean annual precipitation and the ratio of potential evapotranspiration (PET) to precipitation. The life zones are depicted by a series of hexagons in a triangular coordinate system. Two climate variables, biotemperature and annual precipitation, determine the classification. Biotemperature is a temperature sum over a year with the unit temperature values (i.e., average daily, weekly or monthly temperatures) that are used in computing the index set to 0°C if they are less than or equal to 0°C.

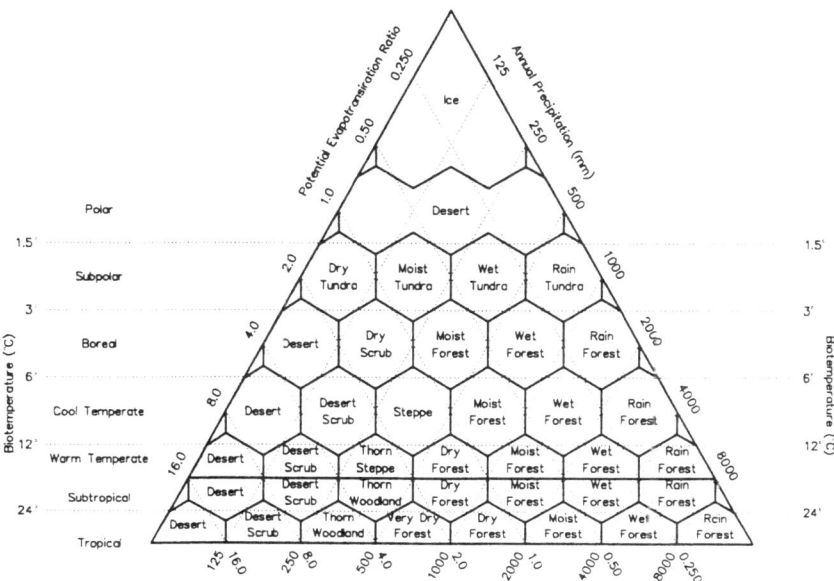

Figure 1. Holdridge climate-vegetation classification scheme (Holdridge, 1967). A detailed description of the model is provided in the text.

Identical axes for average annual precipitation form two sides of an equilateral triangle. The potential evapotranspiration (PET) ratio forms the third side, and an axis for mean annual biotemperature is oriented perpendicular to its base. By striking equal intervals on these logarithmic axes, hexagons are formed that designate the Holdridge Life Zones.

The potential evapotranspiration ratio is the quotient of PET and average annual precipitation. Holdridge (1959) assumes, on the basis of data from several ecosystem types, that PET is proportional to biotemperature (constant of proportionality = 58.93). The PET ratio in the Holdridge Diagram is therefore dependent on the two primary variables, annual precipitation and biotemperature.

One additional division in the Holdridge Classification is based on the occurrence of killing frost. This division is along a critical temperature line that divides hexagons between 12 and 24°C into Warm Temperate and Subtropical Zones. The complete Holdridge Classification at this level includes 37 life zones.

Expected current distributions of life zones were mapped using a global climate data base of mean monthly precipitation and temperature at a 0.5°x0.5° (latitude and longitude) resolution (Leemans and Cramer 1990). Simulations of current ($1XCO_2$) and $2XCO_2$ climates from the Goddard Institute for Space Studies (GISS) (Hansen $et\ al.$, 1988) and Geophysical Fluid Dynamics Laboratory (GFDL, Princeton U.) (Manabe and Wetherald, 1987) general circulation models were used to construct change scenarios. Changes in mean monthly precipitation and temperature were calculated for each GCM scenario for each computational grid element by taking the difference between simulated current and $2XCO_2$ climates. Temperatures were expressed as absolute difference ($2XCO_2$-$1XCO_2$) and precipitation as the ratio of $2XCO_2$ to $1XCO_2$. These data from each GCM were interpolated to 0.5°x0.5° using the same technique as applied in the development of the database for current climate (Leemans and Cramer,1990). The technique used was a triangulation of all datapoints (algorithm developed by Green and Sibson, 1978) followed by a smooth surface fitting (Akima, 1978). Changes in monthly precipitation and temperature were then applied to the global climate data base to provide a change scenario. The altered data bases corresponding to each of the GCM scenarios were then used to reclassify the grid cells (0.5°x0.5°) using the Holdridge Classification.

2.1.2. Carbon Storage

To determine the effects of the climate change scenarios on patterns of global C storage, each of the 37 Holdridge Life Zones was assigned a value for C in both above-ground biomass and soil. Soil C estimates for the Holdridge Life Zones were taken directly from Post *et al.* (1982). Estimates of C in above-ground biomass for each life-zone were calculated using data from Olson *et al.* (1983). Olson *et al.* (1983) does not classify ecosystems according to the Holdridge Life Zones, therefore it was necessary to estimate C values for the life zones directly from the climate variables used in the Holdridge Life Zone Classification (i.e., biotemperature and annual precipitation). Combining the biotemperature and annual precipitation data for each 0.5°x0.5° cell with the C estimates from the corresponding cell of the Olson *et al.* database (also 0.5°x0.5°), an average value of C storage in above-ground biomass was calculated for the climate space corresponding to each Holdridge Life Zone (see Figure 1). This approach provides a single estimate of above- and below-ground C for each life zone, and therefore, changes in C storage under the $2XCO_2$ scenarios are a function of changes in the extent of the various life zones. All estimates of change in C storage are based on the complete Holdridge Classification (i.e., all 37 life zones).

2.2 RESULTS AND DISCUSSION

Changes in the areal coverage of major biome-types under the two climate change scenarios are shown in Table 1. These biome-types are aggregates of the Holdridge Life Zones presented in Figure 1, and the zones comprising the types are defined in Table 1. There is a general qualitative agreement between the scenarios in the directions of change. The extent of tundra and desert decreased, and those of grasslands and forests increased under both scenarios. Despite the agreement in increased forest cover, the scenarios differed in the degree to which the increase is attributable to mesic and xeric forest components. Mesic forest cover increased under the GISS scenario, but decreased in the GFDL scenario. The predicted decrease in mesic forest by GFDL is offset by a larger increase in dry forest, thus forest cover increased overall.

Table 1. Changes in areal coverage of major biome types* under current and changed climate conditions. Values in km²X10⁴.

	Current	GISS	GFDL
Tundra	939	-314	-515
Desert	3699	-962	-630
Grassland	1923	694	969
Dry Forest	1816	487	608
Mesic Forest	5172	120	-402

* Tundra: Polar Dry Tundra, Polar Moist Tundra, Polar Wet Tundra, Polar Rain Tundra
  Desert: Polar Desert, Boreal Desert, Cool Temperate Desert, Warm Temperate Desert,
       Subtropical Desert Bush, Tropical Desert, Tropical Desert Bush
  Grassland: Cool Temperate Steppe, Warm Temperate Thorn Steppe, Subtropical Thorn
       Steppe, Tropical Thorn Steppe, Tropical Very Dry Forest
  Dry Forest: Warm Temperate Dry Forest, Subtropical Dry Forest, Tropical Dry Forest
  Mesic Forest: Moist, Wet and Rain Forest Classes for the Boreal, Cool Temperate, Warm
       Temperate, Subtropical and Tropical Zones

The changes in coverage of the biome-types presented in Table 1 are the outcome of a dynamic process of spatial changes in the climate pattern, and associated spatial changes in the distribution of life zones. These spatial dynamics can be described as a matrix of transitions between types (Table 2). Rows of the matrix show transitions from that biome-type (i.e., aggregated life zones) to the specified type in the column headings. The diagonal elements show the area occupied by the biome-type under current climate which does not change (to another biome-type) under the new climate conditions. Therefore, the sum of the elements in each row is the current coverage for that biome-type, while the sum of the elements in each column is the coverage for that type under the changed climate conditions.

The decline in tundra observed under both scenarios is primarily due to a shift from tundra to mesic forest. This transition is a result of the warming at higher latitudes and the subsequent northward movement of boreal forest into the areas now occupied by tundra. This shift is associated with a slight decrease in soil C, but a large increase in above-ground C. A second major vector of change in the tundra region is the desertification of areas where warming and/or decreases in precipitation occurs, resulting in a decrease in C stored in both soil and above-ground biomass.

Table 2. Matrices of transitions for the GISS and GFDL climate change scenarios. Matrices show the change in coverage between biome types. Values are in km2X104. (T - Tundra; G-Grassland; DF-Dry Forest; MF-Mesic Forest).

GISS

	To: T	D	G	DF	MF	Total
From: T	167.4	114.5	2.1		655.0	939.0
D	456.7	2559.5	617.0	5.4	60.5	3699.1
G		33.8	1645.3	198.2	45.7	1923.0
DF			152.9	1427.2	235.6	1815.7
MF		5.3	200.0	672.0	4295.2	5172.5
Total	624.1	2713.1	2617.3	2302.8	5172.5	

GFDL

	To: T	D	G	DF	MF	Total
From: T	18.5	177.6	29.2	0.2	713.3	938.8
D	405.2	2784.1	337.8	0.3	171.8	3699.2
G		67.9	1663.4	181.0	10.8	1923.1
DF			141.8	1490.8	183.2	1815.8
MF		9.2	720.6	751.9	3690.6	5172.3
Total	423.7	3038.8	2892.8	2424.2	4769.7	

The decrease in the global extent of desert seen in both scenarios is a function of the shift from desert to tundra in the higher latitudes, and from desert to grassland in the temperate and tropical regions. Furthermore, there is a significant conversion from desert to mesic forest under the GFDL scenario. These shifts occur in the northern latitudes where cold desert/dry tundra zones increase in both temperature and precipitation. All shifts from desert to other biome-types represent a large net increase in stored C (both soil and aboveground).

The increased cover of grassland under both scenarios is a function of both shifts from desert to grassland with increased precipitation in areas of the temperate and tropical regions, and the transition of dry and mesic forests to grassland as a result of drying in all forested regions. These shifts represent both an increase and decrease in C storage.

Potential C storage increases in the transition from desert to grassland, but decreases significantly with the transition from forest to grassland.

The extent of dry forest increases with increasing precipitation in grassland regions, and with increased temperatures and/or decreased precipitation in mesic forests. The later transition occurs primarily in the subtropical and tropical regions and represents a large decrease in C in all scenarios. This transition is the largest in the GFDL scenario, resulting in a 30% increase in the global extent of dry forest.

The major transition towards mesic forest is the shift from tundra to boreal forest discussed earlier. This transition is followed in importance by the shift from dry to mesic forest, primarily in the subtropical and tropical regions. The extent of the transition from tundra to boreal forest is similar for both scenarios (GISS - 655.0 and GFDL - 713.3 km^{2}X10^4). The major difference between the scenarios in the predicted areal coverage of mesic forest is the larger degree of mesic forest decline associated with drying in the subtropical and tropical regions (i.e., the shift to dry forest discussed above) under the GFDL scenario.

The transitions discussed above result in a net increase in terrestrial C storage under both climate change scenarios (Tables 3, 4 and 5). The potential for total C storage increased by 2% (37.9 Gt) under the GFDL scenario as compared to 7.7% (146.9 Gt) for GISS (Table 3). Above-ground C increased under both scenarios (Table 4), however, the GFDL scenario shows a slight decrease in soil C while the GISS scenario predicts a 4.7% increase (Table 5).

Both scenarios considered in this study suggest a potential increase in terrestrial C storage with an associated reduction in atmospheric CO_2 levels from concentrations used in the $2XCO_2$ GCM simulations. The potential increase in C storage is primarily due to the; 1) poleward shift of the forest zones, with an increase in the extent of tropical forest and a

Table 3. Changes in C storage (Gt) in above-ground biomass and soils under the GISS and GFDL climate change scenarios. Values in parentheses are percentage change from current. All other values in Gt.

Scenario	Biomass	Soil	Total	Change
Current	737.2	1158.5	1895.7	
GISS	829.6 (12.5)	1213.0 (4.7)	2042.6	146.9 (7.7)
GFDL	782.3 (6.1)	1151.3 (-0.6)	1933.6	37.9 (2.0)

Table 4. Changes in C storage in above-ground biomass for major biome types under GISS and GFDL climate change scenarios. Values in parentheses are percentage change from current. All other values in Gt.

Biome	Current	GISS	GFDL
Tundra	23	15	10
Desert	48	37	44
Grassland	57	79	86
Dry Forest	92	114	120
Mesic Forest	518	585	523
Total	738	830 (12.5)	783 (6.1)

Table 5. Changes in soil C storage for major biome types under GISS and GFDL climate change scenarios. Values in parentheses are percentage change from current. All other values in Gt.

Biome	Current	GISS	GFDL
Tundra	199	129	77
Desert	95	67	86
Grassland	163	193	248
Dry Forest	115	178	176
Mesic Forest	587	646	564
Total	1159	1213 (4.7)	1151 (-0.6)

northern movement of the boreal forest zone into areas currently occupied by tundra, and 2) a decrease in the global extent of desert. However, the results represent equilibrium solutions for both climate (i.e., $2XCO_2$) and vegetation dynamics. In reality, the vegetation would most likely be unable to track the transient climate dynamics. In areas where the potential biomass which can be supported is predicted to decrease due to increased aridity, the changes may occur quite quickly as the environmental conditions become such that the present vegetation can no longer be supported (eg., forest to grassland). In contrast, increases in biomass or soil C may require much longer periods of time. In some cases the present vegetation may exhibit increased growth or recruitment under the more favourable conditions. However, major shifts in vegetation structure such as the transition from grassland or tundra to forest, depend on the immigration of species and the ability of new species to invade existent communities. These changes would operate over timescales relate to the longevity of the component species and the rates of species dispersal. Increases in so C would depend on rates of net primary productivity and the accumulation of soil organic matter. Although a complete analysis of these transient dynamics at a global scale would require detailed models of demographic, landscape and ecosystem processes for the wide array of ecosystems represented in the above analyses (i.e., vegetation or life zones), a first approximation of the transient dynamics of C storage can be estimated by examining the rate at which critical ecological processes controlling these dynamics occur.

3. Estimates of the Transient Dynamics of Terrestrial Carbon Storage

3.1 METHODS

The equilibrium estimates of vegetation transitions and associated C pools can be used to obtain an estimate of the transient response of the terrestrial surface to a step change in climate. The estimates of change in the coverage of major biome types presented in Table 2 are aggregates based on a matrix of possible transitions between the 37 different life zones presented in Figure 1. These transitions between Holdridge Life Zones under current and changed climate were classified as to the processes controlling the changes in vegetation and associated soil C. For vegetation those processes include: 1) dieback of existing vegetation as a result of increased aridity (accompanied by wildfire) followed by secondary succession of species predicted to occupy the area under the new climate conditions, 2) successional replacement of existing vegetation by vegetation predicted for that location under the new climate, and 3) successional replacement limited by the need for immigration of new species Immigration was deemed necessary if the transition implied a major shift in physiognomic structure or life form dominance (eg., tundra or grassland to forest). This definition of immigration dependent vegetation change is very conservative since the majority of vegetation transitions would involve a major taxonomic shift.

Changes in soil C are a result of either increased rates of decomposition and/or increase accumulation of soil organic matter associated with succession.

Carbon pools for both above-ground biomass and soils associated with each of the categories of vegetation change were tabulated and transfer rates associated with these processes were estimated (Table 6). The loss of existing vegetation which will be replaced i categorized as to the processes controlling the net loss of C. Likewise, the C in the new vegetation and soils predicted to replace the existing vegetation is categorized as to the processes controlling rates of replacement and C accumulation.

For those transitions involving dieback of existing vegetation, 0.02 yr^{-1} of the C associated with the original vegetation was computed as a source. When the change was expected to occur over successional time (natural mortality of the original vegetation followed by the establishment and growth of replacing vegetation), 0.004 yr^{-1} of the C in the original vegetation was taken as a source. In symmetry with the C losses from succession-controlled sources, the C gain from replacing vegetation was computed at a rate of 0.004 yr^{-1} times the C for the new vegetation type. For those cases in which the transition in vegetation would be dependent on the rate of immigration of new species, the rate of development of the replacement vegetation was estimated as 0.001 yr^{-1}. These rates were intended to be conservative with the respect to reducing the likelihood of estimating a strong source from the terrestrial surface in the initial stages of the transient C dynamics in the response to climate change.

For below-ground sources of C, the rate at which soil C was lost from soils that had a lower expected soil organic pool under the new climate was 0.02 yr^{-1}. Even though there is likely to be a lag in soil development as succession proceeds, the rate at which soils accumulate C was set at the successional rate (0.004 yr^{-1}). In cases in which the replacement vegetation providing organic matter is limited in its development by immigration, the C accumulation rate for soils was 0.001 yr^{-1}. As is the case for above-ground rates, these below-ground rates were felt to be conservative.

Transfer rates (1/turnover time) were estimated from published values. Estimates of transfer rates for above-ground biomass associated with succession were derived from gap-model simulations (Shugart, 1984; Shugart et al., 1980; Solomon, 1986; Bonan and Hayden, 1990; Smith et al., 1992a) and relate to the demographic limitations on biomass accumulation. Transfer rates associated with vegetation immigration were estimated based on paleo-studies (Davis, 1976, 1984, 1989). Estimates of soil C turnover were based on average parameters from global C models (Peng et al., 1983; Goudriaan and Ketner, 1984; Emanuel et al., 1984; Houghton et al., 1983) . A simple compartment model of C accumulation and release was then applied to the transitions. The compartment model used in this analysis parallels the approach used in current global C models (Peng et al., 1983; Goudriaan and Ketner, 1984; Emanuel et al., 1984; Houghton et al., 1983) . All transfer rates were varied plus or minus 50% to determine the envelope of responses.

3.2 RESULTS AND DISCUSSION

When the transfer rates are applied to the changes in vegetation and associated C pools as mapped by the Holdridge Classification, the terrestrial surface is found to initially behave as a significant source of C in response to climate changes for both the GFDL and the GISS climate change scenarios (Figure 2). For the GFDL scenario, the source strength is at a maximum of 225 Gt C at year 134 after the change in climate. The GISS case has a maximum source strength of 150 Gt occurring at year 120. When the transfers of C losses and gains are varied independently $\pm$ 50%, these results are qualitatively unchanged. In all cases, the terrestrial surface behaved as a source in its initial response. The responses differed with respect to the value of the maximum source strength and the timing of this maximum. For the GFDL scenario, the maximum source strength was between 194 and 240 Gt C with the maximum occurring between 115 to 181 yr. The analogous range of responses for the GISS scenario were 121 to 163 Gt maxima occurring between 101 and 163 yr. The principal difference in the responses of the two GCM's stems from the differences in the associated patterns of vegetation change.

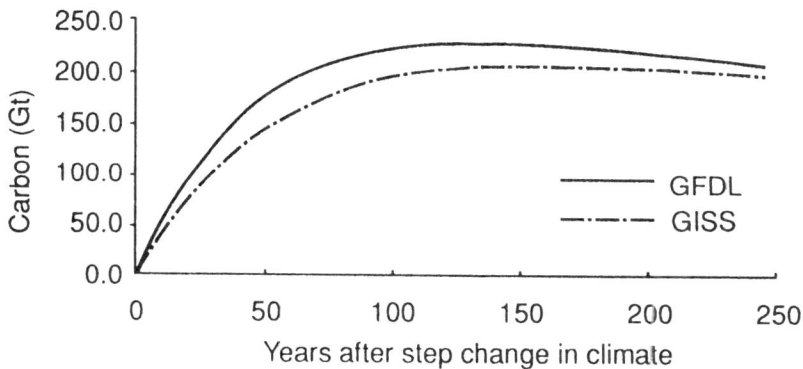

Figure 2. Net C flux to the atmosphere from the terrestrial surface in response to a step climate warming. The fluxes under the two scenarios are derived from the changes in vegetation and soils associated with climate change as predicted by two general circulation models for a doubled-CO_2 atmosphere (from Smith and Shugart, 1993).

In general, the results can be directly interpreted based on the transfer rates and the size of C pools for the three classes of vegetation dynamics included in the analysis (Table 6). The net increase in C storage for vegetation transitions involving successional dynamics is relatively small for both scenarios in comparison to the losses from dieback/disturbance and the gains due to transitions dependent on vegetation immigration. It is the difference in the response times of these two latter pools which drive the temporal dynamics of the atmospheric-terrestrial biosphere transfer. Overall, the robust result that the terrestrial surface behaves as a C source under an abrupt climate warming is a consequence of vegetation changes that result in a net release of CO_2 to the atmosphere (eg., conversion of forests to grassland) being potentially rapid in comparison to those changes which result in a

Table 6. Carbon pools for soils and above ground biomass associated with the changes in vegetation and corresponding soils under the two GCM climate change scenarios. Values for above ground biomass represent either 1) carbon gain from successional replacement of vegetation, 2) carbon loss from vegetation that is being replaced, 3) carbon lost from vegetation impacted by dieback and disturbance as a function of increased aridity, and 4) carbon gain from successional replacement which is rate limited by species immigration. Soil carbon values represent the soil carbon gains and losses associated with the changes in vegetation. Losses of soil carbon would result from increased decomposition rates associated with the new climate and vegetation. Gains are dependent on the buildup of organic material from either 1) succession replacement of vegetation, or 2) development of vegetation that is dependent on species immigration. All values are in Gt. (from Smith and Shugart, 1993)

Above-ground Biomass

Scenario	Loss GFDL	GISS	Gain GFDL	GISS	Net GFDL	GISS
Successional Replacement (Rate: .004 y^{-1})	323.0	292.9	338.7	346.2	+15.7	+53.5
Immigration (Rate: .001 y^{-1})			122.4	171.8	+122.4	+158.5
Dieback/Disturbance (Rate: .02 y^{-1})	124.1	119.4			-124.1	-119.4

Soils

Scenario	Loss GFDL	GISS	Gain GFDL	GISS	Net GFDL	GISS
Increased Decomposition (Rate: .02 y^{-1})	196.8	213.3			-196.8	-213.3
Successional Replacement (Rate: .004 y^{-1})	87.3	111.3			+87.3	+111.3
Immigration (Rate: .001 y^{-1})			101.9	102.0	+101.9	+102.0

TOTAL	GFDL	GISS
	+37.9	+146.9

net increase in terrestrial C storage (eg.,conversion of polar desert to tundra and boreal forest). Two major classes of transitions associated with these processes are: 1) forest decline in areas with predicted increases in aridity, and 2) northern movement of the boreal and tundra zones.

Areas of potential forest decline, resulting from increased aridity, represent a large potential loss of C to the atmosphere. Values of C in above-ground biomass for these areas are 38.4 Gt for GISS and 66.3 Gt for the GFDL scenario.

In the long term these patterns of C loss due to forest decline are largely offset by forest expansion at the northern boundary of the boreal zone. This transition represents a potential increase of 51.2 Gt and 54.1 Gt potential C storage in above-ground biomass under the GISS and GFDL scenarios respectively.   The northern movement of the tundra is an extremely important potential C sink over the long-term and is responsible to a large degree for the net C gain in the equilibrium analyses. The predicted expansion of the tundra and boreal zones into what is currently cold (polar) desert represents a potential for a 102 Gt increase in soil C storage under both scenarios. A large part of the predicted net positive flux from the terrestrial surface in the above analyses is a result of the fact that net positive fluxes of CO_2 from soils resulting from increased decomposition will occur faster than the buildup of organic matter in this region under the changed climate conditions.

The analysis we have presented is simplistic, but clearly shows the importance of considering the temporal dynamics of vegetation in any analysis of the potential response of terrestrial C storage to a changed climate. Although the equilibrium analysis (Smith *et al.*, 1992b) suggests a net increase in potential C storage, there is clearly the potential for a sizeable net transfer of C to the atmosphere when the vegetation dynamics necessary to achieve the equilibrium conditions are examined (Smith and Shugart, 1993). The analyses presented assume a step change in climate in that the vegetation transitions were defined from climate change scenarios based on $2XCO_2$ equilibrium model runs of the two GCM's. However, the rates of C transfer are based on estimates of the time required for those transitions in vegetation and soil properties to occur. In that sense the results represent a transient analysis. The degree to which the results presented would differ if a transient climate change scenario were used would depend on the rate of climate change relative to the rates of transfer associated with the vegetation and soil processes (i.e., species dieback, succession, species immigration and decomposition).   If the climate change is fast relative to these processes, these biological processes will control the rates of CO_2 flux. If the climate change is slow relative to the processes, the rate of climate change will control the flux rates.

4. References

Akima, H.: 1978, *ACM Trans. Math. Software* **4**, 148-159.
Bonan, G.B. and Hayden, B.P.: 1990, *Quaternary Research* **33**, 204-218.
Davis, M.B.: 1976, *Geoscience and Man* **13**, 13-26.

Davis, M.B. 1984. Climatic instability, time lags and community disequilibrium, in: Diamond, J. and Case, T.J. (eds), *Community Ecology*. Harper and Row, New York, pp. 269-284.

Davis, M.B.: 1989, *Climatic Change* **15**, 75-82.

Emanuel, W.R., Killough, G.G., Post, W.M. and Shugart, H.H.: 1984, *Ecology* **65**, 970-983.

Emanuel, W.R., Shugart, H.H. and Stevenson, M.P.: 1985, **Climatic Change 7**, 29-43.

Goudriaan, J. and Ketner, P.: 1984, *Climatic Change* **6**, 167-192.

Green, P.J. and Sibson, R.: 1978, *The Computer Journal* **21**, 168-173.

Hansen, J., Fung, I., Lacis, A., Rind, D., Russell, G., Lebedeff, S., Reudy, R. & Stone, P.: 1988, *J. Geophys. Res.* **93**, 9341-9364.

Holdridge, L.R.: 1959, *Science* **130**, 572.

Holdridge, L.R.: 1967, *Life Zone Ecology*, Tropical Science Center, San Jose.

Houghton, R.A., Hobbie, J.E., Melillo, J.M., Moore, B., Peterson, B.J., Shaver, G.R. and Woodwell, G.M.: 1983, *Ecol. Monogr.* **53**, 235-262.

Lashof, D.A.: 1987, *The Role of the Biosphere in the Global Carbon Cycle: Evaluating Through Biospheric Modeling and Atmospheric Measurement*, Ph.D. dissertation, University of California, Berkley.

Leemans, R. and Cramer, W.: 1990, *The IIASA climate database for land area on a grid of 0.5° resolution*, WP-41, International Institute for Applied Systems Analysis, Laxenburg.

Manabe, S. and Wetherald, R.T.: 1987, *J. Atm. Sci.* **44**, 1211-1235.

Olson, J.S., Watts, J.A., and Allison, L.J.: 1983, *Carbon in live vegetation of major world ecosystems*, ESD Pub. No. 1997, Oak Ridge National Laboratory, TN.

Peng, T.H., Broecker, W.S., Feyer, H.D. and Trumbore, S.: 1983, *J. Geophys. Res.* **88**, 3609-3620.

Post, W.M., Emanuel, W.R., Zinke, P.J., and Stangenberger, A.G.: 1982, *Nature* **298**, 156-159.

Prentice, K.C. and Fung, I.Y.: 1990, *Nature* **34**, 48-51.

Shugart, H.H.: 1984, *A Theory of Forest Dynamics*, Springer-Verlag, New York.

Shugart, H.H., Emanuel, W.R., West, D.C. and DeAngelis, D.L.: 1980, *Mathematical Biosceiences* **50**, 163-170.

Smith, T.M., Shugart, H.H., Bonan, G.B. and Smith, J.B.: 1992a, *Advances in Ecological Research* **22**, 93-113.

Smith, T.M., R. Leemans, and H.H. Shugart: 1992b, *Climatic Change* **21**, 367-384.

Smith, T.M. and Shugart, H.H.: 1993, *Nature* **361**, 523-526.

Solomon, A.M.: 1986, *Oecologia* **68**, 567-569.

MODELING THE EFFECTS OF CLIMATIC AND CO₂ CHANGES ON GRASSLAND STORAGE OF SOIL C

DENNIS S. OJIMA — *Natural Resource Ecology Laboratory, Colorado State University, Fort Collins, CO 80523, USA*

WILLIAM J. PARTON — *Natural Resource Ecology Laboratory, Colorado State University, Fort Collins, CO 80523, USA*

DAVID S. SCHIMEL — *National Center for Atmospheric Research, P.O. Box 3000, Boulder, CO, 80307-3000, USA, and Natural Resource Ecology Laboratory, Colorado State University, Fort Collins, CO 80523, USA*

JONATHAN M.O. SCURLOCK — *Division of Life Sciences, King's College London, Campden Hill Road, London W8 7AH, United Kingdom*

TIMOTHY G.F. KITTEL — *University Corporation for Atmospheric Research, P.O. Box 3000, Boulder, CO, 80307-3000, USA, and Natural Resource Ecology Laboratory, Colorado State University, Fort Collins, CO 80523, USA*

Abstract. We present results from analyses of the sensitivity of global grassland ecosystems to modified climate and atmospheric CO_2 levels. We assess 31 grassland sites from around the world under two different General Circulation Models (GCM) double CO_2 climates. These grasslands are representative of mostly naturally occurring ecosystems, however, in many regions of the world, grasslands have been greatly modified by recent land use changes. In this paper we focus on the ecosystem dynamics of natural grasslands. The climate change results indicate that simulated soil C losses occur in all but one grassland ecoregion, ranging from 0 to 14% of current soil C levels for the surface 20 cm. The Eurasian grasslands lost the greatest amount of soil C ($\sim$1200 g C m^{-2}) and the other temperate grasslands losses ranged from 0 to 1000 g C m^{-2}, averaging approximately 350 g C m^{-2}. The tropical grasslands and savannas lost the least amount of soil C per unit area ranging from no change to 300 g C m^{-2} losses, averaging approximately 70 g C m^{-2}. Plant production varies according to modifications in rainfall under the altered climate and to altered nitrogen mineralization rates. The two GCM's differed in predictions of rainfall with a doubling of CO_2, and these differences are reflected in plant production. Soil decomposition rates responded most predictably to changes in temperature. Direct CO_2 enhancement effects on decomposition and plant production tended to reduce the net impact of climate alterations alone.

Water, Air, and Soil Pollution **70**: 643–657, 1993.
© 1993 *Kluwer Academic Publishers.*

1. Introduction

The seasonal distribution of rainfall is a major determinant of plant production in many semiarid and arid regions, where grasslands naturally occur. In the North American Great Plains, most of the rainfall occurs during the growing season, April through September. This permits greater biological productivity, more evapotranspiration (ET), and less runoff than a rainfall pattern more evenly distributed throughout the year, and accounts for a productive system despite the modest annual input of rain (250-800 mm). Many of the grasslands regions of the world share this characteristic. The Intergovernmental Panel on Climate Change (IPCC, Houghton *et al.* 1990) estimates indicate that potential changes in seasonal rainfall and temperature patterns in central North America and the African Sahel will have a greater impact on biological response and feedback to climate than changes in the overall amount of annual rainfall. Ecosystem model simulations of responses to climate change in the Great Plains demonstrated sensitivity of soil carbon storage and grassland biogeochemistry processes to seasonal distribution of precipitation changes (Schimel *et al.*, 1990; Ojima *et al.* 1991).

Grasslands may be among the earliest systems to exhibit the effects of climatic change (OIES 1991). Sensitivity to climatic change may be a reflection of inadequate reserves of water or soil nutrients. Temperature increases will modify ecosystem process such as ET, decomposition, and photosynthesis. The combined effect of temperature and precipitation changes may alter rates of ecosystem processes in such a way that they may offset each other, or alternatively, they may act synergistically together to amplify a positive or negative effect on system C storage. In addition, grassland ecosystem response to increased atmospheric concentrations of CO_2 will be determined by the changes of increased plant production inputs, increased water use efficiency, and modified nutrient availability.

Biospheric control over terrestrial ecosystem-atmosphere feedbacks is strong in grassland ecosystems because of the relationship between weather variations and biotic responses (Pielke and Avissar 1990). Ecosystem changes in these regions resulting from changes in weather patterns or due to land use can alter processes controlling terrestrial C fluxes, such as plant production, and decomposition (Ojima *et al.*, 1991; Schimel *et al.*, 1990; Burke *et al.*, 1990), as well as other key ecosystem processes, such as evapotranspiration and trace gas production and consumption (Schimel *et al.* 1990, 1991; Pielke and Avissar 1990; Mosier *et al.*, 1991; Ojima *et al.*, 1992). The semiarid nature of the climate regime and the extensive land-use changes which have taken place have accentuated the degree of susceptibility of the natural and managed grassland ecosystems.

The objective of this paper is to characterize the factors controlling carbon dynamics in grassland ecosystems worldwide, and to evaluate the potential future response of plant production and soil organic matter C and N dynamics to change in climate and atmospheric CO_2.

2. Modeling Framework

In order to study the impact of climate and atmospheric perturbations on soil organic matter and ecosystem dynamics, the CENTURY model (Parton *et al.* 1987, Parton *et al.*, in press) was employed. CENTURY is a general model of the plant-soil ecosystem that has been used to represent carbon and nutrient dynamics for different types of ecosystems. The CENTURY model is set up to simulate the dynamics of C and N for different plant-soil systems. The model can simulate the dynamics of grassland systems, and implement land management options which influenced the level of grazing, fire frequency and N deposition (Figure 1). The soil organic matter submodel simulates the flow of C and N through plant litter and the different inorganic and organic pools in the soil. The model runs using a monthly time step and the major input variables include monthly average climate, plant chemistry characteristics (e.g., lignin content, plant N content), soil properties (e.g., soil pH, texture, C and N levels, and bulk density), and atmospheric and soil N inputs.

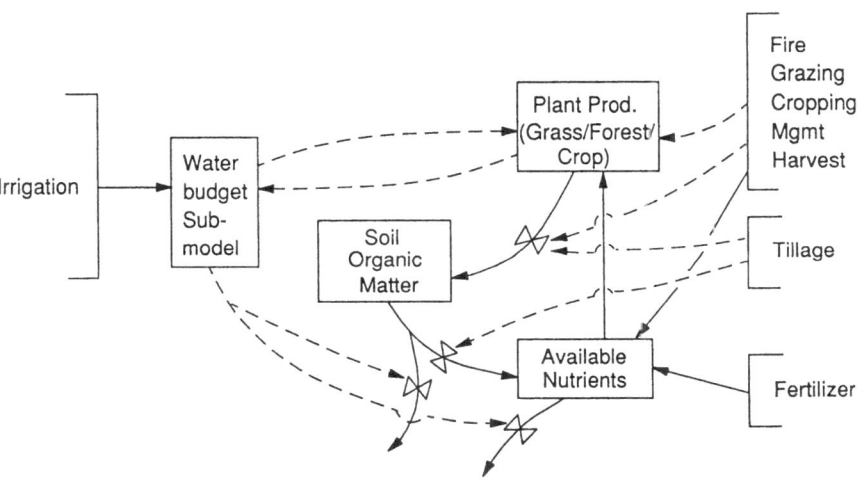

CENTURY
MANAGEMENT PRACTICES

Figure 1. General structure of the CENTURY model. Various management practices (e.g., fire and grazing) were used in the grassland simulations to better represent actual land use practices for the various sites.

The soil organic matter (SOM) submodel simulates the dynamics of C and N in the organic and inorganic parts of the soil system. The soil organic C is divided up into three major components which include active (microbe), slow, and passive soil C. The model also includes a surface microbial pool which is associated with decomposing surface litter. Flows of carbon between these pools are controlled by decomposition rate and microbial respiration loss parameters, both of which may be a function of soil texture. The turnover time of these pools varies with the soil abiotic decomposition parameter (function of monthly precipitation and temperature).

The CENTURY model includes a simplified water budget model which calculates monthly evaporation and transpiration water loss, water content of the soil layers, snow water content and saturated flow of water between soil layers. The potential evapotranspiration rate (PET) is calculated as a function of the average monthly maximum and minimum air temperature and is modified by a user specified multiplier. Interception water loss is a function of aboveground biomass (increases with biomass level), rainfall and PET. Transpiration water loss is a function of the live leaf biomass (exponential function of leaf biomass), rainfall and PET. Water going below the profile can be lost as storm flow or leached into the subsoil where it can accumulate or move into the stream flow at a specified rate.

3. Model Studies of Ecosystem Response

3.1. MODELING CO_2 IMPACT

We modified the plant production parameters for both C_3 and C_4-type grasslands under a doubled atmospheric CO_2 by changing production relative to PET and to nitrogen use efficiency (NUE). The C_3 specific modifications related to CO_2 induced changes of the assimilation efficiency of the rubisco C uptake pathway was not simulated. For the current analysis we uniformly implemented the impact of increased atmospheric CO_2 concentrations on all grasslands. The magnitude of the effect is to cause a maximum of 20% increase in plant production with a doubling of atmospheric CO_2 concentration. The effect of modified NUE and PET on plant production is a simple linear effect on these processes. In addition, changes in NUE affect litter quality so that higher NUE results in a slower decomposition of this material under a given temperature and moisture regime.

3.2. REGIONAL ECOSYSTEM MODELING

We have divided the grassland regions of the world into 7 ecoregions (Figure 2) based on a scheme developed by Bailey (1989). Certain modifications have been made to Bailey's original ecoregion classification to accommodate differences in temperature (T) and rainfall (PPT) of the 31 grassland sites we used in this analysis. Site characteristics for the sites include mean annual T, mean annual PPT, aboveground net primary production (ANNP), and soil C levels to 20cm depth (Table 1).

Table 1. Grassland sites modelled in the present study. Each site is assumed to be representative of its surrounding area on the Bailey (1989) ecoregion map. Soil C and aboveground C values simulated using CENTURY model.

	Ecoregion (Bailey, 1989)	Lat/Long. (approx)	Land Area Represented (10^6 km^2)	Annual Precip. (mm)	Mean Annual Temp.	Soil C (kg/m^2)	Above-ground C (g/m^2)
DRY DOMAIN (300)							
1. Cold continental steppe division (331/333)							
Shortandy, KAZAKHSTAN	(331)	52N 71E	.804	351	1.4 C	7.37	73.1
Tummensogt, MONGOLIA	(333)	46N 113E	.740	269	1.5 C	4.15	35.3
Tuva, RUSSIA	(333)	52N 94E	.109	214	-3.4 C	4.02	46.2
Xilinhot, CHINA	(333)	44N 117E	.441	360	-0.1 C	6.56	84.3
2. Dry steppe division (330) and temperate desert division (340)							
Comodoro Rivadavia, ARGENTINA	(342)	49S 68W	.454	222	12.7 C	1.13	19.1
CPER, USA	(311/315)	40N 105W	.581	300	8.7 C	2.31	45.9
Havre, USA	(331/332)	49N 110W	1.450	312	5.5 C	3.23	33.9
Santa Rosa, ARGENTINA	(311/312)	37S 64W	.335	532	13.5 C	1.80	66.5
Sarmiento, ARGENTINA	(331/332)	46S 69W	.124	141	10.8 C	1.63	15.9
HUMID TEMPERATE DOMAIN (200)							
3. Warm continental division (210) and prairie division (250)							
Khomutov, UKRAINE	(252/332)	47N 38E	.860	441	9 5 C	7.56	112.3
Konza, USA	(251/255)	39N 97W	.781	818	12.9 C	5.78	184.8
Kursk, RUSSIA	(252)	52N 37E	.607	560	5.5 C	9.95	175.8
Otradnoye, RUSSIA	(212)	61N 30E	1.181	543	3.8 C	6.09	106.3
Montevideo, URUGUAY	(254/255)	35S 56W	.529	936	15.1 C	6.48	89.5
4. Mediterranean division (260)							
Bari, ITALY	(262)	41N 17E	.084	574	15.8 C	3.17	107.6
Davis, USA	(262)	39N 122W	.076	420	15.6 C	2.84	50.1

	Ecoregion (Bailey, 1989)	Lat/Long. (approx)	Land Area Represented (10^6 km^2)	Annual Precip. (mm)	Mean Annual Temp.	Soil C (kg/m^2)	Above-ground C (g/m^2)
DRY DOMAIN (300) **5. Tropical/subtropical steppe division (310)**							
Kalgoorlie, AUSTRALIA	(312/315)	31S 122E	1.813	255	17.2 C	3.70	40.1
Khartoum, SUDAN	(314)	16N 33E	1.006	138	25.8 C	2.15	23.2
Menaka, MALI	(314)	16N 2E	.846	290	29.7 C	1.23	35.5
Niamey, NIGER	(415)	14N 2E	1.444	467	23.6 C	2.03	119.5
HUMID TROPICAL DOMAIN (400) **6. Savanna division (410)**							
Charleville/WarraA USTRALIA	(411/416)	26S 146E	1.138	489	210.5 C	4.88	92.6
Cuidad Bolivar, VENEZUELA	(416)	8N 64W	.129	981	27.5 C	4.51	330.6
Kurukshetra Ludhiana, INDIA	(412)	31N 76E	.353	715	24.4 C	3.33	143.4
Marondera, ZIMBABWE	(411/415)	18S 31E	2.995	819	18.3 C	4.60	221.8
Nagpur, INDIA	(413/415)	21N 79E	1.184	1203	26.9 C	5.12	228.6
Nairobi, KENYA	(413/416)	1S 36E	1.565	680	19.0 C	6.24	209.5
Towoomba, SOUTH AFRICA	(314)	25S 29E	.636	630	26.5 C	2.47	119.9
7. Humid savanna (414/415) and rainforest division (420)							
Calabozo/San Fernando, VENEZUELA	(414/415)	9N 67W	.101	1318	28.1 C	7.52	366.5
Carimagua, COLOMBIA	(414)	4N 72W	.185	2338	26.6 C	1.58	226.1
Hat Yai, THAILAND	(423)	6N 101E	.797	1540	27.6 C	2.22	461.6
Lamto, IVORY COAST	(414)	6N 5W	.543	1170	27.9 C	1.78	305.0

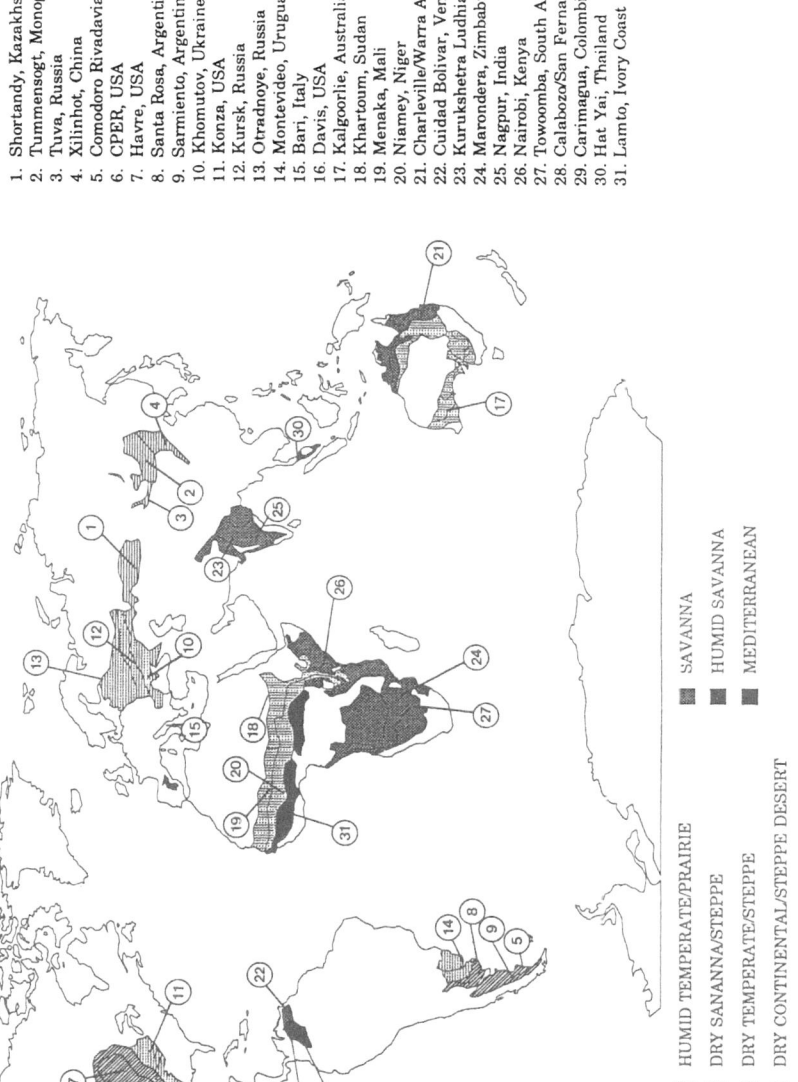

1. Shortandy, Kazakhstan
2. Tummensogt, Mongolia
3. Tuva, Russia
4. Xilinhot, China
5. Comodoro Rivadavia, Argentina
6. CPER, USA
7. Havre, USA
8. Santa Rosa, Argentina
9. Sarmiento, Argentina
10. Khomutov, Ukraine
11. Konza, USA
12. Kursk, Russia
13. Otradnoye, Russia
14. Montevideo, Uruguay
15. Bari, Italy
16. Davis, USA
17. Kalgoorlie, Australia
18. Khartoum, Sudan
19. Menaka, Mali
20. Niamey, Niger
21. Charleville/Warra Australia
22. Cuidad Bolivar, Venezuela
23. Kurukshetra Ludhiana, India
24. Marondera, Zimbabwe
25. Nagpur, India
26. Nairobi, Kenya
27. Towoomba, South Africa
28. Calabozo/San Fernando, Venezuela
29. Carimagua, Colombia
30. Hat Yai, Thailand
31. Lamto, Ivory Coast

HUMID TEMPERATE/PRAIRIE
DRY SANANNA/STEPPE
DRY TEMPERATE/STEPPE
DRY CONTINENTAL/STEPPE DESERT

SAVANNA
HUMID SAVANNA
MEDITERRANEAN

Figure 2. Approximate location of the 31 grassland sites associated with the 7 ecoregions (as modified from Bailey, 1989) used in this study. See Table 1 for the actual locations of the sites.

Summary information of current annual PPT and CGM simulated percent change in annual PPT is depicted Figure 3. The global network of grassland sites had a range of mean annual rainfall of < 100 to > 2000 mm (Figure 3a-f). The mean annual temperatures ranged from less than -2 to greater than 25°C. The dry continental steppes are characterized by a cold and dry climate, in contrast to the humid tropical region which is characterized by a wet and warm climate. The other five regions are intermediate to these two extremes. For regional simulations, we selected representative sites and used site specific climate and soil characteristics to simulate equilibrium current grassland conditions. These initial conditions were then used to simulate ecosystem levels of soil C and plant production under perturbed CO_2 and climate patterns.

Site specific soil texture information was used for the sites which CENTURY had been used for site validation. For the other sites and for extrapolation of the regional estimates of plant and soil C, zonal estimates of soil texture were used. Land management, such as grazing intensity and fire frequency, were applied to a site according to historical land use. The management scenarios are described in the SCOPE grassland modeling paper for the 11 sites used in that study (Parton *et al.* in press), the remaining sites used a similar land management practices characteristic of the region. CENTURY simulations for each site were run to equilibrium (approx. 5000 model years) based on current climatologies.

3.3. CLIMATOLOGY

Ecosystem sensitivity to the temporal and regional resolution of climate change was evaluated by modifying the current monthly weather record for the past 25 years with two different high resolution general circulation model (GCM) climate experiment outputs from Canadian Climate Center (CCC, Boer and Lazare 1988) and the Geophysical Fluid Dynamics Laboratory (GFDL) models as described in the 1990 IPCC report (Houghton *et al.* 1990). We tested the sensitivity of grassland ecosystems to the following global change effects:

- Climate change effects (+CC, i.e., alterations to monthly mean temperature and precipitation);
- Double CO_2 response (+CO_2, i.e., increases in plant production resulting from changes in water use efficiency (i.e., modified PET) and NUE due to elevated atmospheric CO_2 levels;
- Combined effect of climate change and doubling of atmospheric CO_2 (+CC+CO_2).

For each grassland site, a current weather file of monthly precipitation and monthly mean maximum and minimum temperatures was created using existing weather station data from the site itself or from a nearby meteorological station. When data were not available from a researcher at the site, we relied on the "World Weather Disc"(Weather Disk Associates, Inc., National Climate Data Center) for a site. We used a 25-year weather record as the base climate to simulate the equilibrium

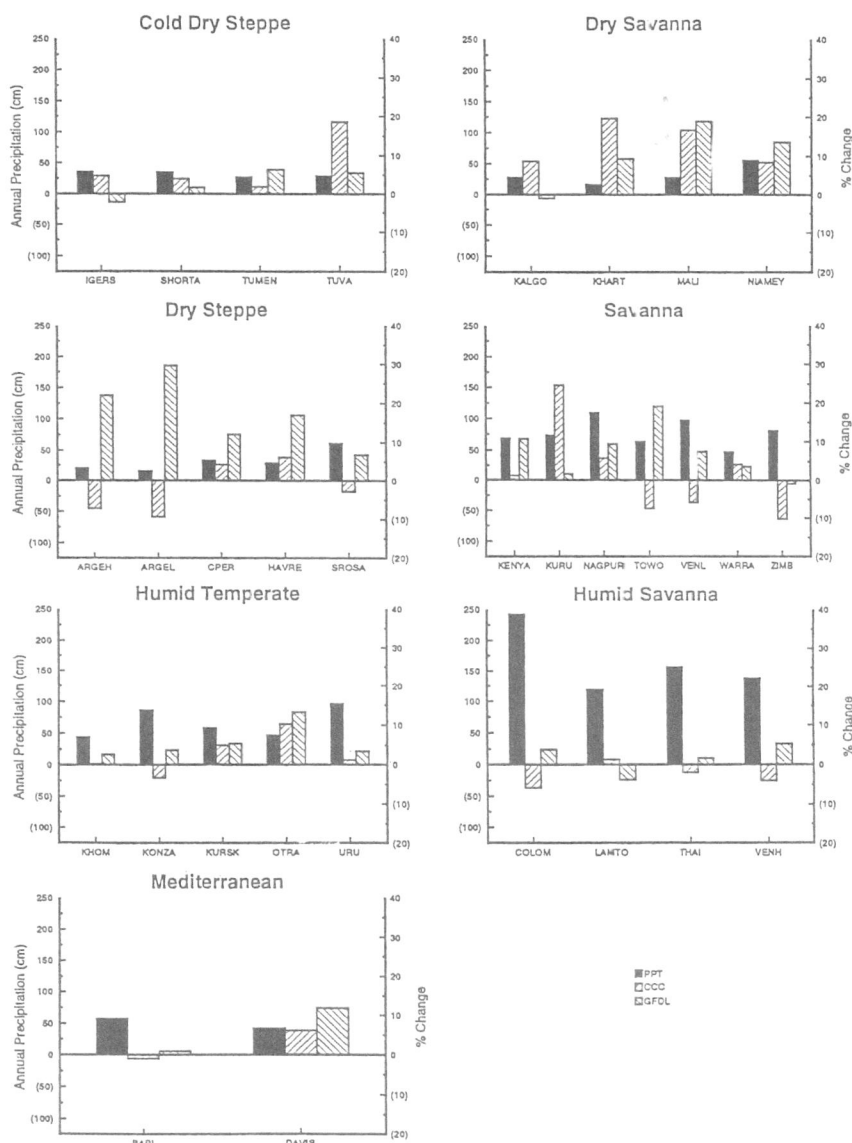

Figure 3. Current mean annual precipitation and percent changes of the two GCM
scenarios for the 31 sites. The solid bar, associated with left axis, is the
current value, and the percent changes of the CCC and GFDL models can be
determined using the right axis of each figure. The site names are associated
with the list of sites found in Table 1.

grassland for each site. In order to generate the double CO_2 climate, we spatially
interpolated GCM grid values of projected double CO_2 climate changes of monthly
temperature (T) and PPT for each site based on the actual model output made for the
1990 IPCC report. The climate change inputs were based on the business-as-usual
scenarios for the CCC and the GFDL high resolution runs (Houghton *et al.* 1990).
We applied these projected monthly values in a linear fashion in a 50 year ramp (e.g.
Figure 4). The 50 yr ramp was generated by taking the projected monthly climate
changes and dividing these by 50, the number of years in the ramp. This incremental
change then was added to each respective month during the 50 year ramp. We
applied the projected monthly temperature changes equally to the minimum and the
maximum mean monthly temperature values used by CENTURY model.

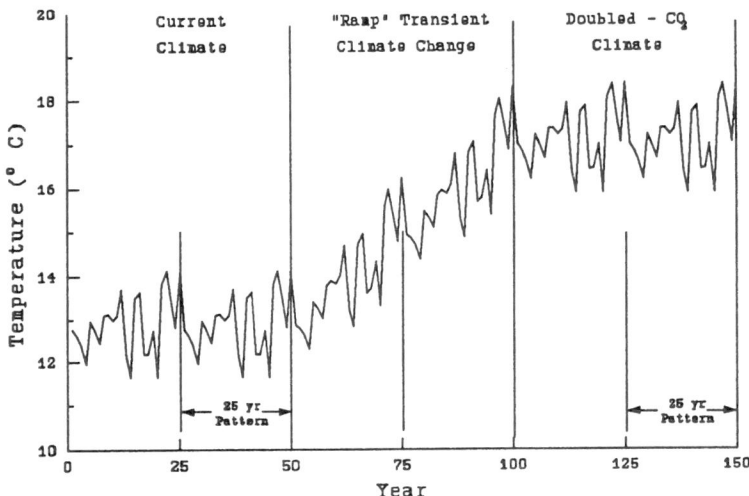

Figure 4. Typical mean annual temperature climate file used in the CENTURY
        climate change runs. These were generated based on current 25 yr weather
        patterns which served as the template to implement the 50 yr transient
        ('RAMP') and the modified doubled CO_2 climate for mean annual
        temperature and precipitation. A repeating 25 yr pattern is used throughout
        the CENTURY simulations.

4. REGIONAL CHANGES TO CLIMATE AND CO_2 PERTURBATIONS

The following section describes the ecosystem results for the grassland ecoregions,
and describes the site variation within each grassland ecoregion. Overall, changes in
aboveground annual net primary production (ANPP) in the climate change runs (CC,
Figure 5) were correlated to changes in rainfall. Cases where rainfall amounts
declined resulted in a lower simulated ANPP. Differences between GCM scenarios
were mostly observed in the differences between projected rainfall amounts.

ANNUAL ABOVEGROUND PLANT C

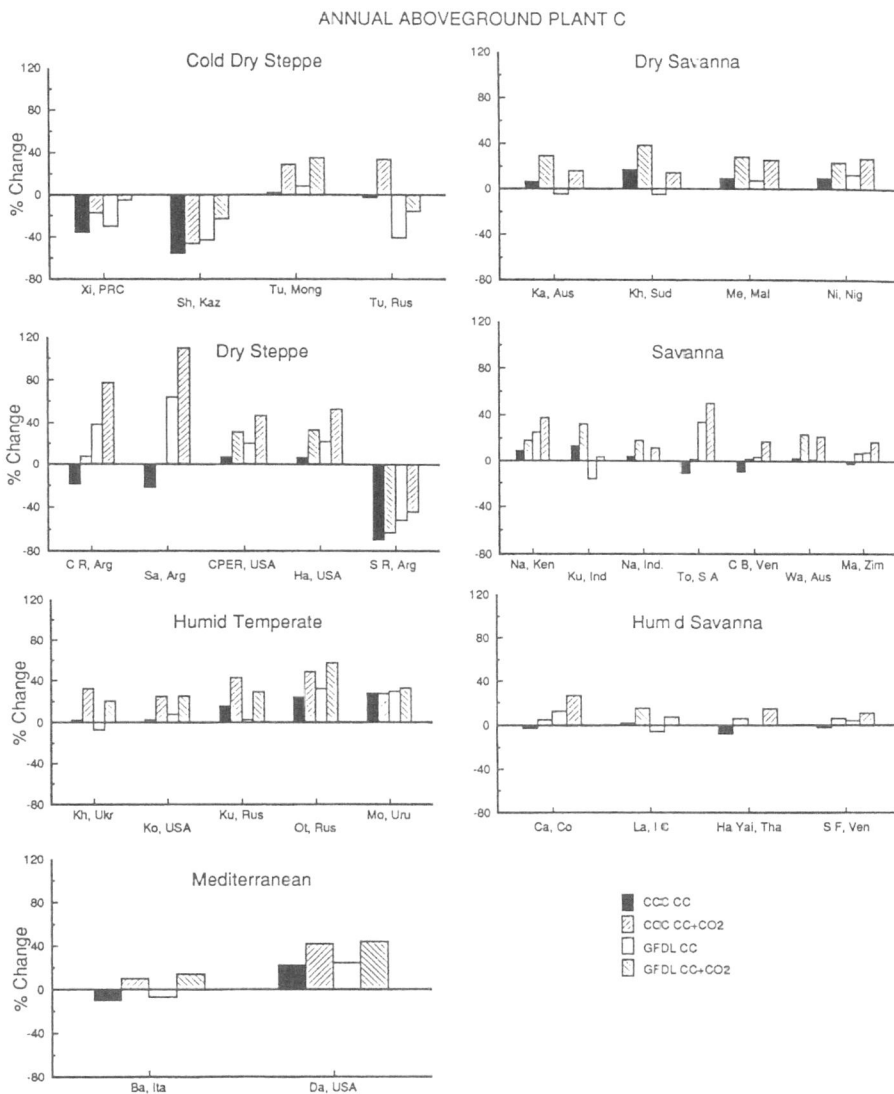

Figure 5. Percent changes in annual aboveground plant C for the 31 sites in the 7 ecoregions. The solid and open bars represent climate change only (CC) for the Canadian Climate Centre (CCC) and the Geophysical Fluid Dynamics Laboratory (GFDL) GCM's, respectively. The hatched bars represent the climate change plus increased atmospheric CO_2 simulations ($CC+CO_2$) for the two GCM's. Site labels are associated with sites listed in Table 1.

% CHANGE IN SOIL C

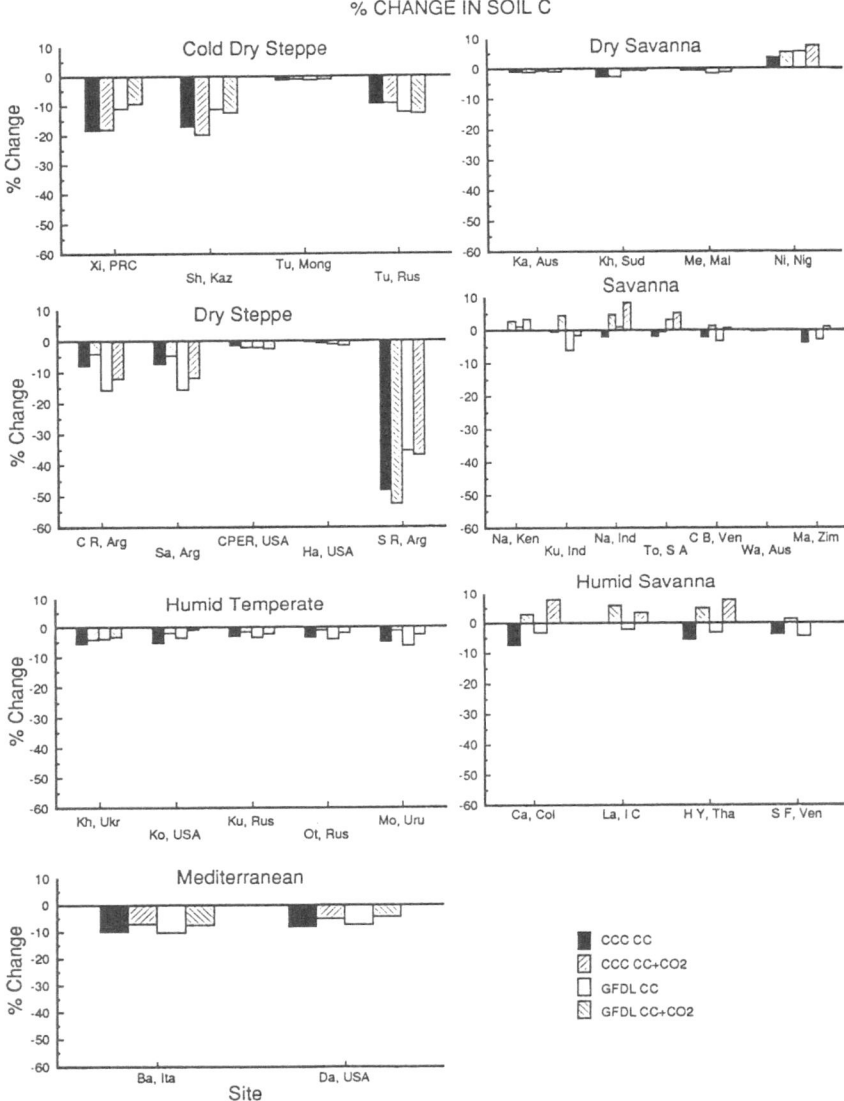

Figure 6. Percent changes in soil organic C in the top 20 cm of soil for the 31 sites in the 7 ecoregions. The solid and open bars represent climate change only (CC) for the Canadian Climate Centre (CCC) and the Geophysical Fluid Dynamics Laboratory (GFDL) GCM's, respectively. The hatched bars represent the climate change plus increased atmospheric CO_2 simulations (CC+CO_2) for the two GCM's. Site labels are associated with sites listed in Table 1.

The impact of modifying both the climate and atmospheric CO_2 ($CC+CO_2$) resulted in enhancement of ANPP relative to CC simulations (Figure 5). The tropical simulations (dry savanna, savanna, and humid savanna) all resulted in increased ANPP with $CC+CO_2$ scenarios, regardless of GCM. Most of the temperate sites, had increased ANPP, except for most of the cold dry steppe sites, as did the Santa Rosa site of the dry steppe region.

The climate change results indicate that simulated soil C losses occur in all but one of grassland ecoregions, losses ranged from 0 to 14% at most sites of current soil C levels for the surface 20 cm (Figure 6). The Santa Rosa site declined in soil C to a much greater extent, losing 48% soil C under the projected CCC climate scenario and 35% under the GFDL scenario. The Eurasian grasslands lost the greatest amount of soil C (approximately 1200 g C m^{-2}). The other temperate grasslands sites losses ranged from 0 to 1000 g C m^{-2}, averaging approximately 350 g C m^{-2}. The tropical grasslands and savannas lost the least amount of soil C per unit area ranging from no change to 300 g C m^{-2} losses, averaging approximately 70 g C m^{-2}. Plant production varies according to modifications in rainfall under the altered climate and to altered nitrogen mineralization rates. The two GCM's differed in predictions of rainfall with a doubling of CO_2, and this difference is reflected in plant production. Soil decomposition rates responded most predictably to changes in temperature. Direct CO_2 enhancement effects on soil C loss and plant production tended to reduce the net impact of climate alterations alone.

Soil C levels declined in the tropical regions, with only a few exceptions under the CC scenarios (Figure 6). In the dry savanna region, the Niamey site produced 3 to 5% more soil C with CC climates. The savanna simulations using the GFDL climate projections, resulted in a net increase in soil C at the Nairobi site (approximately 1%) and the Towoomba site, a 3% increase.

The simulations with the combined CO_2 and CC modifications resulted, in most cases, in a net enhancement of soil C levels relative to the CC simulations. The regions where this trend did not take place were the extremely dry regions in both the temperate and the tropical regions. Specifically, in the cold dry steppe region, dry steppe region, and the dry savanna region, one observes several cases where soil C levels under the $CC+CO_2$ simulations declined more or increased less than the CC simulations. This outcome most likely was the result of greater decomposition of soil C under better soil moisture conditions. The altered soil moisture conditions resulted from modifications in water use efficiency (WUE) under increased CO_2 conditions.

Globally, the impact of climate changes on grassland soil C is estimated to result in a loss of 3 to 4 Pg C over a 50 yr period or a source of .06 to .08 Pg C yr^{-1}. The effect of increasing atmospheric CO_2 concentrations under current climatic conditions on ecosystem productivity and soil C storage is to enhance plant production and to increase soil C storage by 2 Pg under a doubled atmospheric CO_2 environment. The estimate of soil C changes which include the climate and direct CO_2 effects resulted in soil C loss of 1 to 2 Pg C over the 50 yr simulation. These estimates were made based on current grassland areas, and did not include any modifications related to potential land cover or land management changes.

7. Summary

Two major controls related to climate factors are evapotranspiration and decomposition. These processes affect rates of plant productivity and soil C levels. The net result of the CO_2 and climate change impacts will depend on the differential changes in decomposition rates due to climate changes, litter quality changes, and changes to decomposition due to altered soil water dynamics resulting from plant mediated changes in ET rates. Soil C levels will then be determined by the net changes in decomposition rates, plant production changes due to climate and CO_2 mediated effects, and rates of plant residue inputs into the soil system.

Acknowledgements

Data synthesis and model validation was made possible by the many collaborators involved with the SCOPE (Scientific Committee on Problems of the Environment) Project "Effects of Climate Change on Production and Decomposition in Coniferous Forests and Grasslands"; the National Institute for Global Environmental Change (NIGEC) Midwestern Region Center; US-DOE project on "Biological Hysteresis in Climate Change Models for Grasslands: Implications of plant community dynamics on biogeochemical feedbacks". Current model development was funded by the US NASA Earth Observing System project NACW-2662 "Using Multi-Sensor Data to Model Factors Limiting Carbon Balance in Global Grasslands" and the Ecological Research Division, Office of Health and Environmental Research, US Department of Energy (DOE) project, "Rangeland - Plant Response to Elevated CO_2." Model developmenet conducted by Rebecca McKeown and data analysis and integration by Brian Newkirk and Song Bo at NREL, CSU. Access to GCM output datasets was made possible through the help of Dennis Joseph and Roy Jenne, Data Support Section, National Center for Atmospheric Research (NCAR). GCM model data handling was undertaken by Tom Painter and Donna Beller and were supported by Model Evaluation Consortium for Climate Assessment (MECCA). NCAR is supported by the National Science Foundation.

8. References

Bailey, R.G.: 1989, *Environ. Conserv.*, **16**, 307-309.
Boer, G.J. and Lazare, M.: 1988, *J. Clim.*, **1**, 789-806.
Burke, I.C., Kittel, T.G.F., Lauenroth, W.K., Snook, P., Yonker C.M. and Parton, W.J.: 1990, *Bioscience* **41**, 685-692.

Houghton, J.T., Jenkins, G.J. and Ephraums, J.J. (eds): 1990, *Climate Change. The IPCC Scientific Assessment*, World Meteorological Organization (WMO), Cambridge University Press, Cambridge.

Mosier, A. R., Schimel, D., Valentine, D., Bronson, K. and Parton, W.: 1991, *Nature* **350**, 330-332.

Office for Interdisciplinary Earth Studies (OIES): 1991, *Arid Ecosystems Interactions: Recommendations for Drylands Research in the Global Change Research Program*. OIES-Report 6, p. 81.

Ojima, D.S., Kittel, T.G.F., Rosswall, T. and Walker, B.H.: 1991, *Ecological Applications* **1**:316-325.

Ojima, D.S., Valentine, D.W., Mosier, A.R., Parton, W.J. and Schimel, D.S.: 1993, *Chemosphere* **26**, 675-685.

Parton, W.J., Schimel, D.S., Cole, C.V. and Ojima, D.S. 1987, *Soil Science Society of America Journal* **51**, 1173-1179.

Parton, W.J., Schimel, D.S., Ojima, D.S. and Cole, C.V.: *Soil Science Society of America Journal*. (in press).

Pielke, R.A. and Avissar, R.: 1990, *Landscape Ecology* **4**, 133-135.

Schimel, D.S., Parton, W.J., Kittel, T.G.F., Ojima,D.S. and Cole, C.V.: 1990, *Climate Change* **17**, 13-25.

Schimel, D.S., Kittel, T.F.G. and Parton, W.J.: 1991, *Tellus* **43AB**, 188-203.

VEGETATION REDISTRIBUTION:
A POSSIBLE BIOSPHERE SOURCE OF CO$_2$ DURING CLIMATIC CHANGE

RONALD P. NEILSON

*U.S.D.A. Forest Service, Pacific Northwest Research Station,
3200 S.W. Jefferson, Corvallis, Oregon, USA*

Abstract. A new biogeographic model, MAPSS, predicts changes in vegetation leaf area index (LAI), site water balance and runoff, as well as changes in Biome boundaries. Potential scenarios of equilibrium vegetation redistribution under 2 x CO$_2$ climate from five different General Circulation Models (GCMs) are presented. In general, large spatial shifts in temperate and boreal vegetation are predicted under the different scenarios; while, tropical vegetation boundaries are predicted (with one exception) to experience minor distribution contractions. Maps of predicted changes in forest LAI imply drought-induced losses of biomass over most forested regions, even in the tropics. Regional patterns of forest decline and dieback are surprisingly consistent among the five GCM scenarios, given the general lack of consistency in predicted changes in regional precipitation patterns. Two factors contribute to the consistency among the GCMs of the regional ecological impacts of climatic change: 1) regional, temperature-induced increases in potential evapotranspiration (PET) tend to more than offset regional increases in precipitation; and, 2) the unchanging background interplay between the general circulation and the continental margins and mountain ranges produces a fairly stable pattern of regionally specific sensitivity to climatic change. Two areas exhibiting among the greatest sensitivity to drought-induced forest decline are eastern North America and eastern Europe to western Russia. Drought-induced vegetation decline (losses of LAI), predicted under all GCM scenarios, will release CO$_2$ to the atmosphere; while, expansion of forests at high latitudes will sequester CO$_2$. The imbalance in these two rate processes could produce a large, transient pulse of CO$_2$ to the atmosphere.

1. Introduction

Anthropogenic emissions of CO$_2$ and other greenhouse gases are expected to produce a global warming of as much as 1.5 to 4.5° C under the equivalent forcing of double pre-industrial CO$_2$ levels (Houghton et al., 1990). A global warming of such a magnitude could produce large shifts in the distribution of global vegetation (Emanuel et al., 1985; Neilson et al., 1989; Prentice and Fung, 1990; Smith et al., 1992). Extratropical biomes are expected to shift toward the poles, while equatorial biomes (forests) could expand or contract *in situ.*

Regions no longer suitable for forests could produce emissions of CO$_2$, while regions favoring additional vegetation growth could sequester CO$_2$ (Neilson et al., 1989; King and Neilson, 1992). Previous equilibrium estimates of terrestrial C storage under a 2 x CO$_2$ climate are equivocal as to whether the terrestrial biosphere would eventually be a source or a sink (Emanuel et al., 1985; Prentice and Fung, 1990; King and Neilson, 1992; Smith and Shugart, 1993; Smith et al., 1993). However, before equilibrium is ever attained, imbalances in the rates of CO$_2$ release from forest dieback and sequestering of CO$_2$ from forest growth during vegetation redistribution could produce a large, transient pulse of CO$_2$ into the atmosphere (King and Neilson, 1992; Smith and Shugart, 1993). Previous estimates of the transient 'carbon pulse', based on projected 2 x CO$_2$ equilibrium vegetation changes, indicate that the net rate of CO$_2$ release from the terrestrial biosphere could be as high as 40% of current anthropogenic emissions over a period of several decades (King and Neilson, 1992;

Water, Air, and Soil Pollution **70**: 659–673, 1993.
© 1993 *Kluwer Academic Publishers.*

Smith and Shugart, 1993). Large uncertainties exist in these estimates, among the most important being 1) the potential magnitude of vegetation redistribution in the extratropics, and 2) whether or not tropical forests will expand or contract (King and Neilson, 1992; Neilson and King, 1992). These uncertainties are the subjects of this paper and are addressed using a new global biogeography model (Neilson, unpublished).

2. Methods

2.1.   CLIMATE SCENARIOS

Scenarios of double CO_2 climate change, used to drive the vegetation model, were derived from five general circulation model (GCM) 2 x CO_2 equilibrium simulations. Climate scenarios were supplied by the Data Support Section within the Scientific Computing Division of the National Center for Atmospheric Research (NCAR). Model outputs for current and 2 x CO_2 climates were obtained from the following models: GISS (Goddard Institute of Space Studies, Hansen et al., 1988); UKMO (United Kingdom Meteorological Office, Mitchell and Warrilow, 1987); GFDL (Geophysical Fluid Dynamics Laboratory, Wetherald and Manabe, 1988); and OSU (Oregon State University, Schlesinger and Zhao, 1989). Both the GFDL R15 (ca. 4° x 5° grid) and R15 Q-flux versions were used. The Q-flux version of the GFDL model (GFDL-Q) includes a prescribed ocean-atmosphere coupling (Manabe et al., 1991). The coarse grid from each model was interpolated using a 4 point, inverse distance squared algorithm to a 0.5° x 0.5°, lat.-long. grid. The scenarios were applied as per recommended and calculated by the NCAR Data Support Section. Scenarios were constructed by applying ratios (2 x CO_2/1 x CO_2) of all climate variables (except temperature) back to a baseline dataset, the IIASA 0.5° resolution gridded climate dataset (Leemans and Cramer, 1991). Ratios were used to avoid negative numbers, but were not allowed to exceed 5, to prevent unrealistic changes in areas with normally low rainfall. Temperature scenarios were calculated as a difference (2 x CO_2 - 1 x CO_2) and applied to the baseline dataset.

In addition to the published, gridded IIASA climate dataset, Leemans and Cramer have assembled additional datasets for relative humidity, vapor pressure (1771 and 1776 stations, respectively) and windspeed (ca. 3995 stations), and kindly shared these with us (Anonymous 1984a,b; Anonymous 1987a,b; Müller 1982) humidity, vapor pressure and temperature datasets were quality controlled (by the author) by using the temperature data to calculate the saturation vapor pressure, which, in combination with the vapor pressure, was used to calculate the relative humidity (RH). The calculated RH was subtracted from the actual RH (as contained in the separate data file) and the residuals plotted against the temperature. Above zero Celsius, the residuals clustered very close to zero. Residuals exceeding a 10% threshold (there were very few) were either discarded or corrected, if an obvious data coding error was detected. The data were then interpolated as RH values (four point, inverse-distance-squared) over the 0.5° grid and converted back to vapor pressure using the adiabatically-corrected grid temperature. This procedure may slightly underestimate high mountain humidity, but protects against physically impossible values obtained by interpolating vapor pressure directly from low elevations to neighboring mountains.

Winds were interpolated directly to the 0.5° grid using the same four point, inverse distance squared procedure. Mountain winds will be poorly captured by this procedure, but it represents the best available data at present. Future climate scenarios for vapor pressure were constructed using the ratio approach, described above. However, only current winds were used in the present study, since the change in future winds is quite extreme (GCM predictions) and carries many uncertainties (Marks, unpublished).

Site water balance and thermal constraints are thought to be the primary controls on the distribution of most of the world's vegetation (Whittaker, 1975; Box, 1981; Neilson and Wullstein, 1983; Neilson, 1987; Stephenson, 1990). All large-scale biogeography models incorporate these controls in some form or another, ranging from strictly statistical (e.g., Holdridge, 1947) to more fundamental approaches (Box, 1981; Prentice et al., 1992; Neilson et al., 1992; Neilson, unpublished). Site water balance is usually related to the vegetation distribution through some form of index, such as the ratio of actual to potential evapotranspiration, AET/PET, (Prentice et al., 1992), or the difference between potential and actual transpiration, PET-AET (Stephenson, 1990; Lenihan and Neilson, unpublished) integrated over an average annual cycle.

The water balance approach taken here differs from previous approaches by directly coupling the rate of transpiration to canopy conductance, a function of the surface area of leaves (leaf area index, LAI) and their stomatal conductance (Woodward, 1987; Neilson,

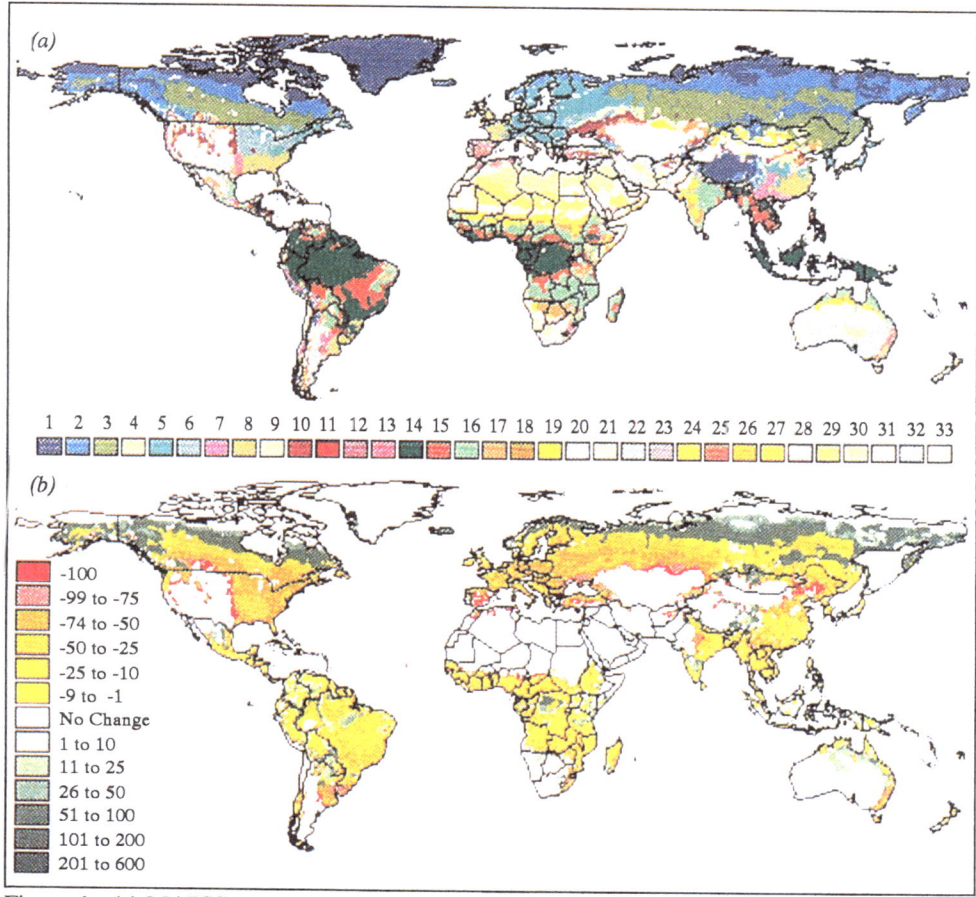

Figure 1. (a) MAPSS control (preliminary calibration, see key in Table 2). (b) Average percent change in LAI from five 2 x CO_2 climate scenarios.

unpublished). The maximum LAI (i.e., maximum rate of transpiration) that can be sustained at a site without depleting soil water is calculated through a process of iteration. The calculated LAI integrates all of the factors that influence transpiration and site water balance, including snow formation and melt, canopy rainfall interception and evaporation, transpiration, soil drainage and runoff. Thus, the influence of site water balance on the distribution of vegetation can be directly calculated from LAI, rather than inferred through a water balance index. For example, the transition from closed forest to open forest or savanna occurs when the forest LAI falls below a certain canopy closure level. The model, described in detail elsewhere (Neilson, unpublished), is termed a Mapped Atmosphere-Plant-Soil System (MAPSS).

Table 1. Definition of thermal zones within MAPSS.

Boundary	Calibrated Temp. Threshold	Physiological Interpretation
Tundra - Taiga/Tundra	735 Growing Degree Days	Short Growing Season (Frost desiccation)
Taiga/Tundra - Boreal	1330 Growing Degree Days	Short Growing Season (No reproduction)
Boreal - Temperate	-16° Celsius (ave. monthly)	Supercooled freezing point (-40° C)
Temperate - Subtropical	1.25° Celsius (ave. monthly)	Annual hard frost (24 hrs. < 0° C)
Subtropical - Tropical	13° Celsius (ave. monthly)	No frost

Maximizing the LAI that can be supported at a site with respect to the site water balance presumes that water is the primary limiting factor of the site carrying capacity. Current theory and empirical analyses tend to support this premise for most of the world's upland ecosystems (Woodward, 1987; Neilson et al., 1989; Stephenson, 1990). However, some systems are limited by available energy (e.g., high latitude systems), soil chemistry (e.g., serpentine soils), or possibly nutrients. It is assumed that north of the closed boreal forest the open taiga/tundra and tundra regions are primarily energy limited (Woodward, 1987; Stephenson, 1990; Lenihan and Neilson, unpublished) and MAPSS does not calculate the water balance for those biomes. The maximum LAI is calculated at all other sites under the assumption that the site is water limited. However, when the calculated LAI reaches the maximum allowed by the model (currently set to 15), it is likely that energy rather than water is the primary limiting factor under the current climate. Nutrient limitations are currently not considered, nor are chemical constraints (e.g., serpentine soils). Poorly drained sites are also not considered in MAPSS.

Three basic vegetation lifeforms (trees, shrubs and grasses) are incorporated in the model. A site can support either trees or shrubs (not both) in competition with grass for light and water (Neilson, unpublished). The woody lifeforms project roots into the top two layers of a three-layer soil, while the grasses reach only into the surface soil layer. The third soil layer is present to accurately simulate unsaturated base flow from the soils. Within the light constraint of the woody vegetation, grasses compete with the woody vegetation for water in the surface soil layer as a function of their relative proportions of LAI (modified by stomatal conductance).

Canopy conductance, used in calculating transpiration is an exponential function of LAI, the form of which is common to many crop and atmospheric models (Abramopolous et al., 1988, Neilson, unpublished). Canopy conductance is also constrained by stomatal conductance, which is a function of vapor pressure deficit and soil water potential. The current simulations are based on a generic sandy-loam soil with the surface layer being 0.5 m thick and the second layer 1.0 m thick. The third layer, providing base flow, is ca. 2 m thick.

Thermal limits on vegetation distribution (Woodward, 1987) define six latitudinal zones within MAPSS, tundra, taiga/tundra (open, boreal woodland), boreal, temperate, subtropical and tropical (Table 1). The tundra-taiga/tundra and taiga/tundra-boreal ecotones are defined by growing degree day limits (base 0° C) of 735 and 1330, respectively (Lenihan and Neilson, unpublished). The boreal-temperate ecotone is defined by winter temperatures that fall below the supercooled freezing point of water (-40° C), determined as the limit for most temperate hardwoods (Burke et al., 1976; Woodward, 1987) and is indexed by mean monthly temperatures below -16° C (Neilson et al., 1989; Neilson et al., 1992). The temperate-subtropical ecotone is defined by the high probability of hard frost (mean daily maximum temperatures below freezing) occurring, on average, in every year, and is indexed by a mean monthly temperature of ca. 1.25° C (Neilson et al., 1989; Neilson et al., 1992). First and last frosts are indexed by mean monthly temperatures less than 13° C (Neilson, unpublished). The subtropical-tropical ecotone is defined by the coolest winter month exceeding the 13° index of last frost.

2.3.   VEGETATION CLASSIFICATION

Vegetation types in MAPSS are classified from a functional perspective (Neilson et al., 1992). I have not attempted to adhere to any particular nomenclature. The classification separates closed forests, tree savannas, dense shrub communities (chaparral), shrub savannas, grasslands and deserts based on woody and grass LAI (Table 2). Above a specified woody LAI threshold, the woody vegetation is of the tree form and below the threshold is of the shrub form. The tree/shrub LAI threshold is set lower in the tropics than in the extratropical zones. I hypothesize that differences in land-use and natural disturbance regimes (fire and large herbivores) preclude widespread open savannas from the extratropical zones compared to those observed in the tropics.

Vegetation types are further split on whether their phenology is evergreen or deciduous, as determined by physiologically-based rules relying on the seasonality of weather (Marshall and Waring, 1984; Neilson et al., 1992; Lenihan and Neilson, unpublished). The tropical zone is defined as evergreen with drought-deciduous forms being indexed by a lower LAI than the closed, evergreen forests. The classification also splits vegetation based on broadleaf or microphyllous leaf form, again based on the seasonality of weather.

Vegetation types are further partitioned by the thermal zone. The subtropical vegetation types are all classified as 'warm mixed' in phenology (evergreen/deciduous) due to the wide variation in frost and cold adaptations. Within the temperate zone, broadleaf deciduous types are subdivided into 'deciduous broadleaf' and 'cool mixed' based on an LAI threshold above which water limitations are reduced (Neilson et al., 1989). The growth of broadleaf, deciduous forms is optimal, but the cool conditions and short growing season also favor cold-adapted conifers, thereby producing a mixture. Within the 'mixed' classes, the leaf form rule is used to designate a potential climax dominant. So, for example, the forests in the northwest coastal mountains (U.S.) are classified as warm mixed (subtropical zone), but microphyllous (conifer) dominant; while, the southeast U.S. is classified as warm mixed, but broadleaf dominant (Fig. 1a). Tropical seasonal and savanna types have been only broadly defined, based on LAI and , in the future, will be better defined using physiologically-based drought-deciduous rules.

Grassland rules have been defined for tall, mixed, and short grass types; however, the grasslands require a fire regime to constrain the woody LAI component (Neilson et al., 1992).

Table 2. Preliminary classification and calibration of vegetation types within MAPSS.[1]

Aggregation	Classification	LAI	Phen.	Leaf	Zone
Tundra	1. Tundra	N/A	N/A	N/A	Boreal
Taiga/Tundra	2. Taiga/Tundra	N/A	N/A	N/A	Boreal
Boreal	3. Forest Evergreen Needle (Taiga)	>3.75	E	M	Boreal
and	4. Forest Evergreen Needle (Temperate)	>3.75	E	M	Temp
Temperate Forest	5. Forest Mixed Cool (Temperate)	> 11	D/E	B/M	Temp
	6. Forest Deciduous Broadleaf	3.75-11	E	B	Temp
	7. Forest Mixed Warm (EN)	>3.75	D/E	B/M	S.Trop
	8. Forest Mixed Warm (DEB)	>3.75	D/E	B	S.Trop
Boreal and	9. Tree Savanna Cool Mixed (EN)	2-3.75	D/E	B/M	Boreal
Temperate	10. Tree Savanna (Evergreen Needle)	2-3.75	E	M	Temp
Savanna	11. Tree Savanna (Deciduous Broadleaf)	2-3.75	D	B	Temp
	12. Tree Savanna Warm Mixed (EN)	2-3.75	D/E	B/M	S.Trop
	13. Tree Savanna Warm Mixed (DEB)	2-3.75	D/E	B	S.Trop
Tropical Forests	14. Tropical Forest Evergreen Broadleaf	>3.75	E	B	Trop
	15. Tropical Seasonal Forest (Moist)	2-3.75	D/E	B	Trop
Tropical Savanna	16. Tropical Dry Forest/Savanna	.65-2	D/E	B/M	Trop
Non-Forest	17. Chaparral	2.1-3.5	D/E	B/M	All
	18. Open Shrubland	<2.1	D/E	B/M	All
	19. Shrub Savanna Cool Mixed (EN)	<2.1	E	B/M	Boreal
	20. Shrub Savanna Evergreen Micro.	<2.1	E	M	Temp
	21. Shrub Savanna Deciduous Broadleaf	<2.1	D	B	Temp
	22. Shrub Savanna Warm Mixed (EN)	<2.1	D/E	B/M	S.Trop
	23. Shrub Savanna Warm Mixed (DEB)	<2.1	D/E	B	S.Trop
	24. Shrub Savanna Tropical	<.65	E	B	Trop
	25. Tall Grass Prairie	0.9-6			All
	26. Mixed Grass Prairie	.5-.9			All
	27. Short Grass Prairie	.4-.5			All
	28. Semi-Desert Grassland	.1-.4			All
	29. Tropical Desert				Trop
	30. Subtropical Desert				S.Trop
	31. Temperate Desert				Temp
	32. Boreal Desert				Boreal
	33. Extreme Desert				All

[1]N/A=Not applicable; E=evergreen; D=deciduous; B=broadleaf; M=microphyllous.
Parenthetical abbreviations in 'Mixed' vegetation types indicate likely climax physiognomy;
EN=evergreen-needleleaf; DEB=deciduous/evergreen broadleaf.

Desert classes are only separated by thermal zone, after having been classified as desert based
on both woody and grass LAI conditions. An extreme desert category is defined where

rainfall is so low that the site water balance calculations cannot be solved.

Most of the classification rules are post-hoc determinations and they do not affect the functional calculation of LAI. The LAI threshold defining closed forest (>3.75, Table 2) is the only critical LAI threshold and is used for defining forest/non-forest boundaries. The fundamental importance of LAI calculations (beyond classification) is as a direct indicator of the vegetation carrying capacity with respect to site water balance.

The 33 vegetation classes serve to demonstrate the qualitative accuracy of MAPSS (Fig. 1a), but are less important from the perspective of global terrestrial carbon balance estimation. The aggregated tundra, taiga/tundra, temperate and boreal forest, temperate and boreal savanna, tropical forest, tropical savanna and non-forest classes (Table 2) are adequate for a first approximation of carbon pool changes and for validation testing of MAPSS.

Above and belowground carbon densities are based on Olson et al. (1983) and Zinke et al. (1984) as presented in Cramer and Solomon (in press). The vegetation classes used in Cramer and Solomon (in press) were aggregated to the levels described above and the carbon density values for each of their classes were averaged for the aggregated classification (Table 2).

Table 3.   Above and belowground carbon density (kg/m^2).

	Vegetation Carbon			Soil Carbon		
	Low	Medium	High	Low	Medium	High
Tundra	0.5	0.8	1.3	15.7	18.2	20.7
Taiga/Tundra	1.0	2.0	5.0	10.0	16.6	23.2
Boreal Forest	3.2	6.9	9.9	6.9	14.8	22.6
Temperate Forest	6.7	10.0	14.0	11.9	13.8	15.8
Boreal & Temperate Savanna	2.0	4.1	7.3	6.7	7.3	7.9
Tropical Forest	10.0	17.0	21.0	9.5	10.4	11.3
Tropical Dry Forest & Savanna	3.2	0.3	6.6	6.3	7.3	8.3
Non-Forest	1.3	2.6	4.9	11.8	14.0	16.3

2.4.   CALIBRATION AND SENSITIVITY ANALYSIS

2.4.1. Calibration over the conterminous United States: Predicted monthly and annual runoff were calibrated on four small clusters of weather stations, generally over Alabama, Illinois, Nebraska and Oregon for a total of about 10 stations within a network of 1211 dispersed weather stations in the U.S. and about 1100 stream gauges (Neilson, unpublished). MAPSS was subsequently validated, using the full meteorological and stream gauge networks, over the eastern U.S. and shown to accurately estimate the annual water balance (runoff/precipitation) to within ± 10% over most of the area and also accurately simulates the monthly runoff at individual sites (Neilson et al., 1989). The residuals (predictions exceeding ± 10%) are largely underpredictions of runoff, and occur in areas where the potential forest vegetation has been replaced by agriculture or other land uses and in mountainous terrain where valley rain gauges underestimate the precipitation catch in mountain watersheds. The predicted vegetation distribution was visually calibrated against a coarse classification based on Küchler (1964) and Dice (1943) (Neilson et al., 1992). MAPSS was then implemented on a dense grid of ca. 78,000 points (10 km cells) over the same spatial extent and visually appears to accurately predict vegetation and runoff in complex terrain (maps on file with the author). The successful transition from 1,211 points to 78,000 represents partial validation of both the model and the distributed climate datasets over the 78,000 points (Daly et al., unpublished; Marks, unpublished).

2.4.2. *Calibration and Validation of MAPSS-Global:* Attempts to calibrate closely to observed or expected vegetation presume that the climate used to drive the models is accurate and that the vegetation maps are accurate. Thus, we have chosen to calibrate MAPSS-global against a wide assortment of vegetation maps, both

Table 4.   Kappa Statistic comparisons for model control runs.

	MAPSS vs. Olson	BIOME vs. Olson	MAPSS vs. BIOME
Tundra	.61	.65	.73
Taiga/Tundra	.41		
Boreal and Temperate Forest	.6	.65	.83
Boreal & Temperate Savanna	0		0
Tropical Forest	.69	.68	.85
Tropical Dry Forest & Savanna	.53	.5	.78
Non-Forest	.59	.64	.77
Total (Temp. Sav. split)	.56	.63	.77
Total (Temp. Sav. lumped)			.78

empirical and simulated. Digital, empirical vegetation maps used are those of Olson et al. (1983) and Matthews (1984) although only intercomparisons with Olson et al. (1983) are presented here. Visual comparison utilized global (Bailey, 1989) and continental vegetation maps (UNESCO, 1981; White, 1983). MAPSS was also compared to the simulated vegetation from BIOME, another global vegetation model (Prentice et al., 1992).

The Kappa statistic, developed for biogeographic intercomparisons, was used to compare MAPSS output to the Olson et al. (1983) map and to the BIOME output (Monserud and Leemans, 1992). The BIOME comparisons are critical because they were produced under the same gridded temperature and precipitation datasets. In addition, the BIOME model utilized a global radiation calculation, a global cloud cover dataset and a spatially variable soil texture dataset for the PET water balance calculations, none of which were used by MAPSS (Leemans and Cramer, 1991; Prentice et al., 1992). However, MAPSS does require winds and humidity for PET calculations, datasets not required by BIOME.

3. Results

3.1.   VALIDATION

The Kappa statistics for control runs of MAPSS and BIOME comparisons with the Olson et al. (1983) dataset, and the MAPSS-BIOME intercomparison are presented in Table 4. MAPSS and BIOME both compare favorably to Olson with a 'Good' rating from the Kappa statistic. The individual biome comparisons are quite parallel between the two models in comparison to Olson et al. (1983). In fact, the two models produce almost the same calibration with an average intercomparison Kappa statistic of .78 (Very Good, boreal and temperate savannas lumped with forests). Most of the temperate forest area in Olson et al. (1983) is occupied by cultivated lands rendering the Kappa statistic comparison ineffective for that class. Therefore, I have lumped temperate and boreal forests for the comparisons to Olson et al. (1983). The boreal/temperate (combined classes) and tropical forest intercomparisons are quite high between the two models (.83), nearing an 'Excellent' rating. The near congruity of the two

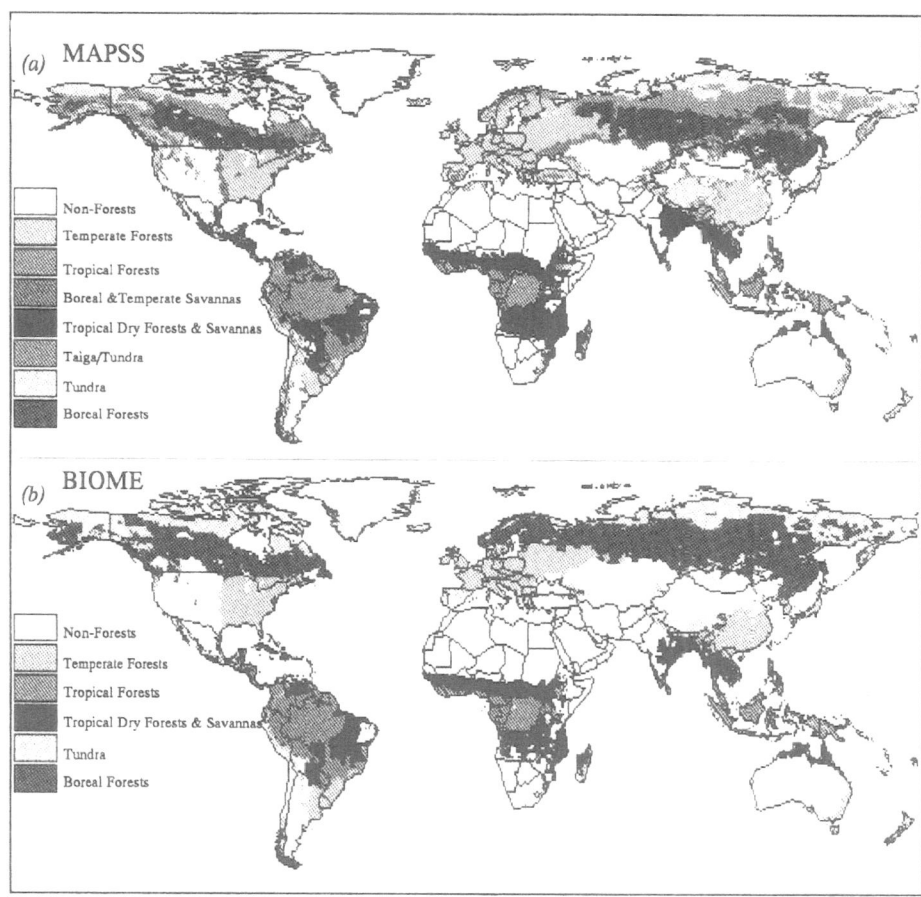

Figure 2. Control runs of MAPSS (a) and BIOME (b), using the coarse vegetation classification (Table 2).

models is most apparent upon visual examination (Fig. 2).

The version of BIOME that we analyzed does not contain a taiga/tundra open woodland, which in MAPSS was lumped with boreal and temperate forests for these comparisons. The poorest showing of MAPSS was in the taiga/tundra comparison with Olson, barely producing a 'Good' rating. This comparison was worst in the Eastern Hemisphere, where larch forests are important, a type that has yet to be defined in MAPSS.

BIOME does not contain temperate and boreal savannas. The definition of these extratropical savanna types in MAPSS is more like a woodland than a savanna with LAI ranging from 2.0 to 3.5 (Table 2). The extratropical savannas, as defined in MAPSS, appear to be contained within the closed forest types of BIOME and produce a slightly improved intercomparison when lumped with forests in MAPSS (Table 4).

Table 5. Area estimates (x 10^6 km^2) of vegetation conversions and areas of stability (MAPSS) under 2 x CO_2 climate scenarios.

Extratropical	OSU	GISS	GFDL	GFDL-Q	UK	Average
Tundra->Taiga/Tundra	4.34	4.63	4.27	4.86	3.68	4.36
Taiga/Tundra->Forest	7.72	8.29	9.17	8.99	9.41	8.72
Forest->Forest	24.29	25.43	18.75	22.93	19.07	22.09
Forest->Savanna	5.37	4.56	8.64	6.52	9.42	6.9
Forest->Non-Forest	.96	.62	3.21	1.16	2.11	1.61
Savanna->Non-Forest	3.37	2.77	4.08	3.57	3.76	3.51
Tropical						
Forest->Forest	15.22	13.93	13.73	12.65	8.43	12.79
Forest->Savanna	.83	2.12	2.32	3.4	7.62	3.26
Savanna->Savanna	17.8	18.36	17.82	17.34	17.87	17.84
Savanna->Non-Forest	.79	.97	1.59	2.11	1.56	1.4

3.2. VEGETATION REDISTRIBUTION

3.2.1. Biome area changes: Both models produce large decreases in the areas of tundra and taiga/tundra (Fig. 3). MAPSS produces decreases in tundra across the five GCM scenarios (range 51% - 72%, $\bar{x}$=62%) and for taiga/tundra (38% - 64%, $\bar{x}$=62%). The two biomes are lumped in BIOME and decrease from 50% to 69% ($\bar{x}$=59%). MAPSS produces large increases in temperate and boreal savannas under the GFDL and UKMO scenarios (36% - 82%). Temperate forest area changes little in MAPSS under 2 x CO_2 climate, but it increases in BIOME. MAPSS generally produces slight decreases in tropical forests (except OSU) and BIOME produces slight increases (Fig. 3).

MAPSS predicts that the temperate forests in the conterminous U.S. could decrease in area by 30% to 94%, while in BIOME they remain virtually unchanged with the GFDL and UKMO scenarios presenting the most extreme shifts (Fig. 4). The differences between

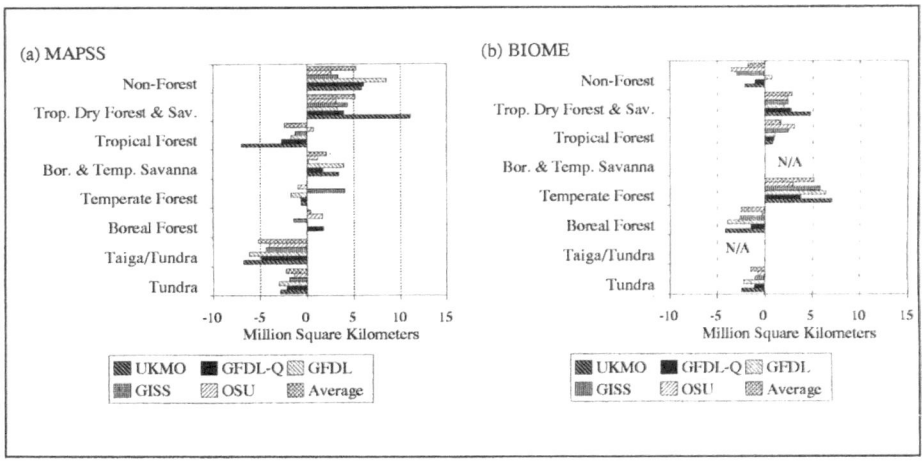

Figure 3. Changes in area predicted by MAPSS (a) and BIOME (b) (average of five 2 x CO_2 climate scenarios).

MAPSS and BIOME in the extratropical forests are much reduced if the temperate and boreal savannas of MAPSS are lumped with closed forest. However, there is a considerable difference in the carbon density between the savanna and forest vegetation types (Table 3).

3.2.2. Vegetation type conversion: Forests shifting toward the poles will colonize treeless areas and sequester carbon through growth. At the opposite boundary, where forests are converted to non-forests, dieback and decay will release carbon to the atmosphere. Intervening areas will remain forested, but could still undergo drought-induced decline. Table 5 presents the area estimates for different types of vegetation conversion under different GCM scenarios (only conversions > 1 x 10^6 km^2 in area are listed).

MAPSS predicts that of the combined boreal and temperate forests of the world, about 72% of the forested area (22 x 10^6 km^2) will remain forested with about 23% (6.9 x 10^6 km^2) being converted to savanna and an additional 5% being converted to non-forest (average over five GCM scenarios). Closed forest expansion into the taiga/tundra could, under 2 x CO_2 equilibrium conditions, add about 8.1 x 10^6 km^2 of newly-forested area (average over five GCM scenarios); while about 8.51 x 10^6 km^2 of closed forest are converted to savanna and non-forest (Table 5). Spatial changes in tropical forests are less remarkable (Table 5). Of the 16 x 10^6 km^2 of forest in the control run about 13 x 10^6 km^2 should remain forested. The UKMO climate was the most severe leaving only 8.4 x 10^6 km^2 as stable, tropical forest (Table 5).

3.2.2. Vegetation Density and Drought Response: Current boreal and temperate forests are predicted by MAPSS to experience a drought-induced reduction in LAI of 33% to 40%, respectively (average over five GCM scenarios, Table 6). The GFDL and UK scenarios are the most extreme with LAI reductions of about 50% each for both boreal and temperate forests (Table 6). Tropical forests average about 10% reduction in LAI across the five scenarios, with the OSU scenario producing the only predicted increase in LAI. Thus, most forested regions in the world are predicted to experience severe drought-induced decline, regardless of whether or not they are expected to remain forested (Fig. 1b). The only areas predicted to increase in LAI are the high latitude areas of forest expansion, a few mountainous regions, and a few small areas in the tropics (Fig. 1b).

Table 6. Average Biome tree LAI (MAPSS control) and predicted change in tree LAI under five, 2 x CO_2 climate scenarios.

	Control	OSU	GISS	GFDL	GFDL-Q	UK	Average
Boreal Forest	12.3	-28%	-15%	-49%	-25%	-50%	-33%
Temperate Forest	11.2	-32%	-24%	-50%	-41%	-53%	-40%
Boreal & Temperate Savanna	2.7	-77%	-65%	-90%	-80%	-85%	-79%
Tropical Forest	10.0	+11%	- 7%	- 7%	-11%	-35%	-10%
Tropical Dry Forest & Savanna	1.8	+ 1%	-12%	-20%	-26%	-24%	-16%

The predicted regional changes in LAI and forest contraction are surprisingly consistent among the five scenarios, excepting the OSU tropical response. The most severe losses of forests are consistently predicted by MAPSS to occur in the eastern United States and Canada and in the western region of the former Soviet Union (Figs. 1b, 4a). The sensitivity of these regions to climatic change appears to result from the nature of the background regional climate, its seasonality and the relative steepness of the regional gradients in the background climate.

Although widespread decline and death of forests is predicted, even in areas that remain forested, forests are predicted to regrow to approximately the original density in new locations. There is, under current climate, a gradient of increasing LAI from warm-temperate to cool-boreal forests. A poleward shift in this gradient will produce a drought-induced LAI decline over the entire gradient, except near the poles where forests will expand into non-forest. Drought-induced forest decline of the magnitude estimated by MAPSS could result in nearly complete forest dieback of extra-tropical forests, if the rate of climatic change is relatively fast, as is being currently predicted (King and Neilson, 1992).

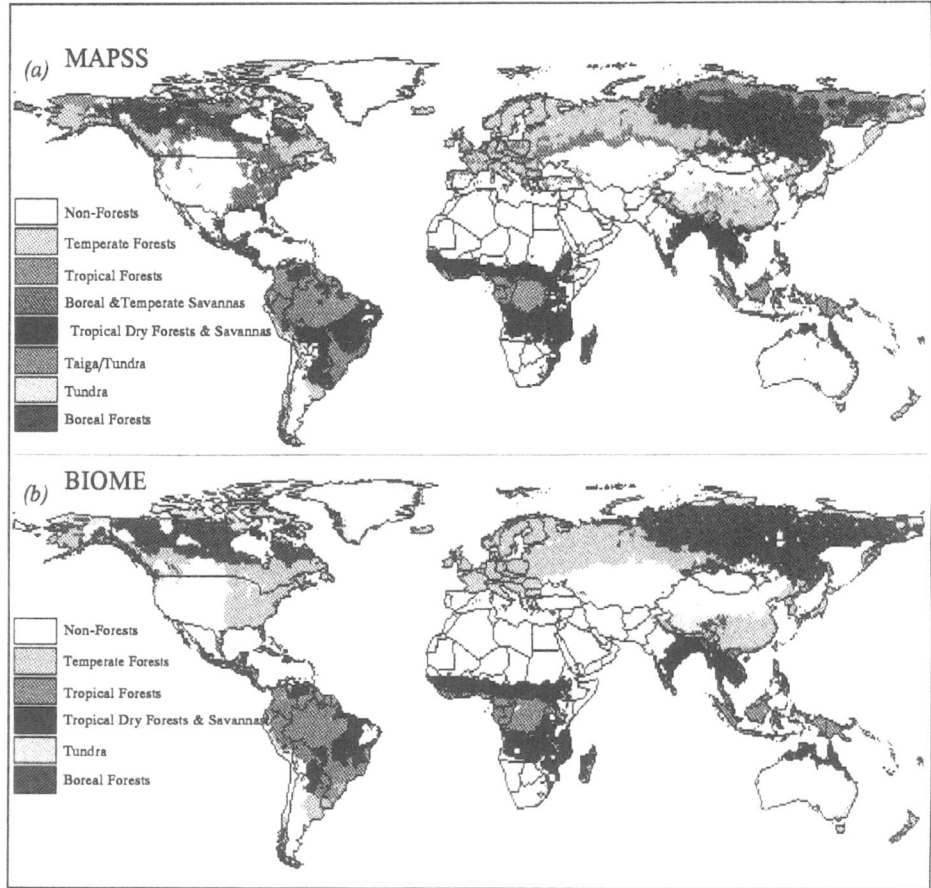

Figure 4. GFDL 2 x CO_2 runs of MAPSS (a) and BIOME (b), using the coarse vegetation classification (Table 2).

3.3.   CARBON BALANCE

The potential equilibrium change in total above and below ground carbon, due to climate-induced vegetation change, was calculated as the product of carbon density (Table 3) and the

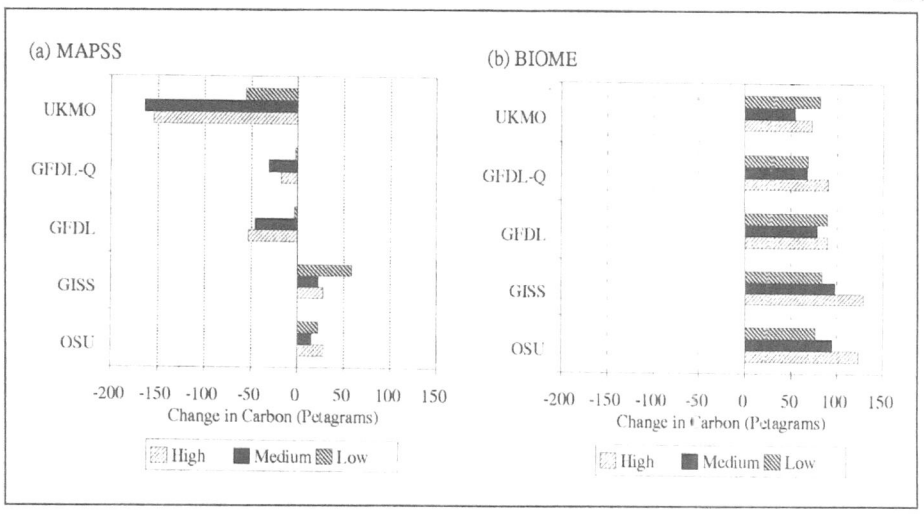

Figure 5. Changes in total terrestrial carbon storage predicted by MAPSS (a) and BIOME (b) under five 2 x CO_2 climate scenarios.

change in area of each biome (Fig. 3). Three of five GCM scenarios under MAPSS produce net losses of carbon from the terrestrial biosphere due to equilibrium 2 x CO_2 climatic change (Fig. 5), in contrast to previous results (Prentice and Fung, 1990; Smith et al., 1993). MAPSS predicts losses of forests in the tropics, where other models predict gains. MAPSS also predicts more severe losses due to drought in extra-tropical forests than do other approaches (Fig. 3). The predicted high-latitude gains in forests by MAPSS are about the same as in other approaches.

Whether the global biosphere is predicted to contain more or less carbon under equilibrium conditions is of less importance than is the temporal trend of carbon fluxes under a rapidly changing climate. Using the equilibrium, area-weighted vegetation conversions (Table 5), the potential transient flux of CO_2 between the biosphere and atmosphere can be estimated (King and Neilson, 1992). The imbalance between decomposition and growth rates, due to forest dieback and regrowth or expansion, could produce a large, net transient "pulse" of CO_2 to the atmosphere. Our previous CO_2 flux estimates and sensitivity analyses indicated that net emissions of CO_2 could be as high as 3.4 Petagrams/year, about a 40% increase over current annual anthropogenic emissions (King and Neilson, 1992; Neilson et al., in press). The projected vegetation changes presented here (Table 5) are in agreement with the earlier estimates of vegetation change (King and Neilson, 1992; Neilson and King, in press). So, for this paper, the transient carbon pulse was not recalculated.

4. Conclusions

A new global vegetation model, MAPSS, has been developed for predictive biogeography under climatic change. Under controlled climate, MAPSS is very similar to another general vegetation model, BIOME (Prentice et al., 1992). However, under 2 x CO_2 climate, the two models diverge considerably with MAPSS exhibiting much greater sensitivity to water stress and loss of extra-tropical forests. MAPSS predicts tropical forest loss, while BIOME predicts tropical forest gains. The reasons for these differences are not fully understood, but are thought to involve the method of calculation of transpiration and the method of relating that transpiration to vegetation properties.

MAPSS predicts almost universal drought stress from climatic change over the world's forests. Drought stress, under five 2 x CO_2 climatic scenarios, would apparently provide the primary mechanism inducing forest decline and dieback. Forest dieback and vegetation conversion to different types could produce a large, net transient pulse of CO_2 to the atmosphere potentially producing a significant positive feedback to the processes of global warming.

These results do not consider the potential mitigating influence of the direct effects of CO_2 on vegetation water-use-efficiency, nor of the aggravating influence of carbon releases from the biosphere due to land use.

5. Acknowledgements

I would like to thank the many people who assisted in background investigations for this research over the past several years. Special thanks go to G. King, J. Lenihan and D. Marks for their continuing input on the issues discussed in this paper. I am particularly indebted to J. Chaney for the skill he brought in rendering this model into computer code and for its implementation over a network of workstations in a distributed computing environment. Additional thanks are accorded G. Koerper for his initial coding of the model. This document has been prepared at the EPA Environmental Research Laboratory in Corvallis, Oregon, USA, through cooperative agreement CR816257 with Oregon State University. It has been subjected to the Agency's peer review and approved for publication. Mention of trade names or commercial products does not constitute endorsement or recommendation for use.

6. References

Abramopolous, F., Rosenzweig, C. and Choudhury, B.: 1988, *J. Climate* 1, 921-941.

Anonymous: 1984a, *Agroclimatological Data for Africa, Volume 1. Countries North of the Equator*, FAO Plant Production and Protection Series 22. Food and Agriculture Organization of the United Nations, Rome.

Anonymous: 1984b, *Agroclimatological Data for Africa, Volume 2. Countries South of the Equator*, FAO Plant Production and Protection Series 22. Food and Agriculture Organization of the United Nations, Rome.

Anonymous: 1987a, *Agroclimatological Data for Asia, Volume 1. A-J*, FAO Plant Production and Protection Series 25. Food and Agriculture Organization of the United Nations, Rome.

Anonymous: 1987b, *Agroclimatological Data for Asia, Volume 2. K-Z*, FAO Plant Production and Protection Series 25. Food and Agriculture Organization of the United Nations, Rome.

Bailey, R.G.: 1989, *Environ. Conserv.* 16, 307-309.

Box, E.O.: 1981, *Macroclimate and Plant Forms: An Introduction to Predictive Modeling in Phytogeography*, Dr. W. Junk Publishers, The Hague.

Burke, M.J., Gusta, L.V., Quamme, H.A., Weiser, C.J. and Li, P.H.: 1976, *Ann. Rev. Plant Physiol.* 27, 507-528.

Cramer, W.P. and Solomon, A.M.: (in press), *Clim. Res.*

Dice, L.R.: 1943, *The Biotic Provinces of North America*, University of Michigan Press, Ann Arbor.

Emanuel, W.R., Shugart, H.H. and Stevenson, M.P.: 1985, *Clim. Change* 7, 29-43.

Hansen, J., Fung, I., Lacis, A., Rind, D., Lebedeff, S. and Ruedy, R.: 1988, *J. Geophys.Res.* 93, 9341-9364.

Holdridge, L.R.: 1947, *Science* 105, 267-268.

Houghton, J.T., Jenkins, G.J. and Ephraums, J.J. (eds.).: 1990, *Climate Change: The IPCC*

Scientific Assessment, Cambridge University Press, Cambridge.

King, G.A. and Neilson, R.P.: 1992, *Water,Air Soil Poll.* **64**, 365-383.

Küchler, A.W.: 1964, *Potential Natural Vegetation*, American Geographical Society, New York.

Leemans, R. and Cramer, W.P.: 1991, *International Institute for Applied Systems Analysis*, RR-91-18, 1-62.

Manabe, S., Stouffer, R.J., Spelman, M.J. and Bryan, K.: 1991, *J. Climate* **4**, 785-818.

Marshall, J.D. and Waring, R.H.: 1984, *Nature* **330**(19), 238-240.

Monserud, R.A. and Leemans, R.: 1992, *Ecol. Model.* **62**, 275-293.

Müller, M.J.: 1982, *Selected climatic data for a global set of standard stations for vegetation science*, Dr W. Junk Publishers, The Hague.

Neilson, R.P.: 1987, *Vegetatio* **70**, 135-147.

Neilson, R.P. and Wullstein. L.H.: 1983, *J. Biogeog.* **10**, 275-297.

Neilson, R.P. and King, G.A.: 1992, Continental Scale Biome Response to Climatic Change, In: McKenzie, D.H., Hyatt, D.E. and McDonald, V.J. (eds), *Ecological Indicators. Volume 2*, Elsevier Applied Science, London, pp. 1015-1040.

Neilson, R.P., King, G.A., DeVelice, R.L., Lenihan, J., Marks, D., Dolph, J., Campbell, W. and Glick, G.: 1989, *Sensitivity of Ecological Landscapes to Global Climatic Change*, U.S. Environmental Protection Agency, EPA-600-3-89-073, NTIS-PB-90-120-072-AS, Washington, D.C.

Neilson, R.P., King, G.A. and Koerper, G.: 1992, *Lands. Ecol.* **7**, 27-43.

Neilson, R.P., King, G.A. and Lenihan, J.: (in press), Modeling forest response to climatic change: the potential for large emissions of carbon from dying forests. In: Kaaninen, M. (ed), *Proceedings of the IPCC Workshop, Carbon Balance of World's Ecosystems: Toward a Global Assessment.* Publications of the Academy of Finland, Helsinki.

Olson, J.S., J.A. Watts, J.A., and Allison, L.J.: 1983, *Carbon in Live Vegetation of Major World Ecosystems*, Oak Ridge National Laboratory, ORNL-5862, Oak Ridge.

Prentice, K.C. and Fung, I.Y.: (1991, *Nature* **345**, 48-50.

Prentice, I.C., Cramer, W., Harrison, S.P., Leemans, R., Monserud, R.A. and Solomon, A.M.: 1992, *J. Biogeog.* **19**, 117-134.

Schlesinger, M.E. and Zhao, Z.C.: 1989, *J. Climate* **2**, 429-495.

Smith, T.M. and Shugart, H.H.: 1993, *Nature* **361**, 523-526.

Smith, T.M., Leemans, R. and Shugart, H.H.: 1992, *Clim. Change* **21**. 367-384.

Smith, T.M., Weishampel, J.F., Shugart, H.H. and Bonan, G.B.: 1993. *Water, Air Soil Poll.* **64**, 307-326.

Stephenson, N.L.: 1990, *Amer. Natural.* **135**, 649-670.

UNESCO.: 1981, *Vegetation map of South America: Explanatory notes*, UNESCO, Paris.

Wetherald, R.T. and Manabe, S.: 1988, *J. Atmos. Sci.* **45**, 1397-1415.

White, F.: 1983, *The vegetation of Africa, a descriptive memoir to accompany the UNESCO, AETFAT, UNESCO vegetation map of Africa*, UNESCO, Paris.

Whittaker, R.H.: 1975, *Communities and Ecosystems.* 2nd Ed. Macmillan Publishing Co., Inc., New York.

Woodward, F.I.: 1987, *Climate and Plant Distribution*, Cambridge University Press, London.

Zinke, P.J., Stangenberger, A.G., Post, W.M., Emanuel, W.R. and Olson, J.S.: 1984, *Worldwide Organic Soil Carbon and Nitrogen Data*, ORNL/TM-8857, Oak Ridge National Laboratory, Oak Ridge.

STRUCTURE OF A GLOBAL AND SEASONAL CARBON EXCHANGE MODEL FOR THE TERRESTRIAL BIOSPHERE

THE FRANKFURT BIOSPHERE MODEL (FBM)

J. KINDERMANN, M. K. B. LÜDEKE, F.-W. BADECK, R. D. OTTO, A. KLAUDIUS,
CH. HÄGER, G. WÜRTH, T. LANG, S. DÖNGES, S. HABERMEHL,
AND G. H. KOHLMAIER

*Institut für Physikalische und Theoretische Chemie, J. W. Goethe-Universität
Frankfurt/M., Niederurseler Hang, D-6000 Frankfurt 50, Germany*

Abstract: Carbon exchange fluxes of terrestrial ecosystems are expected to depend on the internal dynamics of C stocks in vegetation and soils, on nutrient availability, and on the local climatic conditions / weather. The model structure which we present focuses on the internal dynamics in the living vegetation. The mathematical description is derived from two basic hypotheses: 1) vegetation tends to maximize photosynthesizing tissue, and 2) the relative amounts of C in pools with relatively short and long turnover times are given by allometric relations. The model can be calibrated for any vegetation type in a typical climate under the condition to meet mean ecological estimates of e.g. biomass and NPP. For C cycle modeling the FBM yields the net CO_2 flux between the grid elements and the atmosphere in a daily resolution. It is demonstrated that simulations with a $1° \times 1°$ spatial resolution reproduce the response of the time course of C fluxes to local climates.

1. Introduction

A full understanding of the importance of terrestrial biota within the global C cycle has not been achieved, yet. Positive and negative feedbacks must be taken into consideration with respect to the vegetation / climate interaction (Kohlmaier *et al*, 1991). The stimulation of ecosystem production and respiration by atmospheric trace constituents (Kohlmaier *et al*, 1987) as well as the effects of temperature and precipitation change play an important role in the determination of the feedback.

One may also argue that modeling the interaction of internal vegetation dynamics and climatic driving variables will capture some major traits of C exchanges in terrestrial ecosystems. Tight links between climate and vegetation have already laid the basis for numerous scientific approaches in the past (Holdridge, 1947; Box, 1981; Woodward,

Water, Air, and Soil Pollution **70**: 675–684, 1993.
© 1993 *Kluwer Academic Publishers.*

1987; Prentice *et al*, 1992). For these reasons we came to the conclusion that our model has to meet the following requirements:

- The major ecophysiological processes responsible for the reaction of the different ecosystem types to climatic variations must be included.
- The characteristic properties of these ecosystem types should be expressed in terms of (measurable) quantities like biomass, GPP, NPP, soil C, etc.
- A realistic response of the ecophysiological processes to climatic variables requires a time resolution in the range of hours/days.
- The basic structure of the model should be valid for all plant functional types.
- The model structure has to be as simple as possible in order to be manageable at a global scale.

We consider photosynthesis, autotrophic, and heterotrophic respiration as the major ecophysiological processes. Light intensity, temperature, and soil moisture (and in a further state of development: CO_2 concentration), among others, are regarded as the main driving variables.

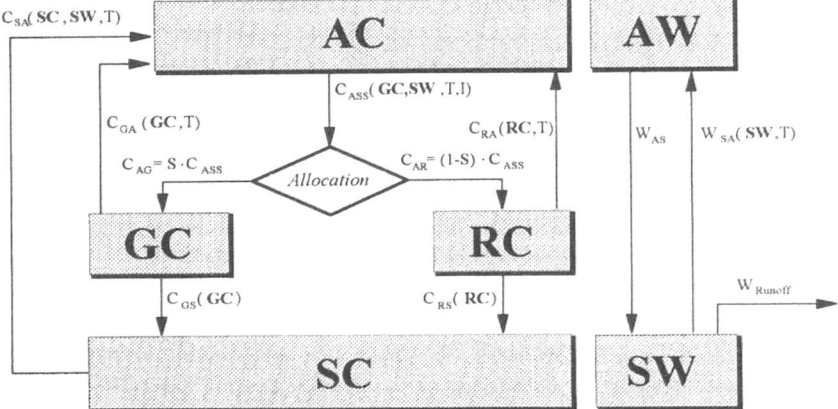

Figure 1: Flow chart and model structure. Bold capital letters represent the reservoirs of C (second letter **C**) and water (second letter **W**): **AC**, atmospheric C; **GC**, C content of green biomass and feeder root biomass plus assimilate store; **RC**, C content of remaining biomass of biota (**GC** + **RC** = **BC**); **SC**, C content of litter, humus and dead biomass; **AW**, water in the atmosphere; **SW**, soil water in the rooting zone. The capital letters C and W represent C and water fluxes. The indices indicate sources and sinks of these fluxes. The functional dependence of the fluxes on the driving variables and pool sizes is given in parentheses (T: hourly air temperature, I: hourly photosynthetic active radiation, PAR). W_{AS} is the daily precipitation, W_{SA} is daily actual evapotranspiration. The factor S represents the fraction of total assimilation C_{ASS} that is allocated to **GC**.

2. Compartmentation

According to Janecek *et al* (1989), for every ecosystem type we propose the same basic model structure, showing the minimal functional subdivision of the total C content of the living biomass (**BC**) into two compartments. Here we distinguish between the parts of the vegetation with a short turnover time (leaves, feeder roots, and stored assimilates, summarized in the **GC**-compartment) and mostly woody, structural material with a long turnover time (**RC**-compartment). To describe the decomposition processes of soil organic matter, a one-compartment model after Fung *et al* (1987) is used (C mass of litter and humus are summarized in the **SC**-compartment). Furthermore, the soil water compartment **SW** is introduced to calculate the actual soil water status. The C and water fluxes between these compartments and the atmosphere generally depend on the climatic variables and pool sizes, both varying in time (Figure 1). This structure is assumed to be sufficient to describe the major C uptake and release processes at any point of the terrestrial surface, the advantage of it being the limited and constant number of parameters required. Hence the local difference of the model source and sink strength is influenced externally by the climatic variables and internally by the vegetation properties, which are reflected by suitable parameter values and developmental state.

Figure 2 illustrates how the C part of the proposed model works for one grid element of a temperate deciduous forest in equilibrium. For C cycle modeling the net C exchange flux between the grid element and the atmosphere is the main output (Heimann and Keeling, 1989). Besides the net exchange flux a number of intermediate results is available for further examinations (Figure 2). A comparison of these intermediate results with available data on standing biomass in selected regions, with documentations of the seasonality of leaf biomass etc. may be used to corroborate the model.

3. Carbon Allocation and Phenology

As shown in Figure 1 it is assumed that the assimilate production C_{ASS} is determined by the mass of compartment **GC**, reflecting the amount of leaves, the actual soil water content **SW** as well as by the external driving variables temperature T and light intensity I. This flux is to be partitioned according to present needs of the plant organs, namely the build-up and maintenance of the photosynthesizing tissue and of the feeder roots, (represented by **GC**) on the one hand, and the build-up and maintenance of stems, branches and coarse roots (represented by **RC**) on the other. Furthermore, assimilates have to be translocated in order to fill particular storage organs, which are included in the C mass of the **GC** compartment as well.

The partitioning of the C assimilation flux C_{ASS} into the **GC** and **RC** compartment is derived in its seasonal and long term patterns from two basic assumptions:

- The vegetation tends to maximize the amount of photosynthesizing tissue (contained in the **GC** compartment).
- It is possible to identify a function **RC** = Ω(**GC**) determining the minimum amount of **RC** needed to support and maintain the given amount of **GC**.

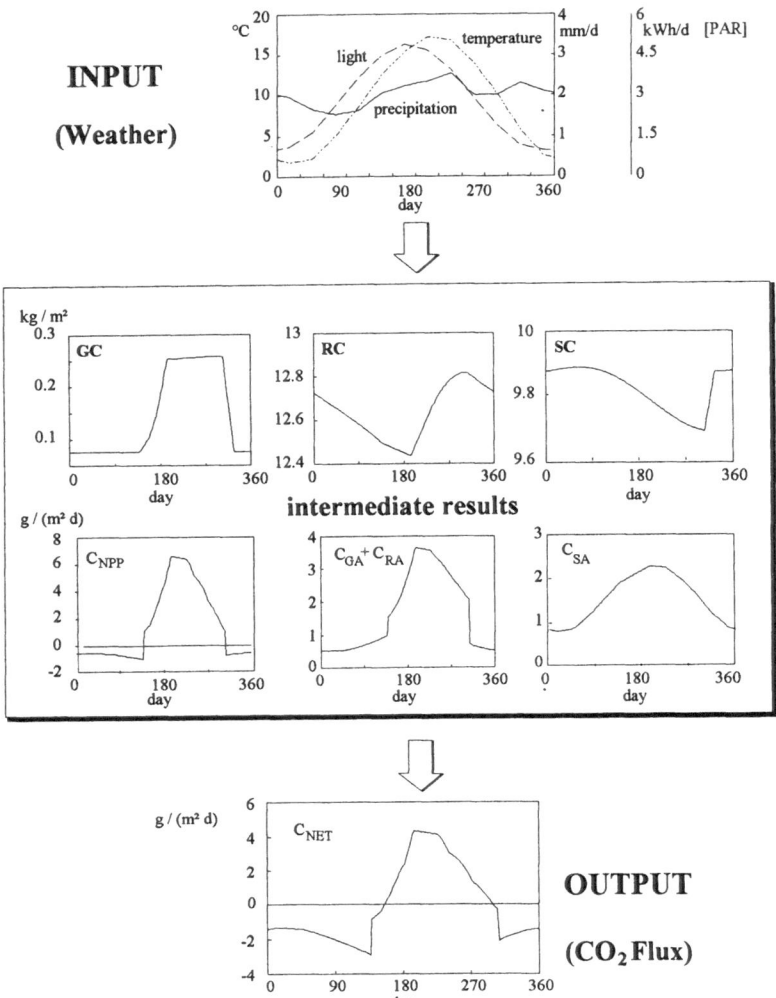

Figure 2: Driving climatic variables (input) and resulting annual course of the net C exchange flux C_{NET} between the whole ecosystem and the atmosphere (output) for a typical grid element of vegetation type 'cold-deciduous forest'. The annual courses of the C compartments and the fluxes $C_{NPP}=C_{ASS}-C_{GA}-C_{RA}$ (exchange flux between living vegetation and atmosphere), $C_{GA}+C_{RA}$ (autotrophic respiration), and C_{SA} (heterotrophic respiration) are displayed as intermediate results.

Data from field measurements (Reichle, 1981) and theoretical considerations suggest that a parabola type of function, the so-called allometric relation, is a suitable parametrisation (see Janecek *et al*, 1989): $RC = \Omega(GC) = \xi \cdot GC^{\kappa}$. For the functional types 'temperate broad leaved forest', 'coniferous forest', 'tropical evergreen forest', and 'grasslands' measurements of woody biomass and maximum leaf mass were used to determine the parameters in this function by least square fitting. For a cold-deciduous forest ecosystem a typical course of the allometric relation is shown in Figure 3. In order to comply with the basic assumptions, the system's development within the **GC/RC** plane is allowed to take place only in the region $RC \geq \Omega(GC)$. Furthermore the system passes through four phases during one annual cycle, as shown in Figure 3.

3.1. Shooting Phase
The C gain from photosynthesis is greater than the C loss. The system allocates most of the assimilates to the **GC** compartment until the trajectory reaches $\Omega(GC)$, maximizing its production ability.

3.2. Secondary Growth
The system is forced to allocate simultaneously into the **GC** and **RC** compartment according to the $\Omega(GC)$ function.

3.3. Leaf Shedding
At the end of the vegetation period, when unfavorable weather conditions do not allow biomass increase (e.g. drought, cold), a leaf abscission phase reduces the **GC** compartment to a remaining amount of feeder roots and assimilate store, which is proportional to the annual maximum of **GC** and characterized by the function $RC = \Theta(GC) = \nu \cdot GC^{\kappa}$.

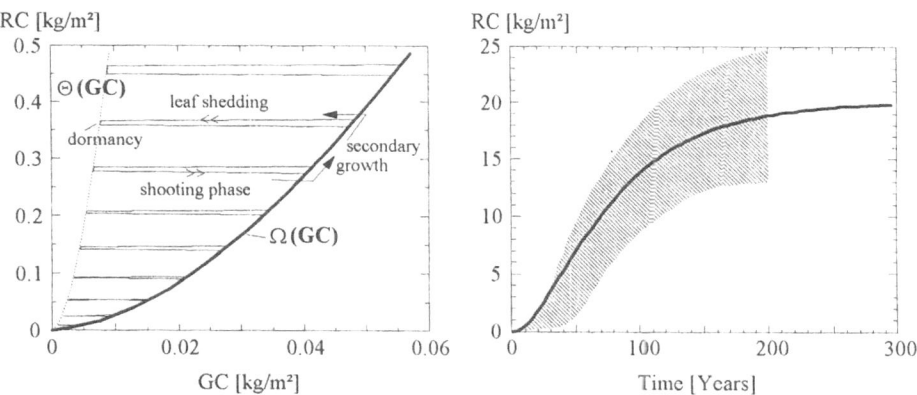

Figure 3: Trajectory in the G-R-space (left hand side) and long term development (right hand side). The hatched area indicates the range of field measurements (Schober, 1975).

3.4.   Dormancy Phase

When the trajectory reaches the $\Theta(GC)$ curve the dormancy phase starts. During this phase the biomass losses, as defined by the **RC** respiration and the total litter production C_{BH}, are distributed among the compartments so that the systems trajectory follows the $\Theta(GC)$ curve. This phase ceases when the weather conditions allow a net biomass increase, assuming a total conversion of stored assimilates into leaf biomass and feeder roots.

These phases are described by the differential equations given in the appendix.

4. Calculation of Fluxes

The net uptake of CO_2 by plants is determined by a balance of two processes: C assimilation, C_{ASS} (i.e. the gross photosynthetic C fixation) and autotrophic respiration, C_{GA} and C_{RA}. As assimilation and respiration show different seasonal courses and different temperature responses, we think it is more convenient to model these processes seperately instead of simulating the NPP directly.

4.1.   Uptake of CO_2

The effective C assimilation rate, C_{ASS}, can be considered as a product function (Richter, 1985) of a term dependent on light and canopy structure, a temperature dependent term and a soil water dependent term. The dependence on incident light intensity and leaf area index, LAI, which is correlated with the **GC** compartment is modeled according to Monsi and Saeki (1953) taking into account the light attenuation in the canopy. For the temperature dependence an optimum curve is used characterized by the minimum, maximum, and optimum temperature of photosynthesis. The dependence of photosynthesis on water availability, represented by the soil water content **SW**, is assumed to follow a saturation curve which is zero at the permanent wilting point and approaches one for field capacity.

4.2.   Release of CO_2 - Autotrophic Respiration

Autotrophic respiration, C_{GA} and C_{RA} respectively, is modeled similarily for both compartments, **GC** and **RC**, depending on the compartment size and an exponential function of the temperature corresponding to a constant Q_{10} value for each ecosystem type (Ryan, 1991).

4.3.   Litter Production

The litter production, C_{GS} and C_{RS}, is assumed to be proportional to the respective compartment size. For the **GC** compartment of the deciduous vegetation types an additional constant rate litter production occurs during the abscission phase.

4.4.   Release of CO_2 - Heterotrophic Respiration

For the climate response of the decomposition of dead organic matter, C_{SA}, we extended the model of Fung et al (1987) taking into consideration a linear dependence on compartment size and introducing a soil moisture factor analoguous to the moisture dependence of photosynthesis.

4.5.   Water Fluxes

Due to the close relation between assimilation and transpiration, the actual evapo-transpiration, W_{SA}, is calculated by the product of potential evapotranspiration after Thornthwaite (1948), W_{PET}, and the soil water dependent function as used in the calculation of assimilation.

W_{Runoff} comprises both surface runoff and drainage. It is taken as the surplus water when the soil water content reaches field capacity (Wilson and Henderson-Sellers, 1985).

5. Results and Discussion

Although the model is constructed to perform simulations on a global scale, in this paper we will only discuss some selected results. For this purpose we have chosen the ecosystem type 'cold-deciduous forest' from Matthews global vegetation map (Matthews, 1983).

Before running the model for a given ecosystem type, one free parameter per flux equation has to be determined. Therefore a calibration procedure is performed based on mean ecological estimates, e.g. of biomass (Matthews, 1984; Rodin et al, 1972), net primary production (Fung et al, 1987), annual respiration (Reichle, 1981) and an averaged climate which was generated conserving the typical phase relations between temperature, precipitation, and light of this vegetation type (data basis: climate atlas of Shea (1986)).

These parameters are equally used for all grid elements of this vegetation type. Yet, taking into consideration the real locally different climate as given in the Shea atlas, different fluxes in the different locations are achieved.

In Figure 4 the driving climatic variables and the corresponding equilibrium results for the annual courses of GC and the net C exchange fluxes for two grid elements of the considered vegetation type are displayed. The model reproduces the expected differences of vegetation dynamics comparing the results of a grid element with moderate (8°E, 50°N) and of a location with continental climate (50°E, 56°N):

In the moderate climate we obtain
* a greater maximum LAI-value
* a greater annual averaged standing biomass
* a longer vegetation period
* a greater amplitude of the annual course of the net C exchange flux.

The following table compares the calculated LAI and standing biomass with the ecological estimates used for the calibration.

	Leaf area index [m² m⁻²]	Standing biomass [kg C m⁻²]
8°E, 50°N (W. Germany)	5.2	13.2
50°E, 56°N (Russia)	4.6	10.9
Ecological estimates (cold-decid. forest)	5	12.6

The evaluation of the whole ecotype results is in preparation and will be published soon. In addition to the discussed equilibrium calculations the model allows for the calculation of all these quantities not only in the steady state, but also under non stationary conditions, i.e. for non climax systems, or systems in transients induced by climate change.

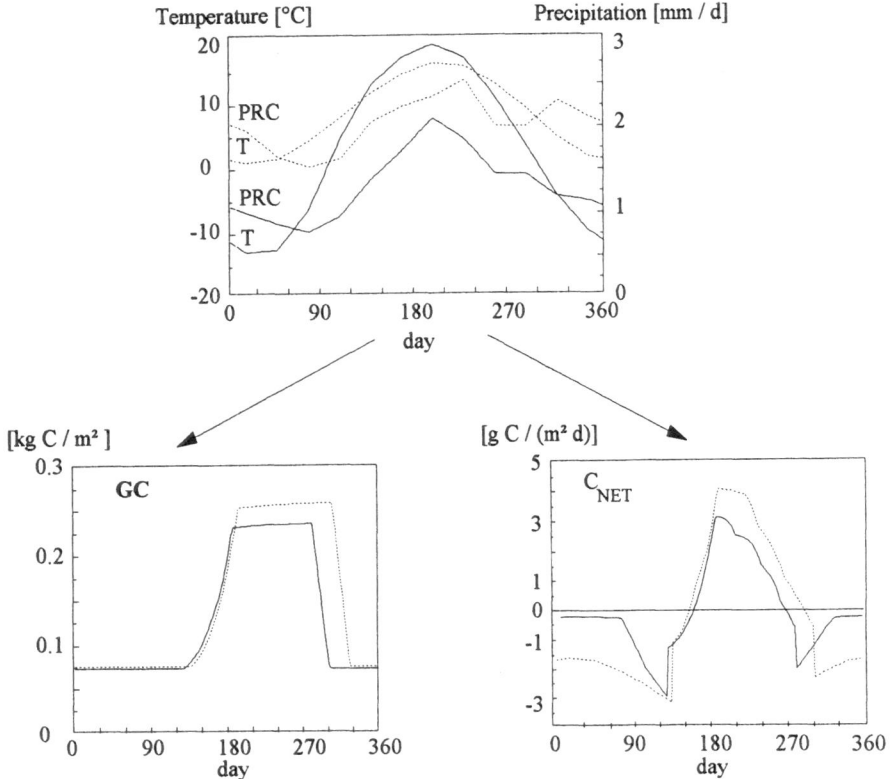

Figure 4: Driving climatic variables and resulting annual courses of the **GC** compartment and the net C exchange fluxes between ecosystem and atmosphere for two grid elements of vegetation type 'cold-deciduous forest'. Solid lines: 50°E, 56°N; dotted lines: 8°E, 50°N.

6. Acknowledgements

The authors thank the Bundesministerium für Forschung und Technologie and the European Communities for financial support. We also acknowledge the helpful comments of M. Bartel.

7. Appendix

DIFFERENTIAL EQUATIONS	DESCRIPTION OF PHASES; DETERMINATION OF S [*]
$$\frac{d\,GC}{dt} = S \cdot C_{ASS} - C_{GA} - C_{GS}$$	**Shooting phase:** All assimilates are allocated to **GC** except for a flux for the compensation of the **RC**-respiration. $$S = \frac{C_{ASS} - C_{RA}}{C_{ASS}}$$
$$\frac{d\,RC}{dt} = (1-S) \cdot C_{ASS} - C_{RA} - C_{RS}$$	**Secondary growth phase:** Simultaneous allocation to the **GC** and **RC** compartment. The systems trajectory equals **RC** = Ω(**GC**). $$S = \frac{C_{ASS} + \frac{d\Omega}{dGC} \cdot (C_{GA} + C_{GS}) - (C_{RA} + C_{RS})}{\left(1 + \frac{d\Omega}{dGC}\right) \cdot C_{ASS}}$$
$$\frac{d\,GC}{dt} = -\Delta$$ $$\frac{d\,RC}{dt} = -C_{RA} - C_{RS}$$	**Leaf shedding phase:** Leaf abscission is performed with a constant rate Δ while leaf metabolism is assumed to be neglectable. The driving variables do not have any impact on the abscission rate. Once started this phase ends when the trajectory reaches the curve **RC** = Θ(**GC**).
$$\frac{d\,GC}{dt} = \frac{1}{1 + d\Theta/dGC} \cdot (-C_{RA} - C_{BS})$$ $$\frac{d\,RC}{dt} = \frac{d\Theta/dGC}{1 + d\Theta/dGC} \cdot (-C_{RA} - C_{BS})$$	**Dormancy phase:** Respiration and litter losses are distributed so that the systems trajectory equals **RC** = Θ(**GC**).

[*] In order to describe the allocation strategy during the vegetation period, we introduce an allocator S, $0 \leq S \leq 1$. As no assimilates are produced out of the vegetation period the allocator is not determined in the leaf shedding phase and in the dormancy phase.

8. References

Box, E.O.: 1981, *Macroclimate and Plant Forms: An Introduction to Predictive Modeling in Phytogeography*, The Hague, Junk.

Fung, I. Y.; Tucker, C. J.; Prentice, K. C.: 1987, *J. Geophys. Res.* **92**(D3), 2999.

Heimann, M. and Keeling, C. D.: 1989, A Three Dimensional Model of Atmospheric CO_2 Transport Based on Observed Winds: Model Description and Simulated Tracer Experiments, in: D. H. Peterson (ed.), *Aspects of Climate Variability in the Pacific and the Western Americas*, Geophysical Monograph 55, Washington DC, pp. 237-275.

Holdridge, L.R.: 1947. *Science*, **105**, 367.

Janecek, A.; Benderoth, G.; Lüdeke, M. K. B.; Kindermann, J.; Kohlmaier, G. H.: 1989, *Ecological Modelling* **49**, 101.

Kohlmaier, G. H.; Bröhl, H.; Siré, E. O.; Plöchl, M.: 1987, *Tellus* **39**B, 155.

Kohlmaier, G. H.; Lüdeke, M.; Janecek, A.; Benderoth, G.: 1991, Land biota, Source or Sink of Atmospheric Carbon Dioxide, in: Schneider, S. H.; Boston, P. J. (eds.), *Scientists on Gaia*, MIT Press, Cambridge, pp 223-239.

Matthews, E.: 1983, *J. Clim. Appl. Meteor.* **22**(3), 474.

Matthews, E.: 1984, *Progress in Biometeorology* **3**, 237.

Monsi, M. and Saeki, T.: 1953, *Jap. Journ. Bot.* **14**, 22.

Prentice, I. C.; Cramer, W.; Harrison, S. P.; Leemans, R.: 1992, *J. Biogeogr.* **19**, 117.

Reichle, D. E. (ed.): 1981, *Dynamic Properties of Forest Ecosystems. International Biological Programm*, **23**, Cambridge University Press, Cambridge.

Richter, O.: 1985, *Simulation des Verhaltens ökologischer Systeme. Mathematische Methoden und Modelle*. VCH, Weinheim.

Rodin, L. E.; Bazilevich, N. I.; Rozov, N. N.: 1972, Productivity of the World's Main Ecosystems, in: *Proc. of World Ecosystems*, Seattle Symp., pp. 13-26.

Ryan, M. G.: 1991. *Ecological Applications* **1**(2), 157.

Schober, R.: 1975. *Ertragstafeln wichtiger Baumarten*, Sauerländer, Frankfurt.

Shea, D. J.: 1986, *Climatological Atlas: 1950-1979, NCAR Technical Note 269+STR*, NCAR, Boulder, Colorado.

Thornthwaite, C. W.: 1948, *Geographical Review* **38**, 55.

Wilson, M. F. and Henderson-Sellers, A.: 1985, *Journal of Climatology* **5**, 119.

Woodward, F. I.: 1987, *Climate and Plant Distribution. (Cambridge Studies in Ecology)*. Cambridge University Press, Cambridge

AN EPILOGUE FOR PERSPECTIVE:
FORESTS ARE MORE THAN CARBON SINKS

WILHELM KNABE

Evironmental Research and Consultancy,
Rumbachtal 69, D-45470 Muelheim an der Ruhr, Germany

Forests are much more than carbon-sinks. They have inspired musicians, painters and poete and may be our teachers and model because trees have developed strategies of survival millions of years before man and many of those strategies proved successful until today in most complicated systems. For example:

*   Trees only depend on solar energy thus forming a solar society.
*   They use renewable resources for prodction CO_2 and water.
*   They live from the site where they stand while they also participate in the worldwide circulation of CO_2 and water, using these renewable resources for production.
*   They carefully deal with water and recycle all limited resources of their debris: Ca, Mg, P, etc.
*   They improve the site where they grow by humus accumulation and do not deliberately poison their surrounding with toxic substances.
*   Trees show us symbiosis, co-operation with other species, and fitting into a system, e.g. by surpressing unlimited growth through negative feed-back.
*   They are masters of a decentralized economy, every leaf contains thousands of solar micro-power plants.
*   Finally trees may teach us patience and communication without words.

Forests are the habitat of indigenous people. They have lived in the forest over centuries without causing its destruction. The Indians in Brazil, Chile, Columbia or Oregon, the Penan in Sarawak, Polynesian tribes, and peasants in the Humalayas had one thing in common: they used nature for their way of living, but they did it with respect and were not restricting the use of a forest to cutting and selling its wood (Heyerdahl, 1976, Jim, 1986). These people could help us to preserve the forests (Colchester et al., 1993).

Forests have great economic importance in various countries of the world. They are needed to produce renewable resources but this could be done without reckless exploitation. The large-scale clear-cut of old growth forests in Oregon or British Columbia, the timber extraction even on steep slopes in Sarawak or the burning and destruction in West Brazil are no sustainable use at all.

Forests are needed for soil and water protection, avalanche control, and outdoor recreation. German forest planners have developed maps where all the different functions of forests are mapped and shown both to convince other planners what value that special woodland has and to adapt the management practice to serve those functions best (Leitfaden, 1975). Other countries have adopted similar programs.

Do you not think that we should make better use of these strategies to countseract the present trends of growth, pollution, destruction, and extinction of life forms?

We need respect towards the subject of our research. The term "manipulation of ecosystems" does not reflect such respect. We should rather adapt natural successions, see them as alternatives to clear-cutting and replanting even-aged stands (Mlinsck, 1991).

There are ways of communication with trees which may help us in our difficult job to care for a sustainable future of mankind (Knabe, 1991). So it is necessary to say to scientists "don't suppress your emotions, they are the roots to life and will help you to a better understanding of nature". It is not enough to plant trees to sequester carbon. We need people who like trees, who protect them and appreciate their benefit and protection.

This workshop dealt with the fluxes of carbon between the atmosphere and vegetation. Quantifying these fluxes and the resulting sinks and sources is a proper programm, well needed for scientific and political decisions. However, the reduction of living organisms to dry matter and C mass at this or any other workshop is far from reality, and needs to be complemented by a broader view of forest ecosystems, to avoid misleading management practices. The respect of the living organisms and their cooperation in systems will help find the optimal solution that best benefits society.

References

Colchester, M., Kirschbaum, S. Schücking, H. and J. Wolters (1993): *Indigene Völker und Wald: Statusbericht, Empfehlungen und Perspektiven für die bundesdeutsche Politik*, Arbeitsgemeinschaft Regenwald (ARA), Klasingstr. 17, D-33504 Bielefeld, Germany, 93 pp.

Hyerdahl, T: 1976, Men and forests. A time perspecive XVI IUFRO World Congress. Norway 1976. Congress report, Aas, Norway: Norwegian Forest Res. Inst., pp. 279–295.

Jim, R: 1986, 18th IUFRO World Congress, Lubljana Congress report, IUFRO Secretariat, Schönbrunn - Tirolergarten, A-1131 Vienna, Austria, pp. 324–335.

Knabe, W.: 1991, What can we learn from the trees? in ; Parkin, S. (ed.), GreenLight on Europe, Heretic Books, London, pp. 116–123.

Leitfaden zur Kartierung der Schutz- und Erholungsfunktionen des Waldes (Waldfunktionenkartierung) WFK, 1974 (=guide for mapping of the protective and recreational functions of forests) Arbeitskreis Zustandservassung und Planung der Arbeitsgemeinschaft Forsteinrichtung, Arbeitsgruppe Landespflege. J.D. Sauerländer's Verlag, Frankfurt am main, 84 S. ISBN 3-7939-0340-0.

Mlinsek, D.: 1991, Is forestry really blind to some facts regarding natural forests? IUFRO-News, 20, 3. p. 20.

LIST OF PARTICIPANTS

**An International Workshop
Terrestrial Biosphere Carbon Fluxes:
Quantification of Sinks and Sources of CO_2**

1–5 March 1993 – Bad Harzburg, Germany

Michael Apps
Forestry Canada
5320 122nd Street
Edmonton, Alberta
Canada T6H 3S5

John Benemann
Private Consultant
343 Caravelle Drive
Walnut Creek, CA 94598 USA

Sandra Brown
Department of Forestry
University of Illinos
W-503 Turner Hall
1102 S. Goodwin
Urbana, IL 61801 USA

Peter Burschel
Silviculture and Forest Management
University of Munich
Hohenbachernstrasse 22
8050 Freising, Germany

C. Vernon Cole
USDA Agricultural Research Service
Natural Resource Ecology Laboratory
Colorado State University
Ft. Collins, CO 80523 USA

Wolfgang Cramer
Potsdam Institute of Climate Impact
    Research
Telegraphenberg
14473 Potsdam, Germany

Ulrich Dämmgen
Bundesforschungsanstalt für
    Landwirtschaft
Institut fur Agrarrelevante
Klimaforschung
Wilhelm-Pieck-Strasse 72
15374 Muncheberg, Germany

Bjorn Dirks
Dept. of Theoretical Prod. Ecology
Wageningen Agricultural University
P.O. Box 430
6700 AK Wageningen
The Netherlands

Robert Dixon
Environmental Research Laboratory
US Environmental Protection Agency
200 SW 35th Street
Corvallis, OR 97333 USA

John Downing
Marine Sciences Laboratory
Pacific Northwes: Laboratories
1529 W Sequim Bay Road
Sequim, WA 98382 USA

Klaus Flach
12601 Builders Road
Herndon, VA 22070 USA

Ed Glenn
Environmental Research Laboratory
University of Arizona
2601 E Airport Drive
Tucson, AZ 85706 USA

Water, Air, and Soil Pollution **70:** 687–691, 1993.

Heinz-Detlef Gregor
Federal Environmental Agency
Bismarckplatz 1
14193 Berlin, Germany

Robert Guderian
Institut fur angewandte Botanik
Universität Essen
Universitätsstrasse 5
45141 Essen, Germany

Charles A. S. Hall
College of Environmental Science and
    Forestry
State University of New York
One Forestry Drive
Syracuse, NY 13210 USA

Linda S. Heath
USDA Forest Service
PO Box 6775
Radnor, PA 19087 USA

Pekka Kauppi
Finnish Forest Research Institute
Unioninkatu 40 A
SF-00170 Helsinki, Finland

John Kinsman
Environmental Affairs
Edison Electric Institute
701 Pennsylvania Avenue, NW
Washington, DC 20004 USA

Wilhelm Knabe
Environmental Research and Consul-
    tancy
Rumbachtal 69
45470 Mülheim an der Ruhr 1,
Germany

Gundolf H. Kohlmaier
Institut für Phys. und theor. Chemie
Universität Frankfurt
Niederurseler Hang
60439 Frankfurt, Germany

Ernst Kürsten
Silviculture and Forest Management
University of Munich
Hohenbachernstrasse 22
84354 Freising, Germany

Werner Kurz
Env. and Social Systems Analysts Ltd.
1765 W 8th Avenue
Vancouver, BC
Canada V6J 5C6

Jeff Lee
Environmental Research Laboratory
US Environmental Protection Agency
200 SW 35th Street
Corvallis, OR 97333 USA

Rik Leemans
Global Change Department
RIVM – P.O. Box 1
3720 BA Bilthoven
The Netherlands

Susanne Lorenz
Institute for World Forestry
Federal Research Centre for Forestry
    and Forest Products
Leuschnerstrasse 91
21031 Hamburg, Germany

Robert Luxmore
Environmental Sciences Division
Oak Ridge National Laboratory
PO Box 2008
Oak Ridge, TN 37831 USA

Billy McCormac
Editor-in-Chief
Water, Air and Soil Pollution
12861 Alta Tierra Road
Los Altos Hills, CA 94022 USA

Michel Meybeck
Laboratorie de Geologie Appliquée
University of Paris
4, place Jussieu
75252 Paris, Cedex 05, France

Ron Neilson
USDA Forest Service
Pacific Northwest Research Station
3200 Jefferson Way
Corvallis, OR 97331 USA

Lars-Owe Nilsson
Dept of Ecology and Environmental
    Research
Swedish University of Agricultural
    Science
S-750 07 Uppsala, Sweden

Dennis S. Ojima
Natural Resource Ecology Laboratory
Colorado State University
Fort Collins, CO 80523 USA

James Orr
LMCE/DSM/CEN Saclay
L'Orme des Merisiers, Bat. 709
91191 Glf-sur-Yvette, France

Dieter Overdieck
Institute of Ecology
TU Berlin
Königin-Luise-Strasse 22
14195 Berlin, Germany

Clinton Owensby
Department of Agronomy
Throckmorton Hall
Kansas State University
Manhattan, KS 66506 USA

James W. Raich
Department of Botany
Iowa State University
353 Bessey Hall
Ames, IA 50011 USA

Jutta Rogasik
Institut für Agrarrelevante
Klimaforschung
BFA fur Landwirtschaft Braunschweig-
    Volkenrode (FAL)
Wilhelm-Pieck-Strasse 72
15374 Müncheberg, Germany

R. Neil Sampson
American Forests
1516 P Street, NW
Washington, DC 20005 USA

Dieter R. Sauerbeck
Bonhoeffer Weg 6
38116 Braunschweig, Germany

Christoph Schlüter
Federal Environmental Agency
Bismarckplatz 1
14193 Berlin, Germany

Ralf Schmidt
Deutscher Bundestag
Enquete Kommission
Schutz der Erdatmosphäre
5300 Bonn 1, Germany

J. M. O. Scurlock
Division of Life Sciences
King's College London
Campden Hill Road
London W8 7AH, UK

H. W. Scharpenseel
Institut für Bodenkunde
Universität Hamburg
Allendeplatz 2
20146 Hamburg, Germany

Florian Scholz
BFA für Forst- und Holzwirtschaft
Institut für Forstgenetik
Sieker Landstrasse 2
22927 Grosshandsorf, Germany

Roger A. Sedjo
Resources for the Future
1616 P Street, NW
Washington, DC 20036 USA

Lloyd Simpson
Department of Biological Sciences
University of California
101 E Victoria Street
Santa Barbara, CA 93101 USA

Lowell Smith
Office of Research & Development
US Environmental Protection Agency
401 M Street, SW
Washington, DC 20460 USA

Tom Smith
Department of Environmental Sciences
University of Virginia
Charlottesville, VA 22903 USA

Al Soloman
Environmental Research Laboratory
US Environmental Protection Agency
200 SW 35th Street
Corvallis, OR 97333 USA

Bobby Stewart
USDA Agricultural Research Service
P.O. Drawer 10
Bushland, TX 79012 USA

Harald Thomasius
Auf der Bismarckhöhe 24
01737 Tharnadt, Germany

Mark C. Trexler
Trexler and Associates
1131 SE River Forest
Milwaukee, OR 97267 USA

Robert R. Twilley
Department of Biology
University of Southwestern Louisiana
P.O. Box 42451
Lafayette, LA 70504 USA

Michael Weber
Lehrstuhl für Waldbau und Forsteinrich-
    tung
Hohenbachernstrasse 22
84354 Freising, Germany

Paul L. Woomer
TSBF c/o UNESCO-ROSTA
UN Complex, Gigiri
P.O. Box 30592
Nairobi, Kenya

Ted Vinson
Department of Civil Engineering
Oregon State University
Corvallis, OR 97331 USA

Jack Winjum
National Council for Air and Stream
  Improvement
US Environmental Protection Agency
200 SW 35th Street
Corvallis, Or 97333 USA

Joe Wisniewski
Wisniewski & Associates, Inc.
6862 McLean Province Circle
Falls Church, VA 22043 USA

Lynn L. Wright
Environmental Sciences Division
Oak Ridge National Laboratory
P.O. Box 2008
Oak Ridge, TN 37831 USA

AUTHOR INDEX

SUBJECT INDEX

Afforestation 55, 139
Agricultural
  C 111, 279, 381
  soil 111
Agroecosystems 3, 357
Agroforestry 533, 545
Anthropogenic CO_2 659
Atmospheric
  C 39, 239
  CH_4 39, 111, 223
  CO_2 3, 19, 39, 55, 71, 95, 123, 163, 357, 413, 425, 431, 465, 533, 595, 629, 643
  N 55, 309, 403, 413, 643
  N_2O 111
  NH_3 187
Atmospheric pollution 177

Biodiesel 499
Bioethanol 499
Biogenic C
  USSR 223
Biogeographical models 19
Biomass
  C 39, 55, 177, 197
  crops 549
  energy 139, 325, 483, 519
  management 3, 139
Biome 19, 39, 71, 95
Biosphere sources of CO_2 659
Biotic C 629
Boreal forest 3, 39, 163, 197

C
  agricultural 111, 279, 381
  atmospheric 39, 239
  biomass 39
  budget 19, 95, 163, 197, 279, 325, 223
  dynamics 675
  emissions 545
  flux 3, 19, 39, 71, 111, 123, 223, 295
  forest impact 187
  fossil fuel emission 19, 55, 123, 139, 483
  mitigation potential 381, 533
  offset projects 561, 579
  pool 19, 39, 55, 95, 123, 177, 357, 403
  radiocarbon 431, 443
  reservoirs 123, 431
  river pollution 123
  sequestration 3, 19, 71, 95, 177, 341, 357, 403, 425, 533, 561
  sinks 3, 55, 71, 163, 177, 207, 295, 309,
  325, 413, 431, 499, 545, 561
  soil 309, 325, 373, 389
  sources 177, 381, 413, 431
  storage 19, 111, 123, 197, 239, 279, 325, 413, 629
  terrestrial storage 19, 71, 545, 595
  use efficiency 413
  wood fuel 139
  wood products 279, 325, 533
CENTURY model 357, 403
CH_4
  atmospheric 39, 111, 223
  ricelands 123
Climate change 19, 39, 55, 71, 95, 187, 309, 357, 373, 403, 579, 595, 529, 643, 659, 675
CO_2
  atmospheric 3, 39, 55, 71, 95, 123, 163, 357, 403, 425, 431, 465, 533, 595, 629, 643
  chemistry 443
  continental shelves 123
  crop residue 373
  emissions 187
  estuaries 123
  fertilization 39, 111, 665
  forest enrichment 309

Deforestation 19, 71
Deserts 3, 341, 629
Drylands 95, 341, 629

Electric utilities
  C 595
Energy
  biomass 483
  conservation 139, 499, 519, 545
  forest 519
EPIC model 389

Fertilization
  CO_2 39, 111, 615
  N 177, 533

Forest
  aboveground C storage 197
  age-class structure 163
  assessing biomass 207
  atmospheric C 239
  biomass 55, 187, 207
  boreal 3, 39, 163, 197